Innovations in Manufacturing Processes and Systems for Sustainable Practices

Innovations in Manufacturing Processes and Systems for Sustainable Practices

Guest Editor

Raul D. S. G. Campilho

MDPI

Basel • Beijing • Wuhan • Barcelona • Belgrade • Novi Sad • Cluj • Manchester

Guest Editor
Raul D. S. G. Campilho
ISEP—School of Engineering
Polytechnic of Porto
Porto
Portugal

Editorial Office
MDPI AG
Grosspeteranlage 5
4052 Basel, Switzerland

This is a reprint of the Special Issue, published open access by the journal *Processes* (ISSN 2227-9717), freely accessible at: https://www.mdpi.com/journal/processes/special_issues/LQ915KB0FQ.

For citation purposes, cite each article independently as indicated on the article page online and as indicated below:

Lastname, A.A.; Lastname, B.B. Article Title. *Journal Name* **Year**, *Volume Number*, Page Range.

ISBN 978-3-7258-4893-5 (Hbk)
ISBN 978-3-7258-4894-2 (PDF)
https://doi.org/10.3390/books978-3-7258-4894-2

Contents

About the Editor

Raul D. S. G. Campilho

Raul D. S. G. Campilho is a researcher at CIDEM, ISEP, the School of Engineering, Polytechnic of Porto, and a collaborator at INEGI, the Institute of Science and Innovation in Mechanical and Industrial Engineering, Portugal. His main research interests include advanced joining technologies, composite materials, manufacturing systems, and structural design for aerospace and mechanical applications. He has coordinated and participated in numerous national and international research projects focusing on structural performance, hybrid bonding techniques, and digital manufacturing. Raul has published extensively in peer-reviewed journals and is involved in the scientific community as an Editor in Chief, Associate Editor and reviewer for scientific journals. He is also a supervisor of MSc and PhD students and regularly collaborates with industry leaders on applied research and innovation initiatives.

Preface

This Reprint compiles a diverse set of contributions originally published in the Special Issue "Innovations in Manufacturing Processes and Systems for Sustainable Practices in the journal Processes." This Reprint intersects technological innovation, sustainable manufacturing, and systems optimization. It captures the current practices in rethinking how products are designed, manufactured, and delivered in ways that are not only efficient but also aligned with environmental and industrial goals.

The scope of this Reprint involves a wide range of topics, including digital transformation, additive manufacturing, green materials, machining innovations, and data-driven production systems. These works show how research and industry converge to develop smarter, cleaner, and more adaptive manufacturing environments. Rather than treating sustainability as a constraint, the studies collected in this Reprint demonstrate how it can serve as an enabler for engineering and sustainability excellence.

The motivation to assemble this Reprint was to provide a reading reference that highlights the progress being made in modern manufacturing. It offers readers a comprehensive view of current challenges, methods, and solutions that support the shift toward responsible production systems. By collecting academic rigor and practical relevance, the Reprint seeks to stimulate new ideas and encourage collaboration across disciplines.

This work is intended for a broad audience of researchers, engineers, industry professionals, and managers who are engaged in advancing sustainable manufacturing. It is also a valuable resource for graduate students and teachers seeking up-to-date examples of applied research in the field.

As Guest Editor, I hope that this Reprint serves as a record of innovation and a source of inspiration for future work that raises the limits of sustainable industrial practices.

Raul D. S. G. Campilho
Guest Editor

processes

MDPI

Editorial

Innovations in Manufacturing Processes and Systems for Sustainable Practices

Raul D. S. G. Campilho [1,2,*] and Flávia B. Barbosa [2]

1 CIDEM, ISEP—School of Engineering, Polytechnic of Porto, 431, 4200-072 Porto, Portugal
2 INEGI—Institute of Science and Innovation in Mechanical and Industrial Engineering, Pólo FEUP, 400, 4200-465 Porto, Portugal; fbarbosa@inegi.up.pt
* Correspondence: raulcampilho@gmail.com

In recent years, manufacturing has undergone significant transformation triggered by pressing societal demand for productivity and sustainability, made possible by accelerating technological innovation [1]. This evolution has redefined the design of industrial systems and has introduced a new vision of production that is smarter, more efficient, and environmentally responsible [2]. A relevant aspect of modern manufacturing is its emphasis on performance and sustainability. The use of the traditional metrics of productivity and quality is now coupled with lifecycle thinking, eco-efficiency, and the implementation of circular economy (CE) principles [3]. Thus, it is necessary to develop innovative approaches that embrace the duality of technologies and methods that not only enhance process capabilities but also reduce environmental impact and support regulatory compliance and societal expectations [4]. The integration of digital tools, particularly those enabling real-time data exchange, intelligent decision-making, and predictive maintenance, has become central to achieving these goals [5]. Currently, digital twins, smart sensors, and cyber–physical systems are enabling more flexible and transparent manufacturing environments [6]. Novel strategies are also being developed for the design and control of sustainable supply chains, exploring how sustainability indicators can be embedded into decision-making from the earliest stages of process development [7]. Another prominent subject in this field is the exploration of additive and micro–nano-manufacturing tools, which are revolutionizing how products are designed and fabricated [8]. These technologies offer unprecedented customization, geometrical complexity, material efficiency, and potential for decentralized and on-demand production, which are vital for a more sustainable industrial model [9]. This Special Issue of *Processes*, titled "Innovations in Manufacturing Processes and Systems Driving Sustainable Practices," presents a collection of 22 original research and review papers that investigate manufacturing innovation. These contributions are grouped into five key subtopics, each representing a fundamental pillar in the transition toward sustainable and intelligent manufacturing: 1—Strategic Approaches to Sustainable Manufacturing and Digitalization; 2—Additive Manufacturing and 3D Printing for Sustainability; 3—Advanced Materials and Green Material Development; 4—Process Innovation and Sustainable Machining; 5—Digital Tools, Simulation, and Production System Optimization. These five subtopics are addressed separately below, and the published papers focusing on each area are briefly described.

Strategic Approaches to Sustainable Manufacturing and Digitalization: Sustainable manufacturing begins with system-level transformation, in which strategy, policy, and digital integration are required [10]. The adoption of CE philosophies, lifecycle thinking, and green design principles is triggering manufacturers to rethink traditional labor-intensive production models. Digitalization and Industry 4.0 technologies, including digital twins, IoT-enabled monitoring, and cyber–physical systems, are redefining how processes are controlled, optimized, and aligned with environmental objectives [11]. In their review, Tiuncika and Bormane [12] identified 19 key criteria for evaluating expertise in sustainable entrepreneurship across environmental, social, economic, and managerial dimensions. Based on a structured literature review, they emphasized the central role of sustainable management within the triple-bottom-line (TBL) framework and proposed a conceptual tool to assess sustainability practices in small- and medium-sized enterprises (SMEs) in manufacturing. Rahmani et al. [13] analyzed the transition to smart, sustainable factories in the automotive sector, emphasizing the integration of additive manufacturing (AM), AI, and human–robot collaboration. Focusing on companies in northern Portugal, they highlighted the move toward Industry 5.0 through lightweight materials, modular robotics, and digitalization strategies aimed at improving efficiency and driving the energy transition. Wiegand and Wynn [14] explored how digital technologies (DTs) support CE practices in Germany's textile and clothing industry. Using qualitative methods and interview-based insights, they identified key drivers of and challenges in the adoption of CE principles. In their research, they proposed an operational framework to help companies align with EU sustainability goals through digital transformation. Cai et al. [15] introduced a fuzzy decision-making method to enhance green design and remanufacturability, integrating environmental, cost, and reliability factors. By applying entropy weighting and hesitation fuzzy sets, the approach supported the selection of optimal design solutions. Validated through an injection mold case study, the method proved effective in guiding sustainable design decisions via a functional prototype system. Alotaibi [16] introduced a structured framework to assess the readiness of micro-, small-, and medium-sized enterprises (MSMEs) to adopt remanufacturing practices. Through identifying and modeling nine key attributes, such as core acquisition, reverse logistics, and design for remanufacturing, and through using graph theory and matrices (GTMs), a Remanufacturing Readiness Index (RRI) was developed. The index quantified the preparedness of enterprises and helped pinpoint intervention priorities. A case study validated the model, confirming its value as a decision-support tool in promoting sustainable practices in MSMEs.

Additive Manufacturing and 3D Printing for Sustainability: AM is rapidly becoming the basis of sustainable production [17]. Its ability to minimize material waste, enable decentralized fabrication, and support lightweight, complex designs makes it a green alternative to traditional subtractive methods [18]. Human-centered design and ergonomic assessments are being integrated into AM workflows, reinforcing its role in future-ready, inclusive manufacturing ecosystems [19,20]. Kwon and Hwang [21] conducted a literature review of recurring challenges in AM, focusing on print failures across the AM workflow. Using the PRISMA method, 126 peer-reviewed articles were analyzed and classified into three main categories: design/pre-processing errors, geometric deviations, and failures in in-process monitoring. The review highlighted key mitigation strategies such as STL optimization, thermal control, and real-time sensing, and the findings offered a structured overview of failure types and solutions to support improved AM reliability. In their review, Fianko et al. [22] analyzed 61 studies to explore mass customization strategies in AM, highlighting key technologies, challenges, and applications across various industries. Four main implementation strategies were identified, supported by digital workflows and adaptive production systems, while determined obstacles included material limitations and quality control issues. A multi-tiered framework was proposed to align personalization with operational efficiency, offering clear guidance for AM adoption and future research directions. Mendoza-Muñoz et al. [23] developed a framework combining axiomatic design and the analytic hierarchy process to evaluate AM processes based on sustainability and human factors. Data from 31 users showed that fused deposition modeling (FDM) was the preferred option, particularly for expert users, due to lower material cost and information demand. The method offered a structured approach to AM process selection, with potential for the future integration of additional sustainability indicators. Hasan et al. [24] investigated the reuse of post-consumer polylactic acid (PLA) blended with virgin PLA to produce sustainable 3D-printed parts. Using a Taguchi design, they evaluated the effects of printing parameters on mechanical properties, revealing that greater layer height and nozzle temperature improved strength. The optimized blend achieved notable gains in tensile and flexural performance, confirming its suitability for low-strength applications and reinforcing the potential of recycled PLA in circular manufacturing. Remache et al. [25] analyzed advanced AM techniques for producing turbomachinery components, focusing on impellers and high-performance materials. Powder bed fusion and FDM were the most utilized methods, offering advantages in complexity, material strength, and production speed. Notably, superalloys such as Inconel 718 demonstrated excellent wear and corrosion resistance, with minimal material loss in harsh environments, highlighting AM's potential to outperform traditional manufacturing methods in demanding applications. Garcia-Llamas et al. [26] investigated the dry milling of Ti6Al4V components produced via selective laser melting (SLM), aiming to enhance machining sustainability. They examined the effects of tool geometry, cutting paths, and feed rates on surface finish, tool wear, and temperature at the tool–workpiece interface using infrared thermography. The results identified optimal conditions that balanced machining performance with reduced energy use and environmental impact, supporting greener manufacturing strategies for titanium parts.

Advanced Materials and Green Material Development: The search for materials that meet performance requirements while reducing ecological impact is the focus of green manufacturing [27]. This includes the development of bioinspired composites, low-impact reinforcements, and recyclable or biodegradable materials tailored for mechanical strength, thermal stability, or multifunctionality [28]. Advances in materials engineering are enabling the production of structures that are both durable and eco-efficient, from construction-grade composites to high-performance coatings and tool materials [29]. In their review, Liu et al. [30] described recent advances in self-renewing polycrystalline diamond compact (PDC) drill bit technologies, highlighting rack-, worm-, and ratchet-driven designs. They detailed their mechanical systems and compared their advantages in deep drilling applications, emphasizing the need for smarter wear detection and multi-technology integration to enhance performance and sustainability. These innovations promise longer bit life and more efficient single-trip drilling in challenging conditions. In another review, Adami et al. [31] assessed hydrogen plasma smelting reduction (HPSR) as a low-carbon pathway for steel production, using hydrogen in various states to reduce iron ore. To improve efficiency, they proposed integrating pre-heating and pre-reduction stages into a planned scale-up of an HPSR demo plant. Through a comparative analysis of reactor types, including fluidized beds, rotary kilns, and cyclone cascades, the authors identified a three-stage cyclone system as the most suitable option for optimizing thermal and hydrogen efficiency in future industrial applications. Estevez and Abdallah [32] optimized a hydrogel composed of sodium alginate, gelatin, and 6% calcium phosphate dibasic (CPDB) through thermally induced cross-linking, aiming to enhance rheological and mineralization properties. The hydrogel demonstrated superior elasticity, hardness, and antimicrobial activity, particularly when incubated in air without cells, where hydroxyapatite crystal formation was most prominent. When combined with osteosarcoma SaOs-2 cells, the bioink promoted dense mineralization. In contrast, fetal bovine serum (FBS)-incubated samples demonstrated less mineralization and greater degradation, indicating the hydrogel's suitability for bone tissue engineering and sustainable material profile. Zhang et al. [33] analyzed the motion of 3D five-directional circular braided fibers and derived angle and stress transformation formulas using fiber node coordinates. A volume average method was applied to calculate the stiffness matrix and fiber volume content, considering matrix effects. The model parameters included a 2 mm knuckle length, a 7 mm inner diameter, and a 0.61 filling coefficient. Comparison with experimental data showed minimal error, confirming the model's accuracy in predicting composite mechanical properties. Fazlali et al. [34] introduced a sustainable method to manufacture rigid printed circuit boards (PCBs) using biodegradable PLA and polyhydroxybutyrate (PHB) biopolymers. A novel overmolding process was used to bond these materials to flexible PI-based circuits after standard PCB fabrication, avoiding high-temperature limitations. Functional test circuits confirmed feasibility, while adhesion and end-of-life options such as debonding and composting were evaluated. The approach resulted in significant reductions in CO_2 emissions, energy use, and installation costs.

Process Innovation and Sustainable Machining: Sustainability in manufacturing is increasingly being driven at the process level. Techniques such as dry or near-dry machining, laser-assisted methods, and robotic finishing reduce the need for harmful lubricants, extend tool life, and lower energy consumption [35]. In parallel, surface engineering techniques are evolving to enhance durability, aesthetics, and functionality without compromising ecological standards [36]. These innovations illustrate the potential of process-level refinements to create high-value products with significantly reduced environmental foot-prints. In their review, Pawanr and Gupta [37] presented recent advances in dry machining, highlighting its benefits regarding sustainability and implementation-related challenges. Key findings included the use of textured tools and coatings to enhance performance and reduce costs and the effectiveness of alumina-based ceramic tools with SiC whiskers. Techniques such as microwave sintering and energy-assisted machining were also noted for improving tool properties and cutting efficiency. Shi et al. [38] examined the robotic polishing of NAK80 mold steel using 3M abrasive disks (P180–P800) on an adaptive hydraulic polishing (AHP)-equipped platform. The effects of polishing force, rotational speed, and feed rate on surface roughness were analyzed using single-factor tests and a response surface methodology. Optimal parameters improved surface finish from 0.4 μm to 0.017 μm (95.75%), and a predictive model for P800 abrasives showed $\leq$7.14% error. Evin et al. [39] applied Six Sigma to optimize the electro-discharge texturing of cold mill work rolls, which directly affects sheet metal surface quality in automotive applications. A predictive model for average roughness was developed using current, voltage, and time as input parameters, and control charts, capability indices, and a modified weighted product method validated the model. Additionally, an algorithm was proposed for process control, aligning with continuous improvement goals.

Digital Tools, Simulation, and Production System Optimization: The digitalization of manufacturing systems has enabled new levels of efficiency, flexibility, and technical insight [40,41]. Simulation-based design, process modeling, and real-time data analytics are driving smarter decision-making and predictive maintenance, thereby minimizing downtime and resource use [42]. Additionally, approaches such as topology optimization, lean mapping, and AI-driven diagnostics allow for process reengineering that aligns productivity goals with sustainability [43]. These advancements are not only improving operational efficiency but also supporting long-term environmental stewardship in manufacturing systems. Zhang et al. [44] presented TopADDPi, a cost-effective Raspberry Pi cluster designed for learning and research in parallel-computing topology optimization (PCTO). They detailed its assembly and configuration, evaluating various setups for computing efficiency, and benchmark tests highlighted its strengths in parallel processing, debugging, energy efficiency, and reduced environmental impact. The results emphasize its value as a sustainable and practical tool for structural engineering education and research. Costa et al. [45] applied lean tools, particularly value stream mapping (VSM), to optimize metal treatment processes. In a case study, value-added and non-value-added activities were identified, revealing a low initial process efficiency (6.29%). After improvements, lead time dropped from 336 to 318 h, and efficiency rose to 7.15%. Despite limitations in discontinuous flows, VSM was proven effective for enhancing performance in complex industrial settings. Wójkowski et al. [46] explored the use of servomotors in industrial equipment control systems, highlighting their advantages over classical mechanical mechanisms. A methodology was proposed for dynamic parameter determination using electronic motion profiles, resulting in an accurate mathematical model. Results showed enhanced system flexibility and adaptability and suggested an optimized angular range (205–270°) to improve performance. The authors emphasized that precise modeling is vital for ensuring system stability and durability.

Collectively, the contributions in this Special Issue offer a comprehensive view of advances redefining the manufacturing sector. These span a wide range of scales and disciplines yet share a common objective: to promote innovation that is technically sound and aligned with environmental standards and social responsibility. As Guest Editors, we are pleased to present this collection of high-quality research and review papers. We thank all the authors for their excellent contributions, the reviewers for their rigorous evaluations, and the editorial team of *Processes* for their continuous support. It is our hope that this Special Issue will serve as a valuable reference for academics, engineers, and policy-makers alike, inspiring further research and collaboration and driving a more sustainable future in manufacturing.

Conflicts of Interest: The authors declare no conflicts of interest.

References

1. Fu, H.; Zheng, C.; Yang, P. Digital inclusive finance and green transformation of manufacturing enterprises: Empirical analysis based on the dual perspectives of demand and supply. *Technol. Forecast. Soc. Change* **2024**, *200*, 123152. [CrossRef]
2. Lenz, J.; MacDonald, E.; Harik, R.; Wuest, T. Optimizing smart manufacturing systems by extending the smart products paradigm to the beginning of life. *J. Manuf. Syst.* **2020**, *57*, 274–286. [CrossRef]
3. Kandpal, V.; Jaswal, A.; Santibanez Gonzalez, E.D.; Agarwal, N. Circular economy principles: Shifting towards sustainable prosperity. In *Sustainable Energy Transition: Circular Economy and Sustainable Financing for Environmental, Social and Governance (ESG) Practices*; Springer: Berlin/Heidelberg, Germany, 2024; pp. 125–165.
4. Qi, Y.; Mao, Z.; Zhang, M.; Guo, H. Manufacturing practices and servitization: The role of mass customization and product innovation capabilities. *Int. J. Prod. Econ.* **2020**, *228*, 107747. [CrossRef]
5. Wellsandt, S.; Rusak, Z.; Ruiz Arenas, S.; Aschenbrenner, D.; Hribernik, K.A.; Thoben, K.-D. Concept of a voice-enabled digital assistant for predictive maintenance in manufacturing. In Proceedings of the 9th International Conference on Through-life Engineering Services (TESConf 2020), Fully Online, 3–4 November 2020.
6. Andreev, V.N.; Charuyskaya, M.A.; Kryzhanovskaya, A.S.; Mursalov, I.D.; Mozharovskaia, A.A.; Chervenkova, S.G. Application of intelligent engineering in the planning of cyber-physical production systems. *Int. J. Adv. Manuf. Technol.* **2021**, *115*, 117–123. [CrossRef]
7. Jemai, J.; Chung, B.D.; Sarkar, B. Environmental effect for a complex green supply-chain management to control waste: A sustainable approach. *J. Clean. Prod.* **2020**, *277*, 122919. [CrossRef]
8. Yu, H.; Han, J.; Li, S.; Zhao, D.; Wang, J.; Tian, Y.; Lin, J. Hierachical micro/nano structures fabrication by a novel tri-axial piezoelectric servo system. *Mater. Des.* **2022**, *224*, 111330. [CrossRef]

9. Mayoussi, F.; Doeven, E.H.; Kick, A.; Goralczyk, A.; Thomann, Y.; Risch, P.; Guijt, R.M.; Kotz, F.; Helmer, D.; Rapp, B.E. Facile fabrication of micro-/nanostructured, superhydrophobic membranes with adjustable porosity by 3D printing. *J. Mater. Chem. A* **2021**, *9*, 21379–21386. [CrossRef] [PubMed]
10. Turan, E.; Konuşkan, Y.; Yıldırım, N.; Tunçalp, D.; İnan, M.; Yasin, O.; Turan, B.; Kerimoğlu, V. Digital twin modelling for optimizing the material consumption: A case study on sustainability improvement of thermoforming process. *Sustain. Comput.* **2022**, *35*, 100655. [CrossRef]
11. Dhivya, K. Harnessing Synergy: IoT and Digital Twins. In *Digital Twins for Sustainable Healthcare in the Metaverse*; IGI Global Scientific Publishing: Hershey, PA, USA, 2025; pp. 107–142.
12. Tiuncika, L.; Bormane, S. Sustainable Management of Manufacturing Processes: A Literature Review. *Processes* **2024**, *12*, 1222. [CrossRef]
13. Rahmani, R.; Jesus, C.; Lopes, S.I. Implementations of Digital Transformation and Digital Twins: Exploring the Factory of the Future. *Processes* **2024**, *12*, 787. [CrossRef]
14. Wiegand, T.; Wynn, M. Circularity and Digitalisation in German Textile Manufacturing: Towards a Blueprint for Strategy Development and Implementation. *Processes* **2024**, *12*, 2697. [CrossRef]
15. Cai, Y.; Ke, C.; Ji, Q. A Fuzzy Decision-Making Method for Green Design for Remanufacturability. *Processes* **2024**, *12*, 911. [CrossRef]
16. Alotaibi, A. Development of Remanufacturing Readiness Index for MSMEs: A Comprehensive Framework. *Processes* **2025**, *13*, 1744. [CrossRef]
17. Yi, L.; Glatt, M.; Sridhar, P.; de Payrebrune, K.; Linke, B.S.; Ravani, B.; Aurich, J.C. An eco-design for additive manufacturing framework based on energy performance assessment. *Addit. Manuf.* **2020**, *33*, 101120. [CrossRef]
18. Moses, J.J.; Muthu, M.F.X.; Vijumon, V. *Additive Manufacturing*; RK Publication: Vellore, India, 2024.
19. Rossi, E.; Di Nicolantonio, M. The Human Integration into Sustainable 3D Printing Systems Part I: Methodological Setting and Human-System Integration. In *Advances in Manufacturing, Production Management and Process Control, Proceedings of the AHFE 2020 Virtual Conferences on Human Aspects of Advanced Manufacturing, Advanced Production Management and Process Control, and Additive Manufacturing, Modeling Systems and 3D Prototyping, San Diego, CA, USA, 16–20 July 2020*; Springer: Berlin/Heidelberg, Germany, 2020; p. 65.
20. Di Nicolantonio, M.; Rossi, E. The Human Integration into Sustainable 3D Printing Systems Part II: Design Experimentation. In *Advances in Manufacturing, Production Management and Process Control, Proceedings of the AHFE 2020 Virtual Conferences on Human Aspects of Advanced Manufacturing, Advanced Production Management and Process Control, and Additive Manufacturing, Modeling Systems and 3D Prototyping, San Diego, CA, USA, 16–20 July 2020*; Springer: Berlin/Heidelberg, Germany, 2020; pp. 71–77.
21. Kwon, S.; Hwang, D. Understanding and Resolving 3D Printing Challenges: A Systematic Literature Review. *Processes* **2025**, *13*, 1772. [CrossRef]
22. Fianko, S.K.; Dzogbewu, T.C.; Agbamava, E.; de Beer, D.J. Mass Customisation Strategies in Additive Manufacturing: A Systematic Review and Implementation Framework. *Processes* **2025**, *13*, 1855. [CrossRef]
23. Mendoza-Muñoz, I.; Montoya-Reyes, M.I.; Maldonado-Macías, A.A.; Jacobo-Galicia, G.; Vargas-Bernal, O.Y. A Hierarchical Axiomatic Evaluation of Additive Manufacturing Equipment and the 3D Printing Process Based on Sustainability and Human Factors. *Processes* **2024**, *12*, 1083. [CrossRef]
24. Hasan, M.R.; Davies, I.J.; Paramanik, A.; John, M.; Biswas, W.K. Fabrication and Characterisation of Sustainable 3D-Printed Parts Using Post-Consumer PLA Plastic and Virgin PLA Blends. *Processes* **2024**, *12*, 760. [CrossRef]
25. Remache, A.; Pérez-Sánchez, M.; Hidalgo, V.H.; Ramos, H.M.; Sánchez-Romero, F.-J. Towards Sustainability in Hydraulic Machinery Manufacturing by 3D Printing. *Processes* **2024**, *12*, 2664. [CrossRef]
26. Garcia-Llamas, E.; Ramirez, G.; Fuentes, M.; Vidales, E.; Pujante, J. Exploring Post-Machining Alternatives Under Dry Conditions for Thin-Walled Additive Manufacturing Components Aided by Infrared Thermography. *Processes* **2025**, *13*, 717. [CrossRef]
27. Gupta, A.; Khanna, A. A holistic approach to sustainable manufacturing: Rework, green technology, and carbon policies. *Expert Syst. Appl.* **2024**, *244*, 122943. [CrossRef]
28. Feng, G.; Xiao, C.; Wang, H.; Zhang, H.; Liu, B.; Wang, C. Enhancing mechanical properties of 3D-printed continuous carbon fibre-reinforced composites via bio-inspired design. *Thin-Walled Struct.* **2025**, *207*, 112737. [CrossRef]
29. Li, Y.; Xiao, K.; Huang, C.; Wang, J.; Gao, M.; Hu, A.; Tang, Q.; Fan, B.; Xu, Y.; Chen, X. Enhanced Potassium-Ion Storage of the 3D Carbon Superstructure by Manipulating the Nitrogen-Doped Species and Morphology. *Nano-Micro Lett.* **2020**, *13*, 1. [CrossRef] [PubMed]
30. Liu, X.; Ji, G.; Yang, L.; Ren, H.; Wu, Q.; Yu, J.; Huang, K. Polycrystalline Diamond Composite Bit Technology for Sustainable Work. *Processes* **2025**, *13*, 421. [CrossRef]

31. Adami, B.; Hoffelner, F.; Zarl, M.A.; Schenk, J. Strategic Selection of a Pre-Reduction Reactor for Increased Hydrogen Utilization in Hydrogen Plasma Smelting Reduction. *Processes* **2025**, *13*, 420. [CrossRef]
32. Estevez, A.T.; Abdallah, Y.K. Biomimetic Approach for Enhanced Mechanical Properties and Stability of Self-Mineralized Calcium Phosphate Dibasic–Sodium Alginate–Gelatine Hydrogel as Bone Replacement and Structural Building Material. *Processes* **2024**, *12*, 944. [CrossRef]
33. Zhang, W.; Li, C.; Li, L.; Wang, W.; Yang, L.; Zhang, C.; Zhang, X. Numerical Analysis on Mechanical Properties of 3D Five-Directional Circular Braided Composites. *Processes* **2025**, *13*, 800. [CrossRef]
34. Fazlali, Z.; Schaubroeck, D.; Cauwe, M.; Cardon, L.; Bauwens, P.; Vanfleteren, J. Polylactic Acid and Polyhydroxybutyrate as Printed Circuit Board Substrates: A Novel Approach. *Processes* **2025**, *13*, 1360. [CrossRef]
35. Maurya, U.; Vasu, V. Ionic Liquids as a Potential Sustainable Green Lubricant for Machining in the Era of Industry 4.0. In *Advances in Sustainable Machining and Manufacturing Processes*; CRC Press: Boca Raton, FL, USA, 2022; pp. 135–155.
36. Zhang, Y.; Peng, L.; Ye, Y.; Chi, Y.; Gao, L.; Zha, X.; Huang, T.; Zhang, Y.; Ding, H.; Ye, C. Exploring the strengthening mechanisms of additive manufactured metals treated by ultrasonic nanocrystal surface modification. *Int. J. Fatigue* **2025**, *190*, 108609. [CrossRef]
37. Pawanr, S.; Gupta, K. Dry Machining Techniques for Sustainability in Metal Cutting: A Review. *Processes* **2024**, *12*, 417. [CrossRef]
38. Shi, D.; Zeng, X.; Wang, X.; Zhang, H. Parameter Optimization and Surface Roughness Prediction for the Robotic Adaptive Hydraulic Polishing of NAK80 Mold Steel. *Processes* **2025**, *13*, 991. [CrossRef]
39. Evin, E.; Kokarda, V.; Tomáš, M. Quality Control of Electro-Discharge Texturing of Rolls Through Six Sigma. *Processes* **2025**, *13*, 450. [CrossRef]
40. Klimant, P.; Koriath, H.-J.; Schumann, M.; Winkler, S. Investigations on digitalization for sustainable machine tools and forming technologies. *Int. J. Adv. Manuf. Technol.* **2021**, *117*, 2269–2277. [CrossRef]
41. Li, H.; Yang, D.; Cao, H.; Ge, W.; Chen, E.; Wen, X.; Li, C. Data-driven hybrid petri-net based energy consumption behaviour modelling for digital twin of energy-efficient manufacturing system. *Energy* **2022**, *239*, 122178. [CrossRef]
42. Goyal, M.K.; Singh, A.P. Real-Time Data Streams: The Future of Predictive Maintenance in Manufacturing. In Proceedings of the 2023 IEEE International Conference on ICT in Business Industry & Government (ICTBIG), Indore, India, 8–9 December 2023; pp. 1–7.
43. de Vito, A.F.; Vicente, W.M.; Xie, Y.M. Topology optimization applied to the core of structural engineered wood product. *Structures* **2023**, *48*, 1567–1575. [CrossRef]
44. Zhang, Z.-D.; Yu, D.-Y.; Ibhadode, O.; Meng, L.; Gao, T.; Zhu, J.-H.; Zhang, W.-H. TopADDPi: An Affordable and Sustainable Raspberry Pi Cluster for Parallel-Computing Topology Optimization. *Processes* **2025**, *13*, 633. [CrossRef]
45. Costa, B.; Varejão, J.; Gaspar, P.D. Development of a Value Stream Map to Optimize the Production Process in a Luxury Metal Piece Manufacturing Company. *Processes* **2024**, *12*, 1612. [CrossRef]
46. Wójkowski, K.; Talaśka, K.; Wilczyński, D. Mathematical Model of the Electronic Cam in Terms of Application in a Dosing Machine. *Processes* **2024**, *12*, 1909. [CrossRef]

processes

MDPI

Review

Sustainable Management of Manufacturing Processes: A Literature Review

Lasma Tiuncika * and Santa Bormane

Faculty of Social Sciences, Riga Stradins University, LV-1007 Riga, Latvia; santa.bormane@rsu.lv
* Correspondence: lasma.tiuncika@rsu.edu.lv

Abstract: Industrial economic activity is one of the primary causes of environmental degradation and a source of issues related to social inequality. Nevertheless, research has shown that the application of the sustainable entrepreneurship model can significantly improve environmental and social conditions while maintaining continuous and sustainable economic growth. Despite that, statistics show relatively low engagement in sustainable entrepreneurship, potentially due to a lack of expertise among entrepreneurs. Due to the low engagement, available sustainability performance metrics are not a relevant measurement tool. Thus, the purpose of this study is to identify a set of criteria that evaluate expertise levels in sustainable entrepreneurship within the manufacturing industry. Through the application of monographic and secondary data analysis methods, the authors carried out a literature review. The findings indicated that sustainable management is a key component to the triple-bottom-line (TBL) framework that is used as a conceptual basis for sustainable entrepreneurship. Additionally, the authors identified 19 criteria of sustainable entrepreneurship within four dimensions of sustainability (environmental, social, economic, and management). Despite the limitations, with a more refined literature review, the criteria could be applied as a comprehensive resource to estimate expertise in sustainable entrepreneurship in future studies on small and medium enterprises (SMEs).

Keywords: sustainability; sustainable development; sustainable entrepreneurship; sustainable management; sustainable production; triple-bottom-line approach

1. Introduction

Modern society is challenged by numerous environmental and social threats such as global warming, climate change, poverty, depletion of natural resources, and other issues that compromise further development. The current problems are attributed to industrial economic activity which continuously fails to respect the limitations of the planetary boundaries in an effort to meet the current needs [1]. As a result, the sustainability concept was introduced to balance the environmental, social, and economic dimensions that altogether are referred to as the triple-bottom-line (TBL) framework [2].

In recent years, the framework has been widely applied to entrepreneurial research due to the extensive benefits that sustainability-oriented entrepreneurship may potentially have in relation to sustainable development efforts. With respect to the theory, a concept of sustainable entrepreneurship was developed. It positions sustainable development as a business opportunity rather than a threat or an obstacle to entrepreneurial activities.

Despite the popularity of the topic, the clarity of the concept remains to be an issue among the researchers. The main reason was pointed out to be the disagreement among the definitions caused by concept complexity, uncertainty of the demands, and contextual differences [3]. Following the TBL framework, the general production procedures in sustainable entrepreneurship must adopt sustainability practices synchronously within each of the three domains—environmental, social, and economical—where performance in each is positioned above profit maximization [2].

In the environmental domain, the central focus is the planet; thus, beyond simply advocating for environmentally friendly behavior, the main goal is to address the negative externalities caused as a result of production. Optimization of production processes through reductions in material waste and carbon footprint, clean technologies, rational and responsible use of resources, and substitution of natural resources with renewables are some examples that a company can use to reverse ecological degradation and benefit environmental sustainability [4–6].

In the social domain, the central element is people; thus, companies are expected to protect human rights and empower social change through various social initiatives and cooperation with stakeholders. Most of the change can already be effected within companies by setting cultural norms through labor-inclusive workspaces, socially conscious treatment, and investment in employees [7,8].

In the economic domain, the central object will largely remain to be profit. Thus, actions must align with sustainability principles while maintaining revenue streams. This can be achieved through the identification of market imperfections and the promotion of innovative market solutions along with the delivery of functional products and respective maintenance services to reduce end-user consumption. At the policy level, companies should target the maximization of material and energy efficiency and transition toward a circular economy in an effort to close the resource loops [2,5].

Meeting TBL values is imperative in sustainable entrepreneurship [2], though not all researchers agree with this concept. For example, some also position culture as a pillar of sustainable entrepreneurship [9]. On the other hand, the ESG (environment, society, and governance) concept is often applied to measure the sustainability performance of enterprises [10,11]. As a revised TBL model, it considers three pillars—environment, social issues, and corporate governance—where governance refers to the top management's decisions on addressing sustainability concerns [12].

This example highlights the management's role in directing the organization's sustainability strategy and overall attitude, which can further impact sustainability efforts and outcomes [13–15]. Although financial and economic aspects are also considered within the ESG framework [16], authors argue that the economic dimension should be separated from environmental, social, or governance aspects. As a result, adding to the basis of the TBL framework, the management dimension must be considered as a separate aspect of the sustainable entrepreneurship concept.

Moreover, researchers have identified numerous advantages of sustainable entrepreneurship. There is evidence that the implementation of sustainability-oriented values at a firm level has positive implications for the environmental performance of employees [13,17]. Proactive engagement with stakeholders can benefit society in terms of sustainable living, health, and overall well-being. Additionally, social initiatives targeted at reducing negative impacts have been shown to deliver positive value for enterprises and society in terms of increased customer loyalty and strengthened competitive advantage [4,15].

However, these findings alone are not sufficient to promote the integration of sustainable entrepreneurship into current business processes. Although about 30% of enterprises undertake due diligence in all three dimensions of sustainability, only 3.7% of established and 3.6% of starting enterprises in the world put forward sustainability goals beyond financial profitability [2,18–20]. Thus, there is no substantial progress in the transition to more sustainable entrepreneurship practices.

Statistics indicate a lack of understanding of how to begin the transition towards sustainable entrepreneurship. Education not only fosters the desire for innovations but also helps detect market opportunities and tools that are necessary for the achievement of entrepreneurial goals [21,22]. Thus, the authors suggest that entrepreneurs have limited expertise in sustainable entrepreneurship, which is essential for the promotion of sustainable behaviors that contribute to a transition towards sustainable business practices.

The education issue also highlights the present research problem—the need for concept clarification and indicators of expertise in sustainable entrepreneurship. Since sustainable entrepreneurship has only been adopted by a small percentage of companies, the assessment of sustainability performance may not be applicable to facilitate the transition process. First and foremost, it is rather important to understand the competence levels before developing a methodology to assist the transformation towards sustainable entrepreneurship. As a result, the main objective of this study is to develop a set of criteria that estimate the expertise levels in sustainable entrepreneurship.

The authors suggest that clarification of criteria would facilitate the adoption of the sustainable entrepreneurship model. This would determine whether the efforts are in line with the sustainability agenda and how radical changes must be made in current business practices. It has not yet been possible to tackle this issue because previous studies have primarily focused on preliminary research on interactions between different factors or the assessment of sustainability performance in general.

The present research is distinctive as it aims to fill the educational gap with the identified criteria that indicate expertise levels in sustainable entrepreneurship. As a result, this study contributes to the current literature by summarizing various factors that have been discovered in previous research in one place and providing future directions for studies on the progress and challenges of sustainable entrepreneurship model application procedures.

2. Materials and Methods

To achieve the objective of this study, monographic and secondary data analysis methods were applied in order to conduct a literature review. The review covered a careful and thorough examination of recent publications in sustainable entrepreneurship. Based on an analysis of the available literature on different aspects and factors associated with sustainable entrepreneurship, the main objective is to identify and group criteria that indicate the level of expertise in sustainable entrepreneurship. The present study serves as a basis for the second research phase, which focuses on the development of a metric system for the evaluation of expertise levels in sustainable entrepreneurship in further studies.

A monographic method was carried out to perform the primary literature review. The studies used in the review were selected based on the relevant keywords, timeframe, type of document, and language. The articles were selected from databases such as ScienceDirect, JSTOR, and ResearchGate according to keyword search terms "sustainability", "sustainable entrepreneurship", "sustainable business", "sustainable management", "green enterprise", "green competencies", "sustainable strategy", "sustainable model", "entrepreneurship", "TBL", "ESG", "circular economy", "resource optimization", and "product functionality". The articles were further filtered based on the year of publication to include articles published from 2019 to 2024.

All available and relevant publications were restricted to only peer-reviewed articles written in English. All irrelevant articles were further eliminated by reading titles and abstracts. After the complete reading of the remaining publications, 73 articles met the review criteria. Most of the publications were extracted from *The Journal of Cleaner Production*, followed by *Sustainability* (Switzerland). The publications were further used for literature analysis in sustainable entrepreneurship, including the review and collection of the latest findings regarding the various factors and their relationship to different aspects of sustainable entrepreneurship.

A secondary data analysis method was used to narrow down the numerous previously discovered factors to a single list of criteria that relate to the expertise level in sustainable entrepreneurship the best. In total, the authors identified 19 criteria and grouped them into four dimensions of sustainability—environment, social, economic, and management. Each of the criteria was defined through 2 to 3 performance-characterizing statements that could serve as a measurement method in further research.

With the present literature review, the authors aim to develop a self-report questionnaire matrix that estimates the expertise levels in sustainable entrepreneurship. The metric system is currently in the development process in the second research phase and will be applied to test the knowledge and expertise in sustainable entrepreneurship. In further research phases, the authors will investigate the underlying conditions that prevent successful integration of the sustainable entrepreneurship model into manufacturing industries.

3. Results

3.1. Sustainable Management

The findings of the literature review indicate that the TBL approach is not sufficient to illustrate the sustainable entrepreneurship model. In theory, sustainable management is a vital contributor to sustainable development efforts in general and, therefore, presents as a key indicator of performance in sustainable entrepreneurship [23].

Research illustrates that corporate governance highly influences an organization's behavior, including sustainability reporting initiatives, and is significantly linked to the overall sustainability performance [23,24]. Overall, governance represents the top management's responsibility towards stakeholder interests, accountability, transparency, and fairness. Corporate governance is also reported to mediate social responsibility efforts and financial performance, in the sense that efforts tend to result in better financial outcomes [25]. Thus, it can be concluded that the top management, as an internal control mechanism, assures that all internal objectives related to sustainability are reached [23].

As a result, the authors suggest that management must be considered as a separate dimension in the sustainable entrepreneurship concept that may potentially regulate the performance in environmental, social, and economic dimensions. Figure 1 illustrates the comparison between the theoretical model presented by Farny and Binder and the authors' proposed model of sustainable entrepreneurship [3].

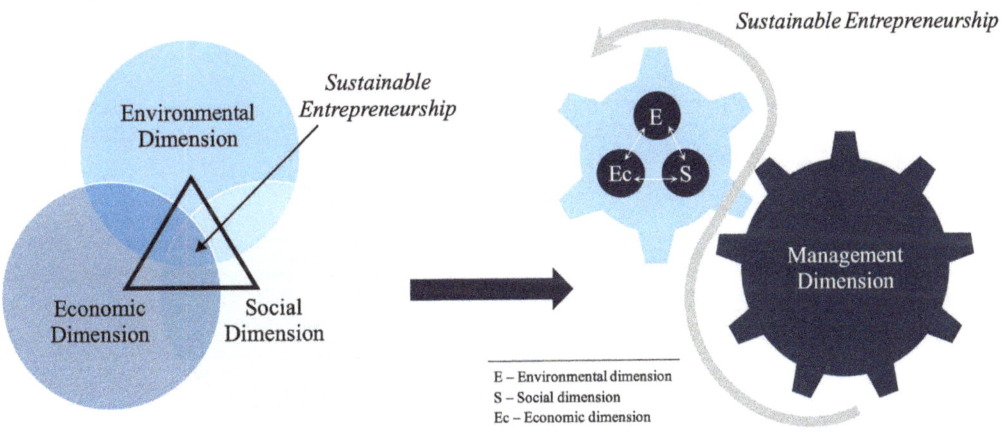

Figure 1. Comparison of the theoretical and proposed models of sustainable entrepreneurship.

The authors suggest that the relationship between sustainable management and general environmental, social, and economic sustainability performance generates the potential and success for sustainable entrepreneurship for three reasons. First, the locus of control of the top management; second, the relationship between the top management's actions and the outcomes at an operational level; and third, the resourcing and cooperative capabilities. Thus, sustainable management should not be merely viewed as a factor but as a separate dimension and a fundamental building block to the sustainable entrepreneurship theory.

3.1.1. The Locus of Control of the Top Management

The locus of control of the top management is expressed through the belief of to what extent the top management is in control of the business processes. By adapting Rotter's (1966) theory, the external locus of control points to the belief that the top management is not in control of consequences and performance resulting from business operations, whereas the internal locus of control indicates that the top management fully controls business operations and therefore the consequences and performance [26].

So far, measuring the environmental, social, and economic performance indicators has been the commonly suggested way to estimate and explain sustainable entrepreneurship. Instead, research shows that there is a necessity to pay closer attention to managerial practices, especially in the manufacturing sector. The prerequisite to any type of action, whether it concerns environmental, social, or economic sustainability, is the management's strategic decision-making process. Any business activity is carried out through a sequence of planned and coordinated strategic actions and arises from deliberate leadership rather than being something that is a byproduct of external events. Following this logic, sustainable management serves as a critical driver of sustainable entrepreneurship, especially when the top management expresses a high internal locus of control [14,17].

As a result, it can be concluded that the integration of the sustainable entrepreneurship model into current business processes entails high levels of long-term commitment and support from the top management [13]. In addition to that, there is no other organizational level within a company that would have enough influence, knowledge, expertise, or action abilities to address and execute such a complex business model. Only the top management facilitates the implementation of strategic goals, allocates the resources, and sets organizational norms and standards that altogether direct the further identity, reputation, and success.

3.1.2. The Relationship between the Top Management Action and the Outcomes at an Operational Level

Corporate governance has been shown to have a critical role regarding the outcomes at an operational level [25]. For example, managerial practice can not only foster but also sabotage the integration of the sustainable entrepreneurship model into regular practice. Top management performance is an indispensable factor in an organization's functioning, and it regulates how well the administrative efforts translate to the operational level or to what extent they will actually be carried out in practice. For example, if the top management does not have an objective outlook on the company's culture and tends to falsely idealize the overall sustainability performance, any policy or sustainability initiatives will likely fail to reach the operational level [12,15].

Poor performance in sustainable management may also be reflected in the form of greenwashing—a misleading and deceptive tactic that is used to present the company and its products as more environmentally friendly than they are in reality. Usually, companies adopt different sustainability attributes and labels that falsely indicate environmental protection efforts [27]. As a result, their overall sustainability is compromised and requires a clear set of transparency and accountability standards specifically within the top management.

3.1.3. The Resourcing and Cooperative Capabilities

Another important aspect that must be considered in the sustainable entrepreneurship model is the management's relation to the overarching governance. If sustainable entrepreneurship is viewed as an entity within an ecosystem, then we can assume that there will be external stimuli that the entrepreneurs must respond to. In this case, the ecosystem consists of the company itself, governmental, international, and non-governmental institutions and organizations, stakeholders, consumers, and other relevant interested parties. Each plays a particular role in the functioning of an enterprise [1,21,28].

The institutional context provides standards and regulations that determine the production process of goods [5]. For example, the government lays normative regulations and implements new targeted policies that not only are legally binding to companies but also provide context to the sustainability demands [21,28,29]. The first responder to such legal demands will be the top management. A clear understanding of the overarching regulations within the top management will therefore determine the following sustainability efforts and proper integration of law into practice [5].

Governmental policy can also be perceived as an assistive guide towards sustainable entrepreneurship, especially for companies that have difficulties with taking proactive action, for example, due to financial or educational restrictions [2,8,29]. Thus, close cooperation with institutions must be maintained at all times and the company's engagement levels may directly point to the importance of sustainability within the top management.

3.2. Sustainable Entrepreneurship Criteria

The secondary data analysis revealed numerous factors that contribute to the understanding of sustainable entrepreneurship. To better understand and illustrate the indicators of sustainable entrepreneurship, the authors compiled the discovered factors and values within 19 criteria split among four subgroups of sustainable entrepreneurship—environmental, social, economic, and management dimensions (Figure 2). This approach helps to sufficiently cover the various aspects of sustainable entrepreneurship in one set of criteria.

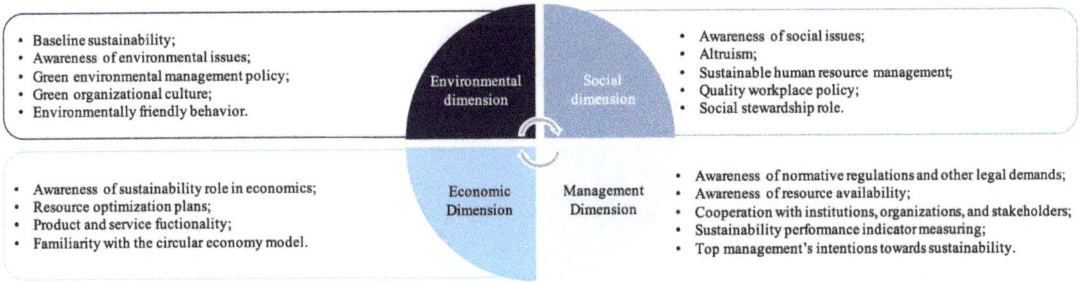

Figure 2. Sustainable entrepreneurship criteria.

3.2.1. Environmental Dimension

The focal point of the environmental dimension is the planet. Therefore, the environmental dimension within sustainable entrepreneurship is centered around the protection of ecological balance to improve environmental sustainability and ensure continuous resource availability for future generations. The aspects that determine the environmental performance in sustainable entrepreneurship can be organized into five criteria:

1. Baseline sustainability;
2. Awareness of environmental issues;
3. Green environmental management policy;
4. Green organizational culture;
5. Environmentally friendly behavior.

Overall, 20 articles (27%) touched upon specific aspects of business and production processes that are associated with the environmental dimension of sustainable entrepreneurship. Of all relevant publications, five articles covered factors associated with "baseline sustainability"; "awareness of environmental issues" as a sustainable entrepreneurship aspect was highlighted in five studies; "green environmental management policy" was the focus of three publications; "green organizational culture" was mentioned as a criterion for sustainable entrepreneurship in four studies; and "environmentally friendly behavior" in entrepreneurship was specifically discussed in three articles.

3.2.2. Social Dimension

The central element of the social dimension is people. Therefore, sustainable entrepreneurship should carry out business activities with respect to social equality, welfare, stability, and protection of human rights. The social responsibility aspects of sustainable entrepreneurship can be highlighted in the following five criteria:

1. Awareness of social issues;
2. Altruism;

3. Sustainable human resource management;
4. Quality workplace policy;
5. Social stewardship role.

With a minor overlap with the publications covering the environmental dimension, 17 articles (23%) covered sustainable entrepreneurship aspects that address social sustainability issues. "Awareness of social issues" as a part of sustainable entrepreneurship was identified in five articles; two articles covered "altruism" in depth as an important business aspect in relation to social sustainability; five articles discussed "sustainable human resource management" policy; "quality workplace policy" in the context of sustainable entrepreneurship was reviewed in two studies; and the importance of the acceptance of the "social stewardship role" in sustainable entrepreneurship was discussed in detail in three articles.

3.2.3. Economic Dimension

The main objective of the economic dimension in sustainable entrepreneurship is profit. Nonetheless, the objective is limited to the condition that economic growth must be sustainable and cannot be achieved on account of the exploitation of environmental or social resources. Sustainable entrepreneurship within the economic dimension can be outlined with the following four criteria:

1. Awareness of sustainability role in economics;
2. Resource optimization plans;
3. Product and service functionality;
4. Familiarity with the circular economy model.

The economic dimension is an essential part of any business activities and the identification of key aspects that contribute to the economic sustainability the strongest is a common consideration for most researchers. Nevertheless, only 17 articles (23%) provided a clear set of factors that serve as indicators for sustainable entrepreneurship. The "awareness of sustainability role in economics" was covered in four articles; "resource optimization plans" as an essential element to business was discussed in six articles; four studies covered "product and service functionality" in detail; and three provided statistical evidence for the "familiarity with the circular economy model" as a predictor of sustainable entrepreneurship.

3.2.4. Management Dimension

The management dimension is centered around strategic objectives and initiatives pushed forward by the top management to facilitate the employment of sustainability values within the company [13]. Although the management already plays a substantial role in each of the previously described criteria, the main distinction is the way in which each of the criteria is acquired and carried out. In other words, fulfilling the criteria of the social dimension results in social sustainability, whereas fulfilling the criteria of the management dimension results in the ability to carry out further actions designed to reach environmental, social, or economic sustainability objectives. As a result, sustainable management serves as a prerequisite for successfully carrying out further sustainability processes that contribute to sustainable entrepreneurship. Based on the literature, the expertise in sustainable management can be expressed through five criteria:

1. Awareness of normative regulations and other legal demands;
2. Awareness of resource availability;
3. Cooperation with institutions, organizations, and stakeholders;
4. Sustainability performance indicator measuring;
5. Top management's intentions towards sustainability.

The criteria of the management dimension were collected from the majority (22) of articles (30% of all). The "the awareness of normative regulations and other legal demands" was covered in four studies; five studies touched upon the "awareness of resource availability" as an aspect of sustainable entrepreneurship; the "cooperation with institutions, organizations, and stakeholders" criterion was highlighted in six articles; the "sustainability performance indicator measuring" and the associated criteria were researched by three studies; and, lastly, "top management's intentions towards sustainability" was mentioned as an important factor for sustainable entrepreneurship in four articles.

4. Discussion

Despite the vast literature on benefits and numerous legal obligations, the lack of a clear set of criteria and standards continues to delay the integration of sustainable entrepreneurship into current business practices. As a result, engagement rates are still relatively low, which calls for a deeper exploration of standards and criteria of sustainable entrepreneurship. Numerous measures exist that evaluate sustainability performance [30]. However, they are constricted by two limitations. First, due to concept clarity issues, measures do not provide a thorough evaluation of sustainable entrepreneurship. Second, in cases where sustainable entrepreneurship is only involved in the strategic planning process, applying such measures would not be entirely relevant as they do not explain what criteria the company is missing specifically to better facilitate the transition process.

To address the research problem, the authors identified 19 criteria among four sustainability dimensions—environmental, social, economic, and management—that can be used as indicative factors pointing to the understanding and expertise levels in sustainable entrepreneurship. With that, this research aims to fill not only the educational gap but also the lack of criteria to evaluate the expertise level in sustainable entrepreneurship.

The management dimension was added to the theoretical matrix based on the available literature-identified limitations of the TBL approach. The authors suggest that as a result of the locus of control effect, the impact that top management has on the business outcomes is high, and due to the resourcing and cooperative abilities, the management role has been overlooked in previous research with regard to TBL theory. Therefore, it must be considered as a separate aspect or value with regard to sustainable entrepreneurship and should be analyzed in relation to TBL values.

The authors propose that for a certain company to be defined as following the sustainable entrepreneurship model, it must fulfill the 19 criteria across the four aforementioned sustainability dimensions. In the theoretical model, the environmental dimension consists of five criteria that cover aspects of entrepreneurship regarding the understanding, beliefs, and action competence towards the environment—baseline sustainability, awareness of environmental issues, green environmental policy, green organizational culture, and environmentally friendly behavior.

Baseline sustainability refers to the overall awareness of the sustainability concept. It encompasses the understanding of sustainability considerations that are critical to the development of further business strategies [31]. Such awareness also translates to the attraction of employees with a similar vision and promotes overall conscientiousness towards sustainability within an organization [17].

Despite the fact that the sustainability concept above the environmental dimension entails performance in social and economic sustainability, the baseline sustainability values tend to be the predictors of environmentally friendly behaviors within a company [17]. That, however, becomes problematic when addressing social or economic sustainability aspects if entrepreneurs tend to associate sustainability more with environmental issues rather than social or economic. In support of the research problem, this also points to the lack of general knowledge about sustainable entrepreneurship.

Awareness of environmental issues specifically indicates the extent to which entrepreneurs understand the current environmental challenges and market flaws or imperfections that amplify environmental degradation risks [1]. It is general environmental knowledge that plays a major role in intention and decision formation [17]. Additionally, it impacts employees' green behavior [32]. The level of awareness tends to indicate the respective level of proactive behavior in environmental protection [31]. In other words, the higher the awareness, the more likely that a company will adjust production policies and engage in environmental protection [33]. Awareness of environmental degradation also promotes innovations and the development of activities necessary to generate environmental gains as a byproduct of entrepreneurial processes [22,34,35]. Thus, it can be concluded that better competence in this area will affect business processes.

Green environmental policy entails planned action, oriented to minimize or reverse negative environmental externalities caused directly by carrying out regular production activities. In many cases, such policies will involve the necessity for innovative interventions [8,13]. As a result, this criterion also points out the level of engagement of innovations and green technology, as well as the overall innovative capacity that is a critical aspect of sustainable entrepreneurship. Nonetheless, a solid corporate environmental policy may reduce financial risks associated with the implementation of a sustainability-oriented strategy and boost overall financial performance through the strengthening of the firm's competitive advantage [36].

Green organizational culture refers to the identity of the company and its link to environmental support. Pro-environmental behavior is most often fostered within the inner culture of a company, and in previous research, it has been linked to better environmental performance [37–40]. It includes different ecological initiatives aimed at, for example, reducing waste, increasing energy efficiency, and reducing CO_2 pollution within the company itself [13]. Studies also indicate an association between green culture and a better dynamic balance between economic and environmental goals [41]. In addition to that, financial and non-financial reward systems can be created to reward environmentally friendly behavior, in an effort to boost the organization's culture [17]. It is necessary that both management and workers have an aligned vision regarding the environment, and this can be achieved with the help of a strong organizational culture.

Finally, environmentally friendly behavior cannot be limited to general support. The exploitation of resources has highly compromised environmental sustainability. As a result, there have to be clear and visible patterns of actions directed at the preservation of the environment, biological diversity, ecosystems, and mitigation of the climate crisis [17]. Pro-environmental behavior also tends to moderate customer attitudes towards sustainable products and is reported to have an impact on green innovation practices [35]. As a result, such behavior is also associated with eco-friendly production [42].

The central aspect of this criterion is the understanding of the relation between short-term costs and long-term goals. This can be expressed through responsible resource management, the use of renewable energy sources, recycling, and the use of recyclable materials in an effort to provide continuous growth capabilities for the next generations [2]. The transition may involve increased costs but would benefit the planet in the long term. This criterion is distinctive from the green environmental policy criterion as it aims to avoid potential environmental risks, whereas green environmental policy deals with already-caused damage and externalities.

The criteria of the sustainable entrepreneurship matrix in the social dimension covers the understanding of the impact that any business activity can have on society and the placement of appropriate policies to improve or at least mediate social issues. According to the literature review, the authors identified five aspects of sustainable entrepreneurship that relate to the social dimension the closest—awareness of social issues, altruism, sustainable human resource management, quality workplace policy, and acceptance of a social stewardship role.

Just like awareness of environmental issues, the awareness of social issues determines the understanding of the relationship between business and society and relates to meeting both social and economic needs [43]. It also moderates the sense of responsibility and overall response to social issues [44]. Social resources catalyze industrial growth since society not only defines the consumer base but is also the source of the workforce. Previous research indicates that higher levels of interest in social sustainability point to a more pronounced interest in sustainable entrepreneurship [2,28]. Therefore, such knowledge within a company can help in identifying the practices that are commonly associated with social problems and make changes to improve not only the general welfare but also the performance of the company.

Altruism is a personal characteristic that epitomizes a person's concern for others' well-being. In business, the characteristic serves as a motivator of social responsibility within a company [45]. In fact, other studies recognize altruism as a precondition to social entrepreneurship oriented at social value creation [46]. In the case of high altruism, personal benefit is not nearly as important as helping others, which aligns with sustainable entrepreneurship's intentions to put forward other non-financial goals above profit maximization. With high levels of empathy-based altruism within a company, there is a greater chance that entrepreneurial activities will be carried out with high regard for social matters and social sustainability [2]. With greater altruism, entrepreneurs will tend to discover more developmental opportunities regarding sustainability [47]. As a result, altruism potentially predicts the intention to engage in sustainable entrepreneurship.

Sustainable human resource management is based on an adapted version of green human resource management theory. The main difference is that the orientation to sustainability entails all TBL framework aspects. Particularly in the social dimension, such practice indicates consideration of employees and their well-being. In other words, resource management not only oversees the establishment of a just and equal workplace but also takes care of employee work–life balance, well-being, and social security [17,48,49]. Studies show that sustainable human resource management practices are highly associated with sustainable innovation [50]. The benefits that arise from a company supporting its employees permeate to the society level and can contribute to better social welfare overall. Nonetheless, sustainable human resource management has been shown to attract talents that further promote sustainability within an organization and increase the overall sustainability performance [51].

Quality workplace policy relates to the company's efforts to establish employment opportunities to support especially the underprivileged part of society and boost employment rates. In the long run, such entrepreneurial initiative contributes to overall social sustainability. Quality refers to stable, long-term positions with competitive salaries that do not discriminate against age, gender, disabilities, ethnicity, religion, or other social factors [7,21]. In consequence, establishing quality workplaces annually indicates concern about employment and social security levels that can be further interpreted as a criterion for sustainable entrepreneurship.

Studies highlight social stewardship goals as an inherent part of sustainability strategy within a business and overall corporate social responsibility [52,53]. A company that accepts a social stewardship role and actively engages in social and human rights protection much more strongly expresses values that align with sustainable entrepreneurship. Often, this role is displayed through marketing activities that in turn help raise awareness of social issues among different groups of society [5]. For best results, companies with a sustainable entrepreneurship model would be expected to work together with stakeholders, as well as governmental and non-governmental organizations, to potentially generate a greater impact through various social value-promoting campaigns and charity events.

Among numerous factors associated with the interaction between economic dimension and sustainable entrepreneurship, the authors propose four criteria that relate to the level of integration of economic sustainability values into ongoing business processes. Those are awareness of sustainability's role in economics, resource optimization plans, product and service functionality, and familiarity with the circular economy model.

Economic issues are some of the primary concerns that have driven efforts towards sustainable development [54,55]. Thus, awareness of sustainability's role in economics refers to the baseline understanding of the interaction between sustainability and economic growth. Primarily, that means diverting the focus from short-term financial gains to long-term economic growth without compromising the company's financial viability [2]. Sustainable entrepreneurship, thus, is concerned with carrying out feasible and profitable business solutions that, in the meantime, support sustainable development. Studies show a direct link between entrepreneurship and economic growth. Thus, the ability to relate both objectives and implement them into practice in a sustainability context demonstrates the potential for the successful application of sustainable entrepreneurship [56].

Resource optimization plans play a critical role in ensuring continuous economic growth in a sustainability context. Research indicates that resource optimization is significantly associated with sustainable development [57]. However, that entails strategic calculations and planning as it may not only have the potential to cut production costs and boost productivity but may also require investing in costly innovations [4]. In the manufacturing sector, resource optimization leads to increased competitiveness and highly contributes to positive environmental performance [58].

Production and resource optimization procedures that fit the sustainability goals will require unique approaches and might not be financially sound, although they are essential in maintaining a competitive edge over other producers [6]. Such considerations are fundamental to sustainable entrepreneurship as they not only concern the possibility of facilitating overall economic growth with respect to sustainable development but also ensure the ability to safeguard the company's position within the market in the long run. Additionally, resource optimization plans may affect the transition towards sustainable entrepreneurship, though the argument calls for more in-depth research to confirm the relationship [59].

Product and service functionality is a key aspect of sustainable entrepreneurship. Goods that serve more than one purpose decrease the need for multiple different items and, thus, reduce unsustainable consumption patterns [5]. The initiative aligns with a frugal business model where value is prioritized over costs. Products with improved functionality not only diversify the supply of a company and optimize costs but also highly contribute to sustainable development [60]. Adding to that, companies can expand the range of market opportunities by offering guarantee and repair services that can help build a solid and loyal customer base. Additionally, customers are already requesting sustainable products that provide endless opportunities for entrepreneurs [4,15]. As a result, the producers can maintain a supply of goods that not only follows sustainability principles but also continuously produces income.

Familiarity with the circular economy model within sustainable entrepreneurship is expressed through thoroughly planned product and project life cycles. Circularity refers to a closed-loop system where the life cycle of materials and products is lengthened through repeated use, repairs, recycling, sharing, and other maintenance activities [5]. A circular economy is associated with positive environmental outcomes and continuous economic growth despite resource constraints, and stimulates sustainability-oriented innovation and technologies [61]. Such a model contributes greatly to economic sustainability and tends to serve as an indicator of sustainable entrepreneurship.

The literature review allowed us to conclude that the final dimension of the sustainable entrepreneurship matrix, management, is most closely associated with the following five criteria—awareness of normative regulations and other legal demands; awareness of resource availability; cooperation with institutions, organizations, and stakeholders; sustainability performance indicator measuring; and, lastly, top management's intentions towards sustainability.

Awareness of normative regulations and other legal demands refers to the understanding of the legal context and binding requirements that entrepreneurs must comply with and put in place within business processes. High levels of sustainability will require high governmental engagement; thus, laws are inevitable, and compliance is mandatory [1]. Institutional context contributes to sustainable development by aiding the strengthening of the competitive advantage of a company [62]. The legal framework may work in favor of sustainable entrepreneurship if the company sees the government as a resource rather than an entity that limits the freedom of entrepreneurship. On the contrary, institutional uncertainty may cause further entrepreneurial uncertainty, which is detrimental to the advancement of sustainability performance within a company [63]. Thus, to that understanding, knowledge of the incentives and legal context behind the pure law is highly important.

Awareness of resource availability illustrates entrepreneurs' knowledge of where to access educational or financial support to carry out sustainability objectives within a business. For example, investment in innovative technologies aids sustainable entrepreneurship; however, lack of knowledge on resource availability can hinder further success [22,64]. As a result, the ability to detect resources and opportunities is an attribute of sustainable entrepreneurship.

Cooperation with institutions, organizations, and stakeholders is essential to success in sustainable entrepreneurship [8]. Previous research presents collaboration with stakeholders and other involved groups as a mechanism for the pursuit of sustainability-oriented business practices [65]. Considering the needs of the interested parties is associated with positive social and environmental outcomes [66]. Thus, such practice undeniably points to sustainable entrepreneurship intentions. The larger the variety of actors, the greater the potential to address different entrepreneurial aspects; for example, cooperation with business schools can share knowledge and skills that are necessary to run sustainable business processes [12]. In addition to that, the strengthening of cooperative ties can ensure the ability to carry out sustainable entrepreneurship in the long run [6]. Therefore, it can be concluded that sustainable management practice entails an outlook on a wide range of cooperative opportunities.

It is not only important to implement different sustainability values but also to measure performance. Currently, a company's performance is measured based on financial indicators; however, sustainable entrepreneurship requires close monitoring of non-financial performance [31]. The assessments can be made internally as well as by inviting an external examiner. Regardless, the top management should be aware of the criteria of sustainable entrepreneurship, know how to interpret the results, and hold a department or at least a position that consecutively supervises the processes. For example, key performance indicator analysis may serve as a tool to estimate success in sustainable entrepreneurship [13,22].

Numerous metric systems, indicators, and measurement methods have been developed to estimate specifically sustainability performance [30]. However, the authors argue that most of them fail to provide a thorough evaluation of all sustainability aspects concerning entrepreneurial activities. Some overly focus on environmental aspects, while others disregard social matters; however, the majority tend to cover the basis of the TBL or ESG concepts [67,68]. Thus, the authors suggest that based on the new concept of four sustainable entrepreneurship dimensions and 19 criteria, a new metric system should be developed and used as a resource for self-evaluation purposes in production companies. This way, companies could reduce reliance on audit firms, boost competence in sustainable entrepreneurship, and compare their annual performance.

Finally, the top management's intentions point to the likelihood that the sustainable entrepreneurship model will be integrated into ongoing business processes [2,69]. Equally important is whether the top management has any goals set to promote sustainability and what the management is ready to do to improve performance in the sustainability area [14,70]. As a result, properly oriented attitudes, incentives, and intentions will determine how successfully the sustainability principles will be adopted into the company. Another key aspect that points to intentions to transform towards sustainable entrepreneurship is sustainability orientation [2].

The present research may be subject to several limitations. First, the research topic requires a methodology that can cover the existing literature comprehensively. For example, a formal systematic literature review would have provided insights into the sustainable entrepreneurship criteria in greater depth compared to the monographic method. Second, despite the exhaustive efforts, not all research could have been covered due to limitations in databases. Third, the findings are subject to the personal bias of the researchers; as a result, the interpretation of findings may be subjective and vary depending on the context.

The limitations may be resolved by conducting a comprehensive systematic literature review to enlarge the scope of articles processed in this review and thoroughly examine the available data. Additionally, further reviews would help refine the criteria in greater depth. Besides that, the present article provides an interesting gap for future research. In other words, more studies should be conducted on sustainable entrepreneurship in relation to small and medium enterprises (SMEs) as they make up a significant portion of all enterprises in total. The authors suggest that raising awareness of sustainable entrepreneurial practices among SMEs in the manufacturing industry could make more impactful changes in the transition towards sustainability. Additionally, future research should focus on the generalizability of the findings in an effort to develop a measure that would estimate not only expertise levels in sustainable entrepreneurship but also sustainability performance regardless of the company size and geopolitical context.

Despite the limitations, the present study contributes to the research by identifying gaps in theory and providing directions for future research. The theoretical implications are currently being used in the second research phase as a basis for a semi-quantitative study that explores the expertise levels in sustainable entrepreneurship in the Latvian manufacturing industry. The main goal is to estimate the key areas that highlight educational gaps or lack of expertise. The results will be further used to develop a theoretical matrix that could be applied as a self-evaluation tool for expertise in sustainable entrepreneurship for enterprises that are considering adopting sustainable practices into present business processes. In such a way, entrepreneurs would be able to identify the lacking areas, compare their performance on a quarterly or annual basis, and transition to sustainable entrepreneurship more successfully.

Nonetheless, the transition to sustainable production calls for an overarching organization that could not only assist companies with theoretical and practical resources specifically related to entrepreneurial activities but also provide statistics on the performance comparison among different manufacturing companies, production sectors, and locations. Most importantly, the implications of the present study could be used by entrepreneurs to evaluate their expertise in sustainable entrepreneurship based on the identified criteria. However, a metric system should be developed and tested in order to serve as a resource to entrepreneurs. Distinctively from other metric systems, the criteria could sufficiently cover and balance different aspects of sustainable entrepreneurship while providing more thorough insights without overly focusing on particular dimensions.

5. Conclusions

This literature review provided valuable insights into current findings regarding sustainable entrepreneurship. The authors primarily identified that definitions of sustainable entrepreneurship among different researchers are inconsistent, though few have attempted to clarify the concept which could be used as a common base for future research [3]. In general, the sustainable entrepreneurship concept illustrates a systemically coordinated set of strategic actions that facilitate the integration of environmental, social, economic, and managerial sustainability practices into ongoing business processes without compromising the developmental capabilities of future generations.

Since research has shown great potential for enterprises to mitigate the environmental and social crisis, without a doubt, the transition to a sustainable entrepreneurship model is highly supported. However, this could be a long and complex process, especially within the current capitalism-oriented traditions and commercial mindset among entrepreneurs. Nevertheless, understanding the expertise position in sustainable entrepreneurship would help to address the key areas that need the most attention in order to successfully facilitate the transition to sustainable business practices in the manufacturing industry.

In conclusion, this study provides a comprehensive summary and analysis of sustainable entrepreneurship criteria. It was concluded that the sustainable entrepreneurship concept should include a sustainable management dimension as an indispensable accessory to the TBL framework. The inclusion rests upon three reasons. First, the locus of control of the top management; second, the relationship between the top management action and the outcomes at an operational level; and third, the resourcing and cooperative capabilities of the top management.

Altogether, the listed factors help better illustrate the role of management in the integration of sustainability pillars into production processes and the overall transition to sustainable entrepreneurship. Thus, sustainable entrepreneurship should be reviewed only with regard to all four dimensions—environmental, social, economic, and managerial—to avoid excessive focus on only certain aspects of sustainability and their role in entrepreneurship.

Based on the present data, a theoretical matrix could function as a resource to help promote the expertise levels and competence in sustainable entrepreneurship among entrepreneurs. On top of that, it presents great potential as a performance metric system for entrepreneurs in quarterly and annual sustainability performance evaluations. Nonetheless, the criteria can be adapted for use in future studies.

Author Contributions: Conceptualization, L.T.; methodology, L.T. and S.B.; formal analysis, L.T.; writing—original draft preparation, L.T.; writing—review and editing, L.T. and S.B.; visualization, L.T.; supervision, S.B. All authors have read and agreed to the published version of the manuscript.

Funding: This research received no external funding.

Conflicts of Interest: The authors declare no conflicts of interest.

References

1. Hummels, H.; Argyrou, A. Planteraty Demands: Redefining Sustainable Development and Sustainable Entrepreneurship. *J. Clean. Prod.* **2021**, *278*, 123804. [CrossRef]
2. Thelken, H.N.; de Jong, G. The impact of Values and Future Orientation on Intention Formation within Sustainable Entrepreneurship. *J. Clean. Prod.* **2020**, *266*, 122052. [CrossRef]
3. Farny, S.; Binder, J.K. Sustainable Entrepreneurship. In *World Encyclopedia of Entrepreneurship*, 2nd ed.; Dana, L.P., Ed.; Edward Elgar Publishing: Cheltenham, UK, 2021; pp. 605–611.
4. Mondal, S.; Singh, S.; Gupta, H. Green Entrepreneurship and Digitalization Enabling the Circular Economy through Sustainable Waste Management—An Exploratory Study of Emerging Economy. *J. Clean. Prod.* **2023**, *422*, 138433. [CrossRef]
5. Bocken, N.; Boons, F.; Baldassarre, B. Sustainable Business Model Experimentation by Understanding Ecologies of Business Models. *J. Clean. Prod.* **2019**, *208*, 1498–1512. [CrossRef]
6. Moya-Clemente, I.; Ribes-Giner, G.; Pantoja-Diaz, O. Configurations of Sustainable Development Goals that Promote Sustainable Entrepreneurship over Time. *Sustainability* **2019**, *28*, 762–788. [CrossRef]
7. Tarnanidis, T. Measuring Sustainable Entrepreneurship in the Food Sector. *Int. J. Sustain. Entrep. Corp. Soc. Responsib.* **2020**, *7*, 1–8. [CrossRef]
8. Calderon, A.Y.C.; Carillo, L.F.L.; Contreras, J.L.G. Sustainable Entrepreneurship in Colombia: Strengths and Opportunities. *J. Manag.* **2020**, *36*, 190–203.
9. Nguyen, H.T.T.; Costanzo, L.A.; Karatas-Özkan, M. Stakeholders' Perceptions of Sustainable Entrepreneurship within the Context of a Developing Economy. *J. Small Bus. Manag.* **2023**, *61*, 441–480. [CrossRef]
10. Gu, W.; Mo, W.; Wang, M. Enterprise-level Sustainable Entrepreneurship Index Construction and its Applications. *Int. Entrep. Manag. J.* **2023**, 1–34. [CrossRef]
11. Mansouri, S.; Momtaz, P.P. Financing Sustainable Entrepreneurship: ESG Measurement, Valuation, and Performance. *J. Bus. Ventur.* **2022**, *37*, 106258. [CrossRef]
12. Lee, S.H.; Zhou, Y. The Outlook for Sustainable Development Goals in Business Management: A Systematic Literature Review and Keyword Cluster Analysis. *Sustainability* **2022**, *14*, 11976. [CrossRef]
13. Ali, M.; Malik, M.; Yaqub, M.Z.; Jabbour, C.J.C.; de Sousa Jabbour, A.B.L.; Latan, H. Green Means Long Life—Green Competencies for Corporate Sustainability Performance: A Moderated Mediation Model of Green Organizational Culture and Top Management Support. *J. Clean. Prod.* **2022**, *427*, 139174. [CrossRef]
14. Dhir, A.; Khan, S.J.; Islam, N.; Ractham, P.; Meenakshi, N. Drivers of Sustainable Business Model Innovations. An Upper Echelon Theory Perspective. *Technol. Forecast. Soc. Change* **2023**, *191*, 122409. [CrossRef]
15. Coffay, M.; Bocken, N. Sustainable by Design: An Organizational Design Tool for Sustainable Business Model Innovation. *J. Clean. Prod.* **2023**, *427*, 139294. [CrossRef]
16. Satyadini, A.; Song, L. Modern Entrepreneurship and the 'Doughnut': Productive or Destructive? *Asian-Pac. Econ. Lit.* **2023**, *37*, 119–141. [CrossRef]
17. Saeed, B.B.; Shakir, H.; Afsar, B.; Khan, I. Promoting Employee's Proenvironmental Behavior through Green Human Resource Management Practices. *Corp. Soc. Responsib. Environ. Manag.* **2018**, *26*, 424–438. [CrossRef]

18. European Commission. *Proposal for a Directive of the European Parliament and of the Council on Corporate Sustainability Due Diligence and Amending Directive (EU) 2019/1937*; European Commission: Brussels, Belgium, 2022.
19. Smit, L.; Bright, C.; McCorquodale, R.; Bauer, M.; Deringer, H.; Baeza-Breinbauer, D.; Torres-Cortés, F.; Alleweldt, F.; Kara, S.; Salinier, C.; et al. *Study on Due Diligence through the Supply Chain*; Publications Office of the European Union: Luxembourg, 2020; pp. 1–572.
20. Korporatīvās Ilgtspējas un Atbildības Institūts. Available online: https://www.incsr.eu/novertejumi/ilgtspejas-indekss/ilgtspejas-indekss-2023/ (accessed on 7 May 2024).
21. Huang, Y.; Li, P.; Bu, Y.; Zhao, G. What Entrepreneurial Ecosystem Elements Promote Sustainable Entrepreneurship? *J. Clean. Prod.* **2023**, *422*, 138459. [CrossRef]
22. Roomi, M.A.; Saiz-Alvarez, J.M.; Coduras, A. Measuring Sustainable Entrepreneurship and Eco-Innovation: A Methodological Proposal for the Global Entrepreneurship Monitor (GEM). *Sustainability* **2013**, *13*, 4056. [CrossRef]
23. Jamil, A.; Ghazali, A.M.; Nelson, S.P. The influence of Corporate Governance Structure on Sustainability Reporting in Malaysia. *Soc. Responsib. J.* **2021**, *17*, 1251–1278. [CrossRef]
24. Wasdani, K.P.; Vijaygopal, A.; Manimala, M.J.; Verghese, A.K. Impact of Corporate Governance on Organizational Performance of Indian Firms. *Indian J. Corp. Gov.* **2021**, *14*, 180–208. [CrossRef]
25. Pekovic, S.; Vogt, S. The Fit Between Corporate Social Responsibility and Corporate Governance: The Impact on Firm's Financial Performance. *Rev. Manag. Sci.* **2021**, *12*, 1095–1125. [CrossRef]
26. Suraya, E.; Azis, N.; Majid, M.S.A. Does Performance Matter for Business Survival Based on the Enhancement of Locus of Control and Business Competence. *Sumerianz J. Bus. Manag. Mark.* **2020**, *3*, 98–106.
27. de Freitas Netto, S.V.; Sobral, M.F.F.; Ribeiro, A.R.B.; da Luz Soares, G.R. Concepts and Forms of Greenwashing: A Systematic Review. *Environ. Sci. Eur.* **2020**, *32*, 19. [CrossRef]
28. Aliabadi, V.; Ataei, P.; Gholemrezai, S. Identification of the Relationships among the Indicators of Sustainable Entrepreneurial Ecosystems in Agricultural Startups. *J. Innov. Knowl.* **2022**, *7*, 100245. [CrossRef]
29. Shakeel, J.; Mardani, A.; Chofreh, A.G.; Goni, F.A.; Klemes, J.J. Anatomy of Sustainable Business Model Innovation. *J. Clean. Prod.* **2020**, *261*, 121201. [CrossRef]
30. Saidani, M.; Yannou, B.; Leroy, Y.; Cluzel, F.; Kendall, A. A taxonomy of circular economy indicators. *J. Clean. Prod.* **2019**, *207*, 542–559. [CrossRef]
31. Mendes, A.C.S.; Ferreira, F.A.F.; Kannan, D.; Ferreira, N.C.M.Q.F.; Correia, R.J.C. A BWM Approach to Determinants of Sustainable Entrepreneurship in Small and Medium-Sized Enterprises. *J. Clean. Prod.* **2022**, *371*, 133300. [CrossRef]
32. Safari, A.; Salehzadeh, R.; Panahi, R.; Abolghasemian, S. Multiple Pathways Linking Environmental Knowledge and Awareness to Employees' Green Behavior. *Corp. Gov.* **2018**, *18*, 81–103. [CrossRef]
33. Fryxell, G.E.; Lo, C.W.H. The Influence of Environmental Knowledge and Values on Managerial Behaviors on Behalf of the Environment: An Empirical Examination of Managers in China. *J. Bus. Ethics* **2003**, *46*, 45–69. [CrossRef]
34. Cojoianu, T.F.; Clark, G.L.; Hoepner, A.G.F.; Veneri, P.; Wójcik, D. Entrepreneurs for a Low Carbon World: How Environmental Knowledge and Policy Shape the Creation and Financing of Green Start-ups. *Res. Policy* **2020**, *49*, 103988. [CrossRef]
35. Khan, M.A.S.; Du, J.; Malik, H.A.; Anuar, M.M.; Pradana, M.; Yaacob, M.R.B. Green Innovation Practices and Consumer Resistance to Green Innovation Products: Moderating Role of Environmental Knowledge and Pro-environmental Behavior. *J. Innov. Knowl.* **2022**, *7*, 100280. [CrossRef]
36. Gangi, F.; Daniele, L.M.; Varrone, N. How Do Corporate Environmental Policy and Corporate Reputation Affect Risk-Adjusted Financial Performance? *Bus. Strategy Environ.* **2020**, *29*, 1975–1991. [CrossRef]
37. Alvesson, M. Organizational Culture: Meaning, Discourse, and Identity. In *The Handbook of Organizational Culture and Climate*, 2nd ed.; Ashkanasy, N.M., Wilderom, C.P.M., Peterson, M.F., Eds.; SAGE Publications, Inc.: Los Angeles, CA, USA, 2011; pp. 11–28.
38. Chatman, J.A.; O'Reilly, C.A. Paradigm Lost: Reinvigorating the Study of Organizational Culture. *Res. Organ. Behav.* **2016**, *36*, 199–224. [CrossRef]
39. Rizvi, Y.A.; Garg, R. The Simultaneous Effect of Green Ability-Motivation-Opportunity and Transformational Leadership in Environment Management: The Mediating Role of Green Culture. *Benchmarking Int. J.* **2021**, *3*, 830–856. [CrossRef]
40. Ogiemwonyi, O.; Harun, A.; Hossain, M.I.; Karim, A.M. The Influence of Green Behavior Using Theory of Planned Behavior Approach: Evidence from Malaysia. *Millenn. Asia* **2022**, *14*, 582–604. [CrossRef]
41. Fallah, M.R.; Soori, M. Presenting a Framework for the Successful Entry of Women Entrepreneurs into Green Entrepreneurship. *J. Sci. Technol. Policy Manag.* **2023**, *14*, 467–486. [CrossRef]
42. Xie, Y.; Chen, Z.; Tang, H.; Boadu, F.; Yang, Y. Effects of Executives' Pro-environmental Education and Knowledge Sharing Activities on Eco-friendly Agricultural Production: Evidence from China. *J. Clean. Prod.* **2023**, *395*, 136469. [CrossRef]
43. Hägg, G.; Jones, C. Educating towards the Prudent Entrepreneurial Self—An Educational Journey Including Agency and Social Awareness to Handle the Unknown. *Int. J. Entrep. Behav. Res.* **2021**, *27*, 82–103. [CrossRef]
44. Dooly, M.; Masats, D.; Mont, M. Launching a Solidarity Campaign: Technology-Enhanced Project-Based Language Learning to Promote Entrepreneurial Education and Social Awareness. *J. Technol. Sci. Educ.* **2021**, *11*, 260–269. [CrossRef]
45. Faust, L.; Kolbe, M.; Mansouri, S.; Momtaz, P.P. The Crowdfunding of Altruism. *J. Risk Financ. Manag.* **2022**, *15*, 138. [CrossRef]
46. Lorenzo-Afable, D.; Lips-Wiersma, M.; Singh, S. 'Social' Value Creation as Care: The Perspective of Beneficiaries in Social Entrepreneurship. *Soc. Enterp. J.* **2020**, *16*, 339–360. [CrossRef]

47. Vallaster, C.; Kraus, S.; Kailer, N.; Bladwin, B. Responsible Entrepreneurship: Outlining the Contingencies. *Int. J. Entrep. Behav. Res.* **2019**, *25*, 538–553. [CrossRef]
48. Faisal, S. Twenty-Years Journey of Sustainable Human Resource Management Research: A Bibliometric Analysis. *Adm. Sci.* **2023**, *13*, 139. [CrossRef]
49. Mishra, P. Green Human Resource Management: A Framework for Sustainable Organizational Development in an Emerging Economy. *Int. J. Organ. Anal.* **2017**, *25*, 762–788. [CrossRef]
50. Grabara, J.; Hussain, H.I.; Szajt, M. Sustainable University Development through Sustainable Human Resources and Corporate Entrepreneurship: The Role of Sustainable Innovation and Work Environment. *Amfiteatru Econ.* **2020**, *54*, 480–495. [CrossRef]
51. Kandukuri, V. Aligning Talent and Business: A Key for Sustainable HRM in SMEs. *J. Entrep. Innov. Emerg. Econ.* **2023**, *9*, 80–91. [CrossRef]
52. Rahim, E. Applying the Triple Bottom Line for Corporate Sustainability toward Zero Environmental, Social, and Economic Footprints in Corporate Practice. In *Applied Research Approaches to Technology, Healthcare, and Business*; IGI Global: Hershey, PA, USA, 2023; pp. 121–139.
53. O'Sullivan, N. Sustainable finance. In *Concise Encyclopedia of Corporate Social Responsibility*; Edward Elgar Publishing; Cheltenham, UK, 2024; pp. 196–201.
54. Kraus, S.; Burtscher, J.; Niemand, T.; Roig-Tierno, N.; Syrjä, P. Configurational Paths to Social Performance in SMEs: The Interplay of Innovation, Sustainability, Resources and Achievement Motivation. *Sustainability* **2017**, *10*, 1828. [CrossRef]
55. Akhtar, S.; Arshad, M.A.; Mahmood, A.; Ahmed, A. Spiritual Quotient towards Organizational Sustainability: The Islamic Perspective. *World J. Entrep. Manag. Sustain. Dev.* **2017**, *13*, 163–170. [CrossRef]
56. Tahir, M.; Burki, U. Entrepreneurship and Economic Growth: Evidence from the Emerging BRICS Economies. *J. Open Innov. Technol. Mark. Complex.* **2023**, *9*, 100088. [CrossRef]
57. Wang, Z.; Deng, Y.; Zhou, S.; Wu, Z. Achieving Sustainable Development Goal 9: A Study of Enterprise Resource Optimization Based on Artificial Intelligence Algorithms. *Resour. Policy* **2023**, *80*, 103212. [CrossRef]
58. Tazhibekova, A.; Shametova, A. Ecological Initiative and Their Influence on the Competitiveness and Sustainability of Companies: 'Green' Strategies of SMEs. *J. Knowl. Econ.* **2024**. [CrossRef]
59. Tandon, A.; Chaudhary, S.; Nijjer, S.; Vilamová, Š.; Tekelas, F.; Kaur, P. Challenges in Sustainability Transitions in B2B Firms and the Role of Corporate Entrepreneurship in Responding to Crises Created by the Pandemic. *Ind. Mark. Manag.* **2024**, *118*, 93–109. [CrossRef]
60. Topleva, S.A.; Prokopov, T.V. Integrated Business Model for Sustainability of Small and Medium-Sized Enterprises in the Food Industry, Creating Value Added Through Ecodesign. *Br. Food J.* **2020**, *122*, 1463–1483. [CrossRef]
61. Foroozanfar, M.H.; Imanipour, N.; Sajadi, S.M. Integrating Circular Economy Strategies and Business Models: A Systematic Literature Review. *J. Entrep. Emerg. Econ.* **2022**, *14*, 678–700. [CrossRef]
62. Woolthuis, R.K.; Hooimeijer, F.; Bossink, B.; Mulder, G.; Brouwer, J. Institutional Entrepreneurship in Sustainable Urban Development: Dutch Successes as Inspiration for Transformation. *J. Clean. Prod.* **2013**, *50*, 91–100. [CrossRef]
63. D'Andrea, F.A.M.C. Entrepreneurship and Institutional Uncertainty. *J. Entrep. Public Policy* **2023**, *12*, 10–31. [CrossRef]
64. Urbaniec, M. Sustainable Entrepreneurship: Innovation-Related Activities in European Enterprises. *Pol. J. Environ. Stud.* **2018**, *27*, 1773–1779. [CrossRef]
65. Karatas-Ozkan, M.; Tunalioglu, R.; Ibrahim, S.; Ozeren, E.; Grinevich, V.; Kimaro, J. Actioning Sustainability through Tourism Entrepreneurship: Women Entrepreneurs as Change Agents Navigating through the Field of Stakeholders. *Cent. Eur. Manag. J.* **2024**, *32*, 31–56. [CrossRef]
66. Cucino, V.; Marullo, C.; Annunziata, E.; Piccaluga, A. The Human Side of Entrepreneurship: An Empirical Investigation of Relationally Embedded Ties with Stakeholders. *Manag. Res. Rev.* **2023**, *47*, 1–21. [CrossRef]
67. Pislaru, M.; Herghiligiu, I.V.; Robu, I.B. Corporate sustainable performance assessment based on fuzzy logic. *J. Clean. Prod.* **2019**, *223*, 998–1013. [CrossRef]
68. Martínez, M.P.; Cremasco, C.P.; Gabriel Filho, L.R.A.; Junior, S.S.B.; Bednaski, A.V.; Quevedo-Silva, F.; Padgett, R.C.M.L. Fuzzy Inference System to Study the Behavior of the Green Consumer Facing the Perception of Greenwashing. *J. Clean. Prod.* **2020**, *242*, 116064. [CrossRef]
69. Kimuli, S.N.L.; Orobia, L.; Sabi, H.M.; Tsuma, C.K. Sustainability Intention: Mediator of Sustainability Behavioral Control and Sustainable Entrepreneurship. *World J. Entrep. Manag. Sustain. Dev.* **2020**, *16*, 81–95. [CrossRef]
70. Meshram, S.A.; Rawani, A.M. Entrepreneurial Success Measures and Factors for Sustainable Entrepreneurship. *Int. J. E-Entrep. Innov.* **2019**, *9*, 15–34. [CrossRef]

processes
MDPI

Article

Implementations of Digital Transformation and Digital Twins: Exploring the Factory of the Future

Ramin Rahmani [1,2,*], Cristiano Jesus [1,3,4] and Sérgio I. Lopes [1,3,5]

1 CiTin—Centro de Interface Tecnológico Industrial, 4970-786 Arcos de Valdevez, Portugal; cristiano.jesus@citin.pt (C.J.); sergio.lopes@citin.pt (S.I.L.)
2 proMetheus—Instituto Politécnico de Viana do Castelo (IPVC), 4900-347 Viana do Castelo, Portugal
3 ADiT-Lab, Instituto Politécnico de Viana do Castelo (IPVC), 4900-347 Viana do Castelo, Portugal
4 ALGORITMI Research Center, School of Engineering, Production and Systems Department, University of Minho, 4800-058 Guimarães, Portugal
5 IT—Instituto de Telecomunicações, Campus Universitário de Santiago, 3810-193 Aveiro, Portugal
* Correspondence: ramin.rahmani@citin.pt or ramin.rahmaniahranjani@gmail.com

Abstract: In the era of rapid technological advancement and evolving industrial landscapes, embracing the concept of the factory of the future (FoF) is crucial for companies seeking to optimize efficiency, enhance productivity, and stay sustainable. This case study explores the concept of the FoF and its role in driving the energy transition and digital transformation within the automotive sector. By embracing advancements in technology and innovation, these factories aim to establish a smart, sustainable, inclusive, and resilient growth framework. The shift towards hybrid and electric vehicles necessitates significant adjustments in vehicle components and production processes. To achieve this, the adoption of lighter materials becomes imperative, and new technologies such as additive manufacturing (AM) and artificial intelligence (AI) are being adopted, facilitating enhanced efficiency and innovation within the factory environment. An important aspect of this paradigm involves the development and utilization of a modular, affordable, safe human–robot interaction and highly performant intelligent robot. The introduction of this intelligent robot aims to foster a higher degree of automation and efficiency through collaborative human–robot environments on the factory floor and production lines, specifically tailored to the automotive sector. By combining the strengths of human and robotic capabilities, the future factory aims to revolutionize manufacturing processes, ultimately driving the automotive industry towards a more sustainable and technologically advanced future. This study explores the implementation of automation and the initial strides toward transitioning from Industry 4.0 to 5.0, focusing on three recognized, large, and automotive companies operating in the north of Portugal.

Keywords: Industry 4.0; collaborative robots; digital transformation; digital twins; factory of future; hybrid vehicles; strategic roadmap

1. Introduction

From AI-powered robots streamlining production lines to 3D printing minimizing waste, the convergence of transformative technologies is revolutionizing industries. This clears the path for a future where factories with unprecedented efficiency, sustainable practices become the norm, and remote collaboration unlocks new possibilities. This paper explores the interconnected domains of digital transformation (DTR) and digital twins (DTWs), exploring their important roles in shaping the factory of the future (FoF). At the heart of this transformation lie the Internet of things (IoT) and the industrial Internet of things (IIoT), offering a conduit to usher in the circular economy (CE). However, the emergence of the Green IoT paradigm, focused on energy reduction and the carbon footprint, finds itself in juxtaposition with the advent of edge artificial intelligence (AI) [1]. Industry 5.0 represents the next phase in manufacturing evolution, fusing automation, digitalization, and human ingenuity to cultivate intuitive, interactive, and sustainable work environments. Moreover, advanced fabrication approaches like additive manufacturing (AM), prominently known as 3D printing, emerge as a cornerstone technology enabling customization, efficiency, and waste reduction compared to subtractive manufacturing [2]. To navigate the complexities of Industry 4.0 in small- and medium-sized enterprises (SMEs), the formulation of tailored maturity models (MMs) assumes significance [3,4].

The global upheaval caused by the COVID-19 pandemic has not only catalyzed a pattern shifts in day-to-day business operations but has also accelerated digital maturity aspirations. Manufacturing entities are expediting their DTR endeavors and raising their digital readiness levels (DRLs) in response to the pandemic's impact [5]. The imperatives of "Digitize or drown!" and "Innovate or die!" underscore the essence of DTR as a driving force. As Industry 5.0 gains traction, the convergence of technology-focused Industry 4.0 with smart logistics and smart factories amplifies research endeavors [6]. This exploration extends to diverse industries' adoption of DTR, unveiling how enterprises embrace novel technologies for their distinct business landscapes [7]. The concept of the DTW emerges as a bridge between virtual simulations and real-world operations. This seamless exchange of data between digital and physical twins (PTWs) in real time empowers operational efficiency, smart manufacturing (SM), and informed decision-making [8]. Exploring the evolution of DTWs from their origins to their projected future is crucial for understanding the benefits they offer to specific industries. This analysis aims to unify various manifestations and definitions of DTWs found across the literature, simplifying the identification of DTWs amid other related terms like "product avatar", "digital thread", "digital model", and "digital shadow" [9]. Furthermore, the symbiosis of building information modeling (BIM) and DTWs holds the potential to revolutionize digital construction across the lifecycle [10].

The evolution of robotic technology introduces new dimensions to manufacturing, addressing productivity and ergonomic concerns. Collaborative robots (cobots) stand as exemplars of Industry 4.0 progress, offering enhanced productivity, adaptability, and safety. However, efficient design remains a challenge, as ergonomic assessment tools are not tailored to these devices [11,12]. A facet of this concept involves human–robot collaboration (HRC), which aims to synergize the finest capabilities of humans and robots to jointly accomplish shared tasks. For instance, an assembly process necessitates a fusion of dexterity, intelligence, and sensing skills. Collaborative robotics emerges as a potential remedy for addressing the issue of musculoskeletal disorders (MSDs) within the industrial domain. The utilization of cobots has the potential to revolutionize manufacturing by addressing efficiency and ergonomic concerns. Exploring their applications and showcasing advancements in human–robot collaboration (HRI) in the coming years hold immense potential [13]. In contrast to traditional industrial robots, collaborative robots offer heightened productivity, flexibility, adaptability, and safety. These user-friendly machines feature intuitive interfaces, allowing even those without formal programming experience to easily adjust settings and functionally [14].

The FoF is designed to embrace circular economy principles and take advantage of sensors and AI from Industry 4.0 to optimize resource usage and minimize waste, leading to a more sustainable production environment. This paper focuses on case studies and assessment tools to extend circular principles across product lifecycles, facilitating the robust evaluation and monitoring of circular efficiency [15]. The urgency for manufacturers to embrace Industry 4.0's transformative potential is undeniable, yet the uniqueness of each business necessitates tailor-made strategic roadmaps (SRs) for a successful transition [16]. The primary facilitative element of the IoT concept lies in the amalgamation of diverse technologies and communication solutions. The next generation of intelligent systems integrates identification and tracking technologies, sensor and actuator networks (both wired and wireless), advanced communication protocols for ultra-fast data exchange, and distributed intelligence for smarter decision-making at the device level. This combination enables real-time monitoring, optimized resource management, and automated processes, and improved communication for the next-generation Internet (NGI), leading to significant improvements in efficiency and functionality [17]. Amid the IoT model, the fusion of diverse technologies shows the approach for novel solutions. Further research is crucial for defining the precise boundaries and application frameworks of Industry 4.0, despite valuable insights gleaned from European strategic roadmaps (ESRs) and the broader scientific literature [18]. The fusion of radio frequency identification (RFID), cyber-physical systems (CPS), IoT, and data mining propels Industry 4.0 towards novel forms of personalization and shared value creation [19]. In this rapidly evolving landscape, DTWs transcend the industrial area and venture into various sectors, heralding efficiencies in terms of time and cost. The convergence of DTR and DTW technologies heralds a revolution with the potential to reshape multiple facets of our lives [20]. This paper undertakes a comprehensive survey of the digital transformation and digital twins paradigms, unveiling their profound influence on the regional and international FoF. The automation of processes emerges as a key enabler, leading to increased productivity, improved production control, and a decrease in difficulty for workers, thereby enabling better process management.

The current case study encompasses various topics, such as Industry 4.0, safety and ergonomics, digital twins, and lightweight additive manufacturing, which contribute to the overarching vision of the FoF. The diverse examples presented herein are drawn from three automotive companies located in northern Portugal, ranging from internal logistics to mobile intralogistics, utilizing intelligent robot and automated vehicles. These include the hierarchical pyramid implementation of DTR and DTW, real-time analysis of physical objects, virtual replication in HRC definition, and adoption of automation and the early steps taken towards transitioning from Industry 4.0 to 5.0. The present investigation has a specific focus on three well-established automotive companies in northern Portugal. These companies have a global reputation for manufacturing car components (Section 2), seat leathers (Section 3), and technology-driven vehicles (Section 4). The focused aims for each of the sections are as follows: internal logistic utilization using mobile robots, the human-centered mechanism design for collaborative robots, and the virtual modeling of factories with simulations and tests.

2. Configuration of Digital Transformation

Digital transformation is the process of using technology to fundamentally change how businesses operate, deliver value to customers, adapt to the evolving digital landscape, and improve efficiency and innovation. On the other hand, a digital twin is a virtual replica or representation of a physical object, system, or process [21]. It is created using digital data and simulations to mimic the real-world counterpart. Digital twins help monitor, analyze, and optimize physical assets and processes, enabling better decision-making and performance improvements in various industries like manufacturing. Within this digital transformation, digital twins play a key role, and they are closely related concepts. Digital transformation is the broader process of using technology to change how organizations operate and deliver value. Digital twins replicate the digital counterparts of physical assets or processes, created as a part of digital transformation efforts. In essence, digital twins are a powerful tool within the digital transformation toolbox, enabling businesses to make more informed decisions, enhance efficiency, and drive innovation by exploiting the insights and simulations they provide.

2.1. Logistic Challenges

The traceability system (TS) presents a promising avenue for potential improvement, employing digital technologies to enhance the monitoring of the production process and internal logistics. It ensures the authenticity and immutability of data, supported by a digital control and monitoring platform. It is essential to note that the company is not responsible for the external transportation of final products to customers, and this falls outside the scope of the supply chain. As illustrated in Figure 1, the focus of traceability lies in individually limited internal logistics for the automobile parts/assembly producer company. Nevertheless, traceability can be extended to cover the entire journey, up to the point of product delivery to the customer. Heavy and bulky components like steel rails, wind turbine blades, and large motors significantly impede the technical challenge of traceability. Tracking the origin, history, and location of these massive parts becomes complex, potentially leading to increased costs, safety hazards, and maintenance difficulties. This challenge also presents an opportunity for innovation in the field. This case study suggests the implementation of collaborative human–robot technologies, which require the following components:

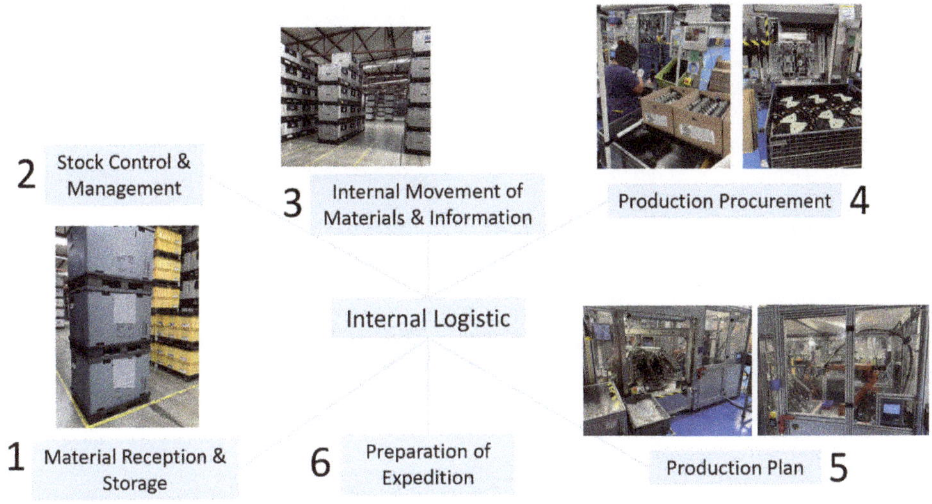

Figure 1. Internal logistic demonstration.

–Utilization of logistics trains equipped with AGV to facilitate the supply of workstations with components and transport lifters to the warehouse.

It is notable that IMR and AGV are both types of mobile robots, but they differ in their functionality and operation. IMRs are more versatile, adaptable, and intelligent, with advanced sensors, cameras, and mapping. They can perform a wider range of tasks in various environments with a higher degree of autonomy and safety features. IMRs can be programmed or trained to perform a variety of tasks, including inspection, logistics, and collaborative work with humans. On the other hand, AGVs are more specialized, fixed-route robots, following a line or path on the ground for navigation, or magnetic strips, wires, or QR codes designed for material transport in factories and warehouses. Moreover, IMRs are often equipped with advanced safety systems to avoid collisions with humans, whereas AGVs are designed to operate in segregated areas and may not have safety features. The development of an autonomous mobile robots with a perception-based navigation system is currently underway, and, coupled with the challenge of designing complex architecture and movement, there is a need for designers to construct it to operate effectively and compactly across various mediums [22,23].

–Enhanced inventory control of both the warehouse and in-process components through the cross-validation of data from the manufacturing execution system (MES), lifters assembled at each station, and transported items in the warehouse. Currently, operators are manually recording production by barcode after the completion of small batches. The integration of the MES with the maintenance system to ensure seamless data exchange and co-ordination is necessary.

–Development of a digital platform that displays real-time information about the internal movement of vehicles, component production, in-process stock levels, machine stoppages due to material shortages, and stoppages caused by breakdowns and malfunctions.

–Implementation of personal protective equipment (PPE) control, including the use of an object detection or RFID algorithm to issue alerts in non-serious cases and control machine operation in mandatory cases. The detection algorithm will generate alerts in non-serious cases when a worker is not wearing the required PPE but will not prevent the worker from continuing work. However, machine actuation control for mandatory cases will be employed when the use of PPE is essential to prevent serious accidents, ensuring that the system prevents a machine from starting if a worker is not wearing the appropriate PPE.

The roadmap for integrating digitalization into the factory of the future and additive manufacturing in Industry 5.0 involves the seamless convergence of advanced fabrication technologies [24]. This journey emphasizes the benefit of automation to transform manufacturing processes. Additionally, additive manufacturing (3D metal/plastic printing) will play a key role in I 4.0 and I 5.0 by enabling on-demand, customized production, reducing waste, decreasing weight, and enhancing product design flexibility [25]. The goal is to create highly efficient and sustainable manufacturing systems that are interconnected and capable of producing complex, personalized products with minimal environmental impact.

2.2. Intralogistics Approach

1. Industry 5.0 exemplifies itself in an innovative internal logistics system for factories. The system utilizes lightweight, retractable, and portable racks integrated with an autonomous mobile robot (AMR) for efficient material movement. This setup highlights allied automation through the collaboration of the AMR and the racks. The use of potentially porous aluminum structures, combined with digitalization, further embodies Industry 5.0's principles. In-house research and development are feasible for smaller-scale simulation. It creates an agile, interconnected, and data-driven internal logistics system that can adapt to the evolving needs of the factory, ultimately leading to increased efficiency and productivity. Mobile intralogistics progress is defined by below terms and shown in Figure 2.

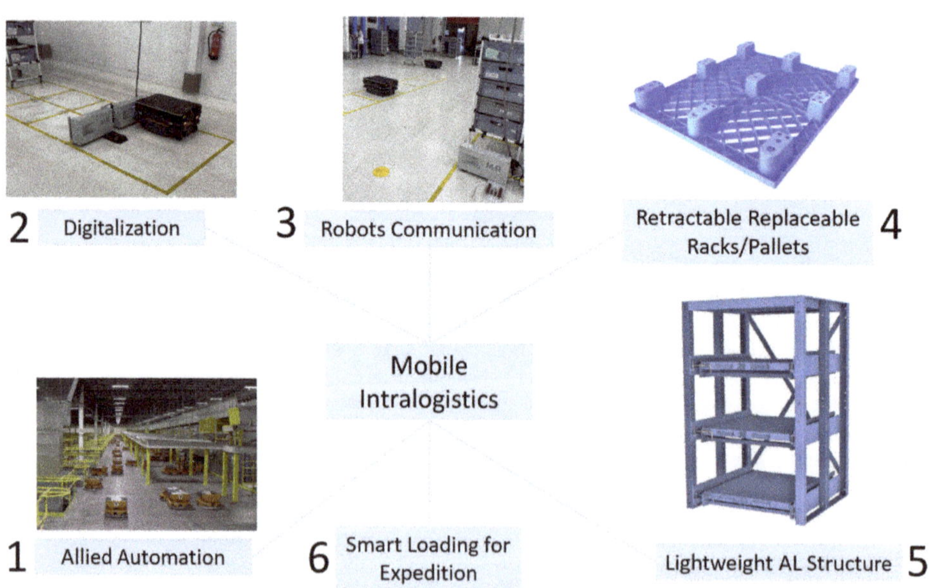

Figure 2. Mobile intralogistics.

Allied automation—This solution prepares allied automation principles to optimize the internal logistics process. By combining a mobile AMR robot with smart racks, the system ensures a seamless and coordinated flow of materials and products within the factory.

Aluminum structures: The lightweight racks and pallets are constructed from advanced aluminum structures (potentially porous), which are not only durable but also exceptionally lightweight. This design minimizes the overall weight of the system, allowing for the greater efficiency and mobility of the AGV/IMR robot.

Digitalization—It is at the core of this setup. Each rack is equipped with sensors and RFID tags for real-time tracking (RTT) and inventory management. The mobile robot/cobot is equipped with sensors and cameras for navigation and safety. All the data from these sensors are integrated into a digital platform, providing real-time visibility into the logistics process. This enables better decision-making, route optimization, communication, and predictive maintenance.

Industry 5.0—This solution aligns with the tenets of Industry 5.0 by creating a highly flexible and interconnected logistics system. AGV/IMR robots, guided by AI and machine-learning (ML) algorithms, can adapt to changing production demands. The porous aluminum racks are modular and can be reconfigured easily to accommodate different products, promoting customization and rapid retooling.

2. The use of metal additive manufacturing (MAM) and wire arc additive manufacturing (WAAM) in lightweight jig and fixture design (JFD) represents a transformative shift in traditional manufacturing practices. This innovative approach allows for the creation of highly customized and complex jigs and fixtures with significant advantages:

Design freedom for complex geometries—MAM offers unparalleled design freedom in situ. Engineers can create intricate, lightweight, and optimized structures that were previously impossible or too costly to produce using conventional methods [26]. The technology enables the production of complex geometries that precisely match the workpiece, ensuring a more secure and accurate fit. This is especially valuable for intricate or irregularly shaped components.

Reduce lead times and increase cost efficiency—Additive manufacturing drastically reduces lead times. Design modifications and iterations can be quickly implemented, making it ideal for rapid prototyping and adapting to design changes. While the initial investment in metal 3D printing equipment can be high, it often results in cost savings in the long run. This is particularly true for low-volume, high-complexity applications where the tooling and setup costs of traditional methods can be substantial. Customized jigs and fixtures result in better alignment and reduced setup times, ultimately improving manufacturing efficiency and accuracy.

3. The process of designing cutting-edge solutions for robot arms or grippers specialized in handling hard or soft materials is targeted. The primary goal is to enable the gentle and damage-free small-space picking and placing of various materials. It can use a combination of compact articulated joints (CAJ) and advanced computer vision. It should employ precise end-effectors with sensors to detect and adapt to object orientation, enabling it to navigate within confined areas while ensuring damage-free manipulation mechanisms for different stiffnesses and hardness ranges of materials. The fabrication of these parts is firmly rooted in the field of digitalization and relies on additive manufacturing techniques. The below outcomes have been followed for implementation in this stage:

- Using compact, collapsible, and easily transportable shelving units integrated with an AMR or AGV for efficient mobile intralogistics operations, facilitating the transportation of stocks and productions.
- Utilizing MAM to create lightweight jigs and fixtures offers advantages in rapid prototyping, on-site fabrication, cost and time savings, and adaptability to evolving design requirements and changes.
- Design of specialized robot arms or grippers tailored for the gentle handling of both hard and soft materials, enabling damage-free pick-and-place operations with the use of compact articulated joints suitable for confined spaces.

2.3. Mobile and Cloud Computing

The hierarchical pyramid of main parameters for DTR from bottom to top is leveled and illustrated in Figure 3. This pyramid illustrates how digital transformation is the overarching concept, with technology infrastructure and data collection as its foundation. DTW, HRC, and mobile intralogistics represent specific areas of focus within the transformation, and all these components collectively contribute to optimization and decision-making at the highest level.

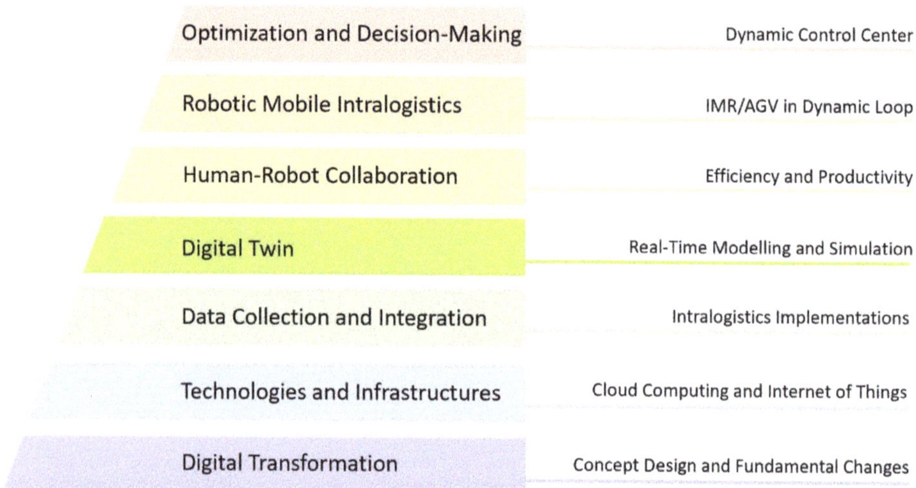

Figure 3. Hierarchy of digital transformation (DTR) based on digital twins (DTWs).

Level 1—At the base of the pyramid, the overarching concept of digital transformation encompasses the entire process.

Level 2—Infrastructures and technologies, such as IoT sensors, communication networks, and cloud computing, form the foundation for digital transformation.

Level 3—Gathering data from various sources, including robots and intralogistics processes, is a crucial step in the digital transformation journey.

Level 4—Digital twins are a key component, enabling real-time modeling and simulation of physical systems [27].

Level 5—Human–robot collaboration is an integral part of the transformation, focusing on how humans and robots work together efficiently [28].

Level 6—Mobile robots play a specific role in intralogistics, representing the application of technology to transport goods within a facility.

Level 7—At the top of the pyramid, optimization and decision-making take center stage, along with data capitalization, digital twins, and collaborative efforts to make informed choices that improve efficiency, productivity, and overall operations.

3. Configuration of Digital Twins

3.1. Physical and Virtual Systems

Key challenges to implementation include the adoption of new technologies such as additive manufacturing (AM), novel materials like metal–plastic composites, the integration of artificial intelligence (AI), the use of IMR/AGV, efficient warehouse space management, and, notably, the transition from human operators to robots with effective human–robot interaction.

Human–robot collaboration (HRC) partnerships are an approach for adaptable workgroups combining human and robot capabilities. These partnerships are leading the way in creating efficient and flexible teams that leverage both human and robot strengths. Robots are enhancing operations by rectifying human errors in factory settings, resulting in increased production speed and uninterrupted precision. The configuration of DTWs in HRC work is depicted in Figure 4, and the terms are explained here as bullet points.

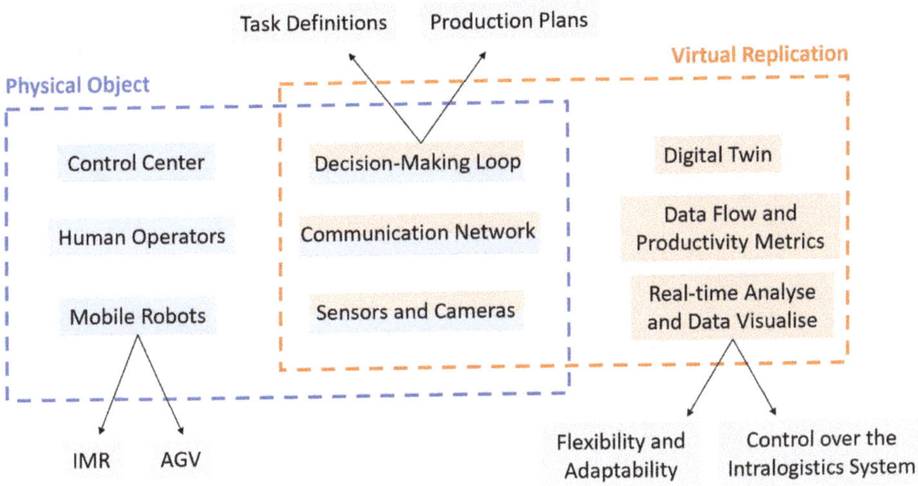

Figure 4. Digital twins (DTWs) in human–robot collaboration (HRC).

Human operator—The human worker actively participates in the intralogistics process, overseeing and interacting with robots and cobots to ensure smooth operations and increase productivity [29].

IMR/AGV robots—IMR is an autonomous mobile unit specialized in intralogistics tasks, capable of navigating and handling materials efficiently, while AGV, another mobile automation solution, is designed for material transport and contributes to streamlined intralogistics operations.

Digital twin—The digital twin is a virtual replica of both IMR and AGV robots, allowing for the real-time monitoring, analysis, and simulations of their actions and surroundings.

Sensors—Sensors on the robots collect data from the physical world, including information on the location, obstacles, and environmental conditions.

Communication network—The network facilitates data exchange between robots and their digital twins, enabling seamless real-time communication.

Data flow and productivity metrics—Data flow from physical robots to the digital twin, providing critical insights for decision-making and control. Moreover, performance metrics and data visualization tools help gauge improvements in efficiency and productivity due to the DTW influence.

Control center—The control center is the interface where operators or AI systems interact with the digital twin, monitoring and managing robot activities [30].

Real-time analytics—This tool processes data from the digital twin, offering actionable insights and performance feedback. The digital twin enables adaptive responses to change conditions, ensuring the intralogistics system remains flexible and efficient.

Decision-making loop—Decisions are made based on digital twin data, influencing the actions and behaviors of the IMR/AGV robots in a dynamic loop. In fact, the decision-making loop acts as the dynamic control center that bridges the gap between production plans and the execution of tasks by robots and humans.

Digital transformation (DTR) can be implemented across various vehicle component manufacturers. For instance, this study focuses on examining a leather company that specializes in crafting car seats and interior designs. The diverse materials used in car seats, such as foams, fabrics, and leathers, pose technical challenges in handling and transport to avoid damages. Implementing the leather defect inspection (LDI) helps reduce waste, especially with expensive leathers. To address these challenges, utilizing data collection technologies like MES, RFID, and computer vision is advised to identify and rectify tracking errors for items and personnel. Utilizing warehouse management systems, including RFID for RTT, aids in efficient material and product storage and shipping. Employing a blend of technologies ensures data accuracy. Collaborative tech implementation, along with intralogistics trains (IMR and AGV), aids in reducing the carbon footprint by facilitating the supply of raw materials, movement between workstations, and transportation to the warehouse. Note that the carbon footprint is the total amount of greenhouse gas, specifically carbon dioxide, emitted directly or indirectly by human activities, impacts in global warming and climate change, and its reduction stands as a key objective within Industry 5.0.

3.2. RFID-Based Smart Factory

MES is a software solution that helps manage and control the production process on the factory area and shop floor of a manufacturing facility. It provides real-time information about manufacturing operations, enabling better decision-making, increased efficiency, and improved overall performance. Therefore, cloud MES processing based on RFID configuration utilizes RFID technology for tracking and managing manufacturing processes [31]. The layers from factory to user are depicted in Figure 5. RFID tags on items communicate with the system through radio waves, allowing for real-time data collection and analysis in a cloud-based environment, promoting efficiency and visibility in the manufacturing process. For MES systems that operate based on the cloud, instead of being hosted on local servers within a factory, the MES software and data are stored and processed on remote servers accessible through the Internet. This allows for more flexibility, scalability, and, often, cost savings. Moreover, RFID technology uses radio waves to identify and track objects. In a manufacturing context, RFID tags are attached to products, components, or equipment. These tags contain data that can be wirelessly read by RFID readers and, consequently, process MES software.

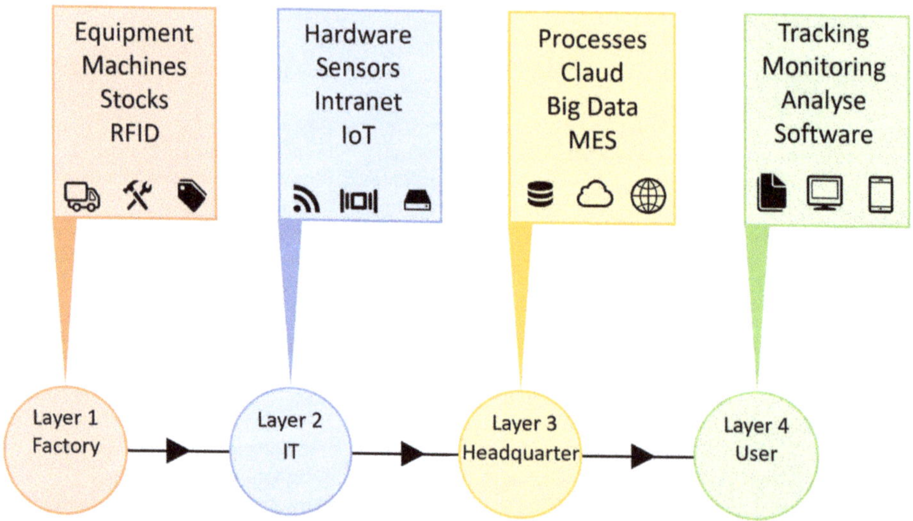

Figure 5. Radio frequency identification (RFID) configuration for manufacturing execution system (MES) processing.

Integrating digital twin technology and Industry 5.0 effectiveness in a company that uses leather for car seat production can significantly enhance efficiency, reduce waste, and improve overall operations. By integrating digital twin technology, RFID-based processes, automated machine vision, and other Industry 5.0 principles, your company can enhance its competitiveness, reduce waste, and improve the overall quality of leather car seat production. Here are some steps and considerations to involve digital twins and digital transformation through Industry 5.0 in car seat manufacturing companies.

DTW implementation—It starts with the creation of a DTW (developing a digital replication) of the manufacturing process, including the entire supply chain, from leather sourcing to the final product. Later, RTT and monitoring encompass implementing sensors and IoT devices to collect real-time data from the manufacturing floor. These data can include temperature, humidity, machine status, cutting design, and other relevant parameters.

RFID-based manufacturing process—Embedding RFID tags on each pack or piece of leather and integrating RFID readers into the manufacturing process help in tracking and monitoring the movement of leather throughout the production line. Integrate RFID data with a Cloud MES to centralize and analyze information. This can provide insights into the production process, reduce errors, and enhance traceability.

Automated machine vision for LDI—Utilize machine vision systems equipped with cameras to inspect leather for defects during the manufacturing process and connect this vision system to the digital twin for real-time defect analysis (based on size and location of defects). This allows for immediate corrective action and reduces the likelihood of defects reaching the final product.

HRC—While the pick-and-place process for leathers is currently carried out by humans, we need to consider introducing robots to this task to reduce the risk of damage during transport and enhance precision. It requires robots to ensure that they work in collaboration with human workers, especially in tasks that require dexterity and attention to detail.

Data analyses and maintenance—Analyze the data collected from sensors and IoT devices to identify patterns, optimize processes, and improve overall efficiency, and, then, implement predictive maintenance based on the data collected to reduce downtime and enhance the lifespan of machinery. Moreover, providing the necessary training to employees for the newly implemented technologies ensures a smooth transition and maximizes the benefits of digital transformation.

Sustainability considerations—Using data from the digital twin and machine vision systems to optimize the cutting process, minimizing waste and maximizing the use of materials, consider the environmental impact of your manufacturing processes and explore ways to reduce the carbon footprint, such as recycling leather scraps [32]. It is worth noting that digital transformation is an ongoing, iterative process. Continuously gather feedback, monitor performance, and adjust to improve efficiency and effectiveness.

For a detailed application and specific example, a case study of the mentioned configuration is employed to address the challenge of manual labor in picking and packing leather, replacing it with robots. The activities and deliverables goals for the leather company in north of Portugal include process identification and material priority in the production lines of the fabrics and handling (pick-and-place) of single and multiple layers. It is a regional automotive company with a global network of partnerships and distribution for car manufacturers worldwide. The *Lectra* machine handles the material cutting, optimization of usage, and positioning of papers and stacks; however, the detachment of cut parts and stacks and the pick-and-place process are performed by human operators. To automate these tasks, the implementation requires artificial vision for detecting part positions, a robotic arm operating independently of the operator, and the ability to recognize and handle single-layer leathers from multilayered ones [33]. The gripper idea of pick-and-place is illustrated in Figure 6, and the defect detection of precious leathers in Figure 7.

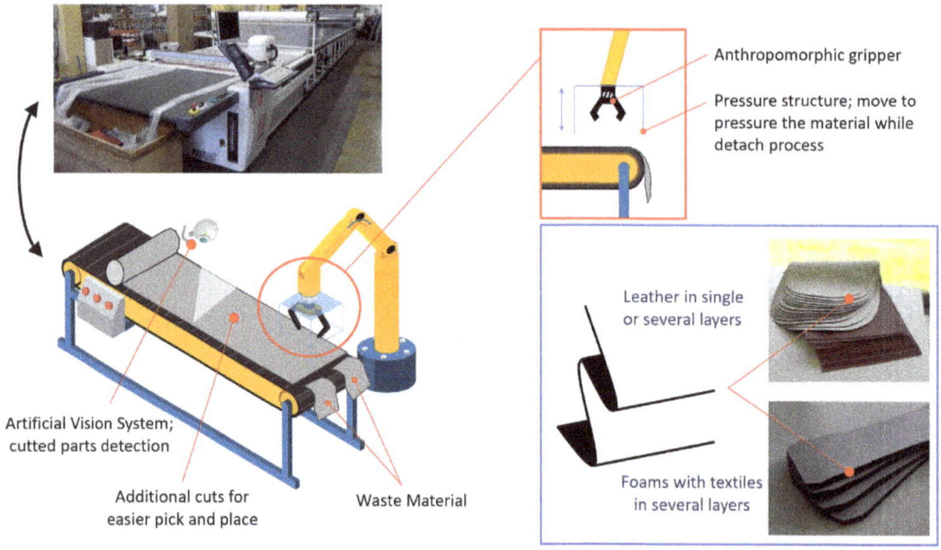

Figure 6. Detach using press and grasp, applied for human-oriented design of collaborative robots in leather-cutting process.

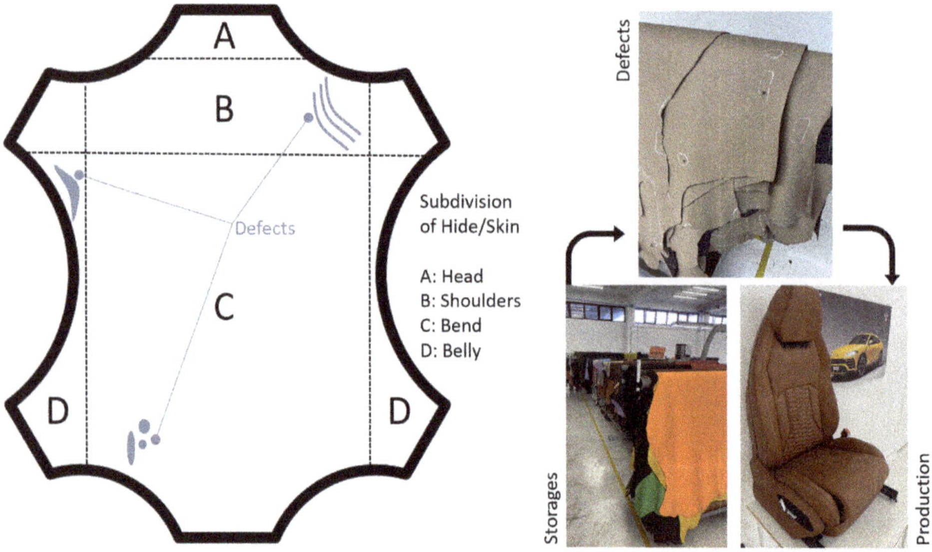

Figure 7. Leather defect detection and classification importance. Visual inspection methods by humans in the beginning can be replaced by HRC and RFID during the process. These effects occur due to different reasons, e.g., raw hides and animal wounds, damage during transport, moisture, folds, wrinkles, and dirt.

The concept of a green factory will function as a research, demonstration, and learning platform, benefiting manufacturing companies, especially those involved in advanced manufacturing such as AM [34]. This can be achieved through the fusion of advanced technologies with traditional ones, as indicated in [35], or it can relate to sustainability in the food and agriculture industries [36]. Green technology aims to achieve smarter, healthier, and more sustainable digitalization, aligning with the promotion of DTR and DTWs. Therefore, it is crucial that we integrate environmentally friendly steps to reduce ecological impact and waste.

4. Technology-Driven in Light Vehicles Industries

4.1. Manufacturing Execution Methodology

The integration of systems, applications, and products (SAP) and MES in industrial management processes is crucial for achieving an efficient manufacturing operation in the third company, focusing on light vehicles, mobility, and supply chain management. Both SAP and MES play distinct but complementary roles in the manufacturing ecosystem [37]. SAP is responsible for managing records related to materials, bills of materials (BOMs), and orders; in contrast, MES oversees the documentation of records like work in progress. There are two developments that can improve the management of the system here. First, SAP extended warehouse management (EWM) with an extended range of functions is a solution for organizing the supply chain, and, second, serving simultaneously enterprise resource planning (ERP) and MES systems for synchronized and efficient manufacturing operations [38]. This development is shown in Figure 8. The key aspects of integrating SAP EWM and MES ERP systems are data flow and communication, master data synchronization, order and production planning, quality and inventory management, real-time monitoring, reporting and maintenance, and user training and support and changes.

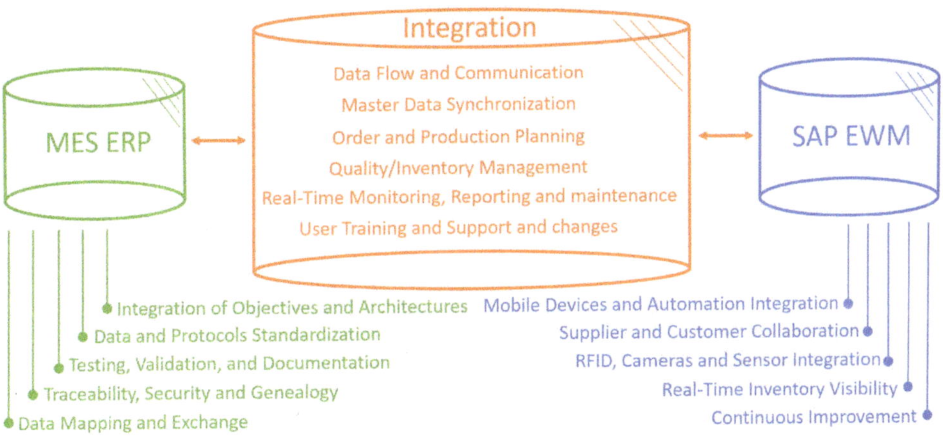

Figure 8. Integration of MES enterprise resource planning (ERP) and SAP extended warehouse management (EWM) systems.

The integration of SAP EWM and MES ERP systems is crucial for seamless and efficient operations in a manufacturing environment [39]. By addressing these key factors, organizations can enhance the integration of SAP EWM and MES ERP systems, leading to improved efficiency, accuracy, and overall performance in both warehouse management and manufacturing execution processes. Here are key factors to consider for a successful integration:

Data flow, consistency, and accuracy—Ensure that the data shared between SAP EWM and MES ERP is consistent and accurate. Inconsistencies can lead to errors, delays, miscommunications, and inefficiencies.

Standardized data formats and protocols—Define standardized data formats and communication protocols for data exchange between SAP and MES to ensure compatibility and smooth integration. Moreover, harmonize or synchronize master data, including product definitions, production schedules, and inventory data, to avoid discrepancies and ensure a single version of truth across systems.

Real-time data exchange—Aim for real-time data exchange to enable quick decision-making and responsiveness to changes in the manufacturing environment. It includes testing and validation, user training, continuous monitoring and optimization, and AGV/cobots movements. Provide training to end-users on the integrated system and implement change management strategies to facilitate a smooth transition. This is crucial for user acceptance and the effective utilization of the integrated solution. Conduct thorough testing, including unit testing, integration testing, and end-to-end testing, to identify and resolve any issues before deploying the integrated solution into a production environment, and regularly optimize the integration process based on feedback and evolving business needs.

Business process alignment—Design the integration solution to be scalable, accommodating future growth and changes in business processes. Ensure that the integration architecture is flexible enough to adapt to evolving business requirements. Regarding the scalability (ability of system to handle a growing amount of work or workload) and flexibility (ability of system to adapt easily to changes, whether they are changes in the business environment, customer requirements, or technological advancements), align business processes between SAP EWM and MES ERP to ensure that workflows are synchronized and there is a flow of information between warehouse management and manufacturing execution.

The digital transformation aims to establish a digital twin, i.e., horizontal and vertical digital processes that function as a virtual mesh adjacent to the real world that is both collaborative and responsive to change and immensely volatile needs. The main objective of a mobility industry player in northern Portugal, as illustrated in Figure 9, is to reduce supply chain bottlenecks. However, despite its advanced technological maturity, achieving this goal hinges on overcoming the challenge of integrating various technologies to ensure seamless interoperability.

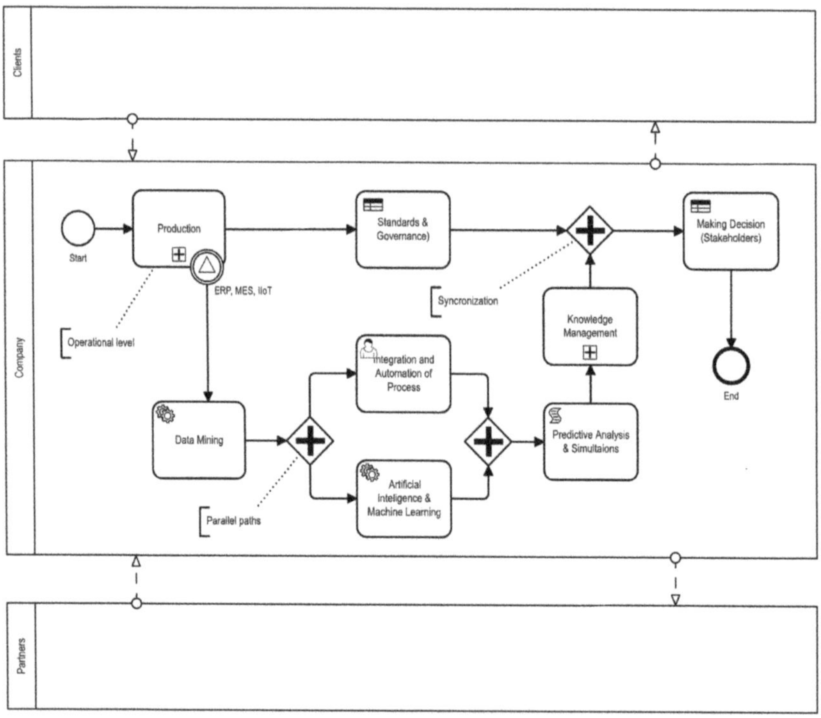

Figure 9. Schematic of future of a light vehicle industry to accelerate mobility solutions.

Therefore, process orchestration is the goal, so that various concepts harmonize to create a symphony of efficiency and productivity. At the center of the factory is production, supported by ERP systems that manage resources and workflow and MES that acts on manufacturing operations, while RPA automates repetitive tasks, freeing up employees for more strategic activities. GenAI injects artificial intelligence into production, optimizing processes and preventing failures. ISO guarantees quality through international standards, while Standards and Governance establish the rules that govern the symphony. Predictive data analysis provides valuable insights for decision-making, while knowledge management preserves and shares accumulated knowledge. With these capabilities, in addition to monitoring and control, the digital twin makes it possible to create a virtual model of the factory, allowing for simulations and tests before implementation in the real world.

The quadruple helix model (QHM) is a concept used in innovation studies and regional development to describe the interaction among four key stakeholders in the innovation process. It builds upon the triple helix model, which includes universities, industries, and governments as the main actors in innovation [40,41]. The QHM adds the fourth helix, civil society. These four helices in the QHM are as follows:

Academia, universities, and research institutions—The first helix represents the traditional knowledge creation and dissemination institutions, such as publications.

Industry and businesses—The second helix relates to the private sector, including businesses and corporations that drive economic development and innovation through commercialization.

Government and public sector—The third helix shows governmental bodies and agencies that create policies, regulations, and infrastructure to support innovation and economic growth.

Civil society—The fourth helix is citizens, non-governmental organizations (NGOs), and community groups who are engaged in the innovation process. They provide input, demand accountability, and contribute to the social context of innovation.

The QHM, specially, emphasizes the importance of collaboration and interaction among the technology institutes and industries for sustainable and innovative digitalization. Then, with the oversight and funding from government agencies, NGOs attend on creating a virtual model of the factory, complete with unit simulations.

4.2. Technology Roadmap and Industry Impact

The phases of industrialization have marked significant shifts in the way societies and economies operate. Here is a brief definition of each phase from Industry 3.0 to Industry 6.0, and it is worth mentioning that our study only emphasizes on satisfying the 4.0 and 5.0 revolution's condition and its transition. These phases of industrialization reflect the evolution of technology and its impact on manufacturing processes, and it is demonstrated in Figure 10. Each phase has brought about transformative changes, leading to increased efficiency, productivity, and innovation in the industrial sector.

Industry 3.0, digital revolution—This phase was characterized by the adoption of computerization and automation in industrial processes [42]. It began in the late 20th century and saw the integration of electronic systems for manufacturing and control. The introduction of computers and programmable logic controllers (PLCs) in this revolution increased efficiency and precision in manufacturing.

Industry 4.0, smart factory—Industry 4.0 builds on the foundation of Industry 3.0 and involves the widespread use of smart technology, sustainability functions, IoT, and CPS [43]. It aims to create smart factories and interconnected systems, integration of IoT, big data, AI, and ML into manufacturing, increased connectivity, real-time data analysis, and customization of products.

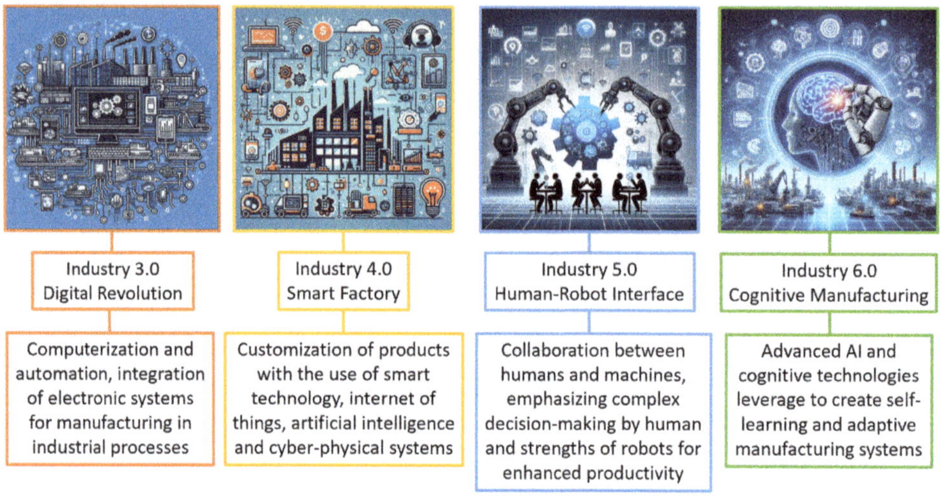

Figure 10. Evolution in industrialization: transition from Industry 4.0 to Industry 5.0.

Industry 5.0, human–robot interface—Industry 5.0 focuses on the collaboration between humans and machines, emphasizing the role of human workers alongside advanced robotic systems. This phase aims to combine the strengths of both humans and machines for enhanced productivity, creating a framework centered around the key enabling technologies driving Industry 5.0 [44]. Advanced robotics with a focus on human–robot collaboration, an emphasis on creativity, problem-solving, and complex decision-making by human workers are key features.

Industry 6.0, cognitive manufacturing—It represents the upcoming wave of industrialization characterized by cognitive manufacturing [45]. This phase utilizes advanced AI and cognitive technologies to create self-learning and adaptive manufacturing systems. Autonomous and self-optimizing systems powered by advanced AI algorithms and machines capable of learning from experience, making decisions, and adapting to dynamic production environments are expected in this revolution.

Based on this brief industrial phase description, currently, we are mostly in Industry 4.0 with an advancement in digital transformation and moving slowly to Industry 5.0 with the utilization of digital twins. Industry 5.0 is an emerging concept that builds upon Industry 4.0 but with a stronger focus on human–robot collaboration and autonomous vehicles.

The three aforementioned sections focused on companies globally renowned for manufacturing and mass-producing car components. These sections, respectively, focus on internal logistic utilization using mobile robots for doors and windows' component assembly factory (Section 2), the human-centered mechanism design for collaborative robots in the seat leather company (Section 3), and the virtual modeling of factories with the ability to conduct test simulations for a multi-technology-driven company (Section 4), with a focus on both theory and implementation. The investigation into automation within worldwide industries located in the northern Portugal region was grounded in local potentials. The graphical presentation of theories and definitions concerning this digitalization has been simplified for process explanation and is depicted in Figures 1–10 herein. In addition to this strength, the research has been limited by the inability to name or publish photos of the companies. Subsequent progress in related factories will be monitored as part of this ongoing work.

5. Conclusions

The survey into the implementation of digital transformation and digital twins in the field of the factory of the future has showcased a remarkable shift in manufacturing paradigms. The foundation of this transformative process extends from internal logistics to the integration of mobile intralogistics, applying the capabilities of IMR and AGV robots. This integration, coupled with the real-time analysis of physical objects and the virtual replication inherent in human–robotic interaction, has ushered in a new era of industrial innovation. The adoption of Industry 4.0 to 5.0 approaches has yielded tangible benefits, leading to a substantial increase in productivity, sustainability, and efficiency across production lines. The reduction in material waste, coupled with improved production control, attests to the positive impact of these advancements. Notably, the transition from traditional robotic systems to advanced technologies, such as detach-and-grasp and pick-and-place robots for a leather company and artificial vision for defect detection, underscores the commitment to enhancing precision and minimizing reliance on human operators. Moreover, the exploration of additive manufacturing, encompassing metal 3D printing and PLA/ABS polymer printing, has emerged as a pivotal subset for rapid prototyping within the manufacturing ecosystem. The versatility and speed offered by these technologies hold the promise of revolutionizing on-site prototyping processes. Finally, in the context of the future factory for the light vehicle industry, artificial intelligence plays a significant role in optimizing production processes by enabling predictive maintenance, efficient resource allocation, and development of autonomous vehicles.

Author Contributions: Conceptualization, R.R. and C.J.; methodology, R.R., C.J. and S.I.L.; investigation, R.R. and C.J.; writing—original draft preparation, R.R.; writing—review and editing, C.J. and S.I.L.; supervision, S.I.L. All authors have read and agreed to the published version of the manuscript.

Funding: This work received funding from Missão Interface, an operation that offers public base funding for Technology and Innovation Centers (CTI) with project code of N.° 03/C05-i02/2022 as part of the Portuguese Plano de Recuperação e Resiliência (PRR). Ramin Rahmani was funded by operation NORTE-06-3559-FSE-000226, Norte Portugal Regional Operational Program (NORTE 2020), under the PORTUGAL 2020 Partnership Agreement, through the European Social Fund (ESF). This work was also developed within the scope of the proMetheus—Research Unit on Materials, Energy, and Environment for Sustainability project, FCT Ref. UID/05975/2020, financed by national funds through the FCT/MCTES.

Institutional Review Board Statement: The authors declare that they have no known competing financial interests or personal relationships that could have appeared to influence the work reported in this paper.

Data Availability Statement: Data are contained within the article.

Conflicts of Interest: The authors declare no conflicts of interest.

References

1. Fraga-Lamas, P.; Lopes, S.I.; Fernández-Caramés, T.M. Green IoT and Edge AI as Key Technological Enablers for a Sustainable Digital Transition towards a Smart Circular Economy: An Industry 5.0 Use Case. *Sensors* **2021**, *21*, 5745. [CrossRef] [PubMed]
2. Rahmani, R.; Karimi, J.; Resende, P.R.; Abrantes, J.C.C.; Lopes, S.I. Overview of Selective Laser Melting for Industry 5.0: Toward Customizable, Sustainable, and Human-Centric Technologies. *Machines* **2023**, *11*, 522. [CrossRef]
3. Jesus, C.; Lima, R.M. Literature Search of Key Factors for the Development of Generic and Specific Maturity Models for Industry 4.0. *Appl. Sci.* **2020**, *10*, 5825. [CrossRef]
4. Mittal, S.; Khan, M.A.; Romero, D.; Wuest, T. A critical review of smart manufacturing & Industry 4.0 maturity models: Implications for small and medium-sized enterprises (SMEs). *J. Manuf. Syst.* **2018**, *49*, 194–214. [CrossRef]
5. Jones, M.D.; Hutcheson, S.; Camba, J.D. Past, present, and future barriers to digital transformation in manufacturing: A review. *J. Manuf. Syst.* **2021**, *60*, 936–948. [CrossRef]
6. Jafari, N.; Azarian, M.; Yu, H. Moving from Industry 4.0 to Industry 5.0: What Are the Implications for Smart Logistics? *Logistics* **2022**, *6*, 26. [CrossRef]
7. Kutnjak, A.; Pihiri, I.; Furjan, M.T. Digital Transformation Case Studies Across Industries—Literature Review. In Proceedings of the 2019 42nd International Convention on Information and Communication Technology, Electronics and Microelectronics (MIPRO), Opatija, Croatia, 20–24 May 2019. [CrossRef]

8. Singh, M.; Srivastava, R.; Fuenmayor, E.; Kuts, V.; Qiao, Y.; Murray, N.; Devine, D. Applications of Digital Twin across Industries: A Review. *Appl. Sci.* **2022**, *12*, 5727. [CrossRef]
9. Singh, M.; Fuenmayor, E.; Hinchy, E.P.; Qiao, Y.; Murray, N.; Devine, D. Digital Twin: Origin to Future. *Appl. Syst. Innov.* **2021**, *4*, 36. [CrossRef]
10. Honghong, S.; Gang, Y.; Haijiang, L.; Tian, Z.; Annan, J. Digital twin enhanced BIM to shape full life cycle digital transformation for bridge engineering. *Autom. Constr.* **2023**, *147*, 104736. [CrossRef]
11. Bilberg, A.; Malik, A.A. Digital twin driven human–robot collaborative assembly. *CIRP Ann.* **2019**, *68*, 499–502. [CrossRef]
12. Maurice, P.; Padois, V.; Measson, Y.; Bidaud, P. Human-oriented design of collaborative robots. *Int. J. Ind. Ergon.* **2017**, *57*, 88–102. [CrossRef]
13. Liu, L.; Guo, F.; Zou, Z.; Duffy, V.G. Application, Development and Future Opportunities of Collaborative Robots (Cobots) in Manufacturing: A Literature Review. *Int. J. Hum.-Comput. Interact.* **2022**, *40*, 915–932. [CrossRef]
14. Sherwani, F.; Asad, M.M.; Ibrahim, B.S.K.K. Collaborative Robots and Industrial Revolution 4.0 (IR 4.0). In Proceedings of the International Conference on Emerging Trends in Smart Technologies (ICETST), Karachi, Pakistan, 26–27 March 2020. [CrossRef]
15. Rossi, J.; Bianchini, A.; Guarnieri, P. Circular Economy Model Enhanced by Intelligent Assets from Industry 4.0: The Proposition of an Innovative Tool to Analyze Case Studies. *Sustainability* **2020**, *12*, 7147. [CrossRef]
16. Ghobakhloo, M. The future of manufacturing industry: A strategic roadmap toward Industry 4.0. *J. Manuf. Technol. Manag.* **2018**, *29*, 910–936. [CrossRef]
17. Atzori, L.; Iera, A.; Morabito, G. The Internet of Things: A survey. *Comput. Netw.* **2010**, *54*, 2787–2805. [CrossRef]
18. Santos, C.; Mehrsai, A.; Barros, A.C.; Araújo, M.; Ares, E. Towards Industry 4.0: An overview of European strategic roadmaps. *Procedia Manuf.* **2017**, *13*, 972–979. [CrossRef]
19. Wang, Y.; Ma, H.-S.; Yang, J.-H.; Wang, K.-S. Industry 4.0: A way from mass customization to mass personalization production. *Adv. Manuf.* **2017**, *5*, 311–320. [CrossRef]
20. Erol, T.; Mendi, A.F.; Doğan, D. Digital Transformation Revolution with Digital Twin Technology. In Proceedings of the 4th International Symposium on Multidisciplinary Studies and Innovative Technologies (ISMSIT), Istanbul, Turkey, 22–24 October 2020. [CrossRef]
21. Qian, C.; Liu, X.; Ripley, C.; Qian, M.; Liang, F.; Yu, W. Digital Twin—Cyber Replica of Physical Things: Architecture, Applications and Future Research Directions. *Future Internet* **2022**, *14*, 64. [CrossRef]
22. Fusic, S.J.; Sugumari, T. A Review of Perception-Based Navigation System for Autonomous Mobile Robots. *Recent Pat. Eng.* **2023**, *17*, 13–22. [CrossRef]
23. Sierra-García, J.E.; Santos, M. Mechatronic Modelling of Industrial AGVs: A Complex System Architecture. *Complexity* **2020**, *2020*, 6687816. [CrossRef]
24. Gunasekaran, J.; Sevvel, P.; John Solomon, I. Metallic materials fabrication by selective laser melting: A review. *Mater. Today Proc.* **2021**, *37*, 252–256. [CrossRef]
25. Suresh, A.; Udendhran, R.; Yamini, G. Internet of Things and Additive Manufacturing: Toward Intelligent Production Systems in Industry 4.0. In *Internet Things Industry 4.0*; Springer: Cham, Switzerland, 2019; pp. 73–89. [CrossRef]
26. Rahmani, R.; Lopes, S.I.; Prashanth, K.G. Selective Laser Melting and Spark Plasma Sintering: A Perspective on Functional Biomaterials. *J. Funct. Biomater.* **2023**, *14*, 521. [CrossRef] [PubMed]
27. Segovia, M.; Garcia-Alfaro, J. Design, Modeling and Implementation of Digital Twins. *Sensors* **2022**, *22*, 5396. [CrossRef]
28. Bonci, A.; Cheng, P.D.C.; Indri, M.; Nabissi, G.; Sibona, F. Human-Robot Perception in Industrial Environments: A Survey. *Sensors* **2021**, *21*, 1571. [CrossRef] [PubMed]
29. Nahavandi, S. Industry 5.0—A Human-Centric Solution. *Sustainability* **2019**, *11*, 4371. [CrossRef]
30. Santhi, A.R.; Muthuswamy, P. Industry 5.0 or industry 4.0S? Introduction to industry 4.0 and a peek into the prospective industry 5.0 technologies. *Int. J. Interact. Des. Manuf.* **2023**, *17*, 947–979. [CrossRef]
31. Wang, C.; Chen, X.; Soliman, A.H.A.; Zhu, Z. RFID Based Manufacturing Process of Cloud MES. *Future Internet* **2018**, *10*, 104. [CrossRef]
32. Kumar, R.; Kariminejad, A.; Antonov, M.; Goljandin, D.; Klimczyk, P.; Hussainova, I. Progress in Sustainable Recycling and Circular Economy of Tungsten Carbide Hard Metal Scraps for Industry 5.0 and Onwards. *Sustainability* **2023**, *15*, 12249. [CrossRef]
33. Aslam, M.; Khan, T.M.; Naqvi, S.S.; Holmes, G.; Naffa, R. On the Application of Automated Machine Vision for Leather Defect Inspection and Grading: A Survey. *IEEE Access* **2019**, *7*, 176065–176086. [CrossRef]
34. Gebbe, C.; Hilmer, S.; Götz, G.; Lutter-Günther, M.; Chen, Q.; Unterberger, E.; Glasschröder, J.; Schmidt, V.; Riss, F.; Kamps, T.; et al. Concept of the Green Factory Bavaria in Augsburg. *Procedia CIRP* **2015**, *32*, 53–57. [CrossRef]
35. Rahmani, R.; Karimi, J.; Kamboj, N.; Kumar, R.; Brojan, M.; Tchórz, A.; Skrabalak, G.; Lopes, S.I. Fabrication of localized dia-mond-filled copper structures via selective laser melting and spark plasma sintering. *Diam. Relat. Mater.* **2023**, *136*, 109916. [CrossRef]
36. Hassoun, A.; Prieto, M.A.; Carpena, M.; Bouzembrak, Y.; Marvin, H.J.P.; Pallarés, N.; Barba, F.J.; Bangar, S.P.; Chaudhary, V.; Ibra-him, S.; et al. Exploring the role of green and Industry 4.0 technologies in achieving sustainable development goals in food sectors. *Food Res. Int. Part B* **2022**, *162*, 112068. [CrossRef] [PubMed]
37. Govindaraju, R.; Putra, K. A methodology for Manufacturing Execution Systems (MES) implementation. *IOP Conf. Ser. Mater. Sci. Eng.* **2016**, *114*, 012094. [CrossRef]

38. Choi, B.K.; Kim, B.H. MES (manufacturing execution system) architecture for FMS compatible to ERP (enterprise planning system). *Int. J. Comput. Integr. Manuf.* **2002**, *15*, 274–284. [CrossRef]
39. Reinkemeyer, L. *Process Mining in Action*; Springer: Berlin/Heidelberg, Germany, 2020. [CrossRef]
40. Craiut, L.; Bungau, C.; Bungau, T.; Grava, C.; Otrisal, P.; Radu, A.-F. Technology Transfer, Sustainability, and Development, World-wide and in Romania. *Sustainability* **2022**, *14*, 15728. [CrossRef]
41. Yun, J.J.; Liu, Z. Micro- and Macro-Dynamics of Open Innovation with a Quadruple-Helix Model. *Sustainability* **2019**, *11*, 3301. [CrossRef]
42. Tantawi, K.H.; Sokolov, A.; Tantawi, O. Advances in Industrial Robotics: From Industry 3.0 Automation to Industry 4.0 Collaboration. In Proceedings of the 2019 4th Technology Innovation Management and Engineering Science International Conference (TIMES-iCON), Bangkok, Thailand, 11–13 December 2019; pp. 1–4. [CrossRef]
43. Ghobakhloo, M. Industry 4.0, digitization, and opportunities for sustainability. *J. Clean. Prod.* **2020**, *252*, 119869. [CrossRef]
44. Leng, J.; Sha, W.; Wang, B.; Zheng, P.; Zhuang, C.; Liu, Q.; Wuest, T.; Mourtzis, D.; Wang, L. Industry 5.0: Prospect and retrospect. *J. Manuf. Syst.* **2022**, *65*, 279–295. [CrossRef]
45. Chourasia, S.; Tyagi, A.; Pandey, S.M.; Walia, R.S.; Murtaza, Q. Sustainability of Industry 6.0 in Global Perspective: Benefits and Challenges. *MAPAN* **2022**, *37*, 443–452. [CrossRef]

processes

Article

Circularity and Digitalisation in German Textile Manufacturing: Towards a Blueprint for Strategy Development and Implementation

Tina Wiegand [1] **and Martin Wynn** [2,*]

[1] Business Department, Hof University of Applied Sciences, Alfons-Goppel-Platz 1, 95028 Hof, Germany; tina.wiegand@hof-university.de
[2] School of Business, Computing and Social Sciences, University of Gloucestershire, Cheltenham GL502RH, UK
* Correspondence: mwynn@glos.ac.uk

Abstract: Textile and clothing (T&C) production and consumption have been steadily increasing for many years, accompanied by significant resource consumption, waste generation, and environmental impact. Germany holds a pivotal position in the industry, both within the EU and globally, as a leading producer, importer, and exporter of T&C. The circular economy (CE) concept has emerged as a promising solution to address the industry's negative impacts by emphasising the reduction, reuse, and recycling of resources across the value chain. Digital technologies (DTs) are increasingly recognised as key enablers and facilitators of this transition, promoting both efficiency and circularity in manufacturing processes. However, the extent to which these are implemented in the German T&C industry remains largely underexplored. This study seeks to investigate the role of DTs in advancing CE practices within German T&C manufacturing. Employing process mapping and technology profiling, this research adopts an inductive, qualitative approach based on primary interview data to explore three key areas: (1) current CE practices, (2) the deployment of supportive DTs, and (3) the challenges in implementing the CE and DTs. The findings reveal that CE adoption is set to accelerate among T&C manufacturers, driven by regulatory compliance and market pressures, with DTs playing a critical role in enabling this transition. This study proposes a new operational framework designed to facilitate the shift towards a more circular textile sector, aligning with the ambitious goals of the European Union and broader environmental and societal objectives.

Keywords: circular economy; CE; digital technologies; digitalisation; textile and clothing industry; manufacturing industry; strategic blueprint; value chain; 9-Rs

1. Introduction

In recent decades, there has been a noticeable surge in textile and clothing (T&C) production and consumption, leading to environmental harm and resource depletion as goods are often discarded after a short use period, with minimal recycling [1]. This trend has positioned the T&C industry as the world's fourth-largest consumer of primary raw materials and water, and the fifth-largest emitter of greenhouse gases [2]. In response, the circular economy (CE) has emerged as a potential way forward that could address these challenges [3]. The CE aims to reduce, reuse, recycle, and recover resources and energy throughout the value chain to minimise waste and pollution [4].

While the CE concept is not new, it has gained momentum in recent years, featuring prominently in sustainability agendas at both governmental and supranational levels, such as in the United Nations Sustainable Development Goals [5] and as an element of the European Union's Green Deal [6]. The European Commission [3] recognised digital technology (DT) as a crucial facilitator for the comprehensive transition from a linear to a more circular economy.

There has also been growing interest in the integration of digitalisation and the CE in recent years [7], with findings increasingly highlighting the significant potential of DTs for companies to reduce resource and energy consumption, as well as waste generation and emissions, while improving efficiency and productivity [8–11].

The T&C industry is a key focus area within the European Union's Circular Economy Action Plan [12]. New regulatory measures include mandatory eco-design specifications, supply chain transparency, the implementation of a digital product passport (DPP), and the reduction in carbon and environmental footprints [13]. European companies are thus facing increasing pressure to reassess business models, products, and processes to align with circularity and sustainability principles, while also addressing competitive advantage and stakeholder interests [14–16].

However, the introduction of sustainable and circular processes, as well as the role of DTs in advancing circularity within the industry, remains largely underexplored in the scientific literature [17–19]. A recent comprehensive literature review by Neri et al. [7] revealed that research on the relationship between the CE and DTs in the manufacturing sector is highly fragmented, highlighting the "need for clearer guidance for industrial practitioners on properly using digital technologies to support the implementation of the circular economy" (page 1). Additionally, contextual factors such as sector-specific characteristics appear to play a significant role, but have scarcely been addressed. Notably, only 2 out of 170 analysed studies explicitly focus on the textile industry.

To address this gap, this article examines the relationship between the adoption of DTs and the transition to the CE within the German T&C manufacturing industry. Specifically, the research questions (RQs) addressed in this paper are as follows:

RQ1. How is the CE currently implemented in German T&C manufacturing companies?

RQ2. How are DTs being used to support and advance CE practices in German T&C manufacturing companies?

RQ3. What are the challenges and critical issues for current and future CE transition and digitalisation in the German T&C manufacturing sector?

The remainder of this paper is organised as follows: In Section 2, the relevant literature is reviewed and a provisional conceptual framework for the primary research is set out. Section 3 then describes the research design and methods applied in this study. Section 4 presents the results of the empirical research, and Section 5 discusses these results and emergent themes. Finally, Section 6 summarises the contribution of this research, discusses its limitations, and further points out some future research directions.

2. Literature Review

2.1. The German T&C Industry

Although the terms textiles and clothing are often used interchangeably, they are delineated and defined here for the purposes of this study. According to the classification of economic activities by the European Union (NACE) [20], the production and processing of textile fibres, yarns, and fabrics, as well as the manufacturing of industrial, technical, and home textiles, are categorised under the textile industry (NACE code 13). The clothing industry, on the other hand, encompasses the manufacturing of apparel such as outerwear, underwear, and workwear (NACE code 14).

The T&C value chain involves numerous actors and complex processes, making it one of the most globalised value chains today [21]. From an end-to-end perspective, eight main processes can be identified [1,22,23] (Figure 1). First, in fibre production, raw materials (whether natural or synthetic) are processed into fibres. This is followed by yarn production, where fibres are spun into yarns and prepared for further processing. The next process is fabric production, where yarns are transformed into woven, knitted, or nonwoven fabrics. The wet processing stage involves treatments such as washing, dyeing, or finishing textile products and materials. Subsequently, these materials are further processed into market-ready end products, representing the final step in manufacturing. The downstream processes, including distribution and retail, use, and end-of-life (EoL) treatment, are covered for completeness here but are not considered part of the core manufacturing processes. Throughout these processes, human and natural resources, as well as energy, are consumed and, as a result, waste is generated and emissions are released into the environment, among other impacts. It is thus evident that a remodelling of the T&C manufacturing processes can potentially make a significant contribution to addressing global environmental and waste challenges.

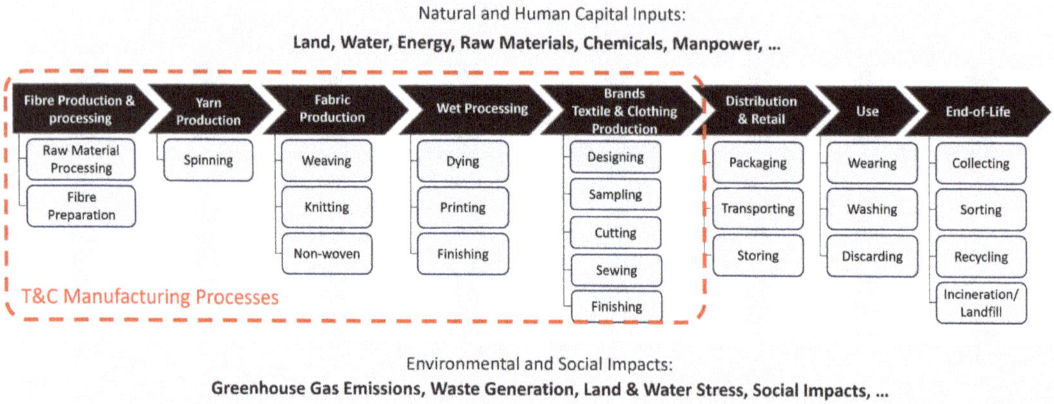

Figure 1. T&C value chain process clusters. Source: Adapted from [1,22,23].

Germany holds a central position both within the EU as the second-largest producer of textile products and the largest importer of clothing, as well as globally being the second-largest importer of clothing and the fifth-largest importer of textiles [24,25]. Moreover, Germany also plays a pivotal role on the global stage as a major exporter of textiles and clothing [25,26]. The emerging social, environmental, and waste problems resulting from these geographically fragmented flows underline the need for a holistic transformation of the T&C industry, shifting from the linear "take–make–dispose" model towards closed-loop textile systems. Gözet and Wilts [1] see Germany as a key player and "change-maker" in this effort.

The T&C industry in Germany had 879 companies in 2023, employing approximately 83,000 people. Of these companies, 689 (78%) are in the textile sector (NACE-13: manufacture of textiles) and 190 (22%) in the clothing sector (NACE-14: manufacture of clothing). The German T&C industry today has regional clusters in North Rhine-Westphalia (224 companies), Bavaria (190 companies), Baden-Württemberg (158 companies), and Saxony (120 companies), of which 94 percent are small and medium-sized enterprises (SMEs) with fewer than 250 employees [27].

2.2. The Circular Economy in T&C Manufacturing

The concept of the CE has evolved over recent decades. Originally focused on waste management, it has evolved and expanded into a systemic approach centred on resource preservation [28]. The CE envisions an industrial system that is restorative and regenerative, aiming to minimise waste and pollution while maximising resource efficiency [29,30]. This vision requires a paradigm shift toward reducing, reusing, recycling, and recovering materials, supported by technological innovation and collaboration among stakeholders along value chains [4]. Two main perspectives for CE implementation can be discerned in the literature, as illustrated in the left and right side of Figure 2. Firstly, from a resource-based perspective, three key strategies can be identified: narrowing, slowing, and closing resource loops [31,32]. According to Bocken et al. [31], "Narrowing" concerns reducing resource consumption related to the product, production, and use process and minimising the environmental impact—also referred to in other sources as resource efficiency [33,34] or eco-efficiency [35]. "Slowing" aims at extending the use and reuse of products and components, which can be achieved by developing durable products and maintaining the value of resources by extending the lifespan of materials, products, and machinery. "Closing" strategies focus on closing resource loops through material recycling and energy recovery. All these strategies emphasise the efficient use and value retention of resources, materials, and products.

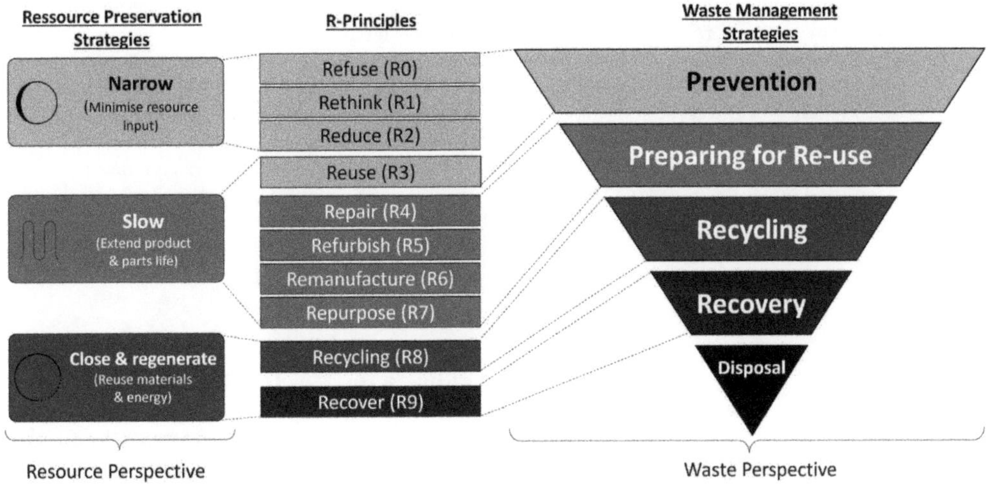

Figure 2. The R-principles and their relationship with resource preservation and waste management strategies. Source: Based on [4,31,32,36].

Secondly, from a waste-based perspective, five waste management strategies are identified that prioritise the prevention and proper treatment of waste [36]. The integration of these perspectives into actionable practices, often referred to as the R-principles, builds the core to guide implementation of the CE concept (centre of Figure 2) [4,28,37,38]. Strategies R0 to R2 primarily focus on the absolute reduction in resource consumption and the avoidance of negative environmental and social impacts through products and processes, making these principles essential for achieving the goals of the CE. Strategies R3 to R7 primarily aim to extend the lifespan of existing products and components. Strategies R8 and R9, known as traditional waste management, involve material recycling and energy recovery from waste, which are seen as the least desirable [28,37]. Furthermore, the use of renewable energy and emissions avoidance are central practices of the CE [29].

Recent research studies have increasingly focused on the practical implementation of CE strategies and practices across various sectors. However, for the T&C industry, existing studies remain fragmented. Franco's study [16] was one of the first to address the T&C industry, adopting a product-based perspective. The author highlights the unique complexity of textile products and value chains as key challenges to implementing circular design strategies, underscoring the need for further industry-specific studies to facilitate a smoother transition for companies in the sector. Country-specific studies from Italy [39] and Sweden [40] emphasise that a combination of diverse strategies and practices among all value chain actors, coupled with strong cooperation and collaboration, are critical for a successful CE transformation. In this context, digital technologies are recognised as enablers of circularity, and yet their role remains largely unexplored. Despite these contributions, there is a notable lack of country-specific studies addressing the CE transformation of European T&C companies. Wiegand and Wynn's analysis [17] provided some insights into the status of CE implementation, but gave an incomplete picture of the industry. Indeed, the authors highlight the need for further research based on industry case studies to investigate strategies and best practices to facilitate the transition to the CE. This study aimed to address that gap by examining the practices implemented in the German T&C manufacturing sector to narrow, slow, and close resource flows, contributing to the development of a more sustainable and circular industry.

2.3. Digital Technology Deployment in T&C Manufacturing

As previously noted, DTs are recognised as facilitators and enablers in CE transformation. A recent review [7] of the relationship between DTs and CE highlights the fragmented nature of the current research landscape. Existing studies tend to focus either on specific technologies or particular CE practices. Furthermore, industry-specific differences have been identified, with very few studies concentrating on the T&C industry. Therefore, Table 1 provides an overview of the current and potential applications of key digital technologies to advance circularity within the T&C industry. The technologies discussed are commonly categorised under the acronyms SMAC (social media, mobile, analytics, cloud) and BRAID (blockchain, robotics, artificial intelligence, Internet of Things, digital fabrication). As shown in the literature sources (Table 1, column 3), all nine SMAC/BRAID technologies are considered relevant for driving circularity in the T&C industry.

Table 1. DTs and their potential role in the CE in T&C manufacturing.

DT	Role of Advancing Circularity in T&C Manufacturing	References
Social media	• provides companies in the T&C industry with new avenues to reach consumers, communicate, inform, and educate them about products, services, innovations, and developments, and influence consumer behaviour.	[41,42]
Mobile computing	• offers diverse organisational opportunities, such as reducing employee travel activities or replacing paper-based processes with digital ones, making them more efficient. • particularly in the T&C industry, mobile applications provide the opportunity to extend product life cycles, for instance, through reselling platforms or to support new business models like rental, leasing, or repairing.	[43,44]
Analytics/Big data	• applications to analyse large and complex datasets, revealing patterns and relationships, fostering informed and data-driven decision-making, and monitoring and controlling CE and sustainability KPIs. • in the T&C value chain, the systematic analysis of data, such as inventory levels, customer activities, product returns, and consumer feedback—but also information regarding resource efficiency, waste generation, or emerging trends and competitor knowledge—enables integration and dissemination into management systems as well as supporting strategic and operational decision-making.	[45,46]
Cloud and edge computing	• provides the basis for networking systems and thereby ensures greater data and information availability and transparency, which enable data-driven decision-making and new business model development. • can contribute to reduced energy consumption and the reduction in physical goods in various locations.	[47,48]
Blockchain technology (BC)	• a distributed ledger database provides tamper-proof recording of transactional data and other information, thus enabling reliable, decentralised data exchange among partners and value chains. • in the complex and highly distributed textile value chain, BC offers various functions such as traceability of product and material data, recording and secure exchange of data between parties, and, within Smart Contract applications, efficient support for procurement management. • forms an essential basis for the development of a digital product passport for textiles.	[18,49–52]
Robotics	• are able to operate autonomously, to interact with each other and to cooperate with humans, e.g., in production or logistics processes. • in the T&C industry, it appears feasible to use robots to support circular strategies by, for instance, leveraging AI systems to undertake sorting or disassembly processes, making these processes economically viable and desirable for companies and contributing to increasing recycling rates.	[48,53,54]

Table 1. *Cont.*

DT	Role of Advancing Circularity in T&C Manufacturing	References
Artificial intelligence (AI)	• is based on the development of machines and systems that are designed to mimic human intelligence by applying machine learning techniques and processing unstructured data such as natural language or images. • some algorithms are used in the context of CE strategies that can help with prediction, decision-making, resource optimisation, and pattern recognition. • in the T&C context, AI could be used to predict trends and customer needs, for dynamic pricing, or in the context of virtual fitting rooms, while other techniques based on image recognition can be used in quality control or for automated textile waste detection and sorting.	[55–59]
Internet of Things (IoT)	• enables physical objects to communicate with each other by using sensors and connecting them with digital applications and systems. • allows for real-time monitoring, tracking, and analysis of product use, condition, or location throughout the life cycle, optimising process and resource efficiency and supporting new service-oriented business models. • in the T&C industry, few concrete real-world use cases aimed at supporting CE strategies could been identified.	[60–63]
Digital fabrication (DF)	• refers to processes where computer-controlled modelling and control tools are used to manufacture physical objects directly from design data. • these processes are often supported by digital, computer-aided design (CAD) tools, which can be directly transferred to manufacturing machines and systems such as milling and cutting machines, automatic assembly systems, or 3D printers (i.e., additive manufacturing). • additive manufacturing seems to be of particular importance for CE strategies, as it can reduce transportation, inventory, and material usage by producing prototypes, spare parts, or products on-demand and at the point of need, as well as implementing circular design strategies through the use of recycled materials in 3D printing. • the implementation of 3D printing processes for textile products is technologically demanding and remains in its early stages, but initial use cases show that utilising 3D-based modelling and manufacturing can enhance communication and collaboration among stakeholders, thereby reducing physical prototypes, incorrect patterns, and overproduction and thus decreasing resource consumption and waste generation. • digital twins—virtual representations of products or facilities—are also included under digital fabrication and allow for the simulation and optimisation of product properties or functions before physical production, as well as providing stakeholders with information on ingredients, manufacturing, disassembly, or repair requirements throughout the product life cycle.	[41,62,64–69]

However, despite the documented potential of DTs to support the CE transition—through, for example, secure database storage, data analytics, AI applications, and digital fabrication—the research lacks specific insights into the practical implementation, challenges, and critical issues faced by German T&C manufacturing companies in effectively adopting and executing these technologies. There is a dearth of detailed information and guidance on how DTs best support each stage of the value chain, and how structural and operational barriers within German T&C production can be overcome. This gap in the literature highlights the need for focused, industry-specific case studies that identify and examine the effective application of specific technologies and associated best practices to confront the challenges in successfully achieving circularity in the sector. This study contributes to closing that gap by seeking to connect the two concepts—the CE and digital technology deployment—and by deriving an operational framework and actionable strategies to support the industry's transition.

2.4. Provisional Conceptual Framework

The provisional conceptual framework (PCF) for this study (Figure 3) comprises three main dimensions identified in the literature review—the five manufacturing processes within the T&C value chain (the 'y'-axis in Figure 3), the nine DTs (SMAC/BRAID) (the 'x'-axis in Figure 3), and the three CE strategies of narrowing, slowing, and closing resource flows (the 'z'-axis in Figure 3). The nine DTs are seen as catalysts and drivers for the introduction and implementation of circular practices within companies and in cross-company processes along the value chain. The PCF was subsequently used as the initial framework for analysing the primary interview data.

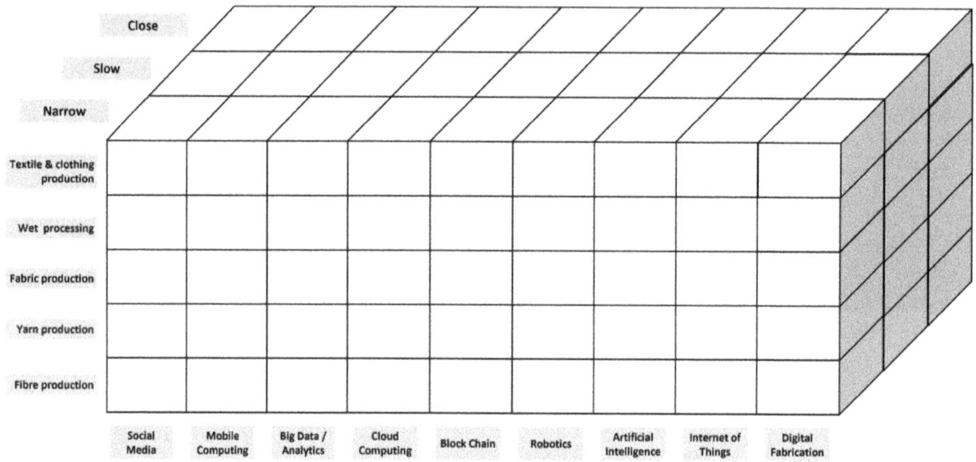

Figure 3. Provisional conceptual framework.

3. Research Method

The research method comprised three distinct phases (Figure 4). Phase one entailed a scoping review of the relevant literature, from which the PCF, outlined above, was developed. A scoping literature review can provide a summary of relevant concepts and pertinent literature, as shown, for example, in Figure 2 and Table 1. Various academic databases, including Web of Science, Google Scholar, and Scopus, were searched to provide a "broad scan of contextual literature" through which "topical relationships, research trends, and complementary capabilities can be discovered" [70] (p. 351). This provided the basis for the development of the PCF, which Miles et al. [71] suggest should identify the key elements to be investigated, as well as any potential relationships between them. It typically serves as the basis for further refinement into a more nuanced framework as the research progresses. In this instance, the PCF involved the mapping of industrial processes, CE strategies, and an initial assessment of which technologies could support the transition towards the CE. This involved mapping the complex manufacturing steps along the textile value chain in a process model and grouping them into process clusters. Relevant DTs enabling and promoting the transition towards the CE were identified. The conceptual framework provided a basis for the subsequent phases of the research. Phase two involved evaluating current circular activities and the use of DTs to promote the CE in German T&C manufacturing companies, through qualitative interviews with industry representatives from the respective T&C process clusters. There are only a few studies on the relationship between CE and DTs to date, but these indicate that DT deployment to support CE practices in manufacturing companies is still rather low, especially in the T&C industry [7,63]. Schöggl et al. [63] (p. 413) point out that "it is important to develop business cases to show/flagship what DTs can be used for and how they can enable better sustainability and circularity management". This study addresses this challenge in phase three of the research by developing a blueprint for strategy development and implementation for the use of DTs to progress the CE transition in the T&C industry.

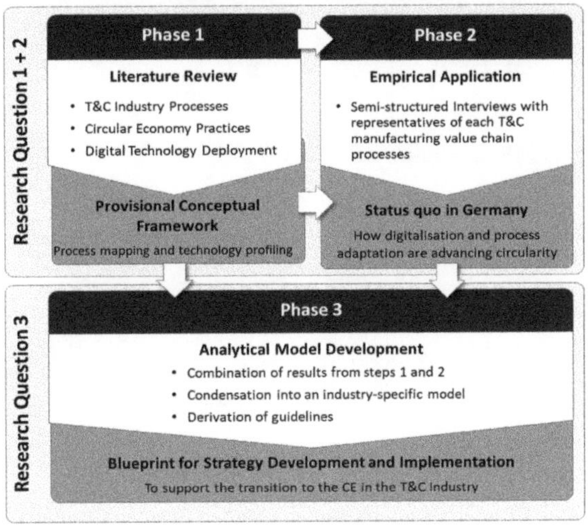

Figure 4. The 3 phases in the research process.

Given the exploratory nature of the research and the fact that knowledge about CE in the T&C industry is still in its infancy [17–19], surveying and interviewing industry experts is considered the most appropriate method for empirical investigation [71] of the current state of knowledge and implementation within the industry. Direct interaction with respondents can also reveal additional relevant areas that were not previously considered but are significant for understanding [72]. Thus, surveying and discussing with industry representatives provide insights into backgrounds, drivers, challenges, and future developments from their perspectives. Since research findings in this field of study are still scarce, an insider perspective also allows for expanding the PCF into areas that were not previously considered but are significant for understanding and developing guidelines for practical application.

To increase the reliability and validity of this study, a research protocol was used for the design of data collection, and for the analysis and formalisation of results. The selection of appropriate interviewees is critical in ensuring rigorous qualitative research [73]. A procedure for selecting company representatives was chosen to enable the researchers to collect sufficient and appropriate data to adequately address the research questions. Two main criteria were considered: firstly, at least one company should be represented from each of the process clusters; secondly, the companies should have incorporated the CE paradigm into their vision or mission statement. This is in line with Yin's [73] approach, which uses a purposive sampling method in case study research. This method is different from the sampling logic of surveys, but rather focuses on investigating specific topics in detail to develop new knowledge. Based on the authors' network, built from previous industry and research projects, a list of potential companies was compiled, and they were contacted and invited to participate in this study. Ten companies, evenly distributed across the process clusters, confirmed their willingness to participate. To ensure confidentiality, the identities of the companies and interviewees are anonymised. Figure 5 shows the allocation of the companies to the process clusters, and in Table 2, a brief description of companies and interviewees is given.

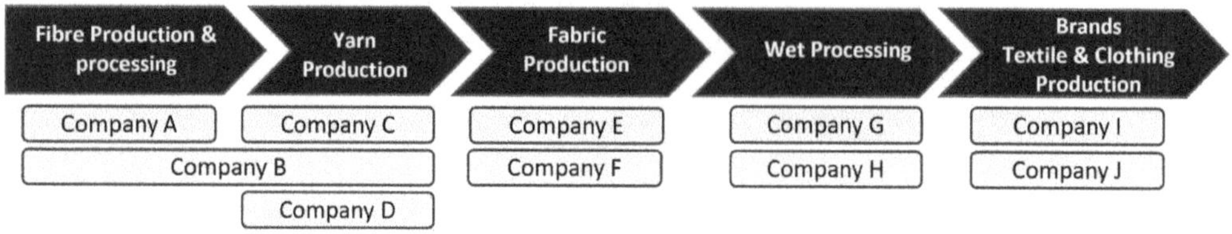

Figure 5. Companies in the T&C value chain process clusters.

Table 2. List of companies and interviewees involved in primary data collection.

Code	Description	Size	Interviewee
A	Manufacturer of manmade fibres for various applications	Large	Project Lead New Business Development
B	Processor of natural fibres and manufacturing of yarns	Medium	Certification and Sustainability Manager
C	Manufacturer of yarns (including dying) from natural fibres	Medium	Management Representative
D	Manufacturer of yarns (including dying) from natural and synthetic fibres	Large	Head of Quality Assurance
E	Manufacturer of nonwovens for various applications	Large	Sustainability Manager
F	Manufacturer of woven fabrics for home textiles	Medium	Head of Sustainability & Marketing
G	Manufacturer of printed fabrics for clothing textiles	Small	CEO
H	Manufacturer of dyed yarns for various applications	Medium	Head of Research and Development
I	Sports and outdoor clothing brand	Medium	Director of Quality and Corporate Responsibility
J	Sports and outdoor clothing brand	Large	Senior Innovation Manager Materials and Manufacturing

An interview guide was created, listing the topics and questions to be addressed [74], and semi-structured interviews were conducted with ten interviewees (one from each company). Outline questions, derived from the PCF, concerned sustainability and CE strategies and activities, and DTs deployed, as well as drivers, challenges, and future development. The interviews were conducted between June and August 2024. Each interview lasted between 50 and 70 minutes and was recorded, transcribed, and coded with the consent of the respondents. In phase three, the findings were integrated into a comprehensive framework with actionable recommendations for practical implementation. Data analysis was conducted using Braun and Clarke's [75] thematic analysis approach as a guide (Figure 6).

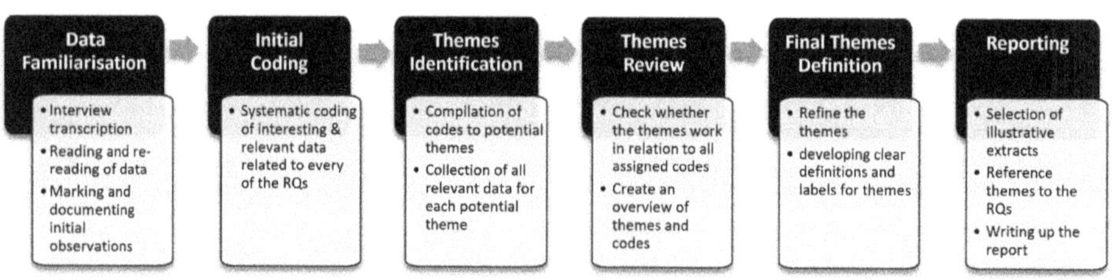

Figure 6. The six steps of thematic analysis. Source: Based on [75].

The first step involved transcribing the interviews and importing the text documents into NVivo, where they were read, and relevant findings were highlighted. In the second step, relevant text passages were coded according to the three RQs and subsequently grouped into themes. Through an iterative process, themes were reviewed, refined, and consolidated based on the assigned codes. Themes were then assigned a definition, and illustrative examples were selected. The final results are reported here in Section 4 below.

4. Results

The results section here is organised around the three RQs. First, in answer to RQ1, the twelve key themes identified in the analysis process outlined above are discussed, based on the current state of CE implementation in the companies studied (Table 2). To address RQ2, CE practices were mapped against the DTs deployed by the companies, while for RQ3, four types of challenges and three critical success factors for digital and circular transformation were identified. These results are detailed in Sections 4.1–4.3 below.

4.1. How Is the Circular Economy Currently Implemented in German T&C Manufacturing Companies? (RQ1)

4.1.1. Overview

In the thematic analysis of the interview responses, twelve CE practices were identified that address the three CE strategies of narrowing, slowing, and closing resource flows (Table 3). An overview of the themes along with corresponding examples of interviewee statements is provided in Tables A1–A3 in Appendix A.

Table 3. Practices from interview data for narrowing, slowing, and closing.

Narrowing	Slowing	Closing
• Reduce negative environmental impact • Reduce waste and scrap • Reduce resource consumption	• Durable products and materials • Extension of product lifespan • Extension of lifespan of machines and equipment • Waste utilisation	• Active recycling of products and materials • Design, development, and use of recyclable products and materials

These three strategies are not mutually exclusive but are complementary, meaning that they are all of relevance in the pursuit of a fully circular system. Table 4 shows in which companies each of the practices was evidenced. This allows for some comparisons between the three CE strategies, which are discussed in turn below.

Table 4. Applied CE strategies and practices in the companies studied.

T&C Process Clusters and Companies (rows) / CE Strategies and Practices (columns)		Narrowing/Eco-Efficiency			Slowing				Closing	
		Reduce Negative Environmental Impact	Reduce Waste and Scrap	Reduce Resource Consumption	Waste Utilisation	Durable Products and Materials	Extend Product Lifespan	Extend Lifespan of Machines and Equipment	Recycling of Products and Materials	Recyclable Products and Materials
Fibre processing	A	x	x	x	x		x			x
	B	x	x	x		x		x	x	
Yarn production	C	x		x	x		x		x	
	D	x	x	x	x					
Fabric production	E	x	x	x	x				x	
	F	x	x	x	x	x	x		x	
Wet processing	G	x	x	x				x		x
	H	x	x	x	x	x	x	x	x	x
T&C brands/production	I	x			x	x	x		x	x
	J	x		x	x	x	x		x	
Total		10	7	9	8	5	6	3	7	4

4.1.2. Narrowing/Eco-Efficiency

Reduce Negative Environmental Impact

Companies are adopting eco-efficiency strategies to meet rising costs and comply with regulatory frameworks like the Corporate Sustainability Reporting Directive (CSRD) [76], which mandates that businesses reduce their negative environmental impacts. All interviewees highlighted measures to reduce these impacts, often following systematic approaches like life cycle assessments (LCAs) to identify key sources of waste and emissions. To reduce these, companies have adopted technologies such as wastewater treatment plants and air filtration systems, while seeking environmental certifications like ISO 14001 [77], GOTS [78], and Oeko-Tex [79].

Reduce Waste and Scrap

Waste reduction and prevention play a significant role in the companies studied, primarily driven by the costs associated with waste management and disposal. Across the value chain processes, various types of waste are generated, and yet the practices mentioned in the interviews appear consistent regardless of the specific process or company size. Waste management systems help effectively categorise waste types, enabling targeted reductions. Quality management systems further contribute by identifying defects early in production, preventing unnecessary material use. Order-based manufacturing minimises overproduction, and innovative production methods and technologies are increasingly employed to reduce process waste. Notably, brand manufacturers did not report engaging in waste reduction practices, likely due to their substantial reliance on outsourced production overseas.

Reduce Resource Consumption

Another key element of the narrowing strategy is reducing resource consumption, particularly water, energy, chemicals, and raw materials. The practices adopted vary depending on the manufacturing processes and products. While nearly all companies report on energy efficiency measures, likely driven by rising energy costs, companies in water-intensive processes place significant emphasis on reducing water consumption. Similarly, in chemical-intensive processes, companies are focusing on reducing the use of dyes and auxiliaries to reduce environmental harm and the volume of chemicals needed. Material use optimisation is particularly achieved by reusing offcuts and by-products in production, lessening the need for new raw materials.

In summary, the practices following the narrowing strategy are widespread in T&C manufacturing, driven by both environmental and economic factors. While effective in reducing resource consumption and waste generation, this approach primarily focuses on internal efficiency gains rather than fully addressing CE goals.

4.1.3. Slowing

This approach encompasses four main themes: waste utilisation, durability of products and materials, extending product lifespan, and extending the lifespan of machines and equipment. Each theme reflects distinct practices adopted by companies across various stages of the value chain. Overall, it is notable that this strategy is pursued far less frequently across the value chain compared to the narrowing strategy.

Waste Utilisation

The majority of the companies (8 of 10) actively engage in waste utilisation, which involves repurposing waste as a resource for other processes or products. Companies have developed systems for separating different types of waste and finding ways to either sell or upcycle these materials to avoid costly disposal. There are even markets for by-products, such as short fibres that cannot be used for yarn production but are valuable to other industries. Brand manufacturers, in particular, focus on upcycling leftover or waste materials into new products to be sold under their brand.

Durability of Products and Materials

Companies are designing their products to last longer, providing them with a competitive advantage, particularly against cheaper, lower-quality imports from Asia. This focus on quality and durability has become a core practice for 5 of the 10 companies. In particular, brands integrate this into their unique selling proposition. But durability and high quality is not only a selling point for finished goods; yarn and fabric manufacturers are also focusing on their material's durability.

Extend Product Lifespan

In addition to designing durable products, companies are actively working to extend the product lifespan through various services and innovations. Both the brands offer repair services, which not only maintain product value but also align with the growing consumer demand for sustainable practices. Additionally, companies in the earlier value chain stages are incorporating reusable packaging into their operations, which helps to extend the life cycle of packaging materials and reduce overall waste.

Extend Lifespan of Machines and Equipment

Lastly, the extension of the machine and equipment lifespan is an important part of the slowing strategy, where companies focus on maintaining and upgrading existing machinery rather than replacing it with new equipment. These measures are more motivated by cost considerations but also contribute to reducing waste and resource consumption.

In conclusion, the slowing strategy is a critical component of CE practices in T&C manufacturing companies. By focusing on maintaining the value of products, materials, and machinery for as long as possible, companies are not only reducing waste but also offering more sustainable and durable products. However, the measures to implement this strategy across the value chain are far less prevalent compared to efficiency strategies, which are often linked to direct cost reduction and potential efficiency improvement.

4.1.4. Closing

Two main clusters of activities emerge related to the closing strategy: process-oriented activities, which involve the recycling of materials and products, and design-oriented activities, which aim to develop materials and products that can be effectively recycled into equivalent products. While these practices are primarily driven by brand manufacturers, some earlier stages of the value chain also engage in relevant activities.

Recycling of Products and Materials

One of the key components of the closing strategy is the recycling of products and materials. Across the value chain, 70% of the companies are using recycled materials. Additionally companies are also looking at ways to recycle their own process waste. This is particularly important in earlier stages of the value chain, where these waste streams can be significant. These efforts help reduce the need for new resource inputs and minimise waste. A critical challenge in closing resource loops is ensuring that post-consumer waste is available for recycling. Companies have limited control over how consumers dispose of products, making it difficult to retrieve materials for recycling. To address this, brand manufacturers are developing take-back systems.

Recyclable Products and Materials

The second major cluster focuses on designing recyclable products and materials. To successfully close resource loops, products must be intentionally designed for ease of recycling. This process often demands collaboration across the value chain: on the one hand, with recyclers to clarify the requirements for recyclable products, and on the other, to ensure these requirements are adhered to at all stages of the value chain. However, only four of the companies studied reported implementing such practices.

In summary, the closing strategy involves both recycling existing materials and designing new products for future recyclability. While challenges remain, particularly with collecting post-consumer waste, companies are making some progress in closing resource loops by integrating recycled materials, improving recycling processes, and designing for recyclability.

4.1.5. Summary

The implementation of the three CE strategies—narrowing, slowing, and closing—reveals distinct patterns across German T&C manufacturing companies (Table 4), with company size appearing to have no significant influence on these variations. Narrowing is more widely adopted across various process clusters due to its focus on resource efficiency and waste reduction, which aligns with immediate regulatory and operational demands i.e., cost reduction. Its integration is relatively straightforward, providing immediate economic benefits with minimal adjustments to existing processes, making it attractive to many manufacturers. In contrast, slowing and closing require more extensive changes in product design, material sourcing, and business models, leading to slower adoption.

4.2. How Are DTs Being Used to Support and Advance CE Practices in German T&C Manufacturing Companies? (RQ2)

4.2.1. Overview

The findings reveal that companies have a similar perception of the role of digitalisation: overall, digital technologies are widely recognised as crucial for enhancing efficiency, productivity, and cost reduction across all stages of the value chain. However, company representatives acknowledge that the full potential of these technologies has yet to be realised. Nevertheless, the nine technologies identified in the literature review as having potential to support and promote circular strategies and practices are being utilised to varying extents, and for different purposes, in the companies interviewed.

Communication technologies, such as social media platforms, are being widely deployed. Similarly, cloud-based systems and data collection and analysis tools are commonly used across the industry. In contrast, technologies that require a higher degree of connectivity and more complex automation—such as blockchain, robotics, AI, the Internet of Things, and digital fabrication—are only being deployed sporadically and to a very limited extent (Table 5).

Table 5. Deployed and planned implementation of DTs in the studied companies.

		Social Media		Mobile Computing		Big Data/ Analytics		Cloud Computing		Blockchain		Robotics		Artificial Intelligence		Internet of Things		Digital Fabrication	
		deployed	planned	deployed	planned	deployed	planned	deployed	planned	deployed	planned	deployed	planned	deployed	planned	deployed	planned	deployed	planned
Fibre production	A	x		x		x		x		x						x			
	B	x		x		x		x								x			
Yarn production	C	x								x									
	D	x		x				x				x				x			
Fabric production	E	x		x		x		x		x			x	x		x		x	
	F	x																	
Wet processing	G	x				x		x						x	x			x	
	H	x		x			x		x			x			x				
Textile and clothing production	I	x		x		x		x							x		x		x
	J	x		x		x		x			x	x					x	x	
Total % of companies		100%	0%	70%	0%	60%	10%	70%	10%	30%	10%	30%	10%	20%	30%	40%	20%	30%	10%

Table 6 provides a systematic mapping of the DTs deployed and their role in supporting CE practices in the T&C companies based on the interview data. The role of each DT is explored in greater detail in the following sections.

Table 6. Digital technologies and their function to support CE strategies in T&C manufacturing.

DT											Other			
CE Strategy	**Related DT Function and CE Practice**	**Social Media**	**Mobile Computing**	**Big Data/ Analytics**	**Cloud Computing**	**Blockchain**	**Robotics**	**Artificial Intelligence**	**Internet of Things**	**Digital Fabrication**	**Management System**	**ERP System**	**PDM System**	**E-Commerce/ Website**
Narrowing / Eco-efficiency	DT Function		Digital collaboration tools for meetings Virtual product presentation	Operational and strategic data analysis	Online platform for chemical management Document sharing with stakeholders	Traceability and transparency	Process automation	Image recognition for quality control Image generation/design	Traceability of materials (physical tracers) Real-time data collection	Digital printing on fabrics Digital colour management 3D design	Centralised data collection for process control and certification	Centralised data collection for operational data and process integration	Centralised product data collection for data and process integration	
	CE Practice		Reduce negative environmental impact Reduce material use and consumption	Reduce negative environmental impact Reduce energy consumption Reduce waste and scrap	Reduce chemical consumption Reduce material use and consumption	Reduce negative environmental impact Reduce resource consumption	Reduce waste and scrap Reduce material use and consumption	Reduce waste and scrap Reduce negative environmental impact Reduce material use and consumption	Reduce negative environmental Reduce waste and scrap	Reduce negative environmental impact Reduce waste and scrap Reduce chemical consumption Reduce water consumption Reduce material use and consumption	Reduce negative environmental impact Reduce chemical consumption	Reduce negative environmental impact Reduce waste and scrap Reduce material use and consumption	Reduce negative environmental impact	
Slowing	DT Function	Consumer education about care and maintenance				Traceability and transparency			Traceability of materials (physical tracers) Digital tags for consumer information					Online marketplace for scrap and secondary raw materials

Table 6. *Cont.*

	DT										Other			
CE Strategy	**Related DT Function and CE Practice**	**Social Media**	**Mobile Computing**	**Big Data/ Analytics**	**Cloud Computing**	**Blockchain**	**Robotics**	**Artificial Intelligence**	**Internet of Things**	**Digital Fabrication**	**Management System**	**ERP System**	**PDM System**	**E-Commerce/ Website**
Slowing	CE Practice	Extend product lifespan				Durable products and materials			Durable products and materials Extend product lifespan					Waste utilisation
Closing	DT Function						Process automation		Digital tags for consumer information					
	CE Practice						Recyclable products and materials		Recycling of products and materials					

4.2.2. Social Media and Digital Collaboration

Social media platforms are widely used in T&C companies (Table 5), primarily for marketing purposes. However, their application for CE-related activities is limited, predominantly employed by brands and, in one case, a fabric manufacturer. These companies utilise social media to communicate with consumers, providing information on care and maintenance for textile and clothing products.

4.2.3. Mobile Computing

Mobile computing tools, including smartphone apps and tablets, are widely used across T&C companies, primarily for communication, project management, and sales presentations. These tools enhance efficiency by reducing the need for travel and improving communication processes. Company J noted that the use of tools like MS Teams had "significantly improved efficiency and reduced travel needs, contributing to resource efficiency". It also contributes to reducing resource use by enabling virtual product presentations. While mobile computing has improved communication and project management processes, its potential to advance circular business models like rental or leasing [43] remains untapped.

4.2.4. Big Data and Analytics

Big data and analytics play a crucial role in supporting CE practices by enabling companies to monitor and optimise resource use, waste generation, and overall efficiency. Data collection occurs primarily through, e.g., spinning, weaving, and dying machines that generate vast amounts of data, although the level of data integration varies between companies. However, many companies are still in the early stages of integrating data analytics into their CE strategies. For example, company F stated, "We can only analyse data retrospectively, but we do not have real-time data showing us error rates or inefficiencies". This lack of real-time data limits the ability to optimise processes and reduce waste proactively. For companies with explicit CE goals, such as reducing environmental impact or increasing the use of secondary raw materials, big data analytics help monitor progress and ensure compliance with sustainability targets.

The systematic collection, integration, and analysis of data are essential for implementing CE strategies. Although the use of big data does not directly drive CE practices, it enables companies to make data-driven decisions and implement measures that contribute to circularity.

4.2.5. Cloud Computing

Cloud computing is primarily used for data sharing, communication, and business operations. The use of cloud applications for data and document sharing contributes to increased efficiency, such as reducing paper consumption.

However, its application in CE practices remains limited.

4.2.6. Blockchain Technology (BC)

BC is viewed as an enabler for ensuring traceability and transparency along the textile value chain. However, only 3 of 10 companies, located in the early value chain processes, currently use blockchain, with one brand manufacturer planning its adoption (Table 5). Specifically, blockchain is employed to track and document the origin of materials, ensuring their authenticity and sustainability credentials. It helps ensure that materials, for instance, originate from organic farming or are free from hazardous chemicals, supporting the reduction in resource consumption and negative environmental impacts. Additionally, blockchain is used to verify the use and application of high-quality materials throughout the value chain, which can positively influence the durability and longevity of final products. At the same time, the companies report that data input on BC platforms is largely manual, raising concerns about the accuracy and reliability of the recorded information. Thus, the potential of the BC to significantly advance CE practices by improving transparency and enabling secure real-time data sharing is only partially utilised.

4.2.7. Robotics

The use of robotics in German T&C manufacturing companies is currently limited to 3 of 10 companies, and within these, to individual process automation tasks, such as packaging or loading and unloading goods. The robotic systems currently deployed are not yet integrated into broader digital ecosystems or circular strategies, and their primary focus is on improving efficiency and productivity in specific sub-processes. Although robotics has not yet been widely adopted for CE purposes, individual process automation solutions can indirectly contribute to circularity by reducing waste and improving efficiency. For example, company G uses laser technology to detect and skip seams during printing, preserving the uncontaminated fabric for recycling.

Company J is beginning to explore the potential of more advanced robotic systems as they invested in robots to optimise production processes, while company E is planning to use robotics to reduce resource consumption: "We are looking to optimise our [logistics] processes with these technologies to make them more efficient, speed up operations [...] and reduce resource consumption, such as packaging materials".

Current applications of process automation aim for increased efficiency and productivity in specific sub-processes but only contribute to CE goals to a very limited extent. Nevertheless, future developments could see these systems integrated with CE strategies to reduce waste and improve resource use.

4.2.8. Artificial Intelligence (AI)

AI applications are currently used at two companies (E and G), while two others (H and I) are planning to implement them. The diverse range of application areas highlights the significant potential of this technology. AI-powered camera systems are increasingly used to detect defects. They thus contribute to increasing product quality and reducing waste. Company E, for instance, uses AI-powered systems to monitor production processes and quality: "We are performing real-time measurements so we can immediately intervene the moment a defect is detected, preventing unnecessary meters of poor-quality material from being produced" (company E). And company H is "working on [systems] that can recognise different product formats using image recognition software [...] for scanning for defects or the quality of the product" (company H).

In addition to improving production processes, AI is being used to optimise design processes. Company G uses AI to generate print designs, significantly reducing the need for physical samples. This not only speeds up the development process but also leads to reduced resource consumption, waste generation, and environmental impacts.

The integration of AI with other systems, such as production or energy management, could lead to major efficiency gains, as company G suggested: "AI could be used to optimise print designs to minimise energy consumption and reduce ink use". Although not yet implemented, this underscores the potential of AI to facilitate CE strategies effectively.

While AI technologies are still in the early stages of adoption in the T&C industry, their potential to drive circular practices by enhancing quality control, design processes, and resource efficiency seems to be significant.

4.2.9. Internet of Things (IoT)

IoT technologies are used in 4 out of 10 companies, all 4 of which can be assigned to the early stages of the value chain (Table 5). The two brand manufacturers are planning to introduce IoT technologies, but no initiatives are planned in the other processes. In companies where IoT is used, sensors for real-time monitoring and error detection (sometimes in conjunction with AI) are used. Company B, for example, uses IoT for machine monitoring: "All our machines are networked, enabling us to conduct analyses and generate reports".

However, the technology's full potential for CE practices remains largely untapped. While the companies collect data from individual machines, the lack of connectivity between systems and companies limits comprehensive data analysis and real-time decision-making. The potential for IoT to enhance resource efficiency and support CE practices lies in its ability to provide real-time, data-driven insights. For example, IoT sensors could track material use and waste generation, helping companies optimise resource use and reduce waste. Tracer systems, such as mentioned by company A, offer the potential to track materials throughout their life cycle, improving transparency and supporting recycling efforts. Digital labels or RFID sensors fulfil a similar function, e.g., in clothing items. However, these technologies were hardly featured in the interviewee discussions; only the brand manufacturers are planning to introduce digital labels to their products. Company I, for example, sees potential in the switch to QR codes instead of physical labels to provide information about their take-back programme because "it can't be washed away [and directs the consumer directly to the relevant website], where it explains what to do".

The integration of IoT systems into broader CE strategies requires investment in networked systems and data analytics. As regulatory demands for transparency increase, IoT technologies could play a crucial role in enabling real-time monitoring and data sharing for CE practices.

4.2.10. Digital Fabrication

Digital fabrication technologies, such as 3D printing and 3D CAD, are not yet widely used in the T&C manufacturing processes; only 3 of 10 companies currently use them. But their potential to reduce resource use and waste is increasingly recognised. Company J noted that digital modelling could replace physical testing and help significantly reduce material and energy consumption: "I believe that not everything needs to be tested physically. Many aspects can be digitally modeled instead". Company G has already transitioned from wet to digital printing, allowing it to manage the entire production process digitally. This shift has improved efficiency and reduced resource consumption and waste generation.

While most companies are still exploring the potential of digital fabrication technologies, their ability to reduce waste, optimise production processes, and enhance resource efficiency makes them promising tools for advancing CE practices.

4.2.11. Other Digital Systems and Platforms

Other digital systems and platforms, notably the core business information software packages—Enterprise Resource Planning (ERP) and Product Data Management/Product Lifecycle Management (PDM/PLM) applications, for example—play a key role in supporting CE practices by ensuring data flows and enhancing traceability. Four of the ten companies interviewed reported plans to upgrade their ERP systems to support sustainability and CE goals. These platforms are essential for enabling seamless data exchange and documenting product recyclability, which is critical for supporting new business models like rental and repair services.

However, manual processes still dominate in some areas, particularly in data collection for sustainability reports. Companies across all process stages continue to rely on spreadsheets to track data for reporting, highlighting the gap between regulatory demands and the current state of digitalisation.

Online marketplaces for secondary raw materials also offer potential for reusing or repurposing surplus or defective products. Company E noted the importance of such platforms: "There's an online market for secondary raw materials where we can sell defective polyester rolls that might still meet the needs of other buyers". However, their existence currently appears to be limited to specific materials.

4.2.12. Summary

DT deployment to support CE practices in the German T&C manufacturing industry varies significantly across processes and companies. A key trend is that later stages in the value chain adopt more DTs, largely aligned with the narrowing strategy. The applications are focused on improving efficiency, reducing resource consumption and waste, and thus cutting costs, which is highly supported by widely used DTs such as mobile computing, big data, and cloud computing, often operating within, or linked to, the corporate ERP system. These technologies enable data integration, facilitating process optimisation and supporting resource efficiency and environmental management.

Although IoT technologies are partly deployed, the collection and analysis of these data are limited. Data are collected automatically from different IoT devices, but are often not available in centralised systems for filtering and analysis to support informed decision-making. ERP systems, for example, which provide the backbone information processing for many companies, are being modified and developed to provide IoT data integration and analysis functions, but this is not straightforward given the range of IoT devices and the need for specific system configurations.

Social media platforms, which are mainly used for marketing and consumer education, indirectly support the slowing strategy by promoting product longevity. However, more complex technologies such as robotics and AI are less common, although they can contribute to quality assurance, process automation, and informed decision making and thus to cost reduction.

For slowing and closing strategies, technologies like digital tags and platform technologies are used sporadically and the use is focused on later stages, where they facilitate traceability and information sharing with consumers and recyclers, essential for reuse, repair, and recycling.

Digital fabrication holds considerable potential for reducing resource use across all processes, but its application is still limited. As investments in digital technologies grow and regulatory pressures increase, the role of these technologies in CE practices will become increasingly vital.

4.3. What Are the Challenges and Critical Issues for Current and Future CE Transition and Digitalisation in the German T&C Manufacturing Sector? (RQ3)

The interview analysis reveals four clusters of challenges and critical issues relating to CE transition and digitalisation. These are (1) product-related, (2) process-related, (3) organisational, and (4) external issues.

4.3.1. Challenges

Product-related challenges arise from the limited availability and complex processing requirements for alternative materials, such as secondary or recycled fibres and materials, which vary in quality, fibre length, and colour consistency. These inconsistencies often require additional testing and adjustments. Company C, for example, tries to source certified fibres from sustainable production, but "it is still the case that these farms, which perform really well, cannot even supply 5% of our wool", and on the processing of recycled wool: "This is a loss of quality because we simply lose fibre length". Company D observed that "recycled polyester has a different quality standard in each batch, sometimes with color variations", which limit its usability in lighter shades, and therefore lead to quality compromises.

Process-related challenges mainly involve the lack of collaboration and standardised data reporting and transparency across the value chain, which results in inefficient and labour-intensive data management practices. For example, Company B pointed out that while their software evolves continually, all data must be manually entered and maintained, creating inefficiencies. Company C echoed this by stressing the lack of uniform data templates, stating, "the data being requested [by different brands and platforms] is similar [...] but it is not the same. Everyone wants something different", which complicates meeting multiple certification and reporting requirements. Furthermore, implementing changes often requires cooperation among multiple stakeholders across the supply chain. For example, company C sought to adopt reusable transport boxes, which would have required participation from transport companies and downstream partners; however, "our customers don't engage sufficiently, and our external service providers don't cooperate enough". This lack of alignment restricts scalable solutions for enhanced transparency and sustainability in processes. The industry-wide absence of uniform digital integration across platforms and reporting systems intensifies these challenges, underscoring the need for standardised, streamlined digital solutions.

Organisational challenges include limited awareness and internal resistance toward new technologies, often stemming from the separate handling of sustainability and digitalisation functions. This separation impedes integrated CE implementation. Company C specifically noted challenges with a newly implemented ERP system because it has "programmed a huge module for sustainability and traceability [in the old system], which will take years to be implemented [in the new system], because too few questions were asked and in the end it was implemented by IT-people who have no idea about all this stuff".

Furthermore, the SMEs in the industry struggle with limited capacity as regards human and financial resources, which slows the innovation process and hampers their ability to meet market demands and regulatory expectations. For example, company F noted, "It's more time-consuming than I would have thought. So it's not that easy. I wish I had more time and resources and manpower".

External challenges encompass fragmented standards, regulatory demands, and market constraints. The varying requirements across industry standards, certification processes, and regional regulations increase operational complexity. Company I cited the varying regulatory demands across European markets, where "France, for example, always does things differently from Germany", complicating compliance. Additionally, the higher costs of sustainable materials, combined with market resistance to price increases, inhibit broader adoption. As company F, who developed a collection of fabrics made of recycled materials, found, "it didn't gain as much popularity as I would have liked [...]. Unfortunately, it's usually one to two euros more expensive per metre of fabric and our customers, especially the German furniture industry, aren't that interested".

4.3.2. Success Factors

To address the challenges discussed above and progress the successful implementation of CE strategies in the German T&C manufacturing industry, three operational areas emerged from the interviews as being of particular importance and having the biggest impact: (1) strengthening collaboration and partnerships; (2) optimising resources and infrastructure; and (3) standardisation and regulatory support.

Strengthening collaboration and partnerships is the most frequently emphasised factor to enable effective transformation. These partnerships are crucial for navigating both organisational and process challenges, as demonstrated by Company A's collaboration with a range of stakeholders, including brands, nonwoven producers, spinners, knitters, and research institutions. Company A emphasises that "having contacts across various levels of the value chain is vital" for product development and innovation, and they actively seek new connections to expand their network. Similarly, company E notes that close communication with supply chain partners is essential, as "suppliers are also continually evolving, more and new recycled materials are being offered [...], which opens up new possibilities". Without such collaboration, circularity initiatives would face significant hurdles due to the complexity of the T&C supply chain, which involves multiple actors and processes.

Efficient resource and infrastructure management are also seen as critical in enabling circular processes and minimising environmental impact. For instance, company G is working with suppliers to prepare fabrics in ways that reduce the need for in-house processing, allowing for "a smoother, energy-efficient production". Furthermore, several companies recognise that using local recycling options could significantly cut environmental costs associated with long-distance transport and handling of recycled materials. However, establishing efficient infrastructure often involves external collaborations. As company E states, "collaborative recycling projects and partnerships with local recyclers" provide viable solutions for reducing waste.

Standardisation and regulatory support: The need for standardised metrics and supportive regulatory frameworks is another significant factor for successful CE implementation. Interviewees stress the need for consistent standards across the value chain to enhance traceability and help companies meet sustainability goals effectively. As noted by company C, the current lack of unified metrics complicates compliance and reporting, as "different brands and platforms require various data formats", which results in inefficiency and an increased workload. A cohesive regulatory framework that establishes clear guidelines for material traceability and reporting would alleviate these issues, as would a central platform for sharing standardised sustainability and CE data.

5. Discussion

Previous studies provide a fragmented picture of the current status and necessary future developments across various T&C value chain stages. This fragmentation arises from the diverse perspectives these studies take in addressing circularity within the T&C industry. Most research adopts a product life cycle perspective, focusing in particular on end-products such as textiles and clothing, to explore potential strategies and practices that could facilitate the transformation from a linear to a circular T&C industry [41,80,81]. Similarly, studies examining the use of digital technologies to support this transformation also rely on the product life cycle as a basis, mapping specific technologies to various phases of the cycle [61,65]. In this context, all manufacturing processes are grouped together under the life cycle phase "production", which does not adequately reflect the complexity of the textile value chain. As outlined in Section 2 of this study, the German T&C sector is highly fragmented, producing both intermediate and final products, as well as working in technical and clothing textiles. These processes involve a wide range of stakeholders across different stages of the value chain. Therefore, to better address the research questions, this study took an organisational and process-focused perspective at the micro level, which allowed for the identification of relevant strategies, practices, and suitable DTs that can support and advance them. It is one of the first studies to focus specifically on T&C companies and their CE practices in the context of DT deployment. The results reveal that the adoption of CE strategies varies across the T&C processes, while the degree of digitalisation is relatively low across the processes, with the primary focus being on efficiency optimisation. A number of related themes emerged that merit further discussion.

Firstly, the companies in this study indicated a particular focus on reducing resource consumption, waste, and the consequent environmental impact, which is driven by a need to reduce high energy costs and labour expenses and comply with regulatory demands in Germany. This focus on internal eco-efficiency reflects broader industry trends but does not necessarily align with a full transformation to the CE. This is consistent with the findings of Barreiro-Gen and Lozano [82], who reported that companies often prioritise internal measures rather than collaborating with external stakeholders to achieve wider-ranging holistic circularity goals. Social media, mobile and cloud computing, and internal systems like ERP, PDM/PLM, and integrated management systems are the most widely used technologies across the value chain. The main goal of digitalisation is to improve process efficiency, as well as reducing resource consumption and waste. While these efficiency measures align with CE principles, companies are not implementing them as part of a holistic CE transformation. It is thus unsurprising that DTs are not primarily adopted to support a broader CE transition or sustainability goals, but rather for the more immediate business benefits they offer.

IoT, big data, and AI have significant potential for real-time process monitoring that could support the adoption of CE practices, but most companies are still in the early stages of implementing these technologies. Blockchain and RFID show promise for enhancing traceability and reducing waste, but a lack of standardisation and high manual data management workloads hinder wider adoption. These technologies are not fully integrated into existing management systems, limiting their effectiveness in supporting CE practices. Digital fabrication (DF) technologies, such as 3D printing, 3D design, and digital twins, are also underutilised, despite their potential to significantly support CE practices. These technologies could eliminate the need for physical samples, reduce resource consumption, and automatically collect data for future applications like the DPP [83]. However, research on the potential of DF technologies to progress the CE is still in its infancy. In summary, the overall level of digitalisation in the industry remains low, which aligns with the findings of Schöggl et al. [63]; and even when DTs are implemented, it is often more for reasons of internal efficiencies rather than transitioning to the CE. In that context, however, this study adds qualitative insights into the motivations and challenges behind the adoption of these technologies, and finds a general lack of strategy concerning the CE transition, in which companies integrate CE principles into their corporate strategy to guide appropriate measures and implement suitable technologies.

Secondly, in terms of the slowing strategy, companies are primarily concerned with high-value waste utilisation. For instance, businesses sell by-products to other actors in the industry, but this practice is largely driven by cost-efficiency rather than a genuine shift toward circularity. Manufacturing companies also focus on product durability to differentiate themselves from low-cost imports from Asia, especially since German manufacturers rely on high-quality and long-lasting products to compete. This strategic focus represents a market niche where durability is valued, particularly in technical textiles and high-performance materials for luxury or functional clothing. However, these efforts fall short, as practices such as repair or reuse do not play a significant role for manufacturing companies, even though there are opportunities for extended business models such as via spare parts supply or repair services [84]. While social media is used by brand manufacturers to engage with consumers, promote sustainability efforts, and advertise repair services, its use remains limited. Social media represents an untapped opportunity for manufacturers to engage directly with consumers, for instance, by educating them about repair practices, and contribute to a more repair-friendly culture. As research has shown, consumers often lack the skills and knowledge to repair products [85,86]. Companies could bridge this gap by sharing knowledge and collaborating with local repair communities.

Thirdly, closing strategies are mainly employed by brand manufacturers, where design and material choices are geared towards recyclability. However, there are contradictions within the value chain. While brand manufacturers favour recycled or mono-materials, upstream companies are hesitant to use them due to quality concerns, favouring material mixes instead to ensure the production of high-quality, durable materials. This mirrors a paradox identified by Colucci and Vecchi [84]. Hedrich et al. [87] note similar issues in their study of textile recycling in Europe, but predict that recycling technologies will soon improve, producing higher-quality materials. Companies could engage with these developments and invest in initiatives to drive innovation in this area, which may strengthen their market position as demand for sustainable materials rises with regulations like the EU's Ecodesign for Sustainable Products Regulation (ESPR) [88]. Additionally, business models aimed at extending the use phase of textile products—such as rental services—are not yet profitable, further complicating the CE transition for brands. The reasons are twofold: consumers do not yet fully embrace these business models; and they involve significant manual effort. Digital support could help by creating effective offerings and efficient processes based on extensive data [89].

Fourthly, this study also highlights the significance of early-stage collaboration between companies across the value chain, particularly in circular innovation projects. Schöggl et al. [90] demonstrated the positive impact of such collaborations on CE strategy implementation. However, these efforts remain project-specific, with little evidence of widespread industry transformation. The lack of collaboration across the value chain represents one of the biggest hurdles to scaling circular strategies. Cooperation is essential for developing circular products and processes [90,91], yet innovation remains siloed within individual processes, limiting CE goals for the end product. For incumbent companies, engaging in innovation ecosystems where various stakeholders such as startups, research institutions, or government actors collaborate can be key to fostering successful innovations [92]. In a similar vein, Colucci and Vecchi [84] see significant potential in collaboration between born-circular firms and established companies to drive a radical shift toward CE, creating a win–win situation for both parties. Business models like repair and resale services also depend on stakeholder cooperation, particularly in terms of decentralised service provision and spare parts supply. While brand manufacturers are key drivers of circular strategies, they cannot achieve these goals without innovation and engagement from upstream and downstream actors. A joint understanding and coordinated efforts across the entire value chain are essential for true CE integration. International processes may impede progress, as studies from Italy [49] suggest local ecosystems are crucial for developing circular and innovative products and business models. More research is needed to explore whether such ecosystems could be established in Germany.

Fifthly, from an organisational perspective, employee and managerial awareness, capabilities, and motivation are critical for advancing digital and CE transformation. The organisational culture plays a crucial role [93], with resistance to change being a common obstacle. However, studies show that this resistance can be overcome by integrating CE into corporate strategies and business practices [94]. Achieving this will require a deeper cultural shift internal to the organisations, making CE a core aspect of the business rather than an external add-on.

Finally, a blueprint for transitioning to circular practices in the German T&C industry can be derived from the research results and the discussion above, which can be customised and applied across the T&C processes studied here (Figure 7). The blueprint revolves around the coordinated adoption of the three CE strategic approaches and related practices, in conjunction with the use of appropriate DTs to enable and support these initiatives.

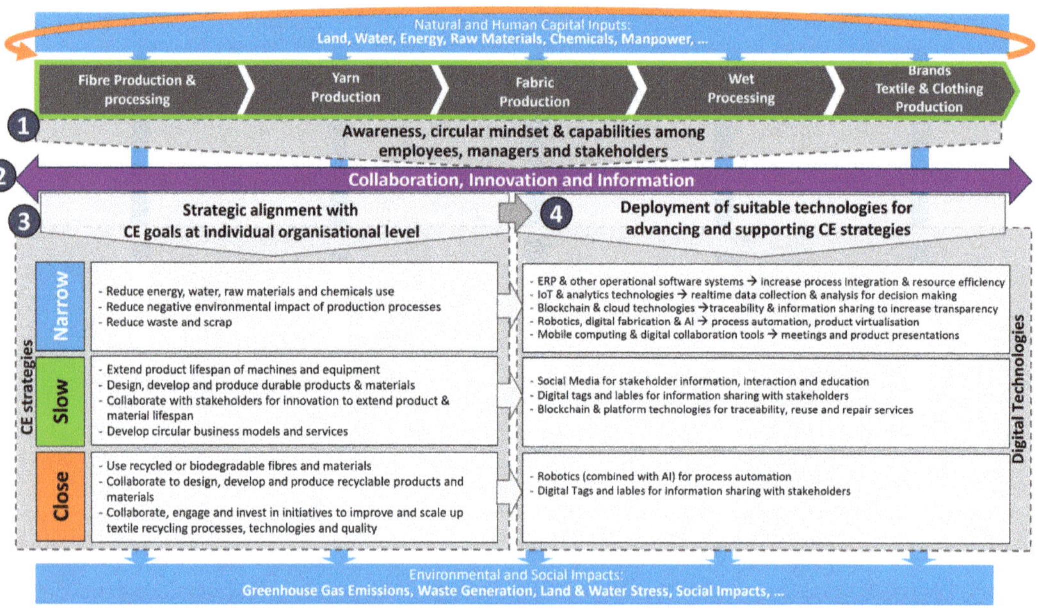

Figure 7. Blueprint for CE strategy development and implementation in T&C manufacturing with appropriate DT deployment.

1. Organisational awareness and capacity building

At the individual organisational level, raising awareness of ecological and social impacts, establishing a mindset for circular thinking, and developing the necessary skills and capabilities among managers, employees, and stakeholders are essential. This includes addressing potential resistance to change, which is common in transitions of this nature. Establishing an organisational culture that values and actively promotes sustainability goals is critical for successful CE implementation. Through training and communication initiatives, organisations can build a foundational understanding of CE principles and practices, creating a workforce that is aligned with circular values.

2. Collaborative networks and ecosystems

Collaboration across the value chain is crucial for developing CE innovations and business models. Strengthening partnerships between organisations within the T&C industry can facilitate knowledge exchange on materials, products, and processes. Establishing regional and innovation ecosystems that also include external stakeholders from other sectors—such as startups, research institutions, and government bodies—can accelerate the creation of circular products and business models. Companies can participate in multiple networks and ecosystems, each dedicated to specific aspects of circularity, to leverage shared resources and drive systemic change in the T&C industry.

3 and 4. Alignment and implementation of strategies, processes, products, and DTs around agreed CE goals

To achieve CE objectives, companies must develop their strategy, processes, and product designs around the principles of narrowing, slowing, and closing resource loops. Each organisation, as well as collaborative initiatives, should work to implement practices associated with these strategies, using digitalisation initiatives to enable and support them.

The narrowing strategy is aimed at minimising resource consumption, waste generation, and the environmental and social impacts associated with manufacturing processes. By leveraging DTs, organisations can optimise their operations to improve efficiency and reduce resource consumption. For example, ERP and other operational software systems can integrate workflows to optimise resource allocation within companies, while IoT devices and analytics provide real-time data on resource consumption and process performance. These insights enable decision-makers to promptly adjust operations to reduce waste. Blockchain and cloud computing further support the narrowing strategy by facilitating seamless data sharing across various stages of the value chain. This integration enhances traceability, allowing companies to monitor and verify resource origins and usage, which is crucial for meeting sustainability standards. Technologies such as robotics, digital fabrication, and AI can also automate processes like quality control and virtual product modelling, minimising physical material usage and ultimately lowering resource requirements. Additionally, mobile computing and collaboration tools help reduce travel by enabling virtual communication, improving cooperation in project-based and operational tasks while reducing the carbon footprint associated with transportation.

The slowing strategy focuses on extending the lifespan and preserving the value of products and materials. This approach aligns with consumer demand for long-lasting, high-quality items and represents a competitive advantage for T&C companies. By engaging stakeholders through social media, companies can educate and encourage consumers to maintain and repair their products, promoting responsible product use. Digital labels and tamper-resistant tags provide reliable product information that is accessible to different stakeholders, aiding in repair, resale, and recycling efforts. Furthermore, blockchain and platform technologies can maintain detailed records of each product's history and condition, facilitating traceability and supporting innovative business models like repair services. This approach not only extends the product life cycle but also strengthens customer trust in the durability and sustainability of companies' offerings.

The closing strategy is essential for enabling effective recycling and end-of-life management, facilitating the development of products that can be reintegrated into production cycles. This strategy encourages companies to design items with recyclable or biodegradable materials, focusing on processes that promote easy recycling at the product's end-of-life stage. Collaborative efforts and investments in recycling methods, technologies, and quality standards are critical for supporting a circular flow of resources. By deploying advanced robotics integrated with AI, companies can improve sorting and separation processes, enhancing productivity and cost efficiency in recycling operations. Digital identifiers, such as tags and labels, further support the closing strategy by aiding reverse flows, for instance, correct disposal and collection, and reprocessing of materials.

As an adaptable roadmap and practical checklist, the blueprint provides a structured approach for companies to strategically adopt and integrate CE principles and DTs across the entire value chain. Through narrowing, slowing, and closing resource loops, the blueprint facilitates a holistic approach that not only enhances operational efficiency but also advances long-term sustainability goals of absolute reduction in resource consumption, as well as negative environmental and social impacts across the value chain. By aligning DTs with CE objectives, companies can transform their practices, creating a more resilient and circular industry. It also recognises the importance of people-related aspects, requiring cultural change and new skills and competencies.

Each company can tailor the framework to specific processes, such as yarn production, textile manufacturing, or garment construction, by selecting appropriate CE practices for narrowing, slowing, and closing resource loops and aligning them with relevant DTs and to meet their strategic goals. The blueprint also highlights how DTs, like IoT and blockchain, can enhance traceability and support compliance with new EU regulations, such as those relating to the DPP. The framework's flexibility enables companies to assess their current status, identify gaps, and outline concrete steps toward transformation. This integrative approach lays the foundation for advancing a circular and digitalised textile sector.

6. Conclusions

This study investigated the relationship between the CE transition and DT deployment in German T&C manufacturing companies. The findings highlight that CE strategies and practices, as well as digitalisation, vary significantly across the value chain. Some key points emerge from the findings, which are briefly discussed below.

It is clear that the focus for many companies remains on internal efficiency measures, driven by economic pressures in the main. While these efforts reduce resource consumption and waste, they are often not aligned with a comprehensive CE transformation. Companies are primarily concerned with improving eco-efficiency within their own processes, rather than pursuing broader collaboration with stakeholders along the value chain, which is essential for achieving circularity. In terms of CE strategies, the slowing strategy is primarily driven by cost-efficiency, with companies focusing on high-value waste utilisation and product durability to differentiate themselves from low-cost imports. While this provides a competitive advantage, practices like repair and reuse are underdeveloped; but there is potential for T&C manufacturers to expand into extended business models, such as repair services and spare parts, to align more closely with CE principles.

The closing strategy, focused on recycling and designing for recyclability, is mainly implemented by brand manufacturers. However, contradictions within the value chain persist, as upstream companies resist using recycled materials due to quality concerns. This reflects broader tensions between sustainability goals and product standards. Advances in recycling technologies, along with regulatory pressure from the EU's ESPR [88], are expected to help overcome these challenges.

DTs, particularly IoT and blockchain, in conjunction with ERP software, are widely adopted to enhance process efficiency. However, their potential to support CE practices remains underutilised. Some of the more advanced technologies, such as big data, AI, and digital fabrication, are still in their infancy in terms of integration into CE strategies. Barriers include a lack of standardisation, manual data management, and limited system integration, which hamper the full utilisation of these technologies for circularity. Collaboration across the value chain remains limited. Although some early-stage projects show promise, a comprehensive transformation toward circularity has not yet emerged. Increased collaboration, especially within circular (innovation) ecosystems that include diverse stakeholders, is crucial to scaling CE strategies and accelerating their adoption. Moving forward, aligning digitalisation with CE goals and fostering greater cooperation across the value chain will be essential for achieving a fully circular T&C manufacturing industry in Germany.

This study has its limitations. It did not encompass all processes of the T&C value chain, with important areas like trade, waste management, and recycling excluded. Future research could encompass these processes to offer a more comprehensive understanding of CE practices and opportunities for collaborative innovations and extending the proposed blueprint accordingly. There were only 10 companies in this study, with two companies per value chain process, which may not accurately represent the industry as a whole. Further research could include quantitative analyses to investigate the variety of actors and processes in the T&C value chain. At the same time, ways to increase the adoption of certain technologies in the industry, notably digital fabrication, could be researched. A further limitation is that this study was conducted solely in Germany, even though the T&C industry is highly globalised. For instance, the analysed brand manufacturers predominantly operate within international networks, rarely collaborating with German manufacturing companies. The findings could be further tested by including companies operating in local supply networks as well as in other regional contexts to validate the blueprint and ensure broader applicability. Expanding the research to different regions could also help to capture the varying dynamics and challenges of CE transition and related DT deployment.

Despite these limitations, the authors believe the research is of value in providing new insights in this field of study and that the blueprint for CE strategy development and implementation put forward here can serve as a starting point for practitioners aiming to implement a systematic CE transformation, and strategically deploy digital technologies to support and enhance this process. The proposed blueprint offers managers a structured and adaptable framework for integrating CE strategies and DTs into their operations. The blueprint's flexibility enables firms to assess current gaps, devise targeted strategies, and foster long-term resilience through collaboration and cultural change. For researchers, the blueprint provides a basis for further development and refinement through additional studies, applying it across diverse regions and contexts to yield new insights. To enhance its effectiveness, the blueprint could be tested, discussed, and extended in further case studies with practitioners, allowing it to be refined in alignment with industry needs. An important next step for expanding the blueprint involves incorporating downstream phases—retail, use, and waste management. Such an expansion would enable companies and stakeholders to devise holistic strategies that engage all relevant actors, promoting a more comprehensive CE transformation. This integrated approach could prove especially valuable for engaging consumers in circular strategies and improving product end-of-life management.

Author Contributions: Conceptualisation, T.W. and M.W.; methodology, T.W. and M.W.; software, T.W.; validation, T.W. and M.W.; formal analysis, T.W. and M.W.; investigation, T.W.; data curation, T.W.; writing—original draft preparation, T.W. and M.W.; writing—review and editing, T.W. and M.W.; visualisation, T.W. and M.W.; supervision, T.W. and M.W.; project administration, T.W. and M.W. All authors have read and agreed to the published version of the manuscript.

Funding: This research received no external funding.

Data Availability Statement: The primary data used in this research article are held in a university environment. Further enquiries can be made via the corresponding author.

Conflicts of Interest: The authors declare no conflicts of interest.

Appendix A

Table A1. Company CE narrowing practices: Interviewee perspectives.

Theme	Company Code	Statement Examples
Reduce negative environmental impact	A	"We have our own wastewater treatment plant on site".
	B	"We have been recording this consistently since 1997. We can systematically record key indicators, and are actually always up to date in terms of the current input and output; this systematic tracking allows us to establish benchmarks and set improvement goals".
	C	"We make significant investments in photovoltaic systems wherever it is feasible".
	D	"We hold various certifications, including Ökotex for our products, as well as GOTS, GRS, and RWS certifications".
	E	"We have a water treatment system that effectively recovers and processes water for reuse. The system employs various process chemicals, ensuring compliance with ZDHC [Zero Discharge of Hazardous Chemicals] standards, to make the treated water suitable for repeated use".
	F	"20% of our energy consumption is covered by our photovoltaic system [. . .]. We calculate and balance our CO_2 footprint annually (Scope 1, 2, and 3) and fully offset the calculated CO_2 emissions through the support of climate projects".
	G	"Regarding emissions, our digital printing processes and the use of certified inks have virtually eliminated exhaust emissions. This also leads to significantly fewer chemicals on the fabric".
	H	"As a textile finisher, energy consumption—both electricity and heat—is a significant factor. For instance, we require 4 million kilowatt-hours of electricity annually, a quarter of which is generated through on-site solar photovoltaics (PV), a practice we began over 20 years ago and have gradually expanded. However, further expansion of renewable electricity generation currently seems unfeasible. In addition to electricity, we also require 8 million kilowatt-hours of thermal energy, which we generate sustainably using biomass".
	I	"We aim to minimize air freight wherever possible and have set specific air freight targets for each department".
	J	"We are collaborating with other brands and partners at our production sites. This initiative focuses on substituting coal with natural gas or biomass and implementing various energy efficiency measures to further reduce our environmental impact".
Reduce waste and scrap	A	"We reduce waste by integrating quality assurance measures to ensure optimal production outcomes".
	B	"When we notice during production checks or inspections that the product quality is deteriorating, adjustments are made at the cotton stage, including adapting the blend to counteract the issue".
	D	"We have system that analyses and categorises waste into different types, aiming to reduce avoidable waste. This effort spans a wide range, starting from copier paper and extending to various other areas [. . .]. We operate on a made-to-order production basis".
	E	"We aim for real-time error detection to minimise waste, so that as soon as a defect is detected, we can intervene before too much low-quality material is produced".
	F	"We produce almost exclusively on-demand, with everything tailored to customer requirements. We maintain only a small stock program, which we constantly adjust by monitoring which colors perform well and to what extent".
	G	We carefully evaluate where we can reduce waste, particularly since we have to pay for disposal. One significant area of improvement has been the reduction in waste from printing ink, which used to be a major issue. Now, thanks to advancements, ink usage has been minimised, and almost no ink needs to be discarded anymore.
	H	"Production is increasingly aligned with actual sales or near-certain demand. This approach minimises overproduction and eliminates surplus stock that would otherwise need to be reprocessed or discarded later".

Table A1. *Cont.*

Theme	Company Code	Statement Examples
Reduce resource consumption	A	"We operate an on-site gas power plant, which not only generates energy but also produces steam and heat that are directly utilized in production processes, ensuring a very high level of efficiency".
	B	"Between 2008 and 2023, we progressively reduced water usage by 30%. For example, in 2008, 132 L of water were needed per kilogram of yarn, whereas by 2023, this figure had dropped to just 96 L".
	C	"We have modernised our dyeing facility in Germany, equipping it with completely new, fully automated dyeing kettles featuring smaller, individually adjustable batch sizes. This allows for a more efficient dyeing process, significantly reducing water consumption and chemical use [...]".
	D	"There are various ongoing projects aimed at energy savings, as is mandatory for everyone, including us. We are certified in energy management".
	E	"Our process involves an important distinction: the precursor to our final product is a fleece [...]. The edges of this need to be trimmed. Since no solidification has yet occurred, these trimmed edges are collected via suction systems and fed back [...]. This allows us to save material that would otherwise need to be added anew".
	F	"Our electricity, gas, heating oil, and diesel consumption are continuously monitored. Through a variety of measures, we have achieved significant energy savings in recent years. For instance, since 2019, our electricity consumption per meter of produced fabric has been reduced by nearly 19%". "Since 1998, we have been monitoring our water consumption and continuously evaluating necessary measures to conserve this resource [...]. We have achieved significant reductions over the past five years, cutting our annual municipal water consumption by nearly 25% and our annual water consumption for climate control in production by almost 47%". "We are reusing leftover yarns by incorporating them into the production process, such as into the selvedge".
	G	"We reuse water up to four times, capturing it with installed pumps and tanks. For instance, we reuse cooling water in the dyeing process, significantly reducing our energy and water consumption [. . .]. We are already forced to do this for energy reasons - we have had this system for over 20 years now and nobody forced us to do it, [but] we are business people and can do the maths".
	H	"We have implemented a rigorous selection process for dyes, chemicals, and auxiliaries, prioritising those with a very high efficacy spectrum. For example, when working with cellulose and reactive dyes, it is possible to purchase inexpensive reactive dyes with yields of only 60–75%, coupled with an extensive need of water. Our approach focuses on sourcing dyes with the highest possible fixation rates, ensuring that 75–95% of the applied dye remains on the material and not in the water".
	J	"We are working intensively with our suppliers, to establish and evaluate new dyeing and finishing processes and implement process optimisation. Particularly in our production countries, water recovery has become a standard practice due to the limited availability of water. We actively support these initiatives and closely monitor developments, ensuring significant progress is made in this area".

Table A2. Company CE slowing practices: Interviewee perspectives.

Theme	Company Code	Statement Examples
Waste utilisation	A	"Defective goods, as long as they are in acceptable condition and not heavily contaminated, are sold as B-grade products".
	C	"Yes, our main revenue driver when it comes to waste utilisation is selling our waste. We have different types of waste with varying lengths and qualities at multiple stages, and even our combings are of interest to those who need shorter fibers. If we simply sell our waste in bulk, we generate significantly more revenue compared to all our recycling efforts, and it doesn't appear that this will change".
	D	"Waste is collected and sold separately as far as possible".
	E	"We try to separate and collect as much as possible—whether it's household waste, production waste like nonwoven scraps, old wood, or plastic packaging".
	F	"We are currently collaborating on a research project with [university], where we produce round-needled tubular structures from our own waste trimmings. These are intended for use in furniture manufacturing. Since 2018, we have also had our own home collection, offering cushions and small furniture, and we aim to incorporate such circular products into this line".
	H	"Almost nothing goes into general waste; we sort and return all cardboard, films, and yarns used in production".
	I	"We repurpose leftover materials for smaller projects. For example, in our own service factory, we create items like bags from these materials. All of our posters and banners, which are made of textile, are collected from our stores and points of sale, then transformed into bags. These are often used as giveaways and feature interesting designs and images. We are actively working on many projects to reuse waste in this area".
	J	"For our products, we actively engage in upcycling. Items that may be faulty, particularly from our manufacturing process, are repurposed and enhanced to create new products. For example, we produce shopper bags or wallets from cutting scraps. We consistently explore ways to make better use of these materials, ensuring they are transformed into valuable items rather than wasted".
Durable products and materials	B	"We aim to explore niche markets to sustain jobs at our location. Technical textiles, for instance, are significantly more profitable than basic yarns used for producing T-shirts. Our focus is consistently on high-quality products".
	F	"Our fabrics are designed to be washable and easy to care for".
	H	"Our focus is on ensuring our product has the highest possible durability", which applies to both final products and intermediary materials.
	I	"As part of our new brand DNA, longevity has become a core focus for us. This means it is undoubtedly an important topic. We're now introducing extended warranties, which further underscores how crucial this is for us [. . .]".
	J	"Our products have a relatively long lifespan. We're talking about 10 years or more, potentially even up to 20 years, depending on the area of application, intensity of use, and maintenance".
	C	"We have already conducted trials using reusable transport boxes for dye cones".
	F	"We offer training and even provide support to end consumers. For example, if someone has a [product] with our fabric, we have a service hotline where they can ask questions like how to deal with pressure marks on velour, or how to properly clean and maintain the fabric. We also advise on removing stains from materials like marker ink or nail polish. While we don't offer repairs, we do provide a care service to ensure the fabric lasts as long as possible, aligning with our commitment to durability".

Table A2. *Cont.*

Theme	Company Code	Statement Examples
	H	"We strive to reuse everything related to packaging whenever possible. For instance, with cardboard boxes, we have agreements with our customers to return the products in the original boxes, keeping them in circulation for as long as possible. Similarly, stretch films and cores are reused whenever feasible, with efforts to employ reusable cores. For materials like plastic films, we ensure they are sorted by type and returned to the manufacturer for recycling. This approach maximizes the lifecycle of packaging materials and aligns with our sustainability goals".
	I	"We try to repair first—we have a Service Factory, and we're developing strategies to expand our repair offerings".
	J	"We have offered a repair service at our company for as long as the company has existed—around 50 years". "We also provide rental models and second-hand offerings. About a month ago, we officially launched these initiatives, which received some media attention [...]. Our aim is to explore how these models can evolve into viable business opportunities [...]. We are also working on designing future products in specific categories to better align them with rental systems. These products may, for example, be heavier or more modular, allowing easier replacement of components, thereby improving their suitability for rental and extending their overall lifespan".
Extend life-span of machines and equipment	B	"In the past, maintenance intervals were based on rough estimates and experience, such as assuming that a carding set would last around three months. However, with modern testing procedures and technologies, these intervals have been extended significantly".
	G	"Our stretching frame is older than I am, but with proper maintenance, it still makes sense to keep it running".
	H	"We have implemented proactive maintenance, [...], based on this knowledge, components are replaced at appropriate intervals".

Table A3. Company CE closing practices: Interviewee perspectives.

Theme	Company Code	Statement Examples
Recycling of products and materials	B	"Using recycled cotton means less water use in farming, fewer chemicals, and reduced transportation impacts. This makes it a sustainable product, and we aim to maximise its potential, particularly in the premium segment".
	C	"We also have our own 'recycled product' [made from recycled fibres], which sells steadily. However, it always includes a portion of virgin fiber to ensure the material can be processed effectively through our spinning machines".
	E	"In our carbon fiber nonwoven production, the process was set up to incorporate waste from carbon fiber nonwoven production. We purchase these materials and produce our own carbon fiber nonwovens from them. This means that these products are essentially made entirely from recycled materials, qualifying as 100% secondary raw materials".
	F	"We have a recycled collection for two years now. However, it's not textile-to-textile recycling; instead, we rely on PET recycled yarns made from PET bottles. Currently, we are also exploring new options and have sourced a new fiber-to-fiber recycled yarn that we are now testing".
	H	"We purchase certain packaging items, such as cartons and foils from recycled materials".
	I	"We are working on a circularity collection, meaning a collection [...] of mono-component products, meaning they are made of 100% recycled polyester wherever possible". "When we use natural fibers, particularly in the wool sector, we exclusively use recycled wool, such as shredded wool or collected wool, to ensure resource efficiency. We aim to use as little conventional [virgin] wool as possible". "The consumer can visit our website to generate and print a free DHL return label, allowing them to send products back to us at no cost. Currently, we handle this process in-house and do not collaborate with an external partner for returns".
	J	"We strive for, and have already largely achieved, the predominant use of non-fossil raw materials in our products. Currently, approximately 90% of our raw materials consist of recycled, bio-based, or biomass-allocated materials. This shift has significantly reduced our carbon footprint". "We also operate a take-back system, allowing customers to return their old products. Depending on their condition, we either repair the items under our manufacturer's warranty or as a paid service [...]. However, some products, due to their condition, cannot be economically repaired. Currently, we collaborate with 'Fairwertung', which oversees the redistribution of these items. This includes second-hand resale, recycling, or appropriate disposal, depending on the product's viability.
Recyclable products and materials	A	"We explore how our fibers can contribute to making textile products more circular. For instance, we assess the potential to replace synthetic fibers with functionalised viscose fibers, enabling products to be designed for biodegradability or improved recyclability".
	G	"We now stop printing at the seam, which means the fabric remains clean and can be more easily recycled".
	H	"We're already working with designers to use only specific dyes and chemicals, so the product can be recycled later".
	I	"We have guidelines from our recycler on what materials can be used, and we integrate these from the beginning of the design process".

References

1. Gözet, B.; Wilts, H. *The Circular Economy as a New Narrative for the Textile Industry: An Analysis of the Textile Value Chain with a Focus on Germany's Transformation to a Circular Economy*; Zukunftsimpuls No. 23; Wuppertal Institut für Klima, Umwelt, Enrgie gGmbH: Wuppertal, Germany, 2022. Available online: https://epub.wupperinst.org/frontdoor/deliver/index/docId/8108/file/ZI23_Textile_Industry.pdf (accessed on 22 March 2024).
2. EEA. Textiles and the Environment in a Circular Economy: Eionet Report—ETC/WMGE 2019/6. Available online: https://www.eea.europa.eu/publications/textiles-in-europes-circular-economy (accessed on 12 April 2024).
3. European Commission. Circular Economy. Available online: https://joinup.ec.europa.eu/collection/rolling-plan-ict-standardisation/circular-economy-0 (accessed on 12 April 2024).
4. Kirchherr, J.; Yang, N.-H.N.; Schulze-Spüntrup, F.; Heerink, M.J.; Hartley, K. Conceptualizing the Circular Economy (Revisited): An Analysis of 221 Definitions. *Resour. Conserv. Recycl.* **2023**, *194*, 107001. [CrossRef]
5. United Nations. Transforming Our World: The 2030 Agenda for Sustainable Development. Available online: https://sustainabledevelopment.un.org/post2015/transformingourworld (accessed on 24 May 2024).
6. European Commission. The European Green Deal: COM (2019) 640 Final. Available online: https://eur-lex.europa.eu/resource.html?uri=cellar:b828d165-1c22-11ea-8c1f-01aa75ed71a1.0002.02/DOC_1&format=PDF (accessed on 13 March 2024).
7. Neri, A.; Cagno, E.; Susur, E.; Urueña, A.; Nuur, C.; Kumar, V.; Franchi, S.; Sorrentino, C. The relationship between digital technologies and the circular economy: A systematic literature review and a research agenda. *R&D Manag.* **2024**. [CrossRef]
8. Neligan, A. Digitalisation as Enabler Towards a Sustainable Circular Economy in Germany. *Intereconomics* **2018**, *53*, 101–106. [CrossRef]
9. Liu, Q.; Trevisan, A.H.; Yang, M.; Mascarenhas, J. A framework of digital technologies for the circular economy: Digital functions and mechanisms. *Bus. Strat. Environ.* **2022**, *31*, 2171–2192. [CrossRef]
10. Das, S.K.; Bressanelli, G.; Saccani, N. Clustering the Research at the Intersection of Industry 4.0 Technologies, Environmental Sustainability and Circular Economy: Evidence from Literature and Future Research Directions. *Circ. Econ. Sust.* **2024**. [CrossRef]
11. Govindan, K. How digitalization transforms the traditional circular economy to a smart circular economy for achieving SDGs and net zero. *Transp. Res. Part E Logist. Transp. Rev.* **2023**, *177*, 103147. [CrossRef]
12. European Commission. A New Circular Economy Action Plan For a Cleaner and More Competitive Europe: COM (2020) 98 Final. Available online: https://eur-lex.europa.eu/resource.html?uri=cellar:9903b325-6388-11ea-b735-01aa75ed71a1.0017.02/DOC_1&format=PDF (accessed on 17 June 2024).
13. European Commission. EU Strategy for Sustainable and Circular Textiles: COM (2022) 141 Final. Available online: https://environment.ec.europa.eu/strategy/textiles-strategy_en (accessed on 17 June 2024).
14. Battesini, T.; Thomas, G.; de Medeiros, J.F.; Kolling, C.; Duarte Ribeiro, J.L.; Morea, D. Redesign in the textile industry: Proposal of a methodology for the insertion of circular thinking in product development processes. *J. Clean. Prod.* **2023**, *397*, 136588. [CrossRef]
15. Jia, F.; Yin, S.; Chen, L.; Chen, X. The circular economy in the textile and apparel industry: A systematic literature review. *J. Clean. Prod.* **2020**, *259*, 120728. [CrossRef]
16. Franco, M.A. Circular economy at the micro level: A dynamic view of incumbents' struggles and challenges in the textile industry. *J. Clean. Prod.* **2017**, *168*, 833–845. [CrossRef]
17. Wiegand, T.; Wynn, M. Sustainability, the Circular Economy and Digitalisation in the German Textile and Clothing Industry. *Sustainability* **2023**, *15*, 9111. [CrossRef]
18. Furferi, R.; Volpe, Y.; Mantellassi, F. Circular Economy Guidelines for the Textile Industry. *Sustainability* **2022**, *14*, 11111. [CrossRef]
19. Rinaldi, F.R.; Di Bernardino, C.; Cram-Martos, V.; Pisani, M.T. Traceability and transparency: Enhancing sustainability and circularity in garment and footwear. *Sustain. Sci. Pract. Policy* **2022**, *18*, 132–141. [CrossRef]
20. European Parliament. Regulation (EC) No 1893/2006 of the European Parliament and of the Council of 20 December 2006 Establishing the Statistical Classification of Economic Activities NACE Revision 2 and Amending Council Regulation (EEC) No 3037/90 as well as Certain EC Regulations on Specific Statistical Domains: EC No 1893/2006. Available online: https://eur-lex.europa.eu/legal-content/EN/TXT/PDF/?uri=CELEX:32006R1893. (accessed on 5 September 2024).
21. European Commission: Directorate-General for Internal Market, Industry, Entrepreneurship and SMEs. In *Data on the EU Textile Ecosystem and Its Competitiveness: Final Report*; Publications Office of the European Union: Luxembourg, 2021.
22. European Environment Agency; Mortensen, L.; Bogdanovic, J.; Reichel, A.; Asquith, M. *Environmental Indicator Report 2014: Environmental Impacts of Production-Consumption Systems in Europe*; Publications Office: Copenhagen, Denmark, 2014. [CrossRef]
23. Schumacher, K.; Forster, A.L. *Facilitating a Circular Economy for Textiles Workshop Report*; NIST: Gaithersburg, MD, USA, 2022. Available online: https://nvlpubs.nist.gov/nistpubs/specialpublications/nist.sp.1500-207.pdf (accessed on 8 August 2024).
24. Statista. Die zehn wichtigsten Importländer für Textilien im Jahr 2022 [In English: The Ten Most Important Import Countries for Textiles in 2022]. Available online: https://de.statista.com/statistik/daten/studie/209862/umfrage/wichtigste-importlaender-fuer-textilien/ (accessed on 17 March 2024).
25. Eurostat. Where Do Our Clothes Come From? Available online: https://ec.europa.eu/eurostat/web/products-eurostat-news/-/edn-20200424-1 (accessed on 17 March 2024).

26. Statista. Die zehn wichtigsten Exportländer für Textilien weltweit nach Exportwert im Jahr 2022 [In English: The Ten Most Important Export Countries for Textiles Worldweide by Ammount in 2022]. Available online: https://de.statista.com/statistik/daten/studie/209828/umfrage/wichtigste-exportlaender-fuer-textilien/ (accessed on 17 March 2024).
27. Federal Statistical Office of Germany. Anzahl der Betriebe in der deutschen Textil- und Bekleidungsindustrie in den Jahren 2008 bis 2022 [In English: Number of Companies in the German Textile and Clothing Industry between 2008 and 2022]. Available online: https://de.statista.com/statistik/daten/studie/6396/umfrage/betriebe-in-der-deutschen-textil-und-bekleidungsindustrie/ (accessed on 17 March 2024).
28. Reike, D.; Vermeulen, W.J.; Witjes, S. The circular economy: New or Refurbished as CE 3.0?—Exploring Controversies in the Conceptualization of the Circular Economy through a Focus on History and Resource Value Retention Options. *Resour. Conserv. Recycl.* **2018**, *135*, 246–264. [CrossRef]
29. Ellen MacArthur Foundation. Towards the Circular Economy: Economic and Business Rationale for an Accelerated Transition. Available online: https://ellenmacarthurfoundation.org/towards-the-circular-economy-vol-1-an-economic-and-business-rationale-for-an (accessed on 22 March 2024).
30. European Commission. Closing the Loop—An EU Action Plan for the Circular Economy. Available online: https://ec.europa.eu/transparency/documents-register/api/files/COM(2015)614_0/de00000000332178?rendition=false (accessed on 17 June 2024).
31. Bocken, N.; de Pauw, I.; Bakker, C.; van der Grinten, B. Product design and business model strategies for a circular economy. *J. Ind. Prod. Eng.* **2016**, *33*, 308–320. [CrossRef]
32. Geissdoerfer, M.; Savaget, P.; Bocken, N.; Hultink, E.J. The Circular Economy—A new sustainability paradigm? *J. Clean. Prod.* **2017**, *143*, 757–768. [CrossRef]
33. Konietzko, J.; Bocken, N.; Hultink, E.J. A Tool to Analyze, Ideate and Develop Circular Innovation Ecosystems. *Sustainability* **2020**, *12*, 417. [CrossRef]
34. Nobre, G.C.; Tavares, E. The quest for a circular economy final definition: A scientific perspective. *J. Clean. Prod.* **2021**, *314*, 127973. [CrossRef]
35. World Business Council for Sustainable Development. *Eco-Efficiency: Creating More Value with Less Impact*; World Business Council for Sustainable Development: Geneva, Switzerland, 2000; ISBN 9782940240173.
36. European Commission. Proposal for a Directive of the Euorpean Parliament and of the Council amending Directive 2008/98/EC on waste: Com (2023) 420 Final. Available online: https://environment.ec.europa.eu/document/download/ca53d82e-a4d3-40b9-a713-93585058f47f_en?filename=Proposal%20for%20a%20DIRECTIVE%20OF%20THE%20EUROPEAN%20PARLIAMENT%20AND%20OF%20THE%20COUNCIL%20amending%20Directive%20200898EC%20on%20waste%20COM_2023_420.pdf (accessed on 1 December 2023).
37. Kirchherr, J.; Reike, D.; Hekkert, M. Conceptualizing the circular economy: An analysis of 114 definitions. *Resour. Conserv. Recycl.* **2017**, *127*, 221–232. [CrossRef]
38. Ghisellini, P.; Cialani, C.; Ulgiati, S. A review on circular economy: The expected transition to a balanced interplay of environmental and economic systems. *J. Clean. Prod.* **2016**, *114*, 11–32. [CrossRef]
39. Coppola, C.; Vollero, A.; Siano, A. Developing dynamic capabilities for the circular economy in the textile and clothing industry in Italy: A natural-resource-based view. *Bus. Strat. Environ.* **2023**, *32*, 4798–4820. [CrossRef]
40. Brydges, T. Closing the loop on take, make, waste: Investigating circular economy practices in the Swedish fashion industry. *J. Clean. Prod.* **2021**, *293*, 126245. [CrossRef]
41. Colucci, M.; Vecchi, A. Close the loop: Evidence on the implementation of the circular economy from the Italian fashion industry. *Bus. Strat. Environ.* **2021**, *30*, 856–873. [CrossRef]
42. UNEP; UNFCCC. *The Sustainable Fashion Communication Playbook*; United Nations Environment Programme: Nairobi, Kenya, 2023. Available online: https://wedocs.unep.org/bitstream/handle/20.500.11822/42819/sustainable_fashion_communication_playbook.pdf?sequence=3 (accessed on 25 May 2024).
43. Faria, R.; Lopes, I.; Pires, I.M.; Marques, G.; Fernandes, S.; Garcia, N.M.; Lucas, J.; Jevremović, A.; Zdravevski, E.; Trajkovik, V. Circular Economy for Clothes Using Web and Mobile Technologies—A Systematic Review and a Taxonomy Proposal. *Information* **2020**, *11*, 161. [CrossRef]
44. Wynn, M.; Jones, P. Digital Technology Deployment and the Circular Economy. *Sustainability* **2022**, *14*, 9077. [CrossRef]
45. Kristoffersen, E.; Mikalef, P.; Blomsma, F.; Li, J. The effects of business analytics capability on circular economy implementation, resource orchestration capability, and firm performance. *Int. J. Prod. Econ.* **2021**, *239*, 108205. [CrossRef]
46. Luoma, P.; Penttinen, E.; Tapio, P.; Toppinen, A. Future images of data in circular economy for textiles. *Technol. Forecast. Soc. Change* **2022**, *182*, 121859. [CrossRef]
47. Bressanelli, G.; Saccani, N.; Perona, M.; Baccanelli, I. Towards Circular Economy in the Household Appliance Industry: An Overview of Cases. *Resources* **2020**, *9*, 128. [CrossRef]
48. Cagno, E.; Neri, A.; Negri, M.; Bassani, C.A.; Lampertico, T. The Role of Digital Technologies in Operationalizing the Circular Economy Transition: A Systematic Literature Review. *Appl. Sci.* **2021**, *11*, 3328. [CrossRef]
49. Abbate, S.; Centobelli, P.; Cerchione, R. From Fast to Slow: An Exploratory Analysis of Circular Business Models in the Italian Apparel Industry. *Int. J. Prod. Econ.* **2023**, *260*, 108824. [CrossRef]

50. Alves, L.; Sá, M.; Cruz, E.F.; Alves, T.; Alves, M.; Oliveira, J.; Santos, M.; Da Rosado Cruz, A.M. A Traceability Platform for Monitoring Environmental and Social Sustainability in the Textile and Clothing Value Chain: Towards a Digital Passport for Textiles and Clothing. *Sustainability* **2024**, *16*, 82. [CrossRef]
51. Bressanelli, G.; Adrodegari, F.; Pigosso, D.C.A.; Parida, V. Circular Economy in the Digital Age. *Sustainability* **2022**, *14*, 5565. [CrossRef]
52. Rusch, M.; Schöggl, J.-P.; Baumgartner, R.J. Application of digital technologies for sustainable product management in a circular economy: A review. *Bus. Strat. Environ.* **2023**, *32*, 1159–1174. [CrossRef]
53. Chauhan, C.; Parida, V.; Dhir, A. Linking circular economy and digitalisation technologies: A systematic literature review of past achievements and future promises. *Technol. Forecast. Soc. Change* **2022**, *177*, 121508. [CrossRef]
54. Lopes de Sousa Jabbour, A.B.; Chiappetta Jabbour, C.J.; Choi, T.-M.; Latan, H. 'Better together': Evidence on the joint adoption of circular economy and industry 4.0 technologies. *Int. J. Prod. Econ.* **2022**, *252*, 108581. [CrossRef]
55. Agrawal, R.; Wankhede, V.A.; Kumar, A.; Luthra, S.; Majumdar, A.; Kazancoglu, Y. An Exploratory State-of-the-Art Review of Artificial Intelligence Applications in Circular Economy using Structural Topic Modeling. *Oper. Manag. Res.* **2022**, *15*, 609–626. [CrossRef]
56. Ellen MacArthur Foundation. Artificial Intelligence and the Circular Economy: AI as a Tool to Accelerate the Transition. Available online: https://emf.thirdlight.com/file/24/GgC25OAGBvwdiFGgtzZGVXuZsz/Artificial%20intelligence%20and%20the%20circular%20economy.pdf (accessed on 8 June 2024).
57. Giri, C.; Jain, S.; Zeng, X.; Bruniaux, P. A Detailed Review of Artificial Intelligence Applied in the Fashion and Apparel Industry. *IEEE Access* **2019**, *7*, 95376–95396. [CrossRef]
58. Rudisch, K.; Jüngling, S.; Carrillo Mendoza, R.; Woggon, U.K.; Budde, I.; Malzacher, M.; Pufahl, K. *Paving the Road to a Circular Textile Economy with AI*; German Informatics Society: Bonn, Germany, 2021.
59. Russell, S.J.; Norvig, P. *Artificial Intelligence: A Modern Approach*, 4th ed.; Pearson: Boston, MA, USA, 2022; ISBN 9781292401171.
60. Li, S.; Da Xu, L.; Zhao, S. The internet of things: A survey. *Inf. Syst. Front.* **2015**, *17*, 243–259. [CrossRef]
61. Bressanelli, G.; Adrodegari, F.; Pigosso, D.C.A.; Parida, V. Towards the Smart Circular Economy Paradigm: A Definition, Conceptualization, and Research Agenda. *Sustainability* **2022**, *14*, 4960. [CrossRef]
62. Hassan, R.; Acerbi, F.; Rosa, P.; Terzi, S. The role of digital technologies in the circular transition of the textile sector. *J. Text. Inst.* **2024**, 1–14. [CrossRef]
63. Schöggl, J.-P.; Rusch, M.; Stumpf, L.; Baumgartner, R.J. Implementation of digital technologies for a circular economy and sustainability management in the manufacturing sector. *Sustain. Prod. Consum.* **2023**, *35*, 401–420. [CrossRef]
64. Agustí-Juan, I.; Habert, G. Environmental design guidelines for digital fabrication. *J. Clean. Prod.* **2017**, *142*, 2780–2791. [CrossRef]
65. Gershenfeld, N. How to make almost anything: The digital fabrication revolution. *Foreign Aff.* **2012**, *91*, 43–57.
66. Huynh, P.H. Enabling circular business models in the fashion industry: The role of digital innovation. *Int. J. Product. Perform. Manag.* **2022**, *71*, 870–895. [CrossRef]
67. Manaia, J.P.; Cerejo, F.; Duarte, J. Revolutionising textile manufacturing: A comprehensive review on 3D and 4D printing technologies. *Fash. Text.* **2023**, *10*, 20. [CrossRef]
68. Preut, A.; Kopka, J.-P.; Clausen, U. Digital Twins for the Circular Economy. *Sustainability* **2021**, *13*, 10467. [CrossRef]
69. Wynn, M.; Irizar, J. Digital Twin Applications in Manufacturing Industry: A Case Study from a German Multi-National. *Future Internet* **2023**, *15*, 282. [CrossRef]
70. Porter, A.L.; Kongthon, A.; Lu, J.-C. Research profiling: Improving the literature review. *Scientometrics* **2002**, *53*, 351–370. [CrossRef]
71. Miles, M.B.; Huberman, A.M.; Saldaña, J. *Qualitative Data Analysis: A Methods Sourcebook*, 3rd ed.; Sage: Los Angeles, CA, USA; London, UK; New Delhi, India; Singapore; Washington, DC, USA, 2014; ISBN 9781452257877.
72. Cooper, D.R.; Schindler, P.S. *Business Research Methods*, 12th ed.; Irwin/McGraw-Hill: Boston, MA, USA, 2014; ISBN 9780073521503.
73. Yin, R.K. *Case Study Research: Design and Methods*, 5th ed.; Sage: Thousand Oaks, CA, USA, 2014.
74. Rheinhardt, A.; Kreiner, G.E.; Gioia, D.A.; Corley, K.G. Conducting and Publishing Rigorous Qualitative Research. In *The SAGE Handbook of Qualitative Business and Management Research Methods*; Cassell, C., Cunliffe, A.L., Grandy, G., Eds.; Sage: Los Angeles, CA, USA; London, UK; New Delhi, India; Singapore; Washington, DC, USA; Melbourne, Australia, 2018; pp. 515–531; ISBN 9781526429261.
75. Braun, V.; Clarke, V. Using thematic analysis in psychology. *Qual. Res. Psychol.* **2006**, *3*, 77–101. [CrossRef]
76. European Parliament. DIRECTIVE (EU) 2022/2464. Available online: https://eur-lex.europa.eu/legal-content/EN/TXT/PDF/?uri=CELEX:32022L2464 (accessed on 18 September 2024).
77. International Organization for Standardization. Environmental Management Systems—Requirements with Guidance for Use (14001:2015). Available online: https://www.iso.org/obp/ui/#iso:std:iso:14001:ed-3:v1:en (accessed on 18 September 2024).
78. Global Standard gGmbH. Global Organic Textile Standard (GOTS)—Key Features. Available online: https://global-standard.org/the-standard/gots-key-features (accessed on 18 September 2024).
79. OEKO-TEX Service GmbH. Our Standards OEKO-TEX Standards Enable Everyone to Make Responsible Decisions and Protect Natural Resources. Available online: https://www.oeko-tex.com/en/our-standards (accessed on 18 September 2024).
80. Dissanayake, D.; Weerasinghe, D. Towards Circular Economy in Fashion: Review of Strategies, Barriers and Enablers. *Circ. Econ. Sust.* **2022**, *2*, 25–45. [CrossRef]

81. De Aguiar Hugo, A.; de Nadae, J.; Da Silva Lima, R. Can Fashion Be Circular? A Literature Review on Circular Economy Barriers, Drivers, and Practices in the Fashion Industry's Productive Chain. *Sustainability* **2021**, *13*, 12246. [CrossRef]
82. Barreiro-Gen, M.; Lozano, R. How circular is the circular economy? Analysing the implementation of circular economy in organisations. *Bus. Strat. Environ.* **2020**, *29*, 3484–3494. [CrossRef]
83. Balchandani, A.; Starzynska, E.; Barrelet, D.; Berg, A.; D'Auria, G.; Rölkens, F.; Amed, I. The State of Fashion 2024. Available online: https://www.mckinsey.com/industries/retail/our-insights/state-of-fashion (accessed on 3 October 2024).
84. Colucci, M.; Vecchi, A. Tackling Climate Change with End-of-Life Circular Fashion Practices—Remade in Italy with Amore. *Br. J. Manag.* **2024**, *35*, 1157–1179. [CrossRef]
85. Fachbach, I.; Lechner, G.; Reimann, M. Drivers of the consumers' intention to use repair services, repair networks and to self-repair. *J. Clean. Prod.* **2022**, *346*, 130969. [CrossRef]
86. Laitala, K.; Klepp, I.; Henry, B. Does Use Matter? Comparison of Environmental Impacts of Clothing Based on Fiber Type. *Sustainability* **2018**, *10*, 2524. [CrossRef]
87. Hedrich, S.; Janmark, J.; Langguth, N.; Magnus, K.-H.; Strand, M. Scaling Textile Recycling in Europe—Turning Waste into Value. Available online: https://www.mckinsey.de/industries/retail/our-insights/scaling-textile-recycling-in-europe-turning-waste-into-value#/ (accessed on 4 May 2024).
88. European Commission. Ecodesign for Sustainable Products Regulation (ESPR). Available online: https://commission.europa.eu/energy-climate-change-environment/standards-tools-and-labels/products-labelling-rules-and-requirements/sustainable-products/ecodesign-sustainable-products-regulation_en (accessed on 18 September 2024).
89. Neligan, A.; Baumgartner, R.J.; Geissdoerfer, M.; Schöggl, J.-P. Circular disruption: Digitalisation as a driver of circular economy business models. *Bus. Strat. Environ.* **2022**, *32*, 1175–1188. [CrossRef]
90. Schöggl, J.-P.; Stumpf, L.; Baumgartner, R.J. The role of interorganizational collaboration and digital technologies in the implementation of circular economy practices—Empirical evidence from manufacturing firms. *Bus. Strat. Environ.* **2023**, *33*, 2225–2249. [CrossRef]
91. Suchek, N.; Franco, M. Inter-organisational Cooperation Oriented Towards Sustainability Involving SMEs: A Systematic Literature Review. *J. Knowl. Econ.* **2024**, *15*, 1952–1972. [CrossRef]
92. Budden, P.; Murray, F. Strategically Engaging With Innovation Ecosystems. *MIT Sloan Manag. Rev.* **2022**, *64*, 38–43.
93. Bressanelli, G.; Perona, M.; Saccani, N. Challenges in supply chain redesign for the Circular Economy: A literature review and a multiple case study. *Int. J. Prod. Res.* **2019**, *57*, 7395–7422. [CrossRef]
94. Dragan, G.B.; Arfi, W.B.; Tiberius, V.; Ammari, A.; Ferasso, M. Acceptance of circular entrepreneurship: Employees' perceptions on organizations' transition to the circular economy. *J. Bus. Res.* **2024**, *173*, 114461. [CrossRef]

processes

Article

A Fuzzy Decision-Making Method for Green Design for Remanufacturability

Yu Cai, Chao Ke * and Qunjing Ji *

School of Art and Design, Wuhan Institute of Technology, Wuhan 430205, China; 21128101@wit.edu.cn
* Correspondence: 22121401@wit.edu.cn (C.K.); 22121402@wit.edu.cn (Q.J.)

Abstract: Designs for remanufacturing (DfRem) consider the remanufacturability of the product in the early stages of product design, which can greatly increase the reusability of the products. However, product design schemes lack reasonable evaluation indicators for remanufacturability, and the decision-makers of the design scheme have subjective preferences and vague hesitation. These result in inaccurate decision making on DfRem schemes that will affect the successful implementation of product remanufacturing. In order to improve the accuracy of the DfRem scheme decision, a fuzzy decision-making method for green design for remanufacturability is proposed. Firstly, an evaluation indicator system for green design schemes was established that takes into account remanufacturability, reliability, cost, and the environment, and the entropy weighting method is used to quantify and weigh the design scheme evaluation indicators. Then, the hesitation fuzzy set is applied to construct the set of evaluations and the optimal design scheme is selected by applying the comprehensive evaluation method. Finally, the feasibility of the above method is verified by using the green design of an injection mold as an example, and the results show that the above method is able to make accurate and effective design scheme decisions. This method has been implemented in a prototype system using Visual Studio 2022 and Microsoft SQL Server 2022. The results show that the fuzzy decision-making system is accurate and effective for rapidly generating a rational green design scheme for remanufacturability.

Keywords: remanufacturing; green design; decision making; entropy weight; hesitant fuzzy set

1. Introduction

Remanufacturing is an industrial process that restores used products to a "new" or "better" than "new" functional state, focusing on material, energy, and cost savings [1–3]. The design for remanufacturing (DfRem) is an important part of the remanufacturing system, which can improve product remanufacturability and facilitate the smooth implementation of product remanufacturing [4,5].

However, the DfRem process can generate multiple design solutions, so designers need to make decisions about the best solution. Unlike traditional designs, DfRem is not only focused on product remanufacturability, but also on the reliability. This makes the design scheme decision very complex. To address this problem, it is necessary to make a comprehensive system of DfRem scheme evaluation indicators. Currently, many scholars have conducted studies on DfRem evaluation indicator system construction. Jiang et al. [6] developed a data-driven ecological performance evaluation method for the remanufacturing process which constructed the evaluation indicator containing the energy-saving rate, remanufacturing process cost, and rate of remanufacturing. Ding et al. [7] proposed an integrated multi-criteria decision-making (MCDM) approach which established the remanufacturability evaluation indicator system for the machine tool guideway. Harivardhini et al. [8] proposed an integrated framework for supporting decision making during early design stages of end-of-life, which can help designers compare and select alternative designs of a product that have better disassembly potential. Peng et al. [9] developed an effective and comprehensive multi-criteria decision-making approach which considered the environmental impact, economic cost, and technical property for the remanufacturing process. Wang et al. [10] proposed a demand-matching multi-criteria decision-making method for reverse logistics, which focuses on four types of evaluation indicators including quality condition, sustainability, economy, and risk. There is no doubt that these studies solved the evaluation indicator system for DfRem, including the remanufacturing process design, design for disassembly, and reverse logistics, and these pieces of research promote the smooth implementation for the DfRem scheme's evaluation.

Unfortunately, this literature only considers remanufacturing in terms of the remanufacturing process, disassembly, or reverse logistics, et al., but seldom considers remanufacturing in terms of the product design phase, and has not constructed a design scheme evaluation indicator system with remanufacturability, which makes it impossible to select a beneficial product design scheme for remanufacturability. Gong et al. [11] established the evaluation criteria for the DfRem scheme, which considered the technical, economic, and environmental factors. Although this method contains the remanufacturing cost and environmentality of the remanufacturing process, the research does not provide an accurate quantitative description of the remanufacturability indicator as a technical indicator. Remanufacturability is an individual characteristic of a product, which is an inherent attribute and should be considered as a technical indicator to guide product design. Moreover, DfRem considers the remanufacturability of the product at the design stage, so the remanufacturability of the product should be evaluated in the design scheme. Most importantly, a design scheme evaluation indicator system for remanufacturability needs to be constructed while mainly including the disassemblability, recyclability, detectability, and reliability of product performance, etc.

Furthermore, designers need to develop evaluation indicator weights in order to make accurate design solution decisions. However, experts determine the importance of indicators by scoring the design scheme evaluation indicators. This is mainly an indicator evaluation based on their personal experience and subjective judgment, which cannot objectively reflect the importance of the indicator. The entropy weighting method is a comprehensive evaluation method that can be used for multiple objects and indicators; moreover, the entropy weighting method is based on the information of each indicator for weight calculation, which is easy to calculate and enable objective weighting based on actual information, thus reducing the interference of subjective factors [12], and many fields have already used the entropy weight method to calculate weights, for example, the weights calculation of nanoparticle evaluation indicators [13], the risk assessment of the tunnel [14], the Groundwater quality assessment [15], and so on. Owing to the advantages of the entropy weight, it can be used to calculate the weights of the DfRem scheme evaluation indicators in order to avoid subjective weighting.

Decision making on DfRem schemes requires the quantification of each evaluation indicator, which is generally scored according to designers or experts. However, experts have hesitation in the face of design scheme evaluation indicators and are unable to give a precise evaluation value, but rather an evaluation interval, which makes the evaluation process ambiguous and uncertain. Hesitant fuzzy set, an effective means to cope with the uncertainty and complexity of decision making [16], and many fields have applied this method for uncertainty decision making, for example, the assessment of a sustainable supplier [17], sustainable city logistics [18], risk decision making [19], etc. Obviously, from the research results of hesitant fuzzy sets in various fields, hesitant fuzzy sets can effectively process fuzzy evaluation information and improve the accuracy of evaluation results.

To improve the accuracy and reliability of a DfRem scheme decision, this paper proposes a fuzzy decision-making method to select the optimal design scheme. The novelties of this paper are listed in the following: (1) The DfRem scheme evaluation system was constructed to comprehensively consider the remanufacturability indicators at the product design stage, thus ensuring product remanufacturability and improving the remanufacturing efficiency of end-of-life products. (2) A DfRem scheme decision-making method was proposed to fully take into account the hesitancy of decision makers in scoring evaluation indicators, thereby reducing the ambiguity of DfRem scheme decision making. (3) DfRem decision-making software 2024 was developed to visualize the design solution decision-making process and to improve the operability of the DfRem decision-making method, thus improving the decision-making efficiency. The remainder of this paper is structured as follows. Section 2 proposes the framework for decision making on DfRem schemes, which contains DfRem scheme evaluation indicator extraction, evaluation indicator system construction, and the DfRem scheme decision. Section 3 constructs a DfRem scheme evaluation indicator system, including reliability, remanufacturability, cost, and environmental indicators. Section 4 proposes a DfRem scheme decision method based on entropy weight and hesitation fuzzy sets. Section 5 validates the method with a DfRem case for the injection mold, and Section 6 draws conclusions for this study.

2. Decision-Making Framework for DfRem

A green design for remanufacturability contains a large amount of design information and constraint information, so it is necessary to extract key evaluation indicators from the above information, and establish an evaluation indicator system for green design schemes, finally adopt a suitable decision-making method to select the optimal design solution. The specific process is shown in Figure 1.

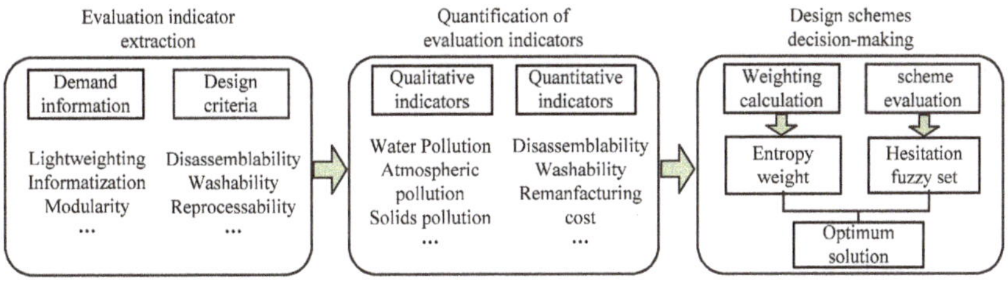

Figure 1. Decision-making framework for DfRem.

The decision-making framework for DfRem includes three main components.

(1) Evaluation Indicator Extraction

The design for remanufacturing needs to take into account remanufacturability requirements, system constraints, and design guidelines. Remanufacturability requirements include connection methods, performance levels, and construction types. System constraints include dimensional constraints, performance constraints, and appearance constraints, etc. System constraints can also be functionally mapped to form functional features of the product. The design guidelines include ease of disassembly, ease of cleaning, and reliability. The green design evaluation indicators need to be extracted from the remanufacturability demand information, and different types of information are described in different ways and quantified in different ways, which can be described by the object element method and transformed into design scheme evaluation indicators. The description is shown in Equation (1).

$$M = (O_r, c_r, v_r, d_r) = \begin{bmatrix} c_{r1} & v_{r1} & d_{r1} \\ \cdots & \cdots & \cdots \\ c_{ri} & v_{ri} & d_{ri} \\ \cdots & \cdots & \cdots \\ c_{rn} & v_{rn} & d_{rn} \end{bmatrix} \tag{1}$$

where M denotes the element information of the product, O_r denotes the r-th target (e.g., product and component), C_{ri} denotes the i-th feature and the corresponding r-th object quantity, respectively, v_{ri} denotes the i-th feature and the corresponding r-th object weight, and d_{ri} denotes the design requirement for the i-th feature of the r-th target.

(2) Quantification of evaluation indicators

Qualitative descriptions are fuzzy evaluations by experts, designers, or clients which are generally described by ambiguous concepts such as degree words and are subjective in nature. Meanwhile, quantitative descriptions are assessed in terms of specific numerical values with objectivity. Since different types of indicators have different levels of measurement, both qualitative and quantitative indicators need to be normalized to ensure the accuracy of the evaluation process.

(3) Design scheme decision making

The evaluation indicator system for green designs containing technical, economic, and environmental factors is constructed from the perspectives of customer demand, enterprise production, and environmental regulations. Meanwhile, the entropy weight method and hesitant fuzzy set (EWM-HFS) are used for the multi-attribute decision making of green design schemes.

3. Evaluation Indicator System Construction

In order to extract and calculate the evaluation indicator values more accurately, the evaluation indicators for green designs for remanufacturability are divided into two levels, the first level indicators include technical, economic, and environmental indicators, the second level indicators are refined based on the first level indicators, for example, technical indicators include easy recyclability, easy disassembly, and reliability, etc. The overall evaluation indicator system is shown in Figure 2.

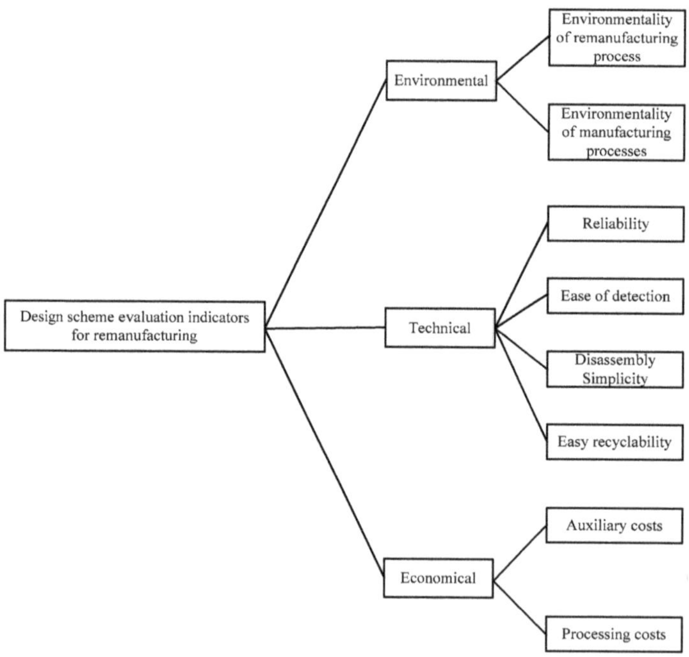

Figure 2. Design scheme evaluation indicator system for remanufacturability.

3.1. Technical Indicators

The technical indicators are evaluated in terms of reliability, disassembly, recyclability, and easy detection, and the process of quantifying each indicator is as follows.

(1) Easy recyclability: Product recycling takes into account factors such as the ease of packaging and transportation and the encapsulation of hazardous materials. The ease

of recycling is mainly measured by the recycling time, and the specific indicators are quantified as follows.

$$T_R = 1 - \frac{T_{R1} - T_{R\min}}{T_{R\max} - T_{R\min}} \quad (2)$$

where T_R denotes the quantified value of the normalized recycling time, T_{R1} denotes the time taken to recycle the product part, $T_{R\min}$ denotes the minimum recycling time, and $T_{R\max}$ denotes the maximum recycling time. All are measured in minutes.

(2) Ease of disassembly: The difficulty of disassembly depends on the number of components, the degree of precision, and the type of connection, etc. The disassembly process minimizes damage to components. The ease of disassembly can be measured by the disassembly time, which is as follows.

$$T_D = 1 - \frac{T_{D1} - T_{D\min}}{T_{D\max} - T_{D\min}} \quad (3)$$

where T_D denotes the quantified value of the normalized product disassembly indicator, T_{D1} denotes the product disassembly time, $T_{D\max}$ denotes the maximum product disassembly time, and $T_{D\min}$ denotes the minimum product disassembly time; the time unit is minutes.

(3) Ease of detection: Since the product is used in different environments and ways, these will affect the quality condition of the product, as well as affecting the difficulty of the parts' quality detection. To simplify the quantification process, the detection time is used to indicate the ease of product detection. Assuming that the product is retired from normal service, the process of quantifying the ease of detection indicator is as follows.

$$T_I = 1 - \frac{T_{I1} - T_{I\min}}{T_{I\max} - T_{I\min}} \quad (4)$$

where T_I is the normalized quantitative value of the product component detectability indicator, T_{I1} is the time required to detect the product component, $T_{I\max}$ is the maximum detection time for the product component, and $T_{I\min}$ is the minimum detection time for the product component; the time unit is minutes.

(4) Reliability: Reliability determines the degree of parts failure during normal service, and also greatly affects the normal operation of products. Generally, the lower the failure degree, the higher the reliability. The reliability of parts can be measured by the normal service time, and then the reliability evaluation indicator is calculated as follows.

$$T_S = \frac{T_{S1} - T_{S\min}}{T_{S\max} - T_{S\min}} \quad (5)$$

where T_{S1} denotes the normal service time of the component, T_S denotes the normalized reliability indicator value, $T_{S\max}$ denotes the maximum service time, and $T_{S\min}$ denotes the minimum service time; the time unit is minutes.

3.2. Economic Indicators

Cost is an important factor to consider in the product design process and determines whether the finished product will be able to circulate in the marketplace, mainly including the processing and auxiliary costs of manufacturing and remanufacturing, and the specific calculation process is as follows.

(1) Processing costs

The green design scheme needs to take into account the manufacturing and remanufacturing costs of the company in order to obtain the maximum economic benefit from the product, which is given by the following formula.

$$C_P = C_M + C_R \quad (6)$$

$$C_Q = 1 - \frac{C_P - C_{P\min}}{C_{P\max} - C_{P\min}} \tag{7}$$

where C_P denotes the total processing cost, C_M denotes the manufacturing cost, C_R denotes the remanufacturing cost, C_Q denotes the normalized processing cost, $C_{P\max}$ denotes the maximum processing cost, and $C_{P\min}$ denotes the minimum processing cost. The unit of cost used is USD.

(2) Ancillary costs

To achieve the proper implementation of manufacturing and remanufacturing, it is necessary to add auxiliary equipment and materials, etc., which are calculated as follows.

$$C_E = C_1 + C_2 \tag{8}$$

$$C_D = 1 - \frac{C_E - C_{E\min}}{C_{E\max} - C_{E\min}} \tag{9}$$

where C_E denotes ancillary costs, C_1 denotes ancillary equipment costs, C_2 denotes ancillary material costs, C_D denotes normalized ancillary costs, $C_{E\max}$ denotes maximum ancillary costs, and $C_{E\min}$ denotes minimum ancillary costs. Here, the cost unit is USD.

3.3. Environmental Indicators

To reduce environmental pollution during the product life cycle, the green design scheme needs to consider the environmental pollution of the manufacturing and remanufacturing process. Product development will produce pollutants during production, and remanufacturing can reduce pollutant emissions by reducing component manufacturing, but the remanufacturing process consumes water, electricity, cutting fluid, and metal materials, which also produce pollutants. The main pollutants in the manufacturing and remanufacturing process are noise pollution, water pollution, dust pollution, and solid pollution. Based on historical manufacturing and remanufacturing data, environmental experts score environmental indicators to determine pollution levels, and the environmental indicators are mainly rated as {very good, good, average, poor, very poor} with a corresponding score value of $\{1.0, 0.8, 0.6, 0.4, 0.2\}$.

4. Fuzzy Decision-Making Methods for Design Schemes

The decision making of green design schemes for remanufacturability mainly consists of three parts. Firstly, the entropy weight method is used to quantify and weigh the design scheme indicators, then the hesitation fuzzy set is used to construct the evaluation value set of each indicator, finally the evaluation value of each solution is calculated according to the comprehensive evaluation function and each solution is ranked in order to select the optimal design scheme [20]. The specific evaluation process is described below.

(1) Entropy weighting method: The entropy weighting method is based on the information of each indicator to set the weight, which is easy to calculate and can be objectively assigned based on the actual information to reduce the interference of subjective factors. The entropy weighting method is calculated as follows.

Firstly, the entropy value of each evaluation indicator is calculated using the following.

$$E_i = -\sum_{j=1}^{n} \frac{x_{ij}}{x_i} \ln \frac{x_{ij}}{x_i} \tag{10}$$

where E_i denotes the entropy value of the evaluation indicator of the i-th design scheme, and x_{ij} denotes the value weight of the i-th indicator in the j-th design scheme.

$$x_i = \sum_{j=1}^{m} x_{ij}, \quad i = 1, 2, \cdots, m \tag{11}$$

where x_i denotes the weight of the i-th indicator value for all design schemes and m denotes the m evaluation indicators.

Then the weight of the i-th evaluation indicator can be calculated as follows,

$$w_i = \frac{d_i}{\sum\limits_{i=1}^{m} d_i} \tag{12}$$

$$d_i = 1 - E_i \tag{13}$$

where d_i indicates the degree of information deviation. Then the weight value of each indicator is $w = (w_1, w_2, \ldots, w_m)^T$.

(2) Hesitant fuzzy sets: In the design scheme decision-making process, decision makers usually hesitate to reach a unified opinion on the decision scheme, which will affect the accuracy of the design schemes' evaluation. In order to obtain reasonable decision results, the hesitant fuzzy set can form a collection of fuzzy evaluation opinions of decision makers and obtain the evaluation level of design schemes through fuzzy operations. The specific process is as follows.

(1) Suppose S is a non-empty set, then the hesitant set S is a function whose every element in the set maps to [0, 1], then the functional expression of the hesitant fuzzy set is as follows:

$$A = \{\langle x, h_A(x)\rangle | x \in S\} \tag{14}$$

where $h_A(x)$ denotes the affiliation of element x with respect to the set A and is also the hesitant fuzzy element of the set.

(2) Given that h_i is a hesitant fuzzy element, the evaluation score of h_{ij} is $s(h_{ij}) = \frac{1}{l_h} \sum\limits_{a \in h} a$, where a is a certain evaluation value in h and l_h denotes the number of evaluation indicators in h.

(3) To calculate the evaluation value of each indicator of the design scheme, the evaluation value can be obtained according to the evaluation value of the indicator. Then the set of evaluation values of each design scheme is described as follows.

$$Y = (y_1, y_2, \cdots, y_n) \tag{15}$$

$$y_i = \{s(h_{i1}), s(h_{i2}), \cdots, s(h_{in})\} \tag{16}$$

(4) The evaluation value of each design scheme is calculated as each indicator value has a corresponding weight, which can be obtained through the entropy weighting method. Moreover, the comprehensive evaluation method is used to calculate the evaluation value of each design scheme, mainly by calculating the multiplication of the weights of each design indicator and the scoring value. The process of calculating the comprehensive evaluation value of each design scheme is as follows.

$$S_i = w * y_i = (w_1, w_2, \cdots, w_n)\begin{vmatrix} s(h_{1i}) \\ s(h_{i2}) \\ \cdots \\ s(h_{in}) \end{vmatrix} \tag{17}$$

where w denotes the weight value of each indicator and y_i represents the assessed value of each design indicator for option i. The evaluation value of all design schemes can be calculated and formed into the set $S = (S_1, S_2, \cdots, S_m)$. The best evaluation scheme can be selected by comparing the evaluation value of each design scheme.

5. Case Study

5.1. Design Scheme Analysis and Information Extraction

Taking injection molding as an example, injection molding is a key device for the production of plastic products, and the structure of the mold is shown in Figure 3. The layout and dimensional accuracy of the mold directly determines the reliability and quality of the product. According to the technical requirements of an electronic product company, a set of injection molds was designed to achieve the mass production of hard disk cover shells while satisfying the quality requirements of a hard disk case. Furthermore, the company hoped that the mold would be able achieve one mold with two cavities and improve the injection precision as much as possible. Meanwhile, due to the high development cost of the injection mold, it was hoped that the injection molding would have a high degree of remanufacturability and reliability. Based on the design requirements, the designer can select three feasible designs from the historical design database, with the specific information shown in Table 1.

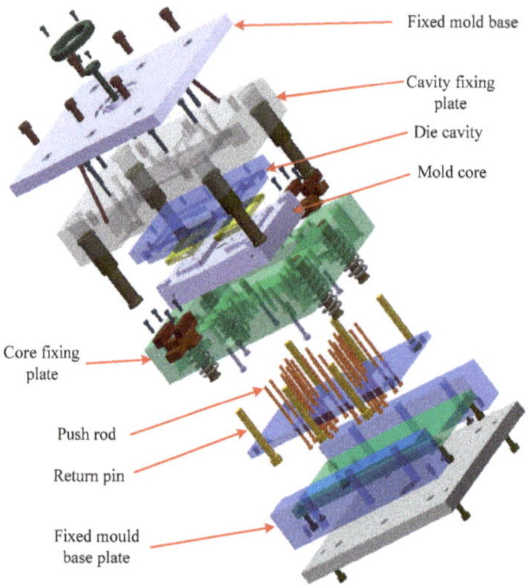

Figure 3. The structure of the injection mold.

Table 1. Design scheme information for injection mold.

Scheme	Material	Total Weight (kg)	Performance	Material Price (USD/kg)	Number of Parts (Pieces)	Number of Modules (Pieces)	Connection Type
A	P20	11	Uniform hardness and good machinability	2.8	98	4	Screws, bolts, pins, welding
	SM45	15	Good machinability, high hardness, good wear resistance	4.9			
	Copper	2	Good toughness and chemical stability	9.8			
B	P20	11	Uniform hardness and good machinability	2.8	90	5	Screws, bolts, pins, welding
	SM45	13.5	Good machinability, high hardness, good wear resistance	4.9			
	Aluminum alloy	1.6	High strength, low toughness, good wear resistance	2.8			
C	P20	12	Uniform hardness and good machinability	2.8	106	3	Screws, bolts, pins
	SM45	15	Good machinability, high hardness, good wear resistance	4.9			
	Copper	2	Good toughness and chemical stability	9.8			

5.2. Design Solution Decisions

Based on the design information in Table 1 and historical manufacturing and remanufacturing data, the design scheme evaluation indicator values for the molds were calculated using Equations (2)–(9), which are shown in Table 2.

Table 2. Quantitative values of evaluation indicators for each design scheme.

Design Scheme	A	B	C
Easy recyclability	$1-\frac{30\text{ min}-27\text{ min}}{37\text{ min}-27\text{ min}}=0.7$	$1-\frac{29\text{ min}-27\text{ min}}{37\text{ min}-27\text{ min}}=0.8$	$1-\frac{33\text{ min}-27\text{ min}}{37\text{ min}-27\text{ min}}=0.6$
Ease of detection	$1-\frac{61\text{ min}-59\text{ min}}{69\text{ min}-59\text{ min}}=0.8$	$1-\frac{62\text{ min}-59\text{ min}}{69\text{ min}-59\text{ min}}=0.7$	$1-\frac{61\text{ min}-59\text{ min}}{69\text{ min}-59\text{ min}}=0.8$
Ease of disassembly	$1-\frac{13\text{ min}-10\text{ min}}{20\text{ min}-10\text{ min}}=0.7$	$1-\frac{12\text{ min}-10\text{ min}}{20\text{ min}-10\text{ min}}=0.8$	$1-\frac{14\text{ min}-10\text{ min}}{20\text{ min}-10\text{ min}}=0.6$
Reliability	$\frac{42367\text{ min}-41667\text{ min}}{42667\text{ min}-41667\text{ min}}=0.7$	$\frac{42467\text{ min}-41667\text{ min}}{42667\text{ min}-41667\text{ min}}=0.8$	$\frac{42267\text{ min}-41667\text{ min}}{42667\text{ min}-41667\text{ min}}=0.6$
Processing costs	$1-\frac{5176\text{ USD}-4896\text{ USD}}{6295\text{ USD}-4896\text{ USD}}=0.8$	$1-\frac{5036\text{ USD}-4896\text{ USD}}{6295\text{ USD}-4896\text{ USD}}=0.9$	$1-\frac{5316\text{ USD}-4896\text{ USD}}{6295\text{ USD}-4896\text{ USD}}=0.7$
Ancillary costs	$1-\frac{1818\text{ USD}-1399\text{ USD}}{2798\text{ USD}-1399\text{ USD}}=0.7$	$1-\frac{1678\text{ USD}-1399\text{ USD}}{2798\text{ USD}-1399\text{ USD}}=0.8$	$1-\frac{1958\text{ USD}-1399\text{ USD}}{2798\text{ USD}-1399\text{ USD}}=0.7$
Environmentality of the manufacturing process	0.8	0.6	0.8
Environmentality of the remanufacturing process	0.8	0.8	0.8

According to the definition of the entropy weight method, MATLAB software 2022 was used to calculate the weight of each evaluation indicator, and the results are as follows.

$$w = [0.0951,\ 0.0835,\ 0.0951,\ 0.0951,\ 0.0951,\ 0.2262,\ 0.0835,\ 0.2262]$$

In order to consider the evaluation indicators comprehensively, industry experts were invited to conduct a subjective evaluation of each indicator. Then, the comprehensive evaluation value of each indicator is shown in Table 3.

Table 3. Score values for each evaluation indicator.

Scheme	A	B	C
Easy recyclability	[0.7, 0.85]	[0.8, 0.85]	[0.6, 0.7]
Ease of detection	[0.8, 0.87]	[0.7, 0.76]	[0.8, 0.84]
Ease of disassembly	[0.7, 0.76]	[0.8, 0.84]	[0.6, 0.68]
Reliability	[0.7, 0.75]	[0.8, 0.86]	[0.6, 0.65]
Processing costs	[0.8, 0.83]	[0.9, 0.95]	[0.7, 0.78]
Ancillary costs	[0.7, 0.75]	[0.8, 0.88]	[0.6, 0.72]
Environmentality of the manufacturing process	[0.8, 0.85]	[0.6, 0.73]	[0.7, 0.75]
Environmentality of the remanufacturing process	[0.738, 0.82]	[0.86, 0.92]	[0.79, 0.83]

Based on the definition of the hesitation fuzzy set, each design scheme is defined as a hesitation fuzzy set. Then, the evaluation indicator and evaluation value of each design scheme can be calculated by Equation (16), and the results are as follows.

$$y(h_A) = \{0.775, 0.835, 0.73, 0.725, 0.815, 0.725, 0.825, 0.779\}$$

$$y(h_B) = \{0.825, 0.73, 0.82, 0.83, 0.925, 0.84, 0.665, 0.89\}$$

$$y(h_C) = \{0.65, 0.82, 0.64, 0.625, 0.74, 0.66, 0.725, 0.81\}$$

According to Equation (17), the overall evaluation value of each design scheme can be calculated, and the results are as follows.

$$S_A = 0.7684,\ S_B = 0.8311,\ S_C = 0.7140$$

From the calculation results, it can be seen that design scheme B has the highest evaluation score, which is better than design scheme *A* and *C*. Therefore, design scheme *B* is chosen as the optimal design scheme, which has the lowest manufacturing cost and can take into account the remanufacturability of the injection molding and ensure the strength of the injection molding.

5.3. Sensitivity Analysis

In order to verify the reliability of the decision-making method, the sensitivity of the comprehensive evaluation results was analyzed by adjusting the weight values of each evaluation indicator. The fluctuation of each indicator weight value was set to 0.1 times, 0.2 times, original value, 5 times, and 10 times of the original weight value, and the experimental results are as follows.

(1) From Figures 4 and 6–9, it can be seen that when the indicators weights including ease of recovery, ease of disassembly, reliability, processing costs, and auxiliary costs are increased, the evaluation values of all three design schemes increase, but design scheme B is always the design scheme with the highest overall rating, which indicates that the evaluation model is stable.

(2) As can be seen from Figure 5, when the weight of the ease of detection indicator increases, the evaluation value of all three design schemes increases. Scheme B is the design scheme with the highest overall rating at 0.1, 0.2, and 5 times the original weight, but at 10 times the original weight, the overall evaluation value is slightly lower than that of design scheme A, which is in the second position and has better overall stability.

(3) As can be seen from Figure 10, when the weight of the manufacturing process' environmental indicator increases, the evaluation value of all three design schemes increases. Scheme B is the design scheme with the highest overall rating at 0.1, 0.2, and 5 times the original weight, but at 10 times the original weight, the overall evaluation value is slightly lower than that of design scheme A and is in the second position, which has better overall stability.

(4) From Figure 11, it can be seen that when the weight of the environmental indicators of the remanufacturing process increases, the evaluation value of all three design schemes increases. Scheme A is the second-highest scoring in terms of overall rating at 0.1, 0.2, and 5 times the original weight, but at 10 times the original weight, the overall evaluation value is slightly lower than that of design scheme C, which is in the third position. Although the evaluation ranking of design scheme A decreases at a multiple of 10, design scheme B is always the highest overall rating, and the stability of the evaluation model is better.

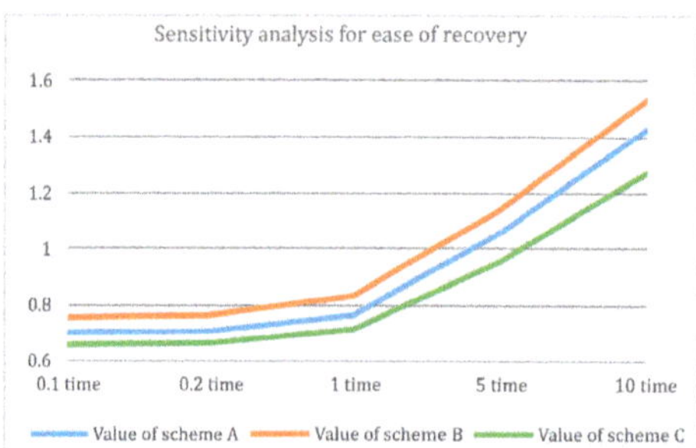

Figure 4. Sensitivity analysis for ease of recovery.

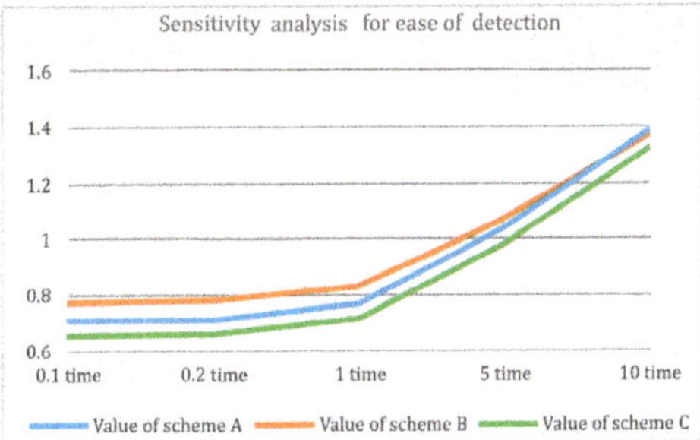

Figure 5. Sensitivity analysis for ease of detection.

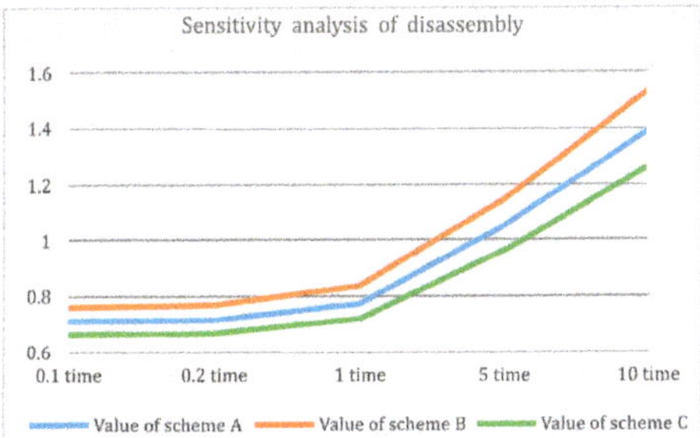

Figure 6. Sensitivity analysis of disassembly.

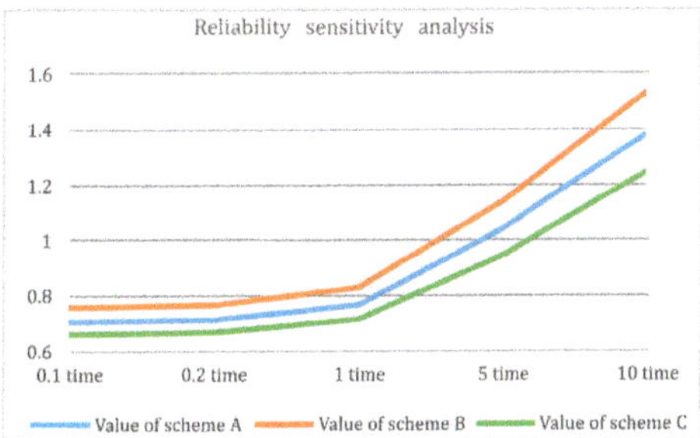

Figure 7. Reliability sensitivity analysis.

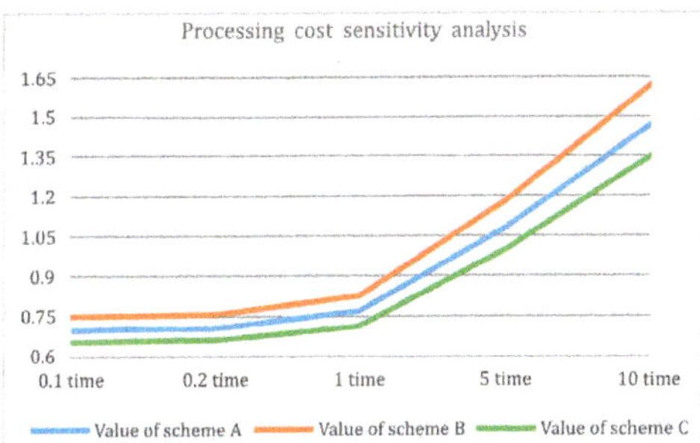

Figure 8. Processing cost sensitivity analysis.

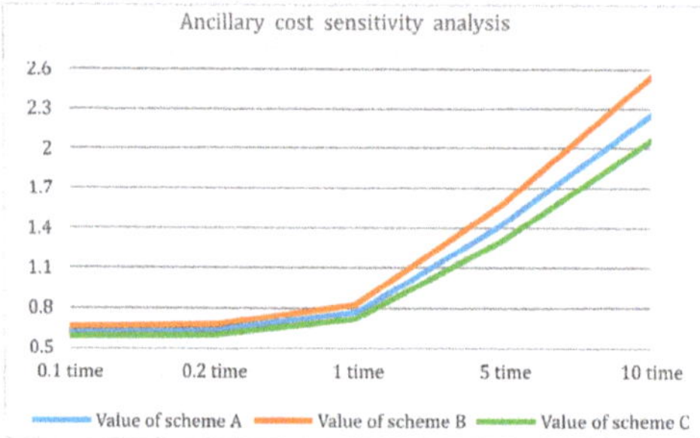

Figure 9. Ancillary cost sensitivity analysis.

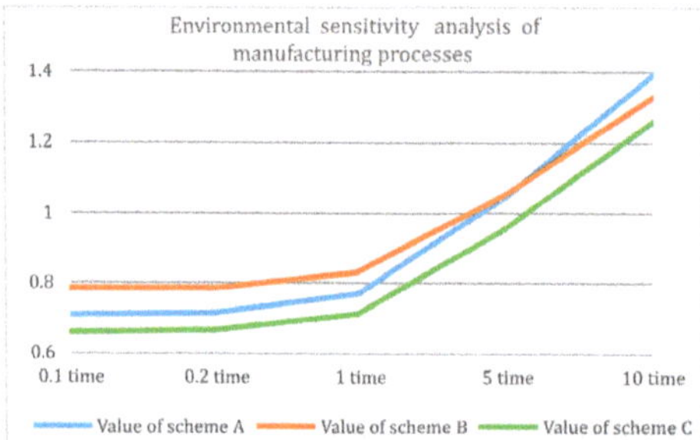

Figure 10. Environmental sensitivity analysis of manufacturing processes.

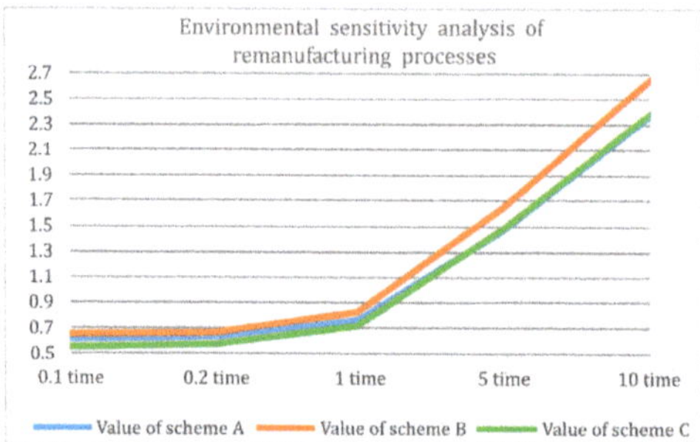

Figure 11. Environmental sensitivity analysis of remanufacturing processes.

From the results of the sensitivity analysis of each indicator, it can be seen that the eight secondary indicators of the design schemes' evaluation system have little influence on the evaluation results when fluctuations occur, and scheme B is always the optimal solution, so the stability of the evaluation model is relatively good.

5.4. Model Implementation on Interactive Interface

To improve the operability of the decision-making method, an interactive interface for green design scheme decision making was developed based on Visual Studio 2022 and SQL Server 2022, as shown in Figure 12. This interface consists of a design search layer, an input and quantification layer, a weight calculation layer, and a decision-making layer to facilitate the implementation of the proposed methodology. The interface allows the user to input information regarding the recycling time, disassembly time, testing time, normal service time, processing cost, and ancillary costs. Then the system performs the similarity calculation, and users can receive a recommended case from the existing case database. The design search interface is used to retrieve green design cases that satisfy the design requirements in the case database. The input and quantization interface are used to input the data for each indicator, and the system backend quantifies the value of each indicator. The weight calculation interface provides the weight values of each indicator and users can choose different visualization forms to generate tables, bar charts, or bar graphs. The decision-making interface is used to input the subjective evaluation value of each indicator by industry experts and the system background calculates the evaluation indicator value of each design scheme.

STEP 1 Design scheme search

STEP 2 Input and quantification

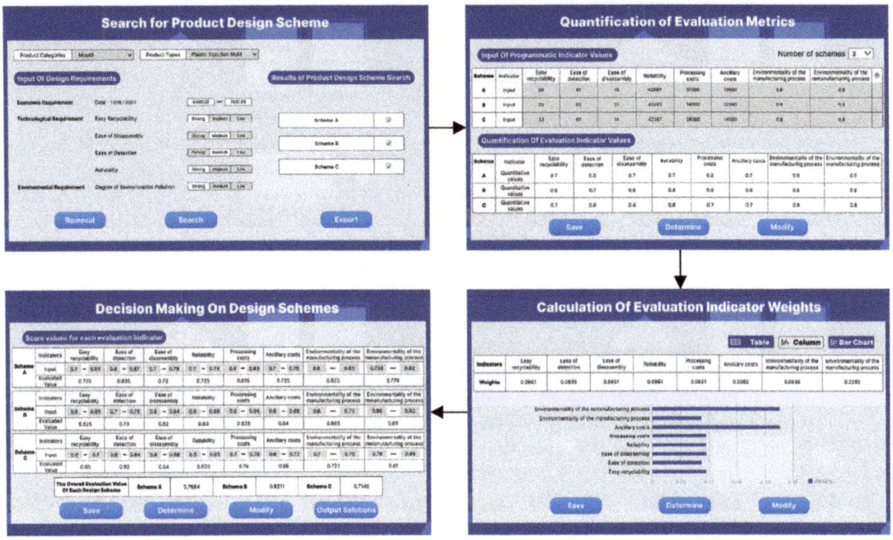

STEP 4 Decision-making

STEP 3 Calculate the weights

Figure 12. Operational process of green design scheme decision-making system.

6. Conclusions and Future Work

This paper proposes a fuzzy decision-making method for green design for remanufacturability, it has constructed a green design evaluation indicator system that includes economic, technical, and environmental aspects of the products, and it fully considers the characteristics of products such as easy disassembly, easy recycling, and reliability. Moreover, entropy weighting and hesitation fuzzy sets are applied to evaluate and decide on alternative design solutions in order to obtain the optimal design solution. Meanwhile, a design scheme decision-making system was developed to improve the user-friendliness of the methodology. Finally, the feasibility of the above-mentioned method is verified by taking the injection molding design as an example.

The results show that experts normally score the design scheme evaluation indicators by giving a point interval rather than an exact value, resulting in ambiguity in the values of the evaluation indicators, which makes it difficult to ensure the accuracy of the decision. In this method, the entropy weighing method can objectively set the weights of design indicators and the hesitation fuzzy set makes it possible to set the scoring intervals, which solves the hesitation and uncertainty of the expert evaluation results, and these contribute to improving the reliability of the design scheme evaluation results.

Future work requires efforts in the following aspects: (1) The proposed decision-making method only utilizes the existing small amount of data for indicator weight calculation, indicator evaluation, and decision making, and in the future, intelligent technologies such as big data, deep learning, and knowledge reuse can be used to make decisions on design solutions for improving the efficiency of DfRem decision making. (2) The decision-making method proposed in this paper is suitable for decision making with a small number of design schemes. When faced with a large number of design schemes, decision-making efficiency is low and the efficiency analysis technique can be considered for large-scale decision making, such as data envelope analysis. (3) Due to different service environments and usage habits, the remanufacturability of electromechanical products is dynamic and the evaluation of remanufactured electromechanical product design solutions only takes into account the current solid remanufacturability, which needs to be taken into account in future research in order to develop a more accurate and comprehensive design solution evaluation indicator system.

Author Contributions: Software, Y.C.; Formal analysis, Q.J.; Resources, Y.C.; Writing—original draft, C.K.; Writing—review and editing, Q.J. All authors have read and agreed to the published version of the manuscript.

Funding: This work was supported by the Wuhan Institute of Technology Research Foundation Project [K2023018] and the Sustainable Design and Product Ecological Innovation Team Project. These financial contributions are gratefully acknowledged.

Data Availability Statement: Data are contained within the article.

Conflicts of Interest: The authors declare no conflicts of interest.

References

1. Ai, X.; Jiang, Z.; Zhang, H.; Wang, Y. Low-carbon product conceptual design from the perspectives of technical system and human use. *J. Clean. Prod.* **2020**, *244*, 118819. [CrossRef]
2. Jiang, Z.; Ding, Z.; Liu, Y.; Wang, Y.; Hu, X.; Yang, Y. A data-driven based decomposition-integration method for remanufacturing cost prediction of end-of-life products. *Robot. Comput.-Integr. Manuf.* **2020**, *61*, 101838. [CrossRef]
3. Ke, C.; Jiang, Z.; Zhang, H.; Wang, Y.; Zhu, S. An intelligent design for remanufacturing method based on vector space model and case-based reasoning. *J. Clean. Prod.* **2020**, *277*, 123269. [CrossRef]
4. Tolio, T.; Bernard, A.; Colledani, M.; Kara, S.; Seliger, G.; Duflou, J.; Battaia, O.; Takata, S. Design, management and control of demanufacturing and remanufacturing systems. *CIRP Ann.-Manuf. Technol.* **2017**, *66*, 585–609. [CrossRef]
5. Ke, C.; Jiang, Z.; Zhu, S.; Wang, Y. An integrated design method for remanufacturing process based on performance demand. *Int. J. Adv. Manuf. Technol.* **2022**, *118*, 849–863. [CrossRef]
6. Chen, D.; Jiang, Z.; Zhu, S.; Zhang, H. A knowledge-based method for eco-efficiency upgrading of remanufacturing process planning. *Int. J. Adv. Manuf. Technol.* **2020**, *108*, 1153–1162. [CrossRef]
7. Jiang, Z.; Jiang, Y.; Wang, Y.; Zhang, H.; Cao, H.; Tian, G. A hybrid approach of rough set and case-based reasoning to remanufacturing process planning. *J. Intell. Manuf.* **2019**, *30*, 19–32. [CrossRef]
8. Jiang, Z.; Ding, Z.; Zhang, H.; Cai, W.; Liu, Y. Data-driven ecological performance evaluation for remanufacturing process. *Energy Convers. Manag.* **2019**, *198*, 111844. [CrossRef]
9. Jiang, X.; Tian, Z.; Liu, W.; Tian, G.; Gao, Y.; Xing, F.; Suo, Y.; Song, B. An energy-efficient method of laser remanufacturing process. *Sustain. Energy Technol. Assess.* **2022**, *52*, 102201. [CrossRef]
10. Ke, C.; Pan, X.Y.; Wan, P.; Jiang, Z.G.; Zhao, J.J. An integrated design method for used product remanufacturing scheme considering carbon emission. *Sustain. Prod. Consum.* **2023**, *41*, 348–361. [CrossRef]
11. Jiang, Z.; Zhou, T.; Zhang, H.; Wang, Y.; Cao, H.; Tian, G. Reliability and cost optimization for remanufacturing process planning. *J. Clean. Prod.* **2016**, *135*, 1602–1610. [CrossRef]
12. Zhao, J.; Xue, Z.; Li, T.; Ping, J.; Peng, S. An energy and time prediction model for remanufacturing process using graphical evaluation and review technique (GERT) with multivariant uncertainties. *Environ. Sci. Pollut. Res.* **2021**, 1–13. [CrossRef]
13. Ramírez, F.J.; Aledo, J.A.; Gamez, J.A.; Pham, D.T. Economic modelling of robotic disassembly in end-of-life product recovery for remanufacturing. *Comput. Ind. Eng.* **2020**, *142*, 106339. [CrossRef]
14. Shi, J.; Xu, H.; Shu, F.; Ren, M.; Lu, Z. A PCA-based method for remanufacturing process optimization from sustainability aspects. *Int. J. Ind. Eng.-Theory Appl. Pract.* **2023**, *30*, 986–998.
15. Peng, H.; Wang, H.; Chen, D. Optimization of remanufacturing process routes oriented toward eco-efficiency. *Front. Mech. Eng.* **2019**, *14*, 422–433. [CrossRef]
16. Liu, C.; Zhu, Q.; Wei, F.; Rao, W.; Liu, J.; Hu, J.; Cai, W. An integrated optimization control method for remanufacturing assembly system. *J. Clean. Prod.* **2020**, *248*, 119261. [CrossRef]
17. Liu, Q.; Shang, Z.; Ding, K.; Guo, L.; Zhang, L. Multi-process routes based remanufacturability assessment and associated application on production decision. *J. Clean. Prod.* **2019**, *240*, 118114. [CrossRef]

18. Cavique, L.; Cavique, M.; Mendes, A.; Cavique, M. Improving information system design: Using UML and axiomatic design. *Comput. Ind.* **2022**, *135*, 103569. [CrossRef]
19. Chen, J.; Tang, J.; Yang, D. Study on sensitivity analysis of tooth surface roughness parameters and contact stress. *J. Northwest. Polytech. Univ.* **2022**, *40*, 883–891. [CrossRef]
20. Zhang, X.; Ao, X.; Jiang, Z.; Zhang, H.; Cai, W. A remanufacturing cost prediction model of used parts considering failure characteristics. *Robot. Comput.-Integr. Manuf.* **2019**, *59*, 291–296. [CrossRef]

processes

MDPI

Article

Development of Remanufacturing Readiness Index for MSMEs: A Comprehensive Framework

Abdulaziz Alotaibi

Industrial Engineering Department, Faculty of Engineering, University of Tabuk, Tabuk 47512, Saudi Arabia; a.aldalbahi@ut.edu.sa

Abstract: Micro-, small-, and medium-sized enterprises (MSMEs) have the opportunity to increase resource efficiency, decrease waste, and promote sustainability by implementing remanufacturing techniques. However, determining whether MSMEs are prepared for the adoption of such techniques requires a methodical approach that considers several aspects of manufacturing readiness, such as core acquisition, the design of remanufacturing, and others. Therefore, this study proposes a framework to measure the readiness for the adoption of remanufacturing practices. Further, the remanufacturing readiness index (RRI) is proposed by combining remanufacturing indicators with a structural approach based on graph theory and matrices (GTM). Nine remanufacturing readiness attributes are identified through the literature and validated by an expert team. These nine attributes include core acquisition, reverse logistics availability, resource availability for remanufacturing initiatives, the design for remanufacturing, enterprise collaboration, remanufactured product positioning, performance measurement, labor skill and availability, and a flexible remanufacturing system. The finalized remanufacturing readiness attributes are modelled using GTM to explore their interdependencies, forming the basis for a quantitative index (RRI) that reflects MSMEs' readiness for the adoption of remanufacturing. It is used to measure the possibility of MSMEs implementing remanufacturing processes. To illustrate the efficacy of the RRI, a case study of a remanufacturing facility was conducted. This RRI acts as a decision-support tool to help MSMEs, industry stakeholders, and governments identify priority areas for intervention, promote resource efficiency, and create sustainable growth. The results highlight the importance of readiness attributes as fundamental components in implementing remanufacturing practices at the MSME level.

Keywords: graph matrix theory; remanufacturing; remanufacturing readiness attributes; end-of-life management; sustainability

1. Introduction

The significance of the environment and sustainable practices has grown significantly with population growth and industrialization across the globe [1,2]. Even though industrial growth makes life easier for people, it also causes issues like ecological damage and environmental degradation, which makes sustainable development challenging [3,4]. In order to address these challenges, countries are taking certain measures such as reducing resource usage [5], the end-of-life management of products [6,7], the adoption of circular practices [8,9], waste reduction [10], and more. End-of-life management practices such as product recovery have gained more attention in the past decades [11]. As a result, there has been a notable increase in product recovery initiatives such as reconditioning, repair, and remanufacturing [12,13]. Remanufacturing is considered one of the facilitators for end-of-life management, waste reduction, value recovery, and circular economy initiatives [14].

Remanufacturing is an industrial process that involves cleaning, dismantling, inspecting, and restoring used products to nearly new conditions [15]. It has gained popularity recently because of its intrinsic benefits to the environment, society, and the economy [16]. Remanufactured goods require only 60% of the energy and 70% of the resources compared to producing new items from virgin materials [2,17]. Remanufacturing procedures also emit fewer emissions than traditional manufacturing methods [18]. In addition, remanufacturing reduces the requirement for energy and virgin materials, which decreases the cost associated with them. These advantages push industries to adopt remanufacturing practices for their sustainable growth [19,20].

Remanufacturing is defined as "an industrial process whereby used products are restored to useful life" [21]. Due to increased consumer awareness, government regulations, and the environment, it has become increasingly important in recent years for both industries and consumers to adopt sustainable practices such as remanufacturing [22,23]. The adoption of remanufacturing practices is influenced by several factors such as core acquisition [21], reverse logistics availability [24], the design for remanufacturing [25], collaboration [24], and others. These factors measure the readiness of an enterprise to adopt remanufacturing practices, and it depends on the type of enterprise: large-, small-, or micro-sized enterprises. The nature of an enterprise can significantly influence its adoption and implementation of remanufacturing practices due to various factors such as resource availability, expertise, market dynamics, and strategic priorities [26]. For instance, large enterprises have greater financial capability to invest in sophisticated remanufacturing facilities and technologies and specialized R&D teams for remanufacturing techniques. Despite this, these enterprises also face some challenges such as higher initial setup costs and complexities involved in the integration of remanufacturing systems. On the other side, MSMEs can implement pilot remanufacturing projects due to simpler organizational structures, but they have limited financial resources and expertise. This suggests that the adoption of remanufacturing varies across enterprises [27].

Several studies focus on the implementation of remanufacturing from various aspects such as barriers [28], determinants [23], and drivers [29]. However, there are few studies that focus on readiness to adopt remanufacturing. Some studies focus on the initiatives that are required to start remanufacturing, such as [30], which explores the initiatives to start remanufacturing, and Ref. [24], which proposes a framework for initiating remanufacturing and identifies the prerequisite factors for remanufacturing. Ref. [31] proposed a maturity model to measure remanufacturing practices. It is evident from the literature that most studies focus on initiatives, prerequisites, and maturity levels for remanufacturing but do not address whether the manufacturing facilities is ready for operations or not?. Additionally, there is a lack of quantitative assessment methods to measure this readiness. Therefore, a significant gap exists in the literature regarding the quantitative evaluation of remanufacturing readiness. This study tries to fill this gap in the literature.

MSMEs face several challenges in preparing to adopt remanufacturing practices [32]. Financial constraints, lack of awareness and expertise, market challenges, and many other factors prevent MSMEs from being ready to adopt remanufacturing practices [33,34]. MSMEs also lack knowledge on how remanufacturing practices can be a game-changer and what factors need to be considered for their readiness to adopt remanufacturing practices [35]. This problem needs to be addressed to prepare MSMEs for the adoption of remanufacturing practices. To address this issue, this study proposes a framework to measure the readiness of an organization for the adoption of remanufacturing practices. Based on this framework, a quantitative measure (i.e., the RRI) is also proposed for measuring the readiness of an organization to adopt remanufacturing practices. This framework can also help assess the organization's progress towards their readiness to adopt remanufacturing practices. Further, this can also be used to compare two organizations in terms of their readiness to use the RRI. The specific objectives of this study are as follows:

- To identify the remanufacturing readiness attributes for MSMEs;
- To propose a framework for measuring their readiness to adopt remanufacturing practices;
- To develop the RRI for MSMEs.

The first objective aims to systematically identify the major critical attributes that determine the readiness of MSMEs to adopt remanufacturing practices. It involves a comprehensive review of the existing literature and engagement with industry experts to capture the attributes that influence readiness. Further, the second objective focuses on designing a structured and quantitative framework to assess remanufacturing readiness. The framework will incorporate the identified attributes and establish a methodological approach for evaluating MSMEs' readiness. It will provide a systematic tool for measuring the readiness levels of the current facility. The third objective involves constructing a quantitative index to measure the degree of readiness for a remanufacturing facility. The RRI will integrate the identified attributes and apply graph matrix theory to generate a comprehensive readiness score. The index will facilitate benchmarking, comparisons between facilities, and informed decision making to support the successful adoption of remanufacturing in MSMEs.

To fulfil the mentioned objectives, a comprehensive literature review was performed to identify the significant remanufacturing readiness attributes of MSMEs. These attributes were validated by an expert team. Further, GTM is applied to develop the interrelationship between the identified attributes. Based on this relationship, GTM is applied to develop the framework for measuring the readiness to adopt remanufacturing practices. This proposed framework is applied to determine the RRI and illustrated through the case study. This study has significant contributions to the existing literature, some of which are highlighted below.

Firstly, this study systematically identifies significant attributes that determine MSMEs' readiness to adopt remanufacturing practices. By filling a literature gap in which no prior work has established these attributes clearly, the research attains a foundational understanding of the most significant attributes influencing remanufacturing readiness. Secondly, this study proposed a novel and structured framework to measure remanufacturing readiness. This framework encompasses identified attributes and assesses readiness to adopt remanufacturing facilities in the context of MSMEs. Thirdly, a significant contribution of this study is the development of the RRI, which quantitatively measures the readiness level of MSMEs. Using GMT, the RRI offers a robust and scalable tool for benchmarking readiness, facilitating objective comparisons between different manufacturing facilities. Finally, the proposed framework and the RRI have been applied to a real-life MSME case study to demonstrate how the RRI (quantitative index) can be utilized to measure readiness. This real-life validation provides decision makers with a concrete illustration of how to apply the framework and interpret results for strategic planning.

The remaining article is structured as follows: Section 2 identifies remanufacturing readiness attributes, Section 3 proposes the remanufacturing readiness assessment framework, Section 4 provides the readiness assessment of a remanufacturing facility, Section 5 demonstrates the application of the proposed framework, Section 6 highlights the implications of the study, and finally, Section 7 concludes the study and outlines its limitations and future research directions.

2. Remanufacturing Readiness Attributes

The integration of a literature review and experts' input is applied to identify the attributes for readiness to adopt remanufacturing practices. A systematic methodology was implemented to identify and validate remanufacturing readiness attributes specific to MSMEs using the PRISMA framework. The PRISMA framework consists of four steps: identifying the relevant literature, screening and selecting studies based on inclusion and exclusion criteria, extracting data from the chosen studies, and synthesizing and summarizing the findings [36]. To identify the relevant literature, this study initiated an extensive search in the Scopus database, employing Boolean operators to combine keywords such as "remanufacturing", "value restore", "refurbishing", "readiness", and "initiative". These keywords are converted into search strings using the Boolean operator (AND/OR). The search was restricted to articles in English published post-2014 to ensure contemporary relevance. The articles were screened through a multi-stage process, starting with title and abstract reviews, followed by an independent evaluation conducted by two authors to minimize selection bias and enhance reliability. The adopted PRISMA framework is illustrated in Table 1.

Table 1. Article selection criteria for identification of remanufacturing attributes.

Stage	Description	Number of Records
Identification	An extensive search in the Scopus database using Boolean operators and keywords such as "remanufacturing", "value restore", "refurbishing", "readiness", "initiative" (TITLE-ABS-KEY ("remanufacturing" OR "value restore" OR "refurbishing") AND TITLE-ABS-KEY ("readiness" OR "initiative")) AND PUBYEAR > 2013 AND PUBYEAR < 2025 AND (LIMIT-TO (LANGUAGE, "English"))	130
Screening	Articles were screened through a multi-stage process: initial title and abstract reviews.	47
Eligibility	Full-text articles were independently evaluated by two authors to minimize selection bias and enhance reliability.	44
Inclusion and Exclusion	Articles were included after meeting all inclusion criteria (English language, post-2014, and relevant to remanufacturing readiness for MSMEs). We excluded articles that were published as a book chapter, editorial, or book series.	21

Following a comprehensive review of the finalized articles, eleven key remanufacturing readiness attributes were identified. To ensure robustness, a validation exercise was conducted through an expert panel comprising ten members: eight industry professionals with over a decade of management-level experience in remanufacturing and sustainability, employed by organizations with a minimum of 18 years of operational history, and two academicians with substantial expertise in sustainability and remanufacturing systems. This integration of industry and academic perspectives provided a rigorous foundation for the validation and applicability of the identified attributes. The experts' information is provided in Table 2.

Table 2. Experts' details.

Expert No.	Organisation Type	Designation	Experience (Year)	Educational Qualification	Expertise Area
1	Small	Manager	12	Postgraduate	Remanufacturing
2	Medium	Operations Manager	13	Graduate	Manufacturing
3	Medium	Procurement Manager	11	Postgraduate	Procurement
4	Micro	Supply chain manager	15	Postgraduate	Supply chain
5	Medium	R&D Head	10	PhD	Remanufacturing and sustainability
6	Micro	General Manager	13	Postgraduate	Remanufacturing
7	Small	Logistics Manager	14	Postgraduate	Logistics
8	Small	Design Engineer	12	PhD	Design
9	Academic Institution	Professor	19	PhD	Remanufacturing and closed-loop supply chains
10	Academic Institution	Associate Professor	15	PhD	Sustainability and circular economy

The identified list of eleven remanufacturing readiness attributes was presented to the expert panel for their feedback. The expert team, comprising industry professionals and academicians, reviewed the remanufacturing readiness attributes and suggested the removal of two attributes deemed no longer relevant in the current MSME context. Based on their recommendations, the two attributes were excluded from the list, and the remaining nine attributes were finalized for subsequent analysis. This iterative refinement process ensured that the finalized remanufacturing readiness attributes were both relevant and aligned with practical and theoretical insights. The finalized remanufacturing readiness attributes are shown in Table 3.

Table 3. Remanufacturing readiness attributes.

Remanufacturing Readiness Attributes	Code	Description	References
Core acquisition	ReM1	Core acquisition refers to the process of collecting used, worn-out, or end-of-life products (called "cores") that can be remanufactured to meet the original quality standards.	[37–39]
Reverse logistics availability	ReM2	Facilitates the collection, transportation, and management of cores across the supply chain, playing a critical role in remanufacturing.	[38,40,41]

Table 3. *Cont.*

Remanufacturing Readiness Attributes	Code	Description	References
Resource availability for remanufacturing initiatives	ReM3	Resource availability such as manpower, capital, and technology to initiate remanufacturing practices	[25,42,43]
Design for remanufacturing	ReM4	Design for remanufacturing focuses on optimizing product design to facilitate efficient disassembly, cleaning, repair, and reassembly, ensuring maximum value recovery.	[31,44–46]
Enterprise collaboration	ReM5	Facilitates the collection, transportation, and management of cores across the supply chain, playing a critical role in remanufacturing	[24,47,48]
Remanufactured product positioning	ReM6	It refers to the strategic placement of remanufactured products in the market to ensure they align with customer expectations and market demand.	[29,49,50]
Performance measurement	ReM7	Performance measurement systems facilitate the evaluation of efficiency and effectiveness of remanufacturing processes, including productivity, quality, environmental impact, and profitability.	[44,51,52]
Labor skill and availability	ReM8	The availability of a qualified workforce is critical for successful remanufacturing operations and maximizing value recovery. It involves ensuring that personnel possess the knowledge, technical skills, and expertise required for the complex processes involved in remanufacturing.	[31,53]
Flexible remanufacturing system	ReM9	A flexible remanufacturing system handles inconsistencies in core quality, fluctuating supply volumes, and variability in product designs, ensuring adaptability to changing demands.	[54–56]

3. Development of the Remanufacturing Readiness Assessment Framework

Remanufacturing attributes and their interdependence are necessary for the assessment of readiness to adopt remanufacturing practices. The identified remanufacturing attributes are modelled using the GTM approach to determine the interdependence among them. GTM allows for the structural modelling of these attributes [57,58]. The digraph model is well suited for understanding the linkages among remanufacturing attributes and the development of the RRI. GTM is an organized and consistent decision-making method that maintains the interdependence between attributes. It can simultaneously accommodate any number of quantitative and qualitative determinants [58]. Unlike conventional MCDM methods, GTM provides a visual representation of the problem, offering insights that other approaches may overlook [59]. Furthermore, GTM mathematically captures the interaction between a problem's components, transforming complex network relationships into simple, comprehensible matrices for systematic and critical analysis [60]. In addition, the matrix representation of the problem is compatible with computer programs, enabling efficient processing and analysis.

A digraph has a limited number of vertices, or nodes, and directed edges that join them [61]. The edges show how they interact, whereas the nodes may stand for different system elements, states, variables, etc. [59]. This notion is utilized in the construction of the RRI and the modelling of remanufacturing attributes. The structure of the system is depicted using a digraph.

The nodes in a digraph represent the remanufacturing attributes, and the connecting edges (or arrow/directed edges) show the interdependence among them. If node i is connected to node j through an arrow from node i to node j then it shows the influence of remanufacturing readiness attributes i over j.

The remanufacturing readiness attribute digraph of the remanufacturing facility consists of nine nodes: core acquisition (ReM1), reverse logistics availability (ReM2), resource availability for remanufacturing initiatives (ReM3), the design for remanufacturing (ReM4), enterprise collaboration (ReM5), remanufactured product positioning (ReM6), performance measurement (ReM7), labor skill and availability (ReM8), and a flexible remanufacturing system (ReM9). The edges also display how these attributes interact with one another. The digraph of remanufacturing attributes is displayed in Figure 1.

The process of creating the digraph of remanufacturing attributes entails establishing the relationships between these attributes. An experienced team was assembled, including academicians and remanufacturing and circular economy professionals, such as managers, operations managers, and logistics managers. The experts were selected based on their managerial positions and years of professional experience, with a focus on those possessing over a decade of expertise in end-of-life management practices, remanufacturing, and sustainable manufacturing and those working at the managerial level. The expert panel included 10 members from the industry and academia, ensuring diverse perspectives and balanced opinions. This selection process ensures that the study benefits from the knowledge of experienced professionals with hands-on experience in remanufacturing and related fields. Table 2 contains the experts' details such as their experience, education, and domain.

As per the experts' input, each remanufacturing attribute impacts the other attributes. For instance, if ReM2 has an impact on ReM1, then an edge exists between nodes 1 and 2. This is depicted by a direct edge originating from node ReM2 and directed to node ReM1. Similarly, the expert team constructed the connections between the other remanufacturing readiness attributes, which can be seen in the digraph in Figure 1.

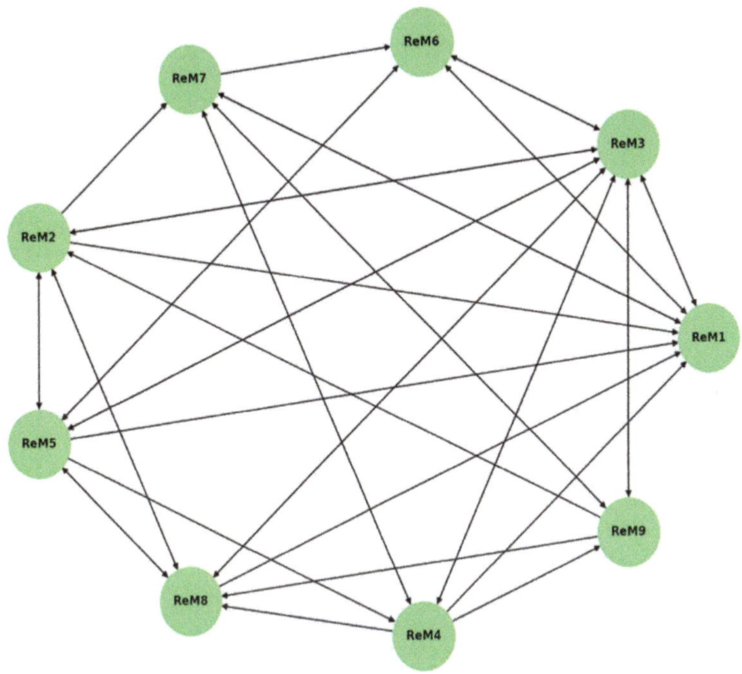

Figure 1. Digraph for the remanufacturing readiness attributes.

Development of Equivalent Matrix of the Digraph

This section explains how to create an equivalent matrix from the digraph of remanufacturing readiness attributes. The equivalent matrix is the matrix version of the digraph of remanufacturing readiness, and it is the asymmetric adjacency matrix. The adjacency matrix will produce a mathematical equation that shows the RRI. Equation (1) represents the remanufacturing readiness attribute matrix (ReM_I) that corresponds to the remanufacturing readiness attribute digraph shown in Figure 1.

$$ReM_I = \begin{bmatrix} M_1 & 0 & m_{13} & 0 & 0 & m_{16} & m_{17} & 0 & 0 \\ m_{21} & M_2 & m_{23} & 0 & m_{25} & 0 & m_{27} & m_{28} & 0 \\ m_{31} & m_{32} & M_3 & m_{34} & m_{35} & m_{36} & 0 & m_{38} & m_{39} \\ m_{41} & 0 & m_{43} & M_4 & 0 & 0 & m_{47} & m_{48} & m_{49} \\ m_{51} & m_{52} & m_{53} & m_{54} & M_5 & m_{56} & 0 & m_{58} & 0 \\ m_{61} & 0 & m_{63} & 0 & m_{65} & M_6 & 0 & 0 & 0 \\ m_{71} & 0 & 0 & m_{74} & 0 & m_{76} & M_7 & 0 & m_{79} \\ m_{81} & m_{82} & m_{83} & 0 & m_{85} & 0 & 0 & M_8 & 0 \\ 0 & m_{92} & m_{93} & 0 & 0 & 0 & m_{97} & m_{98} & M_9 \end{bmatrix} \quad (1)$$

In Equation (1), 'M_i' represents the numerical score of a remanufacturing facility on attribute '*Ai*', using the linguistic score as provided in Table 4. Further, the off-diagonal element 'm_{ij}' represents the degree of influence of attributes i on j using the linguistic score provided in Table 5. The off-diagonal elements with a value of zero in Equation (1) show the absence of any linkages between the corresponding attributes. This matrix represents the remanufacturing readiness attribute matrix for the nine identified attributes; however, it is adaptable and can be extended to include a varying number of attributes based on specific needs as shown in Equation (2).

This matrix is utilized to determine the permanent function to assess the readiness of the facility to adopt remanufacturing. The permanent of the matrix, denoted as *Per (ReM)*, represents the remanufacturing readiness function (RRF) of the remanufacturing facility. Like the determinant of the matrix, the permanent function has a positive sign rather than a negative one [62]. The permanent function was chosen by the researchers over the determined function since the determinant functions lose some information because of negative signs [63]. Several studies have used the permanent function to develop the indices for different measures such as the sustainability index and performance index. For instance, ref. [64] use the permanent function to develop the sustainability index, ref. [58] use it for the sustainability index for logistics, and ref. [57] use it for the circular economy performance index.

Table 4. Attribute assessment for remanufacturing facility.

Linguistic Scale	Crisp Scale
Very High	5
High	4
Medium	3
Low	2
Very Low	1
Unavailable	0

Table 5. Degree of influence among attributes.

Linguistic Scale	Crisp Scale
High	4
Above average	3
Average	2
Minor	1
None	0

To formalize the derivation of a common expression for a remanufacturing readiness matrix (i.e., ReM), having N distinct attributes is critical to remanufacturing readiness. This is represented mathematically in Equation (2).

$$\mathrm{ReM_N} = \begin{bmatrix} M_1 & m_{12} & \cdots & m_{1N} \\ m_{21} & M_1 & & m_{2N} \\ & \vdots & \ddots & \vdots \\ m_{N1} & m_{N2} & \cdots & M_{NN} \end{bmatrix} \tag{2}$$

The corresponding permanent function value of RRF, formulated based on the attributes within the ReM matrix, is expressed as Equation (3).

$$Per\ (ReM_N) = \sum_P \prod_{i=1}^{N} m_{ij},\ P(i) \tag{3}$$

where the sum is calculated for all permutations P.

4. Readiness Assessment of a Remanufacturing Facility

The remanufacturing readiness assessment, incorporating elements M_i and m_{ij}, evaluates the overall readiness of a facility for remanufacturing practices. This quantitative assessment requires numerical values for each of the remanufacturing readiness matrix's M_i and m_{ij} values. The steps involved in performing a quantitative analysis of the elements of the remanufacturing readiness evaluation matrix are outlined below and presented in Figure 2.

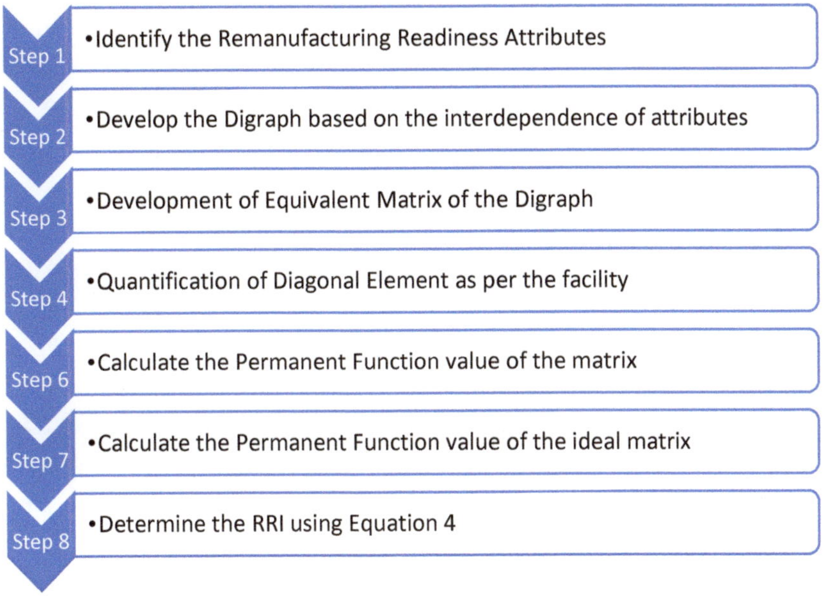

Figure 2. Steps for the determination of RRI.

4.1. *Diagonal Element Quantification*

The diagonal elements of the remanufacturing readiness assessment matrix represent the readiness attributes of the remanufacturing facility. These attributes are shown in Table 1, and the value of these attributes is decided as per the state of the remanufacturing facility. The value of these attributes is assessed by the expert team for the selected remanufacturing facility using the linguistic scale provided in Table 4.

4.2. *Off-Diagonal Element Quantification*

After the quantification of the diagonal elements, the off-diagonal elements are also quantified. The quantification of non-diagonal elements entails the allocation of numerical values to each m_{ij}. In this regard, it is preferable to assign values using a numeric scale from high (4) to none (0). Additional levels are also included as above average (3), medium (2), and minor (1). The degree of influence among a facility's remanufacturing readiness attributes is displayed in Table 6.

Table 6. Remanufacturing readiness assessment matrix.

Remanufacturing Readiness Attributes	Core Acquisition	Reverse Logistics Availability	Resource Availability for Remanufacturing Initiatives	Design for Remanufacturing	Enterprise Collaboration	Remanufactured Product Positioning	Performance Measurement	Labor Skill and Availability	Flexible Remanufacturing System
Core acquisition		0	2	0	0	1	3	0	0
Reverse logistics availability	4		1	0	2	0	2	2	0
Resource availability for remanufacturing initiatives	3	2		4	3	1	0	2	1
Design for remanufacturing	3	0	2		0	0	1	1	2
Enterprise collaboration	3	2	2	1		2	0	1	0
Remanufactured product positioning	1	0	2	0	1		0	0	0
Performance measurement	2	0	0	2	0	3		0	1
Labor skill and availability	3	2	4	0	1	0	0		0
Flexible remanufacturing system	0	1	2	0	0	0	2	1	

4.3. Remanufacturing Readiness Index

The RRI is a quantitative measure of the readiness of a facility for remanufacturing practices. This is derived from the remanufacturing readiness assessment by assigning numerical values to "M_i" and m_{ij}. A higher index value indicates a greater readiness to adopt remanufacturing practices and vice versa. The RRI is the ratio of the permanent function (Per(ReM_K)) value of the remanufacturing readiness assessment matrix to the permanent function (Per(ReM_{Ideal})) of the ideal remanufacturing readiness assessment matrix. The ideal remanufacturing readiness assessment matrix refers to the most desirable and exceptional remanufacturing facility. In the case of the ideal remanufacturing facility, the value of each remanufacturing readiness attribute is the maximum (i.e., four) and the value of m_{ij} is the same as in Table 6. The RRI_k is the remanufacturing readiness index for *'facility k'* and is mathematically expressed as follows:

$$RRI_k = \frac{Per\ (ReM)_k}{Per\ (ReM)_{ideal}} \tag{4}$$

$$\text{Where, } (ReM)_{ideal} = \begin{bmatrix} 4 & 0 & 2 & 0 & 0 & 1 & 3 & 0 & 0 \\ 4 & 4 & 1 & 0 & 2 & 0 & 2 & 2 & 0 \\ 3 & 2 & 4 & 4 & 3 & 1 & 0 & 2 & 1 \\ 3 & 0 & 2 & 4 & 0 & 0 & 1 & 1 & 2 \\ 3 & 2 & 2 & 1 & 4 & 2 & 0 & 1 & 0 \\ 1 & 0 & 2 & 0 & 1 & 4 & 0 & 0 & 0 \\ 2 & 0 & 0 & 2 & 0 & 3 & 4 & 0 & 1 \\ 3 & 2 & 4 & 0 & 1 & 0 & 0 & 4 & 0 \\ 0 & 1 & 2 & 0 & 0 & 0 & 2 & 1 & 4 \end{bmatrix} \tag{5}$$

The value of the RRI lies between zero and one, and a value of the RRI that is close to one represents that the facility is ready for remanufacturing practices. The ideal case value of the RRI is one.

5. A Case Study for RRI Measurement

A case study has been performed to demonstrate the implementation of the proposed RRI. A remanufacturing facility (ABC) located in the National Capital Region (NCR), India is considered for the assessment of the RRI. A case study from India offers an ideal background for the remanufacturing study with its large MSME sector, which can benefit from sustainable and cost-reduction approaches. Government policy support measures such as the National Resource Efficiency Policy and Extended Producer Responsibility promote the use of a circular economy and sustainable manufacturing, thus facilitating remanufacturing practices. Environmental concerns in India and growing market demand further encourage the implementation of remanufacturing to reduce waste, conserve resources, and address a research gap in the developing economy setting. The ABC facility is in the finalization phase and planning to start its operations in the next year (i.e., mid-2025). We have contacted its senior members of management to obtain their input regarding the facility on the identified nine readiness attributes using the linguistic scale as provided in Table 4. As a result, they have provided the input for their remanufacturing facility (ReM_{ABC}) and the response is shown in Table 7.

Based on the above response, the remanufacturing readiness assessment matrix is developed for the *ABC* facility by substituting the value of M_i in the diagonal element of

$\mathrm{ReM_I}$. The off-diagonal element of $\mathrm{ReM_{ABC}}$ is the same as in Table 5. The remanufacturing readiness assessment matrix for ABC ($\mathrm{ReM_{ABC}}$) is as follows:

$$\mathrm{ReM}_{ABC} = \begin{bmatrix} 3 & 0 & 2 & 0 & 0 & 1 & 3 & 0 & 0 \\ 4 & 2 & 1 & 0 & 2 & 0 & 2 & 2 & 0 \\ 3 & 2 & 3 & 4 & 3 & 1 & 0 & 2 & 1 \\ 3 & 0 & 2 & 2 & 0 & 0 & 1 & 1 & 2 \\ 3 & 2 & 2 & 1 & 4 & 2 & 0 & 1 & 0 \\ 1 & 0 & 2 & 0 & 1 & 2 & 0 & 0 & 0 \\ 2 & 0 & 0 & 2 & 0 & 3 & 3 & 0 & 1 \\ 3 & 2 & 4 & 0 & 1 & 0 & 0 & 3 & 0 \\ 0 & 1 & 2 & 0 & 0 & 0 & 2 & 1 & 4 \end{bmatrix} \quad (6)$$

Table 7. Assessment of remanufacturing readiness attributes for ABC facility.

Remanufacturing Readiness Attributes	M_i for ABC
Core acquisition	3
Reverse logistics availability	2
Resource availability for remanufacturing initiatives	3
Design for remanufacturing	2
Enterprise collaboration	4
Remanufactured product positioning	2
Performance measurement	3
Labor skill and availability	3
Flexible remanufacturing system	4

The RRF for the remanufacturing facility ABC is developed from the permanent function of the remanufacturing readiness matrix. ReM_{ABC} is determined as per Equation (3). In order to determine the value of the permanent function, a computer program has been developed in the *R package* R 4.2.1. With the help of this program, the value of the permanent function of the remanufacturing facility is determined as follows:

$$Per(\mathrm{ReM})_{ABC} = 2{,}751{,}357$$

To evaluate the RRI, the permanent function of ideal remanufacturing readiness must be obtained. To do so, the permanent function value is calculated for the ideal matrix that is represented as an equation. The value of the permanent function is as follows:

$$Per(\mathrm{ReM})_{Ideal} = 9{,}581{,}660$$

Now, the RRI for the ABC remanufacturing facility is calculated as per Expression (4).

$$\begin{aligned} \mathrm{RRI_{ABC}} &= \frac{Per(\mathrm{ReM})}{Per(\mathrm{ReM})_{ideal}} \\ &= \frac{2{,}751{,}357}{9{,}581{,}660} \\ &= 0.287148 \end{aligned}$$

The RRI for the remanufacturing facility ABC has been successfully calculated, demonstrating that the RRI value can serve as a reliable metric for decision making in remanufacturing operations. The case result shows that the RRI for the facility is low and it needs certain improvements before operationalizing. This remanufacturing facility can be improved by focusing on the attributes that have low values, such as 'reverse logistics availability', 'design for remanufacturing', and 'remanufactured product positioning'. The framework and RRI can provide such insights to managers before remanufacturing operations start.

6. Implications of Study

This study has several implications for managers, practitioners, and policymakers. The significant implications are outlined below.

6.1. Managerial Implications

The findings of this study provide actionable insights for MSME managers, industry stakeholders, and policymakers to facilitate the adoption of remanufacturing practices, thereby fostering resource efficiency, sustainability, and a competitive advantage. The suggested framework may be used by managers to evaluate how prepared their company is to implement remanufacturing practices. The calculated RRI serves as a measure to assess the extent to which a facility is prepared for remanufacturing operations. Additionally, this index provides decision makers with a valuable tool to compare two or more facilities in terms of their readiness to adopt remanufacturing practices. This comparative analysis can aid in identifying strengths, weaknesses, and areas requiring improvement across different facilities. The findings also suggest that MSMEs may optimize the effectiveness of remanufacturing activities by focusing expenditures on high-impact elements, such as labor skill development, resource availability, and reverse logistics development. Furthermore, the proposed RRI can be utilized to measure a remanufacturing facility's progress towards the adoption of remanufacturing practices by systematically comparing the previous RRI with the current RRI. This quantified measure can also facilitate stakeholders, such as investors and regulatory bodies, in assessing the status of the remanufacturing facility and making informed decisions. In addition, the RRI can also be used to compare the different remanufacturing facilities and set benchmarks for the adoption of remanufacturing.

6.2. Academic Implications

The current study offers a comprehensive addition to the limited studies on remanufacturing practices in MSMEs and offers guidance on how to evaluate readiness. It also quantified a remanufacturing facility's readiness towards the adoption of remanufacturing to embrace sustainability. This study used GTM to outline the relationships between all of the remanufacturing readiness attributes that were found. Further, this study also demonstrates how GTM can be used to develop the index for remanufacturing readiness, and it can be extended for other indices such as the remanufacturing performance index. The proposed framework can be used to measure the readiness to adopt remanufacturing practices. This framework can serve as a foundational reference for future studies exploring readiness assessment in other sectors or geographical contexts. As per the requirements, the proposed framework can be utilized by various researchers in different settings, such as the manufacturing of automobile parts, electronic equipment, and wood furniture. Moreover, the RRI acts as a quantitative decision support tool that provides measurable insights into MSMEs' readiness to adopt remanufacturing practices. Future academic studies can leverage this tool to evaluate readiness across different industries, regions, or levels of business maturity.

7. Conclusions

This study proposes a framework to assess readiness to adopt remanufacturing practices in the context of MSMEs. This study can be considered a step toward fostering sustainable manufacturing practices in MSMEs. Expert feedback and a review of the literature were used to determine the critical readiness attributes for the implementation of remanufacturing practices. Nine readiness attributes for remanufacturing adoption were finalized, including 'core acquisition', 'reverse logistics availability', 'resource availability for remanufacturing initiatives', 'design for remanufacturing', 'enterprise collaboration', 'remanufactured product positioning', 'performance measurement', 'labor skill and availability', and 'flexible remanufacturing system'. Based on these readiness attributes, a framework was established using the GTM methodology to assess readiness to adopt remanufacturing practices. This approach facilitated the measurement of remanufacturing readiness, encapsulated in the form of the RRI, serving as a robust metric for evaluating and benchmarking the readiness of the facility for the adoption of remanufacturing practices. This systematic framework offers a comprehensive tool for MSMEs to evaluate and enhance their readiness to adopt remanufacturing practices. By leveraging this index, MSMEs can systematically identify gaps, prioritize resource allocation, and develop targeted strategies to overcome barriers to the adoption of remanufacturing. To illustrate the proposed framework, a case study was performed, and the result shows that the RRI value for the adopted remanufacturing facility is low. This helps the manufacturing facility to make an informed decision before starting operations. The evaluation of the RRI provides the remanufacturing facility with valuable insights regarding its readiness. Given that the RRI not only encompasses technical aspects but also has managerial implications, manufacturing facilities can leverage the RRI to identify specific areas of deficiency within their operations. This enables informed decision making, allowing managers to implement targeted strategies for improving their readiness towards the adoption of remanufacturing practices.

Certain limitations are also associated with this study. This study is based on a literature review and experts' input; therefore, it has some limitations. One limitation of this study is that the readiness factors were identified through a literature review, and there is a chance some important attributes were not included. In order to eliminate this limitation, a comprehensive review can be performed by broadening the search string and using other databases, such as Web of Science. Further, the interrelationships were identified through the experts' input and their opinions might have been biased. This bias can be reduced in future studies using the fuzzy and grey theories. In addition, this proposed framework was developed for MSMEs, and it can be extended to large-sized enterprises in future studies.

Future research could explore the application of the RRI across diverse industries and geographic regions, refining the framework to account for contextual variations. The proposed framework can be applied to product-specific remanufacturing facilities such as electronic, automobile, or electrical equipment. Future studies need to validate the RRI framework in different sectors and geographical regions to account for industry-specific and cultural variations. Comparative and longitudinal analyses are needed to assess the time adaptability of the framework and its performance compared to other maturity models. The findings can be validated through multiple case studies in future work. Further, the importance of the identified attributes is also considered in the determination of enterprises' readiness to adopt remanufacturing attributes. In addition, the proposed framework can be studied in the context of the Sustainable Development Goals (SDGs). One can explore how remanufacturing readiness helps to achieve specific SDGs in future studies.

Funding: This research received no external funding.

Data Availability Statement: The original contributions presented in this study are included in the article. Further inquiries can be directed to the corresponding author.

Conflicts of Interest: The author declares no conflicts of interest.

References

1. Bulmus, S.C.; Zhu, S.X.; Teunter, R. Competition for Cores in Remanufacturing. *Eur. J. Oper. Res.* **2014**, *233*, 105–113. [CrossRef]
2. Bhatia, M.S.; Srivastava, R.K. Analysis of External Barriers to Remanufacturing Using Grey-DEMATEL Approach: An Indian Perspective. *Resour. Conserv. Recycl.* **2018**, *136*, 79–87. [CrossRef]
3. Xiang, W.; Ming, C. Implementing Extended Producer Responsibility: Vehicle Remanufacturing in China. *J. Clean. Prod.* **2011**, *19*, 680–686. [CrossRef]
4. Khan, S.; Singh, R.; Alnahas, J.; Abbate, S.; Centobelli, P. Navigating the Smart Circular Economy: A Framework for Manufacturing Firms. *J. Clean. Prod.* **2024**, *480*, 144007. [CrossRef]
5. Corsini, F.; Gusmerotti, N.M.; Frey, M. Consumer's Circular Behaviors in Relation to the Purchase, Extension of Life, and End of Life Management of Electrical and Electronic Products: A Review. *Sustainability* **2020**, *12*, 10443. [CrossRef]
6. Godichaud, M.; Tchangani, A.; Pérès, F.; Iung, B. Sustainable Management of End-of-Life Systems. *Prod. Plan. Control.* **2012**, *23*, 216–236. [CrossRef]
7. Golbabaei, F.; Bunker, J.; Yigitcanlar, T.; Baker, D.; Mirhashemi, A.; Paz, A. Sustainable End-of-Life Management of Abandoned and Derelict Vessels through the Lens of Circular Economy. *J. Clean. Prod.* **2024**, *474*, 143559. [CrossRef]
8. Erdiaw-Kwasie, M.O.; Abunyewah, M.; Yusif, S.; Erdiaw-Kwasie, A. Does Circular Economy Knowledge Matter in Sustainable Service Provision? A Moderation Analysis. *J. Clean. Prod.* **2023**, *383*, 135429. [CrossRef]
9. Gajanayake, A.; Ho, O.T.K.; Iyer-Raniga, U. Motivations and Drivers for Adopting Sustainability and Circular Business Strategies in Businesses in Victoria. *Corp. Soc. Responsib. Environ. Manag.* **2024**, *31*, 169–179. [CrossRef]
10. Li, C.; Ahmad, S.F.; Ahmad Ayassrah, A.Y.A.B.; Irshad, M.; Telba, A.A.; Mahrous Awwad, E.; Imran Majid, M. Green Production and Green Technology for Sustainability: The Mediating Role of Waste Reduction and Energy Use. *Heliyon* **2023**, *9*, e22496. [CrossRef]
11. Zailani, S.; Govindan, K.; Shaharudin, M.R.; Kuan, E.E.L. Barriers to Product Return Management in Automotive Manufacturing Firms in Malaysia. *J. Clean. Prod.* **2017**, *141*, 22–40. [CrossRef]
12. Mukherjee, K.; Mondal, S. Analysis of Issues Relating to Remanufacturing Technology—A Case of an Indian Company. *Technol. Anal. Strateg. Manag.* **2009**, *21*, 639–652. [CrossRef]
13. Chong, J.Y.; Mat Saman, M.Z.; Ngadiman, N.H.A. Sustainable Remanufacturing Management Approaches and Applications Model in End-of-Life Vehicles: A Critical Review and Classification. *J. Remanuf.* **2024**, *14*, 169–184. [CrossRef]
14. Singhal, D.; Tripathy, S.; Jena, S.K. Remanufacturing for the Circular Economy: Study and Evaluation of Critical Factors. *Resour. Conserv. Recycl.* **2020**, *156*, 104681. [CrossRef]
15. Sierra-Fontalvo, L.; Polo-Cardozo, J.; Maury-Ramírez, H.; Mesa, J.A. Diagnosing Remanufacture Potential at Product-Component Level: A Disassemblability and Integrity Approach. *Resour. Conserv. Recycl.* **2024**, *205*, 107529. [CrossRef]
16. Mejía-Moncayo, C.; Kenné, J.P.; Hof, L.A. On the Development of a Smart Architecture for a Sustainable Manufacturing-Remanufacturing System: A Literature Review Approach. *Comput. Ind. Eng.* **2023**, *180*, 109282. [CrossRef]
17. Zhang, T.; Chu, J.; Wang, X.; Liu, X.; Cui, P. Development Pattern and Enhancing System of Automotive Components Remanufacturing Industry in China. *Resour. Conserv. Recycl.* **2011**, *55*, 613–622. [CrossRef]
18. Sharma, V.; Garg, S.K.; Sharma, P.B. Identification of Major Drivers and Roadblocks for Remanufacturing in India. *J. Clean. Prod.* **2016**, *112*, 1882–1892. [CrossRef]
19. Zhao, Y.; Zhou, H. Remanufacturing vs. Greening: Competitiveness and Harmony of Sustainable Strategies of Supply Chain under Uncertain Yield. *Comput. Ind. Eng.* **2023**, *179*, 109233. [CrossRef]
20. Barros, M.V.; Salvador, R.; do Prado, G.F.; de Francisco, A.C.; Piekarski, C.M. Circular Economy as a Driver to Sustainable Businesses. *Clean. Environ. Syst.* **2021**, *2*, 100006. [CrossRef]
21. Wei, S.; Tang, O.; Sundin, E. Core (Product) Acquisition Management for Remanufacturing: A Review. *J. Remanuf.* **2015**, *5*, 4. [CrossRef]
22. Karuppiah, K.; Virmani, N.; Sindhwani, R. Toward a Sustainable Future: Integrating Circular Economy in the Digitally Advanced Supply Chain. *J. Bus. Ind. Mark.* **2024**, *39*, 2605–2619. [CrossRef]
23. Khan, S.; Ali, S.S.; Singh, R. Determinants of Remanufacturing Adoption for Circular Economy: A Causal Relationship Evaluation Framework. *Appl. Syst. Innov.* **2022**, *5*, 62. [CrossRef]
24. Vogt Duberg, J.; Kurilova-Palisaitiene, J.; Sundin, E. 5-step Approach for Initiating Remanufacturing (5AFIR). *Bus. Strategy Environ.* **2023**, *32*, 4360–4370. [CrossRef]

25. Kurilova-Palisaitiene, J. On Remanufacturing Readiness Level—An Introduction to a Remometer^{TM}. *Procedia CIRP* **2021**, *98*, 91–96. [CrossRef]
26. Jayashree, S.; Reza, M.N.H.; Malarvizhi, C.A.N.; Gunasekaran, A.; Rauf, M.A. Testing an Adoption Model for Industry 4.0 and Sustainability: A Malaysian Scenario. *Sustain. Prod. Consum.* **2022**, *31*, 313–330. [CrossRef]
27. Fatimah, Y.A.; Biswas, W.; Mazhar, I.; Islam, M.N. Sustainable Manufacturing for Indonesian Small- and Medium-Sized Enterprises (SMEs): The Case of Remanufactured Alternators. *J. Remanuf.* **2013**, *3*, 1–11. [CrossRef]
28. Karuppiah, K.; Sankaranarayanan, B.; Venkatesh, V.G.; Ali, S.M.; Shi, Y. A Causal Sustainable Evaluation of Barriers to Remanufacturing: An Emerging Economy Perspective. *Environ. Dev. Sustain.* **2023**, 1–38. [CrossRef]
29. Jiménez-Parra, B.; Rubio, S.; Vicente-Molina, M.-A. Key Drivers in the Behavior of Potential Consumers of Remanufactured Products: A Study on Laptops in Spain. *J. Clean. Prod.* **2014**, *85*, 488–496. [CrossRef]
30. Duberg, J.V.; Sundin, E.; Kurilova-Palisaitiene, J. Exploring the Incentives for Initiating Remanufacturing: Experiences from Eight European Original Equipment Manufacturers. *Procedia CIRP* **2024**, *122*, 437–442. [CrossRef]
31. Kurilova-Palisaitiene, J.; Matschewsky, J.; Sundin, E. Four Levels of Remanufacturing Maturity as a Circular Manufacturing Indicator: A Theoretical Framework and Practical Assessment Tool. *Resour. Conserv. Recycl.* **2024**, *211*, 107899. [CrossRef]
32. Fatimah, Y.A.; Biswas, W.K. Remanufacturing as a Means for Achieving Low-Carbon SMEs in Indonesia. *Clean. Technol. Environ. Policy* **2016**, *18*, 2363–2379. [CrossRef]
33. Satpathy, A.s.; Sahoo, S.k.; Mohanty, A.; Mohanty, P.P. Strategies for Enhancements of MSME Resilience and Sustainability in the Post-COVID-19 Era. *Social. Sci. Humanit. Open* **2025**, *11*, 101223. [CrossRef]
34. Joshi, K.; Mishra, M.K.; Jamal, M.; Janotra, J. Measuring the Intensity of Challenges and Problems of Small-Scale Entrepreneurs in Uttarakhand, India. *Vilakshan—XIMB J. Manag.* **2024**. [CrossRef]
35. Kerin, M.; Pham, D.T. Smart Remanufacturing: A Review and Research Framework. *J. Manuf. Technol. Manag.* **2020**, *31*, 1205–1235. [CrossRef]
36. Karuppiah, K.; Sankaranarayanan, B.; Lo, H.W. A Systematic Literature Review on the Evolution of Sustainable Manufacturing Practices: Key Findings and Implications. *Clean. Eng. Technol.* **2024**, *22*, 100798. [CrossRef]
37. Sakao, T.; Sundin, E. How to Improve Remanufacturing?—A Systematic Analysis of Practices and Theories. *J. Manuf. Sci. Eng. Trans. ASME* **2019**, *141*, 021004. [CrossRef]
38. Vogt Duberg, J.; Johansson, G.; Sundin, E.; Kurilova-Palisaitiene, J. Prerequisite Factors for Original Equipment Manufacturer Remanufacturing. *J. Clean. Prod.* **2020**, *270*, 122309. [CrossRef]
39. Kleber, R.; Quariguasi Frota Neto, J.; Reimann, M. The Role of Part Failure Rates Asymmetry and Spare Part Proprietariness on Remanufacturing Decision Making. *Eur. J. Oper. Res.* **2023**, *310*, 185–200. [CrossRef]
40. Ren, Y.; Lu, X.; Guo, H.; Xie, Z.; Zhang, H.; Zhang, C.A.; Ren, Y.; Lu, X.; Guo, H.; Xie, Z.; et al. A Review of Combinatorial Optimization Problems in Reverse Logistics and Remanufacturing for End-of-Life Products. *Mathematics* **2023**, *11*, 298. [CrossRef]
41. Guggeri, E.M.; Ham, C.; Silveyra, P.; Rossit, D.A.; Piñeyro, P. Goal Programming and Multi-Criteria Methods in Remanufacturing and Reverse Logistics: Systematic Literature Review and Survey. *Comput. Ind. Eng.* **2023**, *185*, 109587. [CrossRef]
42. Ali, S.; Shirazi, F. The Paradigm of Circular Economy and an Effective Electronic Waste Management. *Sustainability* **2023**, *15*, 1998. [CrossRef]
43. Güsser-Fachbach, I.; Lechner, G.; Ramos, T.B.; Reimann, M. Repair Service Convenience in a Circular Economy: The Perspective of Customers and Repair Companies. *J. Clean. Prod.* **2023**, *415*, 137763. [CrossRef]
44. Kurilova-Palisaitiene, J.; Sundin, E.; Poksinska, B. Remanufacturing Challenges and Possible Lean Improvements. *J. Clean. Prod.* **2018**, *172*, 3225–3236. [CrossRef]
45. Wu, B.; Jiang, Z.; Zhu, S.; Zhang, H.; Wang, Y. A Customized Design Method for Upgrade Remanufacturing of Used Products Driven by Individual Demands and Failure Characteristics. *J. Manuf. Syst.* **2023**, *68*, 258–269. [CrossRef]
46. Gong, Q.; Wu, J.; Jiang, Z.; Hu, M.; Chen, J.; Cao, Z. An Integrated Design Method for Remanufacturing Scheme Considering Carbon Emission and Customer Demands. *J. Clean. Prod.* **2024**, *476*, 143681. [CrossRef]
47. Remanufacturing, L.; Duan, W.; Liu, M.; Xu, D.; Han, L. Remanufacturing Closed-Loop Supply Chain Contract Coordination Considering Quality Control. *Systems* **2024**, *12*, 350. [CrossRef]
48. Cao, J.; Lu, J.; Bian, Z.; Shao, Q. Phase-by-Phase Pricing vs Unified Dual Pricing: How Suppliers Collaborate with Manufacturers in Remanufacturing. *Int. J. Syst. Sci. Oper. Logist.* **2024**, *11*, 2389065. [CrossRef]
49. Kwak, M.; Kim, H. Market Positioning of Remanufactured Products with Optimal Planning for Part Upgrades. *J. Mech. Des.* **2012**, *135*, 011007. [CrossRef]
50. Liu, C.; Chen, J.; Wang, X. Quantitative Evaluation Model of the Quality of Remanufactured Product. *IEEE Trans. Eng. Manag.* **2024**, *71*, 7413–7424. [CrossRef]
51. Liu, Z.; Wan, M.; Yang, X.; Zheng, X.; Chang, C. Ter Operational Decisions and Performance of an Outsourcing Remanufacturing Supply Chain under Different Financing Methods. *Comput. Ind. Eng.* **2024**, *189*, 109946. [CrossRef]

52. Aziz, N.A.; Wahab, D.A.; Ramli, R. Performance Evaluation Approach for the Parametric Upgrade of Automotive Brake Caliper in Remanufacturing. In Proceedings of the 9th International Conference and Exhibition on Sustainable Energy and Advanced Material, Putrajaya, Malaysia, 14 September 2023; Lecture Notes in Mechanical Engineering. Springer: Singapore, 2024; pp. 643–650. [CrossRef]
53. Sharma, V.; Katiyar, R.; Mishra, R. Analyzing the Impact Factors of Remanufacturing Industry on the Indian Economy: A Step towards Circular Economy. *J. Model. Manag.* **2024**, *19*, 2131–2157. [CrossRef]
54. Yao, J.K.; Zhu, S.; Cui, P.Z. Study on Flexible Remanufacturing System and Framework. In Proceedings of the 2010 International Conference on Intelligent Computation Technology and Automation, ICICTA, Changsha, China, 11–12 May 2010; pp. 516–519. [CrossRef]
55. Francas, D.; Öner-Közen, M.; Minner, S. Is It Beneficial to Invest in Flexible Capacity for Hybrid Remanufacturing? *Int. J. Prod. Res.* **2024**, *2024*, 7836–7859. [CrossRef]
56. Zhang, W.; Zheng, Y.; Ahmad, R. The Integrated Process Planning and Scheduling of Flexible Job-Shop-Type Remanufacturing Systems Using Improved Artificial Bee Colony Algorithm. *J. Intell. Manuf.* **2023**, *34*, 2963–2988. [CrossRef]
57. James, A.T.; Kumar, G.; James, J.; Asjad, M. Development of a Micro-Level Circular Economy Performance Measurement Framework for Automobile Maintenance Garages. *J. Clean. Prod.* **2023**, *417*, 138025. [CrossRef]
58. Gupta, A.; Singh, R.K. Developing a Framework for Evaluating Sustainability Index for Logistics Service Providers: Graph Theory Matrix Approach. *Int. J. Product. Perform. Manag.* **2020**, *69*, 1627–1646. [CrossRef]
59. Ali, S.S.; Kaur, R.; Khan, S. Identification of Innovative Technology Enablers and Drone Technology Determinants Adoption: A Graph Theory Matrix Analysis Framework. *Oper. Manag. Res.* **2023**, *16*, 830–852. [CrossRef]
60. Haleem, A.; Khan, S.; Khan, M.I.; Ali, S.S. A Fuzzy-Digraph Matrix-Based Risk Mitigation Framework for Halal Supply Chain Management. *Ann. Oper. Res.* **2024**, 1–33. [CrossRef]
61. Jain, V.; Raj, T. Modeling and Analysis of FMS Performance Variables by ISM, SEM and GTMA Approach. *Int. J. Prod. Econ.* **2016**, *171*, 84–96. [CrossRef]
62. Rao, R.V.; Padmanabhan, K.K. Selection, Identification and Comparison of Industrial Robots Using Digraph and Matrix Methods. *Robot. Comput. Integr. Manuf.* **2006**, *22*, 373–383. [CrossRef]
63. Forbert, H.; Marx, D. Calculation of the Permanent of a Sparse Positive Matrix. *Comput. Phys. Commun.* **2003**, *150*, 267–273. [CrossRef]
64. Beekaroo, D.; Callychurn, D.S.; Hurreeram, D.K. Developing a Sustainability Index for Mauritian Manufacturing Companies. *Ecol. Indic.* **2019**, *96*, 250–257. [CrossRef]

processes

MDPI

Review

Understanding and Resolving 3D Printing Challenges: A Systematic Literature Review

Seulhee Kwon [1] and Dongwook Hwang [2,*]

[1] Graduate School of Metaverse Convergence, Kwangwoon University, Seoul 01897, Republic of Korea; tmf2377@naver.com
[2] School of Media and Communication, Kwangwoon University, Seoul 01897, Republic of Korea
* Correspondence: dongwookkr@kw.ac.kr; Tel.: +82-2-940-8453

Abstract: Additive manufacturing (AM), or 3D printing, enables efficient fabrication of complex and customized components. Despite its growth across industries, users frequently encounter print failures due to design errors, process limitations, and inadequate monitoring. While existing research has explored various aspects of these failures, much of it remains fragmented, with limited consolidated overviews that map common problems, troubleshooting strategies, and guidelines across the AM workflow. This study conducted a systematic literature review using the Preferred Reporting Items for Systematic Reviews and Meta-Analyses (PRISMA) methodology to identify and categorize common 3D printing problems and their solutions. Relevant studies published between 2000 and 2024 were extracted from major databases. A total of 126 peer-reviewed articles were selected and analyzed. Three major categories of recurring challenges were identified: (1) design and pre-processing errors; (2) geometric errors and dimensional deviations; (3) failures in in-process error detection and response. A variety of mitigation strategies have been proposed across the literature, including STL and slicing optimization, thermal management, machine calibration, and sensor-based real-time monitoring. These approaches reflect the multifactorial nature of 3D printing failures, which often arise from the complex interplay of design, material, and process parameters. This review provides a structured summary of failure types and mitigation strategies across the AM workflow.

Keywords: additive manufacturing; guideline; PRISMA; troubleshooting

1. Introduction

3D printing, also referred to as additive manufacturing (AM), is a technology that fabricates physical objects by sequentially depositing material layers based on digital files. Due to its capability to rapidly and cost-effectively produce complex structures, 3D printing has become a core technology in various industries, including the aerospace, healthcare, and automotive sectors [1]. The use of 3D printers for customized production offers significant advantages in small-scale mass production, earning technical recognition from a wide spectrum of users, ranging from individual creators to large-scale manufacturers [2]. Notably, the 3D printing market experienced remarkable growth in 2023, surpassing $20 billion for the first time, driven by advancements in metal 3D printing technologies and a transition toward mass production [3,4].

Despite its innovative potential, users often encounter a variety of challenges when learning and effectively utilizing 3D printing technologies. Issues that arise during the printing process, such as reduced print quality and design and mechanical setting errors, can pose significant difficulties for both novice and experienced users alike [5]. Whilst previous studies have proposed various analyses and solutions to 3D printing issues, most have remained limited to specific case-based or post hoc approaches, lacking comprehensive coverage of problems across the entire process workflow [6]. Additionally, many existing studies tend to address only narrowly defined problems within specific AM processes, which contributes to the persistence of a broader research gap. It is important that knowledge of AM-related errors and their corresponding solutions be made accessible not only to specialists, but also to a wider range of users who interact with these technologies. To support this need, there is value in conducting reviews that take a more comprehensive and structured approach to synthesizing fragmented findings across diverse AM contexts.

While narrative reviews are often sufficient for summarizing research trends, we adopted the Preferred Reporting Items for Systematic Reviews and Meta-Analyses (PRISMA) methodology to enhance the transparency and reproducibility of the literature selection and analysis process. The use of PRISMA helps minimize the influence of researcher bias and provides a logical and standardized framework for systematically identifying and synthesizing relevant studies—even for those who may not be long-standing experts in the field. Building on this foundation, this study aims to propose a proactive approach that systematically identifies and analyzes potential problems that may arise throughout the 3D printing process, categorized by components and stages. Specifically, we apply the PRISMA methodology, which is commonly used to enhance transparency and quality in systematic reviews and meta-analyses [7], to comprehensively review existing literature on 3D printing-related issues and their corresponding solutions. Through this systematic review, we aim to develop practical guidelines that help 3D printing users prevent printing failures and ensure successful output. To guide this investigation, we formulated the following research questions:

Q1: What types of problems are commonly encountered when using 3D printing technologies across different stages of the process?
Q2: What solutions have been proposed in the literature to address these problems, and how might they inform practice and future research?

2. Methodology

2.1. Literature Search Strategy

This study conducted a systematic literature review based on the PRISMA guidelines to identify common issues encountered during the use of 3D printers and their corresponding solutions. PRISMA encourages transparency in the process of identifying, selecting, and reporting research, including the search strategy, classification method, inclusion and exclusion criteria, and the use of automation tools. This methodology enhances the reliability of literature selection and enables researchers to define clear research questions and apply consistent criteria for analysis [8]. Due to its robustness and reproducibility, PRISMA is widely adopted across disciplines to yield trustworthy results.

To ensure comprehensive coverage, the literature search was conducted across major academic databases, including Google Scholar, ScienceDirect, PubMed, and IEEE Xplore. The search targeted publications from 1 January 2000 to 31 July 2024. Keywords used in the search included combinations of terms such as "3D printing", "3D printer", or "additive manufacturing" with "error(s)", "guideline(s)", or "troubleshooting", specifically within the article titles. This ensured that studies addressing errors and guidance related to 3D printing and additive manufacturing were inclusively identified. The keyword combination logic used in this title-based search is summarized in Figure 1.

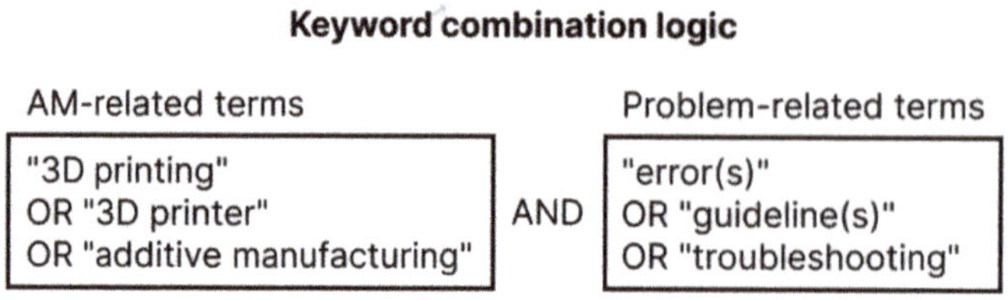

Figure 1. Keyword combination logic used in the title-based search.

2.2. Literature Selection Process

The literature selection process was conducted using the PRISMA flow diagram (Figure 2) and followed specific inclusion and exclusion criteria. Studies were included if they addressed problems encountered during the use of 3D printers and proposed corresponding solutions. Eligible sources included original research articles, experimental studies, review articles, and case studies focusing specifically on technical or methodological aspects of 3D printing.

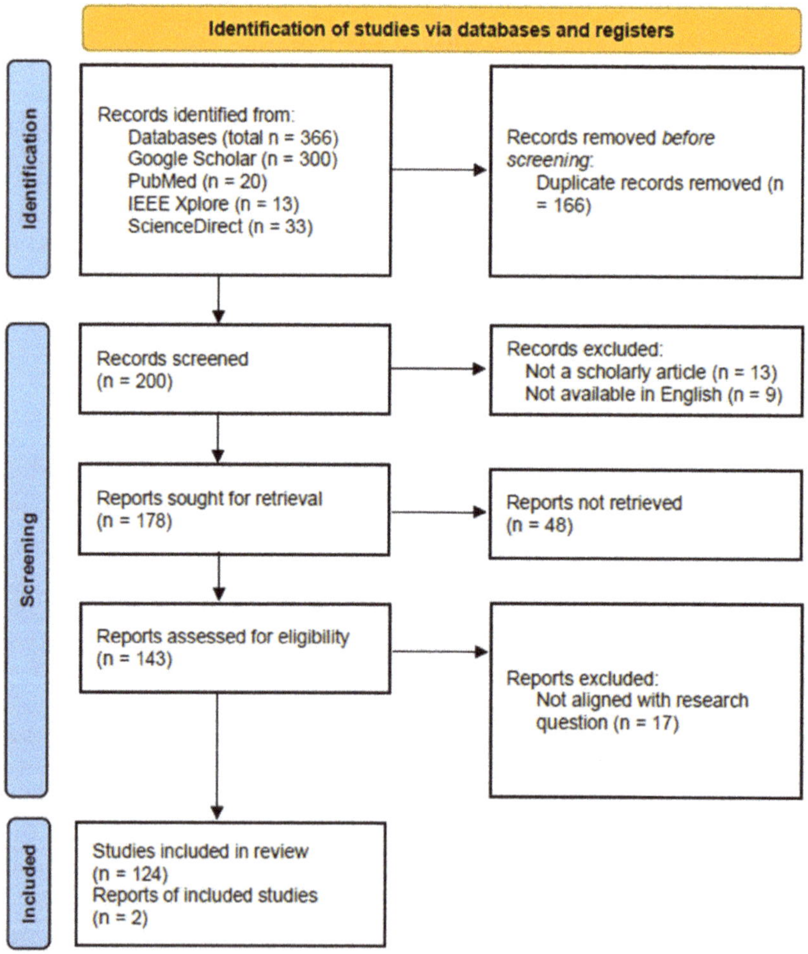

Figure 2. PRISMA flow diagram of the study selection process.

The selection proceeded through a two-stage exclusion process. In the first stage, studies were excluded if only abstracts were available, if full-text access was not accessible, if the papers were not written in English, or if the documents were not in scholarly article format (e.g., books, reports). The second stage excluded studies that focused solely on specific industries, assessed the performance of printed products without addressing failure causes, or provided only general overviews without concrete troubleshooting approaches.

The initial search yielded a total of 366 studies. After removing duplicates, 200 unique studies remained. Applying the first exclusion criteria resulted in 143 eligible papers. A second round of screening was then conducted, and studies that did not align with the research questions were excluded. As a result, a final set of 126 studies was selected for full analysis and data extraction. No automation tools were used during the selection process; all screening and eligibility assessments were conducted manually based on the predefined criteria.

2.3. Data Extraction and Analysis

From the final selection of 126 studies, data were extracted based on predefined criteria, including the study title, publication year, research objectives, key problems encountered during 3D printer use, and corresponding solutions. The data extraction process followed the PRISMA methodology, with two researchers independently conducting the extraction. In cases of disagreement, a third researcher reviewed the discrepancies and facilitated resolution.

In this process, we examined the publication recency, research focus, and objectives of each study, and assessed their methodological rigor and experimental design to ensure validity and reliability. Particular emphasis was placed on identifying recurring issues such as printing errors, quality degradation, and mechanical failures, as well as the solutions proposed to address them. A structured coding process was applied to organize the extracted data, allowing for the consistent categorization of problem types and associated solutions across studies. This analysis enabled us to classify common problem types and extract representative troubleshooting strategies reported across the literature.

Additionally, to evaluate consistency among the findings, similar studies were comparatively analyzed. Major findings and contributions were synthesized according to the research questions. Finally, the limitations of existing research were discussed and future research directions were proposed, providing key insights for the continued advancement of 3D printing studies.

3. Analysis of Troubleshooting Problems in 3D Printing Technologies

AM encompasses a range of process technologies, each characterized by distinct technical attributes depending on the materials used and the specific forming mechanisms involved. The results of the literature analysis revealed that across various AM techniques there are recurring types of issues and corresponding strategies proposed to address them. Accordingly, this study first identifies the common categories of problems that arise throughout AM processes. For each problem type, representative failure cases are outlined along with the specific solutions proposed in the reviewed literature.

To aid interpretation of the following analysis, we refer to the ISO/ASTM 52900 classification, which categorizes AM into seven primary process types [9]. While this study does not aim to conduct a modality-specific comparison, it is acknowledged that certain troubleshooting challenges may vary across different AM processes. A summary of each process, along with its definition and representative technologies, is provided in Table 1.

Table 1. Categories of additive manufacturing processes with definitions and examples.

Process	Definition	Example Technologies
Binder Jetting (BJ)	A liquid bonding agent is selectively deposited to join powder materials.	ExOne, ZPrinting, and VoxelJet.
Directed Energy Deposition (DED)	Focused thermal energy is used to fuse materials by melting as they are being deposited.	LENS, WAAM, and EBAM.
Material Extrusion (ME)	Material is selectively dispensed through a nozzle or orifice.	FDM/FFF, Contour Crafting.
Material Jetting (MJ)	Droplets of feedstock material are selectively deposited.	PolyJet, MJP, and NPJ.
Powder Bed Fusion (PBF)	Thermal energy selectively fuses regions of a powder bed	SLS, SLM, DMLS, and EBM.
Sheet Lamination (SHL)	Sheets of material are bonded to form a part.	LOM, and CBAM.
Vat Photopolymerization (VPP)	Liquid photopolymer in a vat is selectively curedby light-activated polymerization.	SLA, DLP, and CLIP.

3.1. Design and Pre-Processing Errors

In AM the quality of printed objects is influenced not only by post-processing conditions but also by the appropriateness of the design and pre-processing steps performed prior to printing. Key pre-printing procedures include CAD-based modeling, STL file conversion, slicing parameter configuration, build orientation setup, and material and temperature parameter input. These preliminary steps have a direct impact on the dimensional accuracy, surface quality, and mechanical properties of the printed parts. Errors at the design or pre-processing stage often lead to cumulative issues that are difficult to correct in later stages of the process.

According to the literature, the major issues arising during the design and pre-processing stages are as follows. First, during the process of converting CAD models into STL files composed of small triangles, geometric fidelity may not be sufficiently preserved, which can lead to information loss and result in shape distortions [10]. This phenomenon is illustrated in Figure 3a. Such distortions are particularly problematic for models with curves or complex geometries, where chordal errors frequently occur. To reduce these errors, a denser mesh is often used, but this increases the file size and computational load during slicing [11].

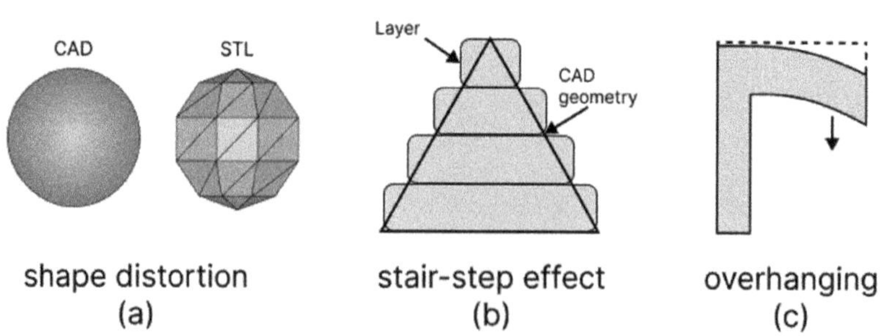

Figure 3. Visual examples of (**a**) shape distortion, (**b**) stair-step effect, and (**c**) overhang deformation.

Second, during slicing, where G-code is generated based on the STL file, the quality and precision of the output are directly influenced by user-defined parameters. A representative example is layer thickness, which affects the trade-off between precision and speed. Thinner layers yield higher resolution but slower print times, whereas thicker layers increase speed at the cost of surface quality. Especially in curved regions, visible layer boundaries can create a phenomenon known as the stair-step effect, which may be partially mitigated by adjusting the build orientation [12]. This effect is illustrated in Figure 3b. To reduce such artifacts, the "Discrete Interpolable-Area Error Profile (DIA-EP)" algorithm has been proposed, applying thinner layers to highly curved areas and thicker layers to flatter surfaces [13]. Additionally, the type of path generation algorithm and its resolution settings are significant factors affecting print quality [14]. Notably, different application domains have adopted tailored strategies to address such fidelity challenges. In medical additive manufacturing, particularly for patient-specific implants or anatomical models, maintaining fine geometric fidelity is critical; hence, adaptive slicing and ultra-fine layer resolution are often prioritized, despite longer build times. In contrast, industrial applications such as tooling or prototyping may tolerate greater dimensional deviation in favor of increased throughput, often leveraging coarser slicing or post-processing techniques. Building on these developments, recent algorithmic approaches such as adaptive slicing and AI-driven design tools have been introduced to further enhance pre-processing accuracy and geometric fidelity.

Third, within the same slicing process, build orientation is another critical factor. Due to the layer-by-layer nature of AM, build orientation influences layer structure, need for and number of support structures, material consumption, print time, and ultimately, durability and post-processing difficulty [15]. One notable issue is the "supporting error". Without adequate support in inclined or overhanging areas, the geometry may collapse or warp. A representative example of such unsupported deformation is illustrated in Figure 3c. In high-precision parts, improper placement or angle of support structures can lead to print failure or deformation [16]. Therefore, build orientation should not be considered merely a visual or esthetic factor but a critical design parameter closely linked to functional quality and manufacturing efficiency.

Since these errors are typically irreversible after printing and often result in significant quality degradation, pre-print simulation-based verification, high-resolution mesh generation, and adherence to standardized design guidelines are essential. Recent research [11] also highlights the importance of pre-evaluating toolpath quality at the G-code level. Techniques such as applying G2/G3 commands for curved path optimization or using post-processing algorithms like Arc Welder have been proposed to enhance toolpath precision. As such, design and pre-processing errors are among the leading causes of print quality degradation and must be considered as both the starting point and a core control node in the overall AM process.

3.2. Geometric Errors and Dimensional Deviations

AM has gained significant attention for its ability to precisely fabricate complex geometries and efficiently produce customized components in small quantities. Despite these advantages, multiple studies have focused on addressing quality and printing errors that occur throughout the AM process [17,18]. During printing, relatively minor issues such as dimensional deviations or surface roughness may arise; in severe cases, geometric distortions and structural instabilities can lead to critical defects such as cracks, porosity, or warping. Unsupported features such as overhangs and bridges can collapse without proper support structures, potentially resulting in severe print failures like the "spaghetti error", which renders the component unusable. These precision-related concerns are particularly critical in medical applications, where AM is widely used for surgical implants, patient-specific prosthetics, and anatomical phantoms [19]. Such components must closely conform to individual anatomical features, often with acceptable dimensional deviations restricted to within a few micrometers as inaccuracies can directly affect patient safety and treatment efficacy [20]. Accordingly, extensive research has been conducted to identify and compensate for dimensional and geometric inaccuracies in medical AM, focusing on process optimization and improving geometric fidelity [21].

The literature review revealed that discrepancies between the designed CAD model and the printed output were among the most frequently observed issues. While design and pre-processing errors have been discussed in detail in Section 3.1, this section focuses on the physical and mechanical factors that contribute to geometric inaccuracies during the actual printing process.

First, thermal variations and the accumulation of residual thermal stress are widely reported as the primary causes of geometric errors. Printed materials tend to shrink during the cooling process, and interlayer temperature gradients can result in asymmetric thermal stress [22]. These stresses may cause overall warping, localized lifting, or tilting of the print, potentially leading to collisions with the recoater or print failures [23]. Although the severity of these effects varies with process and material, they are commonly observed across AM technologies. For example, in material extrusion (ME) processes such as fused deposition modeling (FDM), warping may occur due to filament shrinkage upon cooling [24]. In powder bed fusion (PBF) processes like selective laser melting (SLM) and electron beam melting (EBM), rapid heating and cooling can create steep thermal gradients, resulting in high residual stress [25]. These thermal effects often lead to geometric distortions such as warping or shrinkage. These geometric accuracy concerns tend to be more stringent in medical applications due to patient-specific requirements, whereas industrial contexts may prioritize repeatability and structural performance. Figure 4 illustrates (a) a schematic of layer deformation caused by thermal stress and (b) a simplified shrinkage model comparing the designed geometry to the printed output.

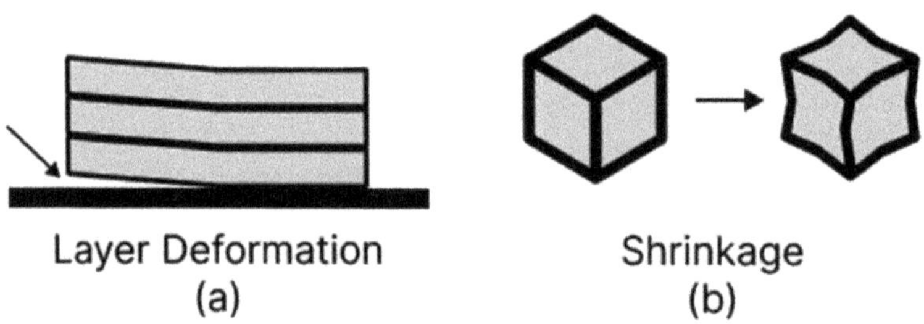

Figure 4. (**a**) Layer deformation and (**b**) shrinkage illustrations.

Second, mechanical vibrations and minor misalignments in the system can accumulate over time and significantly affect print accuracy. Instabilities during machine operation, including vibrations and inconsistent movements, directly influence nozzle positioning and path tracking accuracy, resulting in interlayer misalignment and geometric distortion [26]. Low repeatability in the Z-axis or backlash in the drive mechanisms may cause cumulative misalignment, degrading the consistency of printed layers. Bed leveling errors, where the build plate is not parallel to the reference plane, can also lead to uneven initial layers, poor adhesion, and compromised overall geometric integrity [27]. These issues are often exacerbated by mechanical wear and tear, such as a relaxation of synchronous belts or bearing degradation, causing a gradual decline in print quality over time [28]. In stepper motor-driven systems, excessive vibration during head or platform movement can lead to tracking errors, step skipping, and visible surface artifacts such as ringing or layer registration errors [29].

Third, various process conditions and equipment-related variables have also been reported as key contributors to printing errors. For instance, in PBF processes spatially non-uniform laser energy distribution or suboptimal hatch path planning can lead to reduced flatness and surface accuracy of printed parts [30]. In ME processes, imbalanced material extrusion often results in defects. In particular, temporary interruptions in material flow or excessive self-overlapping in localized regions during extrusion may produce uneven layer thickness, thereby compromising the geometric accuracy of the printed object [22]. In Vat Photopolymerization (VPP), especially in masked stereolithography (mSLA), differences in resin viscosity and curing time can cause variations in surface accuracy and shrinkage behavior, potentially leading to geometric distortion and dimensional mismatch [31]. In Directed Energy Deposition (DED), if the balance between powder feed rate and laser energy input is not properly maintained then irregular melt pool dynamics may occur, resulting in non-uniform layer formation and reduced consistency in deposition [32]. Although the specific causes vary across AM technologies, instability in process parameters universally contributes to degradation in print quality. As such, these process-dependent factors must be systematically considered across all AM platforms.

In conclusion, geometric errors and dimensional inaccuracies in AM are rarely the result of a single cause. Instead, they arise from the complex interplay of multiple factors, including thermal fluctuations, mechanical vibrations, alignment errors, and process parameter variations. Such issues are frequently linked to material shrinkage behaviors, machine repeatability issues, non-uniform deposition paths, and inconsistencies in curing conditions. These problems are not limited to any one AM method but are inherent risks across all additive manufacturing processes. The following table (Table 2) summarizes the issues and proposed solutions related to geometric errors and dimensional deviations. All proposed solutions listed in the table were extracted from the respective referenced studies.

Table 2. Selective literature on the geometric errors and dimensional deviations.

Author(s) (Year)	AM Technology	Issue	Root Cause	Proposed Solution
Auškalnis et al. (2022) [33]	VPP	Photopolymer shrinkage.	Accumulated error during intraoral scanning (IOS); resin shrinkage during printing.	- Shrinkage-compensated design adjustment. - Standardization of post-curing conditions.
Tian et al. (2024) [34]	PBF	Micro-geometry defects.	Node detachment, radius variation, and surface distortion during SLM printing.	- Precision prediction and error quantification. - Geometric parameter modeling and sensitivity analysis.
Pinto et al. (2015) [21]	MJ	Stair-stepping and overbuilding.	STL quality degradation due to an improper tessellation threshold and mesh resolution.	- Layer thickness tuning and material optimization.
Cajal et al. (2013) [35]	VPP	Volumetric dimensional deviation.	Kinematic position error per axis and cumulative mismatch in STL-based build location.	- Real-time probing-based measurement. - Multidimensional compensation model. - Volume correction and reprinting based on STL data.
Das et al. (2015) [36]	PBF	Stair-stepping.	Errors in flatness, verticality, and roundness depending on build orientation and slice thickness.	- Mathematical modeling of geometric errors. - Output orientation optimization. - Visual analysis via quadtree-based support structure calculation.

Table 2. *Cont.*

Author(s) (Year)	AM Technology	Issue	Root Cause	Proposed Solution
Majarena et al. (2017) [26]	ME	Positional and linearity errors.	Positional deviation during axis movement; loss of precision over longer paths.	- Calibrated scanner-based measurement. - G-code correction using positional error matrices. - Reprinting based on modified toolpath.
Li and Anand (2021) [30]	PBF	Flatness error.	Residual stress and asymmetric thermal shrinkage from hatch pattern and scan direction.	- ANN-based deformation prediction. - Backward interpolation for distortion calculation. - GA-based multilayer hatch angle optimization.
Majd et al. (2023) [37]	ME	Infill density deviation.	Discrepancy between target and actual infill, irregular infill geometry, and extrusion profile.	- Neural network for predicting the volumetric percentage error. - Proposed algorithm for infill optimization. - Dataset-based training on infill patterns and parameters.
Zgórniak and Stachurski (2010) [38]	MJ	Dimensional deviation.	Recurrent dimensional deviation depending on print location and orientation.	- Systematic error analysis by XYZ orientation. - Statistical compensation guidelines. - Direction optimization for prototype development.
Moodleah and Kirimasthong (2023) [13]	ME	Geometric error during slicing.	Irregular point cloud and inaccurate inter-layer boundary leading to cumulative shape error.	- Geometric error calculation using triangular interlayer regions. - Adaptive slicing and G-code tuning.

To effectively address these challenges, a comprehensive, end-to-end quality management framework is essential. This should encompass not only path optimization during the design and slicing stages but also post-print 3D scanning-based verification, geometry compensation tailored to process characteristics, shrinkage-aware design compensation, and statistical prediction-based calibration strategies. Only through such holistic control mechanisms can consistent geometric fidelity and dimensional precision be achieved in complex AM workflows.

3.3. *Failures in In-Process Error Detection and Response*

AM is a process that often requires continuous operation over several hours. During this time, the quality of the printed output can be affected in real time by various physical factors such as temperature fluctuations, vibrations, and the wear of mechanical components. Consequently, even when the initial setup is highly precise, unforeseen variables that occur during the printing process can frequently lead to print failures. This issue is particularly critical in long-duration prints, where constant user supervision is not feasible. In such cases, the need for systems capable of detecting and autonomously responding to errors in real time has been increasingly emphasized. To help readers intuitively understand the structure and data flow of such systems, Figure 5 presents a simplified schematic illustration of a typical sensor-integrated real-time monitoring setup in AM.

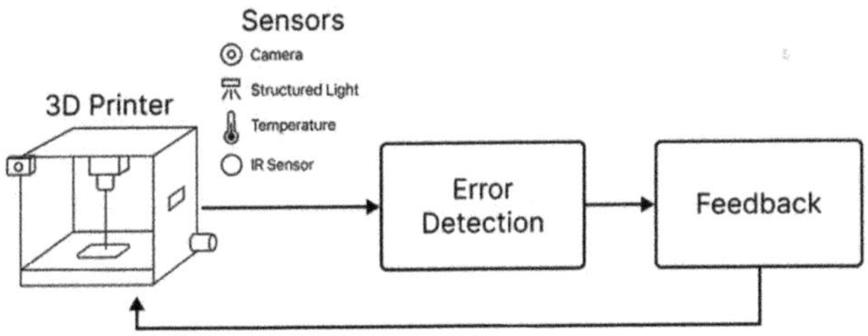

Figure 5. Sensor layout and feedback flow in real-time AM monitoring.

Representative issues observed in this category include the following. First, interlayer deposition errors, such as missing geometry or over-deposition, can compromise the structural integrity of the final product if not immediately detected as the same defect may propagate throughout the remaining layers [39]. Second, minor anomalies such as material extrusion imbalance, machine vibration, or misalignment between axes can significantly degrade the dimensional accuracy of the printed object. However, conventional sensor-based systems often lack the sensitivity or responsiveness required to detect and correct such deviations in real time [29]. Third, if the system fails to quickly identify the exact location and nature of an error after it occurs then a delayed response can result in increased material waste and reduced output quality [40].

In response to these challenges, a few studies have proposed technological approaches that enable real-time error detection and adaptive response during the AM process [41,42]. The following table (Table 3) summarizes selected prior studies and outlines the key techniques and underlying principles proposed to address the problem of in-process error detection and reaction failures. These techniques are designed to detect various abnormalities that may occur during printing, or to enable preemptive control measures. Their ultimate goal is to enhance both the accuracy and speed of error detection and response in additive manufacturing systems.

Table 3. Selective literature on the in-process error detection and response.

Author(s) (Year)	Issue	Technique Used	Technique Description	Advantages
Auškalnis et al. (2022) [33]	Layer deposition error	Structured light-based 2D phase domain error detection.	Detects printing errors in a 2D phase domain, minimizing the need for full 3D reconstruction.	Faster error detection compared to 3D point cloud analysis and enables automatic correction of error regions.
Zhang et al. (2019) [28]	Print anomalies	EFMSAE (error fusion of multiple sparse auto-encoders).	Unsupervised learning-based condition monitoring using multi-sensor data.	Early detection of micro-defects in delta 3D printers using low-cost sensors which enables real-time response.
Duan et al. (2018) [29]	Equipment vibration	LPFBS (limited-preview filtered b-spline).	Feedforward control technique generating predictive B-spline paths for short intervals and adjusting them in real time to correct vibration-induced errors.	Reduces surface roughness and positional misalignment during high-speed printing and enables sensor-free real-time correction on low-cost printers.
Peña et al. (2022) [43]	Geometric errors	CH (conoscopic holography).	Uses non-contact optical interferometry sensors to measure geometric deviation per layer and quantify radial deviations.	Enables fast acquisition of high-density 3D shape data and allows precise, non-contact monitoring of layer-wise deviations in FFF processes.

3.4. Overall Classification and Solution Approaches

In the previous sections, we examined three primary categories of errors in additive manufacturing (AM). Building upon that foundation, this section presents a broader classification by organizing the selected studies according to both the identified problem domains and the types of solution approaches proposed. A total of 126 studies were initially selected based on their relevance to the research questions. However, upon closer examination, 12 of these studies were found to lack a clear alignment with either the defined error categories or the specified solution strategies. To maintain analytical consistency, these studies were grouped under a separate "Other" classification. Table 4 presents the classification of the reviewed literature according to the defined error domains and corresponding solution approaches: (3.1) design and pre-processing errors, (3.2) geometric errors and dimensional deviations, and (3.3) failures in in-process error detection and response. Each study is also classified based on its proposed solution approach, either through the development of software-based algorithms, hardware-based physical systems, or user-controlled operational strategies. For clarity, the table excludes hardware-based interventions during the design/pre-processing stage (3.1) and user-setting-based strategies in the in-process stage (3.3) as these combinations are not logically applicable within the AM workflow. As shown in the table, a substantial number of studies emphasize software-based solutions, such as computational modeling, machine learning-driven parameter optimization, or geometric compensation algorithms aimed at improving accuracy and detecting failures in real time. In contrast, hardware-focused or user-directed solutions appeared less frequently overall; however, a notable number of user-driven strategies were observed within the 3.2 category. This can be attributed to the fact that many studies incorporating the keyword "guideline" proposed user-controllable methods specifically designed to enhance model accuracy.

Table 4. Mapping of representative AM studies by problem domain and solution approach.

Sub-Section	Solution Approach	References
3.1	Development of software-based algorithm.	[13,37,44–49]
	User-controlled operational strategies.	[11,16,20,50–52]
3.2	Development of software-based algorithm.	[12,14,25–27,29,30,32,34–36,53–83]
	Development of hardware-based physical system.	[14,62,74,75,84–87]
	User-controlled operational strategies.	[17,21,23,27,31,33,38,69,78,88–104]
3.3	Development of software-based algorithm.	[18,28,39,105–120]
	Development of hardware-based physical system.	[18,41,42,105,109–112,117,120–124]

From a broader perspective, all studies underscore the importance of systematic error modeling and compensation techniques to improve final-part accuracy and reduce manufacturing defects. For instance, in Section 3.1 (Design and Pre-processing Errors), several algorithms have been proposed to correct slicing and modeling errors by minimizing approximation deviations in model contours [44]. In contrast, Section 3.2 (Geometric Errors and Dimensional Deviations) focuses on geometric compensation of FDM processes, employing methods such as virtual reference points or machine learning regressors to address issues like build platform misalignments or layer-wise shrinkage [34]. In some cases, although the primary focus was on dimensional deviations, several studies have developed algorithms that address these issues by modifying the design and pre-processing steps typically covered in Section 3.1 [53]. Finally, studies corresponding to Section 3.3 (Failures in In-Process Error Detection and Response) incorporate hardware components, such as sensors, to capture critical data in real time, which then serve as inputs for algorithms designed to enable early detection and mitigation of failures [105]. Overall, these findings confirm that software-driven compensation and predictive modeling techniques hold promising potential for improved AM performance. However, to translate these strategies effectively into practice, structured collaboration between design engineers—who ensure that models meet geometric and functional constraints—and control system experts—who develop real-time feedback mechanisms to detect and correct fabrication errors—will be essential.

4. Discussion

This study employed a systematic review methodology following the PRISMA guidelines to ensure a transparent and comprehensive literature survey. A broad search strategy was designed to capture relevant studies on 3D printing failures and mitigation techniques, spanning multiple academic databases. The identification, screening, and eligibility phases were documented using a PRISMA flow diagram, resulting in a final corpus of peer-reviewed articles that directly addressed the research questions. By adhering to PRISMA's structured approach, the review minimized selection bias and provided an overview of the state-of-the-art. This rigorous process underpins the reliability of the findings and allowed for the extraction of key themes. In particular, the included studies converged on three primary categories of challenges in additive manufacturing: (1) design and pre-processing errors; (2) geometric errors and dimensional deviations; (3) failures in in-process error detection and response. These categories formed the basis for organizing the results and are discussed in detail below in relation to the research objectives.

Challenges in the design and pre-processing errors arise from the early stages of the 3D printing workflow, specifically during the design of the 3D model and the preparation of that model for printing, which includes file conversion and slicing. Common problems identified in the literature include a range of design-for-manufacturing issues and setup mistakes that can predispose a print to fail. For example, the accuracy and printability of a part are highly dependent on its design, meaning that 3D printing is not well-suited for certain geometrical features such as unsupported overhangs, extremely thin vertical columns, or large flat horizontal spans without support. Models that ignore these constraints often suffer from collapse, severe sagging, or incomplete layers during printing. Another prevalent issue is the loss of fidelity when converting a CAD model into the triangulated STL format for slicing. The STL approximation represents curved surfaces as a mesh of flat facets, introducing discretization errors that cause subtle geometry distortions. These pre-processing errors can lead to gaps or misalignments between layers, especially for small features or curved surfaces, ultimately affecting the dimensional accuracy and surface quality of the final part. Furthermore, improper slicing configurations such as selecting a suboptimal part orientation, inadequate support structures, or incorrect process parameters are frequently reported for failed builds. Such mistakes can result in issues like poor bed adhesion, internal stresses, or unstable structures during the print.

To minimize errors arising in the design and pre-processing stages, Design for Additive Manufacturing (DfAM) provides a systematic framework that incorporates the physical limitations of the manufacturing process into the design phase itself. This framework is further supported by software tools capable of automatically validating and correcting design features [1]. For instance, in the context of metal powder bed fusion, distortion and buckling in thin-walled structures can be mitigated through the use of design maps that correlate wall thickness and fillet radius, thereby enabling designers to model components with greater confidence [25]. Additionally, Budinoff and McMains (2021) introduced a toolkit that automatically analyzes a 3D model for known problem features such as warping-prone corners, likely weak overhangs, and insufficient support regions, and suggests modifications in geometry or orientation to improve manufacturability [125]. These pre-processing solutions, often integrated into slicing software, have shown success in preventing common failure modes such as part toppling or excessive material deposition in unsupported areas.

The second major category of challenges pertains to the accuracy and fidelity of the printed object's geometry. Even when a part is well-designed and correctly sliced, the physical printing process can introduce geometrical distortions, dimensional inaccuracies, and surface quality issues. A multitude of studies confirm that achieving high dimensional accuracy and repeatability in 3D printing is non-trivial, with errors stemming from both machine limitations and material behavior. For instance, FDM inherently produces layer-by-layer artifacts that can roughen surfaces and slightly alter dimensions. As the extruded filament cools and solidifies, it may shrink unevenly, leading to warping or internal stresses that cause subtle deformations [24]. Such thermally induced deformation means that printed features can deviate from their intended geometry: holes might come out undersized or shafts oversized due to material contraction and the finite extrusion width. Another well-documented source of error is the aforementioned STL model approximation. The conversion of curved CAD surfaces into flat facets introduces a chordal error, which is when the printed object's surface can deviate by a small amount from the true shape and is often noticeable in fine details or circular features. Compounded with the discrete layer thickness, this leads to a "stair-stepping" effect on sloped surfaces and dimensional offsets that accumulate over the height of the part [12]. Consequently, researchers have observed that without compensation even optimally designed parts may not meet tight tolerance requirements. In summary, typical geometric problems include the warping and curling of parts, dimensional shrinkage or expansion, and surface roughness or texture deviations. These issues were frequently highlighted in the surveyed literature as key barriers to using 3D printed components.

Addressing the research question on solutions to improve geometric accuracy, researchers have investigated both hardware- and software-based approaches. On the hardware side, one solution is improved printer calibration and control. High-end machines employ closed-loop controls to ensure each layer is deposited within the correct tolerances. Heating enclosures and controlled cooling systems are used to mitigate warping by maintaining a uniform temperature field around the part, thereby reducing the thermal gradients that cause distortion. On the software side, error compensation algorithms have been proposed. These algorithms pre-emptively modify the print instructions to counteract known sources of deviation. Likewise, some tools adjust the toolpath to account for filament expansion, effectively calibrating out systematic errors. A study on FDM dimensional accuracy noted that tweaking the STL mesh can lessen the chordal discrepancy and improve the match between the printed part and the CAD model. Another set of solutions involves post-processing for dimensional correction; for instance, targeted annealing or vapor smoothing can relieve internal stresses and improve dimensional fidelity after printing. While post-processing veers outside the printing process itself, it is sometimes necessary to meet tight dimensional specs.

The third challenge category identified is the failure to detect and respond to errors during the printing process itself. Even with a perfect design and a well-calibrated printer, unforeseen issues can arise mid-print such as material jams, layer misalignment, support structure failure, or loss of adhesion that could lead to a failed print. A recurring theme in the reviewed studies is that traditional 3D printers operate in an open-loop manner with minimal real-time feedback, making in-process failures disturbingly common. In answer to research question 1, the types of problems often occurring during the print include filament feed problems (e.g., clogged nozzles or filament running out), thermal anomalies like overheating or cooling leading to poor layer bonding, and mechanical issues such as stepper motors skipping. One particular failure mode is the so-called "spaghetti print", where due to a dislodged part or missed layer the printer continues extruding filament into empty space, creating a tangled mess of filament. Without intervention, the print is irrecoverable and results in wasted material and time. Overall, the lack of timely detection means that many prints fail only after significant waste has occurred.

To tackle these in-process issues, researchers have been actively developing solutions centered on real-time monitoring and control systems. One prominent approach is the use of sensor arrays to continuously observe the printing process. Various sensor modalities have been explored, such as visual monitoring with cameras, thermal imaging, acoustic sensors, and vibration/strain sensors attached to the printer. Each of these can capture different signatures of a printing anomaly. For instance, a camera can detect visible defects such as layer delamination, excessive stringing, or part detachment, while a microphone or accelerometer might pick up the sound or vibration pattern of a clogged extruder or stepper motor skip. Once an anomaly is detected, the next step is the response; simpler systems might just alert the operator or pause the print, whereas more sophisticated implementations attempt automated error correction. For example, machine learning algorithms like neural networks have been trained to not only recognize specific failure patterns from sensor data but also to adjust printing parameters on-the-fly to correct them. These types of interventions have shown promise in reducing wasted prints, although they remain largely in experimental stages and are not yet widespread in consumer printers.

Recent studies published since 2023 reflect an emerging trend in AM research toward more integrated, intelligent, and scalable error mitigation strategies. For example, Girard and Zhang (2025) introduced a structured light-based method that detects errors in the 2D absolute phase domain, enabling selective 3D reconstruction and real-time G-code-based correction and thereby demonstrating efficient integration of fast error detection and closed-loop control in AM processes [39]. Similarly, Ntousia et al. (2023) proposed a printability prediction framework that combines neural network-based geometric error estimation with probabilistic modeling of design and process parameters, offering a technology-agnostic tool to support quality assurance across AM platforms from the early design stage [5]. Kwon et al. (2024) further contribute to this trajectory by introducing an AI-driven troubleshooting system that leverages community-annotated data for real-time failure diagnosis and user support [6]. Furthermore, a number of studies emphasize the value of open-source platforms and collaborative ecosystems, which are expected to accelerate the development and democratization of robust AM error-handling solutions. Sani et al. (2024) complement these developments by presenting a comprehensive review of closed-loop AI-augmented additive manufacturing (AI2AM), which integrates real-time monitoring, parameter optimization, and defect correction to enhance the reliability and efficiency of AM systems [126]. Ulkir (2024) further adds to this trajectory by experimentally evaluating the mechanical and thermal behavior of metal-reinforced PLA composites and optimizing FDM process parameters to enhance part quality through the data-driven control of layer thickness, infill density, and nozzle temperature [127]. Additionally, recent advancements in DfAM have led to the development of automated design validation tools and real-time STL repair or slicing optimization techniques, which aim to minimize print failures during the early stages of the AM workflow. Together, these efforts highlight a growing emphasis on real-time, data-driven, and user-supportive frameworks that integrate error detection, prediction, and correction across the AM workflow. This trend reflects a broader shift toward intelligent automation, where adaptive control, predictive modeling, and community-enabled tools work in concert to enhance reliability and accessibility in additive manufacturing.

While the systematic approach of this review provides a high-level understanding of 3D printing challenges and solutions, there are several limitations to acknowledge. First, the scope of the literature survey was constrained to studies available in indexed databases and published in English. It is possible that relevant insights, especially practical knowledge in industry or in non-English publications, were not captured, which may bias the findings toward academic settings. Secondly, the quality and focus of the included studies varied. Some papers provided quantitative data on failure rates or solution effectiveness, whereas others were more anecdotal or conceptual. As a result, our synthesis sometimes had to generalize across different printer technologies and materials. Another limitation lies in the classification of challenges into three categories. In reality, these issues are interrelated as design flaws can lead to in-process failures, and geometric inaccuracies can stem from both design and process. Thus, there is an inherent simplification in our discussion.

5. Conclusions

In conclusion, this PRISMA-guided review has synthesized current knowledge on why 3D prints fail and how to prevent such failures, covering design-stage pitfalls, machine- and material-induced errors, and shortcomings in process monitoring. The challenges of design and pre-processing errors, geometric inaccuracies, and inadequate error detection are significant, but not insurmountable. The literature offers a toolkit of solutions that improve print success rates and part quality. Moving forward, addressing the limitations noted and pursuing the recommended research directions will be critical. By performing this research, the AM community can close the gap between prototype and production, ensuring that 3D printing evolves into a reliably high-quality manufacturing technology. The continued collaboration between design engineers, materials scientists, and control system experts will accelerate progress toward 3D printers that are not only innovative in what they can create, but also consistent and dependable in how they create it.

Future research should aim to address these limitations by expanding the scope of analysis to include non-English literature, gray literature, and practice-oriented case studies from industrial settings. This broader inclusion could enrich the understanding of practical failure modes and solution strategies that are currently underrepresented in academic sources. In addition, more granular classification frameworks may help to capture the interdependencies between design-, geometry-, and process-related errors more accurately. Finally, increased attention to benchmarking datasets, open-source validation tools, and collaborative development environments may facilitate reproducible, scalable, and community-driven progress in AM error detection and correction. In particular, future studies may build upon the PRISMA-based review framework to establish comprehensive troubleshooting guidelines that serve both researchers and practitioners in the AM field.

Author Contributions: Conceptualization, S.K. and D.H.; methodology, S.K. and D.H.; formal analysis, S.K.; writing—original draft preparation, S.K.; writing—review and editing, D.H.; visualization, S.K.; supervision, D.H. All authors have read and agreed to the published version of the manuscript.

Funding: This work was supported by the National Research Foundation of Korea (NRF) grant funded by the Korea government (MSIT) (NRF-2022R1F1A1071767) and also conducted with a Research Grant from Kwangwoon University in 2025.

Data Availability Statement: No new data were created or analyzed in this study.

Conflicts of Interest: The authors declare no conflicts of interest.

References

1. Gibson, I.; Rosen, D.; Stucker, B.; Khorasani, M. *Additive Manufacturing Technologies*; Springer International Publishing: Cham, Switzerland, 2021; ISBN 978-3-030-56126-0.
2. Rinaldi, M.; Caterino, M.; Manco, P.; Fera, M.; Macchiaroli, R. The Impact of Additive Manufacturing on Supply Chain Design: A Simulation Study. *Procedia Comput. Sci.* **2021**, *180*, 446–455. [CrossRef]

3. 3D Printing Market Size and Share | Industry Report, 2030. Available online: https://www.grandviewresearch.com/industry-analysis/3d-printing-industry-analysis (accessed on 31 March 2025).
4. Roland Berger. *Additive Manufacturing: The Money Story*; Roland Berger GmbH: Munich, Germany, 2023.
5. Ntousia, M.; Fudos, I.; Moschopoulos, S.; Stamati, V. Predicting Geometric Errors and Failures in Additive Manufacturing. *Rapid Prototyp. J.* **2023**, *29*, 1843–1861. [CrossRef]
6. Kwon, N.; Sun, T.S.; Gao, Y.; Zhao, L.; Wang, X.; Kim, J.; Hong, S.R. 3DPFIX: Improving Remote Novices' 3D Printing Troubleshooting through Human-AI Collaboration Design. *Proc. ACM Hum.-Comput. Interact.* **2024**, *8*, 1–33. [CrossRef]
7. Liberati, A.; Altman, D.G.; Tetzlaff, J.; Mulrow, C.; Gøtzsche, P.C.; Ioannidis, J.P.; Clarke, M.; Devereaux, P.J.; Kleijnen, J.; Moher, D. The PRISMA Statement for Reporting Systematic Reviews and Meta-Analyses of Studies That Evaluate Healthcare Interventions: Explanation and Elaboration. *BMJ* **2009**, *339*, b2700. [CrossRef] [PubMed]
8. Page, M.J.; McKenzie, J.E.; Bossuyt, P.M.; Boutron, I.; Hoffmann, T.C.; Mulrow, C.D.; Shamseer, L.; Tetzlaff, J.M.; Moher, D. Updating Guidance for Reporting Systematic Reviews: Development of the PRISMA 2020 Statement. *J. Clin. Epidemiol.* **2021**, *134*, 103–112. [CrossRef]
9. *ISO/ASTM 52900:2015*; ISO/ASTM International Additive Manufacturing—General Principles—Terminology. ISO: Geneva, Switzerland, 2015.
10. Cui, K.; Shang, X.; Luo, C.; Shen, Z.; Gao, H.; Xiong, G. A Kind of Accuracy Improving Method Based on Error Analysis and Feedback for DLP 3D Printing. In Proceedings of the 2019 IEEE international conference on service operations and logistics, and informatics (SOLI), Zhengzhou, China, 6–8 November 2019; pp. 5–9.
11. Montalti, A.; Ferretti, P.; Santi, G.M. From CAD to G-Code: Strategies to Minimizing Errors in 3D Printing Process. *CIRP J. Manuf. Sci. Technol.* **2024**, *55*, 62–70. [CrossRef]
12. Oh, Y.; Ko, H.; Sprock, T.; Bernstein, W.Z.; Kwon, S. Part Decomposition and Evaluation Based on Standard Design Guidelines for Additive Manufacturability and Assemblability. *Addit. Manuf.* **2021**, *37*, 101702. [CrossRef]
13. Moodleah, S.; Kirimasthong, K. Adaptive Slicing of Point Cloud Directly with Discrete Interpolable-Area Error Profile in Additive Manufacturing. *SAE Int. J. Mater. Manuf.* **2023**, *16*, 175–188. [CrossRef]
14. Qian, S.; Jiang, X.; Liu, Y.; Wang, S.; Sun, X.; Sun, H. Design and Error Compensation of a 3-Degrees-of-Freedom Cable-Driven Hybrid 3D-Printing Mechanism. *Mech. Sci.* **2023**, *14*, 371–386. [CrossRef]
15. Paul, R.; Anand, S. A Combined Energy and Error Optimization Method for Metal Powder Based Additive Manufacturing Processes. *Rapid Prototyp. J.* **2015**, *21*, 301–312. [CrossRef]
16. Zeng, L.; Zou, X. Error Analysis and Experimental Research on 3D Printing. *IOP Conf. Ser. Mater. Sci. Eng.* **2019**, *592*, 012150. [CrossRef]
17. Scipioni, S.I.; Lambiase, F. Error Introduced by Direct 3D Printing of Compression Samples of PLA Made by FDM Process. *Int. J. Adv. Manuf. Technol.* **2023**, *129*, 4355–4368. [CrossRef]
18. Becker, P.; Gebert, J.; Roennau, A.; Finsterwalder, F.; Dillmann, R. Online Error Detection in Additive Manufacturing: A Review. In Proceedings of the 2021 IEEE 8th International Conference on Industrial Engineering and Applications (ICIEA), Chengdu, China, 23–26 April 2021; pp. 167–175.
19. Kyung, Y.S.; Choi, S.Y.; Kim, G.B.; Song, H.K.; Kim, H.; You, D.; Jeong, I.G.; Homg, J.H.; Kim, N.; Kim, C.-S. Analysis of Errors in 3D Printing Phantoms for Partial Nephrectomy. *Eur. Urol. Suppl.* **2017**, *16*, e1922–e1924. [CrossRef]
20. Martinez-Marquez, D.; Jokymaityte, M.; Mirnajafizadeh, A.; Carty, C.P.; Lloyd, D.; Stewart, R.A. Development of 18 Quality Control Gates for Additive Manufacturing of Error Free Patient-Specific Implants. *Materials* **2019**, *12*, 3110. [CrossRef]
21. Pinto, J.M.; Arrieta, C.; Andia, M.E.; Uribe, S.; Ramos-Grez, J.; Vargas, A.; Irarrazaval, P.; Tejos, C. Sensitivity Analysis of Geometric Errors in Additive Manufacturing Medical Models. *Med. Eng. Phys.* **2015**, *37*, 328–334. [CrossRef]
22. Suzuki, T.; Tateno, T. Tool Path Design of Metal Powder Extrusion in Additive Manufacturing for Suppressing Shape Error Caused During Sintering. *Int. J. Autom. Technol.* **2024**, *18*, 493–502. [CrossRef]
23. Reichwein, J.; Geis, J.; Rudolph, K.; Kirchner, E. Design Guidelines for the Separation of Components to Combine the Potentials of Additive and Conventional Manufacturing Processes. *Procedia CIRP* **2022**, *109*, 592–597. [CrossRef]
24. Zmarzły, P.; Gogolewski, D.; Kozior, T. Design Guidelines for Plastic Casting Using 3D Printing. *J. Eng. Fibers Fabr.* **2020**, *15*, 1558925020916037. [CrossRef]
25. Vastola, G.; Sin, W.J.; Sun, C.-N.; Sridhar, N. Design Guidelines for Suppressing Distortion and Buckling in Metallic Thin-Wall Structures Built by Powder-Bed Fusion Additive Manufacturing. *Mater. Des.* **2022**, *215*, 110489. [CrossRef]
26. Majarena, A.C.; Aguilar, J.J.; Santolaria, J. Development of an Error Compensation Case Study for 3D Printers. *Procedia Manuf.* **2017**, *13*, 864–871. [CrossRef]
27. Wang, Y.; Liu, J.; Guo, M.; Wang, L. Research on the Printing Error of Tilted Vertical Beams in Delta-Robot 3D Printers. *Rapid Prototyp. J.* **2021**, *27*, 1633–1649. [CrossRef]
28. Zhang, S.; Sun, Z.; Long, J.; Li, C.; Bai, Y. Dynamic Condition Monitoring for 3D Printers by Using Error Fusion of Multiple Sparse Auto-Encoders. *Comput. Ind.* **2019**, *105*, 164–176. [CrossRef]

29. Duan, M.; Yoon, D.; Okwudire, C.E. A Limited-Preview Filtered B-Spline Approach to Tracking Control—With Application to Vibration-Induced Error Compensation of a 3D Printer. *Mechatronics* **2018**, *56*, 287–296. [CrossRef]
30. Li, L.; Anand, S. Hatch Pattern Optimization of Powder Bed Fusion Additive Manufacturing Process for Minimizing Flatness Error. *Procedia Manuf.* **2021**, *53*, 456–465. [CrossRef]
31. Junk, S.; Bär, F. Design Guidelines for Additive Manufacturing Using Masked Stereolithography mSLA. *Procedia CIRP* **2023**, *119*, 1122–1127. [CrossRef]
32. Mao, Y.; Wei, H.; Chang, L.; Yue, J.; Liu, T.; Xing, F.; Liao, W. Origin of Deposition Errors and Layer-Wise Control Strategies during Laser Additive Manufacturing. *Virtual Phys. Prototyp.* **2023**, *18*, e2173615. [CrossRef]
33. Auškalnis, L.; Akulauskas, M.; Jegelevičius, D.; Simonaitis, T.; Rutkūnas, V. Error Propagation from Intraoral Scanning to Additive Manufacturing of Complete-Arch Dentate Models: An in Vitro Study. *J. Dent.* **2022**, *121*, 104136. [CrossRef]
34. Tian, W.; Li, Q.; Wang, Q.; Chen, D.; Gao, W. Additive Manufacturing Error Quantification on Stability of Composite Sandwich Plates with Lattice-Cores through Machine Learning Technique. *Compos. Struct.* **2024**, *327*, 117645. [CrossRef]
35. Cajal, C.; Santolaria, J.; Velazquez, J.; Aguado, S.; Albajez, J. Volumetric Error Compensation Technique for 3D Printers. *Procedia Eng.* **2013**, *63*, 642–649. [CrossRef]
36. Das, P.; Chandran, R.; Samant, R.; Anand, S. Optimum Part Build Orientation in Additive Manufacturing for Minimizing Part Errors and Support Structures. *Procedia Manuf.* **2015**, *1*, 343–354. [CrossRef]
37. Majd, Y.F.; Tsuzuki, M.S.; Barari, A. A Machine Learning Approach to Find Density Percentage Error Resulting by Infill Patterns in Additive Manufacturing. *IFAC-Pap.* **2023**, *56*, 4740–4745. [CrossRef]
38. Zgórniak, P.; Stachurski, W. Determination of Systematic Errors of 3D Printer in Order to Ensure Manufacturing Correctness of the Prototype. *Adv. Manuf. Sci. Technol.* **2010**, *34*, 35–45.
39. Girard, J.; Zhang, S. Fast Error Detection Method for Additive Manufacturing Process Monitoring Using Structured Light Three Dimensional Imaging Technique. *Opt. Lasers Eng.* **2025**, *184*, 108609. [CrossRef]
40. Zapico, P.; Peña, F.; Valiño, G.; Rico, J.C.; Meana, V.; Mateos, S. Virtual-Point-Based Geometric Error Compensation Model for Additive Manufacturing Machines. *Rapid Prototyp. J.* **2023**, *29*, 837–849. [CrossRef]
41. Becker, P.; Spielbauer, N.; Roennau, A.; Dillmann, R. Real-Time in-Situ Process Error Detection in Additive Manufacturing. In Proceedings of the 2020 Fourth IEEE International Conference on Robotic Computing (IRC), Taichung, Taiwan, 9–11 November 2020; pp. 426–427.
42. Baumann, F.; Roller, D. Vision Based Error Detection for 3D Printing Processes. *MATEC Web Conf.* **2016**, *59*, 06003. [CrossRef]
43. Peña, F.; Fernández, A.; Zapico, P.; Valiño, G.; Rico, J.C. Conoscopic Holography Feasibility for Form Error In-Situ Monitoring in Additive Manufacturing. *IFAC-PapersOnLine* **2022**, *55*, 1031–1036. [CrossRef]
44. Zeng, L.; Wu, W. Approximation Error Analysis of Contour on 3D Printing. In Proceedings of the International Conference on Mechanisms and Robotics (ICMAR 2022), Zhuhai, China, 25–27 February 2022; Volume 12331, pp. 852–857.
45. Leung, Y.-S.; Kwok, T.-H.; Mao, H.; Chen, Y. Digital Material Design Using Tensor-Based Error Diffusion for Additive Manufacturing. *Comput.-Aided Des.* **2019**, *114*, 224–235. [CrossRef]
46. Brunton, A.; Arikan, C.A.; Urban, P. Pushing the Limits of 3D Color Printing: Error Diffusion with Translucent Materials. *ACM Trans. Graph.* **2015**, *35*, 1–13. [CrossRef]
47. Pritchet, D.; Moser, N.; Ehmann, K.; Cao, J.; Huang, J. Quantifying Discretization Errors in Electrophoretically-Guided Micro Additive Manufacturing. *Micromachines* **2018**, *9*, 447. [CrossRef]
48. Rudolph, J.-P.; Emmelmann, C. Analysis of Design Guidelines for Automated Order Acceptance in Additive Manufacturing. *Procedia Cirp* **2017**, *60*, 187–192. [CrossRef]
49. Simionato, E.; Aldaya, I.; Oliveira, J.A.D.; Jardini, A.L.; Avila, J.; Rosa, G.S.D.; Penchel, R.A. Design Guidelines and Performance Analysis of a Wideband Coaxial Horn Antenna Fabricated via Additive Manufacturing. *IEEE Open J. Antennas Propag.* **2024**, *5*, 1121–1132. [CrossRef]
50. ZHANG, Q.; Lee, B.-C. Research on Solutions to Slicing Errors in FDM 3D Printing of Thin-Walled Structures. *Int. J. Internet Broadcast. Commun.* **2024**, *16*, 176–181.
51. Michaeli, J.G.; DeGroff, M.C.; Roxas, R.C. Error Aggregation in the Reengineering Process from 3D Scanning to Printing. *Scanning* **2017**, *2017*, 1–8. [CrossRef]
52. Naulakha, N.K. Guidelines for Increasing Application of 3D Metal Printing-a Case Study at Equinor ASA. Master's Thesis, UiT Norges Arktiske Universitet, Tromsø, Norway, 2021.
53. Charalampous, P.; Kostavelis, I.; Kontodina, T.; Tzovaras, D. Learning-Based Error Modeling in FDM 3D Printing Process. *Rapid Prototyp. J.* **2021**, *27*, 507–517. [CrossRef]
54. Wu, H.; Luo, H.; Wang, K.; Ye, D.; Huang, Y. Enhanced Geometric Precision of Non-Contact, Conformal 3D Printing via "Error-Transferred" towards Jetting-Direction. *Precis. Eng.* **2021**, *72*, 1–12. [CrossRef]
55. Leirmo, T.L.; Semeniuta, O. Minimizing Form Errors in Additive Manufacturing with Part Build Orientation: An Optimization Method for Continuous Solution Spaces. *Open Eng.* **2022**, *12*, 227–244. [CrossRef]

56. Peña, F.; Rico, J.C.; Zapico, P.; Valiño, G.; Mateos, S. A Layerwise Geometric Error Compensation Procedure for Additive Manufacturing. *Rapid Prototyp. J.* **2024**, *30*, 490–501. [CrossRef]
57. Cajal, C.; Santolaria, J.; Samper, D.; Velazquez, J. Efficient Volumetric Error Compensation Technique for Additive Manufacturing Machines. *Rapid Prototyp. J.* **2016**, *22*, 2–19. [CrossRef]
58. Elkaseer, A.; Mueller, T.; Charles, A.; Scholz, S. Digital Detection and Correction of Errors in As-Built Parts: A Step towards Automated Quality Control of Additive Manufacturing. In Proceedings of the World Congress on Micro and Nano Manufacturing, Portorož, Slovenia, 18–20 September 2018; pp. 18–20.
59. Keaveney, S.; Connolly, P.; O'Cearbhaill, E.D. Kinematic Error Modeling and Error Compensation of Desktop 3D Printer. *Nanotechnol. Precis. Eng.* **2018**, *1*, 180–186. [CrossRef]
60. Ma, X.; Lin, F.; Zhang, L. The Heterogeneous Compensation for the Infiltrative Error of the Binder Jetting Additive Manufacturing Processes. 2015. Available online: https://utw10945.utweb.utexas.edu/sites/default/files/2015/2015-16-Ma.pdf (accessed on 14 October 2024).
61. Bhatia, S. Effect of Machine Positional Errors on Geometric Tolerances in Additive Manufacturing. Master's Thesis, University of Cincinnati, Cincinnati, OH, USA, 2014.
62. Munteanu, A.; Chitariu, D.-F.; Horodinca, M.; Dumitras, C.-G.; Negoescu, F.; Savin, A.; Chifan, F. A Study on the Errors of 2D Circular Trajectories Generated on a 3D Printer. *Appl. Sci.* **2021**, *11*, 11695. [CrossRef]
63. Das, P.; Mhapsekar, K.; Chowdhury, S.; Samant, R.; Anand, S. Selection of Build Orientation for Optimal Support Structures and Minimum Part Errors in Additive Manufacturing. *Comput.-Aided Des. Appl.* **2017**, *14*, 1–13. [CrossRef]
64. Cong, B.; Yoo, H.W.; Pechgraber, D.; Schitter, G. Cross-Scan Error Evaluation of Large Size Polygon Mirror Based Laser Scanning System for Industrial 3D Printing. In Proceedings of the 2024 IEEE International Conference on Advanced Intelligent Mechatronics (AIM), Boston, MA, USA, 15–19 July 2024; pp. 290–295.
65. Wu, Y.; Chiu, G. Error Diffusion Based Feedforward Height Control for Inkjet 3D Printing. In Proceedings of the 2023 IEEE/ASME international conference on advanced intelligent mechatronics (AIM), Seattle, WA, USA, 28–30 June 2023; pp. 125–131.
66. Fu, G.; Gu, T.; Gao, H.; Lu, C. A Postprocessing and Path Optimization Based on Nonlinear Error for Multijoint Industrial Robot-Based 3D Printing. *Int. J. Adv. Robot. Syst.* **2020**, *17*, 1729881420952249. [CrossRef]
67. Song, G.-H.; Lee, C.-M.; Kim, D.-H. Investigation of Path Planning to Reduce Height Errors of Intersection Parts in Wire-Arc Additive Manufacturing. *Materials* **2021**, *14*, 6477. [CrossRef]
68. Xia, Z.; He, Z.; Wang, Q.; Wang, Y. A New Finite Element Model with Manufactured Error for Additive Manufacturing. *Comput. Model. Eng. Sci.* **2020**, *124*, 703–720. [CrossRef]
69. Samperi, M.T. Development of Design Guidelines for Metal Additive Manufacturing and Process Selection. Master's Thesis, The Pennsylvania State University, University Park, PA, USA, 2014.
70. Lei, G.; Lu, L.; Bi, G.; Zhang, L.; Sun, L. Digital Model Reconstruction and Geometric Error Analysis of Additive Manufacturing Blisk Enabled by 3D Scanner. *Proc. J. Phys. Conf. Ser.* **2024**, *2760*, 012037. [CrossRef]
71. Lv, N.; Ouyang, X.; Qiao, Y. Adaptive Layering Algorithm for FDM-3D Printing Based on Optimal Volume Error. *Micromachines* **2022**, *13*, 836. [CrossRef]
72. Jin, Y.; Qin, S.J.; Huang, Q. Out-of-Plane Geometric Error Prediction for Additive Manufacturing. In Proceedings of the 2015 IEEE International Conference on Automation Science and Engineering (CASE), Gothenburg, Sweden, 24–28 August 2015; pp. 918–923.
73. Tian, W. Nonlinear Dynamic Analysis and Quantification of Additive Manufacturing Error on the Stability of Lattice-Core Composite Sandwich Plate. Ph.D. Thesis, UNSW Sydney, Kensington, NSW, Australia, 2023.
74. Qian, S.; Jiang, X.; Qian, P.; Zi, B.; Zhu, W. Calibration of Static Errors and Compensation of Dynamic Errors for Cable-Driven Parallel 3D Printer. *J. Intell. Robot. Syst.* **2024**, *110*, 31. [CrossRef]
75. Hou, L.; Ye, C.; Guo, J.; Chen, Y.; Jing, X.; Li, Y.; Chen, S.; Huang, X.; Li, H. Allowance Optimization and Error Compensation in Hybrid Additive and Subtractive Manufacturing of Complex Parts. *SSRN* **2023**, SSRN 4503092. [CrossRef]
76. Psulkowski, S. Digital Compensation to Mechanical Error: An Investigation into Cyber-Physical Systems with DeXter™ A Hybrid Additive Manufacturing System. Undergraduate Thesis, Florida State University, Tallahassee, FL, USA, 2018.
77. Yoon, D.; Duan, M.; Okwudire, C.E. Software-Based Compensation of Vibration-Induced Errors of a Commercial Desktop 3D Printer. In Proceedings of the Proceedings-32nd ASPE Annual Meeting, Cambridge, MA, USA, 20–22 April 2016; p. 478.
78. Obeidi, M.A. Metal Additive Manufacturing by Laser-Powder Bed Fusion: Guidelines for Process Optimisation. *Results Eng.* **2022**, *15*, 100473. [CrossRef]
79. Plocher, J.; Wioland, J.-B.; Panesar, A.S. Additive Manufacturing with Fibre-Reinforcement–Design Guidelines and Investigation into the Influence of Infill Patterns. *Rapid Prototyp. J.* **2022**, *28*, 1241–1259. [CrossRef]
80. Badini, S.; Regondi, S.; Frontoni, E.; Pugliese, R. Assessing the Capabilities of ChatGPT to Improve Additive Manufacturing Troubleshooting. *Adv. Ind. Eng. Polym. Res.* **2023**, *6*, 278–287. [CrossRef]

81. Alietti, L. Precision Enhancement of the 3D Printing Process Using Automated Visual Error Detection. 2021. Available online: https://www.politesi.polimi.it/retrieve/cec4914a-fd53-4aa1-96d8-85e2ea1a07e1/2022_07_Alietti_02.pdf (accessed on 14 October 2024).
82. Brion, D. Deep Learning Enabled Error Detection and Correction for 3D Printing. Ph.D. Thesis, University of Cambridge, Cambridge, UK, 2023.
83. Plocher, J. Fibre-Reinforced Additive Manufacturing: From Design Guidelines to Advanced Lattice Structures. Ph.D. Thesis, Imperial College London, London, UK, 2022.
84. Kovács11, S.E.; Varga, L. Additive Manufacturing Mistakes: An Overview of Possible Errors in Laser Power Bed Fusion Processes. *Doktorandusz Alm. PhD Stud. Alm.* **2023**, 184–188.
85. Khaleed, H.M.T.; Badruddin, I.A.; Saquib, A.N.; Addas, M.F.; Kamangar, S.; Yunus Khan, T.M. Novel Approach to Manufacture an AUV Propeller by Additive Manufacturing and Error Analysis. *Appl. Sci.* **2019**, *9*, 4413. [CrossRef]
86. Lu, Y. Error Characterization of EOS M280 Additive Manufacturing Machine. Master's Thesis, The Pennsylvania State University, University Park, PA, USA, 2017.
87. Kumaran, S.V.; Abd Wahab, D.; Abd Aziz, N.; Wahid, Z. Guidelines for Automotive Component Repair Using Additive Manufacturing in Industry 4.0: Design and Process Perspective. *J. Inf. Syst. Technol. Manag.* **2024**, *9*, 20–32. [CrossRef]
88. Buttard, M.; Chehab, B.; Josserond, C.; Charlot, F.; Lhuissier, P.; Bataillon, X.; Deschamps, A.; Villanova, J.; Fivel, M.; Blandin, J.-J. Guidelines to Tailor the Mechanical Response of Aluminium Alloys Designed for Additive Manufacturing. *SSRN* **2024**, SSRN 4607360.
89. Islam, M.N.; Pramanik, A.; Slamka, S. Errors in Different Geometric Aspects of Common Engineering Parts during Rapid Prototyping Using a Z Corp 3D Printer. *Prog. Addit. Manuf.* **2016**, *1*, 55–63. [CrossRef]
90. Choi, Y.-H. Influence of Bed Temperature on Heat Shrinkage Shape Error in FDM Additive Manufacturing of the ABS-Engineering Plastic. *World J. Eng. Technol.* **2016**, *4*, 186. [CrossRef]
91. Salamone, F.; Danza, L.; Meroni, I.; Pollastro, M.C. A Low-Cost Environmental Monitoring System: How to Prevent Systematic Errors in the Design Phase through the Combined Use of Additive Manufacturing and Thermographic Techniques. *Sensors* **2017**, *17*, 828. [CrossRef]
92. Šantak, I.; Duspara, M.; Cumin, J.; Kovačević, B.; Maglić, L.; Stoić, A. Simulation of First Layer Adhesion Errors in 3D Printing. In Proceedings of the 10th International Scientific and Expert Conference TEAM2022, Slavonski Brod, Croatia, 21–22 September 2022; pp. 307–315.
93. Liu, L.; Ngo, K.D.; Lu, G.-Q. Guideline for Paste Extrusion 3D Printing of Slump-Free Ferrite Inductor Cores. *Ceram. Int.* **2021**, *47*, 5803–5811. [CrossRef]
94. Di Tore, S.; De Simone, G.; Todino, M.D. Learning by Making. 3D Printing Guidelines for Teachers. In Proceedings of the Makers at School, Educational Robotics and Innovative Learning Environments: Research and Experiences from FabLearn Italy 2019, in the Italian Schools and Beyond; Springer International Publishing, New York, NY, USA; 2021; pp. 181–186.
95. Agrawal, R. Sustainable Design Guidelines for Additive Manufacturing Applications. *Rapid Prototyp. J.* **2022**, *28*, 1221–1240. [CrossRef]
96. Kabbur, N. Design and Manufacturing Guidelines for Additive Manufacturing of High Porosity Cellular Structures. Master's Thesis, University of Cincinnati, Cincinnati, OH, USA, 2017.
97. García-Vigueras, M.; Polo-Lopez, L.; Stoumpos, C.; Dorlé, A.; Molero, C.; Gillard, R. *Metal 3D-Printing of Waveguide Components and Antennas: Guidelines and New Perspectives*; IntechOpen: London, UK, 2023.
98. Shehadeh, A. Evaluation of 316L Stainless Steel Part Fabrication Using Additive and Subtractive Manufacturing: A Guideline for Process Selection. 1 June 2019. Available online: https://qspace.qu.edu.qa/handle/10576/12335 (accessed on 15 October 2024).
99. Klar, R.M.; Cox, J.; Raja, N.; Lohfeld, S. The 3D-McMap Guidelines: Three-Dimensional Multicomposite Microsphere Adaptive Printing. *Biomimetics* **2024**, *9*, 94. [CrossRef] [PubMed]
100. Gokuldoss, P.K.; Kolla, S.; Eckert, J. Additive Manufacturing Processes: Selective Laser Melting, Electron Beam Melting and Binder Jetting—Selection Guidelines. *Materials* **2017**, *10*, 672. [CrossRef] [PubMed]
101. Kranz, J.; Herzog, D.; Emmelmann, C. Design Guidelines for Laser Additive Manufacturing of Lightweight Structures in TiAl6V4. *J. Laser Appl.* **2015**, *27*, S14001. [CrossRef]
102. Lozada, N.; Miguel, J. Evaluation of Carbon Fiber Reinforced Polymer Composites Produced by Additive Manufacturing for Design Guidelines. Master's Thesis, Campus Monterrey, Monterrey, Mexico, 2018.
103. Mihasan, M. A Beginner's Guideline for Low-cost 3D Printing of Macromolecules Usable for Teaching and Demonstration. *Biochem. Mol. Biol. Educ.* **2021**, *49*, 521–528. [CrossRef]
104. Moylan, S.; Slotwinski, J. Assessment of Guidelines for Conducting Round Robin Studies in Additive Manufacturing. In *Proceedings of the ASPE Spring Topical Meeting—Dimensional Accuracy and Surface Finish in Additive Manufacturing*; NIST: Berkeley, CA, USA, 2014; Volume 57, pp. 82–85.

105. Çıklaçandır, S.; Ciklacandir, S.; Çulha, S.; Culha, S.; İşler, Y.; Isler, Y. Design of a Real-Time Tracking System to Eliminate 3D Printing Errors Caused by Thermoplastic Materials. *Akıllı Sist. Ve Uygulamaları Derg.* **2022**, *5*, 82–85.
106. Duman, B.; Süzen, A.A. Modeling of IoT-Based Additive Manufacturing Machine's Digital Twin for Error Detection. *Mühendis. Bilim. Ve Tasar. Derg.* **2023**, *11*, 486–497. [CrossRef]
107. Jin, Y.; Liao, H.; Pierson, H.A. A Multi-Resolution Framework for Automated in-Plane Alignment and Error Quantification in Additive Manufacturing. *Rapid Prototyp. J.* **2020**, *26*, 1289–1303. [CrossRef]
108. Kim, H.; Okwudire, C.E. Combined Servo Error Pre-Compensation and Feedrate Optimization—with Application to a 3D Printer and Precision Motion Stage. *Int. J. Adv. Manuf. Technol.* **2020**, *109*, 809–821. [CrossRef]
109. Engle, J.; Nguyen, R.; Buah, K.; Weaver, J.M. Reducing Computer Visualization Errors for In-Process Monitoring of Additive Manufacturing Systems Using Smart Lighting and Colorization System. 2019. Available online: https://utw10945.utweb.utexas.edu/sites/default/files/2019/125%20Reducing%20Computer%20Visualization%20Errors%20for%20In-Proc.pdf (accessed on 15 October 2024).
110. Straub, J. Physical Security and Cyber Security Issues and Human Error Prevention for 3D Printed Objects: Detecting the Use of an Incorrect Printing Material. In Proceedings of the Dimensional Optical Metrology and Inspection for Practical Applications VI, Anaheim, CA, USA, 13 April 2017; Volume 10220, pp. 90–105.
111. Ramamurthy, R.; Bromberg, V.; Fiorillo, T.; Harding, K. Error Mapping Method for Multi-Axis Additive Manufacturing System. *Dimens. Opt. Metrol. Insp. Pract. Appl. VII* **2018**, *10667*, 99–109.
112. Zeng, L.; Wu, W. Analysis and Prediction of Height Error of FDM 3D Printing. In Proceedings of the 2022 2nd International Conference on Control and Intelligent Robotics, Nanjing, China, 24–26 June 2022; pp. 667–670.
113. Shen, Z.; Shang, X.; Zhao, M.; Dong, X.; Xiong, G.; Wang, F.-Y. A Learning-Based Framework for Error Compensation in 3D Printing. *IEEE Trans. Cybern.* **2019**, *49*, 4042–4050. [CrossRef]
114. Omairi, A.; Ismail, Z.H. Towards Machine Learning for Error Compensation in Additive Manufacturing. *Appl. Sci.* **2021**, *11*, 2375. [CrossRef]
115. Bandari, Y.K.; Charrett, T.O.; Michel, F.; Ding, J.; Williams, S.W.; Tatum, R.P. Compensation Strategies for Robotic Motion Errors for Additive Manufacturing (AM). In Proceedings of the 27th Annual International Solid Freeform Fabrication Symposium, Austin, TX, USA, 8–10 August 2016.
116. Wang, D.; Shen, Z.; Dong, X.; Fang, Q.; Wang, W.; Dong, X.; Xiong, G. Deep Reinforcement Learning for Dynamic Error Compensation in 3D Printing. In Proceedings of the 2023 IEEE 19th International Conference on Automation Science and Engineering (CASE), Auckland, New Zealand, 26–30 August 2023; pp. 1–7.
117. Wylie, B.; Moore Jr, C. Optical Methods of Error Detection in Additive Manufacturing: A Literature Review. *J. Manuf. Mater. Process.* **2023**, *7*, 80. [CrossRef]
118. Paulsen, T. Visual Error Detection on 3D Printing. Bachelor's Thesis, NTNU, Trondheim, Norway, 2019.
119. Brion, D.A.; Pattinson, S.W. Generalisable 3D Printing Error Detection and Correction via Multi-Head Neural Networks. *Nat. Commun.* **2022**, *13*, 4654. [CrossRef]
120. Mehta, P.; Mujawar, M.A.; Lafrance, S.; Bernadin, S.; Ewing, D.; Bhansali, S. Editors' Choice—Review—Sensor-Based and Computational Methods for Error Detection and Correction in 3D Printing. *ECS Sens. Plus* **2024**, *3*, 030602. [CrossRef]
121. Liu, X.; Lv, C.; Liu, B.; Li, X.; Jia, X. Machining Error Analysis of Large-Size Ceramic Additive Manufacturing Based on Scanning Exposure Technology. In Proceedings of the Second International Conference on Advanced Manufacturing Technology and Manufacturing Systems (ICAMTMS 2023), Nanjing, China, 26–28 May 2023; Volume 12744, pp. 44–49.
122. Klingert, J.; Bruder, R.; Schweikard, A. Optical Recognition of Error States to Secure Quality in FDM 3d Printing. *Trans. Addit. Manuf. Meets Med.* **2019**, *1*, 1909S03P04. [CrossRef]
123. Lianghua, Z. Research on Error Analysis and Online Recognition Method of Contour in 3D Printing. *Proc. J. Phys. Conf. Ser.* **2020**, *1605*, 012031. [CrossRef]
124. Gabstur, P.; Pollak, M.; Kocisko, M. Detection of Process Errors of Additive Manufacturing. *MM Sci. J.* **2024**, *2024*, 7306–7311. [CrossRef]
125. Budinoff, H.D.; McMains, S. Will It Print: A Manufacturability Toolbox for 3D Printing. *Int. J. Interact. Des. Manuf. IJIDeM* **2021**, *15*, 613–630. [CrossRef]
126. Sani, A.R.; Zolfagharian, A.; Kouzani, A.Z. Artificial Intelligence-Augmented Additive Manufacturing: Insights on Closed-Loop 3D Printing. *Adv. Intell. Syst.* **2024**, *6*, 2400102. [CrossRef]
127. Ulkir, O. Investigation on the Mechanical and Thermal Properties of Metal-PLA Composites Fabricated by FDM. *Rapid Prototyp. J.* **2024**, *30*, 2113–2122. [CrossRef]

processes

MDPI

Systematic Review

Mass Customisation Strategies in Additive Manufacturing: A Systematic Review and Implementation Framework

Samuel Koranteng Fianko [1,*], Thywill Cephas Dzogbewu [2], Edinam Agbamava [3] and Deon Johan de Beer [4]

1 Department of Business Support Studies, Faculty of Business Management, Central University of Technology, Bloemfontein 9301, South Africa
2 Department of Mechanical and Mechatronics Engineering, Faculty of Engineering, Built Environment and Information Technology, Central University of Technology, Bloemfontein 9301, South Africa; tdzogbewu@cut.ac.za
3 Department of Accounting, Ho Technical University, Poly Road, Ho VH-036, Ghana
4 Centre for Rapid Prototyping and Manufacturing, Central University of Technology, Bloemfontein 9301, South Africa; ddebeer@cut.ac.za
* Correspondence: agapesam4@gmail.com

Abstract: Additive manufacturing (AM) has transformed mass customisation by allowing personalised production with remarkable efficiency. This systematic review compiles findings from 61 peer-reviewed articles (2010–2024) to highlight strategies for implementation, technological facilitators, challenges, industry applications, and evaluation frameworks relevant to mass customisation in AM contexts. Utilising the Preferred Reporting Items for Systematic Reviews and Meta-Analyses (PRISMA) methodology, the review applies stringent inclusion criteria and thematic analysis to create an in-depth understanding of this developing area. Four major strategies for implementation have been identified: combining AM with conventional manufacturing, integrating customer-centred design, establishing flexible manufacturing networks, and creating adaptive production systems. Key technological facilitators include capabilities for multi-material processing, integration of digital workflows, and advanced monitoring of processes, while obstacles consist of limitations in materials, challenges in quality assurance, and complexities related to digital asset management. Industry applications reveal tailored approaches specific to medical, industrial, and architectural sectors. This analysis presents a multi-tiered implementation framework encompassing strategic, tactical, operational aspects and performance evaluation aspects to assist organisations in embracing AM-based mass customisation. This framework fills a notable gap in existing literature by aligning personalisation goals with operational efficiency. This paper also outlines future research priorities, such as creating standardised evaluation methods, improving system reliability, incorporating sustainability, and leveraging emerging tools like AI for process improvement. Ultimately, this review bridges theory and practice, offering a clearer path forward for mass customisation in the era of AM.

Keywords: additive manufacturing; mass customization; additive manufacturing-enabled mass customisation; implementation framework; technological enablers; industry applications

1. Introduction

The intersection of additive manufacturing (AM) and mass customisation marks a transformative shift in modern manufacturing practices [1]. Lacroix et al. [2] claim that traditional manufacturing has always had to select between easy manufacturing of standard goods in a low price range or production of individually customised products in a high price range and low efficiency. This fundamental manufacturing dilemma, however, is being resolved through AM technologies. The AM approach could eliminate the traditional trade-off between customisation and cost efficiency [3]. AM-enabled mass customisation represents systematic approaches that encompass design methodologies, production workflows, software solutions, and organisational structures that allow companies to leverage AM capabilities while balancing customisation with operational efficiency [4]. Through these strategies, manufacturers can develop products with standardised bases but customisable features, implement digital workflows that efficiently process individual specifications, and create production systems capable of handling significant product variation without proportional increases in cost. Implementing AM-enabled mass customisation strategies in organisations has been found to maintain quality while reducing lead times in different customisation scenarios [5,6]. Other benefits include resource efficiency, improved facility operation flexibility, and adding customer value due to personalisation [7,8]. Within the context of the medical field, Wazeer [9] and Shidid [10] indicate that AM-produced patient-specific devices have shown improved treatment outcomes. Paoletti [11] further states that AM-enabled mass customisation has also been used in architectural applications to create complex geometric designs previously impossible through conventional manufacturing methods. Despite the transformative potential of AM for enabling mass customisation, there is a critical absence of comprehensive, cross-industry implementation frameworks that address enablers, barriers, and evaluation methodologies of mass customisation strategies in AM, leaving organisations without adequate guidance to successfully adopt and measure AM-enabled mass customisation strategies across diverse manufacturing contexts.

Given the immense potential benefits and associated challenges with implementing AM-enabled mass customisation, several scholars have explored different aspects of this manufacturing technology from various perspectives. Yu et al. [12], for instance, assess the application of 3D and 4D printing technologies in the fashion industry for mass customisation, and the potential of AM to produce tailored products efficiently was revealed in a review. Pajonk et al. [13] explore the role of multi-material AM in architecture and construction, emphasising its customised building components. In another study, Korne et al. [14], through a systematic review, explore trends in the integration of the AM in the fourth industrial revolution, highlighting the AM to create customisable products in line with the tenets of the 4.0, thus enhancing mass customisation. These studies, however, present several limitations. First, there is a lack of comprehensive understanding of the different customisation implementation strategies in AM and frameworks guiding the effective implementation of AM-enabled mass customisation across various industries. Second, these studies do focus on evaluating the mass customisation strategies and performance metrics, creating obstacles for organisations attempting to determine implementation success. Finally, the absence of a comprehensive analysis of technological enablers and inhibitors of these studies hinders stakeholders from gaining insights into the drivers of success or failure regarding mass customisation strategies in AM. This systematic review therefore seeks to address these knowledge gaps by offering a comprehensive evaluation of implementation strategies, alongside technological enablers and inhibitors. The review aims to explore industry-specific applications and integrative frameworks essential for the successful implementation of AM-enabled mass customisation. This systematic review seeks to answer the following research questions:

RQ1: What are the different mass customisation implementation strategies in AM (2010–2024)?

RQ2: What are the technological ecosystems including technological enablers and barriers in AM-enabled mass customisation?

RQ3: What are the evaluation methodologies and performance metrics for assessing the viability of AM-based mass customisation implementations?
RQ4: What are the industry-specific applications and outcomes of AM-based mass customisation?
RQ5: How can a comprehensive multi-layered implementation framework for AM-based on mass customisation be developed that integrates strategic, operational, and evaluative dimensions while addressing industry-specific considerations and technological factors?

This systematic review makes several significant contributions to the understanding and implementation of AM-based mass customisation. Analysing 61 peer-reviewed studies spanning 2010–2024 addresses a critical knowledge gap in balancing customisation capabilities with production efficiency. The study identifies four key implementation strategies: integration with traditional manufacturing, customer-centric design integration, flexible manufacturing networks, and adaptive production systems, each offering distinct advantages across various industrial contexts. The primary contribution of this research is a multi-layered implementation framework that integrates strategic, tactical, and operational dimensions, providing organisations with a structured approach to AM-based mass customisation. Unlike previous reviews that focused on isolated dimensions such as technological capabilities or business models, this study employs a comprehensive analytical approach that simultaneously addresses technological, organisational, and market factors. Practically, it provides actionable guidance for organisations across healthcare, industrial, and architectural sectors by identifying critical success factors, implementation barriers, and evaluation methodologies. The study also establishes a foundation for future research by identifying key gaps in standardised evaluation methods, long-term reliability assessment, and integrating emerging technologies like artificial intelligence.

The remainder of this paper is organised as follows: Section 2 details the methodological approach employed in this systematic review, including search strategy, selection criteria, and analytical procedures. Section 3 presents the findings around four major themes: implementation strategies, technological enablers and barriers, industry-specific applications, and evaluation methodologies. Section 4 proposes a comprehensive implementation framework for AM-based mass customisation, integrating insights from the systematic review. Section 5 discusses practical implications, limitations, and directions for future research.

2. Materials and Methods

The systematic review followed the PRISMA guidelines, an in-depth methodological approach that ensures transparency, completeness, and reproducibility in systematic review studies [15]. The PRISMA was adopted in this study due to the unique advantages in terms of methodological rigor. First, the PRISMA checklist provides a comprehensive 27-item framework that systematically covers all the critical steps for protocol design and synthesis, thereby reducing selective bias and data omission in the review [16]. Second, the requirement of the PRISMA flow diagram enables comprehensive documentation of the inclusion and exclusion procedures of the study, thus increasing transparency where the reasons for decisions are highlighted [15]. Third, through the use of systematic framework, PRISMA enables other scholars to confirm our methodology and try to repeat our findings [17]. Fourth, for intricate topics such as AM-enabled mass customization that transcend multiple disciplines, the framework provides comprehensive coverage in engineering, business, and technology sectors without selection bias [17]. This systematic review was conducted according to PROSPERO guidelines and registered in the PROSPERO International Prospective Register of Systematic Reviews (registration number: CRD420251058742). The protocol was registered prior to study selection and data extraction.

The systematic review followed the PRISMA guidelines of selecting papers on mass customisation strategies in AM from the Scopus database searched on 31st November 2024, covering publications from 2010 to 2024 [18,19]. Scopus was chosen due to its comprehensive multidisciplinary coverage [20], rigorous quality control processes [21], superior search functionality [22], robust citation analysis tools [23], and excellent coverage of recent technological literature [24]. The search string was as follows: TITLE-ABS-KEY (("Additive Manufacturing" OR "3d Printing") AND ("Mass Customisation" OR "Personalisation" OR "Co-creation") AND ("Strategy" OR "Approach" OR "Method" OR "Framework")). This search strategy defines the main keywords employed in related research papers. The search was temporally constrained to cover technological innovations while being as up to date as possible with modern industry, yielding the first results of 559 articles (see Figure 1). The temporal dimension was beneficial given the dynamism of AM technologies and their uses in mass customisation as noted by Singh et al. [25].

2.1. Study Selection and Eligibility Criteria

The research selected publications through a multi-step selection and screening procedure demonstrated in Figure 1. The initial search yielded 559 articles, which were reduced to 303 after 256 articles were excluded (duplicates). The remaining articles were sorted using the titles and abstracts, and inclusion and exclusion criteria were set. The inclusion criteria specifically focused on studies that addressed the application or evaluation of mass customization initiatives in industries utilizing additive manufacturing technology; provided real-world experiences in industries to support conceptual frameworks; and examined specific sectors, technologies, or business infrastructures. After the screening process, 124 articles were excluded, those not focused on AM mass customization (n = 78) and with insufficient empirical evidence (n = 46), leaving a total of 179 full articles for further consideration. Of the 179 articles sought for retrieval, 38 could not be retrieved, leaving 141 records that were assessed for eligibility. Of these, 77 were excluded for lacking focus on mass customisation (n = 27), quality control details (n = 28), digital infrastructure requirements (n = 10), and technological enablers/inhibitors of AM mass customization (n = 12). The final systematic review included 61 articles that met all criteria.

2.2. Data Extraction and Quality Assessment

A standardized data extraction form was developed to systematically extract relevant information from the selected studies. The form captured bibliographic details, study objectives, methodological approach, key findings, implementation strategies, technological factors, industry context, performance metrics, and limitations. Quality assessment was conducted using adapted criteria from the Critical Appraisal Skills Programme (CASP) [26] modified to accommodate qualitative and quantitative research designs. The synthesis process integrated findings from various studies to identify patterns, relationships, and contradictions in the synthesis theme. This involved a constant comparison between studies, an amalgamation of similar concepts, and identification of overarching themes [27]. In synthesising the research questions, the research was guided, and it focused on developing an understanding of the implementation strategy, technological barriers, and the evaluation methods for mass customisation within an AM environment. All 61 articles were subjected to quality assessment using adapted CASP criteria, which were modified to accommodate both qualitative and quantitative research designs present in the sample.

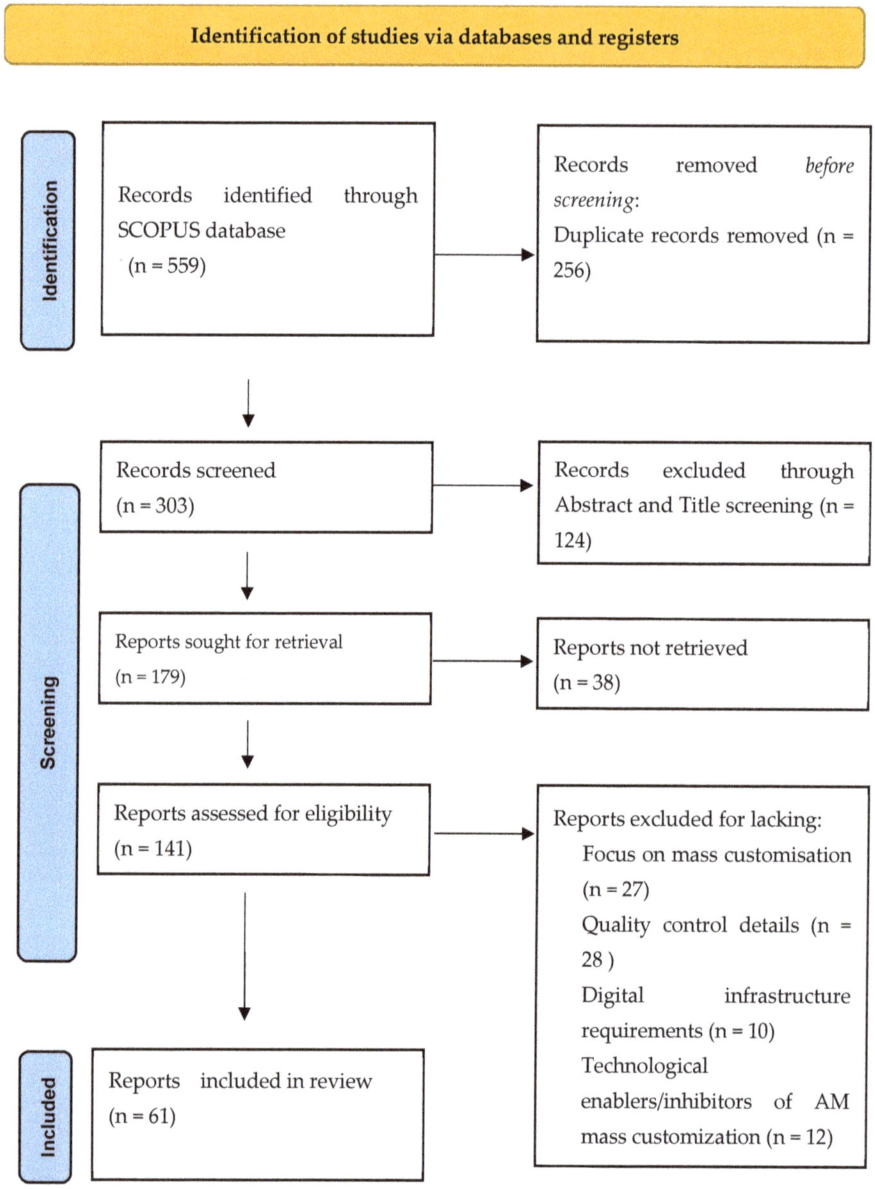

Figure 1. PRISMA flow chart illustrating the systematic selection process of articles for the systematic review. The flow chart follows PRISMA 2020 guidelines, with "n" indicating the number of articles at each stage. Initial database search yielded 559 articles from Scopus database, with 61 articles ultimately included in the final analysis after applying inclusion and exclusion criteria.

2.3. Data Analysis and Synthesis

Qualitative thematic evaluation was combined with a quantitative methodological approach evaluation, integrated research analysis per Creswell and Plano Clark [28]. The qualitative methodological approach incorporated a systematically constructed coding framework for data extraction and thematic organization by predetermined conceptual domains, adhering to the established protocols for thematic analysis as delineated by Braun and Clarke [28]. The data analysis commenced with an initial previously defined code structure based on theoretical requirements and then included new themes which emerged from the information [29]. Two investigators employed coding methodologies on a representative article sample to assess intercoder reliability by reconciling coding discrepancies [30]. The investigators achieved robust reliability, evidenced by a Cohen's kappa coefficient of 0.83 for the initial coding process, indicating substantial agreement, [31]. In addition to intercoder reliability, we utilised a methodological triangulation of cross comparisons of our coding findings among various data sources: empirical papers, conceptual studies, and industry case studies as a means of validating the consistency of identified themes. This triangulation approach showed convergence processes across varying study types, which further confirmed the validity of our thematic framework.

In order to ensure a high degree of methodical rigor and validate the thematic domains, we used a comprehensive multi-stage validation approach. Internal validation involved independent review by two more researchers who were not part of the coding process, resulting in an inter-coder reliability of 0.78 on validation themes. To empirically validate, we cross-validated key themes against alternative coding approaches, as well as performed a sensitivity analysis to ensure the stability of the thematic framework we developed [32]. Practical validation was carried out in the form of validation meetings with professionals from the healthcare, automotive, and construction industries, which confirmed the practical relevance of each thematic domain and established three more sub-themes brought into our final framework. Finally, thematic structure was reviewed by two academic campus experts in manufacturing strategy and systematic review methodology. Table 1 provides the results of the thematic analysis. In the quantitative component of the analysis, descriptive statistical measures were utilised to examine the temporal distribution of the articles and the identification of methodological preferences within the field.

Table 1. Thematic Domains on mass customisation strategies in AM.

Thematic Area	Sub-Dimensions	Papers
Mass customisation implementation strategies in AM	- Integration of AM with - Traditional Manufacturing - Customer-Centric Design Integration - Flexible Manufacturing Networks - Adaptive Production Systems	Zawadzki and Żywicki [33], Jiang et al. [34], Kang and Lee [3], Yuan et al. [35], Hafenecker et al. [36], Paoletti [11], Lacroix et al. [2], Kim and Lee [3], Rajamani et al. [37], Rousseau et al. [38], Rayna et al. [39], Zanetti et al. [40]
Technological ecosystem for AM mass customisation	- Enablers - Barriers and Implementation Challenges - Process Control - Digital Infrastructure	Narendran et al. [41], Lunetto et al. [42], Kumar and Jeong [43], Sun et al. [44], Kumar et al. [45], Lin et al. [46], Pesce et al. [47], Schubert et al. [48], García-Dominguez et al. [49], Turner et al. [50], Shen et al. [51], Bos et al. [52], Gao et al. [53], Lacroix et al. [54], Li et al. [55], da Silva et al. [56], Gulisano et al. [57], Sivabalakrishnan et al. [58], Olakanmi et al. [59], Giunta et al. [60]
Industry-specific applications and outcomes	- Medical and Dental Applications - Industrial and Engineering Applications - Architectural and Construction Applications	Leong et al. [61], Solaimani et al. [62], Bilalis et al. [63], Rana et al. [45], Du et al. [64], Kromoser and Pachner [65], Lacava et al. [66], Bayraklilar [67], Górski et al. [68], He and Han [69], Hauser et al. [70], Alabbasi et al. [71]

Table 1. *Cont.*

Thematic Area	Sub-Dimensions	Papers
Evaluation methodologies and performance metrics	- Cost-Benefit Analysis Frameworks - Quality Assessment Methods - Performance Evaluation Systems	Deradjat and Minshall [72], Huang et al. [73], Basak et al. [74], Martinez-Marquez et al. [75], Frohn-Sörensen et al. [76], Rajamani et al. [37], Zanetti et al. [40], Koop et al. [77], Richter et al. [78], Büscher et al. [79], Sæterbø and Solvang [80], Wang et al. [81], Abdulhameed et al. [82], Bellens et al. [83], Kim and Jeong [43], Huang et al. [73], Lin et al. [84], García-Dominguez et al. [85]

3. Results

3.1. Descriptive Analysis of Included Studies

Analysis of the 61 included studies reveals a clear trajectory of growing research interest in mass customisation strategies in AM across three distinct periods. The temporal distribution shows an initial exploratory phase (2010–2014) with limited output of just two publications, followed by steady development (2015–2019) of 14 publications, and culminating in a significant acceleration phase (2020–2024) that produced 45 publications, representing 73.8% of the total research output (see Figure 2). Rapid research growth directly follows advancement of AM technologies and widespread industrial implementation of mass customisation solutions in fields requiring complex product personalisation [79]. The descriptive analysis reveals that 46 of 61 papers (75.4%) adopt empirical methods, whereas conceptually oriented papers make up only 15 (24.6%). Researchers employ the empirical approach as the dominant methodology within this academic field. Multiple scientific investigations have researched theoretical and practical experiences in mass customisation implementation in various industrial settings. The studies apply mass customisation methodologies to solve genuine manufacturing challenges such as production flow control, quality assurance, and customer involvement.

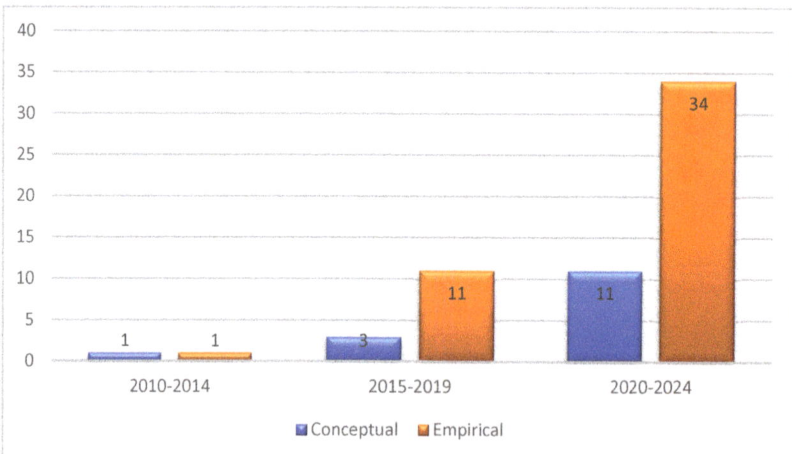

Figure 2. Trends in research methods.

3.2. Thematic Analysis

3.2.1. Mass Customisation Implementation Strategies in AM

Integration of AM with Traditional Manufacturing

Integration of AM with traditional manufacturing emerged as one of the four key mass customisation implementation strategies identified in our systematic review. This strategy represents a sophisticated hybrid production approach that leverages the complementary advantages of both manufacturing paradigms to overcome traditional barriers to mass customisation. AM provides critical customisation capabilities within this strategy through its unique ability to economically produce complex, variable geometries without tooling changes or setup costs. While traditional manufacturing excels at producing standardised components with high efficiency and consistent quality, AM specifically contributes to geometric freedom, tool-less production, and digital-to-physical conversion, enabling customised product aspects [86]. The organisations using this strategy introduce combined manufacturing line configurations that use conventional manufacturing for standardised items alongside additive manufacturing for customised parts [37,75], creating a flexible production ecosystem where mass customisation becomes economically viable. This integration manifests through several key mechanisms. First, digital thread implementation allows direct transmission of data from AM technology to standard processes while Product Lifecycle Management (PLM) systems protect design goals across platforms to perform automated quality assessment [35,36]. This integration is particularly valuable because the digital workflow of AM can accommodate infinite design variations, while traditional manufacturing requires extensive tooling changes for each variant. By reserving AM specifically for the customised components, organisations can achieve mass customisation economics that would be impossible with either manufacturing approach alone. This digital continuity is essential for mass customisation, enabling efficient translation of customer requirements into production specifications without disrupting manufacturing workflows. Advanced simulation tools and digital twins optimise hybrid manufacturing processes by predicting process interactions and potential quality issues pre-production, reducing iterative cycles and material waste while maintaining the efficiency needed for mass customisation at industrial scales. The role of AM is particularly crucial because its layer-by-layer building process enables complex internal structures, conformal features, and integrated functionality that would be impossible to customize with traditional manufacturing methods [87]. Second, hybridization has created distinct post-processing methods that close the performance difference between products produced by AM and conventional manufacturing processes. Organisations use automated finishing operations to exact precision among components while keeping the output quality similar to conventional methods [49]. This ensures that customised components meet the same quality standards as mass-produced parts, a critical requirement for customer acceptance of mass-customised products. Material scientists have formulated composite materials optimised for hybrid manufacturing environments, maintaining consistent mechanical properties across manufacturing techniques [42,87], enabling seamless integration of customised and standardised components in final products.

Organisations have now created intricate supply chains for distributed resource management between various levels of production technology. By strategically deploying AM capabilities closer to customers while maintaining centralised traditional manufacturing for standard components, organisations can provide rapid, localised customisation without sacrificing the economies of scale that make mass customisation economically viable. Companies often establish networks that unite centre-based traditional manufacturing and localised AM technology to offer quick customised resources using standardised component economies [2]. According to Büscher et al. [79] and Song et al. [50], taking into consideration manufacturing challenges such as geometric complexity, production volume, material requirements, and delivery times, advanced production planning systems employ artificial intelligence (AI) systems to maximise work distribution between AM and traditional methods. This is to optimise the balance between customisation capabilities and production efficiency, which is the central challenge of mass customisation. This integration has also created a new group of hybrid manufacturing experts who can handle complex multi-modal production settings by combining traditional manufacturing knowledge with AM-specific skills. This has helped create specialised training programs for the workforce [80].

The hybrid manufacturing paradigm has created the need for sophisticated cost modelling and performance assessment systems. These systems specifically address the unique cost structures of AM (where complexity is virtually free but volume is expensive) and traditional manufacturing (where tooling is expensive but volume is cheap), enabling organisations to optimise which elements to produce with each technology for maximum customisation value at minimum cost. Activity-based costing systems accurately show the economic effects of combining traditional and AM. These systems look at things like how much equipment is used, how many workers are needed, how much material is required, and how quality control is handled in each mode of production [72,74]. Some new performance measures, like process switching efficiency, hybrid material performance indices, and multi-modal quality consistency measures, have been added because they are helpful in hybrid manufacturing settings [73,76]. Research institutes and industry partners are collaborating to establish standardised testing and certification processes for hybrid manufactured components, thereby establishing new quality assurance protocols that address the unique characteristics of combined AM and traditional manufacturing techniques [35,79]. This is anticipated to create the quality infrastructure to scale mass customisation while maintaining regulatory compliance and customer confidence in customised products.

Customer-Centric Design Integration

Customer-centric design integration represents a fundamental shift from traditional customisation approaches, emerging as an AM-enabled mass customisation strategy. This approach is characterised by sophisticated digital platforms that allow real-time co-creation experiences. This approach focuses on placing customer inputs and preferences at the centre of the design process while ensuring manufacturability and structural integrity through AM capabilities [88]. Organisations that adopt this method use advanced parametric design software, which converts customer selection data into production specifications without compromising design requirements or structural stability [34,84], leveraging the geometric freedom of AM to realise customer-specific requirements that would be impossible with conventional manufacturing. In Kim and Lee [3], in their study on personalising the customisation experience, it was shown that customer satisfaction with and likelihood of purchasing the product increased by matching the interface design to customers' culture-specific processing style, a capability uniquely enabled by the tool-less production approach of AM. This strategy is applied in medical applications by developing more advanced biomechanical modelling and simulation tools that take advantage of the ability of AM to produce complex, patient-specific geometries. Automated design platforms that integrate patient-specific anatomical data with finite element analysis are used by healthcare providers to optimise custom implants and prostheses [84,89], creating mass customisation opportunities in medical devices that were previously limited to costly one-off productions. Patient gait analysis, pressure mapping, and dynamic loading patterns are used by machine learning algorithms to generate optimal internal structure for functional performance while satisfying previously constrained manufacturing limitations [61,63], utilising the unique ability of AM to create complex internal structures impossible with traditional manufacturing methods. When machine learning is combined with medical imaging technologies, automated feature detection and design adaptation are made possible design while significantly improving anatomical fit accuracy [75].

In the industrial domain, organisations embrace topology optimisation systems to balance customer requirements and performance constraints. Manufacturing companies use generative design systems to explore multiple design iterations automatically based on customer functional criteria and optimise material distribution and structural efficiency [45,64]. To achieve a Pareto optimal design solution (a set of solutions where no objective can be improved without degrading another objective) for both customer and production capabilities, manufacturing constraints, material properties, and cost are integrated into multi-objective optimisation algorithms [76], enabling economically viable mass customisation through AM. Advanced simulation techniques provide regulatory compliance using real-time evaluation of tailored designs against industry-specific performance criteria, preserving design freedom while assuring regulatory compliance [81].

Advanced encryption techniques and secure data management approaches have enhanced intellectual property protection in collaborative designs. Organisations use distributed ledger solutions to establish transparent modification records, which protect confidential data and intellectual property [69,82], for AM-enabled mass customisation. Organisations can safely handle design rights and manufacturing permissions through smart contracts between customers, designers, and manufacturers working on distributed production networks [80]. This facilitates the complex digital supply chains needed for successful AM-enabled mass customisation.

Flexible Manufacturing Networks

Flexible manufacturing networks represents another AM implementation strategy which enables mass customisation as identified in this systematic review. This strategy provides an organisational format for distributed AM-based mass customisation using coordinated production facilities across different locations. AM technology is central to this strategy because its digital-to-physical production process eliminates the need for physical tooling transfers across locations. Unlike traditional manufacturing, which requires physical moulds, dies, or fixtures to be replicated or transferred between facilities, AM allows digital design files to be instantly transmitted to any location with compatible equipment [2]. This digital nature of AM enables geographically distributed production without sacrificing consistency—the same STL file can be printed at multiple locations while maintaining dimensional accuracy and material properties. Organisations utilising this strategy create decentralised network models with centralised design oversight and production sites positioned close to end-users, particularly when local production results in market advantages [2], allowing mass customisation through AM to be scaled geographically while maintaining consistency. The unique material capabilities of AM systems at each production node allow customisation tailored to local requirements—for example, different material properties for varied climates or regulatory environments—while maintaining the same underlying design architecture [87]. Digital platforms maintain consistent product standards while coordinating design distribution, production scheduling, and quality control across multiple manufacturing sites [45], enabling standardised processes for AM-based mass customisation across diverse locations. These platforms protect intellectual properties through dedicated measures and simultaneously support distributed manufacturing needs, especially when manufacturers handle design changes at various production sites [39], addressing a critical concern in the digital distribution of customisable designs for AM production. This sophisticated digital framework automatically allows companies to allocate their production resources at various distribution centres. Cloud-based information technology systems enable organisations to handle production scheduling, logistics, and quality control within all their facilities through a single system with standardised quality standards [2], ensuring consistent quality of mass customised products regardless of which network node handles the AM production. In customised consumer goods production, flexible manufacturing networks have demonstrated significant value through their ability to rapidly adapt to local market needs and reduce supply chain lead times, resulting in higher customer satisfaction and market responsiveness [90]. Notably, this is a key advantage for AM-enabled mass customisation that can respond to regional preferences while maintaining global design integrity. This adaptability is further enhanced as IoT-enabled inventory management systems track material consumption and quality across multiple manufacturing locations. This enables organisations to automatically replenish and ensure quality consistency [58], critical for maintaining consistent AM material properties across distributed production sites. Digital twin technology applications, which model material movement patterns, assist in identifying upcoming points of congestion so organisations can take preventative steps to optimise their logistics systems [74]. For AM networks specifically, these digital twins simulate material consumption rates for different powder, resin, and filament types across the network. This ensures an adequate stock of specialised materials needed for customised production. Advanced systems for detecting and recovering network faults have contributed to better network resilience. AI-powered monitoring systems in organisations lead to quick responses and production distribution reallocations when they detect quality problems or equipment breakdowns at various manufacturing points [81,86]. These systems capitalise on the digital workflow of AM by redirecting production to alternate facilities with compatible equipment when failures occur, often without requiring physical transfers of tools or equipment—something impossible with traditional manufacturing methods. Digital process twins maintain virtual replicas of production capabilities across the network, enabling dynamic workload allocation in case of equipment failures or capacity limitations [37], allowing mass customisation orders to be rerouted to the most appropriate AM systems based on current network status. Flexible manufacturing networks are associated with reduced delivery times, improved customisation capabilities, and enhanced supply chain resilience [55], making this strategy particularly valuable for scaling AM-enabled mass customisation beyond the limitations of centralised production facilities.

Adaptive Production Systems

Adaptive production systems represent another key implementation strategy for AM-enabled mass customisation identified in our systematic review. This technological strategy makes it possible to provide dynamic adaptation to dynamically varying requirements for customisation through sophisticated sensor networks [91], enabling AM systems to adjust processing parameters in real time to meet diverse customisation requirements. The layer-by-layer building process of AM uniquely suits adaptive production systems in ways traditional manufacturing cannot match. With AM, each layer's processing parameters can be modified during production, adjusting laser power, scan speed, material flow rates, or energy density based on the specific geometry built at that moment [92]. This parametric adaptability allows a single AM system to produce highly varied customised products without the extensive tooling changes, setup adjustments, or production line reconfigurations required in traditional manufacturing. This strategy is typically implemented by organisations that continuously monitor multi-modal sensor arrays for critical process parameters, material properties, and environmental variables for dynamic manufacturing parameters [3,43]. These sensor arrays monitor AM-specific processes like melt pool dynamics, powder bed temperature distribution, recoating uniformity, and layer adhesion, which are factors unique to additive processes that must be controlled precisely to maintain quality across diverse customised parts [93]. According to Giunta et al. [60], machine learning algorithms are used to analyse real-time sensor data and predict potential quality issues, as well as automatically adjust process settings, ensuring consistent product quality even as the number of customised options increases, a critical capability for maintaining quality consistency across diverse AM-produced customised variants. Digital process controllers allow the application of control algorithms that optimise several objectives simultaneously while maintaining process stability, as Buscher et al. [78] described, enabling AM systems to adapt to different customisation requirements without manual reconfiguration dynamically. These controllers specifically manage AM process variables like layer thickness, hatch spacing, and energy density—parameters that would require extensive manual setup in traditional manufacturing but can be continuously adjusted during AM production. Systems of quality have evolved to include automatic fault recovery and predictive maintenance. According to Martinez-Marquez et al. [75] and Rajamani et al. [37], many inspection technologies enabled with AI work in an organisation quality monitoring system to locate and categorise the defects in all product arrangements, ensuring that mass-customised AM products meet quality standards despite their unique geometries. Through the simultaneous combination of staffing and process technology, advanced process control systems that automatically re-optimise setting parameters according to quality data feedback [81], maintain production efficiency, and reduce waste address a key challenge in scaling AM-enabled mass customisation. For AM systems specifically, these controls adjust for material variations and chamber environmental conditions and build plate thermal gradients that would otherwise cause quality variations across different customised designs. Advanced systems that schedule operations and make capacity plans benefit resource optimisation. Different product varieties are distributed optimally across organisation resources, materials inventory, production machine abilities, and delivery deadlines [2,80]. Modern simulation systems assist businesses in real-time estimation of several production scenarios that help them reach optimal production efficiency with quality norms spread among various customisation offerings [71], allowing organisations to predict and optimise AM performance across diverse customisation requirements. Intelligent automation interfaces between people and the systems improve system flexibility. Organisations with automated collaborative robot systems optimise their operational patterns based on product specifications and operator feedback to perform flexible manufacturing operations [50], combining human expertise with automation to handle the complexity of AM-based mass customisation. By employing augmented reality and gesture recognition features, the augmented human machine interfaces can also be used by users to control advanced production systems for safe performance and operational excellence [79], providing intuitive interfaces for managing the complexity of AM-enabled mass customisation production. The adaptive production system strategy is particularly valuable for organisations seeking to accommodate diverse and evolving customisation requirements through AM technologies, as it enables the dynamic process adaptations needed to maintain quality and efficiency across varied product configurations without extensive manual intervention between production runs.

3.2.2. Technological Ecosystem for AM Mass Customisation

Technical Enablers for Mass Customisation

Modern AM systems excel at creating complex geometrical structures that are unbounded by traditional manufacturing constraints. This capability enables the design and production of customised parts with internal features and organic shapes that were previously unattainable through conventional methods [39]. The technological enablers facilitating AM-based mass customisation can be organised into several key categories.

Researchers have significantly expanded options for AM technology in materials, particularly for medical and aerospace applications. These advancements include polymers, metals, and ceramics [42], establishing material diversity as a critical enabler of mass customisation. The broader material portfolio enables manufacturers to precisely align specific customer requirements with suitable material properties, whether biocompatible polymers for personalised medical implants, high-strength aluminium alloys for lightweight aerospace components, or ceramics for heat-resistant custom parts. This material flexibility, combined with the geometric freedom of AM, creates a dual customisation capability where both form and material composition can be adjusted simultaneously. The capability to process such diverse materials through a single manufacturing paradigm overcomes traditional limitations, where material selection often restricts design possibilities, thus fundamentally enabling mass customisation by optimising each product for both function and user needs without prohibitive tooling or process changes [84].

Digital workflow integration is another technological fundamental enabler, as seamless interfaces between CAD systems and AM platforms allow quick design changes and instantaneous production adaptation without tooling changes or major setup periods [43]. Such integration creates a continuous digital thread from customer requirements capture through design iteration to final manufacturing, eliminating communication barriers and reducing translation errors between systems. Data show that companies with completely integrated digital workflows reduce production time while decreasing costs compared to organisations that maintain separate digital systems [84]. This digital continuity is particularly critical for mass customisation, where each product variant must move efficiently from concept to production without manual intervention.

The elimination of traditional tooling requirements has transformed the economics of customised production, particularly in the medical and consumer sectors. Factory flexibility, which allows quick product variation with no extra tooling expenses, readjusts customisation cost structures to support affordable low-quantity manufacturing [47]. Unlike conventional manufacturing, where tooling costs must be amortised across production volumes, AM shifts the economic equation by making the unit cost relatively consistent regardless of quantity. This cost structure democratises customisation by removing the financial penalties traditionally associated with producing unique items, allowing companies to offer personalised products at price points comparable to mass-produced alternatives.

The convergence of multi-material AM techniques with computational design optimisation has created unprecedented opportunities for functional grading and property customisation. When combined with topological optimisation methodologies and multiphysics simulation tools, manufacturers can achieve user-specific optimisation of internal structures. This enables user-specific optimisation of internal structures and the subsequent targeted adjustment of mechanical and thermal properties alongside electromagnetic behaviours per area of the product [41,44]. These advanced design capabilities allow for the creation of components where material composition and structural properties vary continuously throughout the part, optimising performance for specific loading conditions, thermal requirements, or user ergonomics. Such tailored functional properties represent a higher dimension of customisation beyond mere geometry.

In mass customising contexts, the development of digital twin technology has transformed process control and quality assurance. Organisations employ physics-based digital twins to conduct simulations of complex material-to-process relationships, which forecast necessary manufacturing settings for different customised solutions [49,50]. These virtual replicas of physical manufacturing systems enable pre-production validation of customised designs, identifying potential manufacturing issues before committing resources to production. Machine learning algorithms enhance simulation processes across the production line using new data inputs from the manufacturing environment [51]. The closed-loop system created between physical production and digital simulation enables manufacturers to maintain consistent quality across highly varied product offerings, addressing one of the traditional challenges of customisation.

The combination of adaptive slicing methods with multi-axis deposition has produced major enhancements to the ability to create geometries using AM techniques. While early AM systems were limited to simple layer-by-layer construction, modern systems employ sophisticated path planning with rotational freedom and variable layer thicknesses optimised for specific part geometries. Path planning algorithms with advanced capabilities use live process monitoring data to adjust deployment patterns in real time, which enhances mechanical properties alongside surface quality for sophisticated customised designs [55,56]. Organisations that utilise complete technological enablers obtain considerably higher customisation capabilities and operational efficiency than organisations with incomplete implementations [49]. This underscores the importance of adopting a holistic approach to AM-enabled mass customisation rather than focusing on isolated technologies. Companies realising the full potential of AM-enabled mass customisation typically integrate these various enablers into cohesive systems where digital design tools, material capabilities, process controls, and verification technologies work in concert to deliver customised products with consistency, efficiency, and economic viability that rivals traditional mass production approaches.

Technical Barriers and Implementation Challenges

Technology advancements have not eliminated the several ongoing barriers that impede the implementation of mass customization in AM environments. Material property complexity presents a major implementation challenge for AM-enabled mass customization. The final components exhibit anisotropic behaviours and show diverse mechanical properties which vary depending on both construction orientation and process settings [49,50]. This variability complicates quality control and makes it difficult to ensure consistent performance across customized products. Advanced characterization studies reveal complex connections between processing parameters, microstructural evolution, and mechanical properties, particularly in metal AM. The integration of various materials in multi-material AM systems introduces additional difficulties in interface bonding and property matching.

Economic barriers significantly restrict the widespread adoption of AM-enabled mass customisation. The relatively high material expenses coupled with machinery costs prevent small and medium enterprises from feasible implementation. The exact expense of AM technology has been researched extensively through multiple studies, which show that expensive metal printing equipment and precision upgrade systems demand higher investments than traditional manufacturing methods [51,52]. According to Gao et al. [53] and Lacroix et al. [54], there are complex interactions between equipment utilisation, material prices, and production volumes, particularly in cases requiring high-end systems for metal printing or precision applications. Additionally, specialised post-processing complications demand and quality control, which require major capital investment, have become a barrier for many organisations.

Post-processing complexities further challenge mass customisation implementation. These processes demand advanced techniques that create difficulties in maintaining standardised quality across personalised products [53,54]. Post-processing requirements for achieving specific surface finishes or mechanical properties often introduce additional variability into the manufacturing process. The unique characteristics of AM techniques combined with the broad spectrum of possible material combinations complicate the establishment of standardised testing methods and certification processes for customised components.

Maintaining constant quality among tailored components is much challenged by material property anisotropy and process-induced variability. Many factors affect how a material behaves, especially in metal AM [50,51], and advanced characterisation studies show how processing parameters, microstructural evolution, and mechanical properties are all connected in complex ways. While integrating various materials in multi-material AM systems presents difficulties in interface bonding and property matching [52], post-processing needs for achieving specific surface finishes or mechanical properties often introduce additional variability. The special qualities of AM techniques and the broad spectrum of conceivable material combinations challenge the creation of standardised testing methods and certification processes for bespoke components.

Production volume and size limitations constrain mass customisation capabilities in AM environments. Specific obstacles include functional restrictions in build volume capacity and production speed limitations that affect more significant component production and high-volume sectors. These build volume restrictions and manufacturing speed limits directly impact throughput capacities [55], making AM less suitable for specific customisation scenarios requiring large components or high production volumes.

Integration challenges emerge when organisations transition from conventional manufacturing to AM-enabled mass customisation. Companies face exclusive difficulties when integrating AM systems with their existing industrial frameworks due to data management and process control issues [55]. Particularly in companies changing from conventional manufacturing techniques, the integration of AM systems with current industrial infrastructure creates challenges in workflow optimisation and resource allocation [56]. Limited understanding of process–structure–property relationships and the computational requirements for real-time decision-making impede the development of closed-loop control systems that can automatically adjust process parameters based on quality feedback [59,83].

Quality Control and Process Optimisation

Successful implementation of mass customisation through AM requires specialised quality control and process optimisation approaches that differ significantly from traditional manufacturing. The layer-by-layer building process of AM creates unique opportunities for mass customisation by enabling complex geometries and internal features impossible with conventional methods, but it also introduces specific quality challenges that must be addressed. Organisations establish thorough frameworks to implement real-time monitoring systems designed explicitly for AM-based mass customisation, tracking critical parameters like thermal conditions, powder bed uniformity, and layer-by-layer verification [57,60]. These monitoring systems are crucial for mass customisation through AM, as each customised product may require different process parameters despite being produced on the same equipment—a capability that conventional manufacturing cannot match. Prior to production, organisations, especially those operating within controlled sectors such as medical device manufacturers, put in place processes and procedures to ensure that customised designs are accurately verified. The processes include the strict adherence to guidelines and procedures, such as automated checks for manufacturing, compliance with material-specific design, and structural integrity. Several studies [94–96] have established that there is an association between process parameters and the quality of the final product in customised AM applications. According to Giunta et al. [60], this aids in creating predictive models that harness manufacturing parameters in alignment with the needs and preferences of customers. Furthermore, some studies have revealed that adopting machine learning algorithms for optimising the processes, among other things, automatically changes manufacturing parameters to ensure that quality is maintained throughout customisation conditions while reducing waste and production time [59,97]. In order to preserve consistency in mass-customised production environments, quality control systems now integrate multi-modal sensor arrays with real-time analytics systems to offer complete monitoring of the build process, which has the ability to identify and fix deviations in real time.

The integration of the multi-modal sensor arrays and real-time analytics, advanced in situ monitoring systems specifically developed for AM processes, has been found to provide critical transformation to quality control in AM mass customisation. Gulisano et al. [57] and Giunta et al. [98] indicate that the use of thermal imaging systems, machine vision, and acoustic monitoring are combined to track the AM building process for customised components, identifying deviations that would lead to quality issues in complex geometries specific to customised designs. Unlike adaptive control systems which change the processes automatically to maintain ideal manufacturing conditions, multi-sensor data are assessed by machine learning algorithms in real time, aiding in identifying any quality problems before they become severe [99].

Process optimisation frameworks for AM-enabled mass customisation now use multi-dimensional objective optimisation techniques to manage conflicts that specifically balance the geometric complexity enabled by AM with production efficiency. Computational models optimise AM process parameters through thermal–mechanical simulation and microstructural prediction [38,77]. This allows organisations to efficiently produce customised designs with different geometries in the same build—a capability unique to AM that dramatically enhances mass customisation economics. Multiple builds of inspection data undergo statistical process control analysis to expose systematic deviations and spot process drift. Machine learning algorithms develop proper inspection approaches by analysing product features and essential design criteria [74,82]. Automated systems for classification of defects allow quick inspections of unique components to sustain both production throughput and unyielding quality requirements [69,80].

Combining automated inspection systems designed for AM with statistical process control has also improved mass customisation quality assurance. Through the advanced metrology systems that incorporate measuring devices, optical scanning, and computed tomography, it is possible to carry out comprehensive dimensional verification and identification of internal flaws [74,82]. While machine learning models anticipate ideal inspection tactics based on component geometry and essential feature requirements, statistical process control systems examine inspection data over several builds to find consistent variances and process drift. The presence of automated defect classification systems aids in the rapid evaluation of the tailored components, which enhances the effect production flow that confirms that quality standards achieve efficacy in manufacturing [69,80].

Digital Integration and Data Management

Digital integration and data management abilities establish fundamental requirements for efficient mass customisation possibilities within AM operations. The unique layer-by-layer nature of AM processes generates vastly more process data than traditional manufacturing, necessitating sophisticated data handling approaches. Organisations use complete digital thread architectures that transform how they manage data throughout their AM mass customisation structures, from initial capture of customer requirements to final part delivery and lifecycle support. Modern PLM systems unite AM design tools with simulation platforms as well as quality control tools to maintain continuous data flow throughout the product life duration [74,81], enabling seamless transition from customised digital designs to final AM-produced parts without the data translation issues that often occur in traditional manufacturing. These systems incorporate AM-specific parameters such as build orientation, support structures, part nesting, and thermal distortion prediction, which are critical for efficient customisation. Complex systems that control versioning maintain design consistency with production standards, essential when managing thousands of unique part variations, while blockchain deployments protect data authenticity during distributed manufacturing operations. This is a critical capability for AM-enabled mass customisation where design files must be securely transferred to various AM systems without corruption or unauthorised modification. Digital rights management controls how design and process information is shared in controlled ways between organisations to support their collective development of customised products [82,86], protecting the valuable intellectual property in AM build files and parameter settings that enable successful customisation. This allows secure outsourcing of custom AM production while safeguarding intellectual property embedded in designs and process parameters—a critical concern when mass customisation leads to dispersed manufacturing. The analysis of extensive production data by big data analytics systems now enables complete process enhancement and quality improvement. A lot of sensor system data undergo extensive analysis by advanced algorithms, which detect elaborate relationships between AM process variables such as laser power, scanning strategy, and build chamber conditions [81,83].

Real-time sensor data quality issues are automated for identification through pattern recognition systems that use machine learning models with historical production data to predict optimal process parameters for new customisation needs. Data processing through edge computing solutions takes place across distributed AM production networks, which allows rapid decisions to be made and reduces transmission requirements [67,78], enabling local control of AM processes while maintaining centralised design integrity critical for mass customisation. Intelligent knowledge management systems use advanced processes to improve organisational learning and manufacturing improvement in mass customisation applications. By using advanced knowledge bases which include design rules alongside process parameters and quality data obtained from various manufacturing contexts [73,85], new customisation requirements can be addressed speedily. These systems codify critical AM expertise, such as support structure generation rules, optimal part orientation strategies, and post-processing techniques for different geometries and materials, transforming tribal knowledge into digital assets.

The combination of automated documentation systems and machine learning techniques enables comprehensive design decision records, maintenance, and manufacturing parameter tracking alongside automated process best practice definition and improvement identification from production histories. The integration of artificial intelligence tools with knowledge management systems automatically generates AM process specifications for new customisation scenarios, which shortens development periods and upholds quality consistency standards [82,99], a capability that eliminates the extensive manual process planning typically required when switching between different customised designs in traditional manufacturing. Electronic systems supporting AM-enabled mass customisation receive complete security protection through advanced cybersecurity methods to defend design and manufacturing information from sensitive exposure. The combination of cutting-edge encryption methods along with access management protocols protects distributed manufacturing data through network-wide security measures to monitor potential security breaches [57]. Secure architectures in combination with communication technologies enable safe teamwork between multiple participants during custom development, whereas digital signatures alongside audit trails, AM design modifications, and build parameter changes throughout the production process [63,77], ensuring traceability critical for regulated industries adopting AM-enabled mass customisation.

3.2.3. Industry-Specific Applications and Outcomes

Medical and Dental Applications

AM-based mass customisation has revolutionised medical applications by enabling the economical production of highly personalised medical devices at a scale previously impossible with traditional manufacturing methods. Documented evidence demonstrates significant clinical benefits from these customised devices [59,60,83,89], leveraging the unique ability of AM to produce complex patient-specific geometries without the prohibitive tooling costs that would make such mass customisation economically unfeasible with conventional techniques. Advanced orthotics and prosthetic technology have reached new heights with digital scanning technology combined with automated manufacturing approaches. According to Pallari et al. [89], precise digital scanning together with automated manufacturing techniques results in enhanced patient comfort and improved fit accuracy. AM uniquely enables this mass customisation by directly converting digital anatomical scans to physical products without moulds or tools, making each device economically viable despite being one of a kind [100]. Studies demonstrate successful deployment in ankle–foot orthoses, where the capability of AM to produce complex metamaterial structures enables clinicians to integrate patient-specific anatomical data with metamaterial structures to enhance comfort and functionality [61]. Advanced biomechanical modelling and topological optimisation have also transformed patient-specific implant and prosthesis design. Unlike traditional manufacturing, where each design variation requires new tooling, AM enables mass customisation of implants where every device can be unique without cost penalties. These are, however, structures that are impossible to produce with conventional manufacturing but are ideally suited to the layer-by-layer building process of AM. Finite element analysis combined with machine learning algorithms accurately predicts mechanical stress levels and physiological load patterns, enhancing the optimisation of internal lattice structures for optimal osseointegration and mechanical performance [84,89]. Improved material models accurately predict tissue–implant interactions and biological responses, while multi-physics simulation systems combine fluid dynamics with structural analysis to optimise vascular implant designs for individual patient anatomies. These automated optimisation systems develop patient-specific implant designs that modify shape during periods of tissue adjustments by employing individual adaptation models for growth prediction [63,75].

Dental applications have also evolved through the linking of intraoral scanning devices and automated design platforms. AM has transformed dental prosthetics from labour-intensive craftsmanship to digital mass customisation, enabling consistent quality across unique designs. Patented image processing techniques merge different scan technologies, which construct exact digital models of patient teeth structures and plate designs through automated machine learning protocols [61]. The combination of digital design automation with tool-less production of AM creates an economically viable mass customisation workflow where each dental restoration is unique yet produced with industrial efficiency. Real-time visualisation software allows instant confirmation of fitted prostheses along with their tooth relationships, and automated manufacturing sequence programs optimise printing orientation and support frameworks for achieving the best results regarding dimensional precision and surface quality. Printed dental restorations now utilise biocompatible resins to produce custom-made permanent restorations while multi-material printing grants prosthetic manufacturing with artificial structure mechanics suited to mimic natural teeth [63,75].

Orthopaedic medicine has experienced major advancements in bespoke implants, together with specialised surgery instruments for individual patients. AM enables mass customisation of surgical guides and implants that would be prohibitively expensive with traditional manufacturing methods, making patient-specific approaches economically viable for mainstream adoption. Machine learning algorithms analyse medical imaging data to automatically generate optimal cutting guides and positioning tools, while advanced registration systems ensure precise alignment during surgery [45]. This end-to-end digital workflow, combined with AM's geometric freedom, enables mass customisation at a scale and economic feasibility impossible with conventional manufacturing techniques. Computational models simulating postoperative biomechanics and predicting long-term implant performance help optimise implant designs for specific patient activity and lifestyle requirements. Additive manufacturing combined with surface texturing enables precise control of implant–tissue interfaces, resulting in improved biological fixation and reduced infection risk [89]. Employing hybrid manufacturing methods that integrate additive and subtractive techniques, complex internal structures and exact matching surfaces may be produced, therefore maximizing both mechanical and biological performance. Customer-centric design integration has been adopted by the medical and dental fields as the key method of implementing AM, including gait analysis, pressure mapping, and dynamic loading patterns, as well as manufacturing, enabling the production of highly customisable prostheses in both fields. The use of patient-specific anatomical data in combination with finite element analysis allows medical prosthetics to produce custom made prostheses and orthoses [68,75] (see Figure 3a) (Table 2). Healthcare providers take advantage of automated design tools that make use of vast patient data including gait analysis, pressure mapping, and dynamic loading patterns as well as machine learning techniques to formulate the best internal structures for optimum function. Dental prosthetics are a prime example of this approach that uses interactive visualization tools for the fast confirmation of prosthesis fit and dental alignment (see Figure 3b) (Table 2). Medical specialists take advantage of automated production planning tools that tweak build orientations and support design to ensure key functional integration with the underlying dental anatomy. Customer-centric design integration has, to a large extent, been successful in prosthetics with the personalized befit helping the patient feel comfortable, in use, and in treatment at large.

Table 2. AM Strategies (Customer-Centric Design) in Medical and Dental Applications.

Application Area	AM Tools and Method	Specific Products	Clinical Benefits	Cost and Time Benefits	Key Challenges	Supporting Research
Medical Prosthetics	- Digital scanning - Finite element analysis - Machine learning algorithms - Biomechanical modeling	- Ankle-foot orthoses - Custom implants - Patient-specific prostheses - Complex metamaterial structures	- Enhanced patient comfort - Improved fit accuracy - Better osseointegration - Optimal mechanical performance	- No tooling costs for unique designs - Direct digital-to-physical conversion - Economically viable customization	- Complex regulatory approval - Material biocompatibility certification costs - Data privacy protection - Skills gap in clinical settings	[63,89]

Table 2. *Cont.*

Application Area	AM Tools and Method	Specific Products	Clinical Benefits	Cost and Time Benefits	Key Challenges	Supporting Research
Dental Prosthetics	- Intraoral scanning - Automated design platforms - Image processing techniques - Multi-material printing	- Custom dental restorations - Prosthetic crowns - Digital dental models - Biocompatible resin products	- Consistent quality - Precise dimensional accuracy - Natural teeth mechanics - Real-time fit verification	- Labour-intensive to digital workflow - Unique yet industrial efficiency - Optimized printing orientation	- Standard reimbursement models - Technical training requirements - Digital supply chain integration - Quality standardization across sites	[61,63]
Orthopedic Medicine	- Advanced registration systems - Machine learning imaging analysis - Computational biomechanics - Hybrid manufacturing	- Surgical guides - Positioning tools - Patient-specific implants - Surface textured implants	- Precise surgical alignment - Improved biological fixation - Reduced infection risk - Optimized long-term performance	- Economically viable patient-specific tools - End-to-end digital workflow - Complex geometries possible	- Regulatory requirement - Healthcare financing gaps	[75,89,90]

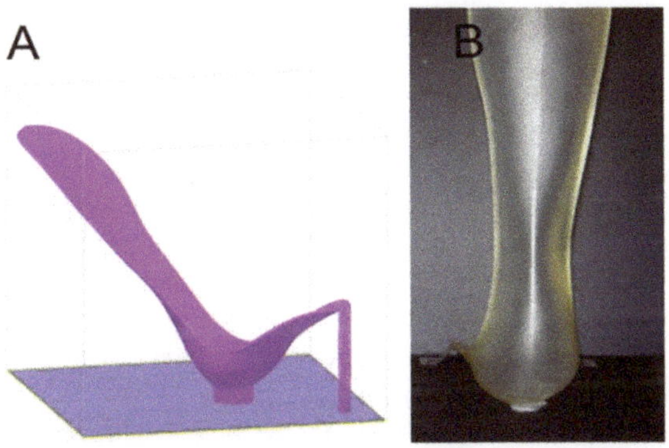

(a)

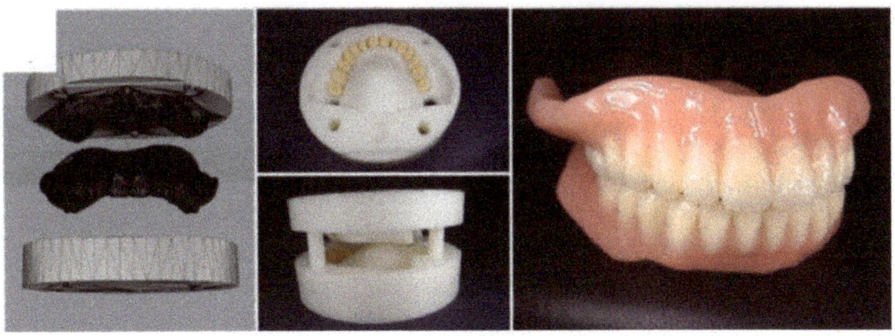

(b)

Figure 3. (**a**): Rigid RP AFO. (**A**) Example of the build platform. (**B**) Completed rigid RP AFO prototype. Source [101,102] (Open Access CC-BY 4.0). (**b**): The CAD design and 3D-printed upper and lower flasks for optimized fitting of the resin teeth and accurate complete dentures fabrication [103] (Open Access CC-BY 4.0).

Industrial and Engineering Applications

The industrial and engineering sectors demonstrate a broad and fast-growing environment for mass customisation through AM practices, which deal with vital performance requirements and intricate operational needs. Unlike traditional manufacturing that requires expensive tooling changes for each design variant, AM uniquely enables mass customisation in industrial applications through its tool-less, direct digital-to-physical production process [84]. Several studies have been conducted on industrial implementations of AM-based mass customisation, mainly in the automotive and aerospace industries. Kumar et al. [45] showcased techniques for producing custom-made components built on the complex geometry with particular functional needs, which demonstrates successful implementations in both prototype and end-use production. This leverages the unparalleled geometric freedom of AM to produce customised industrial components economically unfeasible with conventional manufacturing methods. This employs a mass customisation approach uniquely enabled by the ability of AM to produce parts without tooling investments, making "lot size one" economically viable for industrial components [76]. Through the creation of specific systems for using AM in spare part manufacture, organisations have enabled on-demand manufacturing of tailored components, minimising inventory equipment while enhancing supply chain responsiveness [64]. According to Frohn-Sörensen et al. [76], extant literature has revealed successful instances of rapid tool manufacturing, which has resulted in a significant reduction in setup times and enhancement in production flexibility. This demonstrates how AM enables mass customisation of production tooling itself—an approach impossible with traditional tool-making methods that require extensive machining operations. They further indicate that customised manufacturing aids and fixtures through AM in tooling applications, where the ability of AM to economically produce unique jigs and fixtures for each product variant creates a secondary level of mass customisation support in industrial environments. Additionally, Wang et al. [81] demonstrated how the use of AM provides significant benefits in the development of structural parts. This approach affords manufacturers the capacity to construct highly performing lightweight parts that maintain their mechanical properties while also allowing for customisation, utilising the unique ability of AM to create internal structures and variable densities impossible with traditional manufacturing, enabling mass customised optimisation for specific loading conditions of each component variant. The work by Górski et al. [90] reports that the AM technologies enable the manufacture of cost-effective personalised consumer products via specific methodologies that maintain efficiency in the manufacturing process while customising items. This is fundamentally changing the economics of mass customisation by eliminating the tooling costs that typically make small-batch customisation prohibitively expensive [68]. The relatively high success in the implementation of AM-based mass customisation in industrial and engineering uses has been attributed to advances in material characterisation techniques, process control systems, and design optimisation tools, which aid in achieving efficiency with consistent quality [80].

The aerospace industry now designs components with enhanced methods such as engineering, generative design, and topological optimisation. AM uniquely enables mass customisation in aerospace by making these optimised designs manufacturable regardless of their complexity, unlike traditional manufacturing, where geometric constraints limit customisation options. Multi-objective optimisation software allows engineers to create simple integrated products that keep assembly requirements basic and weight minimal through performance, structural efficiency, and thermal control goal optimisation [63]. Computer simulations of fluid dynamics systems, along with structural models, optimise the design of cooling channels and flow pathways as advanced material models predict mechanical changes due to thermal cycling conditions. The combination of artificial intelligence with design optimisation tools enables users to discover new solutions beyond traditional constraints by exploring complex designs through fast design space exploration and attaining novel solutions that comply with regulations [74,79]. Mass customisation techniques have transformed automotive elements by improving both structural components and attractive features. The integration of advanced design automation systems creates a rapid behavioural generation of personalised interior parts that fulfil crash safety standards and ergonomic comfort needs [90]. In this context, AM serves as the manufacturing technology that makes this level of individualisation economically viable at production scales. Combined material printing, along with topology optimisation, evolves into a process that delivers strong structural components that minimise vibration and suppress sound while maintaining maximum lightness. Special automotive materials allow for the direct manufacturing of final products with superior thermal characteristics and chemical durability through automated quality testing that maintains batch consistency [78,79].

Mass customisation techniques used in structural and cosmetic components have revolutionised automotive applications. Through advanced design automation systems, organisations in the automotive industry are able to rapidly create tailored interior components that meet crash safety criteria while accommodating individual preferences for comfort [43,90]. While the use of topology optimisation techniques builds structures that are lightweight yet very strong for complex loads, multi-material printing creates integrated parts that minimise noise and vibration. Wang et al. [80] and Saeterb and Solvang, ref. [78] further contend that the use of advanced quality control systems ensures consistent production while adopting specialised automotive materials allows direct manufacturing of durable end-use parts with design algorithms optimising customisable components within manufacturing constraints.

Modern industrial tooling applications benefit considerably from conformal cooling implementation, along with topology optimisation implementations. AM uniquely enables mass customisation of industrial tooling by rendering complex internal cooling channels physically producible—features impossible to manufacture with traditional methods, regardless of cost [34]. Complete thermal management in injection moulding equipment relies on computational fluid dynamics, which optimise cooling channel geometry, while structural analysis protects against high-pressure conditions in the equipment [50,74]. The combination of additive and subtractive techniques in hybrid manufacturing systems allows for the production of tools that demonstrate superior thermal performance and precise dimensions by achieving complex internal structures. This enables a level of functional customisation in industrial tooling that is impossible with conventional manufacturing methods alone. Specific moulding conditions prompt machine learning algorithms to analyse process data, which helps predict optimal cooling topologies for automated design methods based on part geometry and material properties [63,78]. This supports the creation of a mass customisation workflow where each tooling design is uniquely optimised for its specific application—an approach that is economically viable only through AM's tool-less production paradigm.

Table 3 shows that four distinct AM strategies enable mass customization across industrial sectors, each finding specific applications based on industry requirements. Integration of AM with traditional manufacturing dominates tooling applications, where hybrid approaches combine AM's geometric capabilities with conventional precision. Injection moulds with conformal cooling channels exemplify this integration, producing superior thermal performance while maintaining dimensional accuracy [39,40] (see Table 3 and Figure 4b). Building on this foundation of integrated manufacturing, customer-centric design integration extends customization possibilities in the aerospace and automotive sectors. While tooling focuses on process improvement, aerospace manufacturers leverage this strategy to create topology-optimized, light-weight components for specific aircraft requirements [45,106]. An example of this is GE Aviation LEAP Engine Fuel Nozzle found in Figure 5. Similarly, automotive applications produce personalized interiors that meet individual preferences while maintaining safety standards [39], with automated design systems bridging customer requirements and manufacturing specifications.

Table 3. AM Strategies Applied to Industrial Applications.

AM Strategy	Industrial Products	Key Applications	Cross-Industry Implementation	Supporting Evidence
Integration with Traditional	- Hybrid manufacturing tooling - Advanced moulds with AM cooling channels - Combined subtractive–additive fixtures	- Conformal cooling implementation - High-precision tooling - Hybrid process optimisation	Tooling: Primary strategy for hybrid manufacturing and optimised cooling channel design Automotive: Critical for safety-critical components requiring high precision	[37,63,75]
Customer-Centric Design	- Personalised automotive interiors - Custom structural components - Tailored safety-critical parts	- Crash safety compliance - Performance optimisation - Individual specifications	Aerospace: Primary strategy for topology optimisation and lightweight high-performance components Automotive: Emphasis on personalised interiors and ergonomic customisation	[45,47]
Flexible Networks	- Distributed spare parts - Regional custom manufacturing - Supply chain components	- Inventory reduction - Rapid response - Distributed capacity	Spare Parts: Primary strategy for distributed on-demand manufacturing All Industries: Enables regional customisation and reduced lead times	[64]
Adaptive Systems	- Variable density structures - Self-optimising components - Quality-adaptive production	- ML-driven optimisation - Real-time quality control - Automated adaptation	Aerospace: Combined with Customer-Centric Design for real-time optimisation Tooling: Used for optimised cooling channel design based on process data	[82]

Moving beyond localized production, flexible manufacturing networks transform spare parts manufacturing through geographically distributed capabilities. Unlike the centralised approach of traditional integration, these networks digitally distribute design files to regional production nodes [102], enabling custom components to be manufactured locally. This strategy eliminates inventory while reducing lead times and maintaining quality consistency across diverse locations. Daimler's Spare Parts On-Demand Network is an excellent example of the flexible manufacturing network AM strategy in action (see Figure 5). Complementing these distributed networks, adaptive production systems provide real-time optimisation capabilities, particularly valuable for aerospace components and advanced tooling. Through dynamic parameter adjustments during manufacturing, these systems enable variable density structures and optimized cooling channels based on specific geometries and materials [82], ensuring consistent quality across diverse customised variants without manual intervention between production runs. Together, these strategic implementations demonstrate how different sectors tactically employ AM capabilities to address specific customisation requirements while balancing performance, economics, and regulatory considerations in delivering the next generation of personalised industrial products.

However, uptake of AM-enabled mass customization in industries is shaped by sector-specific factors and thus limits its potential for widespread use. In mass-producing applications, layer-by-layer AM is still beaten by conventional manufacturing; conventional production typically uses commercial AM machinery that produces 10–50 parts per hour compared with thousands from injection moulding [107]. Material cost, especially in the case of AM grade metal powders, which is three to ten times pricier than traditional materials, prevents mass customization from becoming cost-effective for most industries [73]. The ability to maintain consistent quality in customized variants is particularly problematic in aerospace and automotive sectors, where the certification requirements stipulate flawless manufacture. Achieving 99.9% of consistency during repeated part production for various geometries requires high process control, increasing costs [50]. Traditional supply models still create hurdles because the existing manufacturing infrastructure lacks the ability to support the digital file sharing process as well as networked production required for AM mass customization [39]. Further, the protection of intellectual property often narrows down the scope of collaboration, especially in areas that necessitate the protection of designs while still allowing for the use of distributed production chains [104].

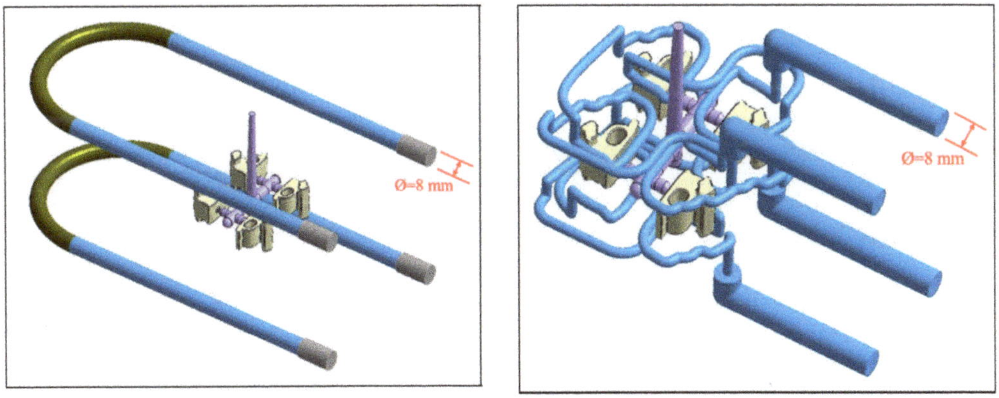

(a) Traditional Cooling (b) AM-enabled conformal cooling

Figure 4. Injection mould with AM-enabled conformal cooling channels demonstrating integration with traditional manufacturing. Source: [108] (Open Access CC-BY 4.0).

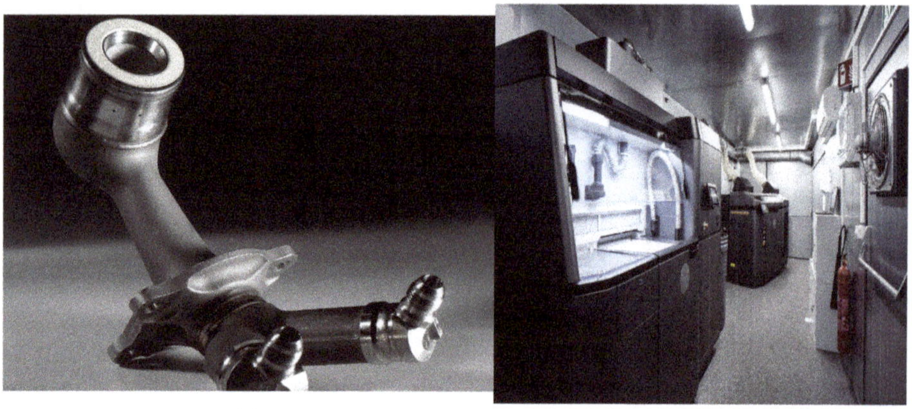

(**a**) (**b**)

Figure 5. (**a**): GE aviation LEAP engine fuel nozzle [109]. (**b**): Daimler spare part on-demand network [110] (customer-centric design integration) (Courtesy GE Aviation) (flexible manufacturing networks). Sample industrial and engineering applications of AM-enabled mass customisation.

Architectural and Construction Applications

Architectural and construction applications represent an emerging domain for AM-based mass customisation, characterised by large-scale components and integration with conventional building systems. Computational design tools adopted by architecture professionals now optimise structures while analysing environmental conditions. The simulation tools combine multiple physics analyses where structural evaluation joins forces with heat and sound models to design optimised building elements across various environmental factors [52,66]. The benefits of this include the creation of customisation opportunities that AM can directly translate to physical components without the economic penalties that traditional construction methods would impose for non-standard elements. Generative design algorithms generate new geometric designs that increase performance potential and aesthetic value, and large-scale printing systems equipped with automatic reinforcement placement systems create structural designs with optimised stress distribution while minimising material requirements. This enhances mass customisation of structural elements where each component can be optimised for its specific loading conditions and location within the building—an approach economically unfeasible with traditional casting methods [76]. Different machine learning algorithms process environmental and occupant behaviour data to develop optimal building components for specific climate regions and usage scenarios [70,71].

Construction applications have undergone transformation through the combination of advanced material science with enhanced process control mechanisms, which resulted from concrete printing technologies. AM enables mass customisation in concrete construction by eliminating the formwork costs that typically make customised concrete elements economically unfeasible [54]. Contemporary material compositions with speciality additives and woven fibres boost rheological attributes as well as mechanical features of printed products. Additionally, in-process property monitoring systems maintain uniform material characteristics across the printing operation [65]. Enhancements in multi-axis printing technologies eliminate formwork constraints, a breakthrough capability that fundamentally enables mass customisation in construction by removing the economic barrier of creating unique moulds for each design variant. Automated tool path generation algorithms optimise stability by optimally distributing material deposition, allowing AM systems to adapt their building strategy for each unique architectural element being produced. Printing parameter adjustments occur in real time based on material and environmental conditions through a combination of process control algorithms and monitoring systems, which lead to high-quality outcomes in large-scale components [52,71]. Modern facade element manufacturing advances because of integrated design processes being combined with manufacturing operations. The integration of automated manufacturing plans increases industrial production effectiveness and waste reduction and enables multi-substance printing to build facade elements that optimise heat and sound resistance, leveraging the unique ability of AM to create functionally graded materials and complex internal structures impossible with conventional facade manufacturing techniques. Existing installation performance data processed by machine learning programs generate optimised design parameters for future projects through digital twin modelling [65,71]. This creates a continuous improvement cycle for AM-enabled mass customisation where each new facade element benefits from the performance data of previously installed customised components.

AM strategies are fundamentally reshaping architecture and construction through diverse implementation approaches that enable unprecedented mass customization. Table 4 illustrates how each strategy addresses specific industry challenges while creating new possibilities. Customer-centric design integration leverages computational power to create responsive building components that would be economically prohibitive using traditional methods [11,52]. Apis Cor's 3D-printed house (Figure 6b) exemplifies the capabilities of adaptive production systems for dynamic adjustment during construction, allowing their mobile technology to respond to changing site conditions while maintaining quality regardless of environment [111]. Harvard GSD's ceramic panels (Figure 6a) showcase integration with traditional manufacturing by combining complex 3D-printed geometries with conventional mounting systems, bridging innovative design with established practices [112]. Flexible manufacturing networks distribute construction capabilities through digital coordination, reducing transportation costs while enabling localization. These contrasting applications demonstrate how AM overcomes traditional constraints, eliminates formwork costs, and enables efficient production of non-standard elements.

(a) (b)

Figure 6. (**a**): Three-dimensional-printed ceramic pattern. Study by Jared Friedman, (**b**): Apis Cor's 3D-printed house [113]. Olga Mesa and Hea Min Kim at Harvard GSD. Instructors: Nathan (Open Access CC-BY 4.0). King, Rachel Vroman [114] (Open Access CC-BY 4.0) (integration with traditional manufacturing). Sample architecture and construction applications of AM-enabled mass customization.

Table 4. AM Strategy Implementation in Architecture and Construction.

AM Strategy	Construction Products	Key Applications	Challenges	Supporting Evidence
Customer-Centric Design Integration	- Functionally graded structures - Optimised load-bearing elements - Customised connection nodes	- Computational design - Multi-physics simulation - Generative algorithms	- Regulatory approval - Performance validation - Digital skills gap	[52,66,71]
Adaptive Production Systems	- Large-scale concrete elements - Multi-axis printed structures - In situ material optimisation	- Real-time parameter adjustment - Property monitoring - Automated tool path generation	- Weather sensitivity - Environmental control - Process monitoring complexity	[52,65,71]
Integration with Traditional Manufacturing	- Large-format structural elements - Reinforced printed components - Hybrid structural systems	- Multi-substance printing - Conventional system integration - Enhanced performance	- Size limitations (10 × 10 × 10m) - Building code integration - System compatibility	[64,70,71]
Flexible Manufacturing Networks	- On-site construction printing - Networked production planning - Iterative optimisation systems	- Performance data collection - Continuous improvement - Digital twin modeling	- Industry fragmentation - Coordination requirements - Digital infrastructure gaps	[65,71]

Architectural and construction industries face significant barriers to AM-enabled mass customization. Regulatory challenges include absent standards for 3D-printed structural parts, requiring expensive custom approvals [11]. Environmental factors like temperature, dust, and moisture compromise print quality and increase equipment costs. Size limitations (10 × 10 × 10 m maximum) necessitate hybrid strategies that reduce customization benefits [112]. Workforce challenges exist with several firms lacking qualified AM personnel. The fragmented nature of construction supply networks complicates implementation, while specialized weather-resistant materials cost up to three times more than conventional alternatives, limiting applications to financially justified projects [104].

3.2.4. Evaluation Methodologies and Performance Metrics

Cost–Benefit Analysis Frameworks

Strategic decision-making depends on complete frameworks to assess the economic feasibility of mass customisation implementations based on AM technologies. This review captures distinctive financial assessment procedures specifically developed for evaluating AM in mass customization contexts, integrating new approaches that suit the unique characteristics of such production environments. Deradjat and Minshall [72] in their work provided a comprehensive model for assessing the direct and indirect costs linked with the implementation of the AM by evaluating the materials, labour, equipment and quality control within the context of mass customization, in ways that account for the unique cost structure of AM where complexity is virtually free but volume carries premium costs. The evaluation of production efficiency in hybrid manufacturing systems uses specific metrics such as the calculation of the use of resources, setup times, and throughput rates across different AM-enabled customization scenarios [73]. Organisations have developed cost models that focus on the cost structure that factors in various production volumes and customization levels, drawing comparisons with that of conventional manufacturing approaches [74]. According to Frohn-Sörensen et al. [76], the use of decision support systems to determine the most appropriate means of using resources in AM-enabled mass customisation settings has demonstrated that this system enables real-time analysis of manufacturing costs and resource utilisation across diverse customisation scenarios.

Economic modelling for AM mass customisation now incorporates comprehensive approaches for multi-dimensional value evaluation. Digital twin simulations use accurate projections to show resource needs and production planning, along with machine learning algorithms that examine historical production information for cost forecasts regarding different customisation approaches [72,73]. Monitoring operations in real time helps activity-based costing systems distribute overhead expenses exactly between different product variations through precise tracking of energy consumption and material usage together with labour resource usages specific to AM processes, while automated quality control systems evaluate cost implications of various inspection strategies for geometrically complex AM-produced customized parts [74,76].

Life cycle cost analysis systems have added sustainability evaluation elements, together with environmental assessment capabilities, to their framework. The combination of advanced simulation tools allows humans to estimate environmental impacts across production scenarios using energy modelling alongside material flow analysis. Machine learning methods discover efficient process parameters and minimise resource usage [39,77]. Open supply chain optimisation methods allow businesses to analyse complete logistics expenses together with inventory needs, which results in well-founded decisions regarding product batches and production sites. The technology enables companies to measure work-related community needs and social benefits through integrated systems that automatically monitor regulatory compliance across multiple geographic regions [11,78].

Mutual integration of dynamic optimization models, along with real options theory, has generated radical changes to investment analysis frameworks. Advanced financial modelling incorporates numerous possibilities for market demand and technological evolution to make complete evaluations about investment timing and scale decisions [79,80]. Digital twin simulations, along with machine learning algorithms, are combined to predict future customisation needs by tracking market trends and competitive dynamics. These methods offer both accurate assessments of equipment specifications and capacity requirements. Investment strategies built with modular structures allow organisations to introduce AM capabilities at different times, and risk management solutions provide complete analysis of market uncertainties and technology expiration risks [81,82].

Quality Assessment Methods

Research has produced specific approaches for assessing product quality in mass customising systems based on AM. With a specific focus on preserving consistency in important characteristics while allowing for varied design needs, Rajamani et al. [37] establish exact methods to evaluate dimensional accuracy and surface quality across many customizing situations. Organisations have put in place complete quality management systems, including in-process monitoring and post-production verification, with particular procedures for matching bespoke components against predetermined quality criteria specifically designed by the layer-by-layer production of customized components in AM [11]. Koop et al. [77] talk about the creation of automated inspection systems for mass-tailored parts. They show how these systems make it possible to quickly confirm important features while keeping the efficiency of manufacturing. Research has revealed particular strategies for applying statistical process control in AM-based mass customization, including real-time monitoring of process parameters and product characteristics to preserve consistent quality across various customization scenarios, enabling consistent quality across all customizing scenarios. Büscher et al. [79] record specific methods for applying quality assurance procedures that address geometric accuracy and material properties in bespoke components, thereby addressing both geometric accuracy and material properties. These methods are particularly relevant in applications that require certain functional performance characteristics.

Quality evaluation in AM mass customization has been transformed by advanced metrology systems which integrate multi-modal measuring methods. While machine learning algorithms automatically identify critical dimensions and geometric features, high-resolution computed tomography combined with laser scanning and structured light systems allows a thorough evaluation of internal features and surface characteristics [37,79]. While automated inspection planning systems optimize measurement tactics based on component geometry and essential feature requirements, real-time monitoring systems track dimensional stability throughout the manufacturing process. Digital twin simulations can estimate measurement uncertainty for complex geometric features [40,77]. However, the development of in situ calibration techniques ensures that measurement accuracy is always checked. Integration of automated defect classification systems helps quickly evaluate tailored components in terms of quality.

Advanced non-destructive testing techniques are now included in material characterisation systems. While machine learning algorithms automatically find possible flaws and material anomalies, integration of ultrasonic testing with thermal imaging and acoustic emission monitoring enables comprehensive evaluation of the internal structure and material properties [75,83]. Automated data analysis tools can track how properties change over several builds. Creating specific test methods for additively manufactured materials ensures an accurate characterisation of uneven features and changes caused by the process. Through the use of digital twin simulations, it is possible to anticipate property distributions for various process parameters [80,81]. However, adopting statistical process control techniques helps in the early identification of material property drift.

The execution of performance validation frameworks became possible by merging accelerated testing methods with computational modelling approaches. Test systems featuring advanced technologies combine equipment that applies mechanical force with environmental simulation to simulate long-term performance under service scenarios [37]. These testing approaches use machine learning processes to optimize test parameter settings according to specific application needs. Data analysis automation monitors how performance shifts during testing, simultaneously with newly developed methodologies that allow testing of additively manufactured components under complicated loading requirements. Test systems operated by computers optimize acceptance methods according to device specifications and digital twin technology projects equipment longevity during different operational conditions [79,82].

Performance Evaluation Systems

System performance assessment frameworks that cover complete mass customization systems based on AM serve as essential methods to boost strategic planning and continuous improvement activities. Widespread frameworks and methodological improvements resulting from the systematic review help organizations to assess comprehensive performance across all dimensions, specifically for these unique manufacturing settings. The implementation of advanced production monitoring systems enables performance assessment through predictive modelling while enhancing real-time analytics capabilities [115]. Several sensor devices that merge process tracking with quality assessment capabilities provide an extensive system for analysing both production operational effectiveness and product quality standing [80,81]. The system autonomously detects operational weaknesses and structural improvement opportunities. Integration produces performance metrics based on productivity standards and flexibility, as well as customizability aspects, using automated data reporting systems that monitor production performance in various scenarios. The use of digital twin technology for the performance prediction in the system simulation allows the automated optimisation algorithms to assess the production system performance across different configuration domains [83,110]. Advanced optimisation methods are employed with sustainability evaluation criteria in resource utilisation models. With the energy monitoring systems in combination with material tracking tools, companies are able to quantify complete resource efficiency measurements, and machine learning algorithms optimise manufacturing parameters to minimise waste production [43,73]. When managing different customised products from manufacturing, automated schedulers use the resources better, whereas digital twins calculate resource use for new variations. They can also be used for enabling waste reduction measurement with material recycling efficiency, along with automated systems enforcing environmental regulations between different control areas [49,84]. Monitoring of customer satisfaction throughout the years has experienced some major changes, largely due to the use of sophisticated analytics in conjunction with a feedback system. Machine learning algorithms can be parallel to evidence from product performance and customer feedback to detect necessary satisfaction factors to measure different product customisation options [3] by means of automated surveys [75]. Digital twin technologies allow companies to estimate customer reaction to design adjustments while stakeholders develop complete quality indicators that use both objective conditions and subjective user perceptions. Customers can receive immediate feedback through time-sensitive systems, and automated analysis platforms track consumer satisfaction through diverse market distributions [79].

4. AM-Enabled Mass Customisation Implementation Framework

In lieu of review, an implementation framework is constructed specifically for AM-enabled mass customisation, comprising four steps that guide organisations from planning through execution (Figure 7). This hierarchical structure addresses the unique challenges and opportunities of using AM technologies for cost-effective, scalable mass customisation. The framework synthesises insights from 61 research studies across diverse industries and technological systems, providing a coherent implementation roadmap that addresses both the technical capabilities unique to AM and the organisational transformations required for successful mass customisation deployment. The framework unites insights from 61 research studies across diversifying industries and technological systems to deliver an organised implementation system that links customisation options to production efficiency.

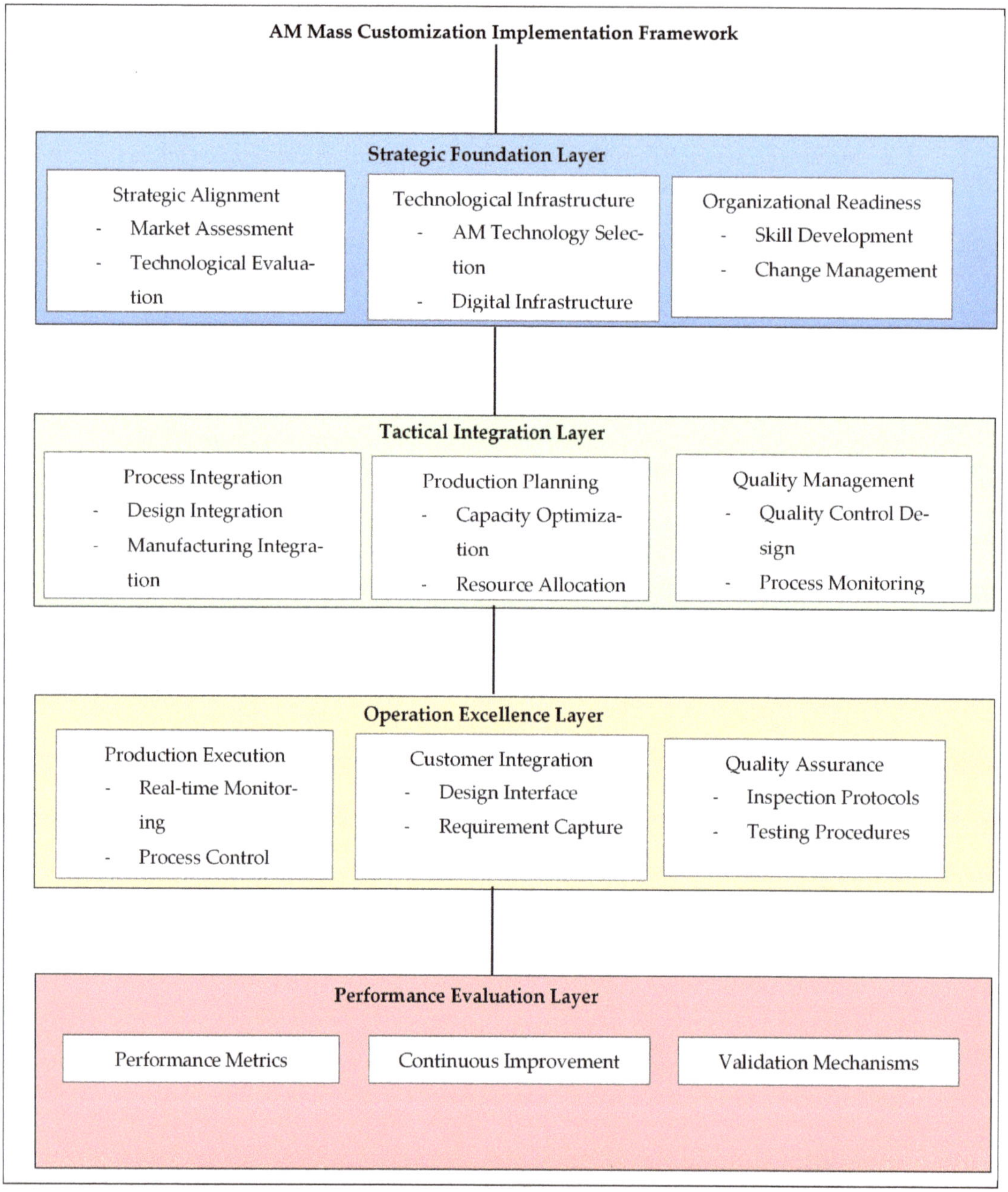

Figure 7. Implementation framework for mass customisation in AM.

This framework differs significantly from current models in the literature in a number of significant ways. Unlike other frameworks, which generally concentrate on single, distinct areas like technology choice [72] or business models [116], our framework is a unique construct integrating the strategic decision-making, operational implementation, and measurement aspects, which are usually described separately in prior research [117]. In addition, whereas current frameworks tend to operationalise theoretical premises in isolation, our integrated perspective offers concrete mechanisms of how these four theoretical perspectives interrelate in the context of AM-supported mass customisation.

This framework builds upon four complementary theoretical foundations, each carefully applied to the specific context of AM-enabled mass customisation. Configuration theory guides organisations to maintain alignment between strategic objectives, AM technological capabilities, organisational structures, and mass customisation market requirements [118]. The resource-based view directs strategic resource allocation and capability development specific to AM technologies, recognising these specialised manufacturing capabilities as sources of competitive advantage in mass customisation markets [119,120]. Dynamic capabilities theory informs the Performance Evaluation Layer [121,122], enabling organisations to continuously adapt their AM processes and customisation offerings as technologies evolve and customer preferences change. The framework adopts a socio-technical systems approach in the Tactical Integration and Operational Excellence Layers [123,124], recognising the critical interdependence between AM technologies and workforce capabilities required for successful mass customisation implementation. A key innovation of this framework is its systematic integration of four theoretical foundations specifically applied to AM-enabled mass customisation, a synthesis not present in existing implementation models. This multi-theoretical approach provides organizations with more robust guidance compared to frameworks that rely on single theoretical perspectives.

The framework organises its four interconnected elements—Strategic Foundation, Tactical Integration, Operational Excellence, and Performance Evaluation—in a progressive structure where each layer addresses specific aspects of AM-enabled mass customisation (Figure 3). Every implementation tier focuses on separate implementation aspects through defined coordination and feedback systems between tiers. This multi-layered approach enables organisations to systematically address both strategic and operational aspects of AM-based mass customisation while maintaining alignment between business objectives and implementation activities. The framework introduces several novel elements not found in previous implementation models: (1) explicit feedback loops and validation checkpoints to ensure continuous verification of alignment during the implementation process, (2) specific considerations for AM processes (build orientation, support structures, nesting efficiency) incorporated into resource planning, as well as (3) adaptive control algorithms that specifically address variable customisation requirements unique to AM environment, and (4) performance metrics specifically tailored to AM-enabled mass customisation that extend beyond traditional manufacturing metrics.

The Strategic Foundation Layer provides a clear direction for organisations with respect to clear targets and key performance indicators for the implementation of their AM-enabled mass customising projects, which align with corporate strategy and the unique capabilities of AM technologies. This involves assessing the practicality and financial feasibility of many mass customisation techniques [71,73] and conduction extensive market research to determine client preferences and customisation demands that specifically leverage the geometric freedom and tool-less production advantages of AM. Based on the particular customisation needs and manufacturing limits of the company, the strategic level should also handle the choice and use of suitable AM technologies and materials. The Strategic Foundation Layer establishes critical organisational infrastructure through three interconnected elements, which comprise the strategic alignment, technological infrastructure, and organisational readiness (see Figure 3). The strategic alignment mechanisms need to ensure synchronisation between mass customisation initiatives and market opportunities [34], establishing clear targets and KPIs aligned with corporate strategy while conducting market research to determine customer preferences and customisation demands [76]. The technological architecture components guide should appropriate AM technology and material selection based on specific customisation requirements [42], evaluating how different AM processes (powder bed fusion, material extrusion, vat photopolymerisation) align with various customisation scenarios [34]. Organisational readiness elements focus on developing AM-specific competencies and managing cultural transformation toward customisation mindsets. The Strategic Foundation Layer provides clear direction for organisations regarding specific targets and key performance indicators for implementing their AM-enabled mass customising projects, which align with corporate strategy and the unique capabilities of AM technologies. This involves assessing the practicality and financial feasibility of various mass customisation techniques [72,74] and conducting extensive market research to determine client preferences and customisation demands that specifically leverage the geometric freedom and tool-less production advantages of AM. Based on the particular customisation needs and manufacturing limits of the company, the strategic level should also address the choice and use of suitable AM technologies and materials. The Strategic Foundation Layer establishes critical organisational infrastructure through three interconnected elements: strategic alignment, technological infrastructure, and organisational readiness (see Figure 3). The strategic alignment mechanisms need to ensure synchronisation between mass customisation initiatives and market opportunities [34], establishing clear targets and KPIs aligned with corporate strategy while conducting market research to determine customer preferences and customisation demands [74]. Technological architecture components should guide appropriate AM technology and material selection based on specific customisation requirements [42], evaluating how different AM processes (powder bed fusion, material extrusion, vat photopolymerisation) align with various customisation scenarios [32], a specificity lacking in previous implementation frameworks. Organisational readiness elements focus on developing AM-specific competencies, managing cultural transformation toward customisation mindsets, and implementing knowledge management systems that capture implementation learning from AM-based customisation initiatives [2].

The Tactical Integration Layer orchestrates relationships between strategic objectives and operational execution of AM-enabled mass customisation (see Figure 3). This layer has three interconnected elements, which include process integration, production planning, and quality management. Process integration should enable seamless collaboration between design, production, and quality control functions [35], while standardised protocols for customer engagement and design validation ensure integrity and traceability of customised product information throughout the AM-enabled customisation workflow [82,83]. Production planning needs to incorporate resource optimisation algorithms and dynamic scheduling systems that balance efficiency with adaptability to varied requirements [31], accounting for AM-specific considerations like build orientation, support structures, and nesting efficiency. Quality management systems incorporate real-time monitoring capabilities specifically designed for AM processes to ensure consistent product quality across different customisation variants [79], addressing the unique quality challenges of producing varied designs through additive processes.

The Operational Excellence Layer focuses on execution optimisation through three interconnected components, which are production execution, customer integration, and quality assurance (see Figure 7). Advanced manufacturing systems need to provide constant performance through real-time process monitoring, adaptive control algorithms that adjust parameters for different customized geometries, and predictive maintenance [36,37]. Customer integration systems facilitate collaborative customisation through advanced requirement capture systems optimised for the design freedom of AM and automated validation tools that ensure manufacturability of customised designs within AM process constraints [50]. Quality assurance combines automated inspection technologies specifically developed for complex AM geometries, testing procedures adapted for AM-produced parts, and comprehensive documentation to maintain product quality while enabling process traceability across varied customisation scenarios [60].

The Performance Evaluation Layer, which is the last in the mass customisation implementation framework, ensures continuous optimisation of the AM-enabled mass customisation system through three components: performance metrics, continuous improvement, and validation mechanisms. Performance measurement systems track key indicators specific to AM-based mass customisation, including production lead times across different customised variants, AM resource utilisation, customer satisfaction with customised products, and quality consistency across varied AM-produced parts [73,124]. This layer is a significant improvement over present frameworks in that it uses machine learning concepts for identifying patterns in the data for the AM process and setting AM-specific metrics for the performance of the mass customisation, something absent in previous implementation models that use traditional manufacturing metrics. Continuous improvement frameworks need to incorporate machine learning techniques for identifying patterns in AM process data, benchmarking studies comparing different AM-enabled customisation approaches, and cost–benefit analyses specific to AM economics [124]. Knowledge-sharing platforms support ongoing learning about AM capabilities and applications, while validation systems ensure continued alignment between implementation outcomes and organisational objectives for AM-enabled mass customisation [39]. In addition, our framework is unique in that it responds to the new technological developments in AM that were not included in previous implementation models, especially in the sphere of multi-material printing and AI-based design optimization that emerged in 2020. This temporal specificity, combined with our systematic review methodology covering 61 studies through 2024, provides organizations with guidance that reflects the current state-of-the-art in AM-enabled mass customisation.

In order to achieve implementation success in the mass customisation implementation framework, there is a need for effective coordination of advancement at every layer to guarantee continuous alignment between elements. Prior to progressing to the next layer, there is a need to build strong foundations in AM technologies and customisation strategies at every level, as early advancement might damage the general system efficacy. It is important to retain alignment with changing organizational objectives, market requirements, and evolving AM capabilities.

5. Conclusions and Prospects

This systematic review of 61 studies provides a comprehensive understanding of mass customization strategies in additive manufacturing environments. We identify four key implementation strategies: integration with traditional manufacturing, customer-centric design integration, flexible manufacturing networks, and adaptive production systems—each offering distinct advantages when aligned with organizational capabilities and market requirements. The analysis reveals critical technological enablers (advanced AM systems, digital workflow integration, elimination of tooling requirements) and barriers (material limitations, post-processing requirements, quality assurance challenges) that significantly influence implementation success. Industry-specific investigations demonstrate how strategies must be adapted across healthcare, industrial/engineering, and architectural contexts to achieve optimal outcomes. The proposed multi-layered implementation framework integrates strategic, tactical, and operational dimensions, providing organizations with a structured yet adaptable approach to AM-based mass customization. Despite these advances, important research gaps remain in standardized evaluation methodologies, long-term reliability assessment, sustainability integration, quality control enhancement, economic modelling, digital workflow management, and emerging technology applications. Addressing these gaps will further advance the potential of AM-based mass customization to transform manufacturing paradigms and create competitive advantages.

5.1. Practical Implications

The implementation framework derived from our systematic review offers significant practical benefits for organizations implementing AM-based mass customization. The multi-layered approach provides a comprehensive roadmap that addresses strategic, tactical, and operational dimensions simultaneously, helping organizations avoid focusing exclusively on technological aspects. By identifying specific technological enablers and barriers, organizations can prioritize technologies aligned with their customization objectives while developing mitigation strategies for common implementation challenges. Industry-specific findings enable organizations in healthcare, industrial, and architectural sectors to adapt implementation approaches to their particular contexts. The evaluation methodologies and performance metrics provide practical tools for assessing implementation effectiveness and guiding continuous improvement. Additionally, the identification of critical success factors supports more effective risk management and change management strategies, increasing the likelihood of implementation success.

5.2. Research Limitations

Despite including 61 studies, several limitations should be acknowledged. The relative newness of AM-based mass customisation means longitudinal studies examining long-term outcomes are limited, with most capturing experiences over only 1–3 years. Geographical biases exist, with studies predominantly from Europe, North America, and parts of Asia limiting understanding of implementation in developing regions where different conditions may influence outcomes. Publication bias may result in over-representing successful implementations. Finally, the rapid evolution of AM technologies means some findings may become outdated as new capabilities emerge.

5.3. Future Research Directions

The systematic review of AM-enabled mass customisation has identified significant knowledge gaps in both theoretical frameworks and practical implementations. This section outlines critical research directions from our analysis, each addressing specific barriers and limitations identified in the literature. These interconnected research priorities collectively form a roadmap for advancing the scientific understanding and industrial adoption of AM-enabled mass customisation strategies:

1. Performance Evaluation Frameworks

Our review revealed a critical absence of standardised metrics for evaluating mass customisation implementations across different industrial contexts [72]. Without consistent evaluation methodologies, organisations cannot reliably assess the effectiveness of their customisation strategies or benchmark against industry standards. Future research should develop and validate comprehensive evaluation frameworks that accommodate industry-specific requirements while enabling meaningful cross-sector comparisons. Such frameworks would facilitate evidence-based decision-making and provide clearer pathways to successful implementation.

2. Long-term Performance and Reliability

While short-term performance characteristics of AM-produced components are increasingly well documented, their long-term durability remains inadequately investigated [50]. This knowledge gap presents a significant barrier to adoption in critical applications requiring extended service life or operation in harsh environments. Longitudinal studies examining ageing behaviours, fatigue characteristics, and maintenance requirements of customised AM components would address a fundamental uncertainty in the literature. Such research is particularly crucial for high-consequence sectors like medical implants and aerospace components, where reliability directly impacts safety and regulatory compliance.

3. Material Development and Characterization

The literature reveals substantial variability in material properties depending on build orientation, process parameters, and post-processing technique [44,51]. This inconsistency undermines the reliability of customised components and limits application scope. Future research should focus on developing specialised materials for AM mass customisation that demonstrate more consistent mechanical, thermal, and chemical properties across varying production conditions. Materials science advancements in this area would significantly expand the practical boundaries of mass customisation and enable more predictable quality outcomes.

4. Sustainable Mass Customisation

Our review identifies a significant gap in understanding how sustainability considerations can be integrated into mass customisation decision frameworks [73]. As environmental regulations tighten globally, research must develop methodologies for systematically evaluating the sustainability impacts of customisation strategies, particularly regarding material utilisation efficiency and energy consumption. Future studies should investigate multi-objective optimisation approaches that balance customisation capabilities with environmental performance and economic viability, creating a more holistic understanding of value creation in AM-enabled mass customisation.

5. Economic Models for Hybrid Manufacturing

The economic implications of various mass customization strategies remain poorly understood, particularly in hybrid manufacturing environments that combine traditional and additive approaches [75]. Current economic models fail to capture the complex relationships between customization levels, production efficiency, and value creation. Research should develop more sophisticated economic frameworks that account for the unique cost structures, resource optimization opportunities, and value propositions in AM-enabled customization. Such models would enable organizations to make more informed implementation decisions based on comprehensive cost–benefit analyses rather than technological determinism.

6. Artificial Intelligence Applications

The literature indicates that artificial intelligence and machine learning technologies hold significant yet largely unrealized potential for advancing mass customization capabilities [81]. Research should systematically explore how these technologies can enhance process optimization, quality control, and design automation in AM environments. Particular attention should be given to developing AI approaches that improve production adaptability and system intelligence while maintaining human oversight of critical decisions. This direction represents a convergence of digital twin technology with computational intelligence that could address multiple barriers identified in our review.

7. Post-Processing Automation

Post-processing operations remain a significant bottleneck in AM-enabled mass customization workflows, often requiring substantial manual intervention [53,54]. This limitation constrains scalability and introduces quality variations that undermine customization advantages. Future research should investigate integrated manufacturing cells that seamlessly combine AM processes with automated post-processing operations. Studies should focus on developing standardized approaches that maintain consistency across customized components while reducing labour intensity and variability, thereby enabling more efficient scaling of mass customization strategies.

8. Cross-Industry Implementation Guidelines

Our systematic review reveals a notable absence of structured implementation guidelines tailored to specific industry contexts [49,56]. Organizations face significant uncertainty when transitioning from traditional manufacturing to AM-enabled mass customization. Research should develop sector-specific implementation frameworks that account for unique industry requirements, constraints, and opportunities. These practical roadmaps would help organizations navigate the complex technological, organizational, and market considerations involved in successful implementation, bridging the gap between theoretical potential and practical realization.

Author Contributions: Conceptualization, S.K.F. and T.C.D.; methodology, S.K.F. and, T.C.D.; software, E.A.; validation, E.A.; formal analysis, S.K.F.; investigation, T.C.D. and D.J.d.B.; resources, T.C.D., D.J.d.B. and E.A.; data curation, S.K.F., T.C.D. and D.J.d.B.; writing—original draft preparation, S.K.F.; writing—review and editing, E.A., T.C.D. and D.J.d.B.; supervision, T.C.D. and D.J.d.B.; project administration, T.C.D. and D.J.d.B.; funding acquisition, T.C.D. and D.J.d.B.; All authors have read and agreed to the published version of the manuscript.

Funding: This work is supported by the South African Research Chairs Initiative of the Department of Science and Technology and National Research Foundation of South Africa (Grant No. SARC 20150101-097994).

Data Availability Statement: Data will be available upon request.

Acknowledgments: We would like to express our sincere gratitude to the anonymous reviewers for their valuable comments and constructive feedback. We also thank Central University of Technology, Bloemfontein, South Africa for providing the necessary resources for this research. Special thanks to Isaac Kwesi Nooni (Wuxi University, Wuxi-China) for his invaluable advice on conducting a systematic review and meticulous proofreading of the paper.

Conflicts of Interest: The authors declare no conflicts of interest.

References

1. Jafar, M.R.; Tripathi, N.M.; Yadav, M.; Nasato, D.S. Additive Manufacturing in the Age of Industry 4.0 and Beyond. In *Advances in Pre- and Post-Additive Manufacturing Processes*; CRC Press: Boca Raton, FL, USA, 2024; pp. 213–230. [CrossRef]
2. Lacroix, R.; Timonina-Farkas, A.; Seifert, R.W. Utilizing additive manufacturing and mass customization under capacity constraints. *J. Intell. Manuf.* **2023**, *34*, 281–301. [CrossRef] [PubMed]
3. Kim, S.; Lee, K. The paradigm shift of mass customisation research. *Int. J. Prod. Res.* **2023**, *61*, 3350–3376. [CrossRef]
4. Renjith, S.C.; Park, K.; Okudan Kremer, G.E. A Design Framework for Additive Manufacturing: Integration of Additive Manufacturing Capabilities in the Early Design Process. *Int. J. Precis. Eng. Manuf.* **2020**, *21*, 329–345. [CrossRef]
5. Jankovics, D.; Barari, A. Customization of Automotive Structural Components using Additive Manufacturing and Topology Optimization. *IFAC-PapersOnLine* **2019**, *52*, 212–217. [CrossRef]
6. Marinho De Brito, F.; Junior, G.D.C.; Frazzon, E.M.; Basto, J.P.T.V.; Alcala, S.G.S. Design Approach for Additive Manufacturing in Spare Part Supply Chains. *IEEE Trans. Ind. Inform.* **2021**, *17*, 757–765. [CrossRef]
7. Stentoft, J.; Wickstrøm, K.A.; Haug, A.; Philipsen, K. Additive manufacturing–enabled innovation in small- and medium-sized enterprises: The role of readiness in make-or-buy decisions. *Ind. Manag. Data Syst.* **2023**, *123*, 1768–1788. [CrossRef]
8. Khorram Niaki, M.; Nonino, F.; Tafakkori, K.; Torabi, S.A.; Kazemian, I. The influence of manufacturing contexts on additive manufacturing-enabled competitive capabilities. *J. Manuf. Technol. Manag.* **2022**, *33*, 1102–1123. [CrossRef]
9. Wazeer, A.; Das, A.; Sinha, A.; Inaba, K.; Ziyi, S.; Karmakar, A. Additive manufacturing in biomedical field: A critical review on fabrication method, materials used, applications, challenges, and future prospects. *Prog. Addit. Manuf.* **2022**, *8*, 857–889. [CrossRef]
10. Shidid, D.; Leary, M.; Choong, P.; Brandt, M. Just-in-time design and additive manufacture of patient-specific medical implants. *Phys. Procedia* **2016**, *83*, 4–14. [CrossRef]
11. Paoletti, I. Mass Customization with Additive Manufacturing: New Perspectives for Multi Performative Building Components in Architecture. *Procedia Eng.* **2017**, *180*, 1150–1159. [CrossRef]
12. Yu, Y.; Kim, G.; Mathur, K. A critical review of additive manufacturing: An innovation of mass customization. *J. Text. Appar. Technol. Manag.* **2020**, *11*, 1–16.

13. Pajonk, A.; Prieto, A.; Blum, U.; Knaack, U. Multi-material additive manufacturing in architecture and construction: A review. *J. Build. Eng.* **2022**, *45*, 103603. [CrossRef]
14. Korner, M.E.H.; Lambán, M.P.; Albajez, J.A.; Santolaria, J.; Corrales, L.D.C.N.; Royo, J. Systematic literature review: Integration of additive manufacturing and industry 4.0. *Metals* **2020**, *10*, 1061. [CrossRef]
15. Moher, D.; Liberati, A.; Tetzlaff, J.; Altman, D.G. Preferred reporting items for systematic reviews and meta-analyses: The PRISMA statement. *BMJ* **2009**, *339*, 332–336. [CrossRef]
16. Liberati, A.; Altman, D.G.; Tetzlaff, J.; Mulrow, C.; Gøtzsche, P.C.; Ioannidis, J.P.A.; Clarke, M.; Devereaux, P.J.; Kleijnen, J.; Moher, D. The PRISMA statement for reporting systematic reviews and meta-analyses of studies that evaluate healthcare interventions: Explanation and elaboration. *BMJ* **2009**, *339*, b2700. [CrossRef]
17. Page, M.J.; Moher, D.; Bossuyt, P.M.; Boutron, I.; Hoffmann, T.C.; Mulrow, C.D.; Shamseer, L.; Tetzlaff, J.M.; Akl, E.A.; Brennan, S.E.; et al. PRISMA 2020 explanation and elaboration: Updated guidance and exemplars for reporting systematic reviews. *BMJ* **2021**, *372*, n160. [CrossRef]
18. Agrawal, S.; Oza, P.; Kakkar, R.; Tanwar, S.; Jetani, V.; Undhad, J.; Singh, A. Analysis and recommendation system-based on PRISMA checklist to write systematic review. *Assess. Writ.* **2024**, *61*, 100866. [CrossRef]
19. Page, M.J.; McKenzie, J.E.; Bossuyt, P.M.; Boutron, I.; Hoffmann, T.C.; Mulrow, C.D.; Shamseer, L.; Tetzlaff, J.M.; Moher, D. Updating guidance for reporting systematic reviews: Development of the PRISMA 2020 statement. *J. Clin. Epidemiol.* **2021**, *134*, 103–112. [CrossRef]
20. Mongeon, P.; Paul-Hus, A. The journal coverage of Web of Science and Scopus: A comparative analysis. *Scientometrics* **2016**, *106*, 213–228. [CrossRef]
21. Baas, J.; Schotten, M.; Plume, A.; Côté, G.; Karimi, R. Scopus as a curated, high-quality bibliometric data source for academic research in quantitative science studies. *Quant. Sci. Stud.* **2020**, *1*, 377–386. [CrossRef]
22. Bramer, W.; Rethlefsen, M.; Kleijnen, J.; Franco, O.H. Optimal database combinations for literature searches in systematic reviews: A prospective exploratory study. *Syst. Rev.* **2017**, *6*, 245. [CrossRef] [PubMed]
23. Zhu, J.; Liu, W. A tale of two databases: The use of Web of Science and Scopus in academic papers. *Scientometrics* **2020**, *123*, 321–335. [CrossRef]
24. Pranckutė, R. Web of Science (WoS) and Scopus: The Titans of Bibliographic Information in Today's Academic World. *Publications* **2021**, *9*, 12. [CrossRef]
25. Singh, S.; Mehla, S.; Bhargava, S.K.; Ramakrishna, S. History and Evolution of Additive Manufacturing. In *Additive Manufacturing for Chemical Sciences and Engineering*; Springer: Berlin/Heidelberg, Germany, 2022; pp. 19–51. [CrossRef]
26. CASP Checklist Website. Available online: https://casp-uk.net/casp-tools-checklists/ (accessed on 4 April 2025).
27. Mishra, S.; Dey, A.K. Understanding and Identifying 'Themes' in Qualitative Case Study Research. *South Asian J. Bus. Manag. Cases* **2022**, *11*, 187–192. [CrossRef]
28. Creswell, J.W.; Plano Clark, V.L. *Designing and Conducting Mixed Methods Research*, 3rd ed.; SAGE Scientific Research Publishing: Thousand Oaks, CA, USA, 2018.
29. Williams, M.; Moser, T. The Art of Coding and Thematic Exploration in Qualitative Research: International Management Review. *Int. Manag. Rev.* **2019**, *15*, 45–55.
30. O'Connor, C.; Joffe, H. Intercoder Reliability in Qualitative Research: Debates and Practical Guidelines. *Int. J. Qual. Methods* **2020**, *19*, 1609406919899220. [CrossRef]
31. Lee, Y.J.; Lee, J.Y.; Cho, J.H.; Choi, J.H. Interrater reliability of sleep stage scoring: A meta-analysis. *J. Clin. Sleep Med.* **2022**, *18*, 193–202. [CrossRef]
32. Allen Forster, C.U.J.T. Validating coding for a theoretical model of information behaviour. *Inf. Res.* **2008**, *13*, 13–14.
33. Zawadzki, P.; Żywicki, K. Smart product design and production control for effective mass customization in the Industry 4.0 concept. *Manag. Prod. Eng. Rev.* **2016**, *7*, 105–112. [CrossRef]
34. Jiang, Z.; Wen, H.; Han, F.; Tang, Y.; Xiong, Y. Data-driven generative design for mass customization: A case study. *Adv. Eng. Inform.* **2022**, *54*, 101786. [CrossRef]
35. Yuan, P.F.; Beh, H.S.; Yang, X.; Zhang, L.; Gao, T. Feasibility study of large-scale mass customization 3D printing framework system with a case study on Nanjing Happy Valley East Gate. *Front. Archit. Res.* **2022**, *11*, 670–680. [CrossRef]
36. Hafenecker, J.; Rigas, N.; Merklein, M. Laser-based ultrasonic measurement of the influence of PBF-LB/M-typical thermal cycles on sheet metal components of hybrid parts. *Prod. Eng.* **2023**, *17*, 861–873. [CrossRef]
37. Rajamani, P.K.; Ageyeva, T.; Kovács, J.G. Personalized Mass Production by Hybridization of Additive Manufacturing and Injection Molding. *Polymers* **2021**, *13*, 309. [CrossRef] [PubMed]
38. Rousseau, M.; Medini, K.; Romero, D.; Wuest, T. Configurators as a means to Leverage Customer-Centric Sustainable Systems—Evidence from the 3D-Printing Domain. *Procedia CIRP* **2020**, *96*, 103–108. [CrossRef]
39. Rayna, T.; Striukova, L.; Darlington, J. Co-creation and user innovation: The role of online 3D printing platforms. *J. Eng. Technol. Manag.* **2015**, *37*, 90–102. [CrossRef]

40. Zanetti, V.; Cavalieri, S.; Kalchschmidt, M.; Pinto, R. The Role of Additive Manufacturing in the B2C Value Chain: Challenges, Opportunities and Models. In Proceedings of the IFIP International Conference on Advances in Production Management Systems, Tokyo, Japan, 7–9 September 2015; pp. 137–145. [CrossRef]
41. Narendran, N.; Perera, I.U.; Mou, X.; Thotagamuwa, D.R. Opportunities and challenges for 3D printing of solid-state lighting systems. In Proceedings of the Sixteenth International Conference on Solid State Lighting and LED-Based Illumination Systems, San Diego, CA, USA, 6–10 August 2017; SPIE: Bellingham, WA, USA, 2017; Volume 35. [CrossRef]
42. Lunetto, V.; Priarone, P.C.; Galati, M.; Minetola, P. On the correlation between process parameters and specific energy consumption in fused deposition modelling. *J. Manuf. Process.* **2020**, *56*, 1039–1049. [CrossRef]
43. Kim, S.; Jeong, B. Mass customization capability planning with additive manufacturing. *IFIP Adv. Inf. Commun. Technol.* **2018**, *535*, 184–192. [CrossRef]
44. Sun, H.; Zou, B.; Wang, X.; Chen, W.; Zhang, G.; Quan, T.; Huang, C. Advancements in multi-material additive manufacturing of advanced ceramics: A review of strategies, techniques and equipment. *Mater. Chem. Phys.* **2024**, *319*, 129337. [CrossRef]
45. Kumar Rana, A.; Kundu, S.; Kayal, S. Digital fabrication with FDM 3D printing on prototype development of toys: Additive manufacturing and sustainability. In Proceedings of the 16th International Conference on Materials Processing and Characterization (ICMPC 2024), Ahmedabad, India, 27–29 June 2024. [CrossRef]
46. Lin, D.; Zhang, Y.; Jiang, B. Identification of Factors Affecting the Application of Additive Manufacturing Technology in Mass Customization Based on Structural Equation Model. *IEEE Int. Conf. Ind. Eng. Eng. Manag.* **2022**, *2022*, 137–141. [CrossRef]
47. Pesce, A.; Hornés, A.; Núñez, M.; Morata, A.; Torrell, M.; Tarancón, A. 3D printing the next generation of enhanced solid oxide fuel and electrolysis cells †. *J. Mater. Chem. A* **2020**, *A8*, 16926–16932. [CrossRef]
48. Schubert, C.; Van Langeveld, M.C.; Donoso, L.A. Innovations in 3D printing: A 3D overview from optics to organs. *Br. J. Ophthalmol.* **2014**, *98*, 159–161. [CrossRef] [PubMed]
49. García-Dominguez, A.; Claver, J.; Sebastián, M.A. Optimization Methodology for Additive Manufacturing of Customized Parts by Fused Deposition Modeling (FDM). Application to a Shoe Heel. *Polymers* **2020**, *12*, 2119. [CrossRef] [PubMed]
50. Turner, R.D.; Wingham, J.R.; Paterson, T.E.; Shepherd, J.; Majewski, C. Use of silver-based additives for the development of antibacterial functionality in Laser Sintered polyamide 12 parts. *Sci. Rep.* **2020**, *10*, 892. [CrossRef]
51. Shen, Z.; Shang, X.; Zhao, M.; Dong, X.; Xiong, G.; Wang, F.Y. A learning-based framework for error compensation in 3D printing. *IEEE Trans. Cybern.* **2019**, *49*, 4042–4050. [CrossRef]
52. Bos, F.; Ahmed, Z.; Rodriguez, C.R.; Figueiredo, S.C. 3D Concrete Printing for Structural Applications. *SPOOL* **2019**, *6*, 5–10. [CrossRef]
53. Gao, M.; Li, L.; Wang, Q.; Liu, C.; Li, X.; Liu, Z. Feature-based energy consumption quantitation strategy for complex additive manufacturing parts. *Energy* **2024**, *297*, 131249. [CrossRef]
54. Lacroix, R.; Seifert, R.W.; Timonina-Farkas, A. Benefiting from additive manufacturing for mass customization across the product life cycle. *Oper. Res. Perspect.* **2021**, *8*, 100201. [CrossRef]
55. Li, Y.; Wu, H.; Tamir, T.S.; Shen, Z.; Liu, S.; Hu, B.; Xiong, G. An efficient product-customization framework based on multimodal data under the social manufacturing paradigm. *Machines* **2023**, *11*, 170. [CrossRef]
56. da Silva, D.P.; Pinheiro, J.; Abdulghani, S.; Lorger, C.K.; Martinez, J.C.; Solano, E.; Mateus, A.; Pascoal-Faria, P.; Mitchell, G.R. Changing the Paradigm-Controlling Polymer Morphology during 3D Printing Defines Properties. *Polymers* **2022**, *14*, 1638. [CrossRef]
57. Gulisano, V.; Papatriantafilou, M.; Chen, Z.; Hryha, E.; Nyborg, L.; Nyborg, L. Towards data-driven additive manufacturing processes. In Proceedings of the 23rd International Middleware Conference Industrial Track, Quebec, QC, Canada, 7–11 November 2022; pp. 43–49. [CrossRef]
58. Sivabalakrishnan, R.; Kalaiarasan, A.; Ajithvishva, M.S.; Hemsri, M.; Oorappan, G.M.; Yasodharan, R. IoT visualization of Smart Factory for Additive Manufacturing System (ISFAMS) with visual inspection and material handling processes. *IOP Conf. Ser. Mater. Sci. Eng.* **2020**, *995*, 012027. [CrossRef]
59. Olakanmi, E.O.; Cochrane, R.F.; Dalgarno, K.W. A review on selective laser sintering/melting (SLS/SLM) of aluminium alloy powders: Processing, microstructure, and properties. *Prog. Mater. Sci.* **2015**, *74*, 401–477. [CrossRef]
60. Giunta, L.; Hicks, B.; Snider, C.; Gopsill, J. A Living Lab Platform for Testing Additive Manufacturing Agent-Based Manufacturing Strategies. *Procedia CIRP* **2023**, *119*, 656–661. [CrossRef]
61. Leong, K.F.; Teng, P.S.P.; Er, B.H.; Tee, C.H. Mass customization and testing of braces using additive manufacturing. *Mater. Today Proc.* **2022**, *70*, 593–598. [CrossRef]
62. Solaimani, S.; Parandian, A.; Nabiollahi, N. A holistic view on sustainability in additive and subtractive manufacturing: A comparative empirical study of eyewear production systems. *Sustainability* **2021**, *13*, 10775. [CrossRef]
63. Bilalis, L.; Canellidis, V.; Papatheodorou, T.; Giannatsis, J. Direct Digital Manufacturing of a Customized Face Mask. *J. Manuf. Mater. Process.* **2022**, *6*, 126. [CrossRef]

64. Du, Y.; Chen, J.; Meng, Q.; Dou, Y.; Xu, J.; Shen, S.Z. Thermoelectric materials and devices fabricated by additive manufacturing. *Vacuum* **2020**, *178*, 109384. [CrossRef]
65. Kromoser, B.; Pachner, T. Optiknot 3D—Free-Formed Frameworks out of Wood with Mass Customized Knots Produced by FFF Additive Manufactured Polymers: Experimental Investigations, Design Approach and Construction of a Prototype. *Polymers* **2020**, *12*, 965. [CrossRef]
66. Lacava, H.; Cherrington, N.; Corrado, A.; Bigdellou, S.; Chen, Q. A Preliminary Study of 3D Printing Home Designs for Improving Efficiency and Sustainability of Indigenous Housing in Canada. *Sustainability* **2024**, *16*, 5781. [CrossRef]
67. Bayraklilar, M.S. Dimensional Accuracy of Acrylonitrile Butadiene Styrene Material Produced by Additive Manufacturing Method. *J. Mater. Eng. Perform.* **2024**, *33*, 2531–2551. [CrossRef]
68. Filip Górski, P.; Zawadzki, P.; Wichniarek, R.; Kuczko, W.; Slupinska, S.; Zukowska, M. Automated Design and Rapid Manufacturing of Low-Cost Customized Upper Limb Prostheses. *J. Phys. Conf. Ser.* **2022**, *2198*, 012040. [CrossRef]
69. He, Q.; Han, L. Research on Adaptive Repair Technology of 3D Printing Based on Convolutional Neural Network. In Proceedings of the 2021 IEEE 21st International Conference on Communication Technology (ICCT), Tianjin, China, 13–16 October 2021.
70. Hauser, M.; King, R.; Wysk, R.; Harrysson, O. Resource Planning for Direct Fabrication of Customized Orthopedic Implants Using EBM Technology. *J. Manuf. Syst.* **2021**, *60*, 500–511. [CrossRef]
71. Alabbasi, M.; Agkathidis, A.; Chen, H. Robotic 3D printing of concrete building components for residential buildings in Saudi Arabia. *Autom. Constr.* **2023**, *148*, 104751. Available online: https://www.sciencedirect.com/science/article/pii/S0926580523000110 (accessed on 5 April 2025). [CrossRef]
72. Deradjat, D.; Minshall, T. Decision trees for implementing rapid manufacturing for mass customisation. *CIRP J. Manuf. Sci. Technol.* **2018**, *23*, 156–171. [CrossRef]
73. Huang, J.; Sun, H.; Kwok, T.-H.; Zhou, C.; Xu, W. Geometric Deep Learning for Shape Correspondence in Mass Customization by Three-Dimensional Printing. *J. Manuf. Sci. Eng.* **2020**, *142*, 061003. [CrossRef]
74. Basak, S.; Baumers, M.; Holweg, M.; Hague, R.; Tuck, C. Reducing production losses in additive manufacturing using overall equipment effectiveness. *Addit. Manuf.* **2022**, *56*, 102904. [CrossRef]
75. Martinez-Marquez, D.; Mirnajafizadeh, A.; Carty, C.P.; Stewart, R.A. Application of quality by design for 3D printed bone prostheses and scaffolds. *PLoS ONE* **2018**, *13*, e0195291. [CrossRef]
76. Frohn-Sörensen, P.; Geueke, M.; Engel, B.; Löffler, B.; Bickendorf, P.; Asimi, A.; Bergweiler, G.; Schuh, G. Design for 3D Printed Tools: Mechanical Material Properties for Direct Polymer Additive Tooling. *Polymers* **2022**, *14*, 1694. [CrossRef]
77. Koop, C.; Grosse Erdmann, J.; Koller, J.; Döpper, F. Circular Business Models for Remanufacturing in the Electric Bicycle Industry. *Front. Sustain.* **2021**, *2*, 785036. [CrossRef]
78. Richter, A.; Kessing, D.; Fischer, F.; Pelzer, L.; Dilger, K. Print-on strategies to bond injection molded parts with structures produced by fused-deposition-modeling. In Proceedings of the Design Society: International Conference on Engineering Design, Delf, The Netherlands, 5–8 August 2019; pp. 5–8. [CrossRef]
79. Büscher, J.; Zajackowski, J.; Rademacher, H.G.; Tillmann, W.; Deuse, J. Trustworthiness of artificial intelligence applications for quality optimisation in metal additive manufacturing. *Procedia CIRP* **2024**, *122*, 497–502. [CrossRef]
80. Sæterbø, M.; Solvang, W.D. A Readiness Model for Facilitating the Implementation of Metal Additive Manufacturing at SMEs. In Proceedings of the 2023 IEEE Conference on Technologies for Sustainability (SusTech), Portland, OR, USA, 19–22 April 2023. [CrossRef]
81. Wang, P.; Li, J.; Yang, J.; Wang, G.; He, L.; Manufacturing, H.Z.-A. Enabling 3D multilayer electronics through the hybrid of vat photopolymerization and laser-activated selective metallization. *Addit. Manuf.* **2023**, *74*, 103717. [CrossRef]
82. Abdulhameed, O.; Mian, S.H.; Moiduddin, K.; Al-Ahmari, A.; Ahmed, N.; Aboudaif, M.K. A multi-part orientation planning schema for fabrication of non-related components using additive manufacturing. *Micromachines* **2022**, *13*, 1777. [CrossRef] [PubMed]
83. Bellens, S.; Guerrero, P.; Janssens, M.; Vandewalle, P.; Dewulf, W. Machine learning based sinogram interpolation for X-ray computed tomography validated on experimental data. *Precis. Eng.* **2024**, *88*, 449–463. [CrossRef]
84. Lin, K.-W.; Chou, L.-W.; Su, Y.-T.; Wei, S.-H.; Chen, C.-S. Biomechanical Effect of 3D-Printed Foot Orthoses in Patients with Knee Osteoarthritis. *Appl. Sci.* **2021**, *11*, 4200. [CrossRef]
85. García-Dominguez, A.; Claver, J.; Sebastián, M.A. Integration of additive manufacturing, parametric design, and optimization of parts obtained by fused deposition modeling (fdm). A methodological approach. *Polymers* **2020**, *12*, 1993. [CrossRef]
86. Gibson, I.; Rosen, D.; Stucker, B. Direct Digital Manufacturing. In *Additive Manufacturing Technologies*; Springer: Berlin/Heidelberg, Germany, 2015; pp. 375–397. [CrossRef]
87. Ngo, T.D.; Kashani, A.; Imbalzano, G.; Nguyen, K.T.Q.; Hui, D. Additive manufacturing (3D printing): A review of materials, methods, applications and challenges. *Compos. B Eng.* **2018**, *143*, 172–196. [CrossRef]
88. Rayna, T.; Striukova, L. Involving consumers: The role of digital technologies in promoting 'prosumption'and user innovation. *J. Knowl. Econ.* **2021**, *12*, 218–237. [CrossRef]

89. Pallari, J.H.P.; Dalgarno, K.W.; Woodburn, J. Mass customization of foot orthoses for rheumatoid arthritis using selective laser sintering. *IEEE Trans. Biomed. Eng.* **2010**, *57*, 1750–1756. [CrossRef]
90. Górski, F.; Suszek, E.; Wichniarek, R.; Kuczko, W.; Żukowska, M. Rapid Manufacturing of Individualized Prosthetic Sockets. *Adv. Sci. Technol. Res. J.* **2020**, *14*, 42–49. [CrossRef]
91. Lu, Y. Industry 4.0: A survey on technologies, applications and open research issues. *J. Ind. Inf. Integr.* **2017**, *6*, 1–10. [CrossRef]
92. Thompson, S.M.; Bian, L.; Shamsaei, N.; Yadollahi, A. An overview of Direct Laser Deposition for additive manufacturing; Part I: Transport phenomena, modeling and diagnostics. *Addit. Manuf.* **2015**, *8*, 36–62. [CrossRef]
93. Dzogbewu, T.C.; Amoah, N.; Fianko, S.K.; Afrifa, S.; De Beer, D. Additive manufacturing towards product production: A bibliometric analysis. *Manuf. Rev.* **2022**, *9*, 1. [CrossRef]
94. Carlone, P.; Palazzo, G.S. Influence of Process Parameters on Microstructure and Mechanical Properties in AA2024-T3 Friction Stir Welding. *Metallogr. Microstruct. Anal.* **2013**, *2*, 213–222. [CrossRef]
95. Maamoun, A.H.; Xue, Y.F.; Elbestawi, M.A.; Veldhuis, S.C. Effect of Selective Laser Melting Process Parameters on the Quality of Al Alloy Parts: Powder Characterization, Density, Surface Roughness, and Dimensional Accuracy. *Materials* **2018**, *11*, 2343. [CrossRef] [PubMed]
96. Gockel, J.; Sheridan, L.; Koerper, B.; Whip, B. The influence of additive manufacturing processing parameters on surface roughness and fatigue life. *Int. J. Fatigue* **2019**, *124*, 380–388. [CrossRef]
97. Tamir, T.S.; Xiong, G.; Fang, Q.; Yang, Y.; Shen, Z.; Zhou, M.C.; Jiang, J. Machine-learning-based monitoring and optimization of processing parameters in 3D printing. *Int. J. Comput. Integr. Manuf.* **2023**, *36*, 1362–1378. [CrossRef]
98. Gudiño Paredes, S.; Jasso Peña, F.d.J.; de La Fuente Alcazar, J.M. Remote proctored exams: Integrity assurance in online education? *Distance Educ.* **2021**, *42*, 200–218. [CrossRef]
99. Frank, S.B.; Christian, A.; Michael, G.; Christian, K.; Gounet, H.P. Towards the Automation of Non-destructive Fault Recognition: Enhancement of Robustness and Accuracy Through AI Powered Acoustic and Thermal Signal Analysis. *J. Fail. Anal. Prev.* **2024**, *24*, 2210–2220. [CrossRef]
100. Baranašić, Z. Additive Manufacturing-Digital Transformation in Manufacturing Business Models. Ph.D. Thesis, University of Zagreb, Zagreb, Croatia, 2022.
101. Mavroidis, C.; Ranky, R.G.; Sivak, M.L.; Patritti, B.L.; DiPisa, J.; Caddle, A.; Gilhooly, K.; Govoni, L.; Sivak, S.; Lancia, M.; et al. Patient specific ankle-foot orthoses using rapid prototyping. *J. Neuroeng. Rehabil.* **2011**, *8*, 1. [CrossRef]
102. Dzogbewu, T.C.; Fianko, S.K.; Amoah, N.; Afrifa, S., Jr.; de Beer, D. Additive manufacturing in South Africa: Critical success factors. *Heliyon* **2022**, *8*, e11852. [CrossRef]
103. Kim, H.; Lee, D.; Lee, S.Y.; Yang, H.; Park, S.-W.; Lim, H.-P.; Yun, K.-D.; Park, C. Denture flask fabrication using fused deposition modeling three-dimensional printing. *Elsevier* **2020**, *64*, 231–234. [CrossRef]
104. Abdulhameed, O.; Al-Ahmari, A.; Ameen, W.; Mian, S.H. Additive manufacturing: Challenges, trends, and applications. *Adv. Mech. Eng.* **2019**, *11*, 1687814018822880. [CrossRef]
105. Lee Ventola, C. Medical Applications for 3D Printing: Current and Projected Uses. *Pharm. Ther.* **2014**, *39*, 704.
106. Gorelik, M. Additive manufacturing in the context of structural integrity. *Int. J. Fatigue* **2017**, *94*, 168–177. [CrossRef]
107. Gibson, I.; Rosen, D.; Stucker, B.; Khorasani, M. Directed Energy Deposition. In *Additive Manufacturing Technologies*; Springer: Berlin/Heidelberg, Germany, 2021; pp. 285–318. [CrossRef]
108. Chung, C.Y.; Liu, C.J. Metal additive manufacturing of conformal cooling channels for improving the quality of metal injection–molded products. *Adv. Mech. Eng.* **2024**, *16*, 16878132241282210. [CrossRef]
109. Metal Additive Manufacturing Magazine. GE Aviation Reaches Milestone with 100,000 Fuel Nozzle Tips. Available online: https://www.metal-am.com/ge-aviation-reaches-milestone-with-100000-fuel-nozzle-tips/ (accessed on 29 May 2025).
110. The Fabricator. Daimler Buses Rolls Out a Mobile Additive Manufacturing Service Center. Available online: https://www.thefabricator.com/additivereport/news/additive/daimler-buses-rolls-out-a-mobile-additive-manufacturing-service-center (accessed on 30 May 2025).
111. Dancel, R. 3D Printed House for Disaster-Affected Areas. In Proceedings of the 2019 DLSU Disaster Risk Reduction and Infrastructure Development (DRRID) Forum, Taft, Manila, 19–21 June 2019. [CrossRef]
112. Stuart-Smith, R. Towards Technically Progressive and Speculative Design Education: Penn MSD-RAS's Collaborative and Interdisciplinary Pedagogy. In Proceedings of the ACSA/EAAE Teachers Conference, Educating the Cosmopolitan Architect, Reykjavik, Iceland, 22–24 June 2023; pp. 181–188. [CrossRef]
113. Suntharalingam, T.; Upasiri, I.; Gatheeshgar, P.; Poologanathan, K.; Nagaratnam, B.; Rajanayagam, H.; Navaratnam, S. Fire resistance of 3D printed concrete composite wall panels exposed to various fire scenarios. *J. Struct. Fire Eng.* **2021**, *12*, 377–409. [CrossRef]
114. Bechthold, M. Ceramic prototypes-design, computation, and digital fabrication. *Inf. Constr.* **2016**, *68*, e167. [CrossRef]
115. Fang, Q.; Xiong, G.; Zhou, M.; Tamir, T.S.; Yan, C.B.; Wu, H.; Shen, Z.; Wang, F.Y. Process monitoring, diagnosis and control of additive manufacturing. *IEEE Trans. Autom. Sci. Eng.* **2022**, *21*, 1041–1067. [CrossRef]

116. Hora, M.; Hankammer, S.; Canetta, L.; Sel, S.K.; Gomez, S.; Gahrens, S. Designing business models for sustainable mass customization: A framework proposal. *Int. J. Ind. Eng. Manag.* **2016**, *7*, 143–152. [CrossRef]
117. Machado, C.G.; Despeisse, M.; Winroth, M.; da Silva, E.H.D.R. Additive manufacturing from the sustainability perspective: Proposal for a self-assessment tool. *Procedia CIRP* **2019**, *81*, 482–487. [CrossRef]
118. Meyer, A.D.; Tsui, A.S.; Hinings, C.R. Configurational Approaches to Organizational Analysis. *Acad. Manag. J.* **2017**, *36*, 1175–1195. [CrossRef]
119. Barney, J. Firm Resources and Sustained Competitive Advantage. *J. Manag.* **1991**, *17*, 99–120. [CrossRef]
120. Robert, M. Porter's Competitive Advantage of Nations: An Assessment. *Strateg. Manag. J.* **1991**, *12*, 535–548.
121. Teece, D.J.; Pisano, G.; Shuen, A. Dynamic capabilities and strategic management. *Strateg. Manag. J.* **1997**, *18*, 509–533. [CrossRef]
122. Eisenhardt, K.M.; Martin, J.A. Dynamic capabilities: What are they? *Strateg. Manag. J.* **2000**, *21*, 1105–1121. [CrossRef]
123. Trist, E. The Evolution of Socio-Technical Systems. 1981. Available online: https://www.lmmiller.com/blog/wp-content/uploads/2013/06/The-Evolution-of-Socio-Technical-Systems-Trist.pdf (accessed on 15 April 2025).
124. Baxter, G.; Sommerville, I. Socio-technical systems: From design methods to systems engineering. *Interact. Comput.* **2010**, *23*, 4–17. [CrossRef]

processes

MDPI

Article

A Hierarchical Axiomatic Evaluation of Additive Manufacturing Equipment and the 3D Printing Process Based on Sustainability and Human Factors

Ismael Mendoza-Muñoz [1,*], Mildrend Ivett Montoya-Reyes [1], Aidé Aracely Maldonado-Macías [2,*], Gabriela Jacobo-Galicia [1] and Olivia Yessenia Vargas-Bernal [1]

[1] Faculty of Engineering, Autonomous University of Baja California, Blvd. Benito Juarez S/N, Mexicali 21280, Baja California, Mexico; olivia.yessenia.vargas.bernal@uabc.edu.mx (O.Y.V.-B.)

[2] Institute of Engineering and Technology, Autonomous University of Ciudad Juarez, Ave. Del Charro 430 Norte, Ciudad Juarez 32310, Chihuahua, Mexico

* Correspondence: ismael.mendoza@uabc.edu.mx (I.M.-M.); amaldona@uacj.mx (A.A.M.-M.)

Abstract: As interest in additive manufacturing (AM) continues to increase, it has become more important to have a robust method to help potential users select the AM process that best suits their technological needs while providing the greatest potential benefits in terms of sustainability and its effect on people. This paper presents the development of a framework for selecting the best AM process for a given application by considering both sustainability and human factors through the combination of axiomatic design and the analytic hierarchy process. Thirty-one participants with varying levels of expertise (novice and advanced users) were involved in the study, considering the frequency of 3D printer usage (novice users: never, rarely; expert users: sometimes, almost always, always) for prototyping parts. They employed fused deposition modeling (FDM) and stereolithography (SLA) (both 3D desktop printers) and collected data on five evaluation criteria. The participation of experts helped establish a novel methodology, with material cost deemed most important (49.8%), followed by cycle time (28%), energy consumption (11.7%), error rate (6.6%), and equipment noise (3.9%). The results showed that FDM was the optimal equipment option for advanced users. By examining the information content of the other options, it was found that FDM demanded less information than SLA, regardless of the user's level of expertise. The proposed method is appropriate to assess the sustainability aspect of FDM and SLA; however, it can be further improved by adding indicators such as environmental impact, recyclability, and ergonomic and occupational health factors.

Keywords: 3D printing; FDM; SLA; human factors (HF); sustainability

1. Introduction

The fourth industrial revolution (I4.0) is transforming production worldwide, leading to the rise of intelligent manufacturing, where humans, machines, and products communicate in virtual environments [1–3]. This has led to the adoption of additive manufacturing (AM) as a key method for producing complex parts and products [1]. The popularity of AM has grown in recent years due to its declining cost and increased accessibility, as well as its innovative applications in fields such as manufacturing and medicine [4]. AM has experienced significant growth and development since its introduction in the 1980s, with machines and processes like stereolithography (SLA) and fused deposition modeling (FDM) being widely used [4].

SLA is considered the world's first 3D printing system and was patented by Charles Hull in 1986. This technique uses light to cure a liquid material and form a 3D object by solidifying thin layers of resin with a laser until a 3D part is fully built. SLA is one of the most popular AM processes due to its ability to produce parts with complex geometries, high accuracy, and good surface finishes at a relatively fast speed [5–7].

FDM is a 3D printing technology invented in 1988 by Scott Crump and is currently the most widely used 3D printer technology [8]. FDM is a cost-effective method for producing end-use polymer products using a computer-controlled extruder that deposits successive layers of semi-melted thermoplastic material in the form of a thin filament to build 3D parts. FDM has been compared to a sophisticated glue gun by Hull [8,9].

3D printing involves designing the desired geometry using CAD software, converting it into a G-code file using a slicing program, and producing the final part using a 3D printer with specific materials [10]. However, the creation of 3D models requires design software, which has become more accessible in recent years. Using design software effectively still requires a certain level of technical knowledge, as it may be challenging for non-expert users to navigate the human–machine interaction [11,12]. Online stores offering 3D designs have reduced the gap between expert and non-expert users, allowing non-experts to access complex designs without advanced technical knowledge [13].

The solution may be helpful for some, but it does not address the DIY trend. Successful and sustainable 3D printing requires users to evaluate capabilities, identify use cases, and make decisions on orientation, process parameters (temperature, layer height, etc.), and support structures, which can be challenging for some [14].

Furthermore, while significant advances have been made in additive manufacturing research, it is critical to acknowledge the existence of unexplored domains, particularly in the application of additive manufacturing in creative industries such as fashion, architecture, sculpture, and cultural heritage preservation [15]. Concurrently, large-scale automated additive construction has issues related to robotic solutions and environmental sustainability. These issues can be solved by improving processes, materials, and large-scale printed structures, which will promote the development of more efficient and sustainable construction technologies [16].

When compared to traditional manufacturing processes, direct metal laser sintering (DMLS) has been shown to offer the highest environmental risk to human health, whereas computer numerical control (CNC) machining has an impact on ecosystem quality. In contrast, AM is more suited for complicated geometries, though at a 91% greater cost than CNC machining. These comparisons highlight the substantial differences in production processes, emphasizing the significance of considering environmental and economic aspects when determining the ideal manufacturing process [17].

Human factors, including ergonomics, psychology, anthropometry, sociology, medicine, safety, and accessibility, are crucial for the use, design, and performance of systems, products, or environments [3,18]. Moreno-Cabezali and Fernandez-Crehuet identified risks that can negatively impact research and development projects focused on automation (AM) in terms of scope, deadline, cost, and quality [19].

Xiong et al. examined the relationship between humans and machines in the context of AM and classified collaborative activities into active–supportive and active–active activities [20]. Reiman et al. reviewed the challenges faced in manufacturing processes in the context of I4.0 in five areas: human aspects, technology, work tasks, work environment, and organizational levels. They concluded that humans will not be eliminated from manufacturing processes due to technological development, but their roles may change and call into question current practices and processes of human factors and ergonomics (HF/E) [21].

With the rise in popularity of 3D printing, there is a need to create easier methods, tools, and frameworks to help users choose the best AM process for their needs, considering technical and economic factors while also considering sustainability and human factors [20,22,23]. Studies have published on the decision making involved in the selection of AM processes, but many focus only on one or two aspects of the decision, such as material, process, and machine combinations [23].

Song and Telenko's research focused on how the decisions and actions of humans affect the environmental impact associated with the use phase of FDM 3D printers. They recognized that novice users often make inadequate design and operation decisions, leading to increased manufacturing failures and increased material and energy consumption [24].

Therefore, the use of an HF-centered approach could be very beneficial for analyzing, understanding, and designing human work in I4.0. The use of this approach to design work systems in I4.0 could improve the overall performance of complex socio-technical systems and workers' well-being.

Zaman et al. introduced a generic decision-making methodology using multiple criteria decision-making (MCDM) tools to consider feasible combinations of additive manufacturing (AM) materials, processes, and machines [25]. Furthermore, Agrawal discussed sustainable materials for FDM, SLA, and selective laser sintering (SLS) technologies [26]. Kadkhoda-Ahmadi et al. proposed a design for an additive manufacturing (DfAM) approach focusing on manufacturing capability, manufacturing process, and resource selection [27].

Accordingly, there is a need for a more comprehensive approach to selecting AM processes [23]. General functionality indices and weights must be considered for each application area, including cost, material strength, energy consumption, environmental impact, and recyclability [25]. Yao et al. solved a multidisciplinary design optimization (MDO) problem for customized AM products, aiming to maximize functionality utility, meet customer performance requirements, and minimize total cost [28].

Mahadik and Masel developed an additive manufacturing cost estimation tool (AM-CET) using a decomposition approach to calculate the total cost of AM by summing machine, material, labor, and post-processing costs [29]. Energy consumption is a relevant factor in the adoption of AM, with Telenko and Seepersad indicating that AM adoption could provide energy savings, especially for small production volumes [30]. The waiting period and preheating time in additive manufacturing can vary between tasks, leading to variability in energy consumption. Errors in printing can increase material consumption and energy consumption and can occur due to insufficient preheating time, the improper geometry of parts, or printer malfunctions [31].

The analytical hierarchy process (AHP) has been utilized in several studies to solve challenges associated with the selection of AM equipment [32–35]. However, while AHP has significance in this context, it may not provide a complete assessment of the AM equipment selection process. This limitation can be effectively addressed by integrating axiomatic design (AD) principles, which can stimulate technical creativity and be seamlessly integrated with the AHP [36]. A hierarchical decision-making model using AD, customer requirements, quality, time, and costs makes it possible to not only give weight to qualities but also systematically select the most suited equipment based on the complete assessment of many design possibilities, thus optimizing evaluation indicators [37].

The main objective of this research is to provide a method for identifying the optimal AM process by adopting a hierarchical, axiomatic evaluation based on sustainability and human factors. This evaluation was applied to undergraduate and graduate engineering students with varying degrees of expertise (novice and advanced users). By enhancing the knowledge base and methodology for selecting and evaluating AM processes, this research can help improve the efficiency, effectiveness, and overall impact of additive manufacturing across a wide range of industries and applications.

1.1. Theoretical Background

1.1.1. Axiomatic Design

Axiomatic design (AD) theory provides a scientific basis for the design and improvement of activities, offering a solid understanding of theories based on logical and rational thinking processes and tools. It is applied in several different fields, such as software, manufacturing, healthcare, and e-commerce [38,39].

AD is a methodology that starts with customer needs (CNs), taking into account the relationship between adjacent domains as shown in Figure 1. For each pair of adjacent domains, the domain on the left represents what we want to achieve, while the domain on the right represents how the designer proposes to achieve it. Functional requirements (FR_s) refer to what customers want to achieve with the system or the main objective of the design. Design parameters (DPs) describe how to meet the functional requirements. Finally, process variables (PVs) encompass the development of the system process based on the established DPs [39,40].

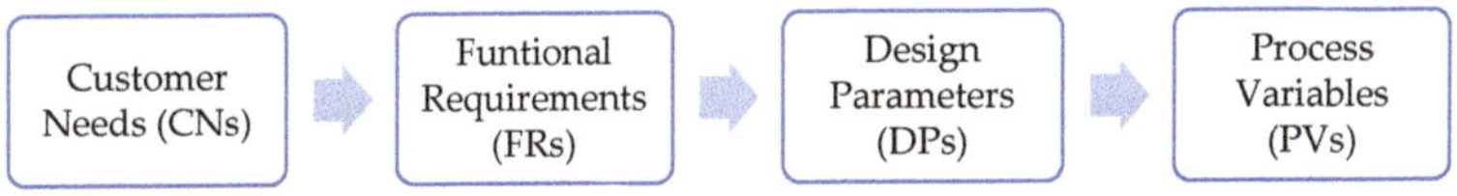

Figure 1. The domains of AD (reproduced with permission from Li, J.; Wu, X.; Zhang, X.; Song, Z.; Li, W., Advanced Engineering Informatics; published by Elsevier, 2022. [40]).

AD establishes that the following axioms must be met to achieve good design: (1) FR_s must be independent (independence axiom) and (2) the information content of the design must be reduced (information axiom) [41].

When the independence axiom is met, a diagonal decoupled design matrix is formed, which demonstrates that each FR corresponds only to one DP [39]:

$$\begin{pmatrix} FR_1 \\ FR_2 \end{pmatrix} = \begin{pmatrix} X & 0 \\ 0 & X \end{pmatrix} \begin{pmatrix} DP_1 \\ DP_2 \end{pmatrix} \quad (1)$$

Typically, the principles of AD are deployed based on probability. The information content is the probability of satisfying the FR_s. For a given FR_i, the information content I_i is defined as follows:

$$I_i = log_2(1/p_i) \quad (2)$$

where p_i expresses the probability of achieving the given FR_i and "log" is a base 2 logarithm (with a unit of bits). If there are n FR_s, the total information content is the sum of all these probabilities [42].

The probability of satisfying the FR_s can be represented by the design range and the system range, as presented in Figure 2. The design range can be considered as the tolerance that the designer is trying to achieve, while the system range is a measure of how well the system is able to comply. Acceptable solutions for design alternatives are determined by the overlapping area between the design range and the system range, called the common range [38].

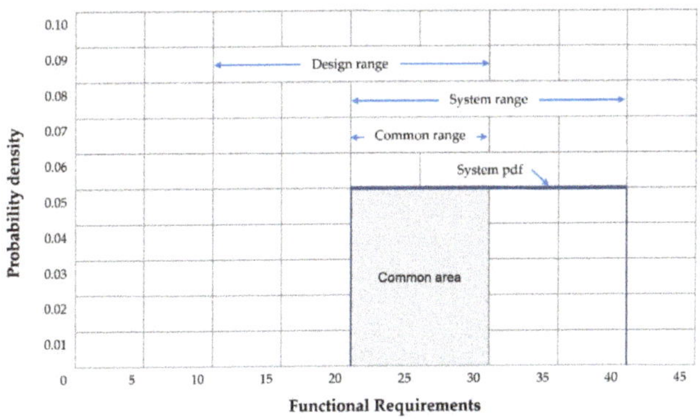

Figure 2. Design range, system range, common range, and probability density function of a FR (Reproduced with permission from Kulak, O.; Kahraman, C., Information Sciences; published by Elsevier, 2005. [43]).

In the simplest case, p_i can be formulated by considering that the system probability is uniformly distributed within its range:

$$p_i = {}^{common\ range}/_{system\ range} \tag{3}$$

A higher p_i denotes that the design presents a high degree of compliance with what the designer wants to achieve; thus, the design is more appropriate in terms of satisfying FR_i. Given this, I_i can be rewritten as follows [44]:

$$I_i = log_2\left({}^{system\ range}/_{common\ range}\right) \tag{4}$$

1.1.2. Analytical Hierarchy Process

The analytical hierarchy process (AHP) is a method used to structure and analyze complex decision problems. It consists of breaking down a problem into a hierarchical structure with objectives, criteria, and sub-criteria and then evaluating the relative importance of each element [45,46]. Currently, AHP is widely used in fields such as operations research, management, engineering, and economics, among others [47].

The process begins by creating a hierarchical structure that represents the aim, criteria based on the goal, and alternatives based on the criteria. Then, comparative judgments are made by constructing a pairwise comparison between the n criteria; thus, a matrix of order n is formed based on these comparisons. It is important to note that the number of comparisons for n factors is given by [45–48]:

$$Number\ of\ comparison = {}^{n(n-1)}/_{2} \tag{5}$$

A pairwise comparison is performed with the help of the fundamental scale of Saaty's relative importance on an absolute scale of 1 to 9. Here, a numerical score of 1 indicates that the two activities contribute equally to the goal, while a numerical rating of 9 represents the extreme order of assertion of one activity over the other [49]. By applying the reciprocal values from the upper diagonal, the lower triangular matrix of the pairwise comparison matrix A is filled, as observed below:

$$A = \begin{bmatrix} a_{11} & \cdots & a_{1n} \\ \vdots & \ddots & \vdots \\ {}^{1}/_{a_{1n}} & \cdots & a_{nn} \end{bmatrix} \tag{6}$$

To assign weights to the parameters, the values in each column of the pairwise comparison matrix A are summed, the matrix is normalized by dividing each element by the total of its column, and then the sum of the elements in each row of the normalized matrix is calculated, obtaining vector w [48]:

$$[w_1, \ldots w_n] \tag{7}$$

In a real-world problem, it is not always possible to find perfect judgments; therefore, a consistent pairwise comparison matrix A is needed after performing pairwise comparisons [36]. The consistency ratio (CR) is an important aspect of AHP, as its value primarily determines optimal decision making in pairwise comparisons [50]. The CR can be verified using the following equation:

$$CR = {}^{CI}/_{RI} \tag{8}$$

where CI is the consistency index and RI is the randomness index. Table 1 presents the RI for different values of n. The CI is calculated as follows:

$$CI = {}^{\lambda_{max} - n}/_{n-1} \tag{9}$$

Table 1. Randomness index (RI) [51].

N	1	2	3	4	5	6	7	8	9	10
RI	0	0	0.58	0.90	1.12	1.24	1.32	1.41	1.45	1.49

λ_{max} is the principal Eigenvalue, which is obtained from the sum of the products between each element of vector w and the sum of the columns of the pairwise comparison matrix A [51]. Judgment entries are not reliable and not accepted if the pairwise comparison matrix A has a high CR value. However, a CR value of less than or equal to 10% is acceptable [52].

2. Materials and Methods

2.1. Materials

The study required the equipment under evaluation, the materials for printing the case study, and formats for collecting data from the participants. In terms of FDM equipment, we used a high-performance LulzBot Mini 1 desktop 3D printer, (Aleph Objects, Loveland, CO, USA) designed to work immediately without a complicated assembly process. It had an open filament system, a fully metal extrusion nozzle that heated up to 300 °C (572 °F), and a bed that heated up to 120 °C (248 °F), which means it could heat up to the temperatures required for industrial-grade materials [53]. To evaluate the SLA technology, a FormLabs Form 2 3D printer (Formlabs Inc., Somerville, MA, USA) was used, which was considered during its launch in the market as a standard in professional and affordable 3D printing, featuring good dimensional accuracy [54]. Additionally, two pieces of auxiliary post-processing equipment (washing and curing) for stereolithographic prints were included: Form Wash and Form Cure (Formlabs Inc., Somerville, MA, USA) [55]. Figure 3 shows the equipment described above.

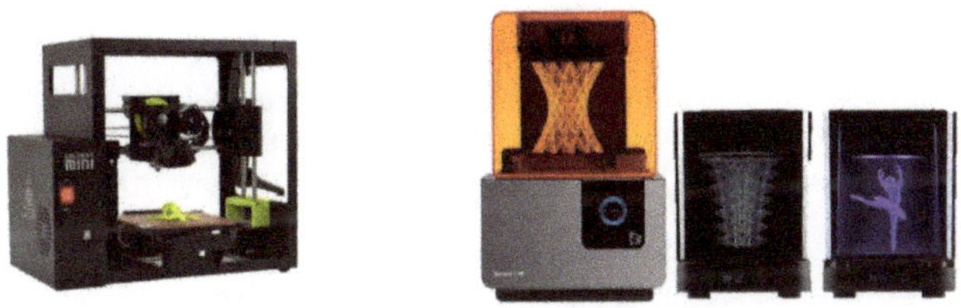

(**a**) (**b**)

Figure 3. Equipment for each AM process: (**a**) Lulzbot Mini 1; (**b**) Form 2 with post-processing equipment.

Furthermore, we used a Microsoft Surface Pro 4 2-in-1 portable computer (Microsoft Corporation, Redmond, Washington, DC, USA) with the Cura LulzBot 3.6.20 Edition (Aleph Objects, Loveland, CO, USA) and PreForm 3.3.0 software (Formlabs, Somerville, MA, USA) to prepare the files for printing and control the operation of the 3D printer [56,57].

Materials such as polylactic acid (PLA) and acrylonitrile butadiene styrene (ABS) were used for prototyping parts, considering the suppliers authorized by LulzBot. The Draft V1 resin was used with the Form 2 3D printer, which handled print resolutions (300 microns) like the Lulzbot Mini (50–500 microns) [53,58]. Information related to the printers and materials used and the process features are presented in Table 2. The study was limited to polymeric materials to focus on specific applications where polymers are commonly used and to provide a starting point for developing their methodology to expand research to include other types of materials in the future.

Table 2. Equipment, materials, and process features.

Feature	Lulzbot Mini	Form2
3D printing technology	Fused deposition modeling	Stereolithography
Material	PLA and ABS	Draft V1 resin
Material consumption	4–7 g	7.13–7.30 mL
Fill density	50–100%	100%
Layer height	250 microns (standard print)	300 microns

A Steren HER-432 120Vca wattmeter and a Steren HER-403 (Steren Electronics, Mexico City, Mexico) sound meter were used to measure the electrical consumption and noise intensity of the equipment used. Likewise, a timer was used to measure the duration of the cycle required to produce a prototyped part.

It is important to note that the participants used a single STL file for a stove knob to simulate a prototyping application for replacement parts. This was established by the research team based on information gathered from different maker communities such as MyMiniFactory, Cults, or Thingiverse. In addition, a series of videos (own elaboration) was provided that broke down and described each subgoal, starting from the quick start guide and user manual of each piece of 3D printing equipment. Finally, an error and doubt log were used to obtain information related to mistakes made and doubts about each of the tasks performed by the participants and estimate their error rate.

2.2. Participants

Given that the study was carried out during the coronavirus (COVID-19) pandemic and considering the pivotal role of 3D printers in manufacturing essential medical models [59], a convenience sampling technique was used, which is a non-probabilistic and non-random sampling technique. A total of 31 subjects were invited to participate according to the following inclusion criteria: gender not specified (45% women and 55% men); over 18 years old (average of 22 years); and students of bachelor's and graduate degrees who also have different levels of expertise (novice and expert user) when considering the frequency of use of a 3D printer. The participants were specifically 1 architecture student, 1 industrial design student, 7 mechatronics engineering students, 19 industrial engineering students, and 3 graduate students (Master of Science). Students from other universities were excluded for health and safety reasons. All the subjects utilized both 3D technologies as part of the study.

To determine the frequency of use of 3D printers, participants were interviewed and categorized as novice users or expert users. The categories were defined as follows: novice users had either never used a 3D printer or used it rarely (a few times in their lives), while expert users used it regularly (at least once per semester). Advanced students, who were classified as expert users, generally created school projects that involved designing and producing prototypes, frequently depending on 3D printing.

In addition, a team of four experts in AM with 3 to 12 years of experience dedicated to the provision of prototyping and design services for parts were also included. These experts comprise materials engineers and project managers with specific roles spanning from product design to prototype manufacturing. Additionally, the team has been involved in relevant projects, such as prototyping automotive components and manufacturing aerospace parts. Their selection was based on their technical skills and expertise to address specific challenges based on their track record of successful projects in the industry, as well as recommendations from colleagues.

2.3. Limitations

The research described has some limitations that permit consideration. Firstly, the study solely focused on polymeric materials, thereby limiting the scope of applications explored and potentially failing to fully capture the diversity of materials used in AM techniques. Additionally, conducting the study during the COVID-19 pandemic may have affected resource availability and the dynamics of the research setting.

Furthermore, the study concentrated on two specific AM processes—FDM and SLA—each represented by a single 3D printer. This narrow focus may restrict the generalization of the findings to other AM techniques and equipment configurations. Additionally, the study's use of a convenience sampling method to select participants could have introduced biases and reduced the sample's representativeness.

The study's cross-sectional design provides a snapshot of the current situation, offering valuable insights into prevailing beliefs and behaviors. However, it is important to note that due to logistical and temporal constraints, following up with the same individuals after the pandemic period was not feasible. Moreover, the lack of access to these participants post study limits the potential for gathering additional information. Therefore, considering the continuation of this study as a longitudinal investigation in the future would be beneficial to examine how attitudes and behaviors evolve over time.

2.4. Methodology

This study adopted an integrated methodology, utilizing both AD and AHP methodologies as they were considered ideal and consistent to create a method for evaluating and selecting 3D printing processes. AD provided the initial structure to clearly define and understand the design requirements and alternatives, while AHP added a layer of quantitative analysis to prioritize these alternatives based on their ability to meet the defined criteria. This integrated approach ensures that the selected 3D printing process not only aligns with the design goals but is also optimized based on a thorough evaluation of all relevant factors. The methodology consists of five stages to evaluate equipment or processes using AD and AHP. Figure 4 illustrates the entire methodological flowchart.

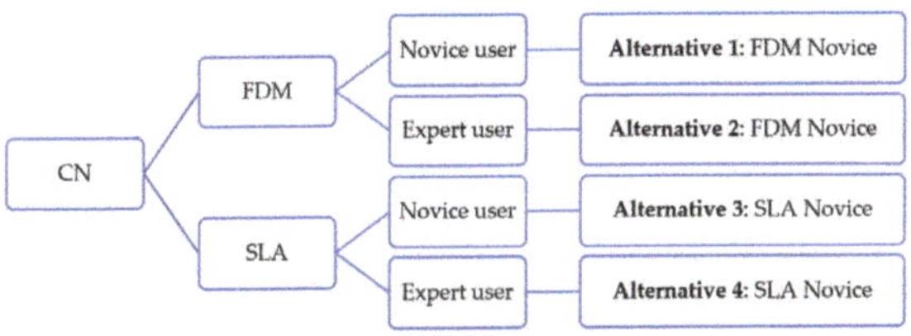

Figure 4. Methodological flowchart of the evaluation process.

In Stage 1, the variables related to the equipment or process are defined. The alternatives are established according to the customer needs (CNs) and the ranges of the functional requirements (FRs), design parameters (DPs), and process variables (PVs) of the AD. With this information, the objectives and criteria are defined, and the experts are identified.

In Stage 2, the process is executed. The equipment is used or the process is developed to obtain and record the data.

In Stage 3, the criteria are evaluated. The information content for each FR is calculated, and AHP is used to make the paired comparisons and generate the normalized matrix to calculate vector w according to the relative weight of the FRs.

In Stage 4, the results are analyzed. The consistency index (*CI*) is calculated, and if it is $\leq$10%, the information content for each FR is adjusted considering the relative weight, and the information content graph is made. If the *CI* is not $\leq$10%, another expert is invited to perform the paired comparisons, and the steps of the methodology continue.

Stage 5 determines the best alternative, which is the option with the lowest information content, and prioritizes the FR.

2.4.1. Stage 1: Define the Variables

Most of the concepts discussed in Stage 1, which focused on establishing the study variables, are defined in Sections 2.1 and 2.2 of the paper. Section 2.1 provides a detailed description of the equipment and materials used, while Section 2.2 outlines the characteristics of the study participants.

A kitchen stove knob was chosen as the piece to produce. This decision was based on its simplicity of printing using the existing setup, minimal material usage, and consideration of equipment limitations such as printing volume. Despite the possibility of selecting a necessary medical item, the stove knob was chosen for its suitability and potential to present a relevant design and functionality challenge. Figure 5 presents the parts that were manufactured using the two examined AM processes.

(a) (b)

Figure 5. Parts manufactured in (**a**) FDM 3D printer and (**b**) SLA 3D printer.

Table 3 lists the tasks necessary to print a part using each AM process studied. As can be seen, the SLA 3D printing process had two additional steps that were associated with the two pieces of post-processing equipment. A series of videos (self-made) was provided to the participants that broke down and described the subgoal of each piece of 3D printing and post-processing equipment.

To define the alternatives (CR) for AD, data collected from the novice and expert participants specified in Section 2.2 were considered, along with the two 3D printing procedures (FDM and SLA) detailed in Section 2.1 Materials. The diagram used to determine the four options that need to be assessed is illustrated in Figure 6.

Table 3. Steps for printing using each AM process.

Lulzbot Mini	Form2
1. Connect and turn on the printer	1. Insert tank of resin, cleaner, and printing platform
2. Turn on the computer and set up printing software	2. Insert resin cartridge
3. Connect and gain control of the printer	3. Connect and turn on the printer
4. Remove the filament	4. Level the 3D printer
5. Insert filament	5. Turn on the computer and set up printing software
6. Print	6. Print
7. Remove part from the printer	7. Remove part from the printer
	8. Wash part
	9. Cure part

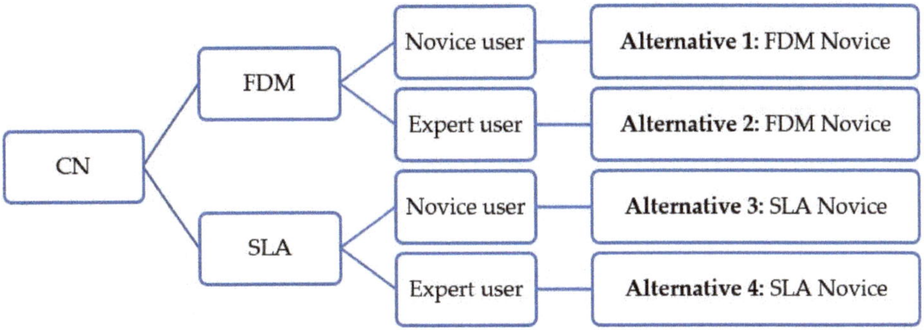

Figure 6. Diagram of alternatives for AD.

Regarding the evaluation criteria, the following were considered to propose a more sustainable selection approach: material cost (MXN), cycle time (h), energy consumption (kwh), error rate (number of events), and equipment noise (dB). These factors can be seen as key sustainability considerations in additive manufacturing. In the current evaluation, sustainability was considered from multiple perspectives, including its economic, environmental, and social impacts. Material cost can have a direct impact on the total cost of production and therefore the economic feasibility of an AM process [60]. Cycle time refers to the time it takes to complete a production cycle and can affect efficiency and production capacity [61]. Energy consumption refers to the use of energy to operate AM equipment and can have a significant effect on the environmental impact of the process [30]. Equipment noise can have an impact on the health and well-being of workers [62], while errors have an impact on both efficiency and safety [63].

In addition, an error and doubt log were given to each participant in which they were asked to record the errors and doubts they had when performing each task; the log was used to estimate their error rate. Because the study included two 3D printing processes (FDM and SLA) and two types of users (novices and experts), a total of four alternatives were evaluated under the five criteria.

The system range data presented in Table 4 were obtained by selecting the minimum and maximum values achieved by the participants for each study alternative. It can be observed that C1 and C5 are the same since they are directly related to the AM equipment. In other words, the values of C1 and C5 are consistent for each equipment, independent of the user type.

The design range can be considered as the tolerance that the designer is trying to achieve. Taking this into account, it is important to consider the information from the provider of each piece of 3D printing equipment. However, the studied AM equipment has been discontinued and there is no technical information related to the proposed criteria. In addition, it is important to note that both pieces of 3D printing equipment were launched on the market in 2015; therefore, it is expected that they have lost efficiency over time. Therefore, the design range data shown in Table 5 are defined as follows: (1) the cost of material commonly represents between 45 and 55% of the total cost of the manufactured part—the design range was estimated considering this statement [64]—and (2) the rest of the criteria were valued based on the mean ± 1 standard deviation of the measurements obtained by the participants in the study.

Table 4. System range data.

Alternatives	C1: Material Cost (MXN/Part)	C2: Cycle Time (h)	C3: Energy Consumption (kwh)	C4: Error Rate (Events)	C5: Equipment Noise (db)
FDM Novice	2.08–4.34	0.732–1.829	0.048–0.268	0–3	49.3–78.0
FDM Advanced	2.08–4.34	0.796–1.309	0.075–0.123	0–2	49.3–78.0
SLA Novice	20.32–20.80	1.589–2.264	0.061–0.139	0–1	57.0–72.3
SLA Advanced	20.32–20.80	1.448–1.709	0.069–0.095	0–2	57.0–72.3

Table 5. Defined design range data.

C1: Material Cost (MXN/Part)	C2: Cycle Time (h)	C3: Energy Consumption (kwh)	C4: Error Rate (Event)	C5: Equipment Noise (db)
2.05–20.50	0.889–1.753	0.066–0.164	0–1	56.8–68.6

2.4.2. Stage 2: Execute the Process

The goal of the study was explained to the participants by the research team meeting with them. All participants were informed of the characteristics of the study in which they were going to collaborate, emphasizing that it was completely voluntary participation and, by request, anonymous.

Then, participants were asked to print the selected part using each of the AM processes in the university facilities. During each session, information about the participant (name, gender, age, and type of user) and the configurations of the 3D printing equipment and pre-processing equipment (predetermined by the providers) were recorded, as well as the criteria to be evaluated.

2.4.3. Stage 3: Evaluate the Criteria

In this stage, the AD methodology was employed, particularly the information axiom to determine the information content for the four alternatives. The main advantage of the information axiom is that it allows decision makers to define the desired characteristics for the considered criteria. In addition, each alternative has its system characteristics [65,66]. On the other hand, the uniform distribution was used for its simplicity, ease of understanding, and compliance with many of the fundamental axioms of probability theory, such as the axiom of independence [67].

The results are presented in Table 6. The option with minimum information content (*) is the most suitable in terms of design requirements. The use of FDM 3D printing equipment by advanced users is marked with an asterisk. Based on the information content of the remaining alternatives, it could be stated that an advanced user requires less information content than a novice, regardless of the AM process used.

Regarding the information content of the error rate (Ier) for an SLA Novice (zero) and an SLA Advanced (one), it indicates that no errors were recorded for novice users, while errors were identified for expert users. This discrepancy could be attributed to the lack of experience and knowledge among novice users, leading to a lower likelihood of error detection. On the other hand, expert users, having a more solid understanding and experience in operating the equipment, might be more aware of potential errors and report them, as reflected in the error rate value of one. Further discussion on this topic will be provided in detail in Section 3.4.

Table 6. Primary information content results.

Alternatives	Imc: Material Cost	Ict: Cycle Time	Iec: Energy Con.	Ier: Error Rate	Ien: Equipment Noise	∑I
FDM Novice	0	0.343	1.169	1.584	1.281	4.379
FDM Advanced	0	0.285	0	1	1.281	2.567 *
SLA Novice	1.432	2.039	0.107	0	0.397	3.976
SLA Advanced	1.432	0	0	1	0.397	2.830

* Is the most suitable in terms of design requirements.

In addition, the opinion of experts was included to derive the relative importance of each criterion using the AHP methodology and thereby estimate a higher level of objective information content for each alternative. For the latter, an instrument was applied using Google Forms that included the following: (1) expert identification data (name, academic background, experience, position, and place of work); (2) a description of the attributes to be evaluated; and (3) pairwise comparisons. Table 7 presents the paired comparisons of the criteria applicable to the study determined by a group of four AM experts.

Table 7. Paired comparison of criteria by experts.

Comparison	Expert A	Expert B	Expert C	Expert D	Average
C1–C2	0.333	3	5	5	3.333
C1–C3	0.142	9	9	0	4.619
C1–C4	9	5	7	7	7
C1–C5	9	9	7	9	8.5
C2–C3	0.142	7	9	0	4.119
C2–C4	0.111	9	7	1	4.277
C2–C5	9	9	7	9	8.5
C3–C4	0.142	5	7	1	3.285
C3–C5	0.111	0.111	9	1	2.555
C4–C5	0.111	0.2	9	1	2.577

With the average of each paired comparison, a paired comparison matrix was formed, as shown in Table 8. With the data from Table 8, the normalized matrix and vector w, which represents the relative weights of each criterion, were obtained (see Table 9).

Table 8. Paired comparison matrix.

Criteria	Material Cost	Cycle Time	Energy Consumption	Error Rate	Equipment Noise
Material cost	1	3.333	4.619	7	8.5
Cycle time	0.3	1	4.119	4.277	8.5
Energy con.	0.216	0.242	1	3.285	2.555
Error rate	0.143	0.233	0.304	1	2.577
Equipment noise	0.117	0.117	0.391	0.387	1

Table 9. Normalized matrix and vector w.

Criteria	Material Cost	Cycle Time	Energy Con.	Error Rate	Equipment Noise	w
Material cost	0.563	0.676	0.443	0.436	0.367	49.8%
Cycle time	0.169	0.203	0.395	0.268	0.367	28%
Energy con.	0.122	0.049	0.095	0.206	0.111	11.7%
Error rate	0.08	0.047	0.029	0.063	0.111	6.6%
Equipment noise	0.066	0.024	0.038	0.024	0.043	3.9%

2.4.4. Stage 4: Analyze the Results

The index and consistency ratio of the answers given by the experts in the paired comparisons were estimated. The results are shown in Table 10, which suggests an acceptable consistency [52].

Table 10. Index and consistency ratio.

Principal Eigenvalue	Consistency Index	Consistency Ratio
5.443	0.111	9.882%

The information content for each criterion concerning the different alternatives—taking the priority vector into account—is presented in Table 11. Figure 7 presents these results as a chart to assist in understanding.

Table 11. Final information content results.

Alternatives	Imc: Material Cost	Ict: Cycle Time	Iec: Energy Con.	Ier: Error Rate	Ien: Equipment Noise	∑I
FDM Novice	0	0.096	0.136	0.105	0.05	0.388
FDM Advanced	0	0.08	0	0.066	0.05	0.196 *
SLA Novice	0.713	0.572	0.013	0	0.016	1.313
SLA Advanced	0.713	0	0	0.066	0.016	0.795

* Is the most suitable in terms of design requirements.

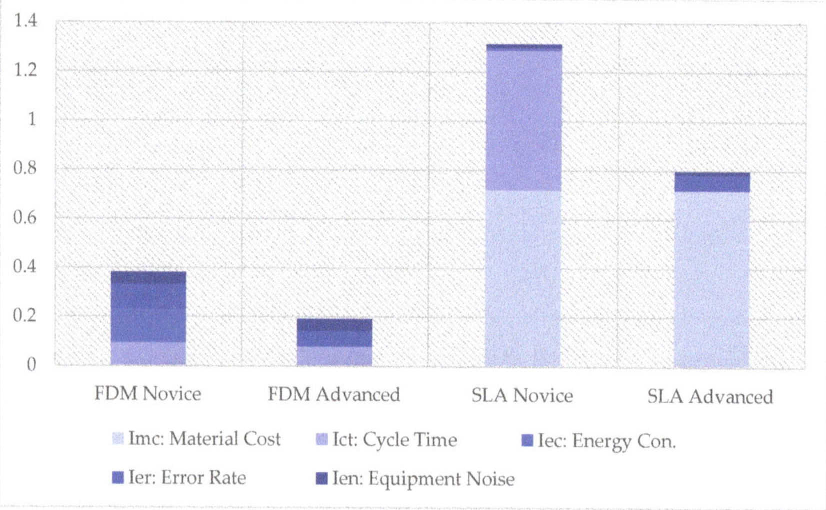

Figure 7. Information content chart. Note: the best result is the alternative with the smallest bar.

2.4.5. Stage 5: Determine the Best Alternative

Considering the relative weight in Table 11, the alternative with the least amount of necessary information is the most suitable in terms of design requirements (*) and is the same as that presented in Table 6.

However, by examining the information content of the other options, we can see that FDM technology requires less information than SLA technology, regardless of the user's skill level. This observation is reinforced by the fact that FDM is one of the most used commercial technologies for 3D-printed parts using thermoplastic polymers [68].

By evaluating each 3D printing technology (FDM and SLA) concerning the user, it can be determined that a novice requires almost double the information for printing a part than that required by an experienced user. This affirmation aligns with the fact that experts tend to perform at faster speeds, exhibit more accurate motor behavior, and operate with greater efficiency than novices [69]. Additionally, expert users can complete a larger proportion of tasks with higher dimensional accuracy and require less effort and revisions compared to proficient users [70].

After comparing the criteria in pairs and reaching vector w shown in Table 9, it was observed that the priorities in the selection of 3D printing equipment were as follows, according to the expert group: (1) material cost (49.8%); (2) cycle time (28%); (3) energy consumption (11.7%); (4) error rate (6.6%); and (5) equipment noise (3.9%). The priorities in selecting 3D printing equipment were primarily associated with economic indicators, as evidenced by the data presented in Table 9, where material cost and cycle time accounted for the majority percentage (77.8%). This shows a clear priority for economic indicators, considering the lack of importance given to HFs (users).

Nevertheless, the relative weight can vary depending on the user's experience, the complexity of the part, and the specific application. Therefore, the results obtained may not be applicable to all scenarios and should be interpreted with caution. Further research could explore the impact of varying indicator weights on the evaluation of different 3D printing technologies to provide a more comprehensive and accurate understanding of the benefits and limitations of each process.

3. Results and Discussions

As previously mentioned, the AD methodology—particularly the information axiom—was used in this study to evaluate the four alternatives (FDM novice, FDM advanced, SLA novice, and SLA advanced). The following sections present a comprehensive analysis of the information content results presented in Table 6.

3.1. Material Cost

Firstly, it is important to note that FDM technology is the most economical option in terms of material cost, regardless of whether the user is a novice or advanced. It is worth noting that the use of filament in FDM 3D printers and its profitability with materials such as ABS or nylon has been demonstrated [71].

However, it is important to keep in mind that 3D printing technologies have different strengths and limitations, and the choice of technology should be based on the specific requirements of the application. For example, there is a wide range of resins available on the market for SLA printing, where the user can choose a resin based on their application and the properties that are of most interest to them. These resins include options for general, engineering, manufacturing, automotive, aerospace, dentistry, jewelry, and medical uses [72]. Consequently, there are very specific applications that require the simple and cost-effective fabrication of devices where the possible use of SLA stands out, even when compared to the FDM 3D printing process [73]. Therefore, it is important to thoroughly evaluate the specific requirements of the application and choose a piece of 3D printing technology that best meets those requirements. By doing so, the resulting 3D-printed parts will not only be cost-effective but also of high quality and reliability for the intended application.

The proposed methodology does not consider some factors such as mechanical properties and geometrical accuracy, which play a significant role in determining the quality and suitability of a 3D-printed part for a specific application. Future research could incorporate these factors to improve our understanding of 3D printing and its potential applications.

3.2. Cycle Time

Although SLA technology requires two post-processing tasks (cleaning and curing), it remains the most advantageous option in terms of cycle time. To achieve these results, both AM processes must be operated by users with a high level of experience and skill. This contrasts with what Finnes reported in a comparison of FDM and SLA printing technologies, where—in general—printing the same part with SLA technology took two-to-four times longer than with an FDM 3D printer [74]. Song and Telenko pointed out that FDM technology must wait for preheating periods that change with each printing task, resulting in variability in energy consumption and cycle time [75].

When comparing the limits of each system in Table 4, it can be observed that the Lulzbot Mini has high variability (wide system ranges) in manufacturing times, which reinforces the notion presented by Song and Telenko [75]. Additionally, a loss of efficiency in the 3D FDM printer time was observed. Although Form 2 has a clear technological disadvantage when compared to the Lulzbot Mini [74], it has narrow system ranges, indicating high precision in its cycle times.

It is important to note that the duration of post-processing tasks, such as cleaning and curing, can vary based on several factors. These factors may include the size and complexity of the part, as well as the specific cleaning and curing methods used. As a result, it is crucial to consider not only the printing and post-processing time but also the part's geometry when evaluating the overall efficiency of different 3D printing technologies. It is, therefore, important to conduct further research on the impact of these factors on the overall efficiency of 3D printing technologies to provide a more accurate evaluation of their suitability for specific applications.

3.3. *Energy Consumption*

When examining the information content results regarding energy consumption, expert users had better results, regardless of which AM process was used. This result is in accordance with Song and Telenko [75].

Regarding the proper functioning and operation of AM equipment, it was observed that novice users obtained better results when operating the SLA 3D printing technology, followed by expert users when using both AM processes. This result was expected since, as Song and Telenko pointed out, experience does not translate into greater expertise or lower error rates [24].

3.4. *Error Rate*

After examining the error and doubt logs, it was found that expert users had issues with (1) leveling the 3D printer (which presented serious unevenness) and (2) operating the SLA washing equipment (poor placement of the parts basket). These problems were not reported initially and were caused by poor operation by a previous user. Upon reviewing the historical records of the tests carried out, it was determined that this user was a novice. Therefore, although novice users were able to successfully print a part, due to their lack of knowledge and skills (even after consulting the descriptive videos of each sub-objective), they were prone to ignoring the negative effects of the incorrect operation of the equipment, which could lead them to making important errors during their subsequent use. It is important to consider that the study did not investigate the impact of training or experience on the performance of novice and expert users, which could be a valuable area for future research to further understand the impact of the learning curve on the use of different AM processes.

3.5. *Equipment Noise*

Finally, it was observed that the SLA 3D printer produced a lower level of noise than its FDM counterpart, regardless of the type of user (novice or expert) that was operating it. This result was expected as all the movements of a 3D FDM printer are controlled by the various stepper motors according to the control program [58]. On the other hand, the SLA AM process uses a stepper motor to control the linear movement of the platform on the Z axis [76]. It is normal to expect stepper motors to generate noise during their operation due to the vibration and rotary movement of their components. The FDM 3D printer uses several stepper motors; hence, it is expected to generate a higher level of noise compared to other equipment that uses fewer motors. However, it is important to note that the level of noise generated by a motor can also depend on other factors, such as the size and type of motor, as well as the conditions under which it is used.

4. Conclusions and Future Work

This paper describes the development of a new method to evaluate and select AM technologies using AD as a scientific basis and the AHP methodology. The proposed framework allows the user to consider equipment conditions, such as efficiency loss over time and the lack of technical information, as well as high variability in cycle time. A novel approach is to also consider human factors like error rate and equipment noise, which are often overlooked in evaluations, without neglecting traditional sustainability criteria such as material cost, cycle time, and energy consumption. Overall, these five aspects provide a good approach to evaluating FDM and SLA 3D printing equipment from a sustainable perspective.

While the presented evaluation criteria may change depending on the specific application, the five factors considered in the proposed methodology are intended to be applicable across a range of potential use cases. The decision to focus on these criteria was based on their potential impact on the overall sustainability and HFs associated with the AM process, as well as their relative ease of measurement and comparison between different technologies and materials. Additionally, by limiting the number of factors considered, the proposed methodology may be more straightforward and accessible to potential users who may not have extensive expertise in the field.

On the other hand, although this study aimed to maintain a sustainable approach in which the proposed indicators were more or less balanced in terms of priority, the experts showed a clear bias towards economic indicators, reflecting the reality that production systems currently face.

For future research, there are multiple alternatives to pursue that could improve the proposed methodology and expand its applicability. Firstly, broadening the evaluation criteria to include additional variables such as environmental impact, recyclability, ergonomics, and occupational health concerns will give a more comprehensive assessment of the sustainability and human factors involved with AM processes. Second, exploring the impact of mechanical characteristics and geometrical precision on the overall applicability and quality of 3D-printed components for various applications may provide useful insights for enhancing component performance. Third, extending the methodology to include other AM methods outside FDM and SLA will require adapting the evaluation criteria and methodology to the specific characteristics of each AM process. Furthermore, a closer examination of user preferences and biases in equipment selection, as well as longitudinal studies to assess the long-term performance and sustainability of selected AM technologies, would contribute to a better understanding of decision-making processes and equipment performance over time. Addressing these areas of future study would improve and broaden the suggested technique, providing more thorough information for evaluating and selecting AM technologies.

Finally, it was demonstrated that the best 3D printing option for a prototyping application for replacement parts is the one that uses FDM technology. Additionally, the user has an important impact on equipment selection, mainly due to their level of skill and experience in using it; in other words, a novice or expert user will define the requirements that they seek to obtain from the manufactured part for its application, such as material properties, print quality, environmental impact, recyclability, effects on individuals, etc., to maximize its benefits in terms of costs, sustainability, and HFs.

Author Contributions: Conceptualization, I.M.-M. and A.A.M.-M.; methodology, I.M.-M., M.I.M.-R. and A.A.M.-M.; software, I.M.-M.; validation, M.I.M.-R., A.A.M.-M. and G.J.-G.; formal analysis, I.M.-M.; investigation, I.M.-M. and A.A.M.-M.; resources, I.M.-M.; writing—original draft preparation, I.M.-M.; writing—review and editing, M.I.M.-R., A.A.M.-M., G.J.-G. and O.Y.V.-B. All authors have read and agreed to the published version of the manuscript.

Funding: This research was funded by the Autonomous University of Baja California at the 22nd internal call for the support of research projects (105/1/C/39/22) and the internal call for research projects 2024-2025 (105/6/C/56/24).

Institutional Review Board Statement: The study was conducted in accordance with the Declaration of Helsinki and approved by the Ethics Committee of the Faculty of Engineering at the Autonomous University of Baja California (028/2023-1, 31 January 2023) for studies involving humans.

Informed Consent Statement: Informed consent was obtained from all subjects involved in the study.

Data Availability Statement: The data presented in this study are available on request from the corresponding author. The data are not publicly available due to the fact that the data contain sensitive information about individuals and organizations, and their release may pose a risk to their privacy or confidentiality.

Acknowledgments: The authors would like to express their gratitude to the participants of the study for their time and effort. They would also like to thank the Autonomous University of Baja California for providing access to the facilities and equipment necessary to carry out this study. Finally, the authors would like to extend their gratitude to the Autonomous University of Ciudad Juárez for providing support with a research stay for the corresponding author (I.M.-M.).

Conflicts of Interest: The authors declare no conflicts of interest.

References

1. Kumar, A. Methods and Materials for Smart Manufacturing: Additive Manufacturing, Internet of Things, Flexible Sensors and Soft Robotics. *Manuf. Lett.* **2018**, *15*, 122–125. [CrossRef]
2. Zhong, R.Y.; Xu, X.; Klotz, E.; Newman, S.T. Intelligent Manufacturing in the Context of Industry 4.0: A Review. *Engineering* **2017**, *3*, 616–630. [CrossRef]
3. Neumann, W.; Winkelhaus, S.; Grosse, E.H.; Glock, C.H. Industry 4.0 and the human factor—A systems framework and analysis methodology for successful development. *Int. J. Prod. Econ.* **2021**, *233*, 107992. [CrossRef]
4. Prabhakar, M.M.; Saravanan, A.K.; Lenin, A.H.; Leno, I.J.; Mayandi, K.; Ramalingam, S. A short review on 3D printing methods, process parameters and materials. *Mater. Today Proc.* **2021**, *45*, 6108–6114. [CrossRef]
5. Zakeri, S.; Vippola, M.; Levänen, E. A comprehensive review of the photopolymerization of ceramic resins used in stereolithography. *Addit. Manuf.* **2020**, *35*, 101177. [CrossRef]
6. Sun, J.; Ye, D.; Zou, J.; Chen, X.; Wang, Y.; Yuan, J.; Liang, H.; Qu, H.; Binner, J.; Bai, J. A review on additive manufacturing of ceramic matrix composites. *J. Mater. Sci. Technol.* **2023**, *138*, 1–16. [CrossRef]
7. Rasiya, G.; Shukla, A.; Saran, K. Additive Manufacturing—A Review. *Mater. Today Proc.* **2021**, *47*, 6896–6901. [CrossRef]
8. Matias, E.; Rao, B. 3D printing: On its historical evolution and the implications for business. In *2015 Portland International Conference on Management of Engineering and Technology (PICMET)*; IEEE: New York City, NY, USA, 2015; pp. 551–558. [CrossRef]
9. Jain, K.; Jain, K. Use of 3D printing for home applications: A new generation concept. *Mater. Today Proc.* **2021**, *43*, 605–607. [CrossRef]
10. Laguna, O.H.; Lietor, F.; Godino, F.J.I.; Corpas-Iglesias, F.A. A review on additive manufacturing and materials for catalytic applications: Milestones, key concepts, advances and perspectives. *Mater. Des.* **2021**, *208*, 109927. [CrossRef]
11. Ashima; Haleem, A.; Bahl, S.; Javaid, M.; Mahla, S.K.; Singh, S. Automation and manufacturing of smart materials in additive manufacturing technologies using Internet of Things towards the adoption of industry 4.0. *Mater. Today Proc.* **2021**, *45*, 5081–5088. [CrossRef]
12. Biedermann, M.; Meboldt, M. Computational design synthesis of additive manufactured multi-flow nozzles. *Addit. Manuf.* **2020**, *35*, 101231. [CrossRef]
13. Patwardhan, A. How 3D Printing Will Change the Future of Borrowing Lending and Spending? In *Handbook of Blockchain, Digital Finance, and Inclusion*; Elsevier: Amsterdam, The Netherlands, 2018; Volume 2, pp. 493–520. [CrossRef]
14. Makanda, I.L.D.; Jiang, P. A Web-based Generative Process Planning System for FDM-based Additive Manufacturing. *IFAC-PapersOnLine* **2020**, *53*, 83–88. [CrossRef]
15. Abisuga, O.A.; De Beer, D. A Systematic Literature Review and Prospects for the Additive Manufacturing in the Creative Industries. In Proceedings of the 13th AHFE International Conference on Additive Manufacturing, Modeling Systems and 3D Prototyping, New York, NY, USA, 24–28 July 2022.

16. Khosravani, M.R.; Haghighi, A. Large-Scale Automated Additive Construction: Overview, Robotic Solutions, Sustainability, and Future Prospect. *Sustainability* **2022**, *14*, 9782. [CrossRef]
17. Mecheter, A.; Tarlochan, F.; Kucukvar, M. A Review of Conventional versus Additive Manufacturing for Metals: Life-Cycle Environmental and Economic Analysis. *Sustainability* **2023**, *15*, 12299. [CrossRef]
18. Sheehan; Joy, A.; Fleming, A.; Vosper, H.; McCarthy, S. Human factors and patient safety in undergraduate healthcare education: A systematic review. *Human Factors Healthc.* **2022**, *2*, 100019. [CrossRef]
19. Moreno-Cabezali, B.M.; Fernandez-Crehuet, J.M. Application of a fuzzy-logic based model for risk assessment in additive manufacturing R&D projects. *Comput. Ind. Eng.* **2020**, *145*, 106529. [CrossRef]
20. Xiong, Y.; Tang, Y.; Kim, S.; Rosen, D.W. Human-machine collaborative additive manufacturing. *J. Manuf. Syst.* **2023**, *66*, 82–91. [CrossRef]
21. Reiman, A.; Kaivo-oja, J.; Parviainen, E.; Takala, E.-P.; Lauraeus, T. Human factors and ergonomics in manufacturing in the industry 4.0 context—A scoping review. *Technol. Soc.* **2021**, *65*, 101572. [CrossRef]
22. Kadir, B.A.; Broberg, O.; da Conceição, C.S. Current research and future perspectives on human factors and ergonomics in Industry 4.0. *Comput. Ind. Eng.* **2019**, *137*, 106004. [CrossRef]
23. Bikas, H.; Porevopoulos, N.; Stavropoulos, P. A decision support method for knowledge-based Additive Manufacturing process selection. *Procedia CIRP* **2021**, *104*, 1650–1655. [CrossRef]
24. Song, R.; Telenko, C. Causes of Desktop FDM Fabrication Failures in an Open Studio Environment. *Procedia CIRP* **2019**, *80*, 494–499. [CrossRef]
25. Zaman, U.K.U.; Rivette, M.; Siadat, A.; Mousavi, S.M. Integrated product-process design: Material and manufacturing process selection for additive manufacturing using multi-criteria decision making. *Robot. Comput. Integr. Manuf.* **2018**, *51*, 169–180. [CrossRef]
26. Agrawal, R. Sustainable material selection for additive manufacturing technologies: A critical analysis of rank reversal approach. *J. Clean. Prod.* **2021**, *296*, 126500. [CrossRef]
27. Kadkhoda-Ahmadi, S.; Hassan, A.; Asadollahi-Yazdi, E. Activity Modeling of Preliminary Additive Manufacturing Process Planning. *Procedia CIRP* **2019**, *84*, 874–879. [CrossRef]
28. Yao, X.; Moon, S.K.; Bi, G. Multidisciplinary design optimization to identify additive manufacturing resources in customized product development. *J. Comput. Des. Eng.* **2017**, *4*, 131–142. [CrossRef]
29. Mahadik, A.; Masel, D. Implementation of Additive Manufacturing Cost Estimation Tool (AMCET) Using Break-down Approach. *Procedia Manuf.* **2018**, *17*, 70–77. [CrossRef]
30. Telenko, C.; Seepersad, C.C. A comparison of the energy efficiency of selective laser sintering and injection molding of nylon parts. *Rapid Prototyp. J.* **2012**, *18*, 472–481. [CrossRef]
31. Jennings, A. 3D Printing Troubleshooting: All Problems & Solutions. All3D. Available online: https://all3dp.com/1/common-3d-printing-problems-troubleshooting-3d-printer-issues/ (accessed on 8 January 2023).
32. Ren, D.; Choi, J.-K.; Schneider, K. A multicriteria decision-making method for additive manufacturing process selection. *Rapid Prototyp. J.* **2022**, *28*, 77–91. [CrossRef]
33. Pancha, U.; Dewa, M. Deployment of the analytic hierarchical process for comparative analysis of additive manufacturing technologies in the manufacture of injection moulds. *South Afr. J. Ind. Eng.* **2022**, *32*, 190–203. [CrossRef]
34. Raja, S.; Rajan, A.J.; Kumar, V.P.; Rajeswari, N.; Girija, M.; Modak, S.; Kumar, R.V.; Mammo, W.D.; Gupta, P. Selection of Additive Manufacturing Machine Using Analytical Hierarchy Process. *Sci. Program.* **2022**, *2022*, 1–20. [CrossRef]
35. Muvunzi, R.; Mpofu, K.; Daniyan, I. An Evaluation Model for Selecting Part Candidates for Additive Manufacturing in the Transport Sector. *Metals* **2021**, *11*, 765. [CrossRef]
36. Laurenţiu, S.L.I.E.N.; Oana, D.O.U.; Margareta, C.O.E.; Valeriu, D.U.G.E.U.; Petru, D.U.; Felicia, B.A.C.U.; Irina, B.E.I. Selection of a Solution When Using Axiomatic Design. *MATEC Web Conf.* **2017**, *127*, 01019. [CrossRef]
37. Weng, F.-T.; Jenq, S.-M. Application integrating axiomatic design and agile manufacturing unit in product evaluation. *Int. J. Adv. Manuf. Technol.* **2012**, *63*, 181–189. [CrossRef]
38. Gölcük, İ. An interval type-2 fuzzy axiomatic design method: A case study for evaluating blockchain deployment projects in supply chain. *Inf. Sci.* **2022**, *602*, 159–183. [CrossRef]
39. Palleti, V.R.; Joseph, J.V.; Silva, A. A contribution of axiomatic design principles to the analysis and impact of attacks on critical infrastructures. *Int. J. Crit. Infrastruct. Prot.* **2018**, *23*, 21–32. [CrossRef]
40. Li, J.; Wu, X.; Zhang, X.; Song, Z.; Li, W. Design of distributed hybrid electric tractor based on axiomatic design and Extenics. *Adv. Eng. Inform.* **2022**, *54*, 101765. [CrossRef]
41. Cavique, L.; Cavique, M.; Mendes, A.; Cavique, M. Improving information system design: Using UML and axiomatic design. *Comput. Ind.* **2022**, *135*, 103569. [CrossRef]
42. Liu, A.-H.; Wan, S.-P.; Dong, J.-Y. An axiomatic design-based mathematical programming method for heterogeneous multi-criteria group decision making with linguistic fuzzy truth degrees. *Inf. Sci.* **2021**, *571*, 649–675. [CrossRef]
43. Kulak, O.; Kahraman, C. Fuzzy multi-attribute selection among transportation companies using axiomatic design and analytic hierarchy process. *Inf. Sci.* **2005**, *170*, 191–210. [CrossRef]
44. Aydoğan, S.; Günay, E.E.; Akay, D.; Kremer, G.E.O. Concept design evaluation by using Z-axiomatic design. *Comput. Ind.* **2020**, *122*, 103278. [CrossRef]

45. Kumar, A.; Pant, S. Analytical hierarchy process for sustainable agriculture: An overview. *MethodsX* **2023**, *10*, 101954. [CrossRef] [PubMed]
46. Singh, R.; Majumder, C.B.; Vidyarthi, A.K. Assessing the impacts of industrial wastewater on the inland surface water quality: An application of analytic hierarchy process (AHP) model-based water quality index and GIS techniques. *Phys. Chem. Earth Parts A/B/C* **2023**, *129*, 103314. [CrossRef]
47. Sharma, A.; Hussain, V.M.S.; Kumar, A.; Pandit, M. Prioritization of forging die design criteria based on failure analysis using fuzzy analytic hierarchy process (FAHP). *Mater. Today Proc.* **2023**, *80*, 925–932. [CrossRef]
48. Ozegin, K.O.; Ilugbo, S.O.; Ogunseye, T.T. Groundwater exploration in a landscape with heterogeneous geology: An application of geospatial and analytical hierarchical process (AHP) techniques in the Edo north region, in Nigeria. *Groundw. Sustain. Dev.* **2023**, *20*, 100871. [CrossRef]
49. Yadav; Yadav, S.; Singh, D.; Kapoor, R.M.; Giri, B.S. An analytical hierarchy process based decision support system for the selection of biogas up-gradation technologies. *Chemosphere* **2022**, *302*, 134741. [CrossRef] [PubMed]
50. Chaisar, M.; Garg, S.K. Selection of Sewage Treatment Technology using Analytic Hierarchy Process. *Mater. Today Proc.* **2022**, *56*, 3433–3440. [CrossRef]
51. Panchal, S.; Shrivastava, A.K. Landslide hazard assessment using analytic hierarchy process (AHP): A case study of National Highway 5 in India. *Ain Shams Eng. J.* **2022**, *13*, 101626. [CrossRef]
52. Shadmaan, S.; Islam, A.I. Estimation of earthquake vulnerability by using analytical hierarchy process. *Nat. Hazards Res.* **2021**, *1*, 153–160. [CrossRef]
53. Lultzbot Mini. Fargo Additive Manufacturing Equipment 3D, LLC. Available online: https://lulzbot.com/store/lulzbot-mini (accessed on 8 January 2023).
54. Form 2. FormLabs. Available online: https://formlabs.com/3d-printers/form-2/ (accessed on 8 January 2023).
55. Post-Processing. FormLabs. Available online: https://formlabs.com/post-processing/ (accessed on 8 January 2023).
56. Cura Lulzbot Edition. Fargo Additive Manufacturing Equipment 3D, LLC. Available online: https://lulzbot.com/support/cura (accessed on 8 January 2023).
57. Formlabs Software. FormLabs. Available online: https://formlabs.com/software/ (accessed on 8 January 2023).
58. FormLabs Draft V1 Resin Cartridge. Source Graphics. Available online: https://sourcegraphics.com/supplies/3d/formlabs/draft-resin/#:~:text=Supports%20print%20resolutions:%20300%20microns (accessed on 8 January 2023).
59. Alsalaet, J.; Munahi, B.S.; Al-Sabur, R.; Al-Saad, M.; Ali, A.K.; Shari, B.A.; Fadhil, H.A.; Laftah, R.M.; Ismael, M. Laminar flowmeter for mechanical ventilator: Manufacturing challenge of COVID-19 pandemic. *Flow. Meas. Instrum.* **2021**, *82*, 102058. [CrossRef]
60. Daher, R.; Ardu, S.; di Bella, E.; Krejci, I.; Duc, O. Efficiency of 3D-printed composite resin restorations compared with subtractive materials: Evaluation of fatigue behavior, cost, and time of production. *J. Prosthet. Dent.* **2024**, *131*, 943–950. [CrossRef]
61. Patel, R.; Desai, C.; Kushwah, S.; Mangrola, M.H. A review article on FDM process parameters in 3D printing for composite materials. *Mater. Today Proc.* **2022**, *60*, 2162–2166. [CrossRef]
62. Sénéchal, O. Maintenance decision support for sustainable performance: Problems and research directions at the crossroads of health management and eco-design. *IFAC-PapersOnLine* **2016**, *49*, 85–90. [CrossRef]
63. Lundstrom, T.; Pugliese, G.; Bartley, J.; Cox, J.; Guither, C. Organizational and environmental factors that affect worker health and safety and patient outcomes. *Am. J. Infect. Control* **2002**, *30*, 93–106. [CrossRef] [PubMed]
64. Raffish, N. How Much Does That Product Really Cost? *Manag. Account.* **1991**, *72*, 36. Available online: https://www.proquest.com/openview/9a7bae00b354b16c6d4f41328f58232f/1?cbl=48426&pq-origsite=gscholar&parentSessionId=X67x5NR0sMTpKuG+ViU/cixBwkOjjNMcxaM61BSvqGk=P (accessed on 8 January 2023).
65. Kahraman, C.; Cebi, S.; Onar, S.C.; Oztaysi, B. A novel trapezoidal intuitionistic fuzzy information axiom approach: An application to multicriteria landfill site selection. *Eng. Appl. Artif. Intell.* **2018**, *67*, 157–172. [CrossRef]
66. Kulak, O.; Cebi, S.; Kahraman, C. Applications of axiomatic design principles: A literature review. *Expert Syst. Appl.* **2010**, *37*, 6705–6717. [CrossRef]
67. Benavides, M. Axiomatic design. In *Advanced Engineering Design*; Elsevier: Amsterdam, The Netherlands, 2012; pp. 77–258. [CrossRef]
68. Arrigo, R.; Frache, A. FDM Printability of PLA Based-Materials: The Key Role of the Rheological Behavior. *Polymers* **2022**, *14*, 1754. [CrossRef]
69. Li, L.; Smith, D.M. Neural Efficiency in Athletes: A Systematic Review. *Front. Behav. Neurosci.* **2021**, *15*, e698555. [CrossRef]
70. Deng, Y.; Chen, J.; Olechowski, A. What Sets Proficient and Expert Users Apart? Results of a Computer-Aided Design Experiment. *J. Mech. Des.* **2024**, *146*, 1–39. [CrossRef]
71. Kariz, M.; Sernek, M.; Obućina, M.; Kuzman, M.K. Effect of wood content in FDM filament on properties of 3D printed parts. *Mater. Today Commun.* **2018**, *14*, 135–140. [CrossRef]
72. Rogosic, R.; Poloni, M.; Marroquin-Garcia, R.; Dimech, D.; Jansen, J.P.; Cleij, T.J.; Eersels, K.; van Grinsven, B.; Diliën, H. Cost-effective, scalable and smartphone-controlled 3D-Printed syringe pump-From lab bench to point of care biosensing applications. *Phys. Med.* **2022**, *14*, 100051. [CrossRef]
73. Xu, X.; Goyanes, A.; Trenfield, S.J.; Diaz-Gomez, L.; Alvarez-Lorenzo, C.; Gaisford, S.; Basit, A.W. Stereolithography (SLA) 3D printing of a bladder device for intravesical drug delivery. *Mater. Sci. Eng. C* **2021**, *120*, 111773. [CrossRef] [PubMed]

74. Finnes, T. High Definition 3D Printing—Comparing SLA and FDM Printing Technologies. *J. Undergrad. Res.* **2015**, *13*, 10–26. Available online: https://openprairie.sdstate.edu/jur/vol13/iss1/3 (accessed on 8 January 2023).
75. Song, R.; Telenko, C. Material and energy loss due to human and machine error in commercial FDM printers. *J. Clean. Prod.* **2017**, *148*, 895–904. [CrossRef]
76. Suryatal, B.K.; Sarawade, S.S.; Deshmukh, S. Fabrication of medium scale 3D components using stereolithography system for rapid prototyping. *J. King Saud. Univ.-Eng. Sci.* **2023**, *35*, 40–52. [CrossRef]

processes

MDPI

Article

Fabrication and Characterisation of Sustainable 3D-Printed Parts Using Post-Consumer PLA Plastic and Virgin PLA Blends

Mohammad Raquibul Hasan [1,*], Ian J. Davies [2], Alokesh Paramanik [2], Michele John [1] and Wahidul K. Biswas [1,*]

[1] Sustainable Engineering Group, School of Civil and Mechanical Engineering, Curtin University, Bentley, WA 6102, Australia; m.rosano@curtin.edu.au

[2] School of Civil and Mechanical Engineering, Curtin University, Bentley, WA 6102, Australia; alokesh.pramanik@curtin.edu.au (A.P.)

* Correspondence: m.hasan21@postgrad.curtin.edu.au (M.R.H.); w.biswas@curtin.edu.au (W.K.B.)

Abstract: Sustainable manufacturing practices are becoming increasingly necessary due to the growing concerns regarding climate change and resource scarcity. Consequently, material recycling technologies have gradually become preferred over conventional processes. This study aimed to recycle waste polylactic acid (PLA) from household-disposed cups and lids to create 3D-printed parts and promote sustainable manufacturing practices. To achieve this, the current study utilised virgin and post-consumer PLA (PC-PLA) (sourced from household waste) blends. The PC-PLA wastes were shredded and sorted by size with the aid of a washing step, resulting in a filament with a 1.70 ± 0.5 mm diameter without fragmentation or dissolution. A 50:50 wt.% blend of virgin PLA (vPLA) and PC-PLA was selected as the standard recycling percentage based on previous research and resource conservation goals. The study investigated the impact of three 3D printing parameters (layer height (LH), infill density (I), and nozzle temperature (NT)) on the quality of 3D-printed parts using a three-level L9 Taguchi orthogonal array. The findings revealed that blending PC-PLA with vPLA led to significant improvements in tensile, flexural, and impact strengths by 18.40%, 8%, and 9.15%, respectively, compared to those of recycled PLA (rPLA). This conclusion was supported by the investigation of the fracture surface area, which revealed fractographic features associated with printing parameters, such as plastic deformation and interfilament debonding. An ANOVA analysis revealed a positive influence of a greater layer height and high nozzle temperature on mechanical properties. Subsequently, the optimal printing parameters (LH: 0.3 mm, I: 100%, and NT: 215 °C) were determined using the S/N ratio, and a confirmation test using the optimum printing parameters exhibited a strong correlation with the statistically predicted outcomes. Finally, the study used optimum printing parameters to fabricate 100% PC-PLA 3D-printed parts, demonstrating their potential for low-strength applications. The findings suggest that employing vPLA and PC-PLA blended filaments for fabricating 3D-printed components presents an effective means of promoting plastic recycling within a closed-loop recycling system and achieving a circular economy.

Keywords: additive manufacturing; post-consumer PLA waste; blended filament; mechanical properties; circular economy

1. Introduction

Plastics, whilst commonly utilised in human life, have been identified as long-lasting pollutants in the environment [1]. Plastics are utilised in several aspects of life, such as clothes, electronics, toys, healthcare supplies, and food packaging [2]. In 2022, approximately 400 million tons of plastic were produced worldwide, using a significant share of crude oil [3]. The majority of current polymer and plastic materials are derived from non-renewable petrochemical resources, which is not a sustainable solution in the long term. The continued increase in plastic production will lead to the depletion of fossil fuels and negatively impact the economy and human livelihoods. In addition, plastic production and disposal release greenhouse gases (GHGs) and contribute to global warming [4]. The decomposition of plastics in landfills takes 10–450 years and significantly contributes to water and environmental pollution [5,6]. Furthermore, the production and disposal of plastics consume a considerable amount of energy [6].

The efficient management of plastic waste in daily life is crucial for environmental conservation and the ecological well-being of future generations. Reusing and recycling plastic products can significantly contribute to achieving a circular economy (CE), offering benefits such as reducing dependence on virgin non-renewable resources, minimising waste, enhancing resource efficiency, and fostering a sustainable recycling ecosystem [7–10]. Integrating plastic recycling into a closed-loop system could enhance waste management. Closed-loop recycling repurposes post-consumer (PC) plastic materials, such as industrial parts or household waste, to create new products with similar properties, thereby ensuring a circular flow of resources [8]. Under these circumstances, PC plastic waste has drawn considerable attention from researchers, governments, and industries. Therefore, recycling PC waste could play a significant role in improving natural resource utilisation efficiency and mitigating the adverse environmental impacts of plastics [11–13].

Various methods for recycling plastic waste, including bioconversion [14], dissolution recycling, injection moulding [15], mechanical recycling [16], additive manufacturing (3D printing) [17], and other similar approaches, have been developed to improve our ability to manage plastic waste and contribute to a sustainable future [1]. Among these technologies, 3D printing has gained widespread acceptance as the optimal solution for producing functional components in various industries, including unmanned aerial vehicles (UAVs), aerospace, civil engineering, agriculture, bioprinting, biomedical engineering, biomedicine, membrane technology, multi-material components, metal matrix composites, and food production [18,19].

The use of recycled polymers in additive manufacturing (AM) has been identified as a potential solution to extend the lifespan of plastic items [7]. AM enables the fabrication of recycled plastic filaments and produces no scraps or solid wastes, thereby establishing a closed-loop supply chain that conforms to the principles of the CE or reduces both sink and source. This approach not only reduces plastic waste but also adds value by reusing PC plastic to create new useable products while simultaneously lowering levels of energy consumption and GHG emissions [20]. Many AM techniques have been developed, and the most popular method is fused deposition modelling (FDM) additive technology. This technology involves the continuous feeding of a thermoplastic filament into the printer, heating it until it melts, extruding the material through a heated nozzle, and depositing it onto a printing platform [21].

Acrylonitrile butadiene styrene (ABS), polylactic acid (PLA), polyethylene terephthalate (PET), and polycarbonate (PC) are commonly used polymer materials for FDM [22,23]. Among these, PLA has gained significant popularity due to its biodegradable nature [24,25]. Specifically, PLA production requires 65% less energy and generates 68% less GHG than traditional plastics. Moreover, PLA is highly recyclable and contains no toxins, making it a potential candidate for replacing traditional plastics [25,26]. The global production capacity of PLA is expected to reach approximately 677,000 metric tons by 2023, reflecting its growing popularity as a sustainable alternative to conventional oil-based plastics [27]. However, the biodegradability of PLA presents challenges for appropriate disposal, as it takes several years to degrade in the environment (between 3 and 5 years) [13,28]. Additionally, PLA made from corn and sugarcane can cause a "food vs. filament" issue as it not only creates pressure on cultivable land but could affect food scarcity unless this biobased filament is recycled. Furthermore, the weaknesses of pure PLA polymers, including their poor mechanical strength, brittleness, low ductility, poor thermal stability, and narrow processing window, have limited their application [29].

Some of these weaknesses can be overcome by selecting optimal printing parameters to obtain the required mechanical properties [30,31]. This is important as the growing market volume of PLA has prompted ongoing efforts to reprocess and recycle PC-PLA. Beltran et al. [32] observed a minor influence of mechanical recycling on the thermal, structural, and mechanical properties of PLA, indicating the reusing capability of recycled PLA in low value-added applications such as packaging. However, the repeated extrusion cycling of virgin PLA could steadily increase the melt flow rate (MFR) of the reprocessed samples, resulting in a decrease in the tensile and impact strengths [33]. Similarly, Pillin et al. [34] found a notable reduction in tensile strength upon repeated injection moulding of PLA. In addition, Zhao et al. [35] found that the recyclability of PLA in 3D printing decreased significantly after only two reprocessing cycles, causing a significant deterioration in the polymer properties and viscosity.

Therefore, blending recycled plastics with virgin plastics could be a straightforward approach to recover the properties of recycled polymers, reduce plastic waste, and save fossil fuels by lessening the conversion of petroleum products to produce virgin plastics [1,9,23,31]. Although the strength of the filament decreased with an increase in the recycled material [36,37], Bergaliyeva et al. [38] found that the strength of the blend specimens increased with an increase in the recycled content. However, further studies are required to analyse the impact of the process parameters on blend-printed specimens, as these studies did not address this research gap.

The fabrication of blend filaments is a fundamental step in FDM 3D printing and requires precise adjustment of extrusion parameters such as screw speed and temperature to ensure a high level of accuracy in the final printed products [38]. These extruder parameters are experimentally fine-tuned and may vary accordingly. In addition, design of experiments (DoE), particularly the Taguchi method, is commonly employed to optimise printing parameters and enhance physical properties [39–42]. Through this method, researchers achieve the desired performance by identifying key influencing factors and conducting experiments and analyses accordingly, thereby demonstrating the reliability and maturity of this approach [29].

The existing literature on 3D printing has primarily focused on limited production orientations, which hinder our understanding of anisotropic behaviour and material degradation. Therefore, a detailed study on the implementation of blended PLA filaments in 3D printing is necessary. This study aimed to bridge the existing gap by examining the implementation of blended PLA filaments, specifically incorporating PC-PLA with vPLA. The resulting physical and mechanical characteristics of 3D-printed components at various printing temperatures were carefully determined. It is noteworthy that relatively few investigations have been carried out regarding the effects of printing temperature on the 3D printability of vPLA/PC-PLA blends. This research aimed to contribute valuable insights into the effects of printing temperature on the quality of 3D-printed parts using a 50:50 wt.% blend of 100% virgin commercial PLA pellets and PC-PLA waste. This research provides in-depth information on material fabrication and characterisation, which will offer additional insights for the manufacturing of blended PLA filaments.

2. Materials and Methods

A systematic approach was employed to manufacture eco-friendly 3D-printed components from PC-PLA, as depicted in Figure 1. This method was implemented to achieve sustainable 3D printing practices by minimising plastic waste and promoting the use of environmentally friendly materials. The detailed methodology employed for the production of the PC-PLA filament included material selection for filament extrusion, printing parameter selection, mechanical property testing, and optimisation of process parameters.

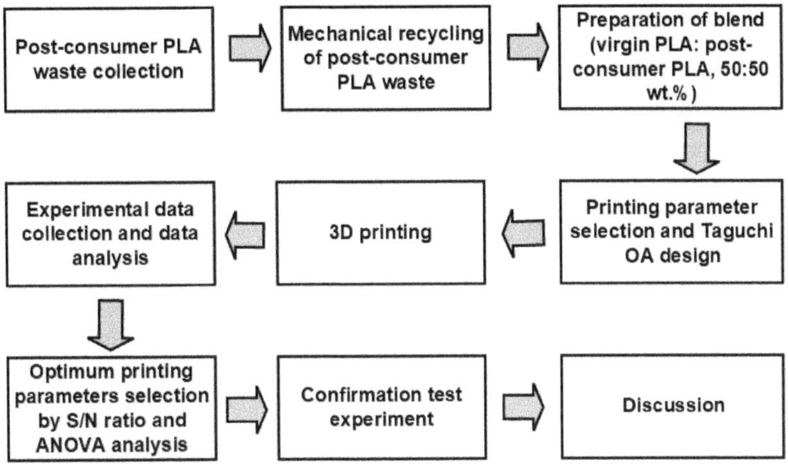

Figure 1. Flow chart of the research methodology.

2.1. Preparation of Raw Materials

Material selection is a critical aspect of fused deposition modelling. This process involves the identification of plastic waste that can be utilised as 3D printing filaments, which requires the selection of specific types of plastic as a source material. Plastic waste can be classified into various categories based on their properties, such as thermoplastics and thermosetting plastics [43]. The aim of this study is to fabricate low-cost blended filaments; therefore, PC-PLA and virgin PLA plastics were selected. Commercially available vPLA pellets from AURARUM (Victoria, Australia) with a 1.24 g/cm^3 density and a printing temperature of 200–230 °C were used as the primary PLA. Waste cups and lids made from PLA plastic were chosen because they are one of the most readily available types of waste. PC-PLA waste was collected from household waste, then washed to remove any visible contaminants, and then dried by exposure to sunlight. Once the waste had been dried, it was cut into pieces that were not larger than 4 mm (between 2 mm and 4 mm) to facilitate its conversion into the intended form and dimensions for the subsequent stage of the procedure. It is essential to cut the waste into smaller pieces for the overall success of the process, which allows the efficient utilisation of recycled plastic in the production of 3D printing filaments [43]. Furthermore, they were then dried in two steps: first in an open atmosphere under hot sunlight (with an atmospheric temperature of 30 °C and humidity of 50%) for one day and then in a hot air oven at 80 °C for 12 h [44,45]. The samples were then stored in a sealed container containing silica gel to avoid moisture adsorption. The processing of the PLA waste used in this study is shown in Figure 2.

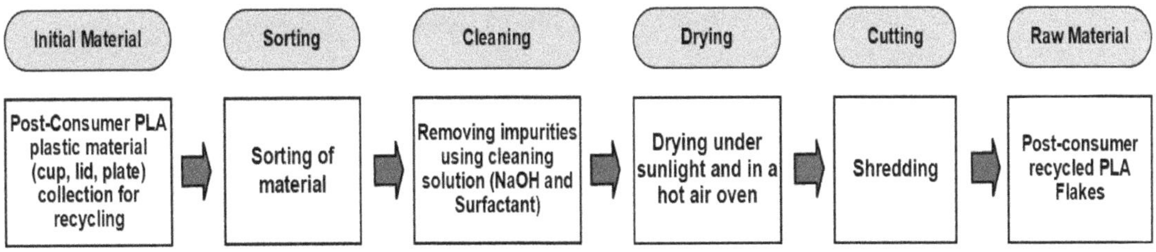

Figure 2. Processing of household PC-PLA waste.

2.2. Filament Extrusion

The next step in the process involved the conversion of PC-PLA and vPLA into 3D printing filaments. Following the preparation of the plastic raw materials, the vPLA pellets were manually combined with PC-PLA with an appropriate weight ratio (50:50 wt.%) to create a blend (Figure 3). The mixture was subsequently dried in a hot air oven at 80 °C for 4 h to eliminate moisture content and prepare it for extrusion [44,45]. The materials were then fed into an extruder (Figure 4). The present research employed a commercial filament manufacturing system, namely FilaBot EX6 (Barre, VT, USA), to produce blend filaments through three main procedures, namely filament extrusion, filament/winding path, and spooling. A pellet feeding hopper, four heating units for separate heating zones, and a motor-driven screw passage characterise the extruder. The material is uniformly heated and sufficiently fluidised for extrusion by means of the four heating units. The extruded material is cooled using fans operating at a steady speed throughout the winding path. A driving wheel and spool holder are the two main components of the spooler. The detailed steps for plastic recycling followed in this study are as follows: the initial step involved setting the temperature of each heating zone to a certain target value and feeding vPLA:PC-PLA into the hopper. Secondly, the hot material (filament) was extruded along the winding path and cooled using fans. Thirdly, the extruded material was wound onto a spooler holder at a constant rate to maintain uniform filament quality (Figure 3) [46,47].

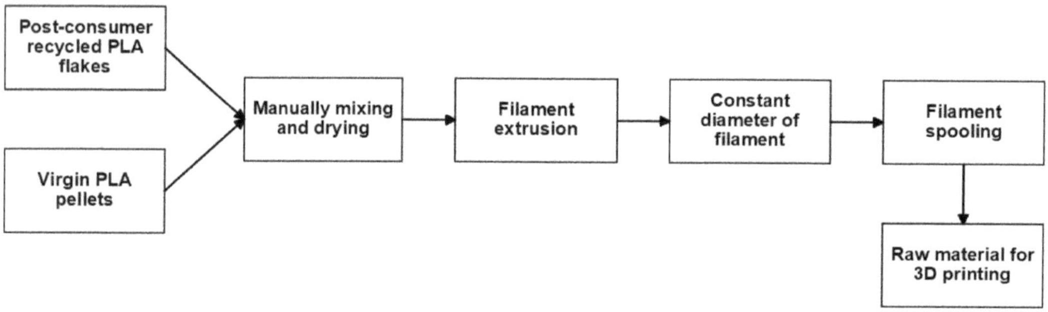

Figure 3. Flowchart of blended filament production.

In this research, the optimal parameters for extrusion were identified through a series of trial-and-error experiments and past studies. The Filabot EX6 configuration for producing the blend filament was assembled in a straight line. The temperature settings for each heating unit were set at 40 °C, 175 °C, 180 °C, and 175 °C, respectively, to ensure that the material inside the extruder barrel was properly heated and melted and then forced through a 1.75 mm diameter nozzle. The extruder motor and spooling speed were then manually adjusted to achieve an extruded filament diameter in the range of 1.65 to 1.75 mm. Once the filament was extruded from the nozzle, it was directed towards the air path, for which the speed of the fan was set at 65 rpm. The filament was carefully wound onto the spool by adjusting the winder speed in accordance with the extrusion speed to ensure that it was securely and uniformly wound, preventing the formation of knots or tangles that could compromise its quality. After the winding process was completed and the spool was filled, the filament was stored and prepared for use in 3D printing. The resulting filament exhibited a consistent diameter and was suitable for use in 3D printers. The process flow for producing the vPLA:PC-PLA (50:50) blend filament is shown in Figure 4.

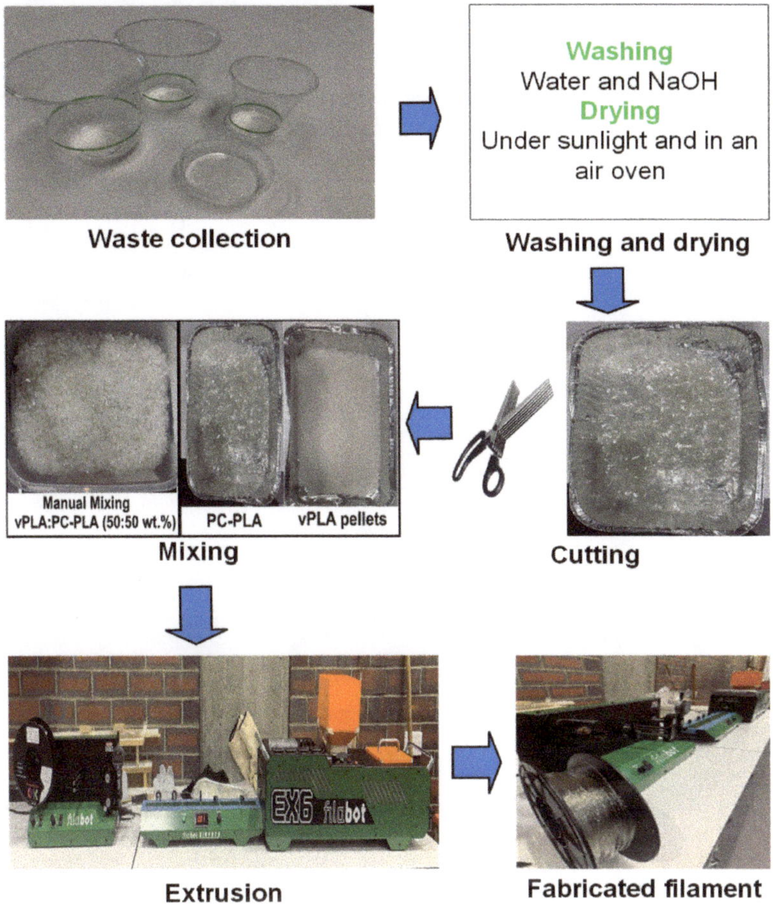

Figure 4. Illustration of the process followed for blended filament production.

2.3. FDM Printing of Specimens

The specimens were fabricated through the FDM method to determine the mechanical properties. To fabricate high-quality samples, this study followed the printing parameter selection criteria proposed by Hasan et al. [48]. In this context, three different layer heights (LH) (0.1 mm, 0.2 mm, and 0.3 mm), infill percentages (I) (60%, 80%, and 100%), and nozzle temperatures (NT) (195 °C, 205 °C, and 215 °C) were determined. The Taguchi L9 orthogonal array (OA) experimental design was used to print the test samples (Table 1). Table 2 lists the parameters for 3D printing that were employed in this investigation.

Following the Taguchi table, the samples were manufactured using a MakerBot Replicator 2X 3D printer purchased from MakerBot Industries, Brooklyn, NY, USA. The 3D printer was equipped with a 0.4 mm nozzle diameter and used MakerBot desktop beta software (version 3.10.1). The test samples were created using SolidWorks 2022 (SP5.0) software, which was subsequently transformed into the standard tessellation language (STL) file format, which is compatible with the 3D printer slicing process. The MakerBot desktop slicing program was used to convert the solid models of the samples into geometric code (G-code) files, which were then used to print the 3D specimens/samples.

Table 1. Experimental layout based on Taguchi L9 OA design [48].

Experimental Run	Coded Matrix			Un-Coded Matrix		
	A	B	C	Layer Height (mm)	Infill (%)	Nozzle Temperature (°C)
1	1	1	1	0.1	60	195
2	1	2	2	0.1	80	205
3	1	3	3	0.1	100	215
4	2	1	2	0.2	60	205
5	2	2	3	0.2	80	215
6	2	3	1	0.2	100	195
7	3	1	3	0.3	60	215
8	3	2	1	0.3	80	195
9	3	3	2	0.3	100	205

Table 2. Fixed printing parameters and their corresponding values.

Parameter	Specific Parameter	Values (Constant)
Layer	Shell thickness	2 mm
Extruder	Extrusion width Retraction speed	0.4 mm 40 mm/s
Infill	Infill pattern	Rectilinear
Speed	Printing speed Infill speed	50 mm/s 50 mm/s
Support	Support structure Platform addition	None None
Temperature	Heat bed temperature	45 °C

2.4. Mechanical Testing and Material Characterisation

The mechanical properties of the 3D-printed specimens were determined using tensile, three-point bending (flexural), impact, and hardness tests. The tests were conducted in accordance with ASTM standards [49–53]. Consequently, a total of 45 specimens were printed for each mechanical test (tensile, three-point bending, and impact), with five samples produced for each experimental run in each of the three tests. In addition, 27 samples were prepared for the hardness tests. Visual inspection of the printed specimens revealed no signs of shrinkage or warping, indicating excellent quality of the samples.

Tensile tests were performed using a Shimadzu Autograph AGS-X universal testing system with a load cell capacity of 50 kN. As depicted in Figure 5a, a dog-bone-shaped Type IV 3D-printed specimen with dimensions of 115 mm × 19 mm × 3.2 mm was tested following the ASTM-D638-22 [49] standard. The specimens were subjected to a crosshead loading speed of 5 mm/min until fracture occurred. Additionally, a 3-point bending setup was utilised on a Shimadzu Autograph AGS-X universal testing machine with a 10 kN load cell capacity to test 3D-printed strip-like samples (127 mm × 12.7 mm × 3.2 mm) in accordance with the ASTM-D790-17 [50] standard (Figure 5b). The samples were examined at a testing speed of 3 mm/min and a support span of 51.2 mm. To calculate the tensile and flexural properties, the steps described by the author Hasan et al. [48] were followed. The tests for each experimental run were repeated a minimum of five times, and the test results are presented as the mean value with standard deviation (SD).

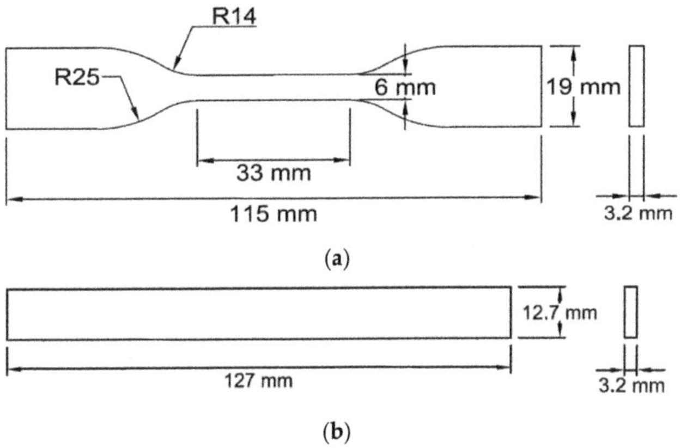

Figure 5. Schematic of (**a**) tensile [49] and (**b**) flexural test specimens [50].

Charpy impact testing was conducted utilising a Zwick 5102 (model D-7900) impact tester fitted with a 6.5 J hammer. The specimens for the examination were fabricated following the ASTM-D6110-18 [51] standard, with measurements of 127 mm × 12.7 mm × 6.2 mm and notched at a 45° angle, as illustrated in Figure 6a. The hardness of the samples was assessed using a Sauter HBD 100-0 Shore D hardness testing device according to the ASTM-D2240-15 [52] standard using dimensions of 40 mm × 40 mm × 8 mm (Figure 6b). The average data, including the standard deviations, were computed using the testing outcomes of a minimum of five samples for each material batch to assess the reproducibility of the tests.

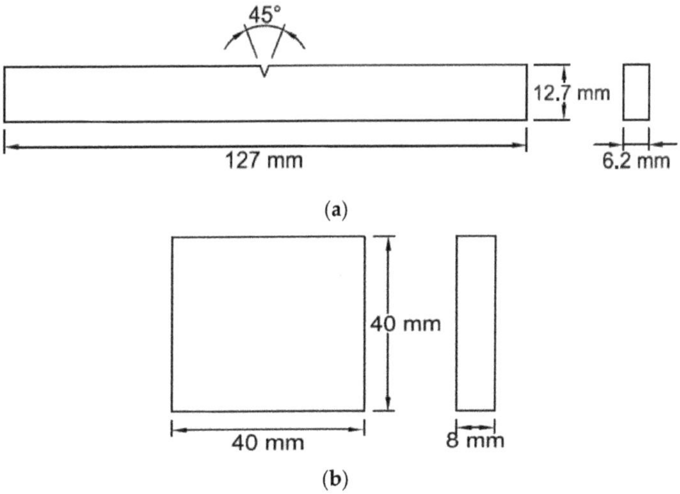

Figure 6. Schematic of (**a**) impact [51] and (**b**) hardness test specimens [52].

The impact of the printing parameters on surface roughness was investigated using a Mitutoyo SJ-210 surface roughness tester. The roughness of the largest surface of the tensile specimens was measured following the ASTM-D7127-17 standard [53]. The arithmetic mean roughness (*Ra*) and root mean square roughness (*Rq*) were assessed quantitatively at the micrometre (μm) scale. The measurements were obtained from five selected locations across the samples (Figure 7), and the mean value with the standard deviation of five measurements was tabulated.

Figure 7. Surface roughness measurement on the selected five points of the top surface of tensile specimens [48].

Fractographic analysis of the fractured cross-sectional structures of the investigated materials was performed using a Pro-MicroScan microscope fitted with an eyepiece camera from Oplenic Corporation, Hangzhou, China. The samples were prepared for analysis according to the procedure described by Hasan et al. [48]. Following the examination, all the test samples were stored in plastic zip bags.

2.5. Optimisation of Process Parameters

The Taguchi method is a well-established approach for analysing quality characteristics in the context of signal-to-noise ratio (S/N). By examining the S/N ratio, researchers can identify the effective process parameters that influence the results and determine the optimal sets for achieving desired outcomes [54]. The S/N ratio can be assessed in terms of three categories: smaller-the-better, larger-the-better, and nominal-is-best [55,56]. In this study, the mechanical properties were evaluated using the "larger-the-better" whereas "smaller-the-better" was used to evaluate surface roughness.

To further refine the analysis, a statistical analysis of variance (ANOVA) was conducted to measure the relative importance of quality attributes. A factor is considered statistically significant if the probability value (p-value) is lower than the predetermined significance level (α) [57]. F-value and percentage contribution were used to represent the relative significance of the factors [54,58,59]. The F-value or percentage contribution is used to quantify the relative importance of these factors. A larger F-value indicates more significant variation in the process parameters, or when the p-value is less than 0.05, the factor is more significant [55,59].

2.6. DoE Confirmation Test

Once the optimum levels of printing parameters were determined, the final step involved the theoretical prediction and experimental confirmation of these parameters. The results obtained are then compared to evaluate the enhancement in the quality characteristics achieved by employing the optimal printing parameters [60]. This comparison aids in evaluating the consistency between the predicted and observed parameters. It is expected that there should be a reasonable agreement of ±5% (with a 95% confidence interval) between the predicted values and the results confirmed through experiments [60].

The predicted S/N ratio and confidence interval ($C.I.$) can be determined using Equations (1) and (2) as follows [60].

$$S/N_{predicted} = \overline{S/N} + \sum_{j=1}^{k}\left(\overline{S/N_k} - \overline{S/N}\right) \quad (1)$$

where $\overline{S/N}$ represents the overall mean of all S/N ratios, $\overline{S/N_k}$ denotes the average S/N ratio derived from the determined significant factors at the optimum level, and k represents the chosen factor.

$$C.I._{predicted} = \sqrt{F_{(\alpha,\vartheta_1,\vartheta_2)} V_e \times \left[\frac{1}{N_e}\right]} \quad (2)$$

where $F_{\alpha,\vartheta_1,\vartheta_2}$ represents the F-ratio obtained from the F-table, α denotes the significance level corresponding to the confidence level of $(1-\alpha)$, ϑ_1 stands for the degree of freedom (DOF) of the mean (which is always set to 1), and ϑ_2 represents DOF of the error term. Meanwhile, V_e denotes the error variance, and N_e signifies the effective number of replications, as mathematically expressed by Equation (3) as follows [61]:

$$N_e = \frac{(Number\ of\ Trials)}{[DOF\ of\ mean\ (always\ 1) + DOF\ of\ all\ factors\ used\ in\ the\ estimate]} \quad (3)$$

Finally, after conducting a confirmation run, the S/N ratio can be used as the value of $S/N_{confirmation}$. The formula for calculating the $C.I.$ for $S/N_{confirmation}$ is presented in Equation (4) as follows, where R is the number of replicas [61].

$$C.I._{confirmation} = \sqrt{F_{(\alpha,\vartheta_1,\vartheta_2)} V_e \times \left[\frac{1}{N_e} + \frac{1}{R}\right]} \quad (4)$$

2.7. Statistical Analysis

Microsoft 365 MSO Excel software (version 2402 build 16.0.17328.20124/32-bit) (a data management tool developed by Microsoft), was employed in this study for the purpose of data management. The results include the average values derived from the material property characterisation tests, along with their corresponding standard deviations. In addition, Minitab 21.4 software was employed to analyse the signal-to-noise (S/N) ratio of the obtained mechanical properties and conduct an analysis of variance to determine the printing parameters that have a significant impact on these mechanical properties.

2.8. Comparative Analysis

In order to assess the improvement or deterioration of the mechanical properties of the virgin PLA and post-consumer PLA blend (vPLA:PC-PLA, 50:50), a comparison was made with the outcomes of 100% virgin PLA (vPLA) and 100% recycled PLA (rPLA) (recycled from residual extrusion waste stream, free from additives, subsequently recompounded and homogenised) filament 3D-printed parts, which were recorded by the author Hasan et al. [48] under the same printing parameters and testing conditions.

3. Results and Discussions

The objective of this study is to determine the optimal printing parameters for fabricating 3D-printed components using a vPLA:PC-PLA (50:50) blended filament consisting of 50 wt.% PC-PLA and 50 wt.% vPLA. To achieve this, 45 samples (five samples for each experimental run) each for tensile, flexural, and impact testing and 27 samples (three samples for each experimental run) were fabricated for hardness testing. Surface roughness measurements were obtained from the tensile samples before conducting the tensile tests.

Following manufacturing, a comprehensive assessment of the surface roughness and mechanical properties, including the hardness, tensile strength, flexural strength, and impact strength, was performed. The performances of the blended filament-produced specimens were compared with those of 100% virgin PLA and 100% recycled PLA (sourced from residual extrusion waste streams) specimens whose results were published by the authors [48]. The same 3D printing parameters were considered in both the current and previous studies. This investigation aimed to gain insights into the performance attributes of specimens printed with blended filaments by assessing selected printing parameters and determining if they exhibit properties comparable to those of vPLA. The following sections present a comprehensive analysis of the experimental data obtained by testing the specimens to determine their material properties.

3.1. Surface Roughness

One primary concern in additive manufacturing is surface quality, which is crucial for ensuring the operation and effectiveness of printed components over time. During the additive manufacturing process, the surface roughness stage plays a critical role in achieving the desired surface quality [62]. The surface roughness, *Ra*, of printed parts has a significant impact on their mechanical behaviour, such as the initiation of cracks, wear resistance, and fatigue life. The requirement for a decrease in surface roughness is contingent upon the specific experimental approach and purpose for which the components are printed [62]. The surface roughness of the vPLA:PC-PLA (50:50) blended printed tensile test samples was analysed to evaluate the impact of the printing parameters on the manufactured samples.

The surface roughness values of all the tested specimens are illustrated in Figure 8a. The results indicate that experimental run 2 (LH: 0.1 mm, I: 100%, and NT: 205 °C) produced the lowest surface roughness of 5.98 μm, followed by experimental run 3 (7.01 μm) and experimental run 1 (8.25 μm). A higher surface roughness was observed when the specimens were printed using experimental run 8 (LH: 0.3 mm, I: 80%, and NT: 195 °C). As shown in Figure 8a, the surface roughness of the blended filaments increases with increasing layer height. The printing temperature also plays a critical role in the production of high-quality prints. The test results indicated that at higher selected temperatures (215 °C) and lower selected temperatures (195 °C), the surface roughness increased, whereas at 205 °C, the surface roughness was found to be at its minimum. A higher printing temperature (215 °C) may cause samples to overheat, whereas a lower printing temperature (195 °C) could lead to inadequate material flow, which can negatively impact the print quality [63]. It was observed that as a result of uneven solidification and possible shrinkage, samples printed at 195 °C had a poorer surface quality than those printed at 205 °C, which might lead to regulated solidification and produce smoother surfaces. Similar trends for the positive correlation between layer height and nozzle temperature were observed in the studies on virgin PLA and recycled PLA [48]. Hasan et al. [48] found that rPLA samples had higher surface roughness values than vPLA samples, indicating rougher surfaces (for experimental run 2). In contrast, the current study observed that the vPLA:PC-PLA (50:50) blended filament had a lower surface roughness than that of rPLA, indicating better surface properties compared to those of rPLA for the same experimental run. The surface roughness of the blended specimen demonstrated a 2.29% lower value compared to rPLA (as shown in Figure 8b). Studies have reported that the blending of virgin material with recycled material positively affects the surface roughness of 3D-printed samples [45,64]. Specifically, an increase in the percentage of virgin material added to the recycled material led to a significant enhancement in surface roughness. This improvement may be attributed to the more pronounced influence of the virgin material on the properties of the recycled material, which are typically more consistent in vPLA than in rPLA. The upgraded surface finish was due to the improved melt flow and cooling behaviour of vPLA, leading to a more consistent 3D printing process [38].

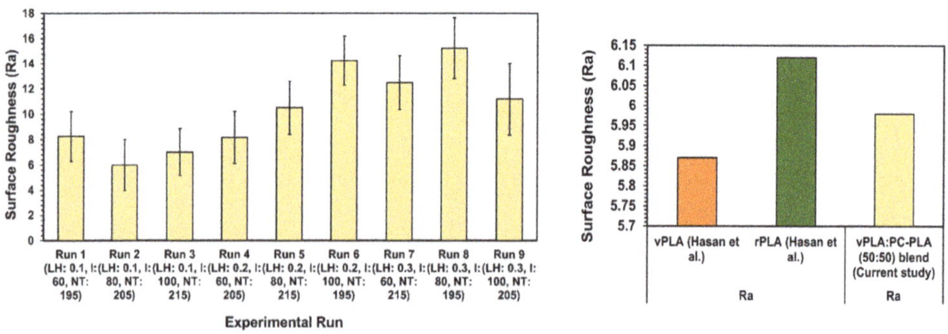

(a) (b)

Figure 8. (**a**) Surface roughness of vPLA:PC-PLA (50:50) blended filament 3D-printed specimens; and (**b**) comparison of surface roughnesses of vPLA [48], rPLA [48], and vPLA:PC-PLA (50:50) blended filament printed specimens (where LH is the layer height, I is the infill percentage, and NT is the nozzle temperature).

3.2. Hardness

The degree of hardness of a material indicates its capacity to withstand local deformation, including the effects of surface indentation or penetration. To attain an optimal number of production runs for an additively manufactured component, a high degree of hardness is required [65]. The detailed results of the hardness of all the vPLA:PC-PLA (50:50) blended prepared samples are shown in Figure 9a. The sample with the maximum infill and layer height exhibited the highest hardness. This may be attributable to the significant influence of the infill density on the output. In experimental run 9 (LH: 0.3 mm, I: 100%, NT: 205 °C), the samples were printed with the highest infill density and layer height, resulting in the highest hardness of 79.51. In contrast to this, experimental run 1 (LH: 0.1 mm, I: 60%, NT: 195 °C) exhibited a lower hardness value of 73.12 for the blended specimen. This increase in hardness may be attributed to a better diffusion of the material in the subsequent layers at higher infill densities, resulting in fewer voids in the interspace of the structure and higher hardness values [66]. Another possible reason is that the cross-sectional area of the material increased as the percentage of filling increased, resulting in a harder internal structure. In contrast, a larger internal air gap with a lower filling percentage facilitates an easier penetration of the hardness tip into the surface of the test specimen [66].

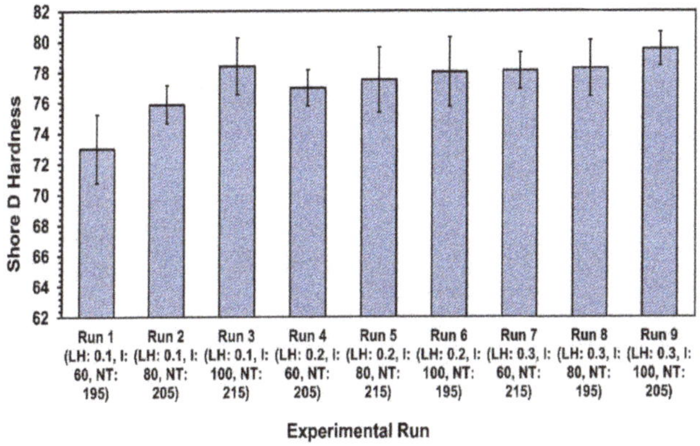

(a)

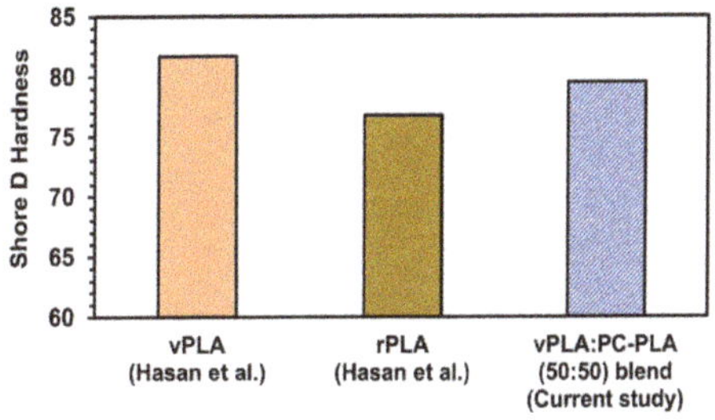

(b)

Figure 9. (**a**) Shore D hardness of the vPLA:PC-PLA (50:50) blended filament 3D-printed specimens; and (**b**) comparison of hardness values for vPLA [48], rPLA [48], and vPLA:PC-PLA (50:50) blended filament printed specimens (where LH is the layer height, I is the infill percentage, and NT is the nozzle temperature).

The printing temperature also plays a significant role in determining the hardness of the material. As the printing temperature decreases, the hardness value also decreases. The reason for this is the incomplete melting of the material, which results in poor layer-to-layer bonding, a larger porosity, and a lower hardness [67]. Conversely, an increase in printing temperature led to a decrease in the rate of change in hardness, which may be attributed to the stronger bonding between the materials caused by the high printing temperature or the work of adhesion between the filaments [67]. In summary, the results of this study illustrate a direct correlation between the nozzle temperature and infill percentage, resulting in an increase in hardness. This observation aligns with the outcomes reported by Hasan et al. [48] for vPLA and rPLA, as well as findings by Mani et al. [68] and Maguluri et al. [69] for vPLA. A comparison of the hardness of the vPLA:PC-PLA (50:50) blended with 100% recycled PLA and 100% virgin PLA showed that the hardness of the blend samples produced using experimental run 9 (LH: 0.3 mm, I: 100%, NT: 205 °C) exhibited a 5% increase compared to that of rPLA, whereas it was 6% lower than that of vPLA (Figure 9b) [48]. Mishra et al. [45] and Abbas et al. [70] noted that blending enhances the compatibility between polymers at the chemical and structural levels, which stimulates the formation of a homogeneous blend. This, in turn, improves intermolecular interactions, thereby enhancing the resistance to abrasion and making it more resilient to plastic deformation, ultimately contributing to an improved hardness [45,64].

3.3. *Tensile Behaviour*

The tensile test results were obtained using the Taguchi OA for the three printing parameters examined in this study, namely, layer height, infill density, and printing temperature. Each Taguchi experiment involved testing five specimens, and consistent repeatability was demonstrated by the tensile results. Therefore, a single stress–strain curve representing each set of five specimens is presented in Figure 10. The stress–strain curve demonstrated that the fracture formed in the vPLA:PC-PLA (50:50) blended 3D-printed specimens exhibited ductile behaviour. In addition, the curves show that after reaching the maximum tensile stress, the tensile stress remained constant with a small increase in the strain values. The raw data obtained from the load and displacement after the tensile test were utilised to calculate the yield strength, ultimate tensile strength (UTS), strain at UTS, elastic modulus, fracture strength, strain at fracture, work until UTS, and work until fracture.

The results of the tensile test showed that the strength decreased when the specimens were printed at a lower temperature (195 °C) and smaller layer height (0.1 mm) and improved when they were printed at a higher temperature (215 °C) and larger layer height (0.3 mm). Studies have shown that specimens with a larger layer height (0.3 mm) exhibit the best tensile properties because they experience minimal distortions [71]. In FDM printing, the interlayer strength typically is lower than the intralayer strength. This is because the temperature gradient within a layer is less pronounced than that between consecutive layers, leading to the formation of weaker bonds between the layers [72]. The bonding strength between the layers is determined by the extruder temperature; a lower temperature restricts the molecular chains of the deposited material from reorganising adequately, thereby reducing the bonding strength between adjacent layers and ultimately lowering the tensile strength [72,73].

The highest tensile strength among the vPLA:PC-PLA (50:50) blended specimens was achieved at 48.88 MPa in experimental run 7 (LH: 0.3 mm, I: 60%, NT: 215 °C) (Table 3). The second highest value for the tensile strength was estimated to be 47.88 MPa for experimental run 9 (LH: 0.3 mm, I:100%, and NT: 205 °C), and the third highest value reached 46.67 MPa in experimental run 5 (LH: 0.2 mm, I: 100%, and NT: 215 °C).

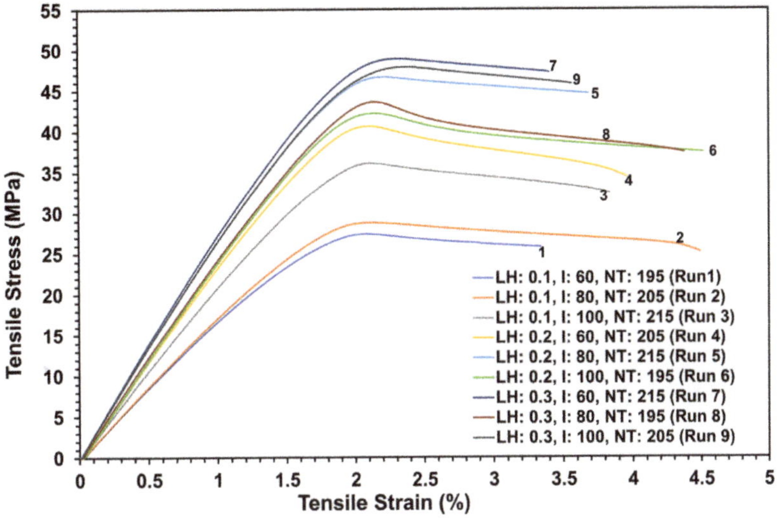

Figure 10. Tensile stress vs. tensile strain curves for vPLA:PC-PLA (50:50) blended filament 3D-printed specimens (where LH is the layer height, I is the infill percentage, and NT is the nozzle temperature).

Table 3. Tensile properties of the vPLA:PC-PLA (50:50) blended filament 3D-printed specimens (where LH is the layer height, I is the infill percentage, and NT is the nozzle temperature).

Experimental Run	Yield Strength (MPa)	Ultimate Tensile Strength (UTS) (MPa)	Strain at UTS (%)	Fracture Strength (MPa)	Strain at Fracture (%)	Elastic Modulus (GPa)	Work until UTS (kJ/m^2)	Work until Fracture (kJ/m^2)
Run 1 (LH: 0.1, I: 60, NT: 195)	25.35 ± 1.98	26.15 ± 2.95	2.09 ± 2.31	25.25 ± 1.56	2.90 ± 1.86	1.98 ± 1.15	21.97 ± 1.58	35.81 ± 1.51
Run 2 (LH: 0.1, I: 80, NT: 205)	28.72 ± 2.25	29.40 ± 2.85	2.13 ± 2.10	27.51 ± 1.85	3.15 ± 2.85	2.02 ± 1.17	23.36 ± 1.02	46.61 ± 1.92
Run 3 (LH: 0.1, I: 100, NT: 215)	28.12 ± 2.10	36.01 ± 1.85	2.14 ± 2.25	34.33 ± 1.93	3.05 ± 2.53	2.33 ± 1.02	28.62 ± 1.09	49.06 ± 1.52
Run 4 (LH: 0.2, I: 60, NT: 205)	31.24 ± 2.05	40.41 ± 1.12	2.12 ± 1.25	37.66 ± 1.83	3.03 ± 2.14	2.65 ± 1.05	32.25 ± 1.19	67.50 ± 1.38
Run 5 (LH: 0.2, I: 80, NT: 215)	41.85 ± 1.95	46.67 ± 1.98	2.20 ± 1.45	44.95 ± 1.84	3.41 ± 1.97	2.85 ± 1.14	39.74 ± 1.14	75.52 ± 1.52
Run 6 (LH: 0.2, I: 100, NT: 195)	32.37 ± 1.91	42.22 ± 2.25	2.17 ± 1.58	38.83 ± 1.79	3.42 ± 2.05	2.70 ± 1.54	34.18 ± 1.27	66.98 ± 1.24
Run 7 (LH: 0.3, I: 60, NT: 215)	43.64 ± 1.68	48.88 ± 1.56	2.37 ± 1.87	46.05 ± 1.89	3.35 ± 2.61	3.20 ± 1.32	42.31 ± 1.34	77.66 ± 1.34
Run 8 (LH: 0.3, I: 80, NT: 195)	34.01 ± 1.81	43.62 ± 2.10	2.14 ± 2.37	38.92 ± 1.87	3.69 ± 2.13	2.75 ± 1.29	34.524 ± 1.51	75.36 ± 1.05
Run 9 (LH: 0.3, I: 100, NT: 205)	43.47 ± 1.88	47.78 ± 2.54	2.28 ± 2.21	47.35 ± 1.94	3.54 ± 2.10	3.40 ± 1.18	42.51 ± 1.24	77.74 ± 1.19

A comparison of the tensile strengths of the vPLA:PC-PLA (50:50) blended specimens with 100% rPLA and 100% vPLA (as shown in Figure 11) [48] showed that, in general, the blended filament specimens exhibited higher tensile properties than those of the 100% rPLA samples, but were lower than those of the vPLA samples. For experimental run 7, the tensile strength of the blended specimens was 18.40% higher than the corresponding value of 39.90 MPa for the rPLA-printed specimens (Figure 11). For experimental runs 9 and 5, the blended specimens demonstrated increases of 20% and 17.81%, respectively, compared to the rPLA specimens. The enhancement in tensile strength observed in the blended specimens may be ascribed to the improved load transfer capacity resulting from the additional cross-linking of polymer chains caused by the blending process while increasing the interlayer bonding [38,74]. Previous research on polymer processing has demonstrated that blending extrusion can cause microstructural changes, which may involve higher shear forces, resulting in the alignment of polymer chains and thereby improving the tensile properties of the material [74,75].

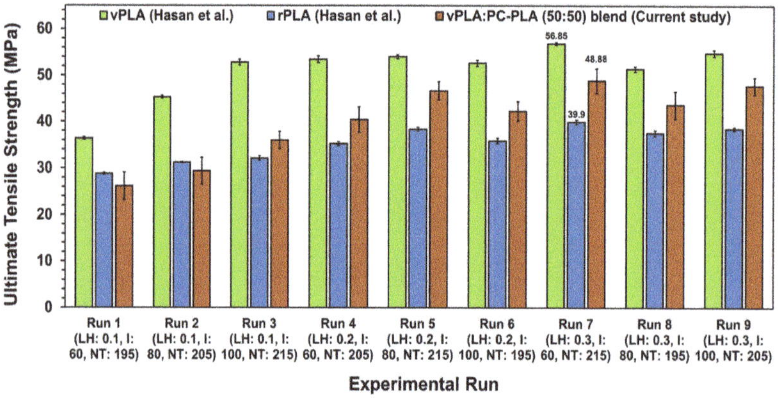

Figure 11. Comparison of UTS for vPLA [48], rPLA [48], and vPLA:PC-PLA (50:50) blended filament printed specimens (where LH is the layer height, I is the infill percentage, and NT is the nozzle temperature).

In contrast, the tensile strengths of the rPLA samples in experimental runs 1 (LH: 0.1 mm, I: 60%, NT: 195 °C) and 2 (LH: 0.1 mm, I: 80%, NT: 205 °C) were higher than those of the blended samples for the same experimental runs by 9.23% and 5.76%, respectively. Notably, the tensile strength of the blended samples decreased with the decrease in the printing temperature and layer height. This decline can be attributed to the incomplete melting of the material caused by a smaller layer height, resulting in reduced layer-to-layer adhesion and subsequent brittleness [67]. In addition, the blended filament specimens showed a comparatively lower strain at fracture of 3.8 to 4.2% for all of the experimental runs, compared to the rPLA specimens, whose strain at fracture was 5.5 to 7% for the same printing parameters; however, this result is consistent with the strain at fracture of vPLA [48]. This phenomenon, which occurred in the blended filament samples, may be attributed to the moisture content of the material [76]. PLA plastic is known to be susceptible to moisture absorption, which can result in hydrolytic degradation during the filament extrusion process. This degradation may cause the breakdown of polymeric chains and decrease the material elongation [77–79]. However, the same moisture content acts as a plasticiser, making the material softer and more ductile [79].

3.4. Flexural Behaviour

The effects of the printing parameters on the flexural properties (yield strength, flexural strength, flexural strain, elastic modulus, and work until flexural strength) of the 3D-printed vPLA:PC-PLA (50:50) blended specimens were obtained and examined through load-deflection data measured during the flexural tests for each set of experimental runs. A total of 45 flexural tests were performed, with five tests for each experimental run. Typical flexural stress–strain curves for each experimental run are shown in Figure 12. The selection of the flexural stress–strain curves in the figure represents the average flexural strength level for each set of experimental runs. The behaviour observed during the flexural tests revealed that the blended samples did not reach the fracture point, which suggests that the samples were ductile and had a high degree of plastic deformation. Examining the stress–strain curves, it is evident that the blended specimens with a larger layer height and high nozzle temperature exhibited the highest flexural stress. Conversely, as the layer height and nozzle temperature were reduced, the flexural stress value tended to decrease, exhibiting a trend similar to that observed for tensile strength.

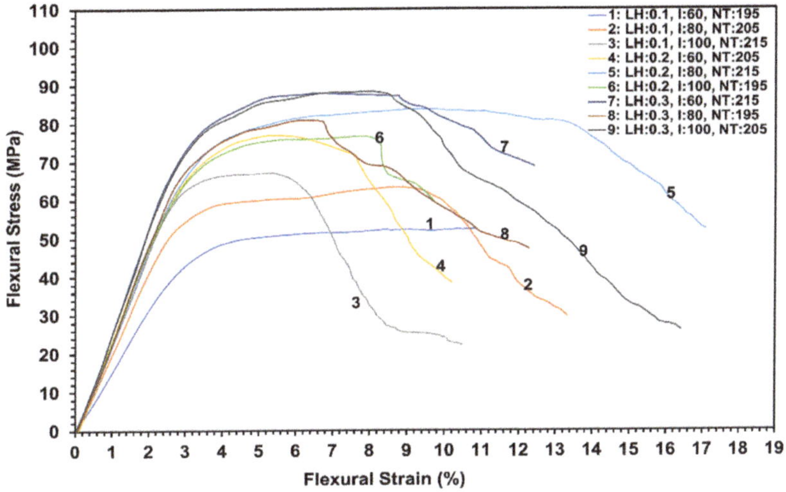

Figure 12. Flexural stress–strain curves for vPLA: PC-PLA (50:50) blended filament 3D-printed specimens (where LH is the layer height, I is the infill percentage, and NT is the nozzle temperature).

A summary of the flexural tests for each experimental run of the vPLA:PC-PLA (50:50) blended specimens is presented in Table 4, which lists the average results of the yield stress, maximum flexural stress, flexural strain, elastic modulus, and work until the maximum flexural stress with the standard deviation. The standard deviation (1.65–4.34%) indicated the excellent repeatability and reliability of the obtained data. The highest flexural stress, 88.36 MPa, was observed in experimental run 7 (LH: 0.3 mm, I: 100%, NT: 205 °C), followed by experimental run 9 (LH: 0.3 mm, I: 100%, NT: 205 °C), which is 88.29 MPa, which showed a similar value. The results clearly show the impact of a larger layer height and nozzle parameters on the flexural properties. The results of this study are consistent with those of other studies that have demonstrated a similar trend with regard to the impact of printing parameters on the flexural properties [48,80,81]. A comparison of the vPLA:PC-PLA (50:50) blended specimens with 100% recycled PLA and 100% vPLA showed a clear increase in their flexural properties compared to those of the rPLA specimens and a slightly lower result than that of the vPLA specimens under the same experimental conditions (Figure 13). The flexural strength of the blended specimen for experimental run 7 was 8% higher than that of the recycled PLA and 4.5% lower than that of vPLA for the same printing parameters [48]. The second highest values of the blended specimens for experimental runs 9 and 5 were 8.41% and 2.3% higher than those of the rPLA specimens and 2.2% and 3.62% lower than those of the vPLA specimens, respectively. Overall, the flexural strength of the blended printed samples for all experimental runs increased by 5–15% compared to that of the recycled PLA-printed samples. This improvement in strength is attributed to the elevated crystallinity and enhanced interfacial bonding among the constituent elements [63]. Research has revealed that a distinct crystal structure is formed when virgin polymers are combined with recycled polymers, which co-crystallise to create stereocomplex crystallites. These crystals increase the melting temperature, and their presence may significantly improve the properties of blended PLA specimens [82,83]. In addition, vPLA acts as a "bridge" within the blended system, reducing the phase boundary between the polymers and enhancing their bonding. This results in improved flexural properties because the phases between the polymers allow for closer connections [84].

Table 4. Flexural test results of the vPLA: PC-PLA (50:50) blended filament 3D-printed specimens (where LH is the layer height, I is the infill percentage, and NT is the nozzle temperature).

Experimental Run	Yield Stress (MPa)	Flexural Stress (MPa)	Flexural Strain (%)	Elastic Modulus (GPa)	Work until Maximum Flexural Stress (kJ/m^2)
Run 1 (LH: 0.1, I: 60, NT: 195)	43.76 ± 3.21	51.01 ± 1.86	6.69 ± 1.89	1.97 ± 1.26	19.09 ± 2.83
Run 2 (LH: 0.1, I: 80, NT: 205)	54.04 ± 2.56	63.81 ± 2.10	7.69 ± 2.67	2.15 ± 2.57	25.86 ± 2.01
Run 3 (LH: 0.1, I: 100, NT: 215)	59.32 ± 2.21	67.23 ± 2.92	6.71 ± 2.15	2.29 ± 2.93	28.78 ± 2.12
Run 4 (LH: 0.2, I: 60, NT: 205)	62.80 ± 2.21	77.20 ± 2.04	6.25 ± 2.65	2.52 ± 2.34	33.49 ± 2.05
Run 5 (LH: 0.2, I: 80, NT: 215)	71.13 ± 2.14	84.23 ± 2.09	8.40 ± 2.39	2.65 ± 2.01	42.34 ± 2.45
Run 6 (LH: 0.2, I: 100, NT: 195)	69.19 ± 2.76	77.18 ± 1.67	7.64 ± 2.95	2.44 ± 1.85	36.30 ± 1.54
Run 7 (LH: 0.3, I: 60, NT: 215)	76.24 ± 2.89	88.36 ± 1.47	7.07 ± 2.28	2.89 ± 2.01	37.92 ± 2.31
Run 8 (LH: 0.3, I: 80, NT: 195)	70.87 ± 3.90	81.37 ± 1.85	6.92 ± 1.12	2.50 ± 1.45	37.19 ± 1.41
Run 9 (LH: 0.3, I: 100, NT: 205)	78.11 ± 2.29	88.29 ± 2.17	7.82 ± 2.56	2.67 ± 2.20	45.19 ± 2.07

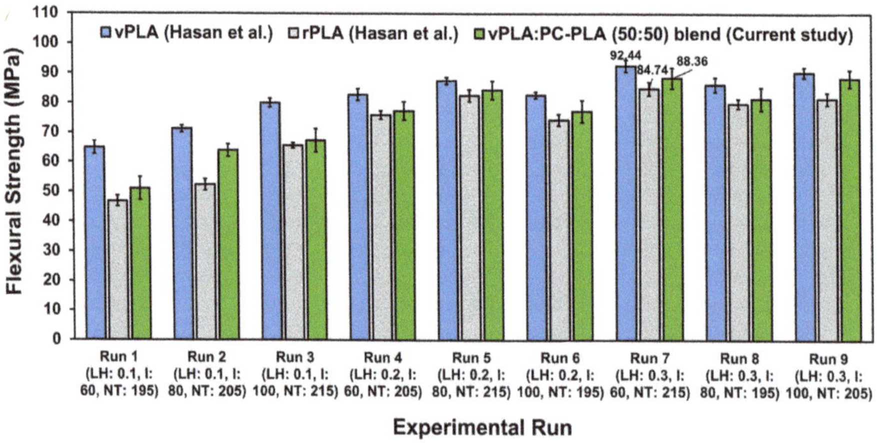

Figure 13. Comparison of flexural strength for vPLA [48], rPLA [48], and vPLA:PC-PLA (50:50) blended filament printed specimens (where LH is the layer height, I is the infill percentage, and NT is the nozzle temperature).

3.5. Impact Behaviour

The results of the Charpy impact tests are depicted in Figure 14. The results indicated that the impact strength of the vPLA:PC-PLA (50:50) blended specimens was influenced by the layer height and infill percentage. The lowest impact strength, measuring 3.69 kJ/m^2, was observed in run 1 (LH: 0.1 mm, I: 60%, NT: 195 °C) when the samples were printed with a low layer height and infill. In contrast, the highest impact strength among all the experimental runs was recorded at a value of 5.37 kJ/m^2 in experimental run 9

(LH: 0.3 mm, I: 100%, NT: 205 °C) when both the layer height and infill were at their maximum. This value was 40% higher than that in run 1. As examined by Syrlybayev et al. [85], an escalation in the infill rate leads to a reduction in the number of voids in the printed substance. A higher amount of material, resulting from the increased infill percentage, is capable of absorbing the force of impact. In addition, a larger layer height offers more resistance to the impact forces. This is because fewer layers are deposited at a larger layer height; therefore, fewer bonding lines and voids exist between the layers, which also helps to improve the impact strength [86].

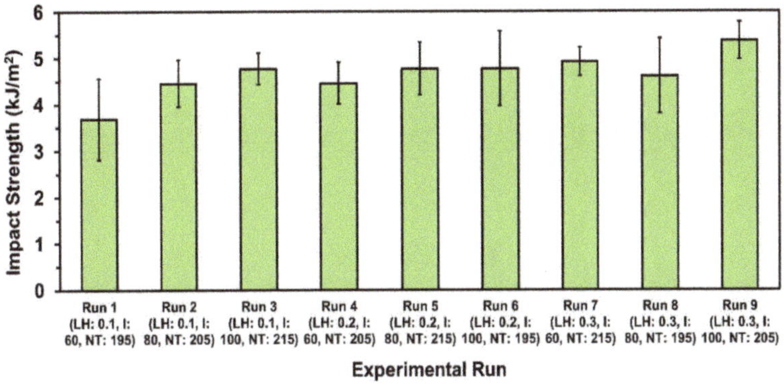

Figure 14. Charpy impact strength of vPLA:PC-PLA (50:50) blended filament 3D-printed specimens (where LH is the layer height, I is the infill percentage, and NT is the nozzle temperature).

A comparison of the impact strength data (Figure 15) for the vPLA:PC-PLA (50:50) blended specimens with those of 100% rPLA and 100% vPLA [48] showed that the vPLA:PC-PLA (50:50) blended specimens exhibited a 9.15% higher impact strength than the rPLA samples and an approximately similar impact strength to that of the vPLA specimens when printed at a larger layer height and high infill (experimental run 9). The impact test results showed that when vPLA was incorporated into PC-PLA, it acted as the centre of the built-in stress, resulting in an increased capacity for absorbing energy during the impact test and enhancing the impact strength [87]. In addition, owing to the increase in infill density, the interlayer adhesion between two consecutive layers increases, which improves the impact strength [88,89]. In contrast, the blended specimens performed poorly compared to rPLA when printed using a smaller layer height and lower temperature, as in experimental runs 1 and 2. The tensile test results also demonstrated a low impact strength for the smaller-layer-height specimens printed at a low temperature (Figure 11). These defects originated from the poor interfacial bonding between vPLA and PC-PLA at a lower nozzle temperature and layer height, which might have reduced the impact strength [86,87]. In addition, a lower layer height results in a higher number of bonding lines between the layers, which leads to a higher probability of void formation. This phenomenon contributes to the premature failure of the specimen, resulting in a lower impact strength [87,88]. Moreover, a lower nozzle temperature can potentially result in the formation of interlayer gaps, facilitating inadequate adhesion between layers during the printing process. Consequently, when the specimens are subjected to loading during impact testing, these interlayer gaps can propagate further cracking, ultimately resulting in material failure [88].

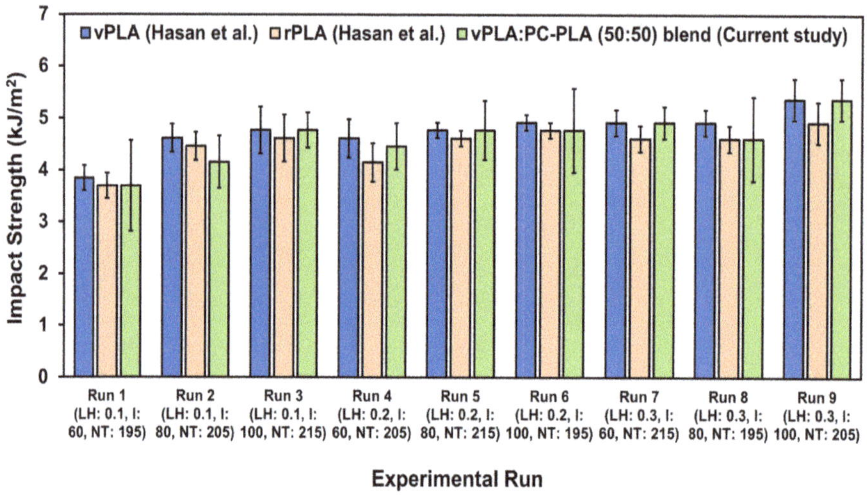

Figure 15. Comparison of the impact strengths of vPLA [48], rPLA [48], and vPLA:PC-PLA (50:50) blended filament printed specimens (where LH is the layer height, I is the infill percentage, and NT is the nozzle temperature).

3.6. Fractography Analysis

To examine the deformation and fracture behaviour of the vPLA:PC-PLA (50:50) blended specimens during mechanical testing, the fracture interface was studied. The micrographs of the specimens were examined using a microscope after performing the tensile tests. One specimen that best represented the failure mode was selected for each batch of five specimens. The micrographs presented in Figure 16a–e depict the cross-sectional images of the blended parts with the selected printing parameter settings.

As shown in Figure 16a, during experimental run 1, the layer height, infill, and nozzle temperature were low (LH: 0.1 mm, I: 60%, NT: 195 °C). As a result, the specimen exhibited noticeable gaps and void areas, which were attributed to filament pull-out. This phenomenon indicates poor layer bonding and ultimately leads to a decrease in strength owing to the separation of the filaments from the material [55]. The cavities in experimental run 1 were also apparent because of the low infill density [90]. Therefore, with an increase in the layer height, infill density, and nozzle temperature, the gaps and voids inside the specimen decreased, as shown in Figure 16b–e. This reduction can be attributed to the increased infill density, which led to a dense structure with a low level of porosity. Nevertheless, the reduction in gaps and improvement of layer bonding in a specimen cannot be achieved solely by increasing the infill density. For instance, when the specimen was printed at a high temperature of 215 °C and a high infill of 100% (experimental run 3), it exhibited a minimal number of or absence of voids (Figure 16b); however, it had a lower tensile strength owing to the low layer height (0.1 mm). This was caused by the higher inner layer porosity of the specimen, which negatively affected its mechanical properties [55,91]. In contrast, an increase in layer height, as observed in experimental runs 7 and 9 (Figure 16d,e), resulted in a noticeable reduction in porosity and improved layer bonding, which led to a significant increase in strength.

The nozzle temperature also played a significant role in determining the strength of the specimens. With an increase in the nozzle temperature from 195 °C to 215 °C, the size of voids in the fractured specimens decreased. For example, in experimental run 7, as illustrated in Figure 16d, it was evident that despite the low infill percentage (60%), the specimens exhibited a higher strength compared to that of the other experimental runs because of the larger layer height (0.3 mm) and higher nozzle temperature (215 °C). The reason for this phenomenon is the increase in the nozzle temperature, which leads to a decrease in the melt viscosity and an improvement in the wettability of the polymers. Consequently, the printed layers remained in a molten state for a longer duration, facilitating increased coalescence within the printed layers [44]. As a consequence, there was a notable enhancement in the adhesion between the printed layers, and a reduction in voids was observed, which improved the strength [90]. In contrast, in experimental run 9, even though there was an increase in interlayer adhesion and a decrease in voids owing to the high infill (100%) and larger layer height (0.3 mm) (Figure 16e), the mechanical strength was lower than that in run 7 because of the lower nozzle temperature (205 °C). In addition, the image of experimental run 6, as depicted in Figure 16c, reveals that although the layer height was increased to 0.2 mm and the infill percentage was set at 100%, the strength of the material was still affected by the low nozzle temperature (195 °C). This delamination was caused by the presence of air gaps and weak interlayer adhesion due to the low printing temperature [44,92]. This finding is supported by the mechanical test results (tensile, flexural, impact, and hardness tests) discussed earlier.

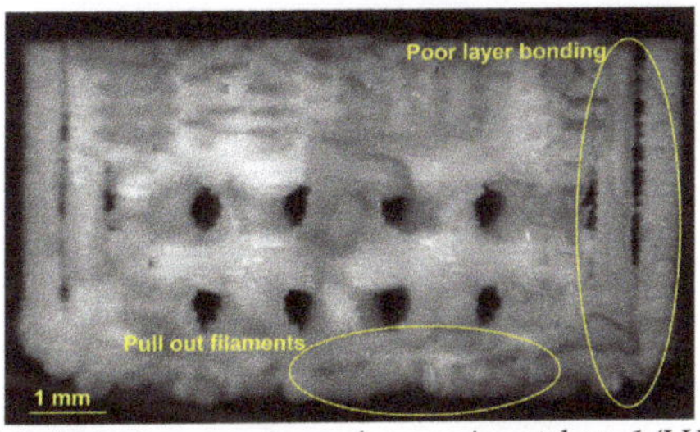

(**a**) Optical microscopy image for experimental run 1 (LH: 0.1 mm, I: 60%, NT: 195 °C)

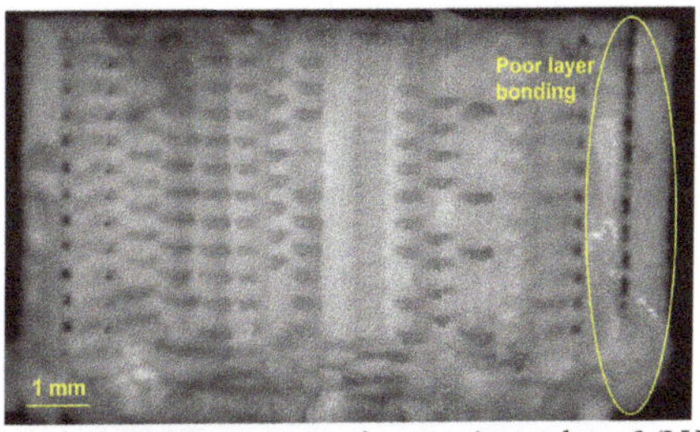

(**b**) Optical microscopy image for experimental run 3 (LH: 0.1 mm, I: 100%, NT: 215 °C)

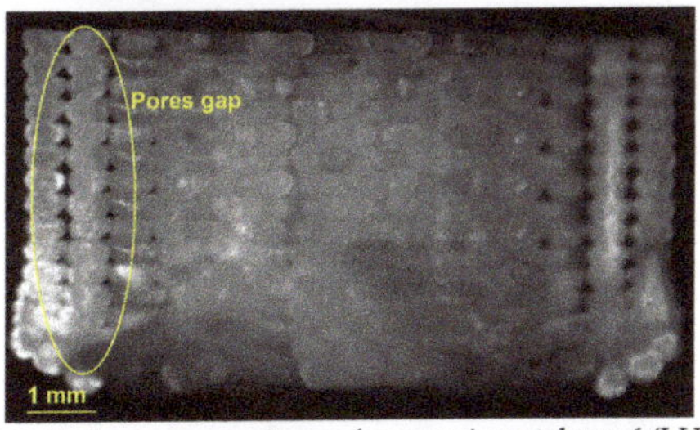

(**c**) Optical microscopy image for experimental run 6 (LH: 0.2 mm, I: 100%, NT: 195 °C)

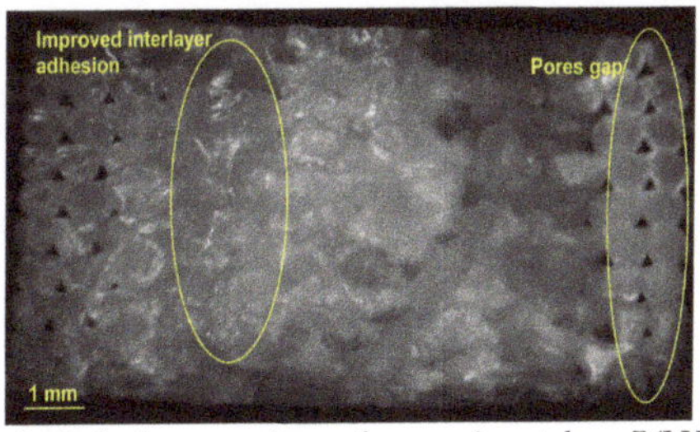

(**d**) Optical microscopy image for experimental run 7 (LH: 0.3 mm, I: 60%, NT: 215 °C)

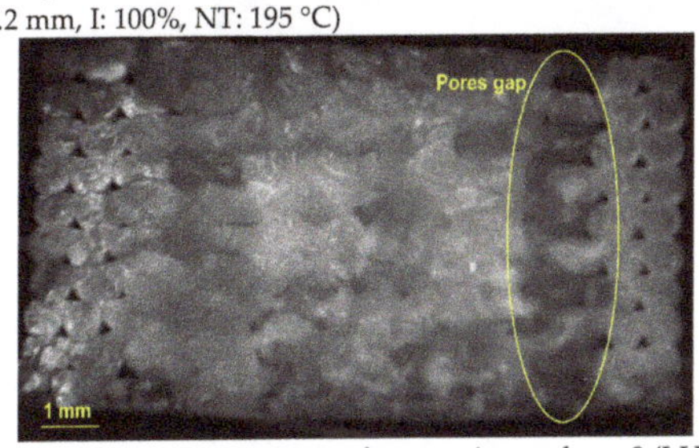

(**e**) Optical microscopy image for experimental run 9 (LH: 0.3 mm, I: 100%, NT: 205 °C)

Figure 16. Micrographs of the fractured vPLA:PC-PLA (50:50) blended filament 3D-printed tensile specimens for (**a**) run 1, (**b**) run 3, (**c**) run 6, (**d**) run 7, and (**e**) run 9.

Overall, an improvement in the layer bonding and reduction in voids were observed for the vPLA:PC-PLA (50:50) blended specimens when the fractured surface was compared with that of recycled PLA, as described by Hasan et al. [48]. This development was attributed to the fact that PC-PLA and vPLA are miscible blends, as both are semicrystalline polymers. Furthermore, both polymers (vPLA and PC-PLA) had identical crystallisation phases, leading to strong interphase adhesion. In addition, more coalescence was observed between the printed layers, indicating a promising level of adhesion strength at the blend interphase [36,37].

3.7. Optimisation of Process Parameters

In this investigation, signal-to-noise (S/N) ratio diagrams were employed to assess the outcomes of the experimental results. The interpretation of the S/N ratio diagram was based on the "larger the better" criterion for mechanical properties (tensile strength, flexural strength, impact strength, and hardness), while the "smaller the better" criterion was applied for surface roughness properties. Specifically, the mechanical properties were analysed using the "larger the better" criterion since the objective of this research is to maximise them. The peak point observed in the S/N ratio diagram signifies the optimal conditions.

Figure 17a–h illustrates the main effects plots of the S/N ratios for the tensile properties. This graph indicates that the S/N ratio is correlated with the layer height, as high S/N ratios were obtained with a larger layer height. The second most significant factor derived from the plot is the nozzle temperature. In contrast, the infill had a minimal effect on the tensile properties compared to the layer height and infill density. This is because the larger layer height and nozzle temperature increased the interlayer adhesion and melt viscosity, which facilitated the bonding of the blend material, ultimately increasing its tensile strength [93]. From the main effects plots, it is evident that the optimal printing parameters for tensile strength are a 0.3 mm layer height, 100% infill, and 215 °C nozzle temperature.

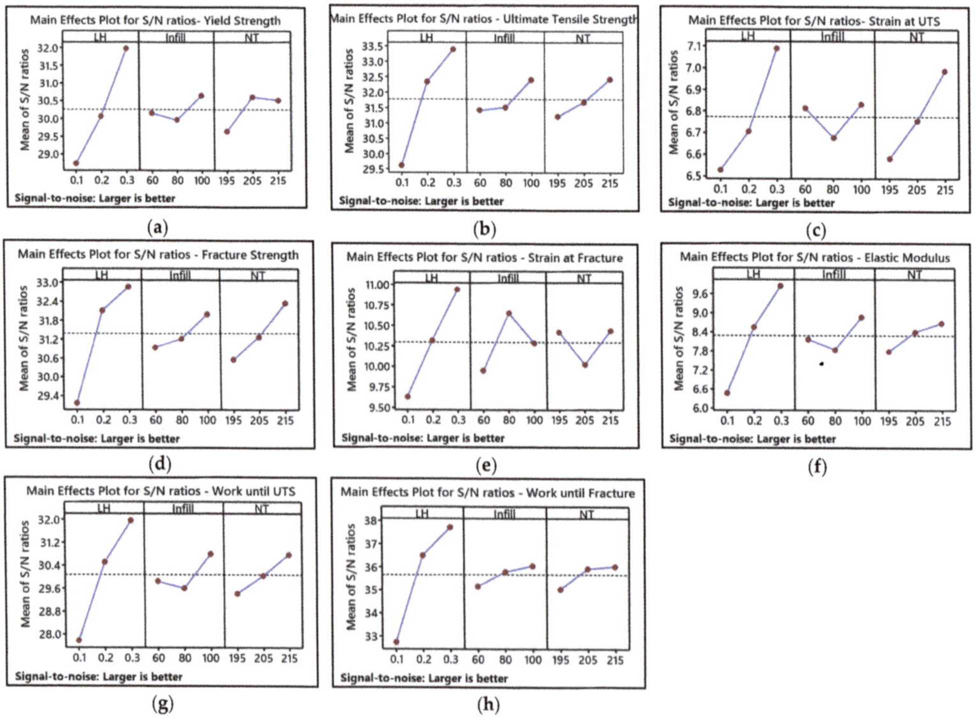

Figure 17. Main effects plots illustrating S/N ratios for the tensile properties of blended specimens: (**a**) yield strength, (**b**) ultimate tensile strength (UTS), (**c**) strain at UTS, (**d**) fracture strength, (**e**) strain at fracture, (**f**) elastic modulus, (**g**) work until UTS, and (**h**) work until fracture.

The flexural properties (Figure 18) exhibited a similar trend for the printing parameters to that of the tensile properties, which also confirmed the impact of the larger layer height (0.3 mm), infill (100%), and nozzle temperature (215 °C) on the flexural properties. This outcome aligns with the findings of Hasan et al. [48], who reported the same trend regarding the optimum printing parameters for 100% virgin PLA and 100% recycled PLA.

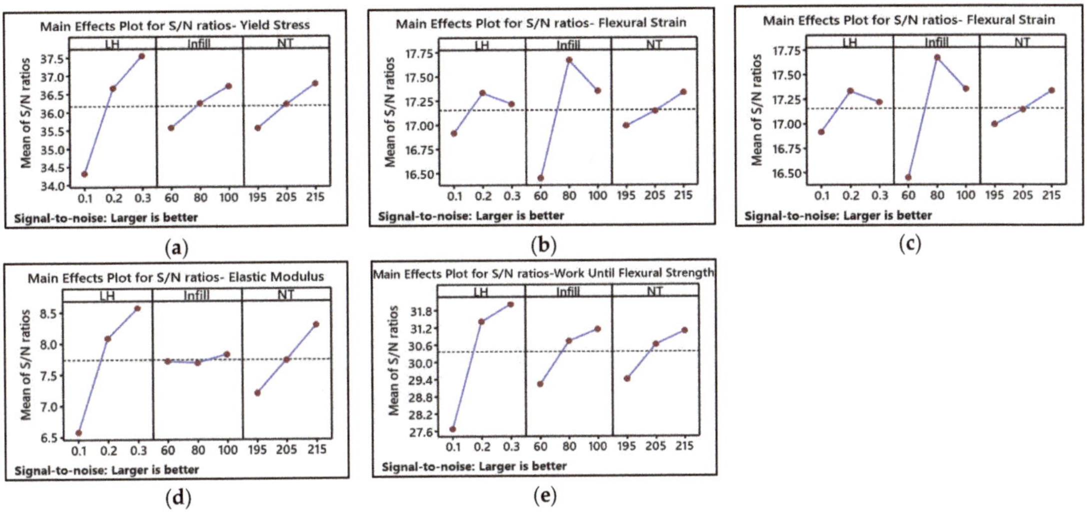

Figure 18. Main effects plots illustrating S/N ratios for the flexural properties of blended specimens: (**a**) yield strength, (**b**) flexural strength, (**c**) flexural strain, (**d**) elastic modulus, and (**e**) work until flexural strength.

The impact strength is mainly dependent on the properties of the material used to resist shock. Therefore, upon analysing the main effects plots of the S/N ratios (Figure 19a), it was observed that compared to the other printing parameters, the infill density exhibited a significant influence on the impact strength. A higher S/N ratio was observed with a larger layer height, high infill, and high nozzle temperature. Consequently, the optimal printing parameters for achieving the maximum impact strength were determined to be a layer height of 0.3 mm, an infill density of 100%, and a nozzle temperature of 215 °C.

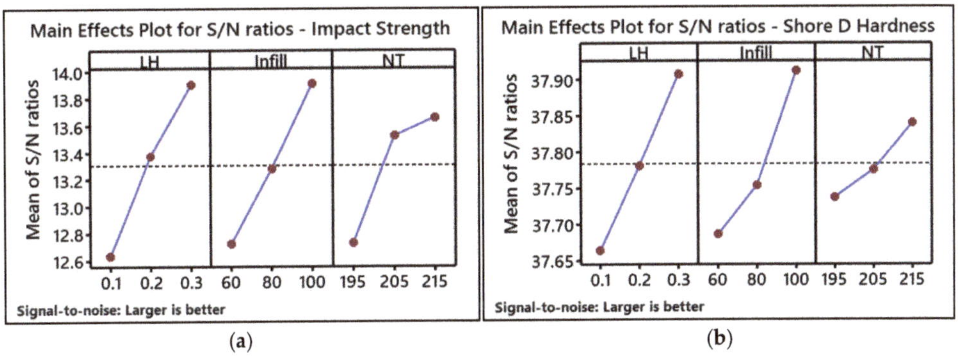

Figure 19. Main effect plots of S/N ratios for (**a**) impact strength and (**b**) hardness of blended specimens.

The hardness properties of the blended samples also demonstrated the optimum printing parameters to be a 0.3 mm layer height, 100% infill, and nozzle temperature of

215 °C (Figure 19b). Therefore, with an increase in the infill density, there is a corresponding increase in the penetration of layers into each other, leading to an improved interlayer strength [94]. Consequently, the impact and hardness properties improved at a high infill density (100%).

The S/N ratio main effects plots of the input process parameter levels for the arithmetic mean roughness (R_a) and root mean square roughness (R_q) are shown in Figure 20a,b, respectively. It can be observed from the main effects plots that the layer height is the most critical and influential parameter in determining the outcome of surface roughness. This is followed by the nozzle temperature and infill in terms of importance. The optimum printing parameter levels for the "smaller the better" criterion for achieving a lower surface roughness were a 0.1 mm layer height, 60% infill, and 205 °C nozzle temperature.

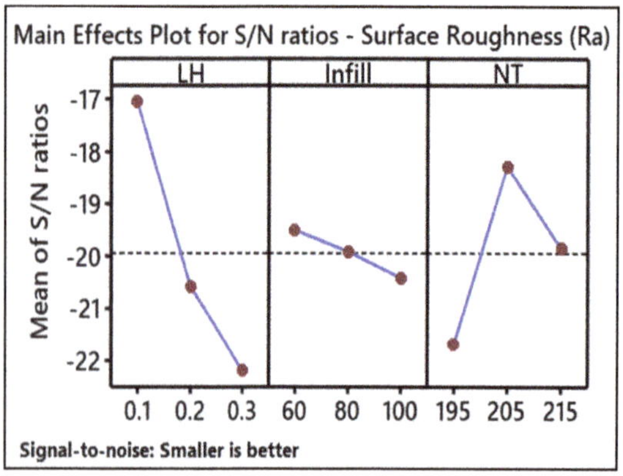

(a)

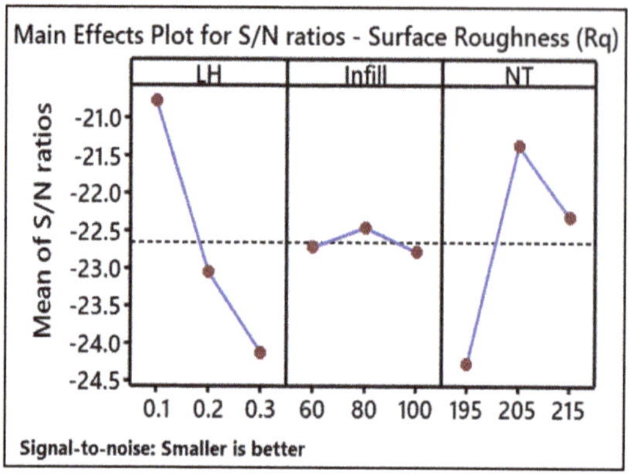

(b)

Figure 20. Main effects plot of S/N ratios for surface roughness of blended specimen (**a**) *Ra*, (**b**) *Rq*.

Analysis of Variance (ANOVA) Assessment

The main objective of an ANOVA is to apply a statistical method to evaluate the impact of individual printing parameters on the material properties. Roy [95] offers a comprehensive explanation of the ANOVA technique. For each quality parameter, the effect of each factor was determined by using a 95% confidence interval. The ANOVA results for tensile strength, flexural strength, impact strength, hardness, and surface roughness are presented in Table 5. A p-value below 0.05 indicates that the variable is considered significant in relation to the response.

In this study, the layer height was found to be the most significant factor influencing the tensile strength, flexural strength, and surface roughness, accounting for approximately 75% of the variation. Conversely, the infill density and nozzle temperature were found to be insignificant because their associated p-values were greater than 0.05. Notably, the infill had a significant influence on the impact strength and Shore D hardness. These findings are consistent with the conclusions reached by Hasan et al. [48], Atakok et al. [41], and Teharia et al. [96]. They also observed the influence of the layer height on the tensile, flexural, and surface roughness properties, as well as the impact of the infill on the impact strength and Shore D hardness.

Table 5. ANOVA responses of various properties examined in this study (where LH is the layer height, and NT is the nozzle temperature).

Response	Source	DF	Seq SS	Adj SS	Adj MS	F-Value	*p*-Value	% Contribution
Tensile strength	LH	2	22.75	22.75	11.37	40.87	0.024	83.24
	Infill	2	1.78	1.78	0.89	3.21	0.238	6.54
	NT	2	2.23	2.23	1.12	4.01	0.199	8.18
	Residual error	2	0.55	0.55	0.28			2.04
	Total	8	27.31					100.00
Flexural strength	LH	2	15.87	15.87	7.93	66.19	0.015	79.89
	Infill	2	1.12	1.12	0.56	4.68	0.176	5.65
	NT	2	2.63	2.63	1.31	10.98	0.083	13.25
	Residual error	2	0.24	0.24	0.11			1.21
	Total	8	19.87					100.00
Impact strength	LH	2	2.13	2.13	1.07	15.89	0.059	34.28
	Infill	2	2.45	2.45	1.22	18.26	0.052	39.19
	NT	2	1.52	1.52	0.76	11.33	0.081	24.38
	Residual error	2	0.13	0.13	0.07			2.15
	Total	8	6.24					100.00
Shore D hardness	LH	2	0.09	0.09	0.04	12.7	0.073	46.29
	Infill	2	0.08	0.08	0.04	11.44	0.08	41.69
	NT	2	0.02	0.02	0.01	2.3	0.303	8.37
	Residual error	2	0.01	0.01	0.00			3.64
	Total	8	0.19					100.00
Surface roughness	LH	2	41.56	41.56	20.78	97.84	0.01	68.47
	Infill	2	1.26	1.26	0.63	2.97	0.252	2.08
	NT	2	17.45	17.45	8.73	41.08	0.024	28.75
	Residual error	2	0.42	0.42	0.21			0.70
	Total	8	60.70					100.00

3.8. Confirmation Test

Based on the S/N ratio main effect plots and ANOVA, the optimum printing parameters were determined for the mechanical properties and surface roughness. A summary of the optimal printing parameters is presented in Table 6.

Table 6. Summary of optimum printing parameters for vPLA:PC-PLA (50:50) blended specimens.

Factors	Tensile Strength		Flexural Strength		Impact Strength		Shore D Hardness		Surface Roughness	
	Level	Description	Level	Description	Level	Description	Level	Description	Level	Description
Layer Height (mm)	3	0.3	3	0.3	3	0.3	3	0.3	1	0.1
Infill (%)	3	100	3	100	3	100	3	100	1	60
Nozzle Temperature (°C)	3	215	3	215	2	205	3	215	2	205

After determining the optimum printing parameters from the Taguchi method, the final step in the process involved confirmation experiments to validate the optimum printing parameters. Therefore, a confirmation experiment was conducted using the optimal printing parameters listed in Table 6. The confirmation test results for the vPLA:PC-PLA (50:50) blended specimens demonstrated an increase in strength compared with that of recycled PLA. In particular, the tensile strength, flexural strength, and impact strength were 51.81 MPa, 94.44 MPa, and 5.54 kJ/m^2, respectively, representing a 12.71%, 5.08%, and 3.17% increase in strength. Additionally, the Shore D hardness increased by 1.59% to 80.78, and the surface roughness decreased by 1.32% to 2.25 µm. These findings illustrate that the specimens fabricated using the optimal printing parameters exhibited a superior strength compared to those produced using L9 OA. Therefore, the S/N ratio of the confirmation test was compared with that of the theoretical prediction of the S/N ratio under the optimal conditions obtained using Equation (1). The confirmation test results for the tensile strengths were 1.14 dB higher than the sum of the predicted S/N ratios (Table 7). The other mechanical properties, namely, the flexural strength, impact strength, and Shore D hardness, were 0.76 dB, 0.35 dB, and 0.09 dB, respectively, higher than those of the predicted S/N values. This result shows that the confirmation test using the optimal printing parameters produces better outcomes than the predicted conditions.

Table 7. Confirmation test summary based on optimum factors.

Output Response	Prediction		Confirmation	
	$S/N_{predicted}$	$C.I._{predicted}$	$S/N_{confirmation}$	$C.I._{confirmation}$
Tensile Strength	34.67	±2.58	35.81	±1.52
Flexural Strength	39.58	±4.19	40.34	±2.51
Impact Strength	14.62	±0.16	14.97	±0.15
Shore D Hardness	38.14	±0.70	38.23	±1.05
Surface Roughness	−14.94	±1.07	−5.90	±1.32

Finally, the confidence intervals (C.I.s) for both the confirmed and predicted results were calculated using Equations (2)–(4) to verify the validity of the optimal printing parameters. A considerable overlap was noted between the C.I.s of the confirmation tests and the predicted results for the mechanical properties and surface roughness of the vPLA:PC-PLA (50:50) blended specimens, as shown in Table 7. The results indicated a significant improvement in both quality characteristics, and the optimum printing parameters were acceptable for use in the optimum manufacturing of blended filament 3D-printed specimens.

4. Comparison between 100% PC-PLA and vPLA:PC-PLA (50:50) Blended Specimens

To assess the tensile properties of 100% PC-PLA, five dog-bone-shaped samples were prepared using the optimal printing parameters and subsequently subjected to testing. The resulting tensile strength of the 100% PC-PLA was then compared with that of the vPLA:PC-PLA (50:50) blended specimens produced under the same printing conditions. The tensile strength of 100% PC-PLA was 32.86 MPa, which was 35% and 28% lower than that of the vPLA:PC-PLA (50:50) blended specimens and 100% recycled PLA, respectively (Figure 21b). The potential cause of the reduction in the tensile strength of 100% PC-PLA could be the presence of moisture during the extrusion process [97]. In addition, the variance in the feeding of PC-PLA flakes in the extrusion process would have induced a greater porosity, which would have contributed to the reduced strength. Moreover, the molecular weight of 100% PC-PLA may decrease because of the recycling process, which breaks the bonds of the original materials and reduces their tensile strength [46]. In addition, the stress–strain curve of the blended filament-printed specimens exhibited a brittle nature, with a lower strain at fracture than that of the vPLA:PC-PLA (50:50) blended specimens and 100% recycled PLA (Figure 21a). During mechanical recycling, plastics are exposed to high temperatures and shear forces, which increases their crystallinity, ultimately leading to a reduction in the strain at break of 100% PC-PLA [93,98]. However, this result demonstrates the potential of using 100% PC-PLA as a feedstock material for 3D printing.

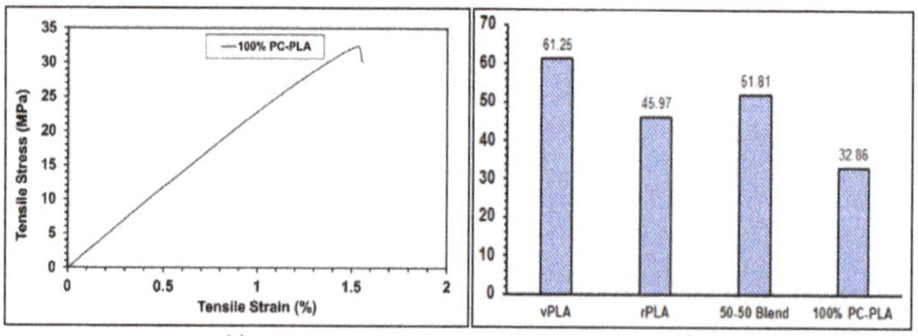

(a) (b)

Figure 21. (**a**) Stress–strain diagram of 100% PC-PLA 3D-printed tensile sample, and (**b**) comparison of UTS for vPLA, rPLA, and vPLA:PC-PLA (50:50) blended specimens and 100% PC-PLA tensile samples.

5. Conclusions and Future Work

This study explored the processing and characterisation of 3D-printed vPLA/PC-PLA blended samples. Sustainable 3D printer feedstock filaments with a 1.7 ± 0.05 mm diameter were made using a vPLA:PC-PLA (50:50) blend. Utilising the Taguchi OA, the samples were subjected to a 3D printing process with three distinct parameters: layer height, infill, and nozzle temperature. These parameters were optimised to generate the desired outcome. The physical and mechanical properties of the printed specimens were evaluated and investigated thoroughly. The incorporation of vPLA in PC-PLA to create blended filaments demonstrated a notable increase in its mechanical properties compared to those of recycled PLA. However, the strength of the blended samples was lower than that of the virgin PLA. Compared to 100% recycled PLA, the blended samples exhibited an improvement of 18.40% in tensile strength, 8% in flexural strength, 9.15% in impact strength, and 5% in hardness. Our fractographic analysis revealed that the blended samples exhibited better interlayer adhesion, contributing to the increase in mechanical properties. The driving agents for this improvement were the increased layer height and nozzle temperature. Specifically, experimental run 7 (LH: 0.3 mm, I: 60%, and NT: 215 °C) demonstrated high tensile and flexural properties. This suggests that a large layer height and high nozzle temperature can lead to increased strength, even with a low infill percentage. On the other hand, experimental run 9 (LH: 0.3 mm, I: 100%, and NT: 205 °C) showed that a high infill percentage can improve impact and hardness properties. The optimisation of the printing parameters showed that for a large layer height (0.3 mm), high infill (100%), and high nozzle temperature (215 °C), the S/N ratio was high. The ANOVA table reveals that the layer height had the most significant influence on the mechanical properties of the three printing parameters. The experimental results of the 3D-printed specimens using optimal printing parameters showed a high tensile strength of 51.81 MPa, which demonstrates the potential of using blended filaments in 3D printing. Future research should concentrate on blending various percentages of vPLA and PC-PLA. Future research could also include value-added product development using blended filaments for industrial applications and the testing of upcycled rPLA-based products. An estimation of the fatigue life and service life of PC-PLA could also be included in future research, which would provide valuable insights into its formulation durability, longevity, and sustainability. Also, it is crucial to consider the sustainability of PC-PLA for various applications. A comprehensive life-cycle assessment (LCA) of PC-PLA could provide valuable information on the environmental impacts of this material. Research in this area can lead to the development of more sustainable and environmentally friendly manufacturing materials.

Author Contributions: Conceptualisation, M.R.H., I.J.D., A.P., M.J. and W.K.B.; methodology, M.R.H., I.J.D., A.P., M.J. and W.K.B.; validation, M.R.H., I.J.D., A.P., M.J. and W.K.B.; formal analysis, M.R.H.; investigation, M.R.H.; data curation M.R.H.; writing—original draft preparation, M.R.H.; writing—review and editing, M.R.H., I.J.D., A.P., M.J. and W.K.B.; supervision, I.J.D., A.P., M.J. and W.K.B. All authors have read and agreed to the published version of the manuscript.

Funding: This research received no external funding.

Data Availability Statement: The research data will be made available at Curtin Research Data Collection.

Acknowledgments: The authors are grateful for the support from the technical staff of Curtin University, Australia.

Conflicts of Interest: The authors declare no conflicts of interest.

References

1. Mishra, V.; Ror, C.K.; Negi, S.; Kar, S.; Borah, L.N. Development of sustainable 3D printing filaments using recycled/virgin ABS blends: Processing and characterization. *Polym. Eng. Sci.* **2023**, *63*, 1890–1899. [CrossRef]
2. Belioka, M.P.; Markozanne, G.; Chrissopoulou, K.; Achilias, D.S. Advanced Plastic Waste Recycling—The Effect of Clay on the Morphological and Thermal Behavior of Recycled PET/PLA Sustainable Blends. *Polymers* **2023**, *15*, 3145. [CrossRef] [PubMed]

3. Bioplastics, E. *Bioplastics Facts and Figures*; European Bioplastics e.V.: Berlin, Germany, 2023.
4. Pinho, A.C.; Amaro, A.M.; Piedade, A.P. 3D printing goes greener: Study of the properties of post-consumer recycled polymers for the manufacturing of engineering components. *Waste Manag.* **2020**, *118*, 426–434. [CrossRef] [PubMed]
5. Björklund, A.; Finnveden, G. Recycling revisited—Life cycle comparisons of global warming impact and total energy use of waste management strategies. *Resour. Conserv. Recycl.* **2005**, *44*, 309–317. [CrossRef]
6. Singh, S.; Ramakrishna, S. Recycling of Thermoplastic Wastes: A Route of Waste to Wealth via Three-Dimensional Printing. In *Encyclopedia of Materials: Plastics and Polymers*; CRC Press: Boca Raton, FL, USA, 2022; pp. 510–515.
7. Fernández, P.; Marqués, R.; Beltrán, N.; Puerto, B.; Blanco, D. Using PETG/rPET Blends in Fused Particle Fabrication: Analysis of Feasibility and Mechanical Behaviour. In *10th Manufacturing Engineering Society International Conference (MESIC 2023), Sevilla, Spain, 28–30 June 2023*; Trans Tech Publications Ltd.: Bäch, Switzerland, 2023; pp. 147–154.
8. Kassab, A.; Al Nabhani, D.; Mohanty, P.; Pannier, C.; Ayoub, G.Y. Advancing Plastic Recycling: Challenges and Opportunities in the Integration of 3D Printing and Distributed Recycling for a Circular Economy. *Polymers* **2023**, *15*, 3881. [CrossRef] [PubMed]
9. Cestari, S.P.; Martin, P.J.; Hanna, P.R.; Kearns, M.P.; Mendes, L.C.; Millar, B. Use of virgin/recycled polyethylene blends in rotational moulding. *J. Polym. Eng.* **2021**, *41*, 509–516. [CrossRef]
10. Suárez-Eiroa, B.; Fernández, E.; Méndez-Martínez, G.; Soto-Oñate, D. Operational principles of circular economy for sustainable development: Linking theory and practice. *J. Clean. Prod.* **2019**, *214*, 952–961. [CrossRef]
11. Hadi, A.J.; Najmuldeen, G.F.; Yusoh, K.B. Dissolution/reprecipitation technique for waste polyolefin recycling using new pure and blend organic solvents. *J. Polym. Eng.* **2013**, *33*, 471–481. [CrossRef]
12. Kamalian, P.; Khorasani, S.N.; Abdolmaleki, A.; Karevan, M.; Khalili, S.; Shirani, M.; Neisiany, R.E. Toward the development of polyethylene photocatalytic degradation. *J. Polym. Eng.* **2020**, *40*, 181–191. [CrossRef]
13. Sirisinha, K.; Wirasate, S.; Sirisinha, C.; Wattanakrai, N. One-Pot Reactive Melt Recycling of PLA Post-Consumer Waste for the Production of Block Copolymer Nanocomposites of High Strength and Ductility. *Polymers* **2022**, *14*, 3642. [CrossRef]
14. Webb, H.; Arnott, J.; Crawford, R.; Ivanova, E. Plastic Degradation and Its Environmental Implications with Special Reference to Poly(ethylene terephthalate). *Polymers* **2012**, *5*, 1–18. [CrossRef]
15. Torres, N.; Robin, J.; Boutevin, B. Study of thermal and mechanical properties of virgin and recycled poly (ethylene terephthalate) before and after injection molding. *Eur. Polym. J.* **2000**, *36*, 2075–2080. [CrossRef]
16. Hopewell, J.; Dvorak, R.; Kosior, E. Plastics recycling: Challenges and opportunities. *Philos. Trans. R. Soc. B Biol. Sci.* **2009**, *364*, 2115–2126. [CrossRef]
17. Huang, Y.; Leu, M.C.; Mazumder, J.; Donmez, A. Additive Manufacturing: Current State, Future Potential, Gaps and Needs, and Recommendations. *J. Manuf. Sci. Eng.* **2015**, *137*, 014001. [CrossRef]
18. Ferreira, R.T.L.; Amatte, I.C.; Dutra, T.A.; Bürger, D. Experimental characterization and micrography of 3D printed PLA and PLA reinforced with short carbon fibers. *Compos. Part B Eng.* **2017**, *124*, 88–100. [CrossRef]
19. Jayawardane, H.; Davies, I.J.; Gamage, J.R.; John, M.; Biswas, W.K. Sustainability perspectives—A review of additive and subtractive manufacturing. *Sustain. Manuf. Serv. Econ.* **2023**, *2*, 100015. [CrossRef]
20. Pavlo, S.; Fabio, C.; Hakim, B.; Mauricio, C. 3D-Printing Based Distributed Plastic Recycling: A Conceptual Model for Closed-Loop Supply Chain Design. In Proceedings of the 2018 IEEE International Conference on Engineering, Technology and Innovation (ICE/ITMC), Stuttgart, Germany, 17–20 June 2018.
21. Hidalgo-Carvajal, D.; Munoz, A.H.; Garrido-Gonzalez, J.J.; Carrasco-Gallego, R.; Alcazar Montero, V. Recycled PLA for 3D Printing: A Comparison of Recycled PLA Filaments from Waste of Different Origins after Repeated Cycles of Extrusion. *Polymers* **2023**, *15*, 3651. [CrossRef] [PubMed]
22. Pulok, M.K.H.; Rahman, M.S.; Chakravarty, U.K. Crack Propagation and Fracture Toughness of Additively Manufactured Polymers. In Proceedings of the ASME 2021 International Mechanical Engineering Congress and Exposition. Volume 4: Advances in Aerospace Technology, Virtual, Online, 1–5 November 2021; ASME: New York, NY, USA, 2021.
23. Jayawardane, H.; Davies, I.J.; Gamage, J.R.; John, M.; Biswas, W.K. Additive manufacturing of recycled plastics: A 'techno-eco-efficiency' assessment. *Int. J. Adv. Manuf. Technol.* **2023**, *126*, 1471–1496. [CrossRef]
24. Al-Mazrouei, N.; Al-Marzouqi, A.H.; Ahmed, W. Characterization and Sustainability Potential of Recycling 3D-Printed Nylon Composite Wastes. *Sustainability* **2022**, *14*, 10458. [CrossRef]
25. Rezvani Ghomi, E.R.; Khosravi, F.; Saedi Ardahaei, A.S.; Dai, Y.; Neisiany, R.E.; Foroughi, F.; Wu, M.; Das, O.; Ramakrishna, S. The Life Cycle Assessment for Polylactic Acid (PLA) to Make It a Low-Carbon Material. *Polymers* **2021**, *13*, 1854. [CrossRef]
26. Balla, E.; Daniilidis, V.; Karlioti, G.; Kalamas, T.; Stefanidou, M.; Bikiaris, N.D.; Vlachopoulos, A.; Koumentakou, I.; Bikiaris, D.N. Poly(lactic Acid): A Versatile Biobased Polymer for the Future with Multifunctional Properties-From Monomer Synthesis, Polymerization Techniques and Molecular Weight Increase to PLA Applications. *Polymers* **2021**, *13*, 1822. [CrossRef] [PubMed]
27. Bioplastics, E. *Bioplastics Market Development Update 2023*; European Bioplastics e.V.: Berlin, Germany, 2023.
28. Taib, N.-A.A.B.; Rahman, M.R.; Huda, D.; Kuok, K.K.; Hamdan, S.; Bakri, M.K.B.; Julaihi, M.R.M.B.; Khan, A. A review on poly lactic acid (PLA) as a biodegradable polymer. *Polym. Bull.* **2022**, *80*, 1179–1213. [CrossRef]
29. Yao, B.; Zhu, Y.; Xu, Z.; Wu, Y.; Yang, L.; Liu, J.; Shang, J.; Fan, J.; Ouyang, L.; Fan, H.J.S. Taguchi design and optimization of the PLA/PCL composite filament with plasticizer and compatibilizer additives for optimal 3D printing. *Polym. Eng. Sci.* **2023**, *63*, 3743–3761. [CrossRef]

30. Alexandre, A.; Cruz Sanchez, F.A.; Boudaoud, H.; Camargo, M.; Pearce, J.M. Mechanical Properties of Direct Waste Printing of Polylactic Acid with Universal Pellets Extruder: Comparison to Fused Filament Fabrication on Open-Source Desktop Three-Dimensional Printers. *3D Print. Addit. Manuf.* **2020**, *7*, 237–247. [CrossRef]
31. Hasan, M.R.; Davies, I.J.; Pramanik, A.; John, M.; Biswas, W.K. Potential of Recycled PLA in 3D Printing: A Review. *Sustain. Manuf. Serv. Econ.* **2024**, *3*, 100020. [CrossRef]
32. Beltran, F.R.; Lorenzo, V.; Acosta, J.; de la Orden, M.U.; Martinez Urreaga, J. Effect of simulated mechanical recycling processes on the structure and properties of poly(lactic acid). *J. Environ. Manag.* **2018**, *216*, 25–31. [CrossRef] [PubMed]
33. Żenkiewicz, M.; Richert, J.; Rytlewski, P.; Moraczewski, K.; Stepczyńska, M.; Karasiewicz, T. Characterisation of multi-extruded poly(lactic acid). *Polym. Test.* **2009**, *28*, 412–418. [CrossRef]
34. Pillin, I.; Montrelay, N.; Bourmaud, A.; Grohens, Y. Effect of thermo-mechanical cycles on the physico-chemical properties of poly(lactic acid). *Polym. Degrad. Stab.* **2008**, *93*, 321–328. [CrossRef]
35. Zhao, P.; Rao, C.; Gu, F.; Sharmin, N.; Fu, J. Close-looped recycling of polylactic acid used in 3D printing: An experimental investigation and life cycle assessment. *J. Clean. Prod.* **2018**, *197*, 1046–1055. [CrossRef]
36. Babagowda; Math, R.S.K.; Goutham, R.; Prasad, K.R.S. Study of Effects on Mechanical Properties of PLA Filament which is blended with Recycled PLA Materials. In Proceedings of the International Conference on Advances in Materials and Manufacturing Applications (IConAMMA-2017), IOP Conference Series: Materials Science and Engineering, Bengaluru, India, 17–19 August 2018; IOP Publishing: London, UK, 2018.
37. Tanney, D.; Meisel, N.A.; Moore, J.D. Investigating material degradation through the recycling of PLA in additively manufactured parts. In Proceedings of the 2017 Annual International Solid Freeform Fabrication Symposium (28th Annual International Solid Freeform Fabrication Symposium)—An Additive Manufacturing Conference, Austin, TX, USA, 7–9 August 2017; Bourell, D.L., Crawford, R.H., Seepersad, C.C., Beaman, J.J., Fish, S., Eds.; University of Texas at Austin: Austin, TX, USA, 2017.
38. Bergaliyeva, S.; Sales, D.L.; Delgado, F.J.; Bolegenova, S.; Molina, S.I. Manufacture and Characterization of Polylactic Acid Filaments Recycled from Real Waste for 3D Printing. *Polymers* **2023**, *15*, 2165. [CrossRef]
39. Kam, M.; İpekci, A.; Şengül, Ö. Taguchi Optimization of Fused Deposition Modeling Process Parameters on Mechanical Characteristics of PLA+ Filament Material. *Sci. Iran.* **2021**, *29*, 79–89. [CrossRef]
40. Kam, M.; İpekçi, A.; Şengül, Ö. Investigation of the effect of FDM process parameters on mechanical properties of 3D printed PA12 samples using Taguchi method. *J. Thermoplast. Compos. Mater.* **2021**, *36*, 307–325. [CrossRef]
41. Atakok, G.; Kam, M.; Koc, H.B. Tensile, three-point bending and impact strength of 3D printed parts using PLA and recycled PLA filaments: A statistical investigation. *J. Mater. Res. Technol.* **2022**, *18*, 1542–1554. [CrossRef]
42. Chen, W.H.; Uribe, M.C.; Kwon, E.E.; Lin, K.Y.A.; Park, Y.K.; Ding, L.; Saw, L.H. A comprehensive review of thermoelectric generation optimization by statistical approach: Taguchi method, analysis of variance (ANOVA), and response surface methodology (RSM). *Renew. Sustain. Energy Rev.* **2022**, *169*, 112917. [CrossRef]
43. Nikam, M.; Pawar, P.; Patil, A.; Patil, A.; Mokal, K.; Jadhav, S. Sustainable fabrication of 3D printing filament from recycled PET plastic. *Mater. Today Proc.* **2023**. [CrossRef]
44. Chu, J.S.; Koay, S.C.; Chan, M.Y.; Choo, H.L.; Ong, T.K. Recycled plastic filament made from post-consumer expanded polystyrene and polypropylene for fused filamant fabrication. *Polym. Eng. Sci.* **2022**, *62*, 3786–3795.
45. Mishra, V.; Ror, C.H.K.; Negi, S.; Kar, S.; Borah, L.N. 3D printing with recycled ABS resin: Effect of blending and printing temperature. *Mater. Chem. Phys.* **2023**, *309*, 128317.
46. Di, L.; Yang, Y. Towards closed-loop material flow in additive manufacturing: Recyclability analysis of thermoplastic waste. *J. Clean. Prod.* **2022**, *362*, 132427. [CrossRef]
47. Naganaboyina, H.P.S.; Nagaraju, P.; Sonaye, S.Y.; Bokam, V.K.; Sikder, P. In-house processing of carbon fiber-reinforced polyetheretherketone (CFR-PEEK) 3D printable filaments and fused filament fabrication-3D printing of CFR-PEEK parts. *Int. J. Adv. Manuf. Technol.* **2023**, *128*, 5011–5024. [CrossRef]
48. Hasan, M.R.; Davies, I.J.; Pramanik, A.; John, M.; Biswas, W.K. Impact of process parameters on improving the performance of 3D printed recycled polylactic acid (rPLA) components. *Int. J. Adv. Manuf. Technol.* **2024**, *131*, 3751–3779. [CrossRef]
49. *ASTM-D638-22*; Standard Test Method for Tensile Properties of Plastics. ASTM International: West Conshohocken, PA, USA, 2022.
50. *ASTM-D790-17*; Standard Test Methods for Flexural Properties of Unreinforced and Reinforced Plastics and Electrical Insulating Materials. ASTM International: West Conshohocken, PA, USA, 2017.
51. *ASTM-D6110-18*; Standard Test Method for Determining the Charpy Impact Resistance of Notched Specimens of Plastics. ASTM International: West Conshohocken, PA, USA, 2018.
52. *ASTM-D2240-15*; Standard Test Method for Rubber Property-Durometer Hardness. ASTM International: West Conshohocken, PA, USA, 2021.
53. *ASTM-D7127-17*; Standard Test Method for Measurement of Surface Roughness of Abrasive Blast Cleaned Metal Surfaces Using a Portable Stylus Instrument. ASTM International: West Conshohocken, PA, USA, 2017.
54. Dehnad, K. *Quality Control, Robust Design, and the Taguchi Method*; Springer: New York, NY, USA, 2012.
55. Ahmad, M.N.; Ishak, M.R.; Mohammad Taha, M.; Mustapha, F.; Leman, Z.; Anak Lukista, D.D.; Irianto; Ghazali, I. Application of Taguchi Method to Optimize the Parameter of Fused Deposition Modeling (FDM) Using Oil Palm Fiber Reinforced Thermoplastic Composites. *Polymers* **2022**, *14*, 2140. [CrossRef] [PubMed]

56. Salam, H.; Dong, Y.; Davies, I.J.; Pramanik, A. Optimization of material formulation and processing parameters in relation to mechanical properties of bioepoxy/clay nanocomposites using Taguchi design of experiments. *J. Appl. Polym. Sci.* **2018**, *135*, 45769. [CrossRef]
57. Greenwald, A.G.; Gonzalez, R.; Harris, R.J.; Guthrie, D. Effect sizes and p values: What should be reported and what should be replicated? *Psychophysiology* **1996**, *33*, 175–183. [CrossRef] [PubMed]
58. Son, J.; Kim, I.; Kim, H.; Kim, I.; Kang, B.; Kim, H. A study on the prediction of bead geometry in the robotic welding system. *J. Mech. Sci. Technol.* **2007**, *21*, 1726–1731. [CrossRef]
59. Muhammad, N.; Manurung, Y.H.P.; Jaafar, R.; Abas, S.K.; Tham, G.; Haruman, E. Model development for quality features of resistance spot welding using multi-objective Taguchi method and response surface methodology. *J. Intell. Manuf.* **2012**, *24*, 1175–1183. [CrossRef]
60. Belavendram, N. *Quality by Design, Taguchi Techniques for Industrial Experimentation*; Prentice Hall: Hoboken, NJ, USA, 1995.
61. Roy, R.K. *A Primer on the Taguchi Method*, 2nd ed.; Society of Manufacturing Engineers: Southfield, MI, USA, 2010.
62. Maier, R.; Bucaciuc, S.G.; Mandoc, A.C. Reducing Surface Roughness of 3D Printed Short-Carbon Fiber Reinforced Composites. *Materials* **2022**, *15*, 7398. [CrossRef] [PubMed]
63. Kumar, S.; Ramesh, M.R.; Doddamani, M. Recycling potential of MWCNTs/HDPE nanocomposite filament: 3D printing and mechanical characterization. *J. Mater. Cycles Waste Manag.* **2023**, *25*, 1168–1178. [CrossRef]
64. Dorigato, A. Recycling of polymer blends. *Adv. Ind. Eng. Polym. Res.* **2021**, *4*, 53–69. [CrossRef]
65. Kim, G.; Barocio, E.; Tsutsui, W.; Wang, P.; Dubikovsky, S.; Pipes, R.B.; Sterkenburg, R. Enhancing surface characteristics of additively manufactured fiber reinforced thermoplastic mold using thermoset coating with ceramic particles. *Surf. Coat. Technol.* **2021**, *422*, 127536. [CrossRef]
66. Kumar, S.; Singh, R.; Singh, T.P.; Batish, A. On investigations of thermal conductivity, circumferential compressive strength, and surface characterization of 3D-printed hybrid blended magnetostrictive PLA composite. *J. Thermoplast. Compos. Mater.* **2020**, *35*, 631–650. [CrossRef]
67. Hsueh, M.H.; Lai, C.J.; Liu, K.Y.; Chung, C.F.; Wang, S.H.; Pan, C.Y.; Huang, W.C.; Hsieh, C.H.; Zeng, Y.S. Effects of Printing Temperature and Filling Percentage on the Mechanical Behavior of Fused Deposition Molding Technology Components for 3D Printing. *Polymers* **2021**, *13*, 2910. [CrossRef] [PubMed]
68. Mani, M.; Karthikeyan, A.G.; Kalaiselvan, K.; Muthusamy, P.; Muruganandhan, P. Optimization of FDM 3-D printer process parameters for surface roughness and mechanical properties using PLA material. *Mater. Today Proc.* **2022**, *66*, 1926–1931. [CrossRef]
69. Maguluri, N.; Suresh, G.; Guntur, S.R. Effect of printing parameters on the hardness of 3D printed poly-lactic acid parts using DOE approach. In Proceedings of the International Conference on Materials Science and Engineering (ICMSE 2022), IOP Conference Series: Materials Science and Engineering, Virtual, Online, 6 November–6 December 2022; IOP Publishing: London, UK, 2022.
70. Abbas, Z.A.; Aleabi, S.H. Studying some of mechanical properties (tensile, impact, hardness) and thermal conductivity of polymer blend reinforce by magnesium oxide. In *Xiamen-Custipen Workshop on the Equation of State of Dense Neutron-Rich Matter in the Era of Gravitational Wave Astronomy*; AIP Publishing LLC: Melville, NY, USA, 2019.
71. Samykano, M.; Selvamani, S.K.; Kadirgama, K.; Ngui, W.K.; Kanagaraj, G.; Sudhakar, K. Mechanical property of FDM printed ABS: Influence of printing parameters. *Int. J. Adv. Manuf. Technol.* **2019**, *102*, 2779–2796. [CrossRef]
72. Farashi, S.; Vafaee, F. Effect of printing parameters on the tensile strength of FDM 3D samples: A meta-analysis focusing on layer thickness and sample orientation. *Prog. Addit. Manuf.* **2022**, *7*, 565–582. [CrossRef]
73. Vanaei, H.R.; Raissi, K.; Deligant, M.; Shirinbayan, M.; Fitoussi, J.; Khelladi, S.; Tcharkhtchi, A. Toward the understanding of temperature effect on bonding strength, dimensions and geometry of 3D-printed parts. *J. Mater. Sci.* **2020**, *55*, 14677–14689. [CrossRef]
74. Sanusi, H.; Mohamad Ridzwan, I.; Mohd Sapuan, S.; Noorfaizal, Y.; Syamir Alihan Showkat, A.; Mohd Sabri, H.; Maliki, I.; Asmawi, S. Tensile Properties of 3D Printed Recycled PLA Filament: A Detailed Study on Filament Fabrication Parameters. *J. Adv. Res. Appl. Mech.* **2023**, *110*, 63–72. [CrossRef]
75. Sam-Daliri, O.; Ghabezi, P.; Steinbach, J.; Flanagan, T.; Finnegan, W.; Mitchell, S.; Harrison, N. Experimental study on mechanical properties of material extrusion additive manufactured parts from recycled glass fibre-reinforced polypropylene composite. *Compos. Sci. Technol.* **2023**, *241*, 110125. [CrossRef]
76. Algarni, M. The Influence of Raster Angle and Moisture Content on the Mechanical Properties of PLA Parts Produced by Fused Deposition Modeling. *Polymers* **2021**, *13*, 237. [CrossRef] [PubMed]
77. Hamid, R.A.; Hamezah, F.H.; Abd Razak, J. Influence of Humidity on the Tensile Strength of 3D Printed PLA Filament. In *Intelligent Manufacturing and Mechatronics*; Ali Mokhtar, M.N., Jamaludin, Z., Abdul Aziz, M.S., Maslan, M.N., Razak, J.A., Eds.; Springer: Singapore, 2022.
78. Aniskevich, A.; Bulderberga, O.; Stankevics, L. Moisture Sorption and Degradation of Polymer Filaments Used in 3D Printing. *Polymers* **2023**, *15*, 2600. [CrossRef] [PubMed]
79. Ror, C.K.; Negi, S.; Mishra, V. Development and characterization of sustainable 3D printing filaments using post-consumer recycled PET: Processing and characterization. *J. Polym. Res.* **2023**, *30*, 350. [CrossRef]
80. Sharma, P.; Dhanopia, A.K.; Joshi, D. An experimental study on carbon fiber thickness and layer thickness of depositing material in fused deposition modeling. *Mater. Today Proc.* **2021**, *44*, 4479–4484. [CrossRef]
81. Nugroho, A.A.R.; Rusita, L.; Larasati, I.L. Effect of layer thickness on flexural properties of PLA (PolyLactid Acid) by 3D printing. *IOP Conf. Ser. J. Phys.* **2018**, *1130*, 012017. [CrossRef]

82. Nofar, M.; Sacligil, D.; Carreau, P.J.; Kamal, M.R.; Heuzey, M.C. Poly (lactic acid) blends: Processing, properties and applications. *Int. J. Biol. Macromol.* **2019**, *125*, 307–360. [CrossRef] [PubMed]
83. Tsuji, H. Poly(lactic acid) stereocomplexes: A decade of progress. *Adv. Drug Deliv. Rev.* **2016**, *107*, 97–135. [CrossRef] [PubMed]
84. Chen, C.C.; Chueh, J.Y.; Tseng, H.; Huang, H.M.; Lee, S.Y. Preparation and characterization of biodegradable PLA polymeric blends. *Biomaterials* **2003**, *24*, 1167–1173. [CrossRef] [PubMed]
85. Syrlybayev, D.; Zharylkassyn, B.; Seisekulova, A.; Akhmetov, M.; Perveen, A.; Talamona, D. Optimisation of Strength Properties of FDM Printed Parts-A Critical Review. *Polymers* **2021**, *13*, 1587. [CrossRef] [PubMed]
86. Rajpurohit, S.R.; Dave, H.K. Impact strength of 3D printed PLA using open source FFF-based 3D printer. *Prog. Addit. Manuf.* **2020**, *6*, 119–131. [CrossRef]
87. Yu, W.; Sun, L.; Li, M.; Li, M.; Lei, W.; Wei, C. FDM 3D Printing and Properties of PBS/PLA Blends. *Polymers* **2023**, *15*, 4305. [CrossRef] [PubMed]
88. Ding, S.; Zou, B.; Wang, P.; Ding, H. Effects of nozzle temperature and building orientation on mechanical properties and microstructure of PEEK and PEI printed by 3D-FDM. *Polym. Test.* **2019**, *78*, 105948. [CrossRef]
89. Gunasekaran, K.N.; Aravinth, V.; Muthu Kumaran, C.B.; Madhankumar, K.; Pradeep Kumar, S. Investigation of mechanical properties of PLA printed materials under varying infill density. *Mater. Today Proc.* **2021**, *45*, 1849–1856. [CrossRef]
90. Maqsood, N.; Rimašauskas, M. Tensile and flexural response of 3D printed solid and porous CCFRPC structures and fracture interface study using image processing technique. *J. Mater. Res. Technol.* **2021**, *14*, 731–742. [CrossRef]
91. Tekinalp, H.L.; Kunc, V.; Velez-Garcia, G.M.; Duty, C.E.; Love, L.J.; Naskar, A.K.; Blue, C.A.; Ozcan, S. Highly oriented carbon fiber–polymer composites via additive manufacturing. *Compos. Sci. Technol.* **2014**, *105*, 144–150. [CrossRef]
92. Kanabenja, W.; Passarapark, K.; Subchokpool, T.; Nawaaukkaratharnant, N.; Román, A.J.; Osswald, T.A.; Aumnate, C.; Potiyaraj, P. 3D printing filaments from plasticized Polyhydroxybutyrate/Polylactic acid blends reinforced with hydroxyapatite. *Addit. Manuf.* **2022**, *59*, 103130. [CrossRef]
93. Gabriel, D.S.; Tiana, A.N. Mechanical Properties Improvement of Recycled Polypropylene with Material Value Conservation Schemes Using Virgin Plastic Blends. *Mater. Sci. Forum* **2020**, *1015*, 76–81. [CrossRef]
94. Maguluri, N.; Suresh, G.; Rao, K.V. Assessing the effect of FDM processing parameters on mechanical properties of PLA parts using Taguchi method. *J. Thermoplast. Compos. Mater.* **2023**, *36*, 1472–1488. [CrossRef]
95. Roy, R.K. *Design of Experiments Using the Taguchi Approach: 16 Steps to Product and Process Improvement*; John Wiley & Sons: Hoboken, NJ, USA, 2001.
96. Teharia, R.; Singari, R.M.; Kumar, H. Optimization of process variables for additive manufactured PLA based tensile specimen using taguchi design and artificial neural network (ANN) technique. *Mater. Today Proc.* **2022**, *56*, 3426–3432. [CrossRef]
97. Little, H.A.; Tanikella, N.G.; Reich, M.J.; Fiedler, M.J.; Snabes, S.L.; Pearce, J.M. Towards Distributed Recycling with Additive Manufacturing of PET Flake Feedstocks. *Materials* **2020**, *13*, 4273. [CrossRef] [PubMed]
98. Cisneros-López, E.O.; Pal, A.K.; Rodriguez, A.U.; Wu, F.; Misra, M.; Mielewski, D.F.; Kiziltas, A.; Mohanty, A.K. Recycled poly(lactic acid)–based 3D printed sustainable biocomposites: A comparative study with injection molding. *Mater. Today Sustain.* **2020**, *7–8*, 100027. [CrossRef]

processes MDPI

Article

Towards Sustainability in Hydraulic Machinery Manufacturing by 3D Printing

Abel Remache [1], Modesto Pérez-Sánchez [2,*], Víctor Hugo Hidalgo [3], Helena M. Ramos [4] and Francisco-Javier Sánchez-Romero [5,*]

1 Industrial Design Department, Facultad de Ingeniería y Ciencias Aplicadas, Universidad Central del Ecuador, Quito 170129, Ecuador; apremache@uce.edu.ec
2 Hydraulic Engineering and Environmental Department, Universitat Politècnica de València, 46022 Valencia, Spain
3 Laboratorio de Mecánica Informática, Facultad de Ingeniería Mecánica, Escuela Politécnica Nacional, Quito 170517, Ecuador; victor.hidalgo@epn.edu.ec
4 Civil Engineering Research and Innovation for Sustainability (CERIS), Department of Civil Engineering, Architecture and Environment, Instituto Superior Técnico, University of Lisbon, 1049-001 Lisbon, Portugal; hramos.ist@gmail.com or helena.ramos@tecnico.ulisboa.pt
5 Rural and Agrifood Engineering Department, Universitat Politècnica de València, 46022 Valencia, Spain
* Correspondence: mopesan1@upv.es (M.P.-S.); fcosanro@agf.upv.es (F.-J.S.-R.)

Abstract: Material wear, maintenance costs, performance, efficiency, and corrosion are some of the issues that turbomachinery impellers may encounter. The optimization of impellers through additive manufacturing (AM) has been the focus of extensive research, aiming to address these challenges in turbine, pump, compressor, fan, and mixer components. This research aims to identify and analyze the main techniques currently being developed to tackle several of these issues. Evaluating the published research, the methodology highlights various AM techniques applied to impellers and related components, as well as the diverse materials used in functional system elements. The analysis revealed that the most commonly used additive manufacturing technologies for the production of turbomachinery components are FDM, with a 22% application rate, and powder bed fusion technology, accounting for 35%, utilized for high-complexity parts and even superalloys. Although more expensive, these technologies employ materials with superior resistance capabilities, surpass the limitations of conventional machining, optimize manufacturing times, and allow for the fine-tuning of multiple parameters. In terms of wear and corrosion resistance, materials such as Inconel 718 exhibited a loss of less than 0.1 mpy (mils per year) in highly corrosive environments, representing a significant improvement over traditional materials.

Keywords: impellers; turbomachinery; additive manufacturing; 3D printing; materials; corrosion

1. Introduction

Turbomachinery encompasses a variety of equipment, including turbines, pumps, compressors, fans, and mixers, and plays a vital role in numerous industrial applications [1]. However, it faces significant challenges, such as material degradation, the need for resource efficiency, maintenance expenses, and the pursuit of optimal performance [2,3]. Failures in turbomachinery are frequently caused by physical factors in the operating environment, with impeller deterioration and damage to adjacent components often attributed to corrosion (related to the chemical attack from the environment) [4,5], erosion (related to the quantity and size of erosive particles or liquid droplets) [6], and cavitation (due to fluid pressure changes) [4]. Studies suggest that the absence of specific classification criteria for material properties suited to their intended applications is a major factor contributing to turbine wear. This issue is compounded by limited research on frictional heating, surface interactions, and contact forces [7]. Expanding the range of available materials and improving knowledge on material selection could significantly alleviate these challenges.

Recent studies have achieved promising results by focusing on critical operational aspects of turbomachinery. For instance, improvements in surface finish and reductions in roughness have been shown to enhance resistance to water droplet erosion (WDE) [8] and to mitigate pressure effects, thereby preserving efficiency [9,10]. Vibrations, which influence lifespan, fatigue failures, and efficiency [11], are frequently caused by cavitation and resonance. Factors such as overspeed and prolonged operation further contribute to wear and are associated with increased vibrations [12]. Currently, efforts are being made to identify failures through algorithms and Industry 4.0 elements, such as machine learning [13], Industrial Internet of Things (IIoT) [14], additive manufacturing (AM), digital twins, among others [15].

Centrifugal compressors often fall short in applications where the process fluid is dense or liquid [16], a crucial consideration for energy systems. Energy systems are fundamental in turbomachinery as they must be reliable; thus, new processes are being optimized. For example, the use of supercritical CO_2 (sCO_2) in energy systems promises efficient, compact, and reliable energy [17]. The impeller, a key component of turbomachinery, primarily faces issues with corrosion [18]. When operating in contact with liquids and solids, the wear from erosion becomes significant, directly affecting the materials used [19].

Erosion degrades materials, leading to losses in performance, productivity, and maintenance efficiency. In centrifugal pumps, the suction surface of the blades is generally most affected by erosion [20]. Erosive wear in turbomachinery can be mitigated by minimizing the quantity and size of erosive particles, though each working environment is unique and presents distinct characteristics. It is crucial to use appropriate materials or coatings resistant to erosion [4]. Particulate characteristics have a direct relationship with erosive damage, reducing pump performance. Larger and more angular particles cause more severe erosion damage. Moreover, particle shape is known to impact turbomachinery performance less significantly than particle size and concentration [21]. In centrifugal pumps, performance diminishes with increased sediment concentrations.

In terms of corrosion, defects become evident over time due to workload and the rotational performance of the impellers [22]. In corrosive and aggressive environments, it is essential to use improved and optimized materials suitable for the specific conditions. Efforts should be made to reduce or eliminate sources or conditions leading to the formation of corrosive compounds [4]. Enhancements in manufacturing methods, chemical compositions, heat treatments, and surface coatings have led to the development of corrosion-resistant turbomachinery components. The mechanical properties of impellers before and after exposure to corrosion have been shown to remain unchanged, indicating that the effects are predominantly external [23].

Corrosion is an electrochemical process that involves oxidation and material loss from metallic surfaces. Thus, polymeric materials and composites are being explored as solutions to this issue. Impellers are critical components that directly influence the performance of turbomachinery; their complex geometries, operational speeds, flow patterns, and fluid compositions typically lead to non-uniform wear [24]. In this context, topological optimization and the use of AM have provided practical solutions. For instance, Laser Metal Deposition (LMD) has enabled the production of optimized parts with complex geometries, maintaining or even exceeding the mechanical properties of conventionally manufactured parts while also offering the required corrosion resistance [25].

Cavitation significantly influences the design and operation of pumps, particularly as operational speeds increase [26]. Research has shown that cavitation typically initiates at the blade's leading edge due to a local pressure drop [27]. Optimizing the shape of the impeller blade's leading edge using AM can enhance performance and mitigate the effects of cavitation [28]. Cavitation often arises from complex flow dynamics, affecting the efficiency, safety, and stability of turbomachinery [29]. In pumps, cavitation is characterized by persistent noise in the suction section, followed by increased vibration and noise levels, along with reduced total head and output capacity [30]. When cavitation ranges increase, pump performance tends to decline significantly [31].

Flow dynamics are vital in the design of turbomachine components. For instance, various types of turbomachinery can be applied, studied, and analyzed according to the Japikse and Baines design window, as illustrated in Figure 1 [1].

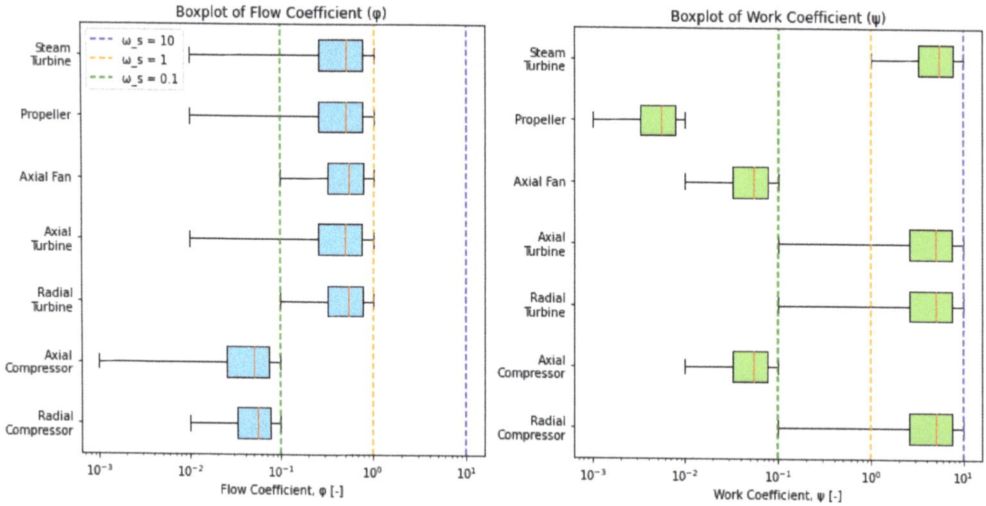

Figure 1. Turbomachines design windows.

Optimizing impellers through advanced manufacturing has become a focal area of research to address the challenges mentioned previously. Material removal manufacturing techniques have shown promising results. The advent of new manufacturing methods and materials has facilitated improvements in addressing specific issues; for instance, the use of unspecified materials is a known cause of impeller failure [24]. Vibration reduction through piezoelectric materials can semi-actively dampen vibrations [32], and the use of carbon nanocomposites such as carbon nanofibers, carbon nanotubes, and graphene has generated positive outcomes in vibration management [33]. Corrosion effects on compressor impellers can be mitigated through surface treatments and coatings [34,35], though certain thermal treatments may induce deviations in blade geometry [36].

Polymeric materials based on epoxy resin with silicon carbide exhibit the highest wear resistance, while silumin shows lower wear resistance. Testing on terrestrial pumps has revealed that parts made from electrocorundum in an epoxy-phenol-formaldehyde bond, compared to parts made from wear-resistant cast iron, can increase the lifespan of pumps by three to eight times [37].

In Computed Numeric Control (CNC) processes, thermoplastics and composites, such as Polyether-ether-ketone (PEEK) and Polyphenylene sulfide (PPS), with and without reinforcements, have been used as suitable options for manufacturing impellers [38]. PEEK and epoxy reinforced with carbon fiber are the most commonly used composites for impeller production via this method, but they have resulted in material waste and excessive energy consumption [38,39].

AM has revolutionized numerous industries. In the medical and dental sectors, SLA and DLP technologies enable the production of customized devices, anatomical models, and prosthetics with high precision and biocompatibility. In the automotive and aerospace fields, AM facilitates the creation of lightweight and functional components using shape-memory polymers and composite materials [40]. It has also allowed the reproduction of discontinued parts through 3D scanning and modeling, generating components that are otherwise unavailable in the market. Additionally, AM proves valuable for low-scale production, rapid tooling, and the development of digital inventories, optimizing supply chains [41]. The production of carbon fiber-specific parts and generative design are highly valued for their durability and adaptability. Furthermore, AM enables the manufacturing of complex tools and molds for injection molding and casting, significantly accelerating production processes [41,42].

In the space sector, 3D printing techniques such as Binder Jetting and FDM are being explored to produce robust structures capable of withstanding extreme temperature fluctuations and meteoroid impacts. These methods ensure that structures maintain adequate thickness and that printing equipment remains resilient under vacuum conditions [42].

From a sustainability perspective, the use of biopolymers and recyclable materials in 3D printing reduces environmental impact. AM also fosters innovation through its integration into education and research. To meet sustainability requirements, AM contributes to reductions in energy consumption, production costs, lead times, material waste, and greenhouse gas emissions while promoting circular economy practices [43]. However, AM does not completely eliminate waste generation, and some evidence suggests that its energy consumption may exceed that of casting and injection molding processes [44,45]. Efforts to make AM more sustainable are ongoing, focusing on areas such as support structure optimization, print orientation, post-processing, and design strategies [46]. These measures can enhance eco-efficiency; for example, in the transportation sector, AM-produced components are lighter, reducing fuel consumption [47].

Most AM processes require post-processing steps, ranging from polishing to heat treatments and coatings, to meet quality standards for aesthetics, functionality, and durability [48]. A significant challenge in the post-processing of metal structures lies in its high cost and time intensity [49]. However, the integration of big data, machine learning, and digital twins has the potential to reshape the AM landscape, driving innovation, sustainability, and efficiency [50].

The application of additive manufacturing has demonstrated enhancements in efficiency and effectiveness in impellers and is recognized as an efficient method for producing components that are challenging to machine [51]. One of the advantages of additive manufacturing is the variety of materials it employs, such as polymers, metals, composites, and ceramics, in addition to generating complex geometries [52]. However, significant differences exist among 3D printing methods; for example, Stereolithography (SLA) and MultiJet (MJ) printing can offer more suitable surface finishes due to their production processes [53]. Studies related to microstructural defects produced by this process are still sparse, though several mention that the microstructure does not undergo radical changes [54–56].

Polymeric Composite Materials (PCMs) have become popular due to their anti-corrosive properties and competitiveness with traditional materials, though market competition remains low. Additive manufacturing (AM) has opened a new window of application in hydraulic machinery, creating more valuable components and materials with reliable performance, addressing many of the Sustainable Development Goals (SDGs) [57].

In the specific context of turbomachinery and additive manufacturing, the current state of the art provides a deep understanding of advanced materials and innovative techniques used in impeller production. The integration of 3D printing in this field not only facilitates the direct creation of parts from digital models but also introduces a range of new possibilities for the design and optimization of critical components. However, this advancement also brings a series of technical challenges and issues that require innovative solutions and a rigorous approach. This review aims to thoroughly examine the different additive manufacturing methods applied in the production of turbomachinery components, evaluating both their advantages and limitations, and highlighting their practical application in real industrial scenarios. This analysis seeks to illustrate the current state of technology and identify emerging trends and future directions in this crucial field.

2. Materials and Methods

The methodology employed in this study is grounded in a comprehensive literature review, adhering to rigorous criteria and methodological recommendations established in the scientific literature. To ensure the validity, clarity, and utility of the analysis, the criteria proposed by [58] and the practical guidelines for methodological reviews outlined by [59] were applied. These methodological approaches are designed to enhance the precision and reproducibility of studies, ensuring that the conclusions drawn are robust and reliable. As a novelty, the methodology addresses an analysis of the mechanical and hydraulic behavior of non-classical materials in the manufacture of impellers.

In constructing the flowchart for the applied methodology, the guidelines established by [60] were followed, which delineate five critical phases in the documentary analysis process: (1) Identification and Initial Selection, where studies potentially relevant for further retrieval are identified and selected; (2) Initial Exclusion, where studies that are neither empirical nor involve both print and digital reading are excluded, ensuring that only research meeting these criteria is considered; (3) Detailed Evaluation, a phase in which a more thorough evaluation of the retrieved studies is conducted, retaining only those that surpass the preliminary selection criteria; (4) Further Exclusion, where studies employing inappropriate methodologies or unverifiable measures are eliminated to maintain scientific rigor; and (5) Final Inclusion, where studies that fulfill all the established methodological requirements are incorporated into the literature review. As a result of the rigorous application of these criteria, the scheme depicted in Figure 2 has been designed, illustrating the workflow and decisions made during the study selection process.

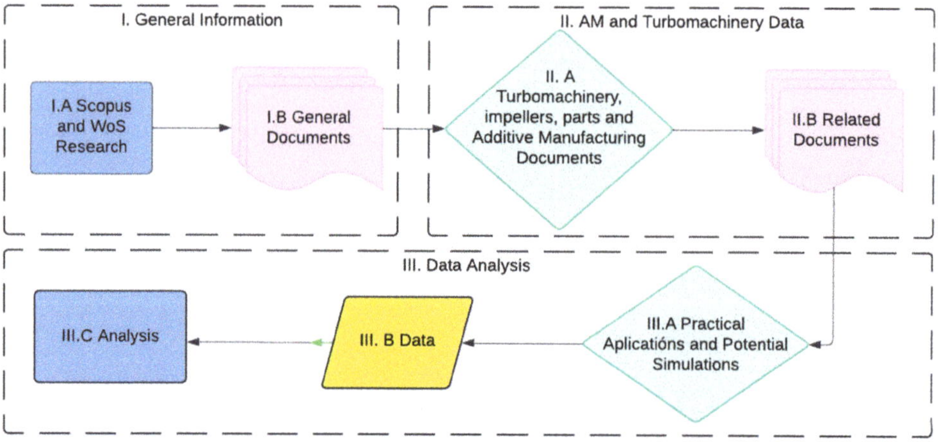

Figure 2. Flowchart of the applied methodology.

To organize and analyze the information gathered, three main groups were structured: (I) General Information, (II) Data on Additive Manufacturing (AM) and Turbomachinery, and (III) Data Analysis. These groups facilitated the systematic analysis and in-depth discussion of the findings from the review.

2.1. I. General Information

I.A Scopus and WoS Research: Initially, a general query related to the topic was conducted using search equations in the Scopus [61] and Web of Science (WoS) [62] databases, as well as high-impact articles. Reviews were excluded, and scientific articles, conference papers, conference reviews, and book chapters were considered.

Keywords and equations with Boolean operators were used as shown in Table 1. The search was conducted as follows: Impellers "AND/OR" 3D Printers, Impellers "AND/OR" 3D Printing, Impellers "AND/OR" Additive Manufacturing.

Table 1. Common materials in impellers used in additive manufacturing.

Keyword	Boolean Operator	Keyword
Impellers	AND/OR	Printers/3D Printing/Additive Manufacturing
Turbomachinery	AND/OR	Printers/3D Printing/Additive Manufacturing
DMLS-LPBF-SLM	AND/OR	Impellers/Turbomachinery/Turbomachines
FDM	AND/OR	Impellers/Turbomachinery/Turbomachines
SLS	AND/OR	Impellers/Turbomachinery/Turbomachines
DLP	AND/OR	Impellers/Turbomachinery/Turbomachines
Polyjet	AND/OR	Impellers/Turbomachinery/Turbomachines
SLA	AND/OR	Impellers/Turbomachinery/Turbomachines
BJP	AND/OR	Impellers/Turbomachinery/Turbomachines
BMD	AND/OR	Impellers/Turbomachinery/Turbomachines
EBM	AND/OR	Impellers/Turbomachinery/Turbomachines
MJF	AND/OR	Impellers/Turbomachinery/Turbomachines
LMD	AND/OR	Impellers/Turbomachinery/Turbomachines
Additive Manufacturing	AND/OR	Impellers/Turbomachinery/Turbomachines

I.B General Document Collection: A detailed collection of documents addressing a wide range of topics related to 3D printing, additive manufacturing, and turbomachinery was undertaken. This involved analyzing trends in scientific production over time and identifying the number of research studies conducted per year, which helped assess the scientific interest in these areas. Additionally, publications by country were reviewed to identify the main focuses of research and international collaboration. Finally, a co-occurrence analysis was conducted to map the relationships and dependencies between research studies, facilitating the identification of patterns and knowledge networks within the field.

2.2. II. AM and Turbomachinery Data

II.A Filtering of Documents Related to Turbomachinery and Additive Manufacturing: Once the general information was gathered, the process of filtering documents specifically relevant to the study commenced. This filtering focused on identifying studies centered on the design, manufacturing, and optimization of key turbomachinery components, such as impellers and blades, manufactured through additive manufacturing. Studies based on unverified simulations or using obsolete technologies were excluded to maintain a focus on recent and relevant empirical studies that reflect the current capabilities of the technology.

II.B Retrieval of Related Documents: From the initial filter, the compilation and organization of documents providing profound insights into the manufacturing and application of turbomachinery components via additive manufacturing were carried out. Information was classified based on the different types of additive manufacturing processes and the mixed processes used, allowing for a comparative evaluation of the various methodologies employed in the industry. This analysis included not only an examination of manufacturing techniques but also an assessment of their practical implementation, costs, challenges, and long-term benefits.

2.3. III. Data Analysis

III.A Filtering of Practical Applications and Potential Simulations: A second thorough filtering of the available documentation was conducted to identify studies presenting practical applications of 3D printing in turbomachinery. This filter particularly focused on those studies that, in addition to documenting the practical use of the technology, included simulations with a high potential for application in real-world scenarios. To ensure the validity and relevance of the information gathered, priority was given to studies published in higher quartile journals (Q1 and Q2), as these tend to offer more rigorous and reliable research. Only those studies that provided clear evidence of performance in real applications were included, ensuring that the derived conclusions were applicable and useful for the industry.

This approach not only allowed for a careful selection of the literature but also a robust validation of the presented simulations, ensuring that any theoretical or experimental model considered had a high likelihood of success in specific industrial applications. This established a solid foundation for analyzing the feasibility and comparative advantages of additive manufacturing in the production of turbomachinery components.

III.B Data Extraction: Subsequently, specific data from the selected studies were extracted. This process involved a meticulous compilation of key information, including the additive manufacturing methods used, the materials employed in the manufacturing, the parts of turbomachinery produced, and the speeds achieved. Comparative data were also gathered, allowing for an evaluation of improvements in the performance of these processes relative to traditional manufacturing methods or other advanced techniques.

Data extraction also focused on identifying significant improvements in terms of efficiency, cost, and quality of the produced components, as well as the durability and resilience of the 3D-printed materials. This enabled a detailed view of the impact of additive manufacturing on the overall performance of turbomachinery, providing critical information for the assessment of its adoption on a larger scale in the industry.

III.C Analysis: The analysis of the extracted data was conducted with a detailed and critical approach, evaluating both the feasibility and advantages of additive manufacturing methods in the production of turbomachinery components. This analysis included a comprehensive review of the selected documents, focusing on the different types of additive manufacturing processes, the materials employed, and specific characteristics such as corrosion resistance and the durability of the manufactured parts.

Possible improvements in process efficiency, as well as the practical applications of the printed materials in real-world scenarios, were discussed. Additionally, specific advantages associated with the different materials and printing technologies used were identified, evaluating their performance under typical operating conditions of turbomachinery. This analysis not only validated the simulations and practical applications identified in earlier stages but also elaborated an integral discussion on the potential improvements that additive manufacturing can offer in this sector. The results of this analysis are crucial for understanding the potential impact of these technologies on the turbomachinery industry and for guiding future research and applications.

The adopted methodology ensures a rigorous and systematic approach in the collection, filtering, and analysis of relevant information, leading to solid and applicable conclusions about the use of additive manufacturing in the production of turbomachinery components. This process, illustrated in Figure 2, ensured that the findings presented are of high relevance to both the scientific community and the industry.

3. Results and Discussion

3.1. I. General Information

The direct search related to impellers and AM technologies yielded 133 documents in Scopus and 96 documents in WoS, while the search linked to turbomachinery and AM technologies resulted in 188 documents in Scopus and 349 documents in WoS. According to Figure 3, it can be observed that from the year 2017, research in the subject increased, reaching its highest peak in 2021 and 2022, although it is notable that there is not a large amount of research in the area.

In the Scopus database, it is possible to verify the top 15 countries with the highest number of publications, with China, the United States, and India being the countries with the most significant contribution and research interest in this topic, as shown in Figure 4.

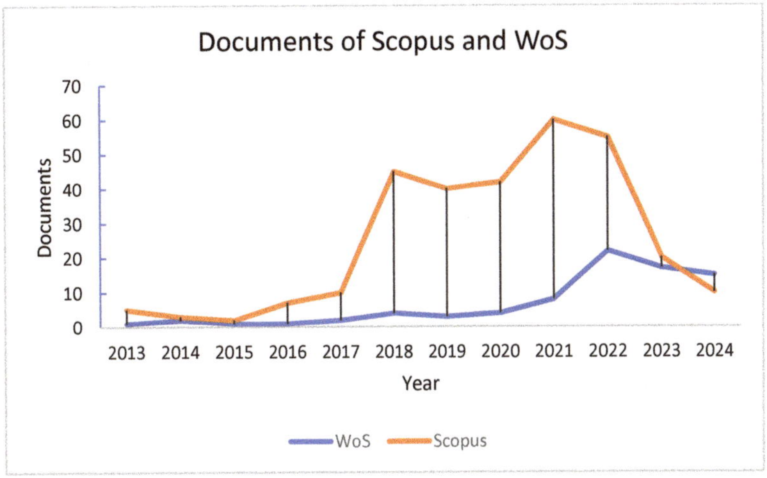

Figure 3. Documents related to the literature review from 2013 to 2024.

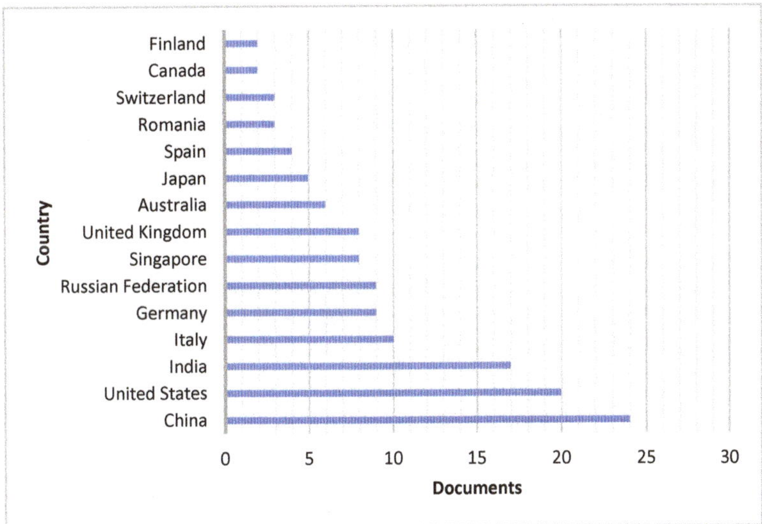

Figure 4. Documents by country related to the search up to 2024.

Using co-occurrence analysis through Vosviewer [63] with Scopus and WoS data, Figure 5 shows that four main clusters of connections have been generated: additive manufacturing, additives, 3D printers, and turbomachinery. The study relations focus on topics related to powder bed fusion technologies, design and topological optimization, characterization and use of materials, alloys, turbine design, among other main areas of study.

The density analysis (Figure 6) shows the most relevant points of study where the aforementioned topics are most evident. Materials and alloys in different 3D printing technologies are key points for researchers, while topology, design, and optimization are hot topics according to the published research.

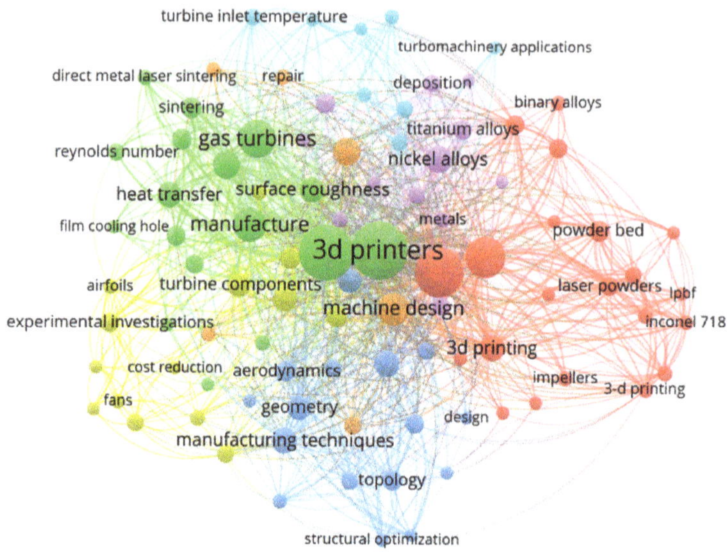

Figure 5. Co-occurrence analysis between documents and keywords.

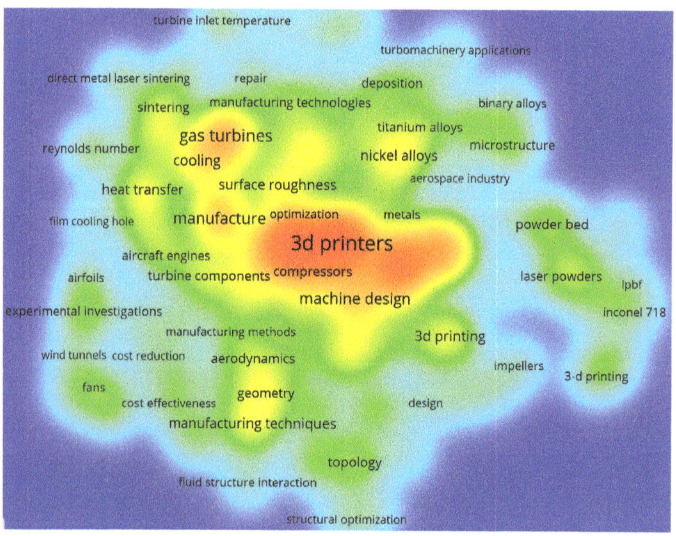

Figure 6. Heat map of scientific documents related to the review.

3.2. II. AM and Turbomachinery Data

Figure 7 shows that the most used technologies in the production of turbomachinery parts are DMLS, LPBF, SLM with an application rate above 3%, and FDM with a usage percentage above 20%. More advanced technologies, such as SLS and DMLS-LPBF-SLM, are increasingly used to produce high-complexity and performance parts. The adoption of additive manufacturing in the production of turbomachinery is limited by the number of materials used, their characteristics, and technological limitations. Despite these limitations, [64] mentions that there are good results such as optimized materials and more sustainable manufacturing processes.

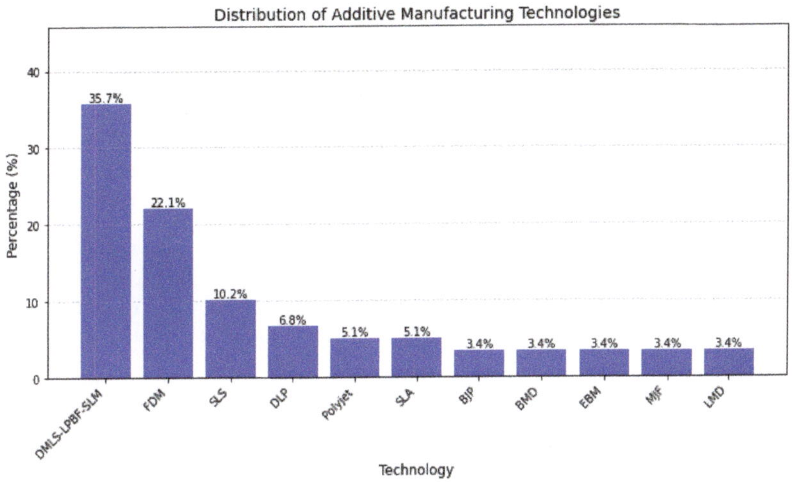

Figure 7. Use of additive manufacturing in turbomachinery.

According to [65], Powder Bed Fusion Technology encompasses various methods, including Direct Metal Laser Sintering (DMLS) or Selective Laser Melting (SLM), Selective Laser Sintering (SLS), Multi Jet Fusion (MJF), and Electron Beam Melting (EBM). The DMLS/SLM process, in particular, has demonstrated the ability to produce impellers with excellent dimensional accuracy, surface finish, and material strength, as shown in prints made from materials such as 17-4 PH, Ti-6Al-4V, Inconel 625, and Inconel 718 [66–68]. This method offers performance comparable to that of forged materials. For example, a study by [69] on impeller design revealed that 3D-printed Inconel 718 (IN718) is a strong alternative to forged components, especially considering the printable dimensions. While the mechanical properties of IN718 are slightly lower than those of forged parts, they are superior to those of cast components [70]. In the case of Ti-6Al-4V applied to impellers, research indicates that residual deformation can be reduced by 20.19% compared to solid compressor impellers, and the residual stress in lattice impellers is lower than in solid ones, with a reduction in the maximum residual stress by 8.72% [71,72]. Additionally, studies on thermal deformation show that the maximum deformation in 3D-printed turbines typically occurs at the base edge [73]. In another case, samples printed using Al-Si-Mg material exhibited a hardness 35% higher than that of the cast alloy [74].

Austenitic steel printing shows that, with a new design, the impellers can operate at higher rotational speeds ranging between 8.2% and 32.73% at an inlet pressure of 5 MPa and different nozzle angles in applications to Pelton turbines [75]. In martensitic steel printing, the rotational speed has a linear influence on the critical crack size and a nonlinear influence on the fatigue crack propagation life. When the external load and the residual stress of 3D printing act at the same time, the crack propagation law is no longer complied with [76]. Martensitic steel shows changes in pore size and difficulty in repeatability of results, but there is no variation in total porosity. The applied heat treatment (solution and aging hardening) showed superior performance and improved tensile strength [77].

The SLM printing process allows the fabrication of high-precision geometries, ensuring imposed constraints [78]. Regarding post-processing operations, results show improvements in surface quality through Abrasive Flow Machining (AFM) [66]. With AlSi10Mg material, it is observed that the component manufactured through SLM has a better surface finish compared to the cast component and meets the impeller specification requirements [74]. Regarding the production of smaller-sized impellers in the future, the results show that the impeller can not only effectively reduce weight [79] but also improve compressor efficiency [80].

In unconventional applications, the material FeSiCr, considered a Soft Magnetic Composite (SMC), is an applicable option for high-efficiency axial flow machines, high-speed airflow, and power density. With this manufacturing method, a 3D-structured motor with high structural strength and low core loss during high-frequency operation could be rapidly fabricated [81]. On the other hand, impellers for cryogenic pumps have also been developed, obtaining improved mechanical properties compared to standard impellers [82].

Regarding new processes, the construction of ceramic cores through selective laser sintering (SLS) was evidenced. Additionally, the organic-inorganic binder conversion process was used to increase the mechanical properties of the prepared core. Ref. [83] used mullite powder with a size of 70 μm and phenolic organic resin as the starting material; the casting of the impeller blade was successful, and the product precision was high in a real application. The ceramics obtained by selective laser sintering (SiC-Si(BN)-Al_2O_3), followed by heat treatment, are three to five times less massive than metallic ones and sufficiently compact for applications such as building materials for the production of microturbines [84]. With polystyrene (PS) after the T5 heat treatment, the tensile strength and Brinell hardness of the cast part increase from 184 MPa and 57 HB to 330 MPa and 109 HB, respectively [85]. Similarly, this technology has been applied with turboexpanders using Nylon [86]. There is also evidence of good performance in closed impellers with stainless steel SS304 [87].

Ceramic Gas Microturbines (μGTU-MTG) produced by SLS technology have enabled increased efficiency, power density, elimination of toxic component emissions from the exhaust, significant reduction in mass and dimensional characteristics, ensuring continuous optimum performance, increasing reliability, and reducing costs and operating costs [84,88]. MTGs have been fabricated in Inconel, but the geometric profiles and finishes are not suitable [89].

Another process used is MJF, which is capable of controlling different properties in its voxels, allowing for the obtaining different mechanical characterizations, different colors, and transparency in the same product. According to [90], when printing parts of a stator, the MJF process turned out to be a reasonable solution in Polyamide, but it is only viable if more exhaustive preliminary tests are carried out. Another study shows that some components in a fan were deformed during the cooling process, generating non-precise tolerances. In addition, only a few materials were allowed in MJF, such as PA 12 or PA11 [91]. Although it is possible to produce elements with suitable geometries and dimensions using this technology, they have microstructural problems, such as porosity between layers or unprocessed powder particles in the printing structure [92].

The Electron Beam Melting (EBM) process can work with materials of superior properties. According to [93], EBM is highly suitable for constructing circular or complex shapes with thin walls. EBM can produce parts with an approximate relative density of 98% in TiAl components, whereas, compared to SLM, the produced parts exhibit poorer mechanical properties and a relative density of 97% due to cracks [94]. This method can also be used to repair parts. When a part is worn or exceeds its anticipated lifespan, it can be cut, designed, and a new feature built onto the existing part [93,95]. In the manufacturing of impellers in Ti-6Al-4V, a decrease in mechanical properties during production due to prolonged annealing has been observed [96], although studies by [97] mention that surface treatments can significantly enhance the quality and reliability of the parts.

For small-scale and low-load applications, Stereolithography (SLA) can prove to be a valuable ally. This method has been tested with resin and polymers on impellers, mostly using Acrylonitrile Butadiene Styrene (ABS). The SLA method applied to turboexpanders with resin results in a glass-like surface with a fine finish, though its fragility poses a problem in practical applications [86]. There are experimental tests in photopolymer demonstrating the possibility for rapid and low-cost construction of rotary compressors, although there are no tests for general characterization [98]. ABS material is capable of producing a significant amount of thrust at extremely low RPMs compared to standard propeller systems in medical applications [99]. Combined applications with casting have also been found, where the wax pattern is generated in high-impact polystyrene (HIPS) using SLS and photosensitive resin using SLA. There are few studies related to turbines used in low RPM and small-scale applications [100].

Furthermore, Digital Light Processing (DLP) technology has proven that it is possible to manufacture a cost-effective ceramic turbine through a turbine rotor with alumina and silicon nitride [101]. Ref. [102] successfully manufactured a complex-shaped silicon nitride impeller suitable for transporting molten mass at high temperatures using this technology. Additionally, Ref. [89] manufactured a ceramic-type Si-Al-O-N rotor under this technology, demonstrating superior efficiency and pressure ratio compared to rotors made with SLS technology and Inconel material due to better geometric features and superior surface finish.

Ref. [103] shows that it is possible to use photopolymers in the manufacturing of impellers, but their operation is at a lower flow pressure than a standard material impeller, due to the absence of sharp edges, although it is acceptable in quick design solutions. It is also known that DLP can be used to obtain lost wax patterns for subsequent casting, achieving lower costs than traditional manufacturing [104].

Regarding Laser Metal Deposition (LMD) technology, it has been shown that combining LMD with multi-axis milling is a viable solution for the production of closed impellers and other complex-shaped parts [105]. Ref. [106] mentions that the cost of components compared to the conventional casting process is slightly higher, and the significant reduction in delivery time can be the deciding factor when parts are needed quickly. In terms of costs, Ref. [107] has proposed an experimental method achieving lower power consumption, higher decomposition performance, and printing efficiency compared to the traditional method. There is not much evidence of the use of this technology on a large scale in turbomachinery, according to the state of the art.

Polyjet technology, on the other hand, allows the design of composite components from different polymeric materials and even medical products or applications [108]. Regarding impellers, Polyjet technology has been used for experiments and evaluations of impeller wear by sludge using VeroGray material, allowing rapid assessment of wear locations and rates in harder materials [109]. In small-scale applications for solar radial turbines, RG525 material was used for experimentation; the maximum allowable deformation was influenced by the rotation speed, the maximum allowable fatigue life was influenced by the rotation speed and the fluid intake temperature, and the tip cover region on the rotor had deflection values greater than 21% [95,110]. In Polycarbonate (PC) material, the hydraulic performance could match the commercial product when manufacturing a monopivot centrifugal blood pump [111]. This additive manufacturing application has enabled functional solutions in renewable energy applications. In applications to small Pelton turbines, two variants were designed: screw-on rotor buckets and a monoblock rotor turbine, allowing for complex designs and customized solutions [112].

Regarding the Fused Deposition Modeling (FDM) process, there are various applications related to turbomachinery offering advantages and disadvantages. In materials such as ABS and Polyactide (PLA), it is evident that costs can be reduced, but there are limitations such as the production of large components and damage in post-processing [113]. In other applications, the use of a large number of support structures leads to high post-production costs [78,114]. It is known that the main disadvantage of filament printers is the formation of gaps when printing intricately shaped objects [115]. In contrast to the above, a mixed process (3D printing and rapid lost wax casting) in centrifugal compressors, processing time and costs were reduced, and casting quality improved [116]. The use of Polyurethane (PU) in impellers for floaters is considerably economical and increases the production process of new parts [117].

In terms of performance [118], it was demonstrated that a pump designed with Polyethylene Terephthalate Glycol-Carbon Filament (PETG-CF) has an efficiency of 26.23%, comparable to a tin-bronze impeller, which has an efficiency of 27.7% in hydrophore pumps. On the other hand, [119] shows that impellers in PLA at speeds not exceeding 1000 rad/s or pressures below 0.22 MPa could operate without compromising their efficiency and structural integrity. The performance characteristics of ABS-P430 pump impellers have similar functional values and very identical trends of the functional curves of total height, energy consumption, and mechanical efficiency [120]. Ref. [121] evidences that among five impeller variants with the same flow section geometry and similar material (ABS-P430), the impellers produced by 3D printing showed slightly higher efficiency compared to cast iron and steel impellers, averaging between 4 and 6%.

Bound Metal Deposition (BMD) currently has limited applications in turbomachinery [122]. Despite this, there are examples of applications in impellers, such as the centrifugal pump impeller printed from SS 316L using BMD. Another clear example is the pump impeller made of SS 316L material, showing similar properties compared to a CNC-machined SS 316L impeller [123].

Binder Jetting Printing (BJP) technology also has few applications in turbomachinery. In another study, the design of an actual impeller blade was successful, and the product precision was excellent [83]. It was discovered that the BJP process prints elements more economically and 20% more resistant than those processed by other additive manufacturing techniques [124]. When comparing SLS printing with BJP, the former offers the advantage of the efficiency of inorganic binder coating, for example, furan [83].

3.3. III. Data Analysis

Regarding manufacturing processes and materials, common subtractive processes such as CNC have been used to design high-performance centrifugal compressors using 2014-T6 aluminum alloy as the manufacturing material [125]. Similarly, materials used for high-performance centrifugal impellers are those with a higher yield strength and minimal specific weight, such as ASTM A3, Hardox 450, and 6061-T6 [126].

In the realm of additive manufacturing, there are ongoing studies and continuous improvements in printable materials. For example, Ref. [127] designed an enhanced agitation impeller printed in polylactic acid (PLA), which was then doped with iron as a raw material to subsequently generate chemical activity through surface etching and oxidation [87]. In an effort to design corrosion-resistant elements, materials used in centrifugal pumps such as cast iron and brass alloy versus polymeric materials like polyamide 6 reinforced with 15% carbon fiber (PA6-CF15) and 15% polyphenylene sulfide reinforced with carbon fiber (PPS-CF15) have been evaluated. Composite materials were suitable for operating at nominal impeller powers but did not withstand the maximum load that metallic materials can bear. The printed impellers were demonstrated to be free of corrosion or related phenomena such as wear, erosion, and fatigue. A low water absorption rate and chemical resistance were also evident [128]. Notably, experimental studies have aided in finding improvements in materials and applications related to turbomachinery.

From the review, various materials used in impellers across different additive manufacturing techniques were observed, as shown in Table 2, with AISI316L, Al_2O_3, Inconel, and Ti-6Al-4V being the most used in metals. It was evidenced that AISI316L is one of the most utilized materials across various processes (LMD, BMD, BJP, DMLS-LPBF-SLM) and is highly used in impellers due to its high mechanical strength, corrosion and wear resistance, longevity, high-temperature resistance, ease of maintenance, and good performance at low temperatures [129,130]. Al_2O_3 is used in ceramic components, rotors, microturbines, ceramic cores, or matrices [84]. Inconel is a difficult-to-machine material used in components subjected to high temperatures and corrosion resistance. It is currently handled by aerospace, oil, gas, nuclear, and chemical industries [131]. Although AM exhibits slightly lower mechanical properties than forged parts [132,133], this process resolves many machining issues. Ti-6Al-4V is another material highly used in AM for its high strength, low density, corrosion resistance, biocompatibility, and high-temperature resistance, although it presents porosity issues, heat treatments, and pressing are performed to correct them [134].

Table 2. Common materials in impellers used in additive manufacturing.

Process	Material	References
SLS	SS 304 (HIPS-Stainlees steel), SiC-Si (BN)-Al_2O_3, Mullite ($3Al_2O_3{\cdot}2SiO_2$), PS, Nylon	[83–86,88,135]
SLA	ABS, Resina	[86,99]
DLP	Si_3N_4, Photopolymer, Al_2O_3, Si_3N_5	[101–103]
LMD	AISI316L	[136]
MJF	PA12GB (Nylon y 40% vidrio)	[90]
Polyjet	PC, FullCure 720 (Clear Resin), RGD 525 (Stratasys material), VeroGray	[109–112]
FDM	ABS-P430, PLA, ABS, PU, Polycast, PET, PLA, PA6-CF15, PPS-CF15, ABS, PETG CF	[78,86,113,115–121,128]
BMD–BJP	AISI316L, Quartz (SiO_2)	[83,123]
EBM	Ti-6Al-4V	[93,96]
DMLS-LPBF-SLM	AlSi10Mg, 17-4 PH Stainless Steel, Ti-6Al-4V, Inconel 718, Inconel 625, AISI316L, FV520B-I, FeSiCr Martensitic Steel, SUS316L	[66–73,75–82,86,137]

For applications in turbomachinery, particularly impellers operating in corrosive fluid environments, it is crucial to select materials with high corrosion resistance, classified under Corrosion Resistance Classes CRC (CRC3 and CRC4), to ensure durability and performance in such environments. Materials such as Inconel 718, Inconel 625, Ti-6Al-4V (Titanium), and Si_3N_4 (Silicon Nitride) are excellent choices for additive manufacturing due to their high corrosion resistance and ability to withstand extreme conditions. Studies show that corrosion can impact the performance of critical components and can form depletion zones in various elements [138]. Materials like alumina (Al_2O_3) can mitigate the effects of erosion and high-temperature corrosion, or remanufacturing processes and coatings with nickel alloys can enhance corrosion resistance [139,140]. Although many materials exhibit good qualities by being stainless, AM alloys may be affected by heat and can contain residual stresses created by the manufacturing process itself [139]. Table 3 shows the materials used in additive manufacturing and their corrosion resistance in various environments.

Table 3. Corrosion resistance of materials used with AM in impellers.

Material	Mils per Year—mpy	Corrosion Resistance	CRCI [1]	CRC [2]
SS 304 (HIPS-Stainless Steel)	2–20 mpy (in seawater)	Medium	0.75–1.0	CRC2
SiC-Si (BN)-Al_2O_3, Mullite	<0.1 mpy	High	>1.0	CRC3
PS (Polystyrene)	Low resistance, not used in corrosive environments	Low	-	CRC1
Nylon	0.1–1 mpy (in diluted acids)	Low	0.8–0.9	CRC2
ABS	1–10 mpy (depending on the medium)	Low	0.6–0.8	CRC1
Resin	Depends on type, generally low	Low	0.5–0.7	CRC2
Si_3N_4 (Silicon Nitride)	<0.1 mpy	High	>1.0	CRC4
Photopolymer	Depends on type, generally low	Low	0.4–0.7	CRC2
Al_2O_3 (Aluminum Oxide)	<0.1 mpy	High	>1.0	CRC3
Si_3N_5	<0.1 mpy	High	>1.0	CRC4
AISI316L (Stainless Steel)	0.5–5 mpy (in seawater)	High	0.9–1.0	CRC3
PA12GB (Nylon and 40% glass)	Similar to Nylon, 0.1–1 mpy	Low	0.8–0.9	CRC2
PC (Polycarbonate)	1–10 mpy	Low	0.7–0.9	CRC2
FullCure 720 (Clear Resin)	-	Low	-	CRC2
RGD 525 (Stratasys material)	-	Low	-	CRC2
VeroGray	-	Low	-	CRC2
ABS-P430	Similar to ABS, 1–10 mpy	Low	0.6–0.8	CRC1

Table 3. *Cont.*

Material	Mils per Year—mpy	Corrosion Resistance	CRCI [1]	CRC [2]
PLA	1–10 mpy (depending on the medium)	Low	0.6–0.8	CRC1
PU (Polyurethane)	1–10 mpy	Low	0.6–0.8	CRC2
Polycast	Similar to ABS	Low	0.6–0.8	CRC2
PET	1–10 mpy (depending on the medium)	Low	0.7–0.9	CRC2
PA6-CF15	Similar to PA6, 1–10 mpy	Low	0.7–0.9	CRC2
PPS-CF15	<0.1 mpy (in acidic media)	Low	0.9–1.0	CRC3
PETG CF	Similar to PET, 1–10 mpy	Low	0.7–0.9	CRC2
Quartz (SiO_2)	<0.1 mpy	High	>1.0	CRC4
Ti-6Al-4V (Titanium)	<0.1 mpy (in seawater)	High	>1.0	CRC4
AlSi10Mg (Aluminum Alloy)	1–5 mpy (in saline environments)	Medium	0.8–0.9	CRC2
17-4 PH Stainless Steel	0.5–5 mpy (in seawater)	High	0.9–1.0	CRC3
Inconel 718	<0.1 mpy (in acidic or saline environments)	Very High	>1.0	CRC4
Inconel 625	<0.1 mpy (in acidic or saline environments)	Very High	>1.0	CRC4
FV520B-I	1–10 mpy	Medium	0.8–0.9	CRC3
FeSiCr (Iron-Silicon-Chromium Alloy)	<0.1 mpy	High	>1.0	CRC3
Martensitic Steel	5–20 mpy (in saline environments)	Medium	0.6–0.8	CRC2
SUS316L (Stainless Steel)	0.5–5 mpy (in seawater)	High	0.9–1.0	CRC3

[1] Corrosion Resistance Cost Index (CRCI); [2] Corrosion Resistance Classes (CRC).

Despite being more costly, the DMLS-LPBF-SLM processes use materials with higher resistance capabilities and can work with superalloys [141]. This method is increasing in use as it can surpass the limitations of conventional machining and can develop complex shapes, new designs, and optimize various parameters [142]. FDM technology is highly utilized for being more accessible, affordable, capable of generating complex geometries, and exhibiting low corrosion [143]. It may present issues with low precision and post-processing costs [144], and is limited by the materials used (polymers) and the final quality of the part [145], although there is evidence of centrifugal compressors constructed from FDM in Ti-6Al-4V material [146]. Figure 8 briefly indicates the materials commonly used in different types of turbomachinery through FDM.

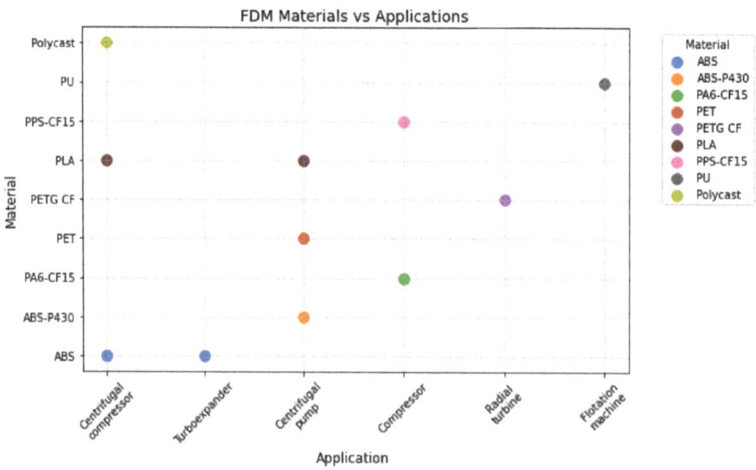

Figure 8. Use of material in FDM in turbomachinery.

Those materials, such as ABS and PLA, are versatile materials used in different types of turbomachinery, like centrifugal compressors, turboexpanders, and centrifugal pumps. Materials like PU and PETG CF are associated with specific applications such as flotation machines and radial turbines, respectively. The concentration of applications is commonly found in compressors and centrifugal pumps.

Regarding the relationship between speed and material, it is known that the value of the stress gradually increases as the speeds of the impeller increase; thus, mechanical properties are influential in the design of turbomachinery [147]. Another important point is related to wear, since accelerated wear in impellers directly affects hydraulic power [148]. Regarding the surface, reducing the roughness of the impeller positively affects the efficiency of the pump and energy requirements, although flow handling tends to be lower with polymeric materials [149]. Ceramic materials like Al_2O_3, Si_3N_5, and metallic materials such as Ti6Al4V printed in 3D have shown good results at high revolutions (Figure 9). Although polymeric materials are not suitable for high-revolution work, in demonstrated applications, they have improved performance parameters, costs, and sustainability [150].

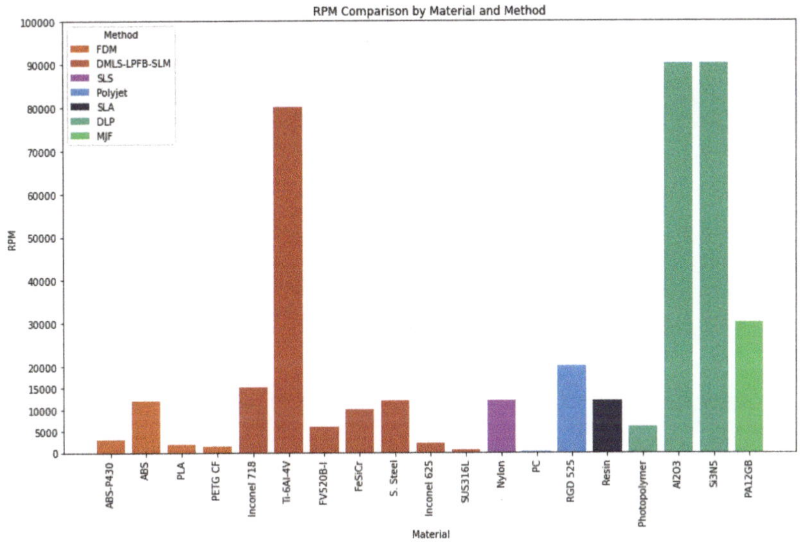

Figure 9. Tested speeds of 3D-printed impellers.

AM compressors could provide significant improvements in cost and weight of the engine for limited-life applications [151]. The introduction of additive manufacturing technologies combined with other standard manufacturing methodologies could yield reasonable benefits regarding production time and component availability. For example, the use of simulation elements, numerical control methods, algorithms, and additive manufacturing, among others, can allow significant advancements in the design and optimization of elements in turbomachinery [152,153]. Table 4 shows some materials, methods, applications, and speeds at which certain materials have been studied, obtaining notable advantages.

Additive manufacturing (AM) has emerged as a versatile and innovative approach for producing turbomachinery components. Different AM techniques, such as DMLS/SLM, SLS, and EBM, offer unique advantages in terms of material optimization, design flexibility, and sustainability. Table 5 summarizes the advantages, disadvantages, and key mechanical/physical defects of various additive manufacturing (AM) techniques applied in turbomachinery.

Table 4. Materials, methods, applications, and speeds applied to turbomachinery related to additive manufacturing.

Material	Method	Application	RPM
PEEK	CNC–3D	High Speed Microturbine	120,000
PPS	CNC–3D	High Speed Microturbine	120,000
Aluminum 2014-T6	Ansys CFD Simulation	Centrifugal Compressor	230,000
ASTM A36	Ansys CFD Simulation	Fan	653
Hardox 450	Ansys CFD Simulation	Fan	1225
Aluminum 6061-T6	Ansys CFD Simulation	Fan	1002
ABS-P430	FDM	Generator	2950
Aluminum 2014-T6	Ansys CFD Simulation	Compressor	210,000
Inconel 718	DMLS-LPFB-SLM	Centrifugal Compressor	14,700
Inconel 718	DMLS-LPFB-SLM	Centrifugal Compressor	15,120
ABS	FDM	Centrifugal Compressor	2800
ABS-P430	FDM	Centrifugal Pump	1450
Ti-6Al-4V	DMLS-LPFB-SLM	Compressor	80,000
FV520B-I	DMLS-LPFB-SLM	Compressor	6000
FeSiCr	DMLS-LPFB-SLM	Fan	10,000
PLA	FDM	Centrifugal Pump	1900
Nylon	SLS	Turboexpander	12,000
ABS	FDM	Turboexpander	12,000
Stainless steel	DMLS-LPFB-SLM	Turboexpander	12,000
PC	Polyjet	Centrifugal Blood Pump	50
Resin	SLA	Turboexpander	12,000
Inconel 625	DMLS-LPFB-SLM	Compressor	2200
RGD 525	Polyjet	Radial Turbine	20,000
PETG CF	FDM	Radial Turbine	1450
SUS316L	DMLS-LPFB-SLM	Pelton Turbine	500
Ti-6Al-4V	DMLS-LPFB-SLM	Compressor	570
Photopolymer	DLP	Centrifugal Pump	5900
Al_2O_3	DLP	Turbine Rotor	90,000
Si_3N_5	DLP	Turbine Rotor	90,000
PA12GB	MJF	Radial Turbine	30,000

AM in turbomachinery offers significant benefits in terms of sustainability, notably optimizing material usage, energy consumption, and production times. Processes such as DMLS-LPBF-SLM have demonstrated substantial weight reductions in components, achieving nearly a 30% decrease in the total mass of impellers manufactured from Inconel 718 while also optimizing operational frequencies [69]. In materials such as AISI316L, BMD technology proves eco-efficient, reducing environmental impacts by 54.6% compared to commercial alternatives [123]. For ABS-P430, the FDM process increases efficiency by 4–6% [121], while PETG CF components achieve efficiencies of 26.23%, comparable to those of bronze materials [118]. Additionally, processes like SLS reduce production costs, with lead times of approximately two weeks compared to 40–50 weeks for traditional methods [135]. In terms of repair and reuse, EBM enables the restoration of worn or out-of-service components, enhancing sustainability by extending their operational lifespan [96]. Furthermore, technologies such as DLP, applied to ceramic materials like Si3N4, meet the demands of high-temperature and molten transfer applications, expanding the possibilities of sustainable manufacturing [102]. Finally, advanced methods, such as those implemented by [81], facilitate the production of 3D motors with high structural strength and minimal core loss, emphasizing the energy efficiency and operational performance of these technologies.

Despite the advancements in AM, key sustainability challenges remain. Processes such as SLM exhibit porosity and structural defects, which affect component durability [91,92]. Similarly, EBM technologies face high energy consumption during production and significant material recovery costs in post-processing [94,96]. Methods like FDM generate considerable waste due to support structures, increasing costs and material losses [114]. Moreover, the reuse of sintered powder in technologies such as SLS and the scalability of processes like BMD for industrial applications remain critical areas for improvement [83,124]. Studies also indicate the need to assess the full life-cycle impact of advanced materials used in Polyjet [71,103]. These limitations highlight the necessity for innovation to achieve truly sustainable AM or to apply recent findings to impeller production. For example, studies on the use of Inconel 625 via wire arc additive manufacturing (WAAM) have shown a threefold reduction in energy consumption and a material waste reduction from 85% to 35% [154].

Table 5. Advantages, disadvantages, and mechanical/physical defects of AM techniques.

AM Technique	Advantages	Disadvantages	Mechanical/Physical Defects
DMLS-LPBF-SLM	High dimensional accuracy, excellent material strength, up to 30% weight reduction, and optimized operational frequencies.	Residual stresses, porosity, lack of fusion, thermal deformation, and high costs.	Residual stresses, porosity, thermal deformation, surface roughness, and lack of fusion.
SLS	Significant cost reductions (up to 99.75%), short production times, lighter and more efficient components.	Low density, high surface roughness, and limited material options.	Low density, high surface roughness, and deformation during cooling.
EBM	Capable of producing and repairing complex components, high mechanical strength, and relative density of 98%.	High energy consumption, susceptibility to cracking in materials such as Ti-6Al-4V.	Cracking in thin components and loss of properties due to prolonged annealing times.
FDM	Cost-effective, efficient for complex geometries, improves performance by 4–6% in materials such as ABS and PETG CF.	Formation of voids, layer adhesion issues, and limitations for high-speed or high-temperature applications.	Delamination between layers, void formation, limited precision, and excessive waste from support structures.
DLP	Excellent surface finish, suitable for high-temperature applications using ceramics such as Si3N4.	Material fragility, limited material compatibility, and flow and pressure challenges.	Material fragility, flow and pressure issues, and limited mechanical strength.
BMD	Highly eco-efficient, reducing environmental impacts by 54.6% compared to conventional methods.	Scalability issues for industrial applications and material density inconsistencies.	Difficulty in achieving repeatability, interlayer porosity, and limited mechanical strength in complex shapes.
MJF	Allows precise control of voxel-specific properties, high resolution, and functional prototypes.	Material restrictions, deformation during cooling, and interlayer porosity.	Interlayer porosity, unprocessed powder particles, and variability in dimensional tolerances.
SLA	Ideal for low-load applications, fast and cost-efficient for rotary compressors using ABS and photosensitive resins.	Fragile parts, roughness in vitreous surfaces, and limited mechanical strength.	Fragility, high roughness on vitreous surfaces, and unsuitability for high-load or high-speed components.
PolyJet	Capable of producing multi-material components, ideal for wear testing, and small-scale energy applications.	Material compatibility limitations, deformation in high-stress conditions, and restricted scalability.	Deformation in high-rotation speed conditions, wear inconsistencies in testing materials.
LMD	Viable for producing closed impellers and other complex shapes, reduces powder consumption, and improves decomposition efficiency.	Higher cost compared to traditional casting processes, limited evidence of large-scale adoption in turbomachinery.	Porosity in deposited layers, residual stresses, and inconsistencies in mechanical properties.

At the policy level, it is crucial to establish financial incentives to support research and industrial implementation of additive manufacturing in turbomachinery, alongside specific regulations to ensure quality and sustainability. Encouraging public-private partnerships is essential to address technical and environmental challenges effectively. Additionally, conducting comprehensive life cycle assessments, providing technical training to professionals in these technologies, and promoting the use of locally sourced materials are fundamental steps to tailor solutions to regional contexts and enhance their practical applicability.

4. Conclusions

A co-occurrence analysis was performed, and it was determined that the number of studies is relatively low, considering that scientific interest, according to publications, has intensified since 2016. It is notable that the areas of interest are related to topics such as topological optimization, simulation, material alloying, performance improvement, and the repair of turbomachinery components.

From the analyzed study, FDM printing and powder bed fusion technology are the most widely used in this field. The advantages and disadvantages of AM depend mainly on the type of process, the material used, post-processing, and its application. For example, the DMLS process produces impellers with adequate dimensional accuracy, surface finish, and good strength in materials such as 17-4 PH, Ti-6Al-4V, Inconel 625, and Inconel 71,800. The hardness of the Al-Si-Mg material produced by this method is 35% higher than that of the cast alloy. In contrast, the SLS process with MTGs printed in Inconel results in inadequate geometric profiles and finishes. Similarly, the results obtained from FDM demonstrate that it is possible to reduce costs and improve performance, but it shows limitations such as the production of large components, load limitations, costs, and damage in post-processing.

In recent years, optimization methods have become more varied, and several techniques have been combined to achieve better results. It is possible to combine manufacturing methods, for instance, using lost wax for casting, resulting in lower costs compared to traditional manufacturing. The same can be done with CNC processes, which offer options for repairing impellers in mixed processes, or by combining various processes to achieve different types of optimizations.

Given the growing scientific interest in 3D printing technologies since 2016, especially in areas like topological optimization, material simulation, material wear, and performance enhancement in turbomachinery, it is clear that additive manufacturing (AM) offers substantial potential for practical applications in pumps. Despite challenges such as size limitations, load capacity, and post-processing costs associated with Fused Deposition Modeling (FDM), the prospects for cost reduction and performance improvement are highly promising. Additionally, integrating AM with traditional manufacturing techniques, such as lost-wax casting or CNC machining, can further enhance process efficiency and cost-effectiveness. This makes the application of 3D-printed materials in pumps not only feasible but also highly advantageous.

This research provides a deeper understanding of the advancements in AM for the production of turbomachinery components, highlighting the use of materials such as Ti-6Al-4V, Inconel 718, and reinforced polymers, which offer enhanced resistance to corrosion, wear, and high temperatures. These developments enable the creation of more complex and optimized geometries, significantly improving the performance and efficiency of impellers and other critical components. Furthermore, the advancements studied establish a solid foundation for mitigating the combined effects of wear, including erosion, cavitation, and corrosion, through the integration of Industry 4.0 digital enablers, such as digital twins and advanced simulations. Future research directions should focus on further optimizing materials, exploring hybrid processes that combine AM with traditional techniques, reducing microstructural defects to enhance component reliability, and advancing algorithms to improve design and operational efficiency in turbomachinery.

Author Contributions: Conceptualization, M.P.-S. and A.R.; methodology, A.R.; formal analysis, F.-J.S.-R. and V.H.H.; investigation, A.R. and M.P.-S.; resources, F.-J.S.-R. and M.P.-S.; writing—original draft preparation, A.R., V.H.H., M.P.-S., H.M.R. and F.-J.S.-R.; writing—review and editing, A.R. and M.P.-S.; supervision, M.P.-S. and V.H.H. All authors have read and agreed to the published version of the manuscript.

Funding: This research received no external funding.

Data Availability Statement: The used data are available in the manuscript.

Conflicts of Interest: The authors declare no conflicts of interest.

Acronyms

ABS	Acrylonitrile Butadiene Styrene
AFM	Abrasive Flow Machining
AM	Additive Manufacturing
BJP	Binder Jetting Printing
BMD	Bound Metal Deposition
CNC	Computed Numeric Control
CRC	Corrosion Resistance Classes
CRCI	Corrosion Resistance Cost Index
DLP	Digital Light Processing
DMLS	Direct Metal Laser Sintering
EBM	Electron Beam Melting
FDM	Fused Deposition Modeling
HIPS	High-Impact Polystyrene
IIoT	Industrial Internet of Things
IN718	Inconel 718
LPBF	Laser Powder Bed Fusion
LMD	Laser Metal Deposition
MJ	MultiJet
MJF	Multi Jet Fusion
MTG	Micro Gas Turbine
PA	Polyamide
PA6-CF15	Polyamide 6 reinforced with 15% carbon fiber
PCMs	Polymeric Composite Materials
PEEK	Polyether-ether-ketone
PET	Polyethylene Terephthalate
PETG	Polyethylene Terephthalate Glycol
PPS	Polyphenylene sulfide
PPS-CF15	Polyphenylene sulfide reinforced with 15% carbon fiber
PU	Polyurethane
SDGs	Sustainable Development Goals
SiC	Silicon Carbide
SLA	Stereolithography
SLM	Selective Laser Melting
SMC	Soft Magnetic Composite
SLS	Selective Laser Sintering
SS	Stainless Steel
WDE	Water Droplet Erosion
sCO_2	Supercritical CO_2
WAAM	Wire Arc Additive Manufacturing

References

1. Lagha, M.; Duplaa, S.; Binder, N.; Carbonneau, X. Unified Classification and Characterization of Axial Turbomachines and Propellers. In Proceedings of the 13th International Symposium on Experimental Computational Aerothermodynamics of Internal Flows, Okinawa, Japan, 7–11 May 2017; p. 1.
2. Singh, A.; Sharma, D. Efficiency Enhancement in Turbomachinery: Bridging Numerical Modeling and Experimental Validation for Fluid Dynamics Advancements. *J. Sustain. Urban Futures* **2023**, *13*, 1–15.
3. Srinivas, G.; Raghunandana, K.; Shenoy, B.S. Recent Developments in Turbomachinery Component Materials and Manufacturing Challenges for Aero Engine Applications. In Proceedings of the IOP Conference Series: Materials Science and Engineering, Melbourne, Australia, 15–16 September 2018; IOP Publishing: Bristol, UK, 2018; Volume 314, p. 012012.
4. Patel, M.; Kumar, A.; Pardhi, B.; Pal, M. Abrasive, Erosive and Corrosive Wear in Slurry Pumps—A Review. *Int. Res. J. Eng. Technol.* **2020**, *7*, 2188–2195.
5. Yonezawa, K.; Nakai, G.; Takayasu, M.; Sugiyama, K.; Sugita, K.; Umezawa, S.; Ohmori, S. Influence of Blade Corrosion on Aerodynamic Characteristics of a Gas Turbine. *Energy* **2021**, *230*, 120665. [CrossRef]
6. Linden, D. Know Your Turbomachinery's Operating Environment! In Proceedings of the 47th Turbomachinery Symposium, Houston, TX, USA, 18–20 September 2018; Turbomachinery Laboratory, Texas A&M Engineering Experiment Station: College Station, TX, USA, 2018.

7. Yang, S.; Gao, S.; Xue, W.; Wu, B.; Duan, D. Phenomenological Modeling of Wear Mechanisms between Nickel-Based Superalloy Blades and Ceramic-Based Abradable Coatings in Turbomachines. *J. Mater. Res. Technol.* **2023**, *27*, 4191–4203. [CrossRef]
8. Di, J.; Wang, S.; Yan, X.; Cai, L.; Xie, Y. Experimental Investigation on Effect of Surface Strengthening Process and Roughness on Water Droplet Erosion Behavior in Turbomachinery. *Tribol. Int.* **2021**, *153*, 106647. [CrossRef]
9. Torner, B.; Wurm, F.-H. Influence of Zonal Surface Roughness on the Pump Characteristics of a Turbopump. In Proceedings of the 15th European Conference on Turbomachinery Fluid dynamics & Thermodynamics, Budapest, Hungary, 24–28 April 2023; European Turbomachinery Society: Florence, Italy, 2023.
10. Alqallaf, J.; Teixeira, J.A. Blade Roughness Effects on Compressor and Engine Performance—A CFD and Thermodynamic Study. *Aerospace* **2021**, *8*, 330. [CrossRef]
11. Radgolchin, M.; Anbarsooz, M. Fatigue Failure of Centrifugal Compressor Impellers: A Comprehensive Review. *Eng. Fail. Anal.* **2023**, *153*, 107592. [CrossRef]
12. Doshi, S.; Katoch, A.; Suresh, A.; Razak, F.A.; Datta, S.; Madhavan, S.; Zanhar, C.M.; Gundabattini, E. A Review on Vibrations in Various Turbomachines Such as Fans, Compressors, Turbines and Pumps. *J. Vib. Eng. Technol.* **2021**, *9*, 1557–1575. [CrossRef]
13. Wang, Y.; Liu, T.; Meng, Y.; Zhang, D.; Xie, Y. Integrated Optimization for Design and Operation of Turbomachinery in a Solar-Based Brayton Cycle Based on Deep Learning Techniques. *Energy* **2022**, *252*, 123980. [CrossRef]
14. Gou, L.; Zeng, X.; Wang, Z.; Han, G.; Lin, C.; Cheng, X. A Linearization Model of Turbofan Engine for Intelligent Analysis towards Industrial Internet of Things. *IEEE Access* **2019**, *7*, 145313–145323. [CrossRef]
15. Xie, R.; Chen, M.; Liu, W.; Jian, H.; Shi, Y. Digital Twin Technologies for Turbomachinery in a Life Cycle Perspective: A Review. *Sustainability* **2021**, *13*, 2495. [CrossRef]
16. Brun, K. Turbomachinery for Carbon Dioxide Transport and Storage. In *Offshore Technology Conference*; OTC: Hollister, MO, USA, 2023; p. D011S010R001.
17. Cetin, T.H.; Zhu, J. Aerodynamic Design and Optimization of SCO2 Turbomachinery: Two-Stage Optimization Approach. *Energy Convers. Manag.* **2023**, *297*, 117711. [CrossRef]
18. Reddy, T.R.; Priyadarsini, C.I.; Navaneetha, R. Design and Analysis of Impeller Using Corrosion Resistant Materials. In *AIP Conference Proceedings*; AIP Publishing: Melville, NY, USA, 2021; Volume 2317.
19. Wang, Y.; Li, W.; He, T.; Liu, H.; Han, C.; Zhu, Z. Experimental Study on the Influence of Particle Diameter, Mass Concentration, and Impeller Material on the Wear Performance of Solid–Liquid Two-Phase Centrifugal Pump Blade. *Front. Energy Res.* **2022**, *10*, 893385. [CrossRef]
20. Bandi, S.; Banka, J.; Kumar, A.; Rai, A.K. Effects of Sediment Properties on Abrasive Erosion of a Centrifugal Pump. *Chem. Eng. Sci.* **2023**, *277*, 118873. [CrossRef]
21. Shen, Z.; Chu, W.; Li, X.; Dong, W. Sediment Erosion in the Impeller of a Double-Suction Centrifugal Pump—A Case Study of the Jingtai Yellow River Irrigation Project, China. *Wear* **2019**, *422–423*, 269–279. [CrossRef]
22. Mirmahdi, E. Defects in Turbine Impeller Blades with Non-Destructive Testing: Modeling, Ultrasonic Waves, Defect Analysis. *J. Inst. Eng. (India) Ser. C* **2021**, *102*, 1395–1401. [CrossRef]
23. Li, X.; Zeng, X.; Zhao, Z.; Zhang, X.; Yan, Q.; Zhang, Y.; Wen, Y.; Yang, X.; Mao, F. Fast-flowing LBE Corrosion of the SiC Pump Impeller: Numerical and Experimental Investigations. *Int. J. Appl. Ceram. Technol.* **2023**, *20*, 436–450. [CrossRef]
24. ElJersifi, A.; Chbihi, A.; Semlal, N.; Bouaouine, H.; Naamane, S. Failure Analysis of a High Chromium Carbon Steel Impeller Operating in Phosphoric Acid Slurry. *Wear* **2021**, *476*, 203660. [CrossRef]
25. Li, L.; Ren, Y.; Chang, S.; Li, M.; Lin, S.; Wang, M. Evaluation of the ATI 718Plus Repaired by Laser Metal Deposition under Heat Treatment Regimes: Microstructures, Mechanical Properties, and Their Heterogeneity. *Mater. Sci. Eng. A* **2023**, *885*, 145540. [CrossRef]
26. Kan, K.; Binama, M.; Chen, H.; Zheng, Y.; Zhou, D.; Su, W.; Muhirwa, A. Pump as Turbine Cavitation Performance for Both Conventional and Reverse Operating Modes: A Review. *Renew. Sustain. Energy Rev.* **2022**, *168*, 112786. [CrossRef]
27. Zhu, D.; Xiao, R.; Liu, W. Influence of Leading-Edge Cavitation on Impeller Blade Axial Force in the Pump Mode of Reversible Pump-Turbine. *Renew. Energy* **2021**, *163*, 939–949. [CrossRef]
28. Zhang, F.; Tao, R.; Zhu, D.; Wu, Y.; Jin, F.; Xiao, R. Anti-Cavitation Leading-Edge Profile Design of Centrifugal Pump Impeller Blade Based on Genetic Algorithm and Decision Tree. *J. Braz. Soc. Mech. Sci. Eng.* **2022**, *44*, 279. [CrossRef]
29. Fei, Z.; Zhang, R.; Xu, H.; Feng, J.; Mu, T.; Chen, Y. Energy Performance and Flow Characteristics of a Slanted Axial-Flow Pump under Cavitation Conditions. *Phys. Fluids* **2022**, *34*, 035121. [CrossRef]
30. Pokharel, N.; Ghimire, A.; Thapa, B.; Thapa, B.S. Wear in Centrifugal Pumps with Causes, Effects and Remedies: A Review. *IOP Conf. Ser. Earth Environ. Sci.* **2022**, *1037*, 012042. [CrossRef]
31. Li, W.; Li, S.; Ji, L.; Zhao, X.; Shi, W.; Agarwal, R.K.; Awais, M.; Yang, Y. A Study on the Cavitation and Pressure Pulsation Characteristics in the Impeller of an LNG Submerged Pump. *Machines* **2021**, *10*, 14. [CrossRef]
32. Kelley, C.R.; Lopp, G.K.; Kauffman, J.L. Optimizing Piezoelectric Material Location and Size for Multiple-Mode Vibration Reduction of Turbomachinery Blades. *J. Vib. Acoust.* **2021**, *143*, 021007. [CrossRef]
33. Jia, X.Q.; Li, S.Y.; Miu, H.J.; Yang, T.; Rao, K.; Wu, D.Y.; Cui, B.L.; Ou, J.L.; Zhu, Z.C. Carbon Nanomaterials: A New Sustainable Solution to Reduce the Emerging Environmental Pollution of Turbomachinery Noise and Vibration. *Front. Chem.* **2020**, *8*, 683. [CrossRef]

34. Nie, D.; Chen, X.; Wu, Q.; Liu, Y. Stress Corrosion Cracking Behaviors of FV520B Stainless Steel Used in a Failed Compressor Impeller. *Eng. Fail. Anal.* **2020**, *116*, 104701. [CrossRef]
35. Fitriyana, D.F.; Anis, S.; Al Qudus, A.R.; Lakuy, M.A.N.; Ismail, R.; Nugroho, S.; Haryadi, G.D.; Bayuseno, A.P.; Siregar, J.P. The Effect of Post-Heat Treatment on The Mechanical Properties of FeCrBMnSi Coatings Prepared by Twin Wire Arc Spraying (TWAS) Method on Pump Impeller From 304 Stainless Steel. *J. Adv. Res. Fluid Mech. Therm. Sci.* **2022**, *93*, 138–147. [CrossRef]
36. Kyaramyan, K.A.; Nochovnaya, N.A.; Zakharova, N.S.; Kashapov, O.S. Microstructure and Mechanical Properties of the Material and the Surface Residual Stresses of Compressor Impeller Blades Made of a VT41 Titanium Alloy Depending on the Initial State and Incomplete Annealing Conditions. *Russ. Metall. (Met.)* **2021**, *2021*, 1587–1593. [CrossRef]
37. Mamazhonov, M.; Shakirov, B.; Matyakubov, B.; Makhmudov, A. Polymer Materials Used to Reduce Waterjet Wear of Pump Parts. *J. Phys. Conf. Ser.* **2022**, *2176*, 012048. [CrossRef]
38. Zirak, N.; Shirinbayan, M.; Deligant, M.; Tcharkhtchi, A. Toward Polymeric and Polymer Composites Impeller Fabrication. *Polymers* **2021**, *14*, 97. [CrossRef] [PubMed]
39. Soori, M.; Jough, F.K.G.; Dastres, R.; Arezoo, B. Sustainable CNC Machining Operations, a Review. *Sustain. Oper. Comput.* **2024**, *5*, 73–87. [CrossRef]
40. Kantaros, A.; Petrescu, F.I.T.; Brachos, K.; Ganetsos, T.; Petrescu, N. Evaluating Benchtop Additive Manufacturing Processes Considering Latest Enhancements in Operational Factors. *Processes* **2024**, *12*, 2334. [CrossRef]
41. Dalpadulo, E.; Petruccioli, A.; Gherardini, F.; Leali, F. A Review of Automotive Spare-Part Reconstruction Based on Additive Manufacturing. *J. Manuf. Mater. Process.* **2022**, *6*, 133. [CrossRef]
42. Hedayati, R.; Stulova, V. 3D Printing for Space Habitats: Requirements, Challenges, and Recent Advances. *Aerospace* **2023**, *10*, 653. [CrossRef]
43. Hegab, H.; Khanna, N.; Monib, N.; Salem, A. Design for Sustainable Additive Manufacturing: A Review. *Sustain. Mater. Technol.* **2023**, *35*, e00576. [CrossRef]
44. Liu, W.; Liu, X.; Liu, Y.; Wang, J.; Evans, S.; Yang, M. Unpacking Additive Manufacturing Challenges and Opportunities in Moving towards Sustainability: An Exploratory Study. *Sustainability* **2023**, *15*, 3827. [CrossRef]
45. Chyr, G.; DeSimone, J.M. Review of High-Performance Sustainable Polymers in Additive Manufacturing. *Green Chem.* **2023**, *25*, 453–466. [CrossRef]
46. Jiang, J.; Fu, Y.-F. A Short Survey of Sustainable Material Extrusion Additive Manufacturing. *Aust. J. Mech. Eng.* **2023**, *21*, 123–132. [CrossRef]
47. Calignano, F.; Mercurio, V. An Overview of the Impact of Additive Manufacturing on Supply Chain, Reshoring, and Sustainability. *Clean. Logist. Supply Chain* **2023**, *7*, 100103. [CrossRef]
48. Kantaros, A.; Ganetsos, T.; Petrescu, F.I.T.; Ungureanu, L.M.; Munteanu, I.S. Post-Production Finishing Processes Utilized in 3D Printing Technologies. *Processes* **2024**, *12*, 595. [CrossRef]
49. Ge, J.; Pillay, S.; Ning, H. Post-Process Treatments for Additive-Manufactured Metallic Structures: A Comprehensive Review. *J. Mater. Eng. Perform.* **2023**, *32*, 7073–7122. [CrossRef]
50. Jin, L.; Zhai, X.; Wang, K.; Zhang, K.; Wu, D.; Nazir, A.; Jiang, J.; Liao, W.-H. Big Data, Machine Learning, and Digital Twin Assisted Additive Manufacturing: A Review. *Mater. Des.* **2024**, *244*, 113086. [CrossRef]
51. Peng, S.; Li, T.; Wang, X.; Dong, M.; Liu, Z.; Shi, J.; Zhang, H. Toward a Sustainable Impeller Production: Environmental Impact Comparison of Different Impeller Manufacturing Methods. *J. Ind. Ecol.* **2017**, *21*, S216–S229. [CrossRef]
52. Andrearczyk, A.; Bagiński, P.; Klonowicz, P. Numerical and Experimental Investigations of a Turbocharger with a Compressor Wheel Made of Additively Manufactured Plastic. *Int. J. Mech. Sci.* **2020**, *178*, 105613. [CrossRef]
53. Parize, H.; Tardelli, J.D.C.; Bohner, L.; Sesma, N.; Muglia, V.A.; Dos Reis, A.C. Digital versus Conventional Workflow for the Fabrication of Physical Casts for Fixed Prosthodontics: A Systematic Review of Accuracy. *J. Prosthet. Dent.* **2022**, *128*, 25–32. [CrossRef]
54. Hernández, F.; Fragoso, A. Fabrication of a Stainless-Steel Pump Impeller by Integrated 3D Sand Printing and Casting: Mechanical Characterization and Performance Study in a Chemical Plant. *Appl. Sci.* **2022**, *12*, 3539. [CrossRef]
55. Yang, Z.; Zhu, L.; Ning, J.; Wang, S.; Xue, P.; Xu, P.; Dun, Y.; Xin, B.; Zhang, G. Revealing the Influence of Ultrasound/Heat Treatment on Microstructure Evolution and Tensile Failure Behavior in 3D-Printing of Inconel 718. *J. Mater. Process. Technol.* **2022**, *305*, 117574. [CrossRef]
56. Popa, I.-F.; Dobromirescu, C.; Vintila, I.S.; Tinca, I.; Mihai, D.; Canache, A.M.; Crunteanu, D.E. Hydraulic Performance Comparison of Centrifugal Closed Impellers Fabricated by Means of Additive Manufacturing and Classical Machining for Space Active Thermal Control Systems of Satellites. *Applied Sciences* **2024**, *14*, 9448. [CrossRef]
57. Makivskyi, O.S.; Kondus, V.Y.; Pitel, J.; Sotnyk, M.I.; Andrusiak, V.O.; Polkovnychenko, V.V.; Mushtai, M.V. The Influence of the Design Features of the Submersible Pump Rotor on the Vibration Reliability. *J. Eng. Sci.* **2024**, *11*, D1–D9. [CrossRef]
58. Cooper, C.; Booth, A.; Britten, N.; Garside, R. A Comparison of Results of Empirical Studies of Supplementary Search Techniques and Recommendations in Review Methodology Handbooks: A Methodological Review. *Syst. Rev.* **2017**, *6*, 1–16. [CrossRef] [PubMed]
59. Aguinis, H.; Ramani, R.S.; Alabduljader, N. Best-Practice Recommendations for Producers, Evaluators, and Users of Methodological Literature Reviews. *Organ. Res. Methods* **2023**, *26*, 46–76. [CrossRef]
60. Alexander, P.A. Methodological Guidance Paper: The Art and Science of Quality Systematic Reviews. *Rev. Educ. Res.* **2020**, *90*, 6–23. [CrossRef]

61. Scopus. Scopus—Document Search. Available online: https://www.scopus.com/standard/marketing.uri (accessed on 27 June 2024).
62. WoS. *Web of Science. WoS.*; Clarivate: London, UK, 2020.
63. VOSviewer. VOSviewer—Visualizing Scientific Landscapes. Available online: https://www.vosviewer.com/ (accessed on 27 June 2024).
64. Jiménez, M.; Romero, L.; Domínguez, I.A.; Espinosa, M.d.M.; Domínguez, M. Additive Manufacturing Technologies: An Overview about 3D Printing Methods and Future Prospects. *Complexity* **2019**, *2019*, 9656938. [CrossRef]
65. Systèmes, D. Introduction to 3D Printing–Additive Processes. *Dassault Syst.* **2018**. [CrossRef]
66. Adiaconitei, A.; Vintila, I.S.; Mihalache, R.; Paraschiv, A.; Frigioescu, T.; Vladut, M.; Pambaguian, L. A Study on Using the Additive Manufacturing Process for the Development of a Closed Pump Impeller for Mechanically Pumped Fluid Loop Systems. *Materials* **2021**, *14*, 967. [CrossRef]
67. Meli, E.; Rindi, A.; Ridolfi, A.; Furferi, R.; Buonamici, F.; Iurisci, G.; Corbò, S.; Cangioli, F. Design and Production of Innovative Turbomachinery Components via Topology Optimization and Additive Manufacturing. *Int. J. Rotating Mach.* **2019**, *2019*, 9546831. [CrossRef]
68. Allison, T.C.; Rimpel, A.M.; Moore, J.J.; Wilkes, J.C.; Pelton, R.; Wygant, K. Manufacturing and Testing Experience with Direct Metal Laser Sintering for Closed Centrifugal Compressor Impellers. In Proceedings of the 43rd Turbomachinery Symposium, Houston, TX, USA, 23–25 September 2014; Texas A&M University, Turbomachinery Laboratories: College Station, TX, USA, 2014.
69. Meli, E.; Furferi, R.; Rind, A.; Ridolfi, A.; Volpe, Y.; Buonamici, F. A General Framework for Designing 3D Impellers Using Topology Optimization and Additive Manufacturing. *IEEE Access* **2020**, *8*, 60259–60269. [CrossRef]
70. Lea, H.G.; Wooding, R.S.; Kuhr, S.; Rotella, J.; Córdova, J.L. Characterization of Properties of Laser Powder Bed Fusion Three-Dimensional-Printed Inconel 718 for Centrifugal Turbomachinery Applications. *J. Eng. Gas Turbine Power* **2023**, *145*, 061017. [CrossRef]
71. Jia, D.; Li, F.; Zhang, Y. 3D-Printing Process Design of Lattice Compressor Impeller Based on Residual Stress and Deformation. *Sci. Rep.* **2020**, *10*, 600. [CrossRef] [PubMed]
72. Zhang, Y.; Li, F.; Jia, D. Residual Stress and Deformation Analysis of Lattice Compressor Impeller Based on 3D Printing Simulation. *Mech. Adv. Mater. Struct.* **2022**, *29*, 717–731. [CrossRef]
73. Guo, J.; Wang, Y.; Wang, Y.; Peng, S.; Wang, F. Simulation Study on the Energy Utilization Efficiency of a Turbine Impeller Based on a Selective Laser Melting Process. *Appl. Sci.* **2023**, *13*, 10657. [CrossRef]
74. Singh, S.P.; Gupta, R.K.; Agarwal, A. Processing and Characterization of Al-Si-Mg Alloy Made through 3D Printing and Comparison with Equivalent Cast Alloy. *Mater. Today Proc.* **2022**, *67*, 422–430. [CrossRef]
75. Auyeskhan, U.; Lee, T.; Park, Y.; Kim, D.; Kim, N.; Kim, C.-S.; Kim, D.-H. Design Study for Performance Improvement of a Hybrid Pico Pelton Turbine and Its Additive Manufacturing Using a Laser Powder Bed Fusion Method. *Int. J. Interact. Des. Manuf. (IJIDeM)* **2024**, *18*, 107–117. [CrossRef]
76. Qu, A.; Li, F. Influence of 3D Printing on Compressor Impeller Fatigue Crack Propagation Life. *Int. J. Mech. Sci.* **2023**, *245*, 108107. [CrossRef]
77. Raghavan, S.; Dzugan, J.; Rzepa, S.; Podany, P.; Soh, N.; Hao, L.J.; Khan, N. Effect of Post-Treatment on Local Mechanical Properties of Additively Manufactured Impellers Made of Maraging Steel. *Rapid. Prototyp. J.* **2023**, *29*, 594–611. [CrossRef]
78. Ponticelli, G.S.; Tagliaferri, F.; Venettacci, S.; Horn, M.; Giannini, O.; Guarino, S. Re-Engineering of an Impeller for Submersible Electric Pump to Be Produced by Selective Laser Melting. *Appl. Sci.* **2021**, *11*, 7375. [CrossRef]
79. Huber, M.; Ess, J.; Hartmann, M.; Würms, A.; Rettberg, R.; Kränzler, T.; Löffel, K. Process Setup for Manufacturing of a Pump Impeller by Selective Laser Melting. In *Industrializing Additive Manufacturing, Proceedings of the Additive Manufacturing in Products and Applications-AMPA2017 1, Zurich, Switzerland, 13–15 September 2017*; Springer: Berlin/Heidelberg, Germany, 2018; pp. 252–263.
80. Zhang, Y.; Li, F.; Jia, D. Lightweight Design and Static Analysis of Lattice Compressor Impeller. *Sci. Rep.* **2020**, *10*, 18394. [CrossRef]
81. Wu, S.-T.; Huang, P.-W.; Chang, T.-W.; Jiang, I.-H.; Tsai, M.-C. Application of Magnetic Metal 3-D Printing on the Integration of Axial-Flow Impeller Fan Motor Design. *IEEE Trans. Magn.* **2020**, *57*, 8201205. [CrossRef]
82. Bauer, D. Additive Manufacturing of Cryogenic Pump Impellers. In Proceedings of the Asia Turbomachinery & Pump Symposium, Lumpur, Malaysia, 23–26 May 2022; Turbomachinery Laboratory, Texas A&M Engineering Experiment Station: College Station, TX, USA, 2022.
83. Kim, E.-H.; Choi, H.-H.; Jung, Y.-G. Fabrication of a Ceramic Core for an Impeller Blade Using a 3D Printing Technique and Inorganic Binder. *J. Manuf. Process.* **2020**, *53*, 43–47. [CrossRef]
84. Chivel, Y.; Sudarev, A.; Konakov, V. Additive Manufacturing Technology of Ceramic Turbomachines. In Proceedings of the European Congress and Exhibition on Powder Metallurgy. European PM Conference Proceedings, Hamburg, Germany, 9–13 October 2016; The European Powder Metallurgy Association: Sandviken, Sweden, 2016; pp. 1–6.
85. Bin, L.I.; Huimin, L.; Yuhang, Z.; Yaxin, C.; Ming, Z.; Qinming, W.U.; Yong, Y. Research on Precision Casting Process and Performance of Impeller Based on SLS. *J. Xihua Univ. (Nat. Sci. Ed.)* **2023**, *42*, 71–77.
86. Špale, J.; Novotny, V.; Novotný, J. Experimental Development of Additively Manufactured Turboexpanders towards an Application in the ORC. In Proceedings of the Heat Powered Cycles Conference, Bilbao, Spain, 10–13 April 2022.
87. Wang, D.; Dong, A.; Zhu, G.; Shu, D.; Sun, J.; Li, F.; Sun, B. Rapid Casting of Complex Impeller Based on 3D Printing Wax Pattern and Simulation Optimization. *Int. J. Adv. Manuf. Technol.* **2019**, *100*, 2629–2635. [CrossRef]
88. Sudarev, A.; Konakov, V.; Chivel, Y. Selective Laser Sintering of Ceramic Turbomachine Components. *Procedia CIRP* **2018**, *74*, 264–267. [CrossRef]

89. Badum, L.; Prochaska, T.; Schwentenwein, M.; Cukurel, B. Ceramic and Metal Additive Manufacturing of Monolithic Rotors From SiAlON and Inconel and Comparison of Aerodynamic Performance for 300W Scale Microturbines. *J. Eng. Gas Turbine Power* **2024**, *146*, 021006. [CrossRef]
90. Stümpfl, D.; Ráž, K.; Streit, P.; Weiß, A.P. Additive manufacturing for fast prototyping of a velocity compounded radial re-entry turbine. In Proceedings of the 7th International Seminar on ORC Power System (ORC2023), Seville, Spain, 4–6 September 2023; pp. 13–22.
91. Liverani, A.; Caligiana, G.; Frizziero, L.; Francia, D.; Donnici, G.; Dhaimini, K. Design for Six Sigma (DFSS) for Additive Manufacturing Applied to an Innovative Multifunctional Fan. *Int. J. Interact. Des. Manuf. (IJIDeM)* **2019**, *13*, 309–330. [CrossRef]
92. Adach, M.; Sokołowski, P.; Piwowarczyk, T.; Nowak, K. Study on Geometry, Dimensional Accuracy and Structure of Parts Produced by Multi Jet Fusion. *Materials* **2021**, *14*, 4510. [CrossRef]
93. Tan, X.; Kok, Y.; Tor, S.B.; Chua, C.K. Application of Electron Beam Melting (EBM) in Additive Manufacturing of an Impeller. In Proceedings of the 1st International Conference on Progress in Additive Manufacturing (Pro-AM 2014), Singapore, 26–28 May 2014.
94. Klocke, F.; Klink, A.; Veselovac, D.; Aspinwall, D.K.; Soo, S.L.; Schmidt, M.; Schilp, J.; Levy, G.; Kruth, J.-P. Turbomachinery Component Manufacture by Application of Electrochemical, Electro-Physical and Photonic Processes. *CIRP Ann.* **2014**, *63*, 703–726. [CrossRef]
95. Mandil, G.; Le, V.T.; Paris, H.; Suard, M. Building New Entities from Existing Titanium Part by Electron Beam Melting: Microstructures and Mechanical Properties. *Int. J. Adv. Manuf. Technol.* **2016**, *85*, 1835–1846. [CrossRef]
96. Wang, P.; Nai, M.L.S.; Lu, S.; Bai, J.; Zhang, B.; Wei, J. Study of Direct Fabrication of a Ti-6Al-4V Impeller on a Wrought Ti-6Al-4V Plate by Electron Beam Melting. *Jom* **2017**, *69*, 2738–2744. [CrossRef]
97. Kahhal, P.; Jo, Y.-K.; Park, S.-H. Recent Progress in Remanufacturing Technologies Using Metal Additive Manufacturing Processes and Surface Treatment. *Int. J. Precis. Eng. Manuf. Green Technol.* **2024**, *11*, 625–658. [CrossRef]
98. Michaud, M.; Milan, P.J.; Vo, H.D. Low-Cost Rotating Experimentation in Compressor Aerodynamics Using Rapid Prototyping. *Int. J. Rotating Mach.* **2016**, *2016*, 8518904. [CrossRef]
99. Derrick, S.; Rauen, J.; Rodriguez, J.; Aller, B. The Dual Impeller Drive System (DIDS). In Proceedings of the ASEE North Central Sectional Conference, Pittsburgh, PA, USA, 27 March 2010; Marshall University: Pittsburgh, PA, USA, 2010; pp. 4C1–4C8.
100. Geiselhart, M. *Design for Manufacturing and Characterization of Small-Scale Turbomachinery Impellers*; Technical University of Denmark: Kongens Lyngby, Denmark, 2021.
101. Leicht, B.T.; Bohan, B.T.; Schauer, F.; Kemnitz, R.; Rueschhoff, L.M.; Lam, B.; Kemp, J.W.; Costakis, W. Manufacturing a Ceramic Turbine Rotor for a Compact Jet Engine. *J. Turbomach.* **2023**, *145*, 081009. [CrossRef]
102. Zhao, J.; Chu, W.; Song, T.; Bi, L.; Liu, J.; Han, Z.; Wang, Y.; Zhou, Q.; Li, L. Digital Light Processing of High-Strength Si3N4 Ceramics: Role of Particle Grading on Slurries Curing and Mechanical Properties. *Ceram Int.* **2024**, *50*, 22722–22732. [CrossRef]
103. Ivanovich, S.M.; Dmitrievich, K.A.; Aleksandrovna, K.I.; Konstantinovich, S.F. Study of Impeller Design Parameters Effect on the Axial Thrust of a Centrifugal Electric Pump Assembly. *Sib. Aerosp. Mag.* **2020**, *21*, 389–399.
104. Zhu, X.; Tong, X.; Wang, F. Rapid Investment Casting Technology of Closed Impeller Based on Digital Light Processing (DLP) 3D Printing. *Spec. Cast. Nonferrous Alloys* **2021**, *41*, 1370–1374.
105. González Barrio, H. Metholology for Hybrid Manufacturing of Turbomachinery Integral Rotary Components in Thermoresistant Super Alloys. Ph.D. Thesis, Universidad del País Vasco-Euskal Herriko Unibertsitatea, Leioa, Spain, 2019.
106. Soo, A.; Ali, S.M.; Shon, H.K. 3D Printing for Membrane Desalination: Challenges and Future Prospects. *Desalination* **2021**, *520*, 115366. [CrossRef]
107. Li, C.; Wu, B.; Zhang, Z.; Zhang, Y. A Novel Process Planning Method of 3+ 2-Axis Additive Manufacturing for Aero-Engine Blade Based on Machine Learning. *J. Intell. Manuf.* **2023**, *34*, 2027–2042. [CrossRef]
108. Pfeil, A.; Siegfarth, M.; Pusch, T.P.; Barbé, L.; Geiskopf, F.; Renaud, P. Design and Experimental Characterization of Hydraulically Actuated Revolute Joint Based on Multimaterial Additive Manufacturing. *J. Mech. Des.* **2022**, *144*, 055001. [CrossRef]
109. Jiang, C.; Fleck, B.A.; Lipsett, M.G. Rapid Wear Modelling in a Slurry Pump Using Soft 3D Impeller Material. *Energies* **2020**, *13*, 3264. [CrossRef]
110. Daabo, A.M.; Hassan, A.A.; Bashir, M.A.; Hamza, H.; Salim, S.; Koprulu, A.; Badawy, T.; Mahmoud, S.; Al-Dadah, R. Experimental Study and 3D Optimization of Small-Scale Solar-Powered Radial Turbine Using 3D Printing Technology. *Machines* **2023**, *11*, 817. [CrossRef]
111. Nishida, M.; Negishi, T.; Sakota, D.; Kosaka, R.; Maruyama, O.; Hyakutake, T.; Kuwana, K.; Yamane, T. Properties of a Monopivot Centrifugal Blood Pump Manufactured by 3D Printing. *J. Artif. Organs* **2016**, *19*, 322–329. [CrossRef] [PubMed]
112. Udroiu, R.; Nedelcu, A.; Deaky, B. Rapid Manufacturing by Polyjet Technology of Customized Turbines for Renewable Energy Generation. *Environ. Eng. Manag. J.* **2011**, *10*, 1387–1394. [CrossRef]
113. Aruna Prabha, K.; Sai Rohit, P.; Nitturi, S.C.; Nithin, B. Manufacturing of 3 D Shrouded Impeller of a Centrifugal Compressor on 3D-Printing Machine Using FDM Technology. In Proceedings of the Materials Science and Engineering Conference Series, Bapatla, Andhra, 7–8 May 2021; Volume 1012, p. 012039.
114. Tamburrino, F.; Barone, S.; Paoli, A.; Razionale, A.V. Post-Processing Treatments to Enhance Additively Manufactured Polymeric Parts: A Review. *Virtual Phys. Prototyp.* **2021**, *16*, 221–254. [CrossRef]

115. Micha Premkumar, T.; Pushpak, V.; Vamsi Krishna, K.; Govardhan Reddy, D.; Sudheer Kumar, N.; Hariram, V.; Seralathan, S.; Nakandhrakumar, R.S. Design and Fusion Deposit Modelling of Radial Flow Centrifugal Pump. *Mater. Today Proc.* **2020**, *33*, 3497–3503. [CrossRef]
116. Choe, C.-M.; Yang, W.-C.; Kim, U.-H.; Ri, B.-G.; Om, M.-S. Manufacture of Centrifugal Compressor Impeller Using FDM and Investment Casting. *Int. J. Adv. Manuf. Technol.* **2022**, *118*, 173–181. [CrossRef]
117. Sedykh, L.V.; Borisov, P.V.; Pashkov, A.N.; Gorbatyuk, N.V.; Surkova, R.Y.; Mamatkulov, Z.K. Development of Flotation Machine Impeller on the Basis of Additive Technologies. *Steel Transl.* **2021**, *51*, 308–313. [CrossRef]
118. Jabłońska, M.; Jurczak, W.; Ozimina, D.; Adamiak, M. Increasing the Operational Reliability of a Ship by Using a Composite Impeller in the Event of Hydrophore Pump Failure. *Eksploat. I Niezawodn.* **2023**, *25*. [CrossRef]
119. Daniyan, I.A.; Muvunzi, R.; Fameso, F.; Mpofu, K.; Ale, F. Computer Aided Modelling and Investigation of the Performance of a Pump Impeller Produced Using Fused Deposition Modelling. In Proceedings of the 2023 14th International Conference on Mechanical and Intelligent Manufacturing Technologies (ICMIMT), Cape Town, South Africa, 26–28 May 2023; pp. 244–249.
120. Sitte, D.; Sutanto, H. Performance Analysis of Hydraulic Machines Using Impellers Made from Conventional Steel and 3D Printing Co-Polymer. In Proceedings of the IOP Conference Series: Materials Science and Engineering, Sanya, China, 12–14 November 2021; IOP Publishing: Bristol, UK, 2021; Volume 1115, p. 012049.
121. Polák, M. Behaviour of 3D Printed Impellers in Performance Tests of Hydrodynamic Pump. In Proceedings of the 7th International Conference on Trends in Agricultural Engineering, Prague, Czech Republic, 17–20 September 2019; pp. 17–20.
122. Do, T.; Bauder, T.J.; Suen, H.; Rego, K.; Yeom, J.; Kwon, P. Additively Manufactured Full-Density Stainless Steel 316L with Binder Jet Printing. In Proceedings of the International Manufacturing Science and Engineering Conference, College Station, TX, USA, 18–22 June 2018; American Society of Mechanical Engineers: New York, NY, USA, 2018; Volume 51357, p. V001T01A017.
123. Jayawardane, H.; Davies, I.J.; Gamage, J.R.; John, M.; Biswas, W.K. Investigating the 'Techno-Eco-Efficiency'Performance of Pump Impellers: Metal 3D Printing vs. CNC Machining. *Int. J. Adv. Manuf. Technol.* **2022**, *121*, 6811–6836. [CrossRef]
124. Sinha, A.; Swain, B.; Behera, A.; Mallick, P.; Samal, S.K.; Vishwanatha, H.M.; Behera, A. A Review on the Processing of Aero-Turbine Blade Using 3D Print Techniques. *J. Manuf. Mater. Process.* **2022**, *6*, 16. [CrossRef]
125. Pakle, S.; Jiang, K. Design of a High-Performance Centrifugal Compressor with New Surge Margin Improvement Technique for High Speed Turbomachinery. *Propuls. Power Res.* **2018**, *7*, 19–29. [CrossRef]
126. Zinchenko, A.; Baiul, K.; Krot, P.; Khudyakov, A.; Vashchenko, S.; Banasiewicz, A.; Wróblewski, A. Materials Selection and Design Options Analysis for a Centrifugal Fan Impeller in a Horizontal Conveyor Dryer. *Materials* **2021**, *14*, 6696. [CrossRef]
127. Sun, X.; Yan, Y.; Zhang, L.; Ma, G.; Liu, Y.; Yu, Y.; An, Q.; Tao, S. Direct 3D Printing of Reactive Agitating Impellers for the Convenient Treatment of Various Pollutants in Water. *Adv. Mater. Interfaces* **2018**, *5*, 1701626. [CrossRef]
128. Mansour, G.; Papageorgiou, V.; Tzetzis, D. Carbon Fiber Polymer Reinforced 3D Printed Composites for Centrifugal Pump Impeller Manufacturing. *Technologies* **2024**, *12*, 48. [CrossRef]
129. Jurković, K.; Schauperl, Z.; Šolić, S.; Bauer, B. Numerical Simulation of Sand Casting of Stainless Steel Pump Impeller. *Metals* **2024**, *14*, 435. [CrossRef]
130. Kennedy, F.E.; Ye, Y.; Baker, I.; White, R.R.; Barry, R.L.; Tang, A.Y.; Song, M. Development of a New Cryogenic Tribotester and Its Application to the Study of Cryogenic Wear of AISI 316 Stainless Steel. *Wear* **2022**, *496–497*, 204309. [CrossRef]
131. De Bartolomeis, A.; Newman, S.T.; Biermann, D.; Shokrani, A. State-of-the-Art Cooling and Lubrication for Machining Inconel 718. *J. Manuf. Sci. Eng.* **2021**, *143*, 050801. [CrossRef]
132. Raddatz, M.; Holländer, D.; Gampe, U. Innovative Material Testing Based on Small-Scale Specimens and Application for Turbomachinery Components. In *Turbo Expo: Power for Land, Sea, and Air*; American Society of Mechanical Engineers: New York, NY, USA, 2017; Volume 50923, p. V07AT31A004.
133. Córdova, J.L. Additive Manufacturing of Centrifugal Impellers for Solid Oxide Fuel Cell Anode Offgas Recycle Blowers. In *Turbo Expo: Power for Land, Sea, and Air*; American Society of Mechanical Engineers: New York, NY, USA, 2022; Volume 86052, p. V007T17A027.
134. Pegues, J.W.; Shao, S.; Shamsaei, N.; Sanaei, N.; Fatemi, A.; Warner, D.H.; Li, P.; Phan, N. Fatigue of Additive Manufactured Ti-6Al-4V, Part I: The Effects of Powder Feedstock, Manufacturing, and Post-Process Conditions on the Resulting Microstructure and Defects. *Int. J. Fatigue* **2020**, *132*, 105358. [CrossRef]
135. Wang, J.; Sama, S.R.; Lynch, P.C.; Manogharan, G. Design and Topology Optimization of 3D-Printed Wax Patterns for Rapid Investment Casting. *Procedia Manuf.* **2019**, *34*, 683–694. [CrossRef]
136. Rettberg, R.; Kraenzler, T. Hybrid Manufacturing: A New Additive Manufacturing Approach for Closed Pump Impellers. In *Industrializing Additive Manufacturing, Proceedings of the AMPA2020, Zurich, Switzerland, 1–3 September 2020*; Springer: Cham, Switzerland, 2021; pp. 146–159.
137. Li, H.; Zhang, Y. Lightweight Design of the Transonic Centrifugal Impeller Based on Fluid–Structure Coupling. *Phys. Fluids* **2022**, *34*, 107109. [CrossRef]
138. Kanesund, J.; Brodin, H.; Johansson, S. High Temperature Corrosion Influence on Deformation and Damage Mechanisms in Turbine Blades Made of IN-792 during Service. *Eng. Fail. Anal.* **2020**, *110*, 104388. [CrossRef]
139. Zhuk, Y. Nanostructured CVD W/WC Coating Protects Steam and Gas Turbine Blades Against Water Droplet Erosion. In *Turbo Expo: Power for Land, Sea, and Air*; American Society of Mechanical Engineers: New York, NY, USA, 2022; Volume 86052, p. V007T17A003.

140. Soffel, F.; Lin, Y.; Keller, D.; Egorov, S.; Wegener, K. Laser Remelting Process Simulation and Optimization for Additive Manufacturing of Nickel-Based Super Alloys. *Materials* **2021**, *15*, 177. [CrossRef] [PubMed]
141. Menon, N.; Mahdi, T.H.; Basak, A. Microstructure of IN738LC Fabricated Using Laser Powder Bed Fusion Additive Manufacturing. *J. Turbomach.* **2022**, *144*, 031011. [CrossRef]
142. Stella, P.; Ceccanti, F.; Cappuccini, F.; Bellacci, M.; Kemble, S.; Biccari, D.V.; Masi, G. Powder Metallurgy and Additive Manufacturing for Compressors and Pumps: A New Manufacturing Paradigm for Corrosion and Heat Resistant Alloys. In Proceedings of the Abu Dhabi International Petroleum Exhibition and Conference, Abu Dhabi, United Arab Emirates, 31 October–3 November 2022; SPE: Kuala Lumpur, Malaysia, 2022; p. D031S082R001.
143. Andrearczyk, A.; Konieczny, B.; Sokołowski, J. Additively Manufactured Parts Made of a Polymer Material Used for the Experimental Verification of a Component of a High-Speed Machine with an Optimised Geometry—Preliminary Research. *Polymers* **2020**, *13*, 137. [CrossRef]
144. Katiyar, P.C.; Singh, B.P.; Chhabra, M.; Kumar, N.; Parle, D. Analysis of Tensile Strength of Thin ABS Specimens 3-D Printed in Various Build Orientations for Turbomachinery Applications. *Int. J. Interact. Des. Manuf. (IJIDeM)* **2024**, 1–13. [CrossRef]
145. Vyacheslav, S.; Iurii, M.; Ilya, K. Prototyping Centrifugal Microcompressors Using Additive Technology. In *AIP Conference Proceedings*; AIP Publishing: Melville, NY, USA, 2023; Volume 2833.
146. Warner, J.; Celli, D.; Scott-Emuakpor, O.; George, T.; Tomlin, T. Fused Deposition Modeling Fabrication Evaluation of a TI-6AL-4V Centrifugal Compressor. *J. Eng. Gas Turbine Power* **2023**, *145*, 031008. [CrossRef]
147. Iyankumar, R.; Mohanavel, V.; Vinoth, N.; Vinoth, T. Modeling and Assessment of Pump Impeller. *Mater. Today Proc.* **2020**, *33*, 3226–3233. [CrossRef]
148. Jiang, C. Investigating Impeller Wear and Its Effect on Pump Performance Using Soft Materials. Master's Thesis, University of Alberta, Edmonton, AB, Canada, 2019.
149. Akdemir, B.; Kayisoglu, B.; Kocoglu, S. Effect of Impeller Materials on Centrifugal Pump Characteristics. *Agric. Mech. Asia Afr. Lat. Am.* **2006**, *37*, 62.
150. Fernández, S.; Jiménez, M.; Porras, J.; Romero, L.; Espinosa, M.M.; Domínguez, M. Additive Manufacturing and Performance of Functional Hydraulic Pump Impellers in Fused Deposition Modeling Technology. *J. Mech. Des.* **2016**, *138*, 024501. [CrossRef]
151. Bauer, A.P. Design, Development, and Testing of a Low Cost, Additively-Manufactured, Centrifugal Compressor, Air Force Institute of Technology. 2020. Available online: https://scholar.afit.edu/etd/3208 (accessed on 21 November 2024).
152. Boccini, E.; Furferi, R.; Governi, L.; Meli, E.; Ridolfi, A.; Rindi, A.; Volpe, Y. Toward the Integration of Lattice Structure-Based Topology Optimization and Additive Manufacturing for the Design of Turbomachinery Components. *Adv. Mech. Eng.* **2019**, *11*, 1687814019859789. [CrossRef]
153. Castro, M.T.; Ocon, J.D. Novel Designs of Blade Mixer Impellers from the Discrete Element Method and Topology Optimization. *Chem. Eng. J.* **2024**, *490*, 151863. [CrossRef]
154. Sword, J.I.; Galloway, A.; Toumpis, A. Analysis of Environmental Impact and Mechanical Properties of Inconel 625 Produced Using Wire Arc Additive Manufacturing. *Sustainability* **2024**, *16*, 4178. [CrossRef]

processes

MDPI

Article

Exploring Post-Machining Alternatives Under Dry Conditions for Thin-Walled Additive Manufacturing Components Aided by Infrared Thermography

Eduard Garcia-Llamas [1,*], Giselle Ramirez [2,*], Miguel Fuentes [3], Eduard Vidales [1] and Jaume Pujante [1]

1 Eurecat, Centre Tecnològic de Catalunya, Unit of Metallic and Ceramic Materials, Plaça de la Ciència, 2, 08243 Manresa, Spain; eduard.vidales@eurecat.org (E.V.); jaume.pujante@eurecat.org (J.P.)

2 CIEFMA, Center for Research in Structural Integrity, Reliability and Micromechanics of Materials, Department of Materials Science and Engineering, Universitat Politècnica de Catalunya, Campus Diagonal Besòs-EEBE, 08019 Barcelona, Spain

3 GSC, Grupo Sevilla Control or Sevilla Control Group, C/Manganeso, 2–P.I. Calonge–Sevilla, 41007 Sevilla, Spain; miguel.fuentes@gsc-aero.com

* Correspondence: eduard.garcia@eurecat.org (E.G.-L.); giselle.ramirez.s1@upc.edu (G.R.); Tel.: +34-938-777-373 (E.G.-L.); +34-934-137-400 (ext. 23382) (G.R.)

Abstract: Additive manufacturing (AM) techniques have transformed the production of parts and components with intricate geometries and customized designs, driving innovation in sustainable manufacturing practices. The additive manufacturing technology used in this work was selective laser melting (SLM), a process that uses laser energy to sinter powdered metals into solid structures. Among the various materials utilized in AM, Ti6Al4V titanium alloys are of particular interest due to their favorable mechanical properties, corrosion resistance, biocompatibility, and potential for reducing material waste. However, the machining of additively manufactured titanium parts presents challenges due to the material's low conductivity, elastic modulus, and chemical affinity with cutting tools, which impact tool wear and surface finish quality. Milling, a commonly employed process for finishing titanium parts, often involves significant energy use and tool wear, highlighting the need for optimized and eco-conscious machining strategies. This study aims to establish correlations among four key aspects: (1) surface finish of machined Ti6Al4V AM parts, (2) cutting tool damage, (3) dry milling parameters including different cutting tools, and (4) variation of temperature at the contact surface of AM parts and tools using infrared thermography. By examining parameters such as feed per tooth (Fz), axial depth of cut (Ap), spindle trajectories (trochoidal, helicoidal, and linear), and cutting tool diameters, this work identifies conditions that enhance process efficiency while reducing environmental impact. Infrared thermography provides insights into temperature variations during milling, correlating these changes to surface roughness and critical machining parameters, thus contributing to the development of sustainable and high-performance manufacturing practices.

Keywords: sustainable machining; additive manufacturing; Ti6Al4V alloy; selective laser melting (SLM); infrared thermography

1. Introduction

Ti-6Al-4V alloy and its variations are extensively utilized in industrial applications, including aerospace, biomedical, and energy sectors, due to their exceptional performance characteristics. These attributes include excellent corrosion resistance, a high strength-to-weight ratio, superior strength retention at elevated temperatures, and outstanding biological compatibility [1]. Additionally, its high potential for material efficiency and waste reduction makes Ti-6Al-4V an attractive material in industries striving to meet sustainability goals. As a result, the demand for Ti-6Al-4V has significantly increased in sectors like aerospace, where weight savings and efficiency are paramount [2].

Additive manufacturing (AM) has emerged as a transformative technology in these sectors, particularly due to its ability to fabricate thin-walled and complex-shaped components with near-net-shape features. This capability significantly reduces material waste, making AM a viable solution for manufacturing high-performance materials such as titanium alloys while minimizing environmental impact and production costs [3,4]. However, AM components often present poor surface quality, which is a critical concern for parts subjected to complex loading conditions, including contact fatigue and sliding wear. In such cases, post-machining operations are essential but become particularly challenging when processing thin-walled surfaces [5].

Milling is a widely used post-machining method for processing thin-walled additive manufacturing (AM) parts. However, when components are made from materials classified as "difficult-to-cut", the cutting process becomes more complex. Titanium alloys, such as Ti-6Al-4V, fall into this category due to their high chemical reactivity and poor thermal conductivity [1,6,7]. Additionally, their low modulus of elasticity results in complex and atypical deformation mechanisms under extreme working conditions, including pronounced thermal softening at elevated temperatures. This distinguishes titanium alloys from conventional metals like steel and aluminum.

These challenges not only hinder productivity but also lead to energy inefficiencies and greater environmental burdens in traditional manufacturing processes. Moreover, the machining of titanium generates significant heat, accelerating tool wear and reducing operational efficiency. This limits production rates and negatively impacts the surface quality of the finished parts. While these phenomena have been extensively studied in the context of milling Ti-6Al-4V blocks (both cast and forged), some of these limitations can be mitigated by incorporating various lubricants into the process [8].

However, the use of lubricants introduces its own set of challenges, such as higher production costs, the contamination of scrap material with metal chips and cutting fluids, and adverse environmental and health effects. Workers may be exposed to harmful aerosols created during the evaporation of lubricants, raising significant health concerns and highlighting the need for more sustainable approaches [9–11]. In response, efforts have been made to develop more environmentally friendly lubricants based on organic oils [9]. Despite these advancements, the most effective solution may lie in minimizing or eliminating the use of lubricants entirely through the adoption of dry machining techniques.

Beyond lubrication, machining AM thin-walled Ti-6Al-4V components for the aerospace industry introduces additional challenges, such as vibrations (chattering) and significant displacements of both the cutter and the workpiece. Milling thin-walled structures is a dynamically unstable process due to cutting and clamping forces acting during chip removal. This phenomenon results in wall deflections, leading to undesirable vibrations that further compromise machining accuracy and surface quality. These issues are exacerbated by the lower stiffness of AM thin-walled structures compared to their bulk counterparts, making dynamic instability a critical factor to consider [12]. For this reason, this study proposes three different spindle trajectories and two cutting tool diameters to investigate the best options for minimizing vibrations and enhancing surface quality.

Finally, it must be noted that these factors do not only affect the efficiency of the machining process itself but also the performance of the finished products. For instance, the role of surface roughness as a limiting factor of fatigue performance is well known and reported in the literature, even with studies on the specific case of additive-manufactured

Ti-6Al-4V [13]. On the other hand, the role of surface finish is more nuanced in medical applications [14]; nevertheless, obtaining a well-controlled and repeatable surface quality is paramount, for instance, to ensure osteointegration in implants.

Given these challenges, this study aims to explore different cutting conditions to reduce heat generation during the milling of thin-walled Ti-6Al-4V components produced via selective laser melting (SLM) [15]. The evaluation will be conducted in terms of tool wear and surface quality of the machined AM component. To accomplish this, infrared thermography is used to correlate heat generation during the milling process with temperature variations and the resulting surface roughness. Previous studies have successfully applied infrared thermography in machining research on materials such as steel, aluminum, and titanium [16–19]. The key advantages of thermographic devices are their non-contact nature and rapid response, allowing measurements in areas where chips could damage the measurement system or in otherwise inaccessible locations. Since no thermocouple needs to be attached to the tool or sample, infrared measurement does not negatively impact industrial processes, making it a reliable technique for monitoring temperature in dry machining operations.

Moreover, this study examines the correlation among four critical aspects: (1) finishing surface quality of machined Ti-6Al-4V AM parts, (2) cutting tool damage, (3) dry milling parameters—including different cutting tools—and (4) temperature variation at the contact surface of the tool and AM part. The milling parameters investigated include feed per tooth (Fz), axial depth of cut (Ap), three spindle trajectories (trochoidal, helicoidal, and linear), and two cutting tool diameters. By focusing on dry machining techniques and their environmental implications, this work contributes to the development of sustainable manufacturing practices [8], and the removal of the lubricant has a positive economic impact, as seen with the most recent 90's versions [20].

2. Materials and Methods

2.1. Materials and Specimens

All samples were obtained using 3D additive manufacturing of Ti-6Al-4V powder. Specifically, the technology employed was Selective Laser Melting (SLM), an advanced version of the Selective Laser Sintering (SLS) process. In SLM, a high-power laser melts the metal alloy powders on the powder bed to form dense, predefined metal objects. The SLM parameters used to produce the samples included a laser power of 200 W, a scanning velocity of 1500 mm/s, a slice thickness of 30 μm, and a hatch distance of 60 μm. Finally, the samples were heat treated by annealing at 850 °C for 2 h.

The chemical composition of the Ti-6Al-4V powder for SLM and its comparison with bulk material specifications are presented in Table 1.

Table 1. Chemical composition of the Ti-6Al-4V powder for SLM, compared to ASTM B265 Ti6Al4V grade alloy [5]; amounts in [wt%].

	Ti	Al	V	Fe	O	C	N	H	Res.
Ti6Al4V powder for SLM	Balance	5.5–6.5	3.5–4.5	≤0.25	≤0.13	≤0.08	≤0.05	≤0.012	≤0.41
ASTM B265 Ti6Al4V alloy	Balance	5.5–6.75	3.5–4.5	≤0.4	≤0.2	≤0.08	≤0.05	≤0.015	≤0.40

The dry milling process was performed on an SLM component, as shown in Figure 1. The component consisted of a platform measuring 70 mm × 48 mm with a thin-wall support of 2.8 mm thickness, attached to a base measuring 64 mm × 64 mm with a thickness of 32 mm.

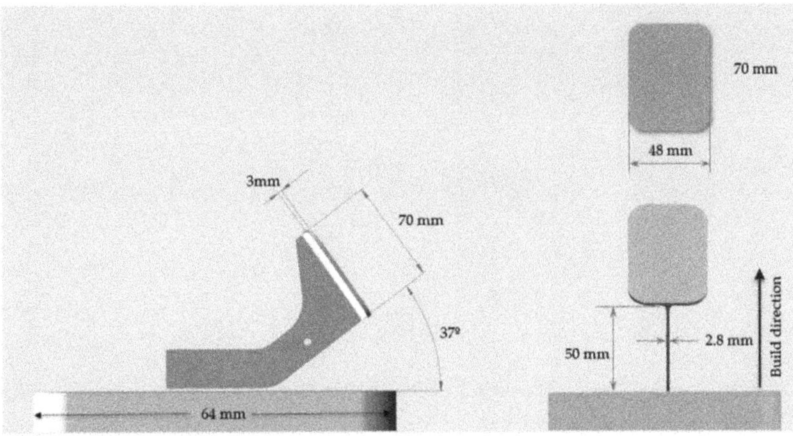

Figure 1. Schematic of the SLM component machined by milling process in this study, along its dimensions.

2.2. Test Methodology

2.2.1. Initial Setup Trials

Initial tests were conducted with the SLM component rigidly fixed in an industrial vise. This setup effectively prevented vibrations caused by the thin-wall support geometry. By ensuring stable clamping, the focus was placed on analyzing cutting tool damage and its correlation with temperature changes, as well as determining the optimal spindle trajectory among trochoidal, helical, and linear paths.

Once the optimal trajectory was determined, the next step was to perform tests on the component without the rigid vise, observing the behavior of the thin-walled support and its inherent geometry, as well as its effects on temperature variations and the surface roughness of the finished part.

All tests were performed using a CNC machining center, the five-axis HAAS UMC-750. Initial trials to determine the best trajectory (trochoidal, helical, and linear) examined milling parameters such as feed per tooth (Fz) and axial depth of cut (Ap). The cutting tool used in these tests was the XDLX 09T308ER-F40 CTC5240, paired with the holder tool GA SD090 C 032 Z3, which had a diameter of 32 mm. This configuration was referred to as Setup 1 (see Figure 2a). Figure 3 illustrates a schematic of the three different trajectories used in the initial trials. The linear trajectory was performed with a total of four milling passes for each component to machine the entire surface. Table 2 presents the variation in feed per tooth (Fz) and axial depth of cut (Ap) for each sample. The fixed milling parameters for all these tests were cutting velocity (Vc) at 60 m/min, radial cutting depth (Ae) at 80%, feed rate (Vf) at 229 mm/min, spindle speed (n) at 764 rpm, and number of effective edges (Zc) at 3. The parameters selected for this study are based on a substantial amount of the literature and references related to titanium machining. A summary can be found in [21], while more specific references for dry conditions, similar 3D printing techniques, and the milling process are provided in [22–24]. Taking into account the reference data, as well as the specifications of our tools and system, the milling parameters and their variations were defined.

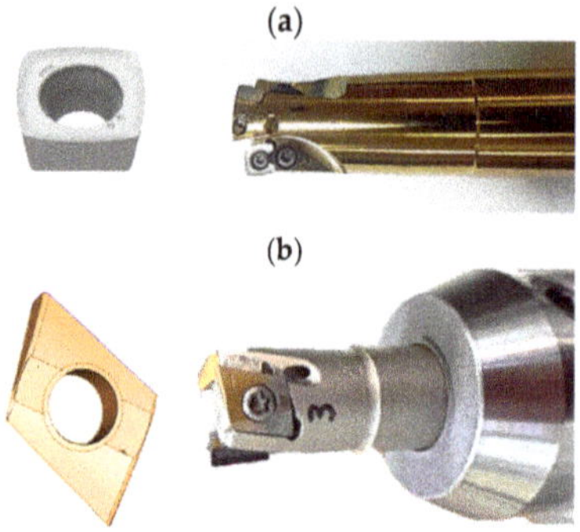

Figure 2. (**a**) Image of cutting tool XDLX 09T308ER-F40 CTC5240 and the tool holder GA SD090 C 032 Z3; (**b**) Image of cutting tool T290 LNMT 100405TR IC808 and the tool holder T290 ELN D20-03-W20-10.

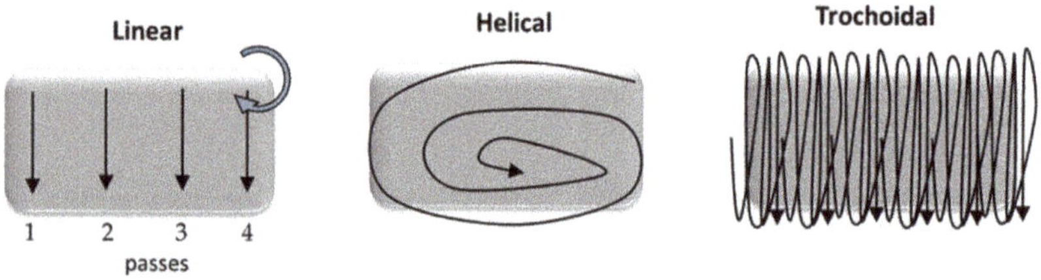

Figure 3. Scheme of the three different trajectories used in the tests.

Table 2. Summary of the variation parameters feed per tooth (Fz) and axial depth of cut (Ap) and trajectory for each test.

Test Number	Ap (mm)	Fz (mm)	Trajectory
1	0.1	0.1	Linear
2	0.1	0.5	Linear
3	0.25	0.1	Linear
4	0.1	0.1	Trochoidal
5	0.1	0.1	Helical

New-edge tools were used for all tests under a dry cutting environment to eliminate the influence of tool wear on cutting temperature and surface roughness. This ensured a reliable correlation among the samples machined under different test conditions.

2.2.2. Setup Variations

Once the optimal parameters and trajectory were identified, a different set of cutting tools and holder was used. The second cutting tool was a T290 LNMT 100405TR IC808 with two cutting edges and a grade IC808, offering a submicron substrate with excellent chipping resistance, combined with a "SUMO TEC" PVD coating of TiAlN/AlTiN+TiN. The corresponding holder was the T290 ELN D20-03-W20-10, with a diameter of 20 mm. This configuration was referred to as Setup 2 (see Figure 2b). The tests conducted with this cutting tool and holder utilized a trochoidal trajectory. In these tests, the feed per tooth (Fz) and axial depth of cut (Ap) were fixed at 0.1 mm. The fixed milling parameters included a cutting velocity (Vc) of 120 m/min, a radial cutting depth (Ae) of 20%, a feed rate (Vf) of 573 mm/min, a spindle speed (n) of 1910 rpm, and a number of effective edges (Zc) of 3. With the new set of tools and holders, the objective was to evaluate how damage to the cutting tool impacts the surface finish and temperature variations.

2.2.3. Surface Roughness Determination

The surface quality of the machined components was assessed using Ra and RSM values, both of which represent surface roughness but are calculated differently. Ra, or Roughness Average, is determined as the arithmetic average of a surface's measured microscopic peaks and valleys. RSM is the arithmetic mean value of the widths of the profile elements of the roughness profile. The roughness images were acquired using an infinite focus microscopy (Alicona Infinite Focus SL, Alicona Imaging GmbH, Graz, Austria) and data analysis was performed with MountainsMap 5.1 software, adhering to ISO 4287 and ISO 25178 standards [25,26]. The analysis involved the evaluation of the absolute ordinate Z(x) within the sampling length. Figure 4 illustrates an example of a component surface area measured by the optical profilometer, including the 3D image obtained and the graph showing the variation of the roughness.

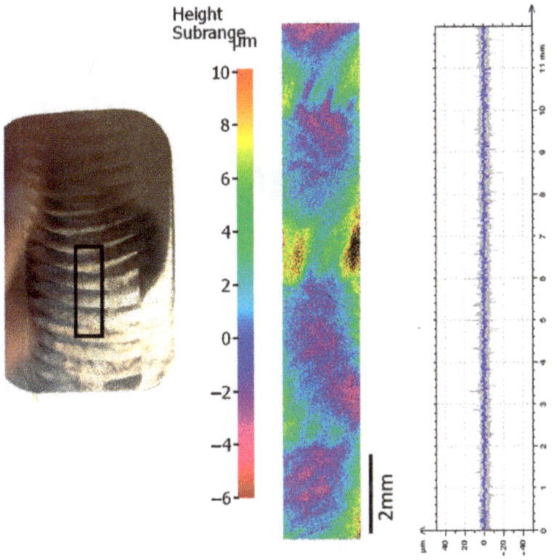

Figure 4. From left to right, the figure shows the machined component with the analyzed surface highlighted by a black rectangle, followed by the extracted 3D image and the graph illustrating the surface roughness variation.

Worn and adhered material at the cutting edge of tools were analyzed using the same microscopy technique, and they were expressed in terms of volume adhered in mm3. The software of the infinite focus microscope allows us to overlap the initial and final topographic images to subtract the adhered or removed material during the testing.

2.2.4. Temperature Measurement by Means of Thermal Imaging

Thermal imaging was used to measure workpiece temperature under the different machining trials. This allowed correlating the different cutting configurations with the resulting workpiece heating generated.

Figure 5a shows the experimental setup with the five-axis CNC machining center, the HAAS UMC-750. On one side, the data acquisition computer can be seen, while inside the HAAS UMC-750, the SLM component and the black square indicating the location of the infrared camera for capturing infrared images are visible. The window glass of the five-axis CNC machining center was removed to ensure clear IR imaging without any interference. The FLIR SC645 high-resolution LWIR science-grade infrared camera (Figure 5b) features an uncooled microbolometer detector with a resolution of 640 × 480 pixels, a 17-micron pixel size, and a spectral detection range of 7.5–13 µm. The thermal camera operates at a frame rate of 25 Hz and within a temperature range of 120 °C to 650 °C. Figure 5c shows an example of an infrared image obtained using the FLIR SC645. To determine the emissivity of the titanium samples during milling trials, laboratory tests were conducted. A machined titanium sample with an attached thermocouple was heated to 650 °C. Then, the thermographic camera was used to measure temperature variations and the cooling process, which were correlated with the appropriate emissivity values to match the thermocouple readings. The 0.30 emissivity for the Ti6Al4V was determined, and it is consistent with references, which place it in the range of 0.20 and 0.35 [19,27]. At the start of the experiments, the distance from the sample to the IR camera, as well as the environmental temperature and humidity, were measured and assumed to remain constant throughout all tests. The reflected temperature was determined using the reflection methodology specified in ASTM E1862-14.

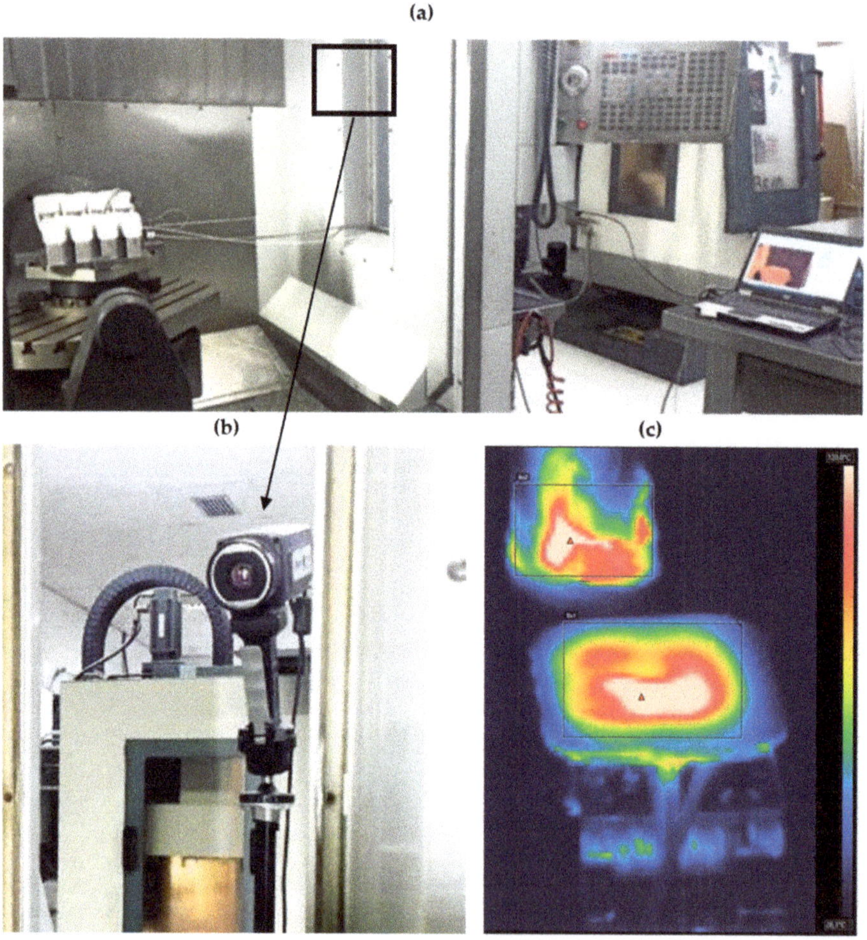

Figure 5. (**a**) Experimental setup with the five-axis CNC machining center, the HAAS UMC-750. Inside, the SLM component is visible, and the black square indicates the location of the infrared camera; (**b**) FLIR SC645 high-resolution LWIR infrared camera; (**c**) Example of an infrared image acquired by the infrared camera. The two registered temperature areas are represented by two rectangles. The bottom rectangle focuses on the component temperature, while the top rectangle focuses on the tool temperature. The red triangles indicate the maximum temperature in each area.

The infrared camera measurements were conducted in situ without interference from chips or tool-workpiece interaction, as the machined area and cutting tool were sufficiently large to ensure accurate temperature registration (see Figure 5c). In this work, infrared images were used to measure and present the maximum temperature, as the thermal effect plays a crucial role in metal machining. Tool life and surface quality largely depend on the maximum temperature reached in the cutting region [16,19].

3. Results and Discussion

3.1. Setup 1: Effect of Cutting Parameters on Surface Roguhness and Temperature

Surface finish is a critical aspect of machining and milling processes, not only because it affects the final aesthetic appearance of the component and its functionality, but also because it provides reliable insights into the condition of the cutting tool. Furthermore, changes in the cutting tool's condition can lead to significant variations in the temperature reached during the milling process, both on the surface of the component and the tool itself. The results of the experimental Setup 1 (cutting tool XDLX 09T308ER-F40 CTC5240 and the tool holder GA SD090 C 032 Z3), previously described, are presented below. In this setup, the component was rigidly fixed in a vise to analyze how variations in feed per tooth (Fz) and axial depth of cut (Ap) affect surface roughness and temperature changes. The most reliable temperature data was recorded during the second and third passes, where the cutting tool maintained full contact with the component. These two passes were used to draw conclusions and establish correlations between surface roughness, temperature, and cutting tool damage.

Figure 6 and Table 3 show how the roughness parameters Ra and RSM, as well as the maximum temperature changes during the four milling passes, are affected by the increase in Fz and Ap. It was observed that increasing the Fz parameter 0.1 and 0.5 mm had no significant effect on either surface roughness or temperature. In contrast, increasing Ap from 0.1 to 0.25 mm caused a marked degradation, particularly in the RSM roughness parameter. The measured roughness parameters of Ra were consistent and comparable with other references [24,28], where milling was performed under dry conditions. Moreover, the maximum temperature obtained with the Fp and Az 0.1 values is similar to the ones obtained in [19], but for this reference, the machining was the turning, not the milling, as it is focused on work.

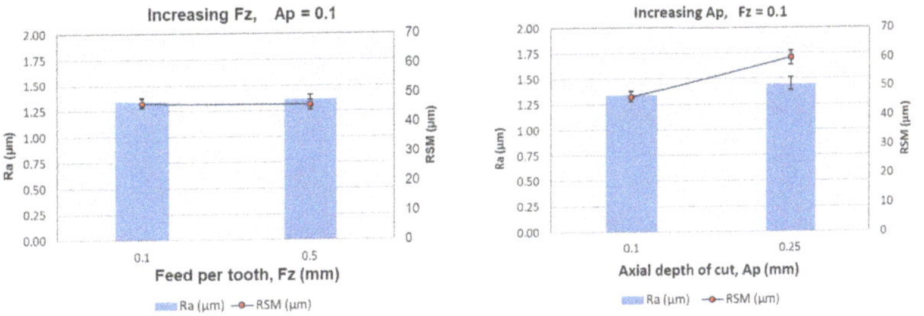

Figure 6. Variation of Ra and RSM roughness parameters while keeping Ap fixed at 0.1 and increasing Fz with values of 0.1 and 0.5 (**left chart**), and while keeping Fz fixed at 0.1 and increasing Ap with values of 0.1 and 0.25 (**right chart**).

Table 3. Maximum temperatures recorded during the four milling passes for the material and tool under varying feed per tooth (Fz) and axial depth of cut (Ap).

Test Number	Ap (mm)	Fz (mm)	Ra (μm)	RSM (μm)	Passes	Max. Temperature (°C)	
						Material	Tool
1	0.1	0.1	1.34 ± 0.04	46 ± 1.7	1st	203	93
					2nd	552	184
					3rd	615	217
					4rt	414	189
2	0.1	0.5	1.37 ± 0.04	46 ± 1.7	1st	423	90
					2nd	471	159
					3rd	567	159
					4rt	462	155
3	0.25	0.1	1.45 ± 0.06	60 ± 2.5	1st	650	280
					2nd	<660	325
					3rd	<660	388

During test number 3, infrared temperature recordings revealed that fixing Ap at 0.25 mm led to a sudden rise in temperature on both the component and the cutting tool. This temperature surge was primarily attributed to the accumulation of titanium swarf or chips, which in turn accelerated tool wear. As a result, the fourth milling pass could not be completed, and the test was halted after the third pass.

3.1.1. Trajectory Changes Affectation

Next, the effect of trajectory changes on the machining process was specifically analyzed. Different milling trajectories, including linear, trochoidal, and helicoidal, were evaluated to determine their impact on surface roughness, temperature, and tool wear.

Figure 7a illustrates the maximum temperatures recorded during the milling process for different trajectories. These temperatures significantly influenced the distribution of cutting forces and the heat generated during machining. Specifically, lineal and helicoidal trajectories were found to provide better cooling and reduced tool wear, whereas trochoidal trajectories led to higher temperatures and increased tool damage under similar cutting conditions. Consequently, the machining process using the trochoidal trajectory had to be halted to prevent catastrophic damage to the cutting tool, holder, and component.

These increased temperatures appeared to correlate with higher resulting workpiece roughness. Figure 7b presents the roughness parameters measured on the component surface for the different milling trajectories, showing a major increase for the trochoidal path. Both Ra and RMS increase by an approximate 50%, a significant deterioration in surface finish.

Similar observations could be drawn for the degradation of the workpiece.The Heat-Affected Zone (HAZ) on the cutting tool edge corresponding to each trajectory is illustrated, along with a chart representing the HAZ extent, in Figure 7c,d. Results show a clear increase in the extent of the HAZ for the machining bits used in the trochoidal path experiment (almost 6 mm), with the linear path keeping the minimal HAZ (1 mm), and spiral path an intermediate value (2 mm).

As a conclusion, the trochoidal trajectory can be dismissed as a feasible option for achieving the most efficient and effective machining process in Setup 1, as it appears to result in the most heat being released and the worst workpiece surface quality.

(a)

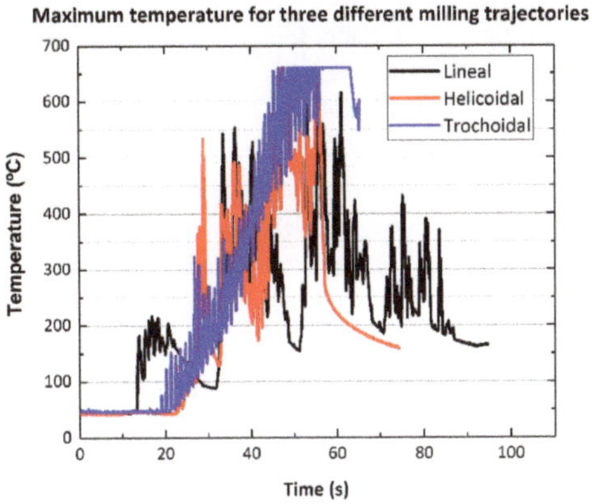

(b)

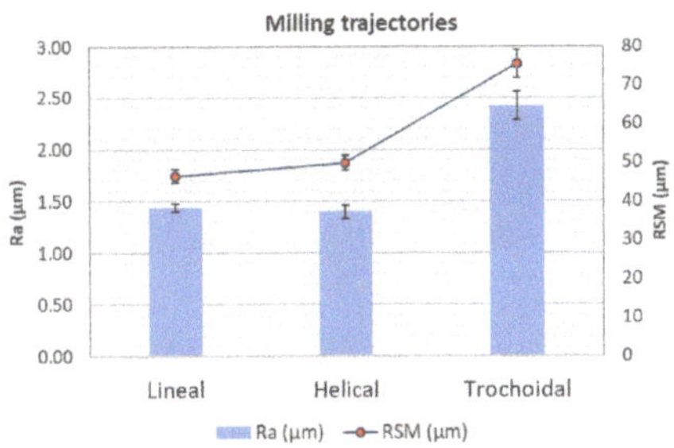

(c)

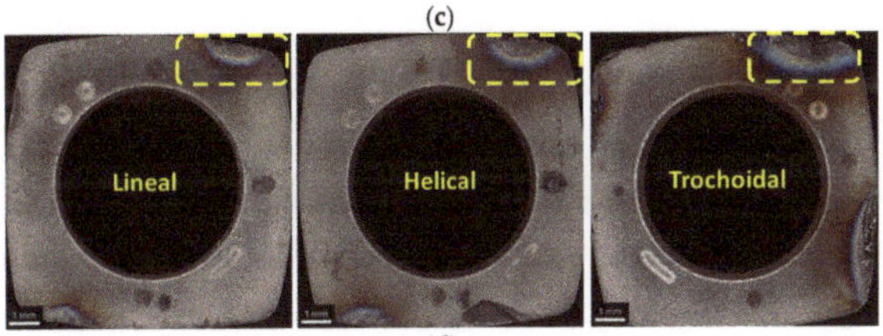

(d)

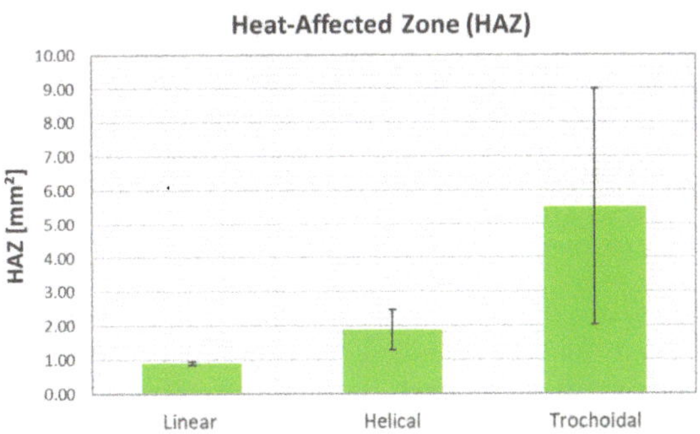

Figure 7. (**a**) The chart shows the maximum temperatures recorded during the milling process for different trajectories. (**b**) Surface roughness parameters measured on the component for different milling trajectories. (**c**) Image of the heat-affected zone (HAZ) on the cutting tool edge (highlighted in a yellow square) for the various trajectories, along with a chart (**d**) illustrating the extent of the HAZ.

3.1.2. Influence of Cutting Tool Wear on Recorded Temperature

The wear of the cutting tool significantly impacts the temperature measurements recorded during the machining process. As the cutting tool experiences wear, its efficiency in material removal decreases, leading to higher friction and increased heat generation at the tool-workpiece interface.

Worn tools tend to produce inconsistent temperature data due to uneven heat distribution and reduced cutting performance. These inconsistencies can compromise the reliability of thermal analysis and its correlation with surface roughness and machining efficiency. In contrast, the thermal gradients were accurately measured by the infrared camera even as the cutting tool wore down. Figure 8a shows the correlation between three different levels of cutting tool wear and the corresponding temperature changes. The chart illustrates three levels of cutting tool wear, labeled as Worn 1, 2, and 3, where a higher number indicates more severe wear on the tool. To clarify, Worn 1 indicates that the milling was performed using a cutting tool that had already undergone wear from one previous round of milling, Worn 2 refers to milling performed with a tool that had experienced wear from two previous rounds of milling, and Worn 3 corresponds to milling performed with a tool that had already undergone wear from three previous rounds of milling. Additionally, Figure 8b illustrates the increase in roughness parameters (Ra and RSM), particularly in the final stage of tool wear (Worn 3). It can be observed that increased cutting tool wear leads to higher temperatures recorded by the infrared camera, accompanied by a rise in roughness parameters due to poorer surface finish quality.

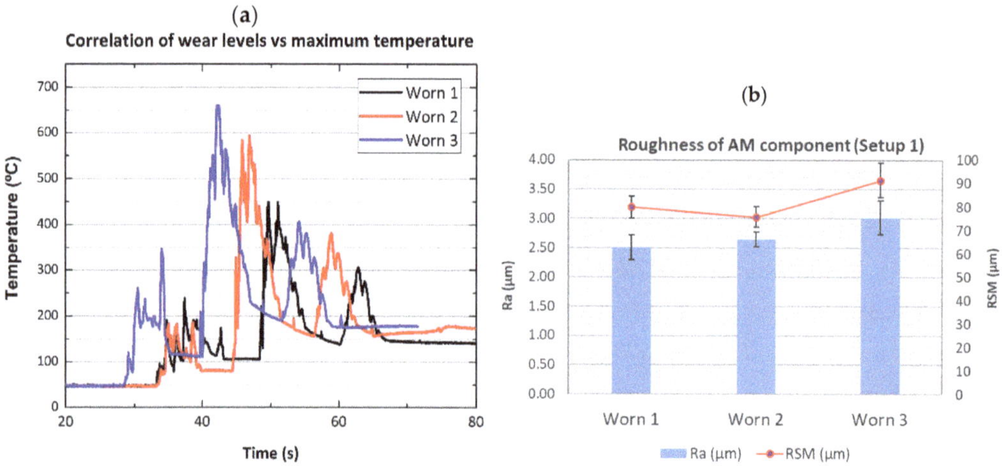

Figure 8. (**a**) Correlation between three levels of cutting tool wear and the corresponding maximum temperature changes measured by the infrared camera. (**b**) Evolution of Ra and RSM roughness of the AM component obtained with the three levels of tool wear.

This highlights that the infrared measurements provide consistent values, enabling effective monitoring of tool wear throughout machining operations, which directly affects the surface finish of the component. This behavior showed a strong correlation with the literature, which indicates that excessive wear on the cutting tool directly contributes to an increase in maximum temperature during the machining process [29].

3.2. Setup 2: Effect of Cutting Parameters on Surface Roughness and Temperature Compared with Setup 1

At this stage, it became clear that an improvement in the cutting tool was necessary to achieve an adequate milling process and ensure high surface quality in the thin-wall component. Although the linear trajectory is a widely used technique in the industry and has been validated in multiple studies for block samples and solid plane surfaces—showing correct results in terms of surface finish, tool wear, and heat-affected zone (HAZ) in the previous section—it was insufficient for milling the thin-wall component.

The need for an improved setup (Setup 2) arose due to the complexities associated with milling a component with thin-wall support. As mentioned in the introduction, additional challenges such as vibrations (chatter) and significant displacements of both the cutter and the workpiece posed major obstacles. Chatter, in particular, proved to be a critical drawback that made milling with a linear trajectory unfeasible.

To address this issue, alternative approaches were explored in the literature. Studies [30–34] highlight trochoidal trajectory as an effective path-planning strategy. This technique has the potential to increase the material removal rate per unit of tool wear, thereby enhancing productivity while reducing cutting energy consumption and improving tool performance. Based on these references, the trochoidal trajectory was evaluated and found to be optimal for minimizing vibrations in the 3D-printed thin-wall support component without rigid clamping during the milling process, ultimately improving process stability and surface quality.

Setup 2 utilized the cutting tool T290 LNMT 100405TR IC808 with the tool holder T290 ELN D20-03-W20-10. This cutting tool featured a combined TiAlN+TiN coating, selected because this multilayer coatings generally deliver superior performance across various applications compared to monolayer and uncoated tools [35,36]. Tests were conducted on both a rigidly fixed component using a vise and a 3D-printed thin-wall support without rigid clamping, using a trochoidal trajectory. The feed per tooth (Fz) and axial depth of cut (Ap) were fixed at 0.1 mm throughout.

The primary change in cutting parameters compared to Setup 1 was an increase in cutting velocity (Vc) to 120 m/min—double that of Setup 1—necessitated by the cutting tool's minimum requirements as specified by the supplier. The tool holder in Setup 2 also supported three effective edges (Zc = 3).

This section includes a comparison of the results from Setup 2 and Setup 1. Initially, the analysis focused on the impact of reducing the tool diameter from 32 mm to 20 mm. Subsequently, the relationship between tool damage, temperature changes, and their effects on surface roughness was examined for both setups. Finally, Setup 2 was used to evaluate tool wear and surface quality differences between a rigidly fixed component, and one supported by the 3D-printed thin-wall structure.

3.2.1. Diameter Change: Setup 1 vs. Setup 2

Figure 9 illustrates the maximum temperature measurements recorded by the infrared camera for the SLM component rigidly fixed with a vise. The results demonstrate that Setup 2 exhibited significantly better thermal performance compared to Setup 1 when employing a trochoidal trajectory. Setup 1 displayed a critical temperature increase exceeding 660 °C, whereas Setup 2 maintained temperatures under 250 °C throughout the milling process. To determine whether the key improvement in Setup 2 was primarily due to the reduction in tool diameter or the advanced coating in the cutting tool, a literature review was conducted. A study [29] demonstrated that using a 32 mm diameter tool with advanced coatings improved machinability, which has a direct impact on the maximum temperature during the milling process. Additionally, another study [37] showed that changes in tool diameter had no significant effect on average surface roughness. This suggests that while diameter reduction may have influenced the process, the primary factor in lowering maximum temperatures was the correct selection of the advanced coating of the cutting tool in Setup 2, which reduces the coefficient of friction. Maintaining lower temperatures ensures improved surface roughness, as detailed in the following subsection.

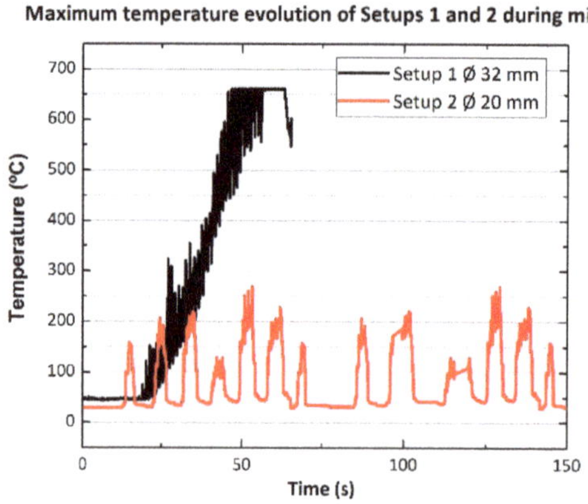

Figure 9. Temperature evolution recorded by the infrared camera for Setup 1 and Setup 2 during the milling process with a trochoidal trajectory.

3.2.2. Influence of Cutting Tool Wear: Setup 1 vs. Setup 2

Figure 10 shows the correlation between three different levels of cutting tool wear and the corresponding roughness parameters (Ra and RSM) for Setup 1 and Setup 2. While the roughness parameters increase with the progression of tool wear in Setup 1, Setup 2 maintains consistent Ra and RSM values. This is because the cutting tools in Setup 2 experienced less wear, despite both setups having the same workload. The comparison of cutting tool damage is provided in detail in the following section.

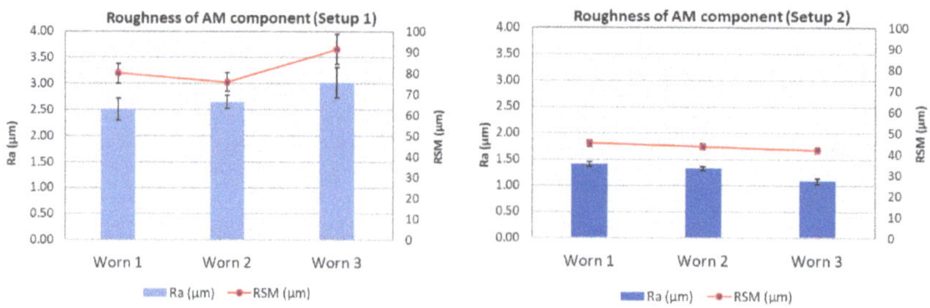

Figure 10. Evolution of Ra and RSM roughness parameters with the three levels of tool wear for Setup 1 and Setup 2.

3.2.3. Damage in the Cutting Tools: Setup 1 vs. Setup 2

Figure 11 displays an example of the 3D overlapped images used to measure the adhered and removed material of the cutting edge of tools. The color thresholding technique is used to represent the differences in terms of picks (adhered material) or valleys (removed material). The yellow and aqua colors indicate the initial surface of tools corresponding to Setup 1 and Setup 2, respectively. In both cases, the valley areas are represented with blue/magenta colors, where tool of Setup 1 achieved deeper values (~600 mm) than Setup 2 (~80 mm). The bar graph shown in Figure 11 corresponds to the measurements of all worn tools evaluated in this work for both setups. In this case, the quantification of removed and adhered material is expressed in terms of volume (mm3). Results indicate that tools used in Setup 1 presented a higher volume of worn material, which is visible in 3D images as chipping and spalling of large particles at the cutting edge. Even more, the volume of adhered material was also higher than the tools of Setup 2. These phenomena can be attributed to the severe machining conditions, where high temperatures were achieved (around 600 °C), as shown in Figure 9, promoting the tool brittleness.

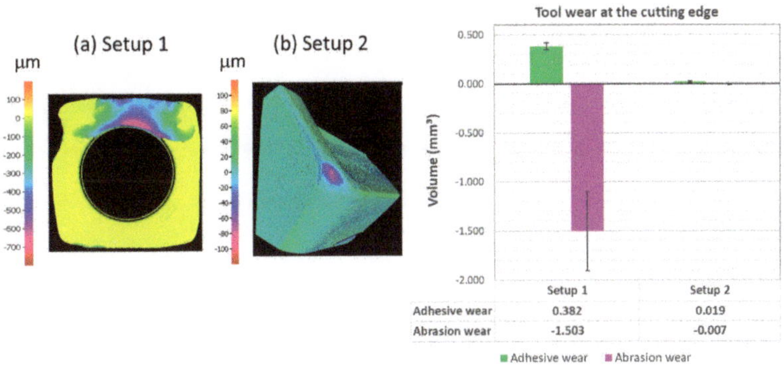

Figure 11. At (**left**), 3D images of the worn cutting tools for Setup 1 and for Setup 2 are shown. At the (**right**), charts depicting tool damage, including adhesion and wear, are presented.

3.2.4. Comparison of Vise Fixation vs. 3D-Printed Thin-Wall Support for Setup 2

In this final subsection, the effect of fixing the component with a vise versus using its own 3D-printed thin-wall support was analyzed. Figure 12 shows 3D images of the cutting edge for both conditions using Setup 2. As observed, both tools exhibit good performance compared to Setup 1 (as shown in Figure 11), whether fixed with the vise or the thin-wall support. However, the second condition, using the 3D-printed thin-wall support, showed an increased wear on the cutting tool. In this case, magenta colors show the deepest values (~100 mm) for the 3D-printed thin-walled supported component. Even with the use of coated tools, these dry machine conditions induced premature damage, and it is attributed to the presence of higher vibrations.

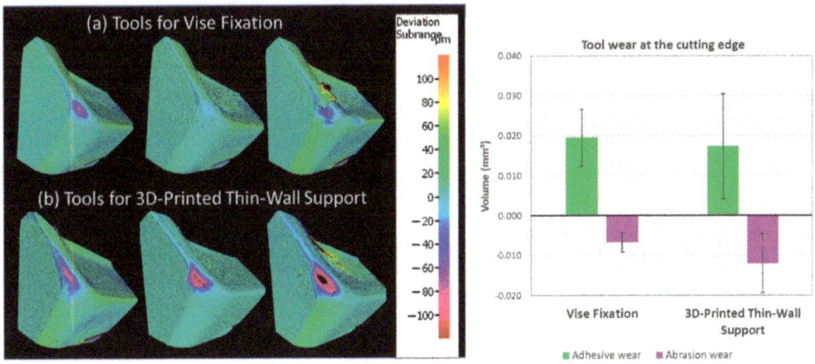

Figure 12. On the (**left**), images of the cutting tools for the compound fixed with a vise and the compound with 3D-printed thin-wall support, both for Setup 2, are shown. On the (**right**), a chart depicting tool damage, including adhesion and wear, is presented.

The vibrations and early wear of the cutting tool were also reflected in the surface roughness of the milled component (Figure 13). In this case, the differences were discerned in the surface patterns and machining patterns (3D profilometer images) and in terms of the mean width of profile elements (RSM), since it reflects the arithmetic mean value of the widths of the profile elements of the roughness profile, where a profile element is a peak and valley in the roughness profile. This value tends to increase in the thin-wall-supported component due to the increment of vibrations attributed to the fixing system, as also seen in the patterns of the surface roughness.

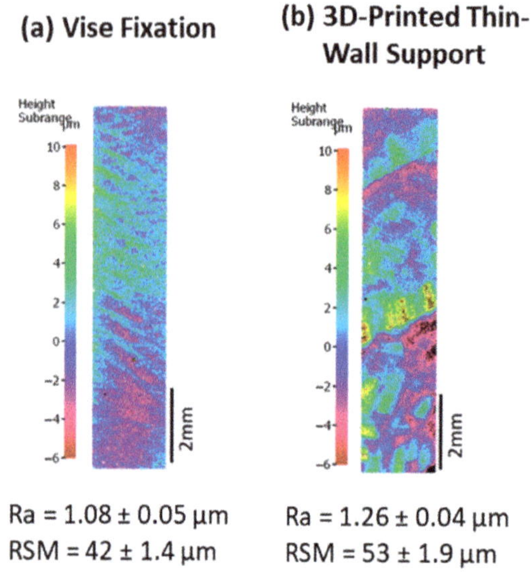

Figure 13. Three-dimensional surface images of the milled, 3D-printed Ti-6Al-4V surface fixed with the (**a**) vise and (**b**) thin-walled supported for Setup 2 and the Ra and RSM roughness values.

The stability of the system is a critical factor to consider in any machining process, as it helps prevent undesired damage to the tool and ensures a high-quality surface finish.

Despite the complexity of machining a thin-wall support 3D AM component, the final Ra parameter of 1.26 ± 0.04 µm met aeronautical industry requirements. While the general requirement specifies a surface roughness average (Ra) of 3.2 µm, more demanding areas require an Ra of 1.6 µm. These standards are based on the internal Airbus specification CAN 16056.

4. Conclusions

This study explored post-machining alternatives for thin-walled additive-manufactured components, utilizing an infrared thermographic monitoring system to correlate and identify tool damage and poor surface finishes in machining processes. This analysis was performed in the context of sustainable manufacturing, as the post-machining was performed under dry conditions. More importantly, this work demonstrated that the complexity of thin-wall support in AM components determines both the machining trajectory and the achievable surface roughness quality.

It has been demonstrated that a cutting tool with a coating designed for machining titanium in Setup 1 produced valid results when used with a linear milling trajectory for the additive-manufactured component fixed in a vise. However, when the component was no longer solidly fixed in the vise and was instead supported only by the thin-walled structure, the linear and helical trajectories were no longer effective. This required a new approach, and by using a trochoidal trajectory combined with a multilayer-coated cutting tool, reliable results were achieved. Additional and more detailed conclusions drawn from this work include:

- Changes in feed per tooth (Fz) and axial depth of cut (Ap) significantly influence tool wear and temperature. While varying Fz and change in diameter of the cutting tool did not have a major impact, increasing Ap led to more noticeable tool degradation, especially in terms of surface roughness. In future works, optimizing these parameters can lead to more energy-efficient and environmentally friendly machining processes.
- Among the two roughness parameters, Ra (arithmetic average roughness) and RSM (root mean square roughness), the RSM parameter proved to be more sensitive and reliable in detecting surface finish differences. As a result, it is considered more dependable for comparison with infrared camera temperature readings, offering a precise method for monitoring sustainability-driven improvements in surface finish quality.
- For Setup 1, it was shown that the linear and helicoidal trajectories provided better cooling and reduced tool wear compared to trochoidal trajectories. Trochoidal paths resulted in higher temperatures and more severe tool damage, indicating that optimizing cutting trajectories can lead to more energy-efficient and sustainable production processes.
- In Setup 2, it was demonstrated that the method of fixing the component, whether with a vise or a 3D-printed thin-wall support, had an effect on tool performance. The 3D-printed support led to increased tool wear, potentially due to vibrations during the milling process. However, both methods performed better than Setup 1, emphasizing the role of material and system stability in optimizing energy use and process sustainability.
- As cutting tool wear/damage increased, infrared temperature readings showed a rise in temperature, which correlated with increased RSM roughness value of the Ti-6Al-4V surface parameters due to the increment of vibrations to worsened surface finishes. These findings underscore the importance of minimizing tool wear to enhance operational efficiency and reduce material waste.
- The use of infrared cameras has been confirmed as an effective tool for monitoring tool condition and surface quality, especially in dry machining conditions, where traditional lubrication is absent. This technique aids in detecting temperature increases that signal potential issues, such as tool wear or surface roughness degradation, while promoting more sustainable practices by eliminating the need for chemical lubricants.

The stability of the machining system is crucial in preventing tool damage and achieving optimal surface finishes. Any instability, such as vibrations, can lead to increased wear and poor component quality. Ensuring system stability not only improves tool life and part quality but also supports more sustainable, cost-effective manufacturing practices.

Future studies at pilot plants or industrial levels will be necessary to evaluate the economic feasibility of the dry milling process, considering the savings from not using lubricants, as well as the increased wear on the tool and the surface finish for thin-wall support 3D AM complex structures presented in this work.

Author Contributions: Conceptualization, G.R., M.F. and E.G.-L.; Methodology, G.R., E.V. and E.G.-L.; Validation, G.R. and E.G.-L.; Formal Analysis, G.R., E.V. and E.G.-L.; Investigation, G.R., M.F. and E.G.-L.; Resources, G.R. and M.F.; Data Curation, G.R., E.V. and E.G.-L.; Writing—Original Draft Preparation, E.G.-L.; Writing—Review & Editing, G.R., J.P. and E.G.-L.; Visualization, G.R., J.P., M.F., E.V. and E.G.-L.; Supervision, J.P. and G.R.; Project Administration, G.R.; Funding Acquisition, G.R. and J.P. All authors have read and agreed to the published version of the manuscript.

Funding: This research was funded by Ministry of Science and Innovation through the 2019 CIEN call as well as by the Catalan Government via the ACCIÓ-Eurecat TRAÇA-IMPULSENS funding grant.

Data Availability Statement: The original contributions presented in this study are included in the article. Further inquiries can be directed to the corresponding authors.

Acknowledgments: This work was performed inside the project LUBRINTEL, which received fundings from CDTI and supported by the Ministry of Science and Innovation through the 2019 CIEN call. Moreover, this work was financially supported by the Catalan Government through the funding grant ACCIÓ-Eurecat TRAÇA-IMPULSENS.

Conflicts of Interest: The authors declare no conflicts of interest.

References

1. Peters, M.; Hemptenmacher, J.; Kumpfert, J.; Leyens, C. *Structure and Properties of Titanium and Titanium Alloys*; Wiley-VCH Verlag GmbH & Co., KGaA: Weinheim, Germany, 2003; pp. 1–36. [CrossRef]
2. Srivastava, M.; Jayakumar, V.; Udayan, Y.; Sathishkumar, M.; Muthu, S.M.; Gautam, P.; Nag, A. Additive manufacturing of Titanium alloy for aerospace applications: Insights into the process, microstructure, and mechanical properties. *Appl. Mat. Today* **2024**, *41*, 102481. [CrossRef]
3. Nguyen, H.D.; Pramanik, A.; Basak, A.K.; Dong, Y.; Prakash, C.; Debnath, S.; Shankar, S.; Jawahir, I.S.; Dixit, S.; Buddhi, D. A critical review on additive manufacturing of Ti-6Al-4V alloy: Microstructure and mechanical properties. *J. Mater. Res. Technol.* **2022**, *18*, 4641–4661. [CrossRef]
4. Al-Rubaie, K.S.; Melotti, S.; Rabelo, A.; Paiva, J.M.; Elbestawi, M.A.; Veldhuis, S.C. Machinability of SLM-produced Ti6Al4V titanium alloy parts. *J. Manuf. Process.* **2020**, *57*, 768–786. [CrossRef]
5. Ni, C.; Zhu, L.; Zheng, Z.; Zhang, J.; Yang, Y.; Hong, R.; Bai, Y.; Lu, W.F.; Wang, H. Effects of machining surface and laser beam scanning strategy on machinability of selective laser melted Ti6Al4V alloy in milling. *Mater. Des.* **2020**, *194*, 108880. [CrossRef]
6. Boyer, R.R. An overview on the use of titanium in the aerospace industry. *Mater. Sci. Eng. A* **1996**, *213*, 103–114. [CrossRef]
7. Kishawy, H.A.; Hosseini, A. *Machining Difficult-to-Cut Materials Basic Principles and Challenges*; Springer: Oshawa, ON, Canada, 2019. [CrossRef]
8. Pimenov, D.Y.; Mia, M.; Gupta, M.K.; Machado, A.R.; Tomaz, I.V.; Sarikaya, M.; Wojciechowski, S.; Mikolajczyk, T.; Kapłonek, W. Improvement of machinability of Ti and its alloys using cooling-lubrication techniques: A review and future prospect. *J. Mater. Res. Technol.* **2021**, *11*, 719–753. [CrossRef]
9. Shokoohi, Y.; Khosrojerdi, E.; Rassolian Shiadhi, B.H. Machining and ecological effects of a new developed cutting fluid in combination with different cooling techniques on turning operation. *J. Clean. Prod.* **2015**, *94*, 330–339. [CrossRef]
10. Samukov, A.; Cherkasova, M.; Kuksov, M.; Dmitriev, S. Metal chips preparation for utilization using advanced reagents. In *IOP Conference Series: Earth and Environmental Science*; IOP Publishing: Bristol, UK, 2021; Volume 938, p. 012009. [CrossRef]
11. Lei, Y.; Zhang, Y.; Newkirk, J.W.; Liou, F.; Thomas, E.; Baker, A. Investigation of Machining Coolant Residue Cleaning Methods for Ti6Al4V Part Fabrication through Hybrid Manufacturing Process. *Manuf. Lett.* **2018**, *16*, 10–13. [CrossRef]
12. Isaev, A.V.; Grechishnikov, V.A.; Petr, P.; Mihail, K.; Yuriy, I.; Andrey, V. Machining of Thin-walled Parts Produced by Additive Manufacturing Technologies. *Procedia cIRP* **2016**, *41*, 1023–1026. [CrossRef]
13. Pegues, J.; Roach, M.; Williamson, R.S.; Shamsaei, N. Surface roughness effects on the fatigue strength of additively manufactured Ti-6Al-4V. *Int. J. Fatigue* **2019**, *116*, 543–552. [CrossRef]
14. Bin Anwar Fadzil, A.F.; Pramanik, A.; Basak, A.K.; Prakash, C.; Shankar, S. Role of surface quality on biocompatibility of implants—A review. *Ann. 3D Print. Med.* **2022**, *8*, 100082. [CrossRef]
15. Rahulan, N.; Sreekala Sharma, S.; Rakesh, N.; Sambhu, R. A short review on mechanical properties of SLM titanium alloys based on recent research works. *Mater. Today.* **2022**, *56*, A7–A12. [CrossRef]
16. Zgórniak, P.; Stachurski, W.; Ostrowski, D. Application of Thermographic Measurements for the Determination of the Impact of Selected Cutting Parameters on the Temperature in the Workpiece During Milling Process. *J. Mech. Eng.* **2016**, *62*, 657–664. [CrossRef]
17. Ramirez-Nunez, J.A.; Trejo-Hernandez, M.; Romero-Troncoso, R.J.; Herrera-Ruiz, G.; Osornio-Rios, R.A. Smart-sensor for tool-breakage detection in milling process under dry and wet conditions based on infrared thermography. *Int. J. Adv. Manuf. Technol.* **2018**, *97*, 1753–1765. [CrossRef]
18. Armendia, M.; Garay, A.; Villar, A.; Davies, M.A.; Arrazola, P.J. High bandwidth temperature measurement in interrupted cutting of difficult to machine materials. *CIRP Ann.* **2010**, *59*, 97–100. [CrossRef]
19. De Maddis, M.; Lunetto, V.; Razza, V.; Russo Spena, P. Infrared Thermography for Investigation of Surface Quality in Dry Finish Turning of Ti6Al4V. *Metals* **2022**, *12*, 154. [CrossRef]
20. Sales, W.F.; Diniz, A.E.; Machado, A.R. Application of cutting fluids in machining processes. *J. Barz. Soc. Mech. Sci.* **2001**, *23*, 224–240. [CrossRef]

21. Khanna, N.; Zadafiya, K.; Patel, T.; Kaynak, Y.; Rashid, R.A.R.; Vafadar, A. Review on machining of additively manufactured nickel and titanium alloys. *J. Mater. Res. Technol.* **2021**, *15*, 3192–3221. [CrossRef]
22. Campos, F.O.; Araujo, A.C.; Munhoz, A.L.J.; Kapoor, S.G. The influence of additive manufacturing on the micromilling machinability of Ti6Al4V: A comparison of SLM and commercial workpieces. *J. Manuf. Process.* **2020**, *60*, 299–307. [CrossRef]
23. Khaliq, W.; Zhang, C.; Jamil, M.; Khan, A.M. Tool wear, surface quality, and residual stresses analysis of micro-machined additive manufactured Ti–6Al–4V under dry and MQL conditions. *Tribol. Int.* **2020**, *151*, 106408. [CrossRef]
24. Zhang, H.; Dang, J.; Ming, W.; Xu, X.; Chen, M.; An, Q. Cutting responses of additive manufactured Ti6Al4V with solid ceramic tool under dry high-speed milling processes. *Ceram. Int.* **2020**, *46*, 14536–14547. [CrossRef]
25. *X1. UNE-EN ISO 4287*; Geometrical product specifications (GPS). Surface texture: Profile method. Terms, definitions and surface texture parameters (ISO 4287:1997 + Technical Corrigendum 1). International Organization for Standardization (ISO): Geneva, Switzerland, 1997.
26. *X2. UNE-EN ISO 25178-2*; Geometrical product specifications (GPS)—Surface texture: Areal—Part 2: Terms, definitions and surface texture parameters (ISO 25178-2:2021). International Organization for Standardization (ISO): Geneva, Switzerland, 2021.
27. González-Fernández, L.; Risueño, E.; Pérez-Sáez, R.B.; Tello, M.J. Infrared normal spectral emissivity of Ti–6Al–4V alloy in the 500–1150 K temperature range. *J. Alloys Compd.* **2012**, *541*, 144–149. [CrossRef]
28. Veiga, F.; Del Val, A.G.; Suárez, A.; Alonso, U. Analysis of the Machining Process of Titanium Ti6Al-4V Parts Manufactured by Wire Arc Additive Manufacturing (WAAM). *Materials* **2020**, *13*, 766. [CrossRef]
29. Jaffery, S.H.; Khan, M.; Nadeem A Sheikh, N.A.; Mativenga, P. Wear mechanism analysis in milling of Ti-6Al-4V alloy. *Proc. Inst. Mech. Eng. Part B J. Eng. Manuf.* **2013**, *227*, 1148–1156. [CrossRef]
30. Niaki, F.A.; Pleta, A.; Mears, L.; Potthoff, N.; Bergmann, J.A.; Wiederkehr, P. Trochoidal milling: Investigation of dynamic stability and time domain simulation in an alternative path planning strategy. *Int. J. Adv. Manuf. Technol.* **2019**, *102*, 1405–1419. [CrossRef]
31. Luo, M.; Hah, C.; Hafeez, H.M. Four-axis trochoidal toolpath planning for rough milling of aero-engine blisks. *Chin. J. Aeronaut.* **2019**, *32*, 2009–2016. [CrossRef]
32. Dong, Y.; Li, S.; Zhang, Q.; Li, P.; Jia, Z.; Li, Y. Modeling and Analysis of Micro Surface Topography from Ball-End Milling in a Trochoidal Milling Mode. *Micromachines* **2021**, *12*, 1203. [CrossRef]
33. García-Hernández, C.; Garde-Barace, J.-J.; Valdivia-Sánchez, J.-J.; Ubieto-Artur, P.; Bueno-Pérez, J.-A.; Cano-Álvarez, B.; Alcázar Sánchez, M.-Á.; Valdivia-Calvo, F.; Ponz-Cuenca, R.; Huertas-Talón, J.-L.; et al. Trochoidal Milling Path with Variable Feed. Application to the Machining of a Ti-6Al-4V Part. *Mathematics* **2021**, *9*, 2701. [CrossRef]
34. Zhou, X.; Zhou, J.; Qi, Q.; Zhang, C.; Zhang, D. Effects of Toolpath Parameters on Engagement Angle and Cutting Force in Ellipse-Based Trochoidal Milling of Titanium Alloy Ti-6Al-4V. *Appl. Sci.* **2023**, *13*, 6550. [CrossRef]
35. Sousa, V.F.C.; Silva, F.J.G. Recent Advances on Coated Milling Tool Technology—A Comprehensive Review. *Coatings* **2020**, *10*, 235. [CrossRef]
36. Zhao, J.; Liu, Z.; Wang, B.; Hu, J.; Wan, Y. Tool coating effects on cutting temperature during metal cutting processes: Comprehensive review and future research directions. *Mech. Syst. Signal Process.* **2021**, *150*, 107302. [CrossRef]
37. Harun, S.; Burhanuddin, Y.; Ibrahim, G.A. The Effect of Cutting Parameters on Surface Roughness and Morphology of Ti-6Al-4V ELI Titanium Alloy during Turning with Actively Driven Rotary Tools. *J. Manuf. Mater. Process.* **2022**, *6*, 105. [CrossRef]

processes

MDPI

Article

Biomimetic Approach for Enhanced Mechanical Properties and Stability of Self-Mineralized Calcium Phosphate Dibasic–Sodium Alginate–Gelatine Hydrogel as Bone Replacement and Structural Building Material

Alberto T. Estevez * and Yomna K. Abdallah *

IBAG, Institute for Biodigital Architecture and Genetics, School of Architecture, Universitat Internacional de Catalunya, 08017 Barcelona, Spain
* Correspondence: estevez@uic.es (A.T.E.); yomnaabdallah@uic.es (Y.K.A.)

Abstract: Mineralized materials are gaining increased interest recently in a number of fields, especially in bone tissue engineering as bone replacement materials as well as in the architecture-built environment as structural building materials. Until the moment, there has not been a unified sustainable approach that addresses this multi-scale application objective by developing a self-mineralized material with minimum consumption of materials and processes. Thus, in the current study, a hydrogel developed from sodium alginate, gelatine, and calcium phosphate dibasic (CPDB) was optimized in terms of rheological properties and mineralization capacity through the formation of hydroxyapatite crystals. The hydrogel composition process adopted a three-stage, thermally induced chemical cross-linking to achieve a stable and enhanced hydrogel. The 6% CPDB-modified SA–gelatine hydrogel achieved the best rheological properties in terms of elasticity and hardness. Different concentrations of epigallocatechin gallate were tested as well as a rheological enhancer to optimize the hydrogel and to boost its anti-microbial properties. However, the results from the addition of EPGCG were not considered significant; thus, the 6% CPDB-modified SA–gelatine hydrogel was further tested for mineralization by incubation in various media, without and with cells, for 7 and 14 days, respectively, using scanning electron microscopy. The results revealed significantly enhanced mineralization of the hydrogel by forming hydroxyapatite platelets of the air-incubated hydrogel (without cells) in non-sterile conditions, exhibiting antimicrobial properties as well. Similarly, the air-incubated bioink with osteosarcoma SaOs-2 cells exhibited dense mineralized topology with hydroxyapatite crystals in the form of faceted spheres. Finally, the FBS-incubated hydrogel and FBS-incubated bioink, incubated for 7 and 14 days, respectively, exhibited less densely mineralized topology and less distribution of the hydroxyapatite crystals. The degradation rate of the hydrogel and bioink incubated in FBS after 14 days was determined by the increase in dimensions of the 3D-printed samples, which was between 5 to 20%, with increase in the bioink samples dimensions in comparison to their dimensions post cross-linking. Meanwhile, after 14 days, the hydrogel and bioink samples incubated in air exhibited shrinkage: a 2% decrease in the dimensions of the 3D-printed samples in comparison to their dimensions post cross-linking. The results prove the capacity of the developed hydrogel in achieving mineralized material with anti-microbial properties and a slow-to-moderate degradation rate for application in bone tissue engineering as well as in the built environment as a structural material using a sustainable approach.

Keywords: bone replacement materials; mineralized materials; biomineralization; calcium phosphate reinforced hydrogels; multi-scale applications; multi-disciplinary; biomimetic structural materials; bone hierarchical structure motifs

1. Introduction

Biomineralized materials are gaining increased interest in various fields, extending from regenerative medicine applications in bone tissue engineering to optimized structures, with large industrial applications such as architectural building blocks (bricks) [1–4] thanks to their strength and enhanced mechanical properties. By definition, biomineralization is the process by which living organisms produce minerals to compose hardened mineralized tissues [5]. For example, calcium phosphates and carbonates are involved in forming structures of shells and the bone in vertebrates. These biominerals, namely phosphate and carbonate salts of calcium, which are usually crystalline, often exist in combination with organic polymers like collagen and chitin to perform as structural support for bones and shells [6]. These bio composite materials are organized in multi-scale hierarchical structures from the nanometer to the macroscopic level, providing intricate architectural aesthetics as well as multifunctional applications. Including materials and tissue engineering applications in regenerative medicine as well as structural efficiency is feasible thanks to their material–structure function and multi-scale complexity. This has triggered research interest in analyzing and mimicking the biological mechanisms of biomineralization [7,8].

Typically, a biomineralization process signifies mineralization induced under biological control of mineral morphology, growth, composition, and site as ruled by the cellular processes of a specific organism, such as the mineralization of collagen granting bone, cartilage, and teeth with the adequate compressive strength and mechanical performance [9]. However, the definition of biomineralization induction in bone replacement material engineering can extend to involve the biomimetic approach of synthesizing or inducing the mineralization reaction to form these biogenic minerals of calcium phosphates and carbonates mineralized crystals [10,11]. Thus, in the current study, the authors focused on developing a biometric approach of biomineralization induction of calcium-phosphate-based hydrogels without the use of cells vs. with the use of embedded cells.

The main marker of the successful biomineralization induction is the formation of mineralized hydroxyapatite crystals. Hydroxyapatite is the most common biogenic form of calcium phosphate, with the chemical formula ($Ca_{10}(PO_4)_6(OH)_2$), and it is a naturally occurring form of apatite and is also the primary constituent of bone: 65 to 70% of bone mass is HA [12]. The biomimetic mineralization induction reaction tries to mimic the natural mechanism of remineralization, where the mineral ions are introduced to reinstate the structure of the hydroxyapatite crystals [13].

To achieve this goal, the current study introduces an optimized and sustainable hydrogel composition based on calcium phosphate by proposing a three-stage, thermally catalyzed cross-linking of the developed 3D-printed hydrogel and detecting its mineralization by the formation of mineralized hydroxyapatite crystals across multiple time intervals and in different incubation conditions. Furthermore, the mechanical properties of the developed hydrogel were tested to draw a picture about its rigidity vs. elasticity. In this paper, these tests are described, along with the same hydrogel composition with embedded osteosarcoma cells, to compare the mineralization efficiency and duration between the passively mineralized hydrogel (without cells) and the actively mineralized bioink (with cells).

To understand the criteria of selection of the hydrogel constituents, the following introduction exhibits the most popular adopted hydrogels in bone tissue engineering, which are based on alginate–gelatine composition, and their compatibility with calcium phosphate for inducing mineralization.

Alginate–Gelatine–Calcium Phosphate Trio-Hydrogels for Bone Tissue Regeneration

Bone tissue engineering is a promising discipline to provide solutions for bone defects [14]. Bone tissue regeneration is controlled by intracellular functions as well as the biophysical traits of the extracellular matrix (ECM) and is mainly affected by these mechanical indications in the surrounding microenvironment. Natural bone regeneration signifies a stiffening physiological process combining viable cells, extracellular matrix, specific cytokines, and growth factors that formulate sufficient mechanical and vascular environments. This is usually tackled in bone tissue bioengineering research by employing bioinks encapsulating specific bioactive cells and molecules for application as grafts in the damaged area. During the bone regeneration, the ECM is dynamically remodeled with structural and compositional changes, resulting in dynamic stiffness alteration. This alteration has been proven to induce different biological behaviors of cells [15,16]. For example, it was reported that a matrix stiffness of 10–30 kPa induces MSCs differentiation into osteoblasts in two-dimensional (2D) microenvironment. Also, a matrix stiffness of 134 kPa endorses the excretion of alkaline phosphatase (ALP), osteopontin (OPN), and osteocalcin (OC) from osteoblasts when cells are cultured on 2D substrates [17]. Therefore, it is more relevant to design biomaterials that are dynamically changing in terms of stiffness to simulate the mechanical microenvironment in vivo while analyzing the effect of ECM dynamic hardening on bone regeneration to promote adopting such biomaterials of special biophysical properties in tissue regeneration in vivo [14]. This necessitates that bone substitute materials (BSM) possess an osteo-inductive 3D structure to encompass osteogenic cells and osteo-inductive factors, to offer sufficient mechanical properties, and to promote vascularization [18].

Hydrogels as high-water-content bio composite materials are vastly applied in tissue engineering scaffolds thanks to their efficient performance in diffusing nutrients and oxygen into their structure, simulating the biological tissues. This is enabled by their soft nature, elasticity, and permeable structure [18–20]. In addition to their viscoelastic traits and injectability, these hydrogels are compatible with additive manufacturing technologies (AMT) and 3D bioprinting. These hydrogels incorporating living cells form the construction blocks of three-dimensional scaffolds in bioprinting, which are called bioinks. Furthermore, hydrogels are ideal for developing scaffolds that mimic the ECM conditions thanks to the controllability of their mechanical properties by the dynamic tuning of their cross-linking degree [21]. It was reported that some cross-linking strategies dynamically changed hydrogels' stiffness under external stimuli. For example, temperature was applied to change hydrogel stiffness through controlling the viscoelasticity of hydrogels [14].

Commonly, interpenetrating network hydrogels are broadly employed to construct biomaterials with dynamic matrix stiffness for 2D and 3D cell culture, especially alginate-based hydrogels, which are the most tested in bone tissue engineering (BTE) applications and 3D bioprinting, thanks to their gelling property, low toxicity, high obtainability, and low cost. They provide an appropriate niche for cell embedding, which, in addition to their intrinsic ionic cross-linking, qualifies them as injectable hydrogels and makes them more favorable than other solid-scaffolds materials. Alginate polysaccharide is safe, according to the FDA (Food and Drug Administration) [22], for application in humans [23]. It is also compatible with 3D bioprinting techniques, which widens the opportunity to expand its applications in bone regeneration [18].

Alginate is chemically composed from a linear polysaccharide of homopolymer units of 1,4-linked (-D-mannuronic acid) (M) and (-L-guluronic acid) (G) [24]. M block segments possess a lined and flexible configuration, whereas the (1→4) linkages to guluronic acid introduce a steric disruption around the carboxyl groups. Therefore, the G block parts deliver folded and stiff structural configuration, giving the molecular chains their stiffness. High-M-content alginates promote the production of cytokine more than the high-G-content alginates [25]. Alginates originating from different sources possess different M and G contents and varied lengths of each block, affecting the final properties of the material [24]. Aqueous alginate solutions are non-Newtonian fluids [26] and become more viscose under lower pH values, obtaining maximum values at pH of 3.0–3.5. This response of alginates to different values of pH is attributed to the carboxyl groups on the alginate backbone. This pH-sensitive behavior is proven by the higher swelling ratios observed at higher pH values [18].

Alginate-based materials also enjoy the capacity of in situ gelation [27], solubility in water [28], cytocompatibility [29], mucoadhesive properties [30], release of active agents [31], and acting as a protective shield for cell and particle release systems [32]. However, their low degradation caused by the use of high-molecular-weight alginates to obtain mechanical properties similar to those of hard tissues remains a challenge for their use in vivo as bone replacement materials [33]. For example, higher-molecular-weight alginates produce stiffer gels, while medium- and low-molecular-weight alginates produce more degradable hydrogels with higher rates of cell proliferation. To moderate the molecular weights of alginate-based hydrogels, there are various methods, such as enzymatic preparation [34], ultrasonic irradiation [35], ultraviolet photolysis [36], oxidative-reductive depolymerization, and thermal degradation [18,37]. Through the past decade, multiple studies have attempted to achieve equilibrium between the mechanical traits and the degradation kinetics (and the resulting potential toxins from the degradation process) of alginate hydrogels by varying cross-linking methods, alginate molecular weight, chemical structure and processing technologies by tuning alginate hydrogels' chemical composition and structure to tailor specific properties for their applications in tissue engineering to improve degradability, mechanical properties, cell adhesion, and special features such as printability. Usually, this tuning occurs either through the carboxyl or the hydroxyl groups. For example, phosphorylated alginates were developed through the hydroxyl modification to induce hydroxyapatite nucleation for bone tissue regeneration [38].

Recently, oxidized alginates (OAs) are becoming widespread thanks to their highly reactive groups and faster degradation rates than the non-modified alginates. Alginate oxidation is usually considered a simple process with easy purification and non-toxic effects. The oxidation is conducted by reacting alginate with sodium periodate to produce two aldehyde groups in each oxidized unit by breaking the carbon–carbon bonds in the alcohol groups; consequently, a lower molecular weight and the oxidized uronate residues make OA susceptible to alkali-catalyzed elimination [39]. However, OA is less efficient in mechanical behavior than the unmodified alginate hydrogels. For example, Ref. [40] conducted the oxidation of sodium alginate at 4.9% to evaluate its degradability and in vivo reactions, where they reported the decrease in the molecular weight of the OA from 390 kDa to 255 kDa and a decrease in the compressive modulus from 150 $\pm$ 13 kPa to 754 $\pm$ 21 kPa in comparison to the non-oxidized gels. This alteration in the compressive strength during degradability implies an increase in the degradation rate, starting at the 6th day of incubation. However, the OA hydrogels maintained their gelling capacity in the presence of calcium ions via ionic bonding. Nevertheless, the ionic cross-linking of the reported OA hydrogels through the interaction between the polymer chains and the divalent ions decreased due to the reduced GG blocks in OA chains. Gomez, C.G. etc., 2007 [41] reported that alginate oxidation is selective towards the guluronate units, decreasing the ionically gelation at oxidation values higher than 10%. In addition, the oxidation degree is limited by the formation of hemiacetal groups and their interference in the reaction [42]. Due to the several aldehyde groups present in the backbone of OA hydrogels, they are highly active in covalent bonding, which alters the final polymer properties. The oxidation degree of covalently bonded hydrogels controls the network cross-linking mark, density, mechanical characteristics, degradation rate, and swelling behavior [39].

The ionic cross-linking of alginate with divalent cations is realized by the cooperative interaction with G monomers blocks to form ionic bridges, while M monomers blocks formulate weak junctions with divalent cations, unlike the tight junctions between G monomers blocks and divalent cations. The cross-linking mechanism occurs through the organization of divalent ions with four-carboxyl groups to form an egg-box like arrangement [43]. Thus, ionic gelation of alginate hydrogels has two options: internal or external gelation. Internal gelation develops in situ gelling hydrogels and is suitable for injectable alginate applications. In internal gelation, divalent cationic salts of low solubility at neutral pH are used, followed by a subsequent acidification of the medium to release the cations to further control the gelation kinetics and result in homogeneous gels. For example, Ref. [44] evaluated the composition of alginate microspheres with calcium chloride and zinc sulphate salts for drug release applications, where Zn cations interacted with alginate at different sites with respect to Ca cations and were less selective but allowed a higher degree of cross-linking. This proves that the higher specificity of calcium ions in alginate cross-linking produces a harder and more rigid polymeric network. On the other hand, external gelation ionic cross-linking is employed to synthesize materials for bone tissue engineering applications within a simple, low-cost process thanks to the fast interaction between divalent ions and alginate. However, it results in non-homogeneous hydrogels with low mechanical properties due to the non-cross-linked alginate zones within the cross-linked alginate layers. Currently, several researches are attempting to alleviate the disadvantages of this gelling mechanism. Figure 1 represents the chemical structures of alginate and the gelation of alginate in the presence of calcium ions as a cross-linking process, showing the famous egg-box gelation model.

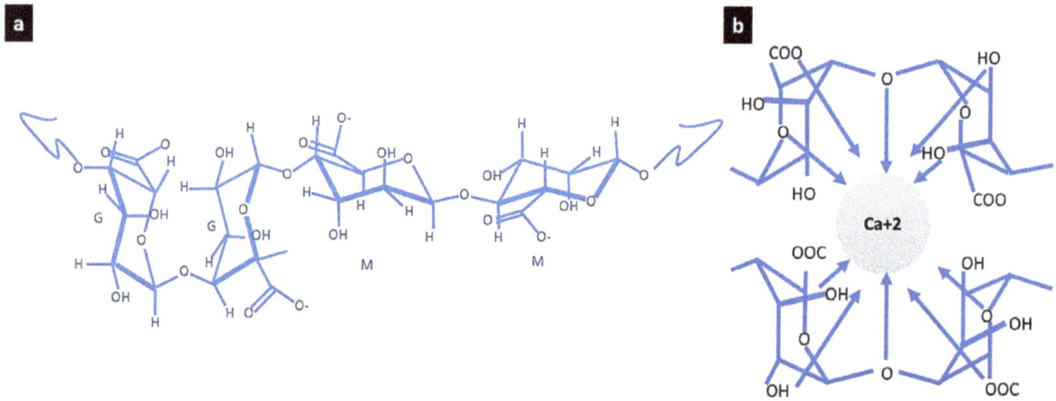

Figure 1. The chemical structure of alginate and its gelation by ionic cross-linking: (**a**) the chemical structure of alginate following [45] and (**b**) the egg-box model of the alginate ionic cross-linking by calcium ions following [18].

Sodium alginate (SA) is widely used to prepare biomaterials with dynamic mechanical properties. SA usually reacts differently to different concentrations of $CaCl_2$ or calcium ion chelating agents where the SA-based hydrogels are either stiffened or softened. Consequently, some studies recommended the continuous supplementation of calcium ions in SA-based hydrogels to sustain the cross-link degree, which could dynamically improve the stiffness of the obtained gel [14].

The covalent cross-linking of hydrogels results in strong and non-reversible chemical bonds that control the mechanical properties of alginate-based hydrogels [46]. On the other hand, these covalently bonded hydrogels usually suffer from lower injectability for in situ scaffolds, which hinders their application. Thus, another line of research to solve the in-situ gelation is by the generation of shear thinning viscoelastic hydrogels, including solutions of certain proteins, colloidal systems, peptides, and polymer mixtures with a self-assembly behavior that results in gelation of the material when it is not under shear stress [47]. Based on this property, the highly viscous polymer solution flows when a shear force is applied through injecting and forms a hydrogel when it stops. Alginates are an excellent base material for the generation of this type of materials due to their polyelectrolytic properties.

Alginates suffer from low cellular attachment due to their lack of any specific signal receptors for binding proteins, which makes them susceptible to unwanted interactions, immunological response, and proteolytic degradation [48]. To avoid this deficiency in alginates, researchers have combined oxidized alginates with proteins like collagen, and gelatine [49,50]. Collagen-derived gelatines enjoy unique gelling traits due to the physical cross-linking of the triple-helix structure of the native collagen. The gelatines retain informational signals, and they are highly degradable in vivo while being controllable in terms of tuning their physicochemical properties. These advantages added to their high cellular adhesion, pliability, availability, and low cost have promoted the use of gelatines for cross-linking with OA, exhibiting high cytocompatibility with a wide range of cells strains [51,52]. Still, the end characteristics of the oxidized alginate-based hydrogels depend largely on the oxidation degree of the alginate since it was reported that a greater oxidation degree induces faster gelling, higher cross-linking degree, and lower swelling [18].

However, raw alginate hydrogels are limitedly applied in tissue engineering due to their lack of specific signal receptors for binding proteins, their poor cellular adhesion, and the difficult control of their degradation kinetics. However, they can be tuned chemically by modifying the alginate hydrogel composition and/or the covalent bonding design. On the other hand, human bone-like biomimetic mechanical properties have not been reached with alginate-based composites, nor have they been sufficiently explored in the injectable hydrogels field. Thus, more research is required to overcome these disadvantages.

Most of the literature focuses on studying how the chemical composition and architecture influence cellular phenotypical expression, differentiation, proliferation, migration, and ECM composition during in vitro and in vivo assays [53]. However, the used cell line and its interaction with the material should also be considered. For example, Ref. [54] developed insitu gelling alginate hydrogels encapsulating MC3T3 cells under controlled gelation kinetics. In terms of materials science, a single material can hardly provide all of the required properties for biomimetic bone replacement materials. Thus, the designing of composite materials emerged to combine the merits of different types of materials [55,56]. One biomimetic approach is the combination of organic and inorganic components for manufacturing materials similar to bone tissue. Bio glasses (BG) and calcium phosphates are the most used inorganic compound for the reinforcement of hydrogels.

Bioactive glasses, which are silica-based materials, have a high surface reactivity, which enhances promoting the nucleation and subsequent formation of calcium phosphate crystals, especially bonelike apatite crystals, on their surface when incubated in simulated body fluids [57,58]. Calcium phosphate ceramics are also highly employed in bone tissue engineering thanks to their bio conductive nature and osteoconductive and biodegradation properties depending on their chemical and physical parameters. The bioactivity of calcium phosphates can be explained by their dissolution releasing calcium and phosphorus ions into the microenvironment of the cells, promoting the formation of new bone, which is also supported by the rough surface of the particles [18,59].

For example, Ref. [60] prepared composite spongy scaffold from alginate and hydroxyapatite through a phase separation method, significantly improving the mechanical strength (compressive and elastic modulus) and enhancing the osteosarcoma cell adhesion and proliferation on the scaffolds. These scaffolds had an average pore size of 150 μm and over 82% porosity. Another study by [61] fabricated oxidized alginate–gelatine–biphasic calcium phosphate hydrogel scaffolds loaded with different contents of spherical HAp particles, where the scaffolds without HAp had higher porosity and interconnected pore structure than materials loaded with HAp (pore size average ≈100–250 μm). The scaffolds with higher granules resulted in a higher compressive strength (dry/compressive modulus 2.45 ± 0.19 MPa; wet/compressive modulus 51 ± 0.39 MPa). HAp-enhanced hydrogels exhibited the high cell viability, cell attachment, proliferation, and differentiation behavior of rat bone marrow-derived stem cells (BMSC). Perez, R.A. etc., 2012 [62] developed a core shell design of fibrous scaffolds made of alginate with α-tricalcium phosphate for in situ cytochrome C protein loading and controlled delivery, which was produced by the injection of the alginate-α-tricalcium phosphate and posterior cross-linking with $CaCl_2$. The hydrogel exhibited an initial drug delivery burst at lower cross-linking time and $CaCl_2$ concentration, and the scaffolds showed more elastic than viscous behavior. Scaffolds with higher amounts of α-TCP yielded higher mechanical properties, with the storage modulus increasing from ~80 kPa without α-TCP to 800 kPa with 75 wt % α-TCP, signifying that the α-TCP phase promotes solidifying in the alginate matrix.

However, calcium phosphate bone cements still have some limitations that halt their clinical applications. For example, their mechanical strength is usually insufficient to provide adequate mechanical support to the defect site, having poor washout resistance and handling properties [63,64] reported that the addition of alginate increases the stiffening rate of calcium phosphate cements, resulting in quicker setting times. The compressive strength increased with incubation time, obtaining the highest value at 3 days and maintaining it up to 7 days of incubation. Particularly, compressive strength significantly increased from ~15 MPa to 60–70 MPa after 3–7 days of incubation at higher contents of alginate. Also, significant rat bone marrow-derived stromal cell adhesion, proliferation, and increased bone-associated gene markers such as collagen type-1, osteopontin, and bone sialoprotein were observed.

Thus, it can be concluded that developing sodium alginate hydrogels enhanced with gelatine and calcium phosphate can provide an integral solution balancing the required mechanical strength and cytocompatibility in bone tissue engineering, which is the main approach in the current study.

Furthermore, tissue engineering requires the development of scaffolds with a complex, customized geometry in combination with a precise control over the internal architecture and maintaining cytocompatibility. Bioprinting offers an integrated solution to the direct manufacturing of constructs with the possibility of incorporating cells and substrate materials including different kinds of additive manufacturing technologies (AMT) [65]. However, the type of AMT depends to a great extent on the processability of the bioink. Ribeiro, A.; etc., 2017 [66] defined printability as the possibility to extrude a hydrogel and dispense it in a pattern with a satisfactory degree of shape fidelity and also identify the accuracy of the printed structure in matching the original design. Printability is ruled by the rheological properties of materials and must be adjusted to the specific fabrication process to generate constructs with high shape fidelity [67]. Injectable alginate bioinks are one of the most used and successful materials for bioprinting due to their shear thinning character, rapid cross-linking ability, and feasibility of printing viable cells. They are particularly effective for bioprinting with nozzle systems due to the ability to protect the encapsulated cells through the process [27]. However, tuning alginate bioinks to meet specific requirements of printability, shape fidelity, and viability of encapsulated cells still requires further development to be used in the fabrication of effective 3D-printed constructs.

For instance, alginate bioinks must have sufficient viscoelasticity to achieve injectability during the printing process (rheological properties) and good shape fidelity to maintain the overall shape of the fabricated scaffold after printing [68]. Since the viscosity of alginate bioinks depend on the alginate concentration, the alginate molecular weight, and the cell density loading, printability can be promoted by controlling these parameters [27,69] Freeman and Kelly, 2017). A first approach is the control of alginate concentration. However, it can negatively influence the long-term biological performance at high concentrations. Park, J.; etc., 2018 [70] studied the influence of alginate concentration M/G 1.6 with high and low molecular weight (high 3.5×10^5 g/mol; low 1.43×10^5 g/mol) on printing fidelity and cell viability of a nozzle-based 3D printing system. Greater printing capacity, stability, and fidelity were obtained for hydrogels of high alginate of 3 wt% alginate and a 1:2 ratio (Low to High alginate). The metabolic activity and proliferation of the cells increased for lower molecular weights. Thus, high molecular weights produce dense networks and hinder mass transfer between the medium and the scaffolds, reducing the viability and proliferation of encapsulated cells.

Izadifar, M.; etc., 2017 [71] analyzed alginate bioinks' viscosity as function of temperature and polymer viscosity for low-concentrated alginate formulations. It was reported that bioinks prepared with 2% concentration of "medium-viscosity alginate" exhibited higher viscosity than formulations having a higher concentration of "low-viscosity alginate" in a range of temperature from 10 °C to 25 °C. Hybrid constructs reported cell viability >80%

for embryonic chick cartilage cells at 14 days. Hence, using alginate bioinks with high concentrations (i.e., 3.5–10 wt/v%) improves resolution printability and the material structural stability but limits cytocompatibility [70,71]

Another study by [72] developed alginate–gelatine blends by varying the individual constituent concentrations for study of printability, print accuracy, compressive behavior, and viability of encapsulated mesenchymal stem cells in bio-printed constructs. Higher concentrations of both alginate and gelatine resulted in printable bioinks with an optimal cross-linking time of 15 min in calcium chloride to improve stability per layer. In addition, the blends with 7% alginate–8% gelatine yielded high printability, mechanical strength, stiffness, and cell viability after printing. However, the compressive behavior of the hydrogel decreased rapidly over time and especially at 37 °C.

Added to the challenge of tuning rheological and mechanical properties of the hydrogel in balance with injectability, cell viability, and degradation, mineralization induction remains the biggest challenge in bone tissue engineering when developing bone replacement materials. Organic matrices offer this possibility based on tunable organic–inorganic interactions [73,74]. Therefore, further studies are required to optimize the osteo-inductive behavior of BRM. The conjunction of growth factors, bone cells, or bioactive materials where hydroxyapatite is used to promote the attachment of cultured osteoblasts in multiple polymers and improve their mechanical properties is a promising approach [75]. However, the alginate–hydroxyapatite system suffers from some limitations, including the difficulties of hydroxyapatite dispersion, the viscosity of the alginate, and the negative charge density of both that partially prevent their interaction. This hinders their processing together, especially for injectable materials. Alternatively, studies have reported the use of surface functionalization of hydroxyapatite with alginate or oxidized alginate since hydroxyapatite can be used as an effective sorbent and carrier for polymers thanks to its ion exchange capacity, adsorption capacity, and acidic properties [76]. These properties together with the alginate's ability to encapsulate cells and growth factors could encourage osteo-induction from simultaneous fronts. In addition, this overcomes the alginate hydrogels' limitation in reaching the structural properties required to function as mechanical support in bone tissue engineering, where the reported alginate–cement compositions achieved higher mechanical properties than alginate hydrogels (compressive strength 25–70 MPa; Young's modulus 90–130 MPa), although still far from the average mechanical properties of human bone (cortical bone: compression strength 200 ± 36 MPa; elastic modulus on compression 23 ± 4.8 GPa; cancellous bone-compression strength 1.5–38 MPa; elastic modulus on compression 10–1570 MPa) [77].

Thus, this alginate–gelatine cement still requires further experimentation to optimize the rheological and mechanical properties of the hydrogel, especially in the current study, where the indicated application is not only directed towards developing bone replacement materials but also developing self-mineralized material with a slow and low degradation rate that can be as well employed as a structural material in architectural and design applications. Since bioprinting can facilitate the tailoring of the properties of oxidized alginate-based hydrogels to the specific rheological properties of bone regeneration thanks to the ability to designing structures layer by layer and the ability to vary their processing conditions with a finer control over the principal alginate manufacturing conditions (addition of components, cross-linking agent concentration, and temperature), it was used in the current study to test and 3D print the developed hydrogels.

2. Materials and Methods

2.1. *Rheological Properties of Alginate–Gelatine Hydrogel Enhanced with Different Concentrations of Calcium Phosphate*

A hydrogel composed from 8% gelatine from bovine skin, type B in powder, and of pH 7 at 100 g/L at 60 °C was purchased from (G9391-Merck Life Science S.L. St. Louis, MO, United States), and + 4% sodium alginate E401 $(C_6 H_7 NaO_6)_n$ with molecular weight of 10,000–600,000 Daltons, purchased from (Quimics Dalmau SL, Barcelona, Spain), was prepared in 60 °C distilled water stirred for 20 min before adding different concentrations of 4, 4.5, 5, 5.5, 6, 6.5, 7, and 7.5 g of calcium phosphate dibasic $CaHPO_4$ (CPDB) (C7263-Sigma Aldrich. St. Louis, MO, United States) and stirring the mixture for 4 h at 60 °C until reaching a homogenous gel, respectively. Each gel was tested by the syringe test before testing it for 3D printing using a Felix Bio dual-head extruder bio printer with a 5 mL syringe with a 0.5 mm nozzle. The objective was to print a $3 \times 3 \times 0.5$ cm^3 cuboid form with a square grid infill pattern with cell size of 0.5×0.5 mm^2. The 3D-printing process settings were 60 °C printing head temperature, 4 °C bed temperature, flow 105%, speed 100%, layer height 0.5 mm, and printing pressure 12 PSI. The 3D printing was conducted in sterile conditions. Post printing, each print was cross-linked with a 5 mL prepared solution of 5 mg calcium chloride $CaCl_2$ (449709-Sigma Aldrich. St. Louis, MO, United States) dissolved in 5 mL sodium carbonate concentrate Na_2CO_3 (72784-Sigma Aldrich. St. Louis, MO, United States). Three different hydrogels of the three different concentrations of 5.5, 6, and 6.5 g/100 mL calcium phosphate were selected to undergo the rheological tests and chosen based on comparison of their printability and post-printing shape retention and fidelity with the other gels containing the different concentrations of calcium phosphate. Three different thicknesses of 1, 3, and 6 mm of the 3D-printed geometry per each concentration were tested. The printed samples were left to dry in ambient temperature for a week before conducting the rheological properties tests. Each sample contained 10 specimens tested under the DISCOVERY HR-2 Rheometer manufactured by TA Instruments at the Biomaterials, Biomechanics, and Tissue Engineering division (BBT), Dept. of Material Science and Engineering (CEM), Universitat Polytechnic de Catalunya. The parameters G′ and G″ were estimated per each sample.

The geometry employed in the analysis consisted of a rugose plate with a diameter of 20 mm. The geometry was adjusted to specific gap heights for each sample height, specifically 3000 μm for samples with a height of 6 mm, 500 μm for samples with a height of 1 mm, and 700 μm for samples with a height of 3 mm. The testing sequence was then applied to each sample, employing a frequency of 1 Hz and a displacement ranging from 1×10^{-5} rad to 10 rad at room temperature 25 °C. The values of G′ and G″ were recorded for each individual sample.

After that, the sample height of 1 mm was standardized for testing the rheological properties of the same three selected concentrations of 5.5, 6, and 6.5 g calcium phosphate. Each sample contained 10 specimens, and the test geometry employed in the analysis consisted of a rugose plate with a diameter of 20 mm. The test geometry was adjusted to 500 μm for all the samples, which had a height of 1 mm. The testing sequence was then applied to each sample, employing a frequency of 1 Hz and a displacement ranging from 10^{-5} rad to 10 rad, at room temperature 25 °C. The values of G′ and G″ were recorded for each individual sample.

2.2. Rheological Properties of Alginate–Gelatine–Calcium Phosphate Hydrogel Enhanced with Different Concentrations of Epigallocatechin Gallate

Three different concentrations of epigallocatechin gallate $C_{22}H_{18}O_{11}$ (E4143-Sigma Aldrich. St. Louis, MO, United States) of 4, 6, and 8 μm were added to the 6 g calcium phosphate-modified SA–gelatine hydrogel, respectively, to test the effect of EGCG on the printability, shape retention, and rheological properties of the developed hydrogel. The 3D-printing process employed identical settings as the previous test, and the tested geometry and the incubation conditions post printing were the same. Each sample corresponding to each concentration contained 10 specimens. The specimens were analyzed using the DISCOVERY HR-2 rheometer manufactured by TA Instruments. The parameters G′ and G″ of the three materials were examined. The test geometry employed in the analysis consisted of a rugose plate with a diameter of 20 mm and 1 mm height. The test geometry was adjusted to a force of 0.2 N for all the samples to determine the gap. The testing sequence was then applied to each sample, employing a frequency of 1 Hz and a displacement ranging from 10^{-5} rad to 10 rad. The values of G′ and G″ were recorded for each individual sample.

2.3. Mineralization Test of the CPDB-Modified SA–Gelatine without Cells, Incubated in Fetal Bovine Serum and Air, Respectively, for 7 and 14 Days, Respectively

The developed hydrogel with optimum concentrations of rheological enhancers was then tested for mineralization without the embedding of living cells (osteosarcoma SaOs-2 cells) to examine the hydrogel capacity of mineralization, being incubated in 10 mL of fetal bovine serum FBS each 3D bio printed sample following the same dimensions as described above and the same 3D printing settings (FBS was kindly provided from the GRC—Research Group in Regulation of Lipid Metabolism in Obesity and Diabetes at the International University of Catalunya UIC) or in the open air in non-sterile conditions. The hydrogel samples incubated in air included three specimens, which were incubated in the open air in ambient room temperature 25 °C for 7 and 14 days, respectively, while the incubated FBS samples included three specimens incubated in 25 °C for 7 and 14 days, respectively, with the FBS solution changed every 48 h. The samples and corresponding specimens were 3D printed in duplicates per each incubation condition (FBS and Air). The samples were tested using scanning electron microscopy (SEM) to detect the formation of hydroxyapatite crystals across these different time intervals. The sample preparation procedures for the SEM included preservation in glutaraldehyde and storing at 4 °C. Later, they were mounted on steel stubs with a bio adhesive layer and a glass cover and scanned under low vacuum for the SEM scanning.

2.4. Mineralization Test of CPDB-Modified SA–Gelatine with Osteosarcoma Cells Incubated in Fatal Bovine Serum and Air for 7 and 14 Days, Respectively

For the mineralization test, 8×10^6 Saos-2 osteosarcoma cells were kindly provided by the Bioengineering Institute of Technology at the International University of Catalunya (UIC). The cell culture was prepared following the reference of the supplier ATCC (USA) and the method of [78] in DMEM containing McCoy's 5 medium supplement with 15% FBS and 1% penicillin/streptomycin (Sigma-Aldrich. St. Louis, MO, United States). Cells were cultivated in T25 flasks (Nunc, USA) in a humidified incubator at 37 °C, using a standard mixture of 95% air and 5% CO_2. After trypsinization, the cells in suspension were morphometrically analyzed. A phase-contrast microscope was used to measure the area and diameter of attached cells (n = 100) by using an image analysis system Eclipse TS2 (Nikon, NY 11747-3064, U.S.A). Detached cells were analyzed by staining 10 µL of the cell suspension with 10 µL Trepan Blue solution. The cells were then counted with a Neuberger chamber. For cell proliferation analysis, Saos-2 cells were seeded at 2.5×10^5 cells/T25 flask. On days two, four, eight, and fourteen, cells were detached, and their number was determined following the previous method. All assays were performed in triplicate, using the early log-phase to calculate the population replication time. The following procedures were followed first to detach the SaOs-2 cells to prepare them for encapsulation; old cell culture media were removed, and the cells were washed twice with 5–10 mL PBS to remove any residues from the old media; 5–10 mL Trypsin was added to the cells before incubation for 5 min at 37 °C to detach the cells. After incubation, the cells in Trypsin were centrifuged for 5 min at 300× g, and then, 500 µL of McCoy medium was added to the three Falcon flasks to suspend the cells. Next, 20 µL of cell suspension was added to 80 µL of Trepan Blue to be counted by the Neuberger chamber. After counting the cells in each Falcon flask, the cells were placed again in three T25 flasks, and 8.1 mL of McCoy medium was added to suspend the cells at 37 °C in each flask, respectively. The mixture was incubated for five minutes at 37 °C. All procedures were conducted in a UV sterile hood. After incubation, 5 mL of the cells with media mixture was added to 10 mL of the optimized hydrogel, and 5 mL of the bioink was loaded to a 5 mL sterile syringe and fixed into the printing head of the Felix Bio bio printer. The printing temperature was adjusted to 4 °C at the printing bed and 37 °C at the printing head (nozzle), while the printing speed was kept at 100% and the flowrate at 100% as well. The 3D digital design file was exported as STL file from Rhinoceros 3D and adjusted to printer settings using Simplify 3D software (https://www.simplify3d.com/) and then transferred to the printer's built-in software.

The bio-printed samples were cross-linked with the cross-linking solution of reacting calcium chloride and sodium carbonate. Then, the cross-linked 3D-printed bioink patches were incubated in FBS (fetal bovine serum) and ambient air for 7 and 14 days, respectively. Each sample contained three specimens that were examined by the SEM to detect the formation of hydroxyapatite and biomineralization. Similarly to the hydrogel samples, the bioink samples (Saos-2 embedded in the hydrogel) were prepared by fixing in glutaraldehyde and kept at 4 °C. Later, the samples were mounted on steel stubs with a bio-adhesive layer and a glass cover and scanned under low vacuum.

3. Results and Discussion

The current study aimed to attain the sustainable synthesis of a self-biomineralized material with possible applications in tissue engineering as a bone replacing material as well as in architecture as a mineralized material that hardens by time. These characteristics are ruled by these main necessities: (1) the minimized and sustainable composition of the material by limiting the components of the material while selecting the most efficient components to achieve mineralization; (2) competent mechanical properties of the developed material balancing elasticity with plasticity to offer a rigid yet elastic material that is adequate for both applications as BRM and a structural material in the built environment; (3) autonomous mineralization without cells in non-sterile conditions, which indicates the antimicrobial trait of the developed material since the developed material is designed to be used in the built environment and to be mineralized by the ambient air, which opens wide potentials for the large-scale 3D printing of sustainable architectural applications with minimum material consumption while contributing to mitigating CO_2 by integrating it in this mineralization reaction; (4) slow-to-moderate degradation by achieving equilibrium between the material resistance to atmospheric conditions (resistance to water, air, and microbial contamination) while being degradable by using available buffers such as, in the case of this study, phosphate-buffered saline. This would enable the customized use of the developed material for the proposed two applications as a BRM or as a building material in the built environment.

However, the topic of optimized hydrogels for bone tissue engineering employing sustainable and green processes have been proposed in a limited number of recent studies in the bone tissue engineering field. For example, Ref. [79] offered a review on cheap and sustainable polymers that are used in bone tissue engineering, including alginate. Similarly, one study [80] proposed the use of cheap and sustainable polymers such as chitin, cellulous, and starch combining with hydroxyapatite for application in bone tissue engineering materials. Lacroix, J.; etc., 2013 [81] also proposed the use of bioactive glasses as green and safe in situ scaffolds for bone tissue engineering. One study [82] proposed self-healing sustainable hydrogels for guided bone regeneration (GBR), and a few other pieces of literature proposed the use of self-mineralized materials as building material in architecture applications either from existing cementous mineralized materials based on the CO_2 to calcium carbonate reactions [83,84] or by inducing mineralization through a bioactive agent as bacteria or fungi [4,85]. However, the current study goes beyond the literature by not only proposing a sustainable composed hydrogel for self-mineralized material but also by providing sustainable minimized materials and processes in preparing this self-mineralized material. Moreover, the current study novelty in the proposed hydrogel/material is that it is antimicrobial and mineralizes in ambient, open air in non-sterile conditions. Furthermore, the minimized materials used in the hydrogel composition and the cheap and relatively easy processes promote its scaling to multi-scale applications ranging from bone replacement material to construction and building material, which offers for the first time this dual use of the developed material in the field of bone tissue engineering and architecture construction, which has not been reported in the literature until now, to our knowledge.

Thus, in this section, the results and discussion show the rheological properties for the developed and optimized hydrogel in two phases, including hydrogels enhanced with different concentrations of calcium phosphate dibasic, followed by the CPDB-modified SA–gelatine hydrogels enhanced with different concentrations of EGCG, revealing the best reached rheological properties corresponding to the best hydrogel composition and concentrations. This is followed by discussion of the mineralization detection results of the optimized hydrogel incubated in different conditions without embedding cells compared to the biomineralization results in the same varied conditions when embedding osteosarcoma cells in the bioink. Finally, the results of degradation duration and conditions are also detailed.

3.1. Rheological Properties of SA–Gelatine Hydrogel Enhanced with Different Concentrations of Calcium Phosphate Dibasic

The current study aimed to develop a sustainable, optimized, self-mineralized hydrogel from available and efficient components to synthesize a bone replacement material. Three-dimensional polymeric hydrogels have been used widely for bone tissue regeneration for their excellent biocompatibility through high water uptake, space for drug or nutrient storage, ability for cell immobilization, and creation of a microenvironment for cell culturing, allowing nutrients and metabolites to diffuse to and from the cells, as well as biodegradability, attachment, proliferation, and differentiation due to their innate structural and compositional similarities to the extracellular matrix. Hydrogel scaffolds have proven suitable mechanical properties for promoting cell attachment and tissue formation [86]. In addition, their degradation products are not toxic in vivo. [87]. However, physically cross-linked hydrogels suffer from low mechanical strength and minute changes in the external environment (pH, temperature, and ionic strength), which affects the stability of the hydrogel system. Thus, in the current study, the author designed a three-stage chemical cross-linking reaction to cross-link the used sodium alginate by gelatine, calcium phosphate, and by the cross-linking solution composed from calcium chloride and sodium carbonate since it was reported that chemically cross-linked hydrogels are mechanically stronger and more stable because the interactions between the chemical linkages are stronger than physical linkages [88].

Tuning the mechanical properties of a hydrogel plays an important role in tissue regeneration because they create and maintain space for cell proliferation [86,89]. Generally, hydrogels suffer from insufficient mechanical strength for load bearing in hard tissue implant applications, especially for skeletally mineralized elements of the body, which indicate that the implanted site has to bear the whole amount of the applied load to maintain the mechanical stability of the implant site, which is not present in most of the common hydrogel systems. Thus, this study adopted a composite hydrogel approach that incorporates a mineral-reinforcing phase within the hydrogel to enhance its mechanical properties as a bone replacement material and as a structural load-bearing material as well. In the current study, we focused on enhancing the mechanical strength and stability of one of the most popular bio composite hydrogels, which is alginate–gelatine enhanced with different concentrations of calcium phosphate dibasic.

This ideal mechanical strength is balanced with a low-to-moderate degradation rate since the degradation of hydrogel scaffolds should be adequate for enabling new tissue formation and eventual bio-integration, and this should be neither too high, which could endanger the physical structure, nor too low as to hinder tissue regeneration and integration. The degradation rate of hydrogels is controlled by tuning the type of backbone polymer and cross-linking characteristics such as the amount of cross-linker, physical or chemical type, and size of cross-linking molecules [86,89]. The current study employed a three-phase, thermally induced chemical cross-linking to ensure the stability of the hydrogel and the low degradation rate even in aqueous environment in the absence of chemical buffers to promote its application as a self-sustaining and self-mineralizing material that can be used as a building material in the built environment and as a bone replacement material in bone tissue regeneration applications where the in vivo conditions, including body fluids, will act as the necessary buffers for its customized degradation.

The current study adopted sodium alginate, gelatine, and calcium phosphate as the main three components of the hydrogel. These materials can mimic the properties of bone. As gelatine is derived from collagen, it exhibits biocompatibility in tissue engineering application, including biodegradability, plasticity, and adhesiveness. Thanks to its excellent solubility in the aqueous medium, gelatine can be modified without any changes in the physiochemical properties, promoting the use of gelatine widely in tissue engineering scaffolds as 3D foams or hydrogels. Furthermore, gelatine contains RGD peptide sequences, which induce excellent cell proliferation and differentiation properties [90–92]. However, gelatine cannot sustain a mechanical load, which necessitates the chemical cross-linking to overcome this deficiency in its mechanical stability [93]. On the other hand, alginate can mimic the extracellular matrix, as it is a biodegradable polysaccharide, is biocompatible, and is a less toxic hydrophilic polymer that is highly suitable for drug delivery applications [94–96]. Sodium periodate is usually used to chemically modify the polymeric chain of native alginate through oxidation reactions on the −OH groups at the C-2 and C-3 positions of the uronic units of sodium alginate [97] to improve its reactive properties since oxidized alginates possess many reactive groups and a faster degradation profile than alginate alone [98,99]. Thus, the integration of sodium alginate with gelatine solves the deficiencies in both of them while augmenting their advantages [100,101].

Calcium phosphates such as hydroxyapatite and calcium phosphate have osteogenic properties and can stimulate natural bone regeneration. For example, tricalcium phosphate β-TCP is used widely for hard tissue repair applications due to its osteo-inductivity; it also has a faster dissolution rate. Some studies have reported the combination of HAp, α-TCP, and β-TCP to obtain improved mechanical properties, controlled biodegradation behavior, better osteointegration response, and increased bioactivity in the hydrogel material system. However, this indicates the use of more components under complicated processes such as a dissolution–precipitation process under physiological conditions, which does not align with the sustainable and minimized processes and material consumption approach of the current study. On the other hand, calcium phosphate dibasic is a biocompatible, bioactive, and osteoconductive reagent with a sustainable release of calcium and phosphate ions by its gradual dissolution, inducing bone formation and mineralization [102]. For example, porous calcium phosphate microspheres were used as a promising bone defect filler [103,104]. Recently, the composites of biopolymers and bio-ceramics are becoming especially popular because of their role in increasing the mechanical stability of bone substitutes and facilitating the interaction between host tissue and scaffold [105]. For instance, spherical hydroxyapatite granules encapsulated into oxidized alginate–gelatine–biphasic calcium phosphate hydrogel with different concentrations up to 35 wt% granules-loaded hydrogel proved their capacity to contribute to cell migration of rat bone marrow-derived stem cells and extracellular matrix growth [103]. Moreover, their porous bio-ceramic structure permits the infiltration and enhancement of implant tissue attachment and resorbability, resulting in faster bone ingrowth due to larger surface area [106,107]. This is due to the self-cross-linking properties and good biocompatibility of the oxidized alginate–gelatine as the hydrogel matrix for the encapsulation of granules [61,108]. Another study reported the use of alginate–hydroxyapatite composites for a drug delivery system for enzymes, antibiotics, and antifungal drugs [15]. These studies proved that the presence of hydroxyapatite along with alginate increased the cell attachment in the inner parts and offered a suitable choice of scaffold for bone tissue engineering [109]. Similarly, some literature reported that mixing gelatine with hydroxyapatite is beneficial for orthopedic applications by combining the osteo-conductivity and bioactivity of hydroxyapatite and the morphological features of gelatine [110]. Hence, incorporating gelatine into alginate and hydroxyapatite, inducing a composite, enhances the crystallinity of the composite.

In the current study, hydrogel composites consisting of sodium alginate–gelatine and enhanced by different concentrations of calcium phosphate dibasic CPDB were prepared by cross-linking the gelatine and the sodium alginate as a first-stage cross-linking, followed by cross-linking by the addition of calcium phosphate dibasic, and finally, a cross-linking solution of calcium chloride and sodium carbonate. The increased interactions between the functional groups had several effects on the materials properties, physical behaviors, and formation of hydroxyapatite crystals upon mineralization. The various concentrations of calcium phosphate affected the physical, mechanical, and morphological properties and characteristics of this hydrogel. Thus, the current study employed detailed morphological and material characterizations of the developed hydrogel for printability and its rheological properties, mineralization through formation of hydroxyapatite crystals, and degradability rate.

The sol-to-gel phase transition was completed within 4 h of mixing, as confirmed by the mixture turning to white color and by applying the syringe test, which exhibited the enhanced coherency and shape retention of the viscous gel thanks to the two-step cross-linking as well as the high molecular weight of the sodium alginate, as supported by [111]. This justifies the printing temperature at 60 °C at the printing end to facilitate the flow of the sol–gel through the printing nozzle, to avoiding clogging, and to promote layer-to-layer adhesion, which is supported by [112], who adopted around a 55 °C printing temperature to achieve 3D printing of tough hydrogels with enhanced mechanical properties by employing a two-step method to prepare the hydrogel as well by mixing at a high temperature of 95 °C, then cooling to 65 °C before printing. This is also congruent with [113], who used an up to 80 °C printing temperature to print a thermally modified alginate–di-aldehyde–gelatine hydrogel with gelatine modified at 80 °C for three hours and printed in the hierarchical, complex structures of cartilage and confirmed via μCT analysis. This is contrary to older methods reported in previous literature that usually employed from 37° to 40 °C as the printing temperature of the alginate–gelatine-based hydrogels, achieving lower mechanical properties and shape retention [114].

In the current study, a thermal shock to the printed hydrogel was provided by keeping the printing bed temperature at 4 °C to regain the gelation capacity of the viscous gel. The post-printing, third cross-linking phase of the 3D-printed CPDB-modified SA–gelatine hydrogel by the prepared solution of $CaCL_2$ + Na_2CO_3 played a crucial role in enhancing the hydrogel stability and antimicrobial property since this material is developed for non-sterile condition applications as building material in the built environment. This is congruent with [115,116], who used a 30 mg/mL $CaCl_2$ solution to cross-link SA to form strong bonding. This also helped in retaining its stability for low degradability in the absence of buffer solutions to enable its dual use as a bone replacement material in tissue engineering with customized degradation rate in vivo and as a resistant building material in the built environment in the current study.

The results of the printability test of the different tested concentrations of calcium phosphate dibasic-enhanced SA–gelatine hydrogels revealed that concentrations of 5.5, 6, and 6.5% achieved the best printability of the gel in terms of viscosity, layer coherence, structural stability, and contact angle of 35° ± 3, which is ideal for a hydrogel for moderate wettability property, as reported by [117]. Moreover, the tested different hydrogels with the three varied CPDB concentrations mentioned above exhibited minor swelling post printing and pre-cross-linking with the Na_2CO_3 + $CaCl_2$ solution, especially in lower CPDB concentrations. This was due to the temperature-catalyzed cross-linking reaction between gelatine and alginate [118], followed by the second-phase cross-linking by the addition of calcium phosphate dibasic, leading to the in situ formation of the coherent hydrogel, congruent with [119,120], and then finalized by the third cross-linking with the calcium carbonate solution. However, all the developed hydrogels suffered from shrinkage post cross-linking from 1.5 to 3 mm in all the three directions of the prints (X, Y, and Z), which is almost 0.05% to 0.1% shrinkage from the original size of the full prints. However, the shrinkage effect was more pronounced in the 5.5% CPDB-enhanced

SA–gelatine hydrogel with 3 mm shrinkage from all directions, while the least recorded shrinkage was in the 6.5% followed by the 6% CPDB hydrogel with a 1.5 mm shrinkage from all directions post cross-linking. The shrinkage rate achieved in the current study beats the 3–5% shrinkage of phosphate-enhanced alginate–gelatine hydrogels post cross-linking as reported in [121]. This shrinkage achieved in the current study in the cross-linked CPDB-modified SA–gelatine hydrogel is attributed to the hydrogen bonding in the third-phase cross-linking with the calcium carbonate solution, as supported by [122]. However, the higher concentrations of CPDB in the 6% and the 6.5% limited the effect of the shrinkage post cross-linking, which was caused by the calcium carbonate in the cross-linking solution (calcium chloride + sodium carbonate) and by the cross-linking realized by the increased calcium ions in the CPDB corresponding to its increased concentration in the hydrogel composition in the second-stage cross-linking (Figure 1). This proves that in the current study, the addition of the calcium phosphate dibasic acted as an active cross-linking agent to the sodium alginate–gelatine mixture, enhancing the hydrogel's stability and coherence.

The obtained gel exhibited dense and coherent texture with low-to-moderate irregular and interconnected porous microstructure structure with $\leq$20% porosity and with pore size between 20 μm to 60 μm. This is due to the thermally induced, three-stage cross-linking of the developed hydrogel adopted in the current study (Figures 2 and 4). This is within the pore size rate of 10–500 μm reported in [123] of gelatine enhanced with different concentrations of cements. This is nearly 40% less than the reported pore size in a similar calcium phosphate-enhanced alginate–gelatine hydrogel in [124], where the pores were interconnected, and the porosity ranged from 44.45 to 67.89%, with pore size ranging from 100 to 300 μm. This can be attributed to the different preparation methods as well as the higher molecular weight of the used alginate and the three-stage cross-linking process adopted in the current study. Although even higher porosity is favorable for cytocompatibility since the pores enable the circulation of air and nutrients, the pore size is relevant to the type, scale, and morphology of the embedded cells in the bioink, with moderate pore size being more cyto-compatible [125]. However, in the current study, the material coherence balance with porosity distribution and pore size is essential to deliver both designated functions: as a hard tissue bone replacement material that is load-bearing, which indicates more coherency of the gel to enable higher mineralization density in the material, and as a stand-alone self-mineralized material for structural applications in the built environment. In the current study, the morphological and rheological characterization was conducted after drying the hydrogel in the open air for 7 days to test the actual morphology of the gel planned for use in non-sterile conditions in the built environment in addition to its possible application as bone replacement material due to its anti-microbial traits that are crucial for in vivo post-operative infection control in bone replacement grafts.

The rheological characterization of the developed gel aimed to identify the hydrogel properties dependent on the different concentrations of CPDB in balancing between elasticity and plasticity and exposing and concluding the intrinsic behavior of the material itself regardless of the thickness. Therefore, three different thicknesses from the three different concentrations of 5.5, 6, and 6.5% CPDB-modified hydrogels were examined for the rheological properties for promoting the sustainability purposes of using the least amount of materials to deliver the same mechanical qualities and also proving that the rheological properties tested translate the material properties better than the geometry characteristics. This rheological identification was decided based on the G-Values: G' (storage" or "elastic" modulus), G'' ("loss" or "plastic" modulus), and $\tan\delta = G''/G'$, which measures the elastic ($\tan\delta < 1$) or plastic ($\tan\delta > 1$) characteristics of the developed hydrogel.

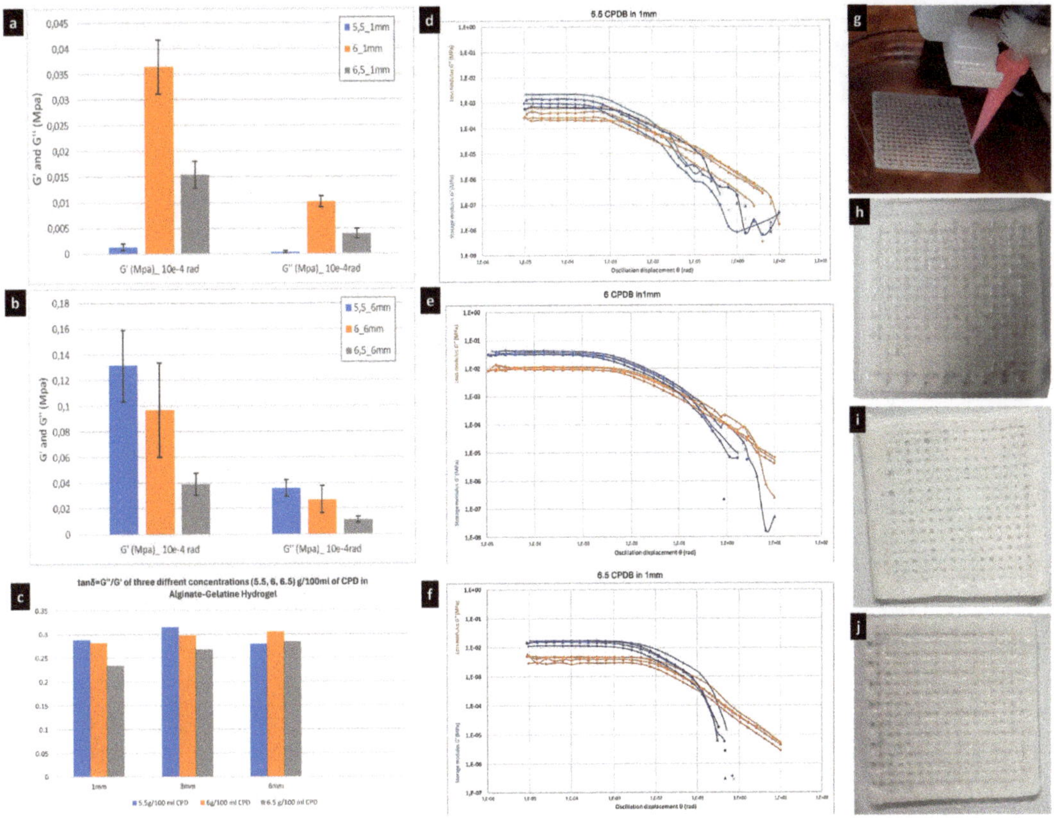

Figure 2. The rheological properties comparison between the three different hydrogels of different CPDB concentrations of 5.5, 6, and 6.5% enhanced SA–gelatine hydrogels in three different thicknesses of 1 mm, 3 mm, and 6 mm. (**a**) Comparison of the G values (G′ storage modulus) and G″ (loss modulus) between the three different 3D-printed hydrogels of 1 mm thickness; (**b**) comparison of G values between the three different 3D-printed hydrogels of 6 mm thickness; (**c**) a total comparison of the tan theta values of the three different hydrogels of different thicknesses of 1 mm, 3 mm, and 6 mm. (**d**–**f**) Each plot exhibits the G values (storage and loss modulus) in relation to the oscillation displacement of the tested three samples (ten replica per each): (**d**) the 5.5% CPDB rheological behavior, (**e**) the 6% CPDB properties, and (**f**) the 6.5% properties. Duplicate values obtained from overlapping in the ten replica behaviors per each sample were culled for easier reading. (**g**) The printability of the developed hydrogel of 6% CPDB-enhanced SA–gelatine hydrogel. (**h**,**j**) The 3D-printed hydrogel coherency and stability post printing and pre-third-stage cross-linking, where (**h**) corresponds to the 5.5% exhibiting swelling and low coherency; (**i**) the 6% exhibiting high coherency, layer-to-layer adhesion, and no swelling; and (**j**) the 6.5% showing minor swelling pre-cross-linking.

This indicates that the optimum criterion for deciding on the best rheological properties in the case of our study is tanδ, as it is in the middle between elasticity (tanδ < 1) and plasticity (tanδ > 1), which ensures the sufficiency of the developed hydrogel to act as a phase-transformative material, i.e., from soft to hard tissue, for application in bone tissue engineering as well as a self-mineralized material in architectural structural applications, where the initial phase of material before the mineralization is in equilibrium between elasticity and plasticity.

This justifies the selection and design of the mechanical properties' characterization of the hydrogel in the current study, where the standard rheological properties of the material expressed in its G values is the main method used to examine the behavior of the different CPDB-enhanced hydrogels [126,127]. This was done instead of testing the compressive strength as reported in other literature, where it does not reflect only the material properties but also depends to a great extent on the geometry, which is not the case in the current study that focuses in particular on studying the material itself [128].

The initial screening of the rheological properties of the different hydrogel composites with the different CPDB concentration and different thicknesses of 1, 3, and 6 mm revealed that the 6% CPDB-modified SA–gelatine hydrogel achieved this equilibrium between elasticity and plasticity in comparison to the two other best printable, coherent, and stable hydrogels of CPDB concentrations 5.5 and 6.5%. Table 1 and Figure 2a–f show the comparison between the G values and tan theta of the three different thicknesses of 1, 3, and 6 mm of the 3D-printed geometry per each concentration of 5.5, 6, and 6.5% of calcium phosphate-enhanced hydrogels.

Table 1. The G values and the tan theta of the tested rheological values of the three different concentrations of 5.5, 6, and 6.5 g/100 mL in three different thicknesses of 1, 3, and 6 mm each. All the results were obtained from the statistical estimation of the mean of 10 replicas per each value. The estimated standard deviation ranged from 0.003 to 0.012 for all values.

Sample	G' (MPa) 10^{-4} rad	G'' (MPa) 10^{-4} rad	$\tan\delta = G''/G'$
5.5 g/100 mL CPD in 1 mm	0.00143814	0.000414447	0.288
6 g/100 mL CPD in 1 mm	0.0384162	0.0108194	0.281
6.5 g/100 mL CPD in 1 mm	0.0156726	0.00367397	0.234
5.5 g/100 mL CPD in 3 mm	0.110418	0.0348274	0.315
6 g/100 mL CPD in 3 mm	0.0162533	0.0048435	0.298
6.5 g/100 mL CPD in 3 mm	0.0223588	0.00598952	0.267
5.5 g/100 mL CPD in 6 mm	0.148103	0.0413324	0.279
6 g/100 mL CPD in 6 mm	0.089164	0.027222	0.305
6.5 g/100 mL CPD in 6 mm	0.0410406	0.0116671	0.284

As exhibited in Table 1 and Figure 2, all the tested samples of the different concentrations CPDB–SA–gelatine hydrogels exhibited a higher storage modulus G' than the loss modulus G'' ratio, which indicates that the mixture of sodium alginate, gelatine, and calcium phosphate under the current methods is mainly elastic. This promotes its use in the designated described applications as a bone tissue replacement material and a structural material. The results also reveal an overall increase in both the storage modulus and loss modulus values congruent with the increase in the 3D prints' thicknesses. This is pronounced clearly when comparing the results between the 1 mm thickness prints of the different concentrations to the 6 mm prints of the same concentrations, respectively, although there were some fluctuations in this relativity between thickness and increase in rheological properties at the 3 mm thickness, as exhibited in Table 1, which is mainly justified by the differentiated rate of drying during the 7-day duration and the mineralization between the different concentrations, which is more augmented in this intermediate size. In the 1 mm thickness, the 6% CPDB concentration recorded the higher values of both storage modulus at 0.0384162 and loss modulus at 0.0108194 in comparison to the same thickness of 1 mm in the 5.5% and 6.5% CPDB concentrations. This might be because the 6% CPDB concentration falls in the middle between not being too-light for post drying, as in the case of the 5.5%, and not too dense for slow drying, as in the 6.5% CPDB.

However, when increasing the thickness to 3 mm, the 6% CPDB came last in the storage modulus value with 0.0162533 MPa compared to 0.110418 MPa in the 5.5% CPDB and 0.0223588 MPa in the 6.5% CPDB as well as having the smallest loss modulus of 0.0048435 MPa compared to 0.0348274 MPa in the 5.5% CPDB and 0.00598952 MPa in the 6.5% CPDB. This is justified by the uneven drying of the printed layers in the 3 mm thickness in the 6 CPDB concentration, as explained above. Nevertheless, at the 6 mm thickness, the 6% CPDB concentration recorded an intermediate storage modulus value of 0.089164 MPa between 0.148103 MPa in the case of 5.5% CPDB and 0.0410406 MPa in the case of 6.5% CPDB as well as recording the intermediate value of loss modulus of 0.027222 MPa in the case of 6% CPDB and between 0.0413324 MPa in 5.5% CPDB and 0.0116671 MPa in 6.5% CPDB.

Thus, it can be deducted that the 5.5% CPDB recorded the highest storage modulus with the 3 mm and 6 mm thicknesses while also recording the highest loss modulus in these two thicknesses, respectively. The 6% CPDB recorded the highest storage modulus in the 1 mm thickness while coming in the second place with the 6 mm thickness and third with the 3 mm thickness. Similarly, it came first in the loss modulus values with the 1 mm thickness and second with the 6 mm thickness. Finally, the 6.5% CPDB came second in the storage modulus values with the 1 mm thickness and the 3 mm while coming last with the 6 mm thickness.

Although it is obvious that increasing the thickness was congruent to the increase in both the storage and loss modulus, which reflects an overall increase in the rheological properties of the material, the main concern in the current study is the behavior of the material itself. This is why the 1 mm thickness was tested twice, and the authors thus focused on the results obtained at the 1 mm thickness to select the best CPDB concentration, which is 6% CPDB, to proceed to the next step of testing the different concentrations of EGCG as a rheological enhancer, which provides the anti-microbial effect of the developed material.

Although it was possible to compare the reached results of the developed hydrogels in the current study with the previous literature in terms of comparing swelling and shrinkage behavior, this was not the case when trying to compare the reached rheological indicators of the tested hydrogels in the current study with the previous literature because the materials and methods adopted in the current study comprise a novel, simplified, and minimized process that depends mainly on the thermal catalysis of the three-stage cross-linking process of the hydrogel developed from only three main components: sodium alginate, gelatine, and calcium phosphate dibasic, which enables its antimicrobial properties for application in built environments, as mentioned before. However, when comparing to the most relevant literature [14], it was found that a reached stiffness starting from 0.0102 MPa of GeLMA hydrogel is congruent with the reached hydrogel stiffness in the current study pertaining to the 6% CPDB-modified SA–gelatine hydrogel of 1 mm height and 0.0108194 MPa, which proves the efficiency and sustainability of the developed hydrogel in the current study in which we employed fewer materials and processes to reach to the same rheological properties of the GeLMA hydrogel reported in [14]. This is thanks to the calcium phosphate reinforcement effect [129]. Moreover, the reached rheological properties in the current study are attributed to the increased concentrations of gelatine (8%) and alginate (4%), which surpass the results obtained by [72] by employing blends of 7% alginate–8% gelatine to achieve high printability, mechanical strength, and stiffness.

3.2. Rheological Properties of CPDB-Modified SA–Gelatine Hydrogel Enhanced with Different Concentrations of Epigallocatechin Gallate

A polyphenolic compound, (–)-epigallocatechin-3-gallate (EGCG), which is a major catechin found in green tea, was tested in the current study as an addition to the CPDB-enhanced SA–gelatine hydrogel because of its reported health benefits, including antioxidant effects, cancer chemoprevention, improving cardiovascular health, enhancing weight loss, and protecting the skin from the damage caused by ionizing radiation. Moreover, EGCG has been shown to regulate a number of disease-specific molecular targets. These properties give EGCG an importance in bioengineering and regenerative medicine research and applications [130]. Furthermore, it enhances the degradability of a hydrogel since it becomes rapidly degraded in both acidic (pH below 2) and neutral conditions [131,132].

EGCG, as other tea polyphenols, exhibits antibacterial properties. As was reported in the literature, epigallocatechin gallate (EGCG)'s disinfectant efficiency was measured in *Escherichia coli* cultures under different concentrations of calcium ion (Ca^{2+}) with reported alteration in the tested bacterial growth and structures, where higher concentrations of Ca^{2+} (6–10 mM) enhanced the EGCG disinfectant effects by facilitating the entry of EGCG into the bacteria. This opens wide potentials in applying EGCG in sustainable bioactive building materials, as it can perform purifying antimicrobial effects in the built environment, and especially as an additive for the disinfection of groundwater and other raw waters containing calcium ions [133]. This also gives EGCG higher value in tissue engineering in bone replacement materials when incorporating it in calcium-based hydrogels, triggering their disinfectant effect. Applying EGCG lessens the need to incorporate additional antibiotics for infection control of the developed hydrogel grafts. For example, it was reported that (EGCG)-modified gelatine sponges incorporating β-TCP granules were applied in bone regeneration. They were prepared under vacuum-heating treatment to induce thermal cross-linking, and the prepared hydrogels exhibited high stability, β-TCP granule retention, operability, and cytocompatibility, with significant higher bone-forming ability than β-TCP alone. The treatment increased the number of osteoclasts in defects treated with the hydrogels compared with that around β-TCP alone, which indicates that the combination of EGCG, gelatine, and calcium phosphates through thermal cross-linking can be a promising process to enhance bone-forming ability [134].

This justifies the incorporation of EGCG in the current research as a disinfectant and cytocompatibility enhancer component for the SA–gelatine–CPDB hydrogel for the two proposed applications in sterile (bone replacement material) and non-sterile (building material) condition, either without or with cells.

Additionally, some research has reported the effect of EGCG in boosting biomineralization with amorphous calcium phosphate (ACP) for defect restoration by preventing and restoring demineralization, combining the antibacterial property of EGCG and the remineralization effect of ACP in both neutral and acidic conditions. In acidic conditions, EGCG exerted a strong antibacterial effect, and the ACP release rate doubled within 4 h, resulting in the prevention of demineralization in the presence of cariogenic bacteria. It also exhibited good biocompatibility with L-929 cells and human gingival fibroblasts [135].

In addition to providing support for infiltrating cells, EGCG could assure the delivery of drugs or biologically active molecules, enhancing bone formation. A study reported that developed hydrogel from gellan gum (GG), gelatine (Gel), and epigallocatechin gallate (EGCG)-loaded $CaCO_3$ microparticles is subjected to enzymatic mineralization with calcium phosphate (CaP) and recorded the release of EGCG for up to 14 days, where EGCG reduced the cytotoxicity of the calcium carbonate particles and increased the viability of the MG-63 cells exhibiting a pore size similar to that of the spongy bone when examined under optical and scanning electron microscopy [136].

Other studies have reported the use of EGCG as a safer alternative treatment for bone disorders by designing a 3D-printed tricalcium phosphate scaffold for localized EGCG delivery to enhance in vitro osteogenic ability, anti-osteoclastogenic activity, vascularization formation, and chemoprevention. It was reported that EGCG release enhances the osteogenic differentiation of hMSCs. However, it was also reported that EGCG significantly downregulates the receptor activator of nuclear factor-κB ligand (RANKL) expression by 7.0-fold, indicating that EGCG suppresses RANKL-induced osteoclast maturation as well as reducing human osteosarcoma MG-63 cell viability by 66% compared to the control at day 11. These results indicate that EGCG might be considerable only for low-load-bearing scaffolds. Nevertheless, further research on the effect of EGCG on the mineralization and its incorporation for development as load-bearing bone tissue replacement material is still needed, especially in terms of its effect on the rheological and mechanical properties of these bone replacement scaffolds [137].

Some studies have reported the effect of EGCG on gelatine-based hydrogels' stability and mechanical behavior, indicating that the addition of EGCG improved the gel strength and thermal stability of gelatine with an optimal final concentration of 1.0 g L epigallocatechin gallate −1 in a gelatine solution (66.7 g L^{-1}) thanks to the hydrogen bonds as the main molecular interactions [138]. Moreover, some recent studies have tested the effect of EGCG on the rheological properties of active edible films prepared from tea polyphenols (TP), gelatine, and sodium alginate. They evaluated the effect of 0.4%–2.0% TP (*w*/*w*, TP/gelatine) on the physical, antioxidant, and morphological properties of gelatine–sodium alginate films. The results revealed that the tensile strength (Ts), contact angle (CA), and cross-linking degree were enhanced congruently with increasing TP concentration in the film, while the water vapor permeability (WVP) and light transmittance of the film were decreased. Antioxidant capacity was improved by increasing TP content in the film, and the interactions between gelatine sodium alginate and TP were confirmed by Fourier transform infrared spectroscopy analyses (FTIR). The resulting surfaces of the developed gel were smooth, continuous, and dense in internal structure as observed by scanning electron microscopy (SEM). This promotes the use of TP in gelatine and sodium alginate film solutions to improve the physical properties and antioxidant activity of the films [139].

Nevertheless, testing the effect of EGCG on the SA–gelatine–calcium phosphate hydrogel has not been sufficiently presented in the literature before, especially for load-bearing bone replacement structural material applications. Thus, in the current study, three different concentrations of EGCG were tested to evaluate their effect on the rheological properties and stability of the developed SA–gelatine–CPDB hydrogel, mainly to enhance stretchability (elasticity) and the antibacterial traits of the prepared material. The three different concentrations of EGCG, namely 4, 6, and 8 µm, were added to the optimum concentration of 6% CPDB-modified SA–gelatine hydrogel, following the recommendations of [139] for the optimum concentration of EGCG for SA–gelatine hydrogels film applications. However, the hydrogel composition used in the current study is designed for hard tissue engineering applications as a bone replacement material. This justifies reducing the tested concentrations of the EGCG for two reasons: (1) to avoid cell inhibition effects in case of embedding the osteosarcoma SaOs-2 cells in the developed hydrogel and (2) to avoid increasing the elasticity of the final material and hindering its rigidity.

The rheological tests of the three different EGCG concentrations hydrogel printed in 1 mm thick prints revealed the following: (1) All the tested concentrations resulted in reduced storage modulus and loss modulus values of the hydrogel (G′ of 0.0275944 MPa, G″ of 0.00809548 MPa in 4 µm EGCG; G′ of 0.0216016 MPa, G″ of 0.00548179 MPa in 6 µm EGCG; and G′ of 0.0277438 MPa, G″ of 0.00765127 MPa in 8 µm EGCG) in comparison to the control group (6% CPDB–SA–gelatine hydrogel without the addition of the EGCG) (G′ of 0.0384162 MPa and G″ of 0.0108194 MPa), as exhibited in Table 2 and Figure 3. Although tanδ values of the EGCG-modified gels witnessed minor changes in comparison to the non-modified gel, however, an overall reduction to the rheological properties of the hydrogel across the different tested concentrations of the EGCG is obvious, the ratio of which is ~0.001% to each of the hydrogel components' dry weight.

Table 2. The G values and the tan theta of the tested rheological values of the three different concentrations of 4, 6, and 8 µm of EGCG-modified hydrogel printed in 1 mm thick prints. Al l the results were obtained from the statistical estimation of the mean of 10 replicas per each value. The estimated standard deviation is 0.003 for all values.

Concentration (µm)	G′ (MPa) 10^{-4} rad	G″ (MPa) 10^{-4} rad	tanδ = G″/G′
4 EGCG	0.0275944	0.00809548	0.2933
6 EGCG	0.0216016	0.00548179	0.2537
8 EGCG	0.0277438	0.00765127	0.2757

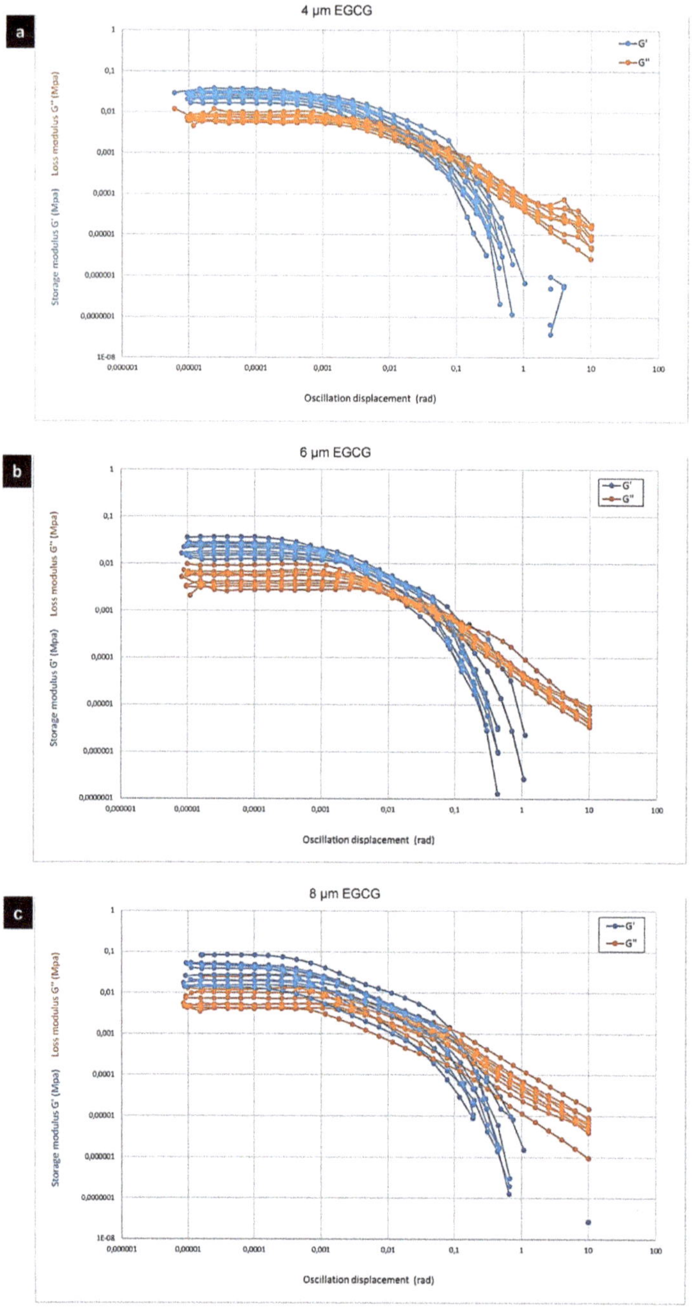

Figure 3. The rheological performance of the three tested EGCG-modified hydrogels using different concentrations of (4, 6, and 8 µm). Each plot exhibits the G values (storage and loss modulus) in relation to the oscillation displacement of the tested three samples (ten replicas each): (**a**) the 4 µm EGCG-modified hydrogel rheological behavior, (**b**) the 6 µm EGCG-modified hydrogel properties, and (**c**) is the 8 µm EGCG-modified hydrogel properties.

It is noteworthy that the rheological tests were always performed on the samples after 1 week of drying in ambient air (non-sterile conditions), which indicates that the EGCG-modified hydrogel either suffered from reduced mineralization due to the reduced ability to interact with the ambient CO_2 or that it suffered slower interaction and mineralization in the ambient air. Thus, further testing of the rheological properties of the EGCG-modified hydrogel will be conducted in the future to test a wider array of different concentrations as well as the change in their rheological properties chronologically since this would exceed the scope and capacity of the current research, which is focused on testing the effect of the EGCG of the developed hydrogel only as a possible antibacterial and stretchability booster.

The G values in all the three tested concentrations of EGCG revealed minor differences and fluctuations, which indicates that the used concentrations might have needed to be significantly varied to sense their effect or that EGCG has a trivial effect on the current hydrogel composition due to the three-stage thermally and chemically induced cross-linking, which might have affected the behavior of the EGCG.

Based on the reached results, EGCG was not found to efficiently promote the rheological properties of the developed hydrogel, especially since the rheological properties in the current study were measured after one week of drying in ambient air, which implies the interaction and mineralization of the EGCG modified hydrogel become unfavorable for the designated application as bone replacement material as well as a structural building material in non-sterile conditions. Furthermore, the EGCG as a polyphenolic tea extract compound is considered an expensive reagent, which hinders its sustainability and alignment with the current study objectives of developing an affordable and available hydrogel for multi-scale applications. Thus, the authors maintained the optimum CPDB concentration-modified SA–gelatine hydrogel for testing its mineralization and microstructure when incubated in different conditions to measure the feasibility of applying the material both inside the human body as a bone replacement material as well as a building material self-mineralized in non-sterile conditions.

3.3. Mineralization of the 6% CPDB-Modified SA–Gelatine Hydrogel Incubated in Fetal Bovine Serum and Air

The design of this experimental procedure was based on the objective of the multi-scale application of the developed material as a load-bearing, self-mineralized, bone replacement material and as a structural building material that can be adopted in open-air, non-sterile conditions. This indicates that the main criteria is the mineralization of this CPDB-modified SA–gelatine hydrogel by detecting the formation of hydroxyapatite $[Ca_{10}(PO_4)_6(OH)_2]$ crystals, which is the characteristic mineral phase of bones, on the 1 mm thick prints of the developed hydrogel.

In the design of the experimental procedures in this phase, we chose incubation in FBS and air, respectively, due to the following reasons; (1) Fetal bovine serum (FBS) is a common media constituent used in tissue engineering cultures [140] thanks to its nourishing composition containing proteins and growth factors, allowing maintenance and proliferation of eukaryote cells in vitro It has proven to maintain and boost the growth of different type of cells, which justifies its standard use in cell culture and tissue engineering protocols [141]. Although in the current phase of the experimental study, the optimized hydrogel does not contain any embedded osteosarcoma cells as planned, the use of FBS for incubating the hydrogel was intended not only to standardize the experimental study procedures by providing a control group where the hydrogel is incubated in FBS without embedding cells but also to test the effect of the FBS as a standard simulation of body fluids [142,143] on the degradation rate and mineralization behavior of the optimized hydrogel in the current study. (2) Incubation in open-air, non-sterile conditions is considered the pivotal point of novelty in the current study since it has not been reported in the previous literature before, especially for the novel proposed application in the current study as a structural building material, to our knowledge. The standard tissue engineering practices have always focused on in vivo testing of the developed hydrogels scaffolds, which necessitates their sterilization. This is not the case in the current study, which takes a different approach to prove that under non-sterile conditions, the incubated, optimized hydrogel exhibits high antimicrobial properties, as exhibited in Figure 4, which shows the SEM results of the incubated, optimized hydrogel in open-air, non-sterile conditions.

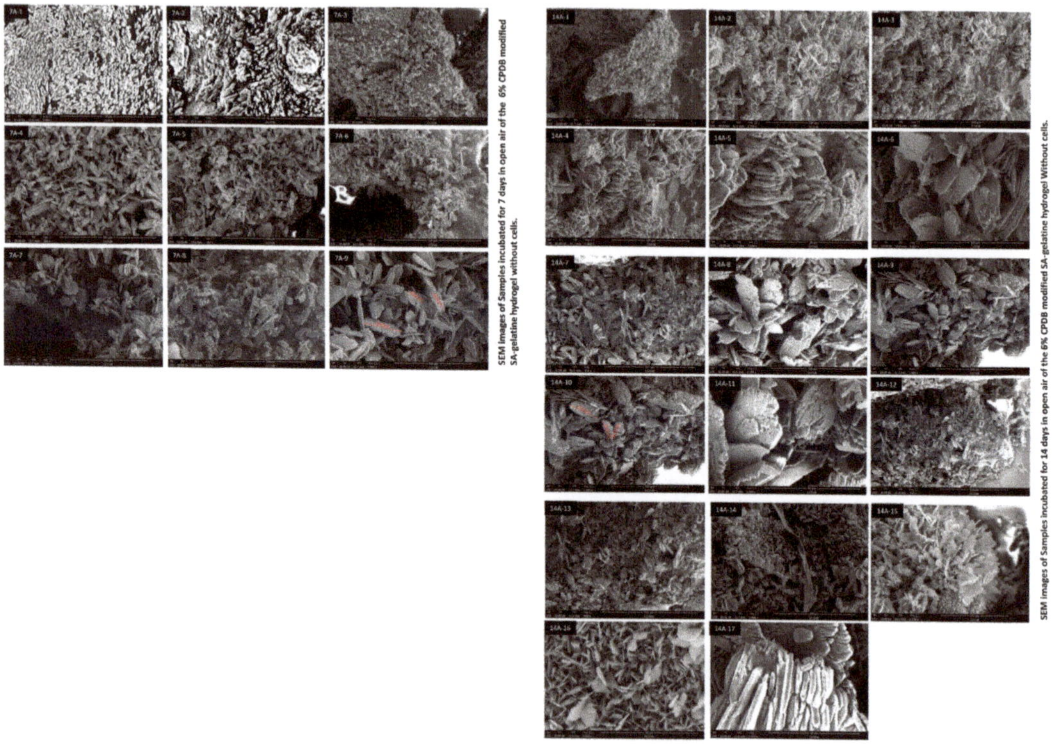

Figure 4. Scanning electron microscopy results of the optimized hydrogel samples (without cells) incubated in open-air, non-sterile conditions for 7 and 14 days, respectively. (7A-1 to 7A-9) The mineralization detection results of the first group of the optimized hydrogel samples incubated in air for 7 days, where obvious and dense formation of hydroxyapatite platelets is manifested in all the SEM images. (14A-1 to 14A-17) The second group of the scanned hydrogel samples incubated in air for 14 days, where all the samples exhibit a significant increase in the density, projection, and orientation of the hydroxyapatite platelets, proving the mineralization of the developed optimized hydrogel in open-air, non-sterile conditions.

The detection of the formation of the hydroxyapatite crystals was conducted by scanning electron microscopy (SEM) on the surface of the hydrogel, using the low-vacuum mood method without further fixative solutions to avoid damaging the samples or altering their chemical structure, as congruent with the methods of [144,145]. This is justified because the surface of the 6% CPDB-modified SA–gelatine hydrogel is a reactive interface in each of the cases, i.e., either in FBS or with the surrounding air containing CO_2 and O_2, which induces the mineralization of the hydrogel. A similar green approach of low cost and low environmental impact material for environmental applications and medical use that induces the formation of hydroxyapatite was tackled by [146]. However, it was conducted in aqueous media, where the synthesis of the hydroxyapatite was achieved by reacting phosphoric acid with calcium carbonate in a water suspension to form a Ca–HA gel of fine particles. Moreover, another recent study [147] proposed using a catalyst of an active hydroxyapatite (HAp) and brushite (Bru) mixture to fixate carbon dioxide from the ambient air as a green approach to produce formic acid or acetic acid as well as to alleviate the carbon dioxide foot print. However, in the current study, we go beyond the current literature by providing a self-mineralized material in ambient air that hardens chronologically, depending on the available minimized components and using thermally induced chemical reactions from the outer layer of the material as the reactive interface to produce hydroxyapatite on the surface of the hydrogel while maintaining the elasticity of the interactive material in its inner layers, following an exact biomimetic approach of the bone formation and mineralization process, i.e., following the anatomical level of long bone (e.g., femur), where the harder, mineralized layer of the bone is the outer layer (cortical bone), while moving inwardly, there is less dense and more of cancellous bone [148,149].

The reaction cycle of hydroxyapatite calcination in native mammalian bone tissue was illustrated in an equation presented in [150], where organic matter, hydroxyapatite, and water, through the calcination phase, turned to calcium phosphate and carbon dioxide. In the current study, we propose a reaction of mineralization of the trio-component hydrogel of sodium alginate, gelatine, and calcium phosphate dibasic cross-linked with calcium chloride and sodium carbonate to formulate hydroxyapatite crystals. This approach is supported by an increasing number of recent studies mainly focused on the synthesis of hydroxyapatite [151]. The standard synthesis of hydroxyapatite is usually achieved by reacting soluble calcium salts [152]. Also, different sources of calcium in the solid state can improve the reaction as well [153]. However, such a reaction is not a simple, straightforward reaction but rather a group of sequent complex reactions, especially in the current study, with three stages of thermally induced chemical cross-linking and incubation in open air. However, the following equation (Equation (1)) draws an overall view of the proposed mineralization reaction, exhibiting the chemical reactive groups in the input components of the hydrogel and the outputs mainly exhibiting the hydroxyapatite crystals, confirming the mineralization of the hydrogel with some expected trace elements of Na, N, and Cl ions that can be present as salts, for example, NaCl.

$(NaC_6H_7O_6)n + (C_{31}H_{27}NO_4)n + (CaHPO_4)n$	→ Drying in open-non sterile conditions →	$(Ca_{10}(PO_4)_6(OH)_2)n$
Thermally induced cross-linking at 60 °C for 4 h. Chemical cross-linking by $CaCl_2 + Na_2CO_3$.		

(1)

Equation (1): The proposed reaction of hydroxyapatite formation from mineralization of the three-stage, thermally induced, chemical cross-linked hydrogel of 6% CPDB-modified SA–gelatine by incubation in open air (CO_2, O_2) and in non-sterile conditions. The equation exhibits the chemical composition of the hydrogel developed from sodium alginate, gelatine, and calcium phosphate dibasic, cross-linked thermally and chemically by the cross-linking solution prepared from reacting calcium chloride and sodium carbonate. The mineralization occurs after incubating the hydrogel in open air for 7 and 14 days, respectively.

The detected mineralization of the air-incubated hydrogel samples obviously proved the formation of hydroxyapatite platelets in all the SEM-analyzed samples, as exhibited in Figure 4 congruent with [152], showing the two groups of tested samples of the optimized hydrogel of 6% CPDB–SA–gelatine incubated in open-air, non-sterile conditions for 7 days and for 14 days, respectively. The figure shows that the density, size, and spatial extension of the hydroxyapatite crystals increased significantly to fill all the hydrogel interfaces (14A-1–14A-6,14A-14–14A-17), with scale bar ranging from 500 μm to 30 μm, as in Figure 4 (14A-17). The first group of SEM images in Figure 4 exhibit the results of the mineralized samples of the optimized hydrogel incubated for 7 days in open air, where the samples were still in the active mineralization process, and some empty regions in the hydrogel interface can be detected, as exhibited in Figure 4 (7A-1,7A-2,7A-6,7A-7,7A-8). And the overall lower density of the hydroxyapatite platelets in this first group (7-day air incubation) is pronounced along the different magnifications used, ranging from 500 μm to

30 μm scale bar, when compared to the second group's mineralization density (14-days air incubation). This can be seen when comparing (7A-1 to 7A-9) scans from the first group with (14A-1 to 14A-17) in the second group. This is mainly justified by the drying of the hydrogel in the open air in room temperature, where it loses its water content, almost reversing the first calcination equation proposed by [150], which is congruent with the chemical reactions involved during the calcination and precipitation methods exhibited in in the same reference [150]. However, the current study employed lower temperature in initiating the mineralization reaction, instead adopting chemical pathways where breaking the calcium carbonate (3rd-stage cross-linking) occurred as a result of adjusting the reactive groups between the hydrogel interface and the surrounding air, as exhibited in Equation (1). Furthermore, the presence of calcium carbonate facilitated the mineralization reaction and the formation of the hydroxyapatite by providing more reactive sites between the hydrogel interface containing the calcium carbonate ($CaCO_3$) and water (H_2O) and the surrounding air containing CO_2 and O_2 [4,150] to facilitate the hydroxyapatite platelets formation. Although the $CaCO_3$ catalyzed the mineralization by air reaction in the current study, the formation of the detected hydroxyapatite crystals is attributed mainly to the high concentrations of the calcium phosphate dibasic in the reaction, as supported by [154], who proved the self-setting and hardening of calcium phosphate cements and their ability for forming hydroxyapatite crystals. Furthermore, the formation of the hydroxyapatite platelet morphology proved by the SEM scans in the current study is congruent with [150,155], who proved the transformation of calcium phosphate particles to platelet-like apatite crystals.

Unlike the mineralization morphology and density detected in the optimized hydrogel samples (without cells) incubated in FBS (Figure 5), the identical hydroxyapatite platelet morphology is not visible in the scanned hydrogel interface, but rather, multiple irregular, varied-sized faceted spheres are detected in all the scanned hydrogel samples incubated in FBS for 7 or 14 days. This still signifies the presence of hydroxyapatite; however, in another morphology of faceted spheres, as reported in [156], the SEM images of the encrusted crystals in artificial urine solutions (AUS) exhibited hydroxyapatite spheroid-like structured crystals observed in the calcium phosphate/struvite AUS. This proves that the detected faceted spheres in the SEM images of the hydrogel incubated in fetal bovine serum in the current study (Figure 5) are of hydroxyapatite since both FBS and urine are a simulation of body fluids. Moreover, Refs. [150,157], proved the varied scales and shapes of faceted, spherical hydroxyapatite crystals using scanning electron microscopy. Furthermore, the varied morphology of the hydroxyapatite crystals between incubation in air (sharp-edged platelets) and incubation in FBS (faceted spheres) is justified by the different chemical structure of the media that affected the crystalline structure of the formed hydroxyapatite [150,156]. The fetal bovine serum is a rich source of mineral ions such as sodium, chlorine, potassium, calcium, and phosphate as well as proteins such as albumin, bilirubin, creatinine, and globulin and lipids such as cholesterol and monosaccharides such as glucose [158,159]. This vivid chemical composition enables more complex reaction pathways that result in the faceted sphere form of the detected hydroxyapatite crystals in the scanned samples of the hydrogel incubated in FBS, as exhibited in Figure 5.

However, the density of these faceted spheres of hydroxyapatite and the overall texture of the hydrogel surface proves the reduced mineralization of the hydrogel incubated in FBS. This was more pronounced in the 7-day period of incubation in FBS, where 7 FBS-2, 7 FBS-4, and 7 FBS-8 yielded many areas of the hydrogel surface without any detected hydroxyapatite crystals, even in the faceted spheres morphology, while the projection (textures) of the second group of the hydrogel samples incubated in FBS for 14 days exhibited increased density of the faceted hydroxyapatite spheres, as exhibited in Figure 5 (14FBS-1,14 FBS-4). This proves that the longer incubation period in FBS facilitated interaction with the hydrogel interface to produce more hydroxyapatite crystals.

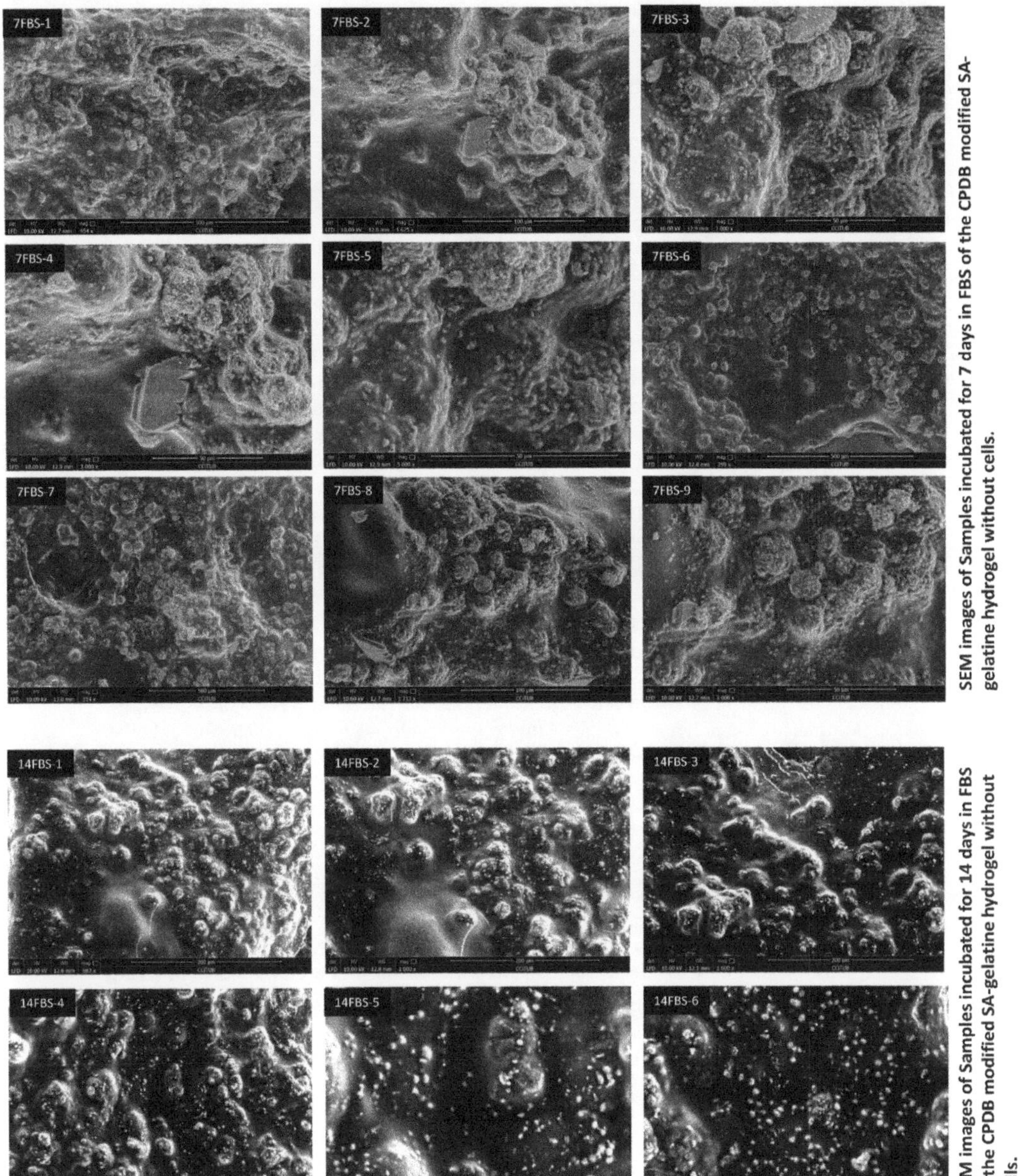

Figure 5. Scanning electron microscopy results of the optimized hydrogel samples (without cells) incubated in FBS for 7 and 14 days, respectively. (7 FBS-1 to 7 FBS-9) First group of the optimized hydrogel incubated in fetal bovine serum for 7 days, with different magnification ranging from 50 to 500 μm scale bars, exhibiting the formation of hydroxyapatite crystals with a faceted spheres morphology. (14 FBS-1 to 14 FBS-6) SEM results of the second group of samples of the optimized hydrogel incubated in FBS for 14 days, with different scalebars ranging from 50 to 200 μm, which exhibits enhanced texture of the mineralized hydrogel with increased density and projection of the faceted hydroxyapatite spheres.

Comparing the results of the hydroxyapatite density, morphology, and distribution, over all the developed hydrogel surfaces between the samples incubated in air and the samples incubated in FBS, it can be concluded that the incubation in air was more potent in accelerating the mineralization of the samples and densifying the resultant hydroxyapatite platelets across the topology of the SEM-scanned hydrogel samples.

3.4. Mineralization Test of CPDB-Modified SA–Gelatine Encapsulating Osteosarcoma Cells Incubated in Fatal Bovine Serum and Air for 7 and 14 Days, Respectively

The SEM results of the hydrogel samples incubated in air and in FBS, respectively, proved the capacity of the proposed optimized hydrogel and the three-stage cross-linking process in achieving mineralization of the hydrogel and forming hydroxyapatite crystals in various media. It was also mandatory to detect the mineralization capacity of the hydrogel with the embedded osteosarcoma cells within it (the bioink). In the current study, the main objective was to prove the hydrogel's capacity without cells to achieve self-mineralization, especially in air, as the main sustainable approach of developing this multi-scale application material. On the other hand, the capacity of the bioink including the embedded osteosarcoma SaOs-2 cells to achieve mineralization came second in importance in the current study. This is because the employment of SaOs-2 cells, culturing, and maintenance in bone replacement material is not a sustainable approach for two reasons: (1) An optimized hydrogel (without cells) applied in bone tissue engineering scaffolds or grafts is designated to be applied in vivo, where the living, healthy cells responsible for bone mineralization are already existing (patient cells), which indicates there is no compatibility to use the osteosarcoma SaOs-2 cell lines for embedding in the hydrogel as bone replacement material for active mineralization and bone remodeling; (2) usually, testing the effect of the hydrogel on cell viability in the literature is conducted for short periods (21 days) due to the sensitivity of the cell culture and the complex maintenance procedures that it requires, added to the complex and inaccurate statistically based estimations of the toxicity tests, which give only an approximated idea of how the cells act under the developed hydrogel. Therefore, in the current research, the authors designed the experimental study to detect the mineralization of the hydrogel under the effect of embedded osteosarcoma cells as an indicator of the cells' viability, bioactivity, and achieved metabolic pathways to realize the mineralization, aligning with the objective and methodology of the current study of developing optimized hydrogel with sustainability in its composition (materials) and operations (processes). Thus, the mineralization of the bioink (optimized hydrogel+ the osteosarcoma SaOs-2 cells) was detected using scanning electron microscopy of the samples of the bioink incubated in air in controlled but non-sterile conditions or incubated in FBS at 37 °C.

In the current study, the osteosarcoma SaOs-2 cell line was used thanks to its availability and reliable reproducibility in unlimited numbers without the need for isolation or ethical approval. The human osteosarcoma cell line SaOs-2 is a phenotypically mature osteoblast with high levels of ALP activity, exceeding other osteosarcoma cell lines such as MG-63 and SaOs-1. These characteristics serve synthesizing autonomous biomineralized materials with greater spatial dimensions and viability. Additionally, it was reported that the ALP activity by the SaOs-2 cell line was similar to human primary osteoblasts at early time points but achieved 120-fold higher results after 14 days of culturing under the same conditions [160]. The capacity of SaOs-2 cells to form a calcified matrix typical of woven bone was also reported [161] (Saldaña, et al., 2011) as well as the high-detailed similarity between the collagen structure synthesized by SaOs-2 and the collagen formed by primary human osteoblast cells but with a higher level of lysyl hydroxylation in the SaOs-2 cells [162]. Finally, the similarity between the cytokine and growth factor expression of SaOs-2 cells and primary normal human osteoblast cells was also reported [163,164]. The SaOs-2cells were employed in the current work due to their strong similarity to human isolated osteoblasts, giving the possibility of application on a wide spatial scale due to its enhanced ability of proliferation. This was proven after only 14 days of incubation, achieving an 8×10^6 cell count/T25 flask using the Neuberger chamber. The bioink composed from the optimized hydrogel and the cell culture was printed in cell-sustaining conditions of 37 °C and 100% printing speed and flow to minimize the post-printing pre-cross-linking time. This is congruent with [165]. After cross-linking, the 3D-printed bioink samples were incubated in FBS at 37 °C as a simulation of the in vivo conditions of using these developed grafts as a bone replacement material [114,166]. Usually, in the literature, a toxicity test is conducted to measure the cell viability under the effect of the developed hydrogel, which is usually conducted by immunofluorescence microscopy methods that employ cell staining as a differentiating marker between the living and the dead cells [167–169]. However, these methods imply exposing the cells to the toxicity effect of the stains and radiation used in these procedures, which hinders the sustainability of the procedure and reduces the accurate prediction of the hydrogel toxicity and the viability of cells under its effect. Thus, the authors adopted a sustainable approach in the current study to conclude the effect of the hydrogel on the cells' viability and behavior through measuring the osteosarcoma cells' effect on the mineralization process of the hydrogel (Figure 6) and comparing it to the control group, which is the hydrogel incubated in FBS (without cells) (Figure 5). The detection of the mineralization of the bioink incubated in FBS was conducted by fixing the samples first, then scanning them under low-vacuum scanning electron microscopy, as discussed before.

The results of the SEM revealed enhanced mineralization of the bioink (the hydrogel with the embedded osteosarcoma cells) in terms of density and the size of the formed faceted, spherical hydroxyapatite crystals, as exhibited in Figure 6 (7FBS-SaOs-2-1–7 FBS-SaOs-2-5), in comparison to their density, size, and distribution in the case of the hydrogel without cells incubated in fetal bovine serum for 7 days (Figure 5 (7FBS-1–7FBS-9)). This is attributed to the bioactivity of the osteosarcoma cells and their metabolic pathways being incubated in FBS, which induced the mineralization of the material and the composition of the hydroxyapatite crystals with larger size, density, and distribution along the bioink prints' topology. Similarly, a mild increase in the density of the formed hydroxyapatite crystals was detected in the bioink prints incubated in FBS for 14 days, as exhibited in Figure 6 (14 FBS-SaOs-2-1–14 FBS-SaOs-2-7), in comparison to the control group (hydrogel without cells incubated in FBS for 14 days) (Figure 5 (14 FBS-1–14 FBS-6)). This proves that the hydrogel has cytocompatibility, by encapsulating, and maintaining the osteosarcoma SaOs-2 cells' viability and enabling their bioactivity in inducing mineralization and hydroxyapatite formation.

The role of the SaOs-2 cells in inducing mineralization was further confirmed by analyzing the SEM results of the bioink prints incubated in air in controlled but non-sterile conditions for 7 and 14 days, respectively, as exhibited in Figure 7. As shown in both groups of SEM images, the first group (from 7Air-SaOs-2-1 to 7Air-SaOs-2-9) incubated in air for 7 days and the second (from 14Air-SaOs-2-1 to 14Air-SaOs-2-9) incubated for 14 days exhibit significantly increased density, distribution, sharpness, and projection of the formed faceted hydroxyapatite spheres, starting at crystal of 2 μm size, when comparing them to the bioink samples (with cells) incubated in FBS (Figure 6). Although this proves the bioactivity of the osteosarcoma cells in boosting the mineralization of the hydrogel incubated in air as well as the biocompatibility of the developed hydrogel for maintaining cell viability and boosting its pathways, when comparing these results (Figure 7) with the control group of the mineralized hydrogel samples incubated in air without cells (Figure 4), it can be inferred that the calcium phosphate dibasic's reaction with air as exhibited in Equation (1) is more responsible for the achieved results of the dense mineralization of the bioink incubated in air (Figure 7). This is due to two reasons: (1) The exhibited well-defined and sharp morphology of the faceted hydroxyapatite spheres in the bioink incubated in air is sharper and denser than the bioink incubated in FBS (Figure 6), which proves the mineralization and hardening of the hydrogel's outer interface that is encapsulating (coating the embedded cells) by the reaction with air, as described above in Section 3.3, although incubating the bioink with the living cells is more compatible with FBS to maintain motile conditions for cells' maintenance and circulation of nutrients and oxygen [170]. This proves that incubating the hydrogel and the bioink with embedded cells in air induces higher mineralization and formation of the hydroxyapatite crystals. (2) This is thanks to the high concentration of calcium phosphate dibasic in the developed hydrogel, where in addition to its biocompatibility, bioactivity, and osteoconductive properties [171], it gradually dissolves to release calcium and phosphate ions, which are beneficial in the hydroxyapatite formation [172]. The dissolution of the calcium phosphate depends on the surface exposed to the physiological environment and cellular activity. Therefore, the mineralization of the bioink (with the embedded cells) incubated in air was denser than the mineralization of the bioink incubated in FBS, where the particles of the calcium phosphate embedded inside the hydrogel will slowly interact with the environment at the outer peripheral surfaces as the hydrogel degrades, showing that when incubating the bioink samples in air, the outer surface reacts faster with the surrounding carbon dioxide and oxygen from the ambient air to form the hydroxyapatite crystals, with slower degradation of the hydrogel matrix that is reinforced by the calcium phosphate dibasic. This is unlike incubation in FBS, where a slower interaction rate and faster degradation rate of the hydrogel occurs, leading to a slower mineralization rate and lower density of the hydroxyapatite formation [124]. Thus, it can be concluded that the optimized hydrogel is the 6% CPDB-modified SA–gelatine mineralizes in open air with anti-microbial properties, as exhibited in the SEM images of Figures 4 and 7, both with and without cells. And it mineralizes with a slower rate in FBS either with cells or without cells, with more detected mineralization in the case of the embedded osteosarcoma cells, which proves the compatibility of the hydrogel with the SaOs-2 cells since they together boosted the mineralization of the hydrogel and the formation of the hydroxyapatite crystals. The effect of the mineralized hydrogel on the embedded osteosarcoma cells enabled maintaining the cells incapsulated within the mineralized outer interface of the hydrogel, isolating them from the open-air conditions yet still enabling their access to oxygen and nutrients circulation gained either from the original carrier media of the SaOs-2 cells added to the hydrogel or from the mineralization reaction products from the outer to the inner layers of the hydrogel. A further detailed analysis of the mineralization reaction kinetics of the bioink and the reciprocal effect of the hydrogel on the cell viability and proliferation will be exhibited in future studies due to the tight scope and objective of the current work.

The degradability of the hydrogel was evaluated by measuring the increase in its dimensions after the two weeks of incubation in FBS, without or with cells, as a function of incubation time in FBS at 37 °C. The authors chose to test the degradation of the hydrogel under the same media of incubation for the following reasons: (1) FBS is a closer simulation to body fluids, where the proposed application of the optimized hydrogel will be conducted as a bone replacement material (graft or scaffold); (2) FBS is used to sustain 3D cell cultures of different bone cell lines, including the SaOs-2 cells used in the current study; and (3) lastly, the test maintains the sustainable approach of developing the optimized hydrogel for multi-scale applications, as mentioned throughout the manuscript. The degradation process occurs initially by breaking a small number of cross-linked bonds, which cannot damage the whole hydrogel network but rather increases the lattice size of the networks, resulting in a large water absorbance [173] until reaching a critical value where the whole cross-linking network dissociates, leading to the loss of hydrogel. After incubation in FBS for 14 days, all the hydrogel and bioink samples displayed an increase in the dimensions of the prints with a 5 to 20% ratio, with higher increase in dimensions in the case of bioink (hydrogel with embedded cells). This can be justified because the viscosity of the hydrogel is already diluted by the effect of embedding the SaOs-2 cells with their liquid media in the hydrogel. In other words, the addition of the cells with their medium already dilutes the concentration of the hydrogel and affects its cross-linking network, leading to an increased swelling effect and subsequent increase in dimensions. However, this reached degradation rate after 14 days of incubation in FBS renewed every 2 days is a promising result since a hydrogel should be maintained for a suitable time in order to enable regeneration of new tissue in case of application as a bone graft or bone replacement material.

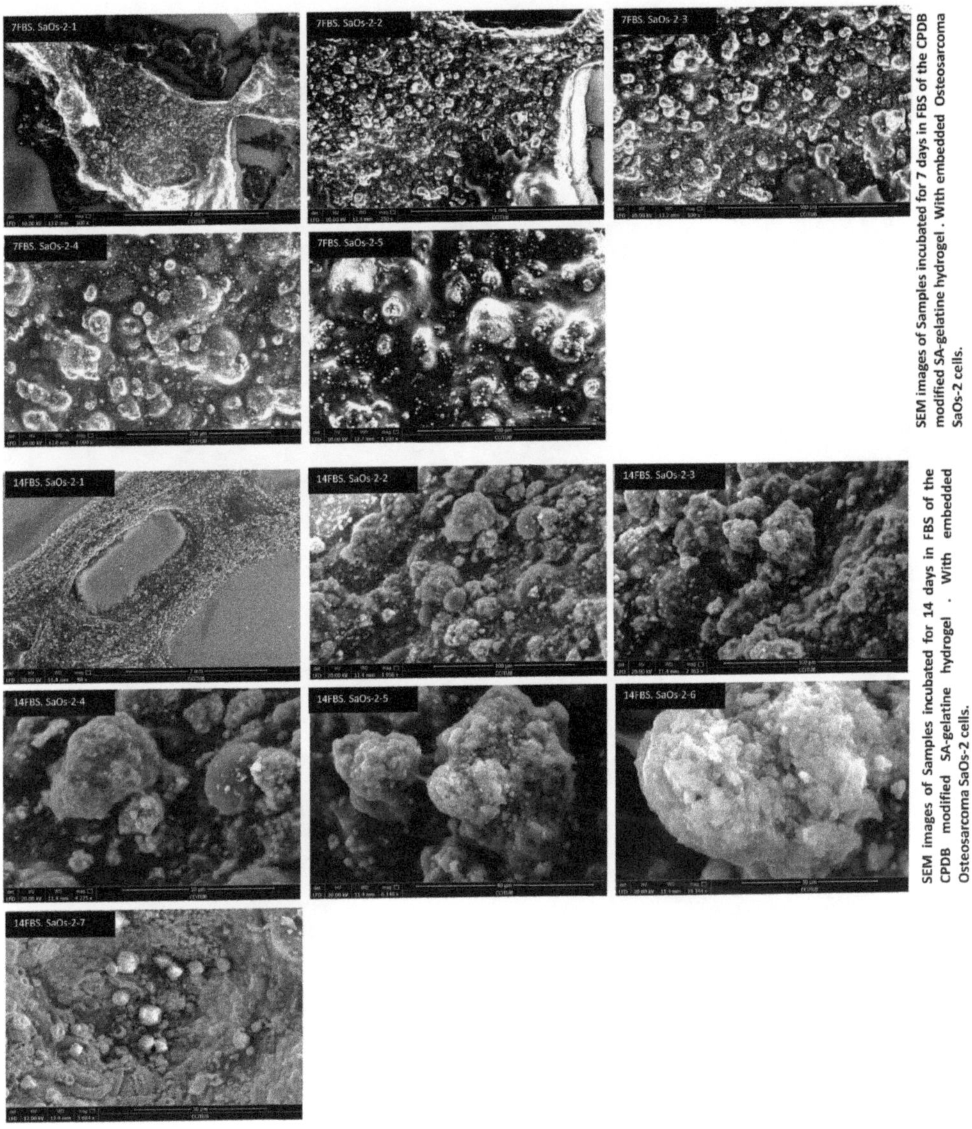

Figure 6. SEM images of the detected mineralization of the bioink prints (optimized hydrogel of 6% CPDB-modified SA–gelatine with embedded osteosarcoma SaOs-2 cells) incubated in FBS for 7 and 14 days, respectively, at 37 °C. (7FBS-SaOs-2-1 to 7FBS-SaOs-2-5) The first group of the SEM images of the scanned bioink prints samples incubated in FBS for 7 days. (14 FBS-SaOs-2-1 to 14 FBS-SaOs-2-7) The second group of the SEM images of the scanned bioink prints incubated in FBS for 14 days.

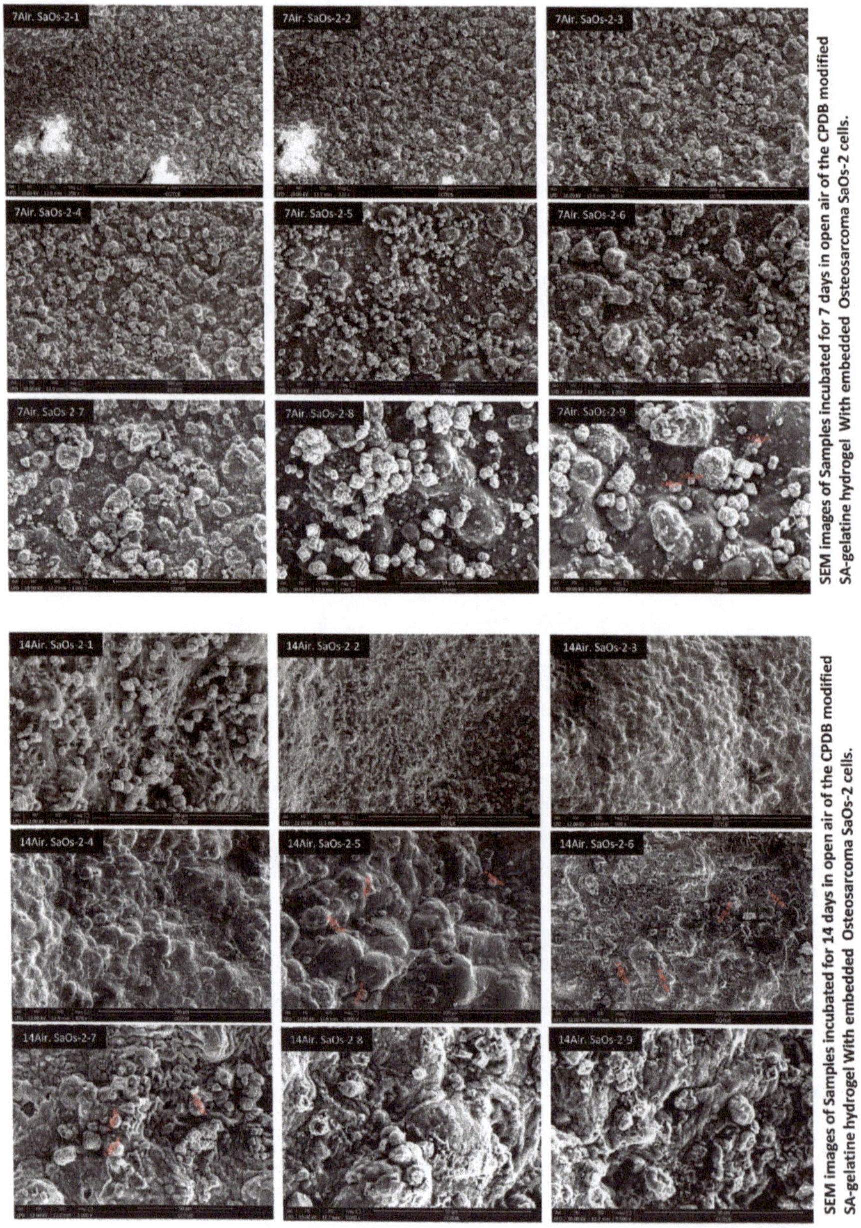

Figure 7. The SEM images of the bioink (optimized hydrogel with embedded osteosarcoma SaOs-2 cells) mineralization being incubated in open air in controlled but non-sterile conditions for 7 and 14 days, respectively. (7Air-SaOs-2-1 to 7Air-SaOs-2-9) The first group of the bioink samples incubated in air for 7 days at 37 °C, with varied scales ranging from 50 µm to 1 mm and with hydroxyapatite crustal size starting from 2 µm. (7Air-SaOs-2-1 to 7Air-SaOs-2-4) The dense distribution of the formed hydroxyapatite crystals across the hydrogel surface. (14Air-SaOs-2-1 to 14Air-SaOs-2-9) The bioink samples incubated in air for 14 days at 37 °C, with varied scales ranging from 50 µm to 500 µm, and with hydroxyapatite crystals size starting from 2 µm.

On the other hand, the hydrogel/bioink incubated in air (without and with cells, respectively) exhibited mild shrinkage in the dimensions of the 3D-printed samples after incubation in air for 14 days with a shrinkage of 2% from all directions from their size (the 3D-printed samples) post-cross-linking and pre-incubation in air for 14 days. However, another phenomenon was observed, which is the skewing or uneven rotation of the prints' upper surfaces, giving the shape of a hyperbolic paraboloid after complete mineralization, which will be exhibited and analyzed in detail in a future study.

4. Conclusions

In the current study, an optimized hydrogel was developed following a sustainable approach to minimize the materials and processes involved in developing self-mineralized hydrogel for multi-scale applications: as a bone replacement material and as a structural building material. The hydrogel developed mainly from three components of sodium alginate, gelatine, and calcium phosphate dibasic underwent a three-stage cross-linking process by cross-linking sodium alginate, and gelatine together; then cross-linking with calcium phosphate dibasic; and lastly, cross-linking the solution of calcium chloride and sodium carbonate. These thermally induced cross-linking reactions generated a viscose and stable hydrogel that has excellent injectability and shape retention as well. Three different concentrations of calcium phosphate dibasic showed higher rheological properties than the other tested concentrations. However, the 6% CPDB concentrations achieved the best rheological properties in terms of elasticity and rigidity. Although three different concentrations of epigallocatechin gallate were tested for enhancing the hydrogel elasticity and anti-microbial properties, the tested concentrations did not show any significance in these aspects. Thus, further investigations were conducted on the 6% CPDB-modified SA–gelatine hydrogel to detect its mineralization capacity in different environments and without and with cells, respectively. The SEM images proved the significant potency of the hydrogel to mineralize through forming hydroxyapatite platelets when incubated in the open air in non-sterile conditions for a duration of up to 14 days, with a congruent increase in mineralization over time. Similarly, the bioink with the embedded cells incubated in air in controlled but non-sterile conditions exhibited a similarly dense, mineralized material of hydroxyapatite crystals, which proves that the hydrogel is biocompatible and enables the cell activity of the embedded osteosarcoma SaOs-2 cells. Although the recorded mineralization in the case of the hydrogel incubated in FBS without and with cells recorded less mineralization density, the SEM images still exhibited hydroxyapatite crystals, showing the mineralization of the hydrogel when incubated in FBS. The degradability of the hydrogel was measured by the increase in its dimension when incubated in FBS for 14 days without and with cells. The results revealed a 5 to 20% increase in the dimensions of the hydrogel and bioink samples from all directions, especially in the case of bioink (with embedded cells). However, this ratio and time rate proves the slow degradation of the hydrogel, which enable its use in bone grafts and as a resistance structural building material. Furthermore, the hydrogel and bioink samples incubated in air did not show any swelling in dimensions but rather a shrinkage of 2% ratio from all directions after 14 days, with a detected skew in their surface morphology that will be evaluated and analyzed in a future study.

Author Contributions: Conceptualization, A.T.E. and Y.K.A.; Methodology, A.T.E. and Y.K.A.; Software, A.T.E. and Y.K.A.; Validation, A.T.E. and Y.K.A.; Formal analysis, A.T.E. and Y.K.A.; Investigation, A.T.E. and Y.K.A.; Resources, A.T.E. and Y.K.A.; Data curation, A.T.E. and Y.K.A.; Writing—original draft, A.T.E. and Y.K.A.; Writing—review & editing, A.T.E. and Y.K.A.; Visualization, A.T.E. and Y.K.A.; Supervision, A.T.E. and Y.K.A.; Project administration, A.T.E. and Y.K.A.; Funding acquisition, A.T.E. and Y.K.A. All authors have read and agreed to the published version of the manuscript.

Funding: This research received no external funding.

Data Availability Statement: All mandatory data are exhibited in the study, further raw data can be given upon request to the authors.

Acknowledgments: The authors would like to extend their sincerest thanks to Nuria Casals (head of the Pharmacology Unit in the Faculty of Medicine and Health Sciences at Universitat Internacional de Catalunya and the group leader of NeuroLipid Group at Universitat Internacional de Catalunya), and Samuel Bru Rullo for their kind and generous support in the logistic aspects of the current study. Equally, the authors would like to express their sincere thanks to the Bioengineering Institute of Technology at Universitat Internacional de Catalunya for their kind support in training and providing the used cell line; it would not have been possible to complete this research without their kind help. Last but not least, the authors extend sincere thanks to David Artiaga Torres and the CCiTUB at the University of Barcelona for their kind and professional support in the microscopy study in this work.

Conflicts of Interest: The authors declare no conflict of interest.

References

1. Wen, B.; Dai, Y.; Han, X.; Huo, F.; Xie, L.; Yu, M.; Wang, Y.; An, N.; Li, Z.; Guo, W. Biomineralization-inspired mineralized hydrogel promotes the repair and regeneration of dentin/bone hard tissue. *NPJ Regen. Med.* **2023**, *8*, 11. [CrossRef] [PubMed]
2. Wu, X.; Walsh, K.; Hoff, B.L.; Camci-Unal, G. Mineralization of Biomaterials for Bone Tissue Engineering. *Bioengineering* **2020**, *7*, 132. [CrossRef] [PubMed]
3. Parra-Torrejón, B.; Jayawarna, V.; Rodrigo-Navarro, A.; Gonzalez-Valdivieso, J.; Dobre, O.; Ramírez-Rodríguez, G.B.; Salmeron-Sanchez, M.; Delgado-López, J.M. Bioinspired mineralization of engineered living materials to promote osteogenic differentiation. *Biomater. Adv.* **2023**, *154*, 213587. [CrossRef]
4. Abdallah, Y.K.; Estévez, A.T. Biowelding 3D-Printed Biodigital Brick of Seashell-Based Biocomposite by Pleurotus ostreatus Mycelium. *Biomimetics* **2023**, *8*, 504. [CrossRef]
5. Definition of Biomineralization. Available online: https://www.merriam-webster.com/dictionary/biomineralization (accessed on 28 September 2022).
6. Vinn, O. Occurrence, Formation and Function of Organic Sheets in the Mineral Tube Structures of Serpulidae (Polychaeta, Annelida). *PLoS ONE* **2013**, *8*, e75330. [CrossRef]
7. Boskey, A.L. Biomineralization: Conflicts, challenges, and opportunities. *J. Cell. Biochem.* **1998**, *30–31*, 83–91. [CrossRef]
8. Sarikaya, M. Biomimetics: Materials fabrication through biology. *Proc. Natl. Acad. Sci. USA* **1999**, *96*, 14183–14185. [CrossRef]
9. Sherman, V.R.; Yang, W.; Meyers, M.A. The materials science of collagen. *J. Mech. Behav. Biomed. Mater.* **2015**, *52*, 22–50. [CrossRef]
10. Yokoi, T.; Kawashita, M.; Ohtsuki, C. Biomimetic mineralization of calcium phosphates in polymeric hydrogels containing carboxyl groups. *J. Asian Ceram. Soc.* **2013**, *1*, 155–162. [CrossRef]
11. Nie, L.; Li, X.; Wang, Z.; Hu, K.; Cai, R.; Li, P.; Han, Y.; Sun, M.; Yuan, H.; Suo, J.; et al. In vitro biomineralization on poly(vinyl alcohol)/biphasic calcium phosphate hydrogels. *Bioinspired Biomim. Nanobiomater.* **2020**, *9*, 122–128. [CrossRef]
12. Onozato, H.; Watabe, N. Studies on fish scale formation and resorption. *Cell Tissue Res.* **1979**, *201*, 409–422. [CrossRef] [PubMed]
13. Pepla, E.; Besharat, L.K.; Palaia, G.; Tenore, G.; Migliau, G. Nano-hydroxyapatite and its applications in preventive, restorative and regenerative dentistry: A review of literature. *Ann. Stomatol.* **2014**, *5*, 108–114. [CrossRef]
14. Chen, Z.; Lv, Y. Gelatin/sodium alginate composite hydrogel with dynamic matrix stiffening ability for bone regeneration. *Compos. Part B Eng.* **2022**, *243*, 110162. [CrossRef]
15. Yan, J.; Miao, Y.; Tan, H.; Zhou, T.; Ling, Z.; Chen, Y.; Xing, X.; Hu, X. Injectable alginate/hydroxyapatite gel scaffold combined with gelatin microspheres for drug delivery and bone tissue engineering. *Mater. Sci. Eng. C* **2016**, *63*, 274–284. [CrossRef] [PubMed]
16. Cai, Z.; Saiding, Q.; Cheng, L.; Zhang, L.; Wang, Z.; Wang, F.; Chen, X.; Chen, G.; Deng, L.; Cui, W. Capturing dynamic biological signals via bio-mimicking hydrogel for precise remodeling of soft tissue. *Bioact. Mater.* **2021**, *6*, 4506–4516. [CrossRef] [PubMed]
17. Engler, A.J.; Sen, S.; Sweeney, H.L.; Discher, D.E. Matrix Elasticity Directs Stem Cell Lineage Specification. *Cell* **2006**, *126*, 677–689. [CrossRef] [PubMed]
18. Hernández-González, A.C.; Téllez-Jurado, L.; Rodríguez-Lorenzo, L.M. Alginate hydrogels for bone tissue engineering, from injectables to bioprinting: A review. *Carbohydr. Polym.* **2020**, *229*, 115514. [CrossRef]
19. Buwalda, S.J.; Boere, K.W.M.; Dijkstra, P.J.; Feijen, J.; Vermonden, T.; Hennink, W.E. Hydrogels in a historical perspective: From simple networks to smart materials. *J. Control Release Off. J. Control Release Soc.* **2014**, *190*, 254–273. [CrossRef]
20. Dragan, E.S. Design and applications of interpenetrating polymer network hydrogels. A review. *Chem. Eng. J.* **2014**, *243*, 572–590. [CrossRef]

21. Li, J.; Yan, J.F.; Wan, Q.Q.; Shen, M.J.; Ma, Y.X.; Gu, J.T.; Gao, P.; Tang, X.Y.; Yu, F.; Chen, J.H.; et al. Matrix stiffening by self-mineralizable guided bone regeneration. *Acta Biomater.* **2021**, *125*, 112–125. [CrossRef]
22. Xu, Z.; Lam, M.T. Alginate Application for Heart and Cardiovascular Diseases. In *Alginates and Their Biomedical Applications*; Springer Series in Biomaterials Science and Engineering; Springer: Berlin/Heidelberg, Germany, 2017; pp. 185–212. [CrossRef]
23. de Vos, P.; Lazarjani, H.A.; Poncelet, D.; Faas, M.M. Polymers in cell encapsulation from an enveloped cell perspective. *Adv. Drug Deliv. Rev.* **2014**, *67–68*, 15–34. [CrossRef]
24. Pina, S.; Oliveira, J.M.; Reis, R.L. Natural-Based Nanocomposites for Bone Tissue Engineering and Regenerative Medicine: A Review. *Adv. Mater.* **2015**, *27*, 1143–1169. [CrossRef] [PubMed]
25. Otterlei, M.; Østgaard, K.; Skjåk-Bræk, G.; Smidsrød, O.; Soon-Shiong, P.; Espevik, T. Induction of Cytokine Production from Human Monocytes Stimulated with Alginate. *J. Immunother.* **1991**, *10*, 286–291. [CrossRef]
26. Becker, T.A.; Kipke, D.R. Flow properties of liquid calcium alginate polymer injected through medical microcatheters for endovascular embolization. *J. Biomed. Mater. Res.* **2002**, *61*, 533–540. [CrossRef] [PubMed]
27. Axpe, E.; Oyen, M. Applications of Alginate-Based Bioinks in 3D Bioprinting. *Int. J. Mol. Sci.* **2016**, *17*, 1976. [CrossRef] [PubMed]
28. Rhim, J.-W. Physical and mechanical properties of water resistant sodium alginate films. *LWT Food Sci. Technol.* **2004**, *37*, 323–330. [CrossRef]
29. Lee, K.Y.; Mooney, D.J. Alginate: Properties and biomedical applications. *Prog. Polym. Sci.* **2012**, *37*, 106–126. [CrossRef] [PubMed]
30. Gonzalez-Pujana, A.; Orive, G.; Pedraz, J.L.; Santos-Vizcaino, E.; Hernandez, R.M. Alginate Microcapsules for Drug Delivery. In *Alginates and Their Biomedical Applications*; Springer Series in Biomaterials Science and Engineering; Springer: Berlin/Heidelberg, Germany, 2017; pp. 67–100. [CrossRef]
31. Osmokrovic, A.; Jancic, I.; Vunduk, J.; Petrovic, P.; Milenkovic, M.; Obradovic, B. Achieving high antimicrobial activity: Composite alginate hydrogel beads releasing activated charcoal with an immobilized active agent. *Carbohydr. Polym.* **2018**, *196*, 279–288. [CrossRef] [PubMed]
32. Qiu, M.; Chen, D.; Shen, C.; Shen, J.; Zhao, H.; He, Y. Preparation of in situ forming and injectable alginate/mesoporous Sr-containing calcium silicate composite cement for bone repair. *RSC Adv.* **2017**, *7*, 23671–23679. [CrossRef]
33. Szekalska, M.; Puciłowska, A.; Szymańska, E.; Ciosek, P.; Winnicka, K. Alginate: Current use and future perspectives in pharmaceutical and biomedical applications. *Int. J. Polym. Sci.* **2016**, *2016*, 7697031. [CrossRef]
34. Falkeborg, M.; Cheong, L.-Z.; Gianfico, C.; Sztukiel, K.M.; Kristensen, K.; Glasius, M.; Xu, X.; Guo, Z. Alginate oligosaccharides: Enzymatic preparation and antioxidant property evaluation. *Food Chem.* **2014**, *164*, 185–194. [CrossRef] [PubMed]
35. Feng, L.; Cao, Y.; Xu, D.; Zhang, D.; Huang, Z. Influence of chitosan-sodium alginate pretreated with ultrasound on the enzyme activity, viscosity and structure of papain. *J. Sci. Food Agric.* **2016**, *97*, 1561–1566. [CrossRef]
36. Burana-osot, J.; Hosoyama, S.; Nagamoto, Y.; Suzuki, S.; Linhardt, R.J.; Toida, T. Photolytic depolymerization of alginate. *Carbohydr. Res.* **2009**, *344*, 2023–2027. [CrossRef]
37. Kelishomi, Z.H.; Goliaei, B.; Mahdavi, H.; Nikoofar, A.; Rahimi, M.; Moosavi-Movahedi, A.A.; Mamashli, F.; Bigdeli, B. Antioxidant activity of low molecular weight alginate produced by thermal treatment. *Food Chem.* **2016**, *196*, 897–902. [CrossRef]
38. Coleman, R.J.; Jack, K.S.; Perrier, S.; Grøndahl, L. Hydroxyapatite Mineralization in the Presence of Anionic Polymers. *Cryst. Growth Des.* **2013**, *13*, 4252–4259. [CrossRef]
39. Reakasame, S.; Boccaccini, A.R. Oxidized Alginate-Based Hydrogels for Tissue Engineering Applications: A Review. *Biomacromolecules* **2018**, *19*, 3–21. [CrossRef] [PubMed]
40. Bouhadir, K.H.; Lee, K.Y.; Alsberg, E.; Damm, K.L.; Anderson, K.W.; Mooney, D.J. Degradation of Partially Oxidized Alginate and Its Potential Application for Tissue Engineering. *Biotechnol. Prog.* **2001**, *17*, 945–950. [CrossRef]
41. Gomez, C.G.; Rinaudo, M.; Villar, M.A. Oxidation of sodium alginate and characterization of the oxidized derivatives. *Carbohydr. Polym.* **2007**, *67*, 296–304. [CrossRef]
42. Painter, T.; Larsen, B.; Sjövall, J.; Kääriäinen, L.; Rasmussen, S.E.; Sunde, E.; Sørensen, N.A. Formation of Hemiacetals between Neighbouring Hexuronic Acid Residues during the Periodate Oxidation of Alginate. *Acta Chem. Scand.* **1970**, *24*, 813–833. [CrossRef]
43. Grant, G.T.; Morris, E.R.; Rees, D.A.; Smith, P.J.C.; Thom, D. Biological interactions between polysaccharides and divalent cations: The egg-box model. *FEBS Lett.* **1973**, *32*, 195–198. [CrossRef]
44. Chan, L.W.; Jin, Y.; Heng, P.W.S. Cross-linking mechanisms of calcium and zinc in production of alginate microspheres. *Int. J. Pharm.* **2002**, *242*, 255–258. [CrossRef] [PubMed]
45. Mollah, M.Z.I.; Zahid, H.M.; Mahal, Z.; Faruque, M.R.I.; Khandaker, M.U. The Usages and Potential Uses of Alginate for Healthcare Applications. *Front. Mol. Biosci.* **2021**, *8*, 719972. [CrossRef] [PubMed]
46. Zhao, Y.; Gao, S.; Zhao, S.; Li, Y.; Cheng, L.; Li, J.; Yin, Y. Synthesis and characterization of disulfide-crosslinked alginate hydrogel scaffolds. *Mater. Sci. Eng. C* **2012**, *32*, 2153–2162. [CrossRef]
47. Radhakrishnan, J.; Subramanian, A.; Krishnan, U.M.; Sethuraman, S. Injectable and 3D Bioprinted Polysaccharide Hydrogels: From Cartilage to Osteochondral Tissue Engineering. *Biomacromolecules* **2017**, *18*, 1–26. [CrossRef]
48. Tønnesen, H.H.; Karlsen, J. Alginate in Drug Delivery Systems. *Drug Dev. Ind. Pharm.* **2002**, *28*, 621–630. [CrossRef] [PubMed]

49. Wright, B.; De Bank, P.A.; Luetchford, K.A.; Acosta, F.R.; Connon, C.J. Oxidized alginate hydrogels as niche environments for corneal epithelial cells. *J. Biomed. Mater. Res. Part A* **2013**, *102*, 3393–3400. [CrossRef]
50. Zhou, H.; Xu, H.H.K. The fast release of stem cells from alginate-fibrin microbeads in injectable scaffolds for bone tissue engineering. *Biomaterials* **2011**, *32*, 7503–7513. [CrossRef]
51. Baniasadi, H.; Mashayekhan, S.; Fadaoddini, S.; Haghirsharifzamini, Y. Design, fabrication and characterization of oxidized alginate–gelatin hydrogels for muscle tissue engineering applications. *J. Biomater. Appl.* **2016**, *31*, 152–161. [CrossRef]
52. Sarker, B.; Singh, R.; Silva, R.; Roether, J.A.; Kaschta, J.; Detsch, R.; Schubert, D.W.; Cicha, I.; Boccaccini, A.R. Evaluation of Fibroblasts Adhesion and Proliferation on Alginate-Gelatin Crosslinked Hydrogel. *PLoS ONE* **2014**, *9*, e107952. [CrossRef]
53. Lee, J.-H.; Kim, H.-W. Emerging properties of hydrogels in tissue engineering. *J. Tissue Eng.* **2018**, *9*, 204173141876828. [CrossRef]
54. Kuo, C.K.; Ma, P.X. Ionically crosslinked alginate hydrogels as scaffolds for tissue engineering: Part 1. Structure, gelation rate and mechanical properties. *Biomaterials* **2001**, *22*, 511–521. [CrossRef] [PubMed]
55. Rodriguez-Lorenzo, L.M.; Saldaña, L.; Benito-Garzón, L.; García-Carrodeguas, R.; de Aza, S.; Vilaboa, N.; Román, J.S. Feasibility of ceramic-polymer composite cryogels as scaffolds for bone tissue engineering. *J. Tissue Eng. Regen. Med.* **2011**, *6*, 421–433. [CrossRef] [PubMed]
56. Sánchez-Téllez, D.; Téllez-Jurado, L.; Rodríguez-Lorenzo, L. Hydrogels for Cartilage Regeneration, from Polysaccharides to Hybrids. *Polymers* **2017**, *9*, 671. [CrossRef] [PubMed]
57. El-Rashidy, A.A.; Roether, J.A.; Harhaus, L.; Kneser, U.; Boccaccini, A.R. Regenerating bone with bioactive glass scaffolds: A review of in vivo studies in bone defect models. *Acta Biomater.* **2017**, *62*, 1–28. [CrossRef] [PubMed]
58. Sarker, B.; Li, W.; Zheng, K.; Detsch, R.; Boccaccini, A.R. Designing Porous Bone Tissue Engineering Scaffolds with Enhanced Mechanical Properties from Composite Hydrogels Composed of Modified Alginate, Gelatin, and Bioactive Glass. *ACS Biomater. Sci. Eng.* **2016**, *2*, 2240–2254. [CrossRef] [PubMed]
59. Fedorovich, N.E.; Alblas, J.; Hennink, W.E.; Öner, F.C.; Dhert, W.J.A. Organ printing: The future of bone regeneration? *Trends Biotechnol.* **2011**, *29*, 601–606. [CrossRef]
60. Lin, H.-R.; Yeh, Y.-J. Porous alginate/hydroxyapatite composite scaffolds for bone tissue engineering: Preparation, characterization, andin vitro studies. *J. Biomed. Mater. Res.* **2004**, *71B*, 52–65. [CrossRef] [PubMed]
61. Sarker, A.; Amirian, J.; Min, Y.K.; Lee, B.T. HAp granules encapsulated oxidized alginate–gelatin–biphasic calcium phosphate hydrogel for bone regeneration. *Int. J. Biol. Macromol.* **2015**, *81*, 898–911. [CrossRef] [PubMed]
62. Perez, R.A.; Kim, H.-W. Core-shell designed scaffolds of alginate/alpha-tricalcium phosphate for the loading and delivery of biological proteins. *J. Biomed. Mater. Res. Part A* **2012**, *101A*, 1103–1112. [CrossRef]
63. Ishikawa, K.; Takagi, S.; Chow, L.C.; Ishikawa, Y. Properties and mechanisms of fast-setting calcium phosphate cements. *J. Mater. Sci. Mater. Med.* **1995**, *6*, 528–533. [CrossRef]
64. Lee, G.-S.; Park, J.-H.; Won, J.-E.; Shin, U.S.; Kim, H.-W. Alginate combined calcium phosphate cements: Mechanical properties and in vitro rat bone marrow stromal cell responses. *J. Mater. Sci. Mater. Med.* **2011**, *22*, 1257–1268. [CrossRef] [PubMed]
65. Derakhshanfar, S.; Mbeleck, R.; Xu, K.; Zhang, X.; Zhong, W.; Xing, M. 3D bioprinting for biomedical devices and tissue engineering: A review of recent trends and advances. *Bioact. Mater.* **2018**, *3*, 144–156. [CrossRef] [PubMed]
66. Ribeiro, A.; Blokzijl, M.M.; Levato, R.; Visser, C.W.; Castilho, M.; Hennink, W.E.; Vermonden, T.; Malda, J. Assessing bioink shape fidelity to aid material development in 3D bioprinting. *Biofabrication* **2017**, *10*, 014102. [CrossRef] [PubMed]
67. Paxton, N.; Smolan, W.; Böck, T.; Melchels, F.; Groll, J.; Jungst, T. Proposal to assess printability of bioinks for extrusion-based bioprinting and evaluation of rheological properties governing bioprintability. *Biofabrication* **2017**, *9*, 044107. [CrossRef] [PubMed]
68. Chung, J.H.Y.; Naficy, S.; Yue, Z.; Kapsa, R.; Quigley, A.; Moulton, S.E.; Wallace, G.G. Bio-ink properties and printability for extrusion printing living cells. *Biomater. Sci.* **2013**, *1*, 763–773. [CrossRef] [PubMed]
69. Freeman, F.E.; Kelly, D.J. Tuning Alginate Bioink Stiffness and Composition for Controlled Growth Factor Delivery and to Spatially Direct MSC Fate within Bioprinted Tissues. *Sci. Rep.* **2017**, *7*, 17042. [CrossRef]
70. Park, J.; Lee, S.J.; Lee, H.; Park, S.A.; Lee, J.Y. Three dimensional cell printing with sulfated alginate for improved bone morphogenetic protein-2 delivery and osteogenesis in bone tissue engineering. *Carbohydr. Polym.* **2018**, *196*, 217–224. [CrossRef] [PubMed]
71. Izadifar, M.; Babyn, P.; Kelly, M.E.; Chapman, D.; Chen, X. Bioprinting Pattern-Dependent Electrical/Mechanical Behavior of Cardiac Alginate Implants: Characterization and Ex Vivo Phase-Contrast Microtomography Assessment. *Tissue Eng. Part C Methods* **2017**, *23*, 548–564. [CrossRef] [PubMed]
72. Giuseppe, M.D.; Law, N.; Webb, B.; Macrae, R.A.; Liew, L.J.; Sercombe, T.B.; Dilley, R.J.; Doyle, B.J. Mechanical behaviour of alginate-gelatine hydrogels for 3D bioprinting. *J. Mech. Behav. Biomed. Mater.* **2018**, *79*, 150–157. [CrossRef]
73. Palmer, I.; Nelson, J.; Wolfgang Schatton Dunne, N.J.; Buchanan, F.; Clarke, S.A. Biocompatibility of calcium phosphate bone cement with optimised mechanical properties: An in vivo study. *J. Mater. Sci. Mater. Med.* **2016**, *27*, 191. [CrossRef]
74. Busch, A.; Jäger, M.; Mayer, C.; Sowislok, A. Functionalization of Synthetic Bone Substitutes. *Int. J. Mol. Sci.* **2021**, *22*, 4412. [CrossRef] [PubMed]
75. You, F.; Chen, X.; Cooper, D.M.L.; Chang, T.; Eames, B.F. Homogeneous hydroxyapatite/alginate composite hydrogel promotes calcified cartilage matrix deposition with potential for three-dimensional bioprinting. *Biofabrication* **2018**, *11*, 015015. [CrossRef] [PubMed]

76. Russo, L.; Taraballi, F.; Lupo, C.; Poveda, A.; Jiménez-Barbero, J.; Sandri, M.; Tampieri, A.; Nicotra, F.; Cipolla, L. Carbonate hydroxyapatite functionalization: A comparative study towards (bio)molecules fixation. *Interface Focus* **2014**, *4*, 20130040. [CrossRef] [PubMed]
77. An, Y.H.; Draughn, R.A. (Eds.) *Mechanical Testing of Bone and the Bone-Implant Interface*, 1st ed.; CRC Press: Boca Raton, FL, USA, 1999. [CrossRef]
78. Pautke, C.; Schieker, M.; Tischer, T.; Kolk, A.; Neth, P.; Mutschler, W.; Milz, S. Characterization of osteosarcoma cell lines MG-63, Saos-2 and U-2 OS in comparison to human osteoblasts. *Anticancer Res.* **2004**, *24*, 3743–3748. [PubMed]
79. Jouyandeh, M.; Vahabi, H.; Rabiee, N.; Rabiee, M.; Bagherzadeh, M.; Saeb, M.R. Green composites in bone tissue engineering. *Emergent Mater.* **2021**, *5*, 603–620. [CrossRef]
80. Pesode, P.; Barve, S.; Wankhede, S.V.; Ahmad, A. Sustainable Materials and Technologies for Biomedical Applications. *Adv. Mater. Sci. Eng.* **2023**, *2023*, 6682892. [CrossRef]
81. Lacroix, J.; Lao, J.; Jallot, E. Green and safe in situ templating of bioactive glass scaffolds for bone tissue engineering. *J. Mater. Chem. B* **2013**, *1*, 1782. [CrossRef] [PubMed]
82. Li, J.; Li, W.; Kong, M.; Li, Z.; Yang, T.; Wang, Q.; Teng, W. Self-healing hybrid hydrogels with sustained bioactive components release for guided bone regeneration. *J. Nanobiotechnology* **2023**, *21*, 62. [CrossRef] [PubMed]
83. Jia, Z.; Deng, Z.; Li, L. Biomineralized Materials as Model Systems for Structural Composites: 3D Architecture. *Adv. Mater.* **2022**, *34*, 2106259. [CrossRef]
84. Beatty, D.N.; Williams, S.L.; Srubar, W.V. Biomineralized Materials for Sustainable and Durable Construction. *Annu. Rev. Mater. Res.* **2022**, *52*, 411–439. [CrossRef]
85. Reinhardt, O.; Ihmann, S.; Ahlhelm, M.; Gelinsky, M. 3D bioprinting of mineralizing cyanobacteria as novel approach for the fabrication of living building materials. *Front. Bioeng. Biotechnol.* **2023**, *11*, 1145177. [CrossRef] [PubMed]
86. Tan, H.; Marra, K.G. Injectable, Biodegradable Hydrogels for Tissue Engineering Applications. *Materials* **2010**, *3*, 1746–1767. [CrossRef]
87. Tran, N.Q.; Joung, Y.K.; Lih, E.; Park, K.M.; Park, K.D. Supramolecular Hydrogels Exhibiting Fast In Situ Gel Forming and Adjustable Degradation Properties. *Biomacromolecules* **2010**, *11*, 617–625. [CrossRef] [PubMed]
88. Taaca, L.M.; Prieto, E.; Vasquez, M.R. Current Trends in Biomedical Hydrogels: From Traditional Crosslinking to Plasma-Assisted Synthesis. *Polymers* **2022**, *14*, 2560. [CrossRef] [PubMed]
89. Hoare, T.R.; Kohane, D.S. Hydrogels in drug delivery: Progress and challenges. *Polymer* **2008**, *49*, 1993–2007. [CrossRef]
90. Ponticiello, M.S.; Schinagl, R.M.; Kadiyala, S.; Barry, F. Gelatin-based resorbable sponge as a carrier matrix for human mesenchymal stem cells in cartilage regeneration therapy. *J. Biomed. Mater. Res.* **2000**, *52*, 246–255. [CrossRef] [PubMed]
91. Pesáková, V.; Stol, M.; Adam, M. Comparison of the influence of gelatine and collagen substrates on growth of chondrocytes. *Folia Biol.* **1990**, *36*, 264–270.
92. Kim, A.Y.; Kim, Y.; Lee, S.H.; Yoon, Y.; Kim, W.-H.; Kweon, O.-K. Effect of Gelatin on Osteogenic Cell Sheet Formation Using Canine Adipose-Derived Mesenchymal Stem Cells. *Cell Transplant.* **2017**, *26*, 115–123. [CrossRef] [PubMed]
93. Liao, H.; Zhang, H.; Chen, W. Differential physical, rheological, and biological properties of rapid in situ gelable hydrogels composed of oxidized alginate and gelatin derived from marine or porcine sources. *J. Mater. Sci. Mater. Med.* **2009**, *20*, 1263–1271. [CrossRef]
94. Boguń, M.; Rabiej, S. The influence of fiber formation conditions on the structure and properties of nanocomposite alginate fibers containing tricalcium phosphate or montmorillonite. *Polym. Compos.* **2009**, *31*, 1321–1331. [CrossRef]
95. Sakai, S.; Yamaguchi, S.; Takei, T.; Kawakami, K. Oxidized Alginate-Cross-Linked Alginate/Gelatin Hydrogel Fibers for Fabricating Tubular Constructs with Layered Smooth Muscle Cells and Endothelial Cells in Collagen Gels. *Biomacromolecules* **2008**, *9*, 2036–2041. [CrossRef] [PubMed]
96. Kong, H. Designing alginate hydrogels to maintain viability of immobilized cells. *Biomaterials* **2003**, *24*, 4023–4029. [CrossRef] [PubMed]
97. Boontheekul, T.; Kong, H.-J.; Mooney, D.J. Controlling alginate gel degradation utilizing partial oxidation and bimodal molecular weight distribution. *Biomaterials* **2005**, *26*, 2455–2465. [CrossRef] [PubMed]
98. Kong, H.J.; Kaigler, D.; Kim, K.; Mooney, D.J. Controlling Rigidity and Degradation of Alginate Hydrogels via Molecular Weight Distribution. *Biomacromolecules* **2004**, *5*, 1720–1727. [CrossRef] [PubMed]
99. Boanini, E.; Rubini, K.; Panzavolta, S.; Bigi, A. Chemico-physical characterization of gelatin films modified with oxidized alginate. *Acta Biomater.* **2010**, *6*, 383–388. [CrossRef] [PubMed]
100. Łabowska, M.B.; Cierluk, K.; Jankowska, A.M.; Kulbacka, J.; Detyna, J.; Michalak, I. A Review on the Adaption of Alginate-Gelatin Hydrogels for 3D Cultures and Bioprinting. *Materials* **2021**, *14*, 858. [CrossRef]
101. Tareq, A.; Hussein, M. Sodium Alginate-Gelatin Cross-Linked Microspheres for Releasing Diltiazem Hcl. *Sci. J. Univ. Zakho* **2016**, *4*, 226–235. [CrossRef]
102. Xu, H.H.; Wang, P.; Wang, L.; Bao, C.; Chen, Q.; Weir, M.D.; Chow, L.C.; Zhao, L.; Zhou, X.; Reynolds, M.A. Calcium phosphate cements for bone engineering and their biological properties. *Bone Res.* **2017**, *5*, 17056. [CrossRef] [PubMed]

103. Fernández, E.; Vlad, M.D.; Gel, M.M.; López, J.; Torres, R.; Cauich, J.V.; Bohner, M. Modulation of porosity in apatitic cements by the use of α-tricalcium phosphate—Calcium sulphate dihydrate mixtures. *Biomaterials* **2005**, *26*, 3395–3404. [CrossRef]
104. del Real, R.P.; Wolke, J.G.C.; Vallet-Regí, M.; Jansen, J.A. A new method to produce macropores in calcium phosphate cements. *Biomaterials* **2002**, *23*, 3673–3680. [CrossRef]
105. An, J.; Wolke, J.G.; Jansen, J.A.; Leeuwenburgh, S.C. Influence of polymeric additives on the cohesion and mechanical properties of calcium phosphate cements. *J. Mater. Sci. Mater. Med.* **2016**, *27*, 58. [CrossRef]
106. Hesaraki, S.; Zamanian, A.; Moztarzadeh, F. The influence of the acidic component of the gas-foaming porogen used in preparing an injectable porous calcium phosphate cement on its properties: Acetic acid versus citric acid. *J. Biomed. Mater. Res. Part B Appl. Biomater.* **2007**, *86B*, 208–216. [CrossRef]
107. Ginebra, M.P.; Driessens, F.C.M.; Planell, J.A. Effect of the particle size on the micro and nanostructural features of a calcium phosphate cement: A kinetic analysis. *Biomaterials* **2004**, *25*, 3453–3462. [CrossRef] [PubMed]
108. Félix Lanao, R.P.; Leeuwenburgh, S.C.G.; Wolke, J.G.C.; Jansen, J.A. Bone response to fast-degrading, injectable calcium phosphate cements containing PLGA microparticles. *Biomaterials* **2011**, *32*, 8839–8847. [CrossRef] [PubMed]
109. Ilie, A.; Ghiţulică, C.; Andronescu, E.; Cucuruz, A.; Ficai, A. New composite materials based on alginate and hydroxyapatite as potential carriers for ascorbic acid. *Int. J. Pharm.* **2016**, *510*, 501–507. [CrossRef] [PubMed]
110. Hillig, W.B.; Choi, Y.; Murtha, S.; Natravali, N.; Ajayan, P. An open-pored gelatin/hydroxyapatite composite as a potential bone substitute. *J. Mater. Sci. Mater. Med.* **2007**, *19*, 11–17. [CrossRef] [PubMed]
111. Devina, N.; Eriwati, Y.K.; Santosa, A.S. The purity and viscosity of sodium alginate extracted from Sargassum brown seaweed species as a basic ingredient in dental alginate impression material. *J. Phys. Conf. Ser.* **2018**, *1073*, 052012. [CrossRef]
112. Wang, J.; Liu, Y.; Zhang, X.; Rahman, S.E.; Su, S.; Wei, J.; Ning, F.; Hu, Z.; Martínez-Zaguilán, R.; Sennoune, S.R.; et al. 3D printed agar/ calcium alginate hydrogels with high shape fidelity and tailorable mechanical properties. *Polymer* **2021**, *214*, 123238. [CrossRef]
113. Kreller, T.; Distler, T.; Heid, S.; Gerth, S.; Detsch, R.; Boccaccini, A.R. Physico-chemical modification of gelatine for the improvement of 3D printability of oxidized alginate-gelatine hydrogels towards cartilage tissue engineering. *Mater. Des.* **2021**, *208*, 109877. [CrossRef]
114. Bociaga, D.; Bartniak, M.; Grabarczyk, J.; Przybyszewska, K. Sodium Alginate/Gelatine Hydrogels for Direct Bioprinting—The Effect of Composition Selection and Applied Solvents on the Bioink Properties. *Materials* **2019**, *12*, 2669. [CrossRef]
115. Zhang, N.; Zhao, D.; Liu, N.; Wu, Y.; Yang, J.; Wang, Y.; Xie, H.; Ji, Y.; Zhou, C.; Zhuang, J.; et al. Assessment of the degradation rates and effectiveness of different coated Mg-Zn-Ca alloy scaffolds for in vivo repair of critical-size bone defects. *J. Mater. Sci. Mater. Med.* **2018**, *29*, 138. [CrossRef]
116. Kim, J.-A.; Lim, J.; Naren, R.; Yun, H.; Park, E.K. Effect of the biodegradation rate controlled by pore structures in magnesium phosphate ceramic scaffolds on bone tissue regeneration in vivo. *Acta Biomater.* **2016**, *44*, 155–167. [CrossRef] [PubMed]
117. Rodríguez-Rego, J.M.; Mendoza-Cerezo, L.; Macías-García, A.; Carrasco-Amador, J.P.; Marcos-Romero, A.C. Methodology for characterizing the printability of hydrogels. *Int. J. Bioprinting* **2023**, *9*, 667. [CrossRef] [PubMed]
118. Pettignano, A.; Häring, M.; Bernardi, L.; Tanchoux, N.; Quignard, F.; Díaz, D.D. Self-healing alginate–gelatin biohydrogels based on dynamic covalent chemistry: Elucidation of key parameters. *Mater. Chem. Front.* **2017**, *1*, 73–79. [CrossRef]
119. Cohen, B.; Panker, M.; Zuckerman, E.; Foox, M.; Zilberman, M. Effect of calcium phosphate-based fillers on the structure and bonding strength of novel gelatin–alginate bioadhesives. *J. Biomater. Appl.* **2013**, *28*, 1366–1375. [CrossRef] [PubMed]
120. Fu, P.-S.; Wang, J.-C.; Lai, P.-L.; Liu, S.-M.; Chen, Y.-S.; Chen, W.-C.; Hung, C.-C. Biodegradable Hydrogel Beads Combined with Calcium Phosphate Bone Cement for Bone Repair: In Vitro and In Vivo Characterization. *Polymers* **2022**, *14*, 505. [CrossRef] [PubMed]
121. Wang, S.; Wang, X.; Neufurth, M.; Tolba, E.; Schepler, H.; Xiao, S.; Schröder, H.C.; Müller, W.E.G. Biomimetic Alginate/Gelatin Cross-Linked Hydrogels Supplemented with Polyphosphate for Wound Healing Applications. *Molecules* **2020**, *25*, 5210. [CrossRef] [PubMed]
122. Gurikov, P.; Smirnova, I. Non-Conventional Methods for Gelation of Alginate. *Gels* **2018**, *4*, 14. [CrossRef] [PubMed]
123. Orshesh, Z.; Hesaraki, S.; Khanlarkhani, A. Blooming gelatin: An individual additive for enhancing nanoapatite precipitation, physical properties, and osteoblastic responses of nanostructured macroporous calcium phosphate bone cements. *Int. J. Nanomed.* **2017**, *12*, 745–758. [CrossRef]
124. Nguyen, T.-P.; Lee, B.-T. Fabrication of oxidized alginate-gelatin-BCP hydrogels and evaluation of the microstructure, material properties and biocompatibility for bone tissue regeneration. *J. Biomater. Appl.* **2011**, *27*, 311–321. [CrossRef]
125. Yao, Y.; Yang, Y.; Ye, Q.; Cao, S.; Zhang, X.; Zhao, K.; Jian, Y. Effects of pore size and porosity on cytocompatibility and osteogenic differentiation of porous titanium. *J. Mater. Sci. Mater. Med.* **2021**, *32*, 72. [CrossRef] [PubMed]
126. Herrada-Manchón, H.; Fernández, M.; Aguilar, E. Essential Guide to Hydrogel Rheology in Extrusion 3D Printing: How to Measure It and Why It Matters? *Gels* **2023**, *9*, 517. [CrossRef]
127. Maciel, B.R.; Baki, K.; Oelschlaeger, C.; Willenbacher, N. The Influence of Rheological and Wetting Properties of Hydrogel-Based Bio-Inks on Extrusion-Based Bioprinting. *Chem. Ing. Tech.* **2022**, *94*, 393–401. [CrossRef]
128. Anwar, F.; Amin, K.F.; Hoque, M.E. Tools and techniques for characterizing sustainable hydrogels. In *Sustainable Hydrogels*; Elsevier: Amsterdam, The Netherlands, 2023; pp. 47–77. [CrossRef]

129. Sen, K.S.; Duarte Campos, D.F.; Köpf, M.; Blaeser, A.; Fischer, H. The Effect of Addition of Calcium Phosphate Particles to Hydrogel-Based Composite Materials on Stiffness and Differentiation of Mesenchymal Stromal Cells toward Osteogenesis. *Adv. Healthc. Mater.* **2018**, *7*, 1800343. [CrossRef] [PubMed]
130. Nagle, D.G.; Ferreira, D.; Zhou, Y.-D. Epigallocatechin-3-gallate (EGCG): Chemical and biomedical perspectives. *Phytochemistry* **2006**, *67*, 1849–1855. [CrossRef] [PubMed]
131. Du, G.-J.; Zhang, Z.; Wen, X.-D.; Yu, C.; Calway, T.; Yuan, C.-S.; Wang, C.-Z. Epigallocatechin Gallate (EGCG) is the most effective cancer chemopreventive polyphenol in green tea. *Nutrients* **2012**, *4*, 1679–1691. [CrossRef] [PubMed]
132. Chu, C.; Deng, J.; Man, Y.; Qu, Y. Green Tea Extracts Epigallocatechin-3-gallate for Different Treatments. *BioMed Res. Int.* **2017**, *2017*, 5615647. [CrossRef]
133. Xu, Z.; Liu, X.; Feng, C.; Zhu, S.; Li, Y.; Guo, Z.; Chen, X. Disinfection properties of the tea polyphenol epigallocatechin gallate in the presence of calcium ions. *AQUA—Water Infrastruct. Ecosyst. Soc.* **2021**, *70*, 731–740. [CrossRef]
134. Gao, B.; Honda, Y.; Yamada, Y.; Tanaka, T.; Takeda, Y.; Nambu, T.; Baba, S. Utility of Thermal Cross-Linking in Stabilizing Hydrogels with Beta-Tricalcium Phosphate and/or Epigallocatechin Gallate for Use in Bone Regeneration Therapy. *Polymers* **2021**, *14*, 40. [CrossRef]
135. Dai, D.; Wang, J.; Xie, H.; Zhang, C. An epigallocatechin gallate–amorphous calcium phosphate nanocomposite for caries prevention and demineralized enamel restoration. *Mater. Today Bio* **2023**, *21*, 100715. [CrossRef]
136. Pietryga, K.; Panaite, A.; Pamuła, E. Composite scaffolds enriched with calcium carbonate microparticles loaded with epigallocatechin gallate for bone tissue regeneration. *Eng. Biomater.* **2022**, *25*, 12–21. [CrossRef]
137. Jo, Y.; Sarkar, N.; Bose, S. In vitrobiological evaluation of epigallocatechin gallate (EGCG) release from three-dimensional printed (3DP) calcium phosphate bone scaffolds. *J. Mater. Chem. B* **2023**, *11*, 5503–5513. [CrossRef] [PubMed]
138. Liu, H.; Jiao, Z.; Guo, S. Effect of Epigallocatechin Gallate on the Properties of Gelatin. *Int. J. Food Prop.* **2014**, *17*, 2119–2130. [CrossRef]
139. Dou, L.; Li, B.; Zhang, K.; Chu, X.; Hou, H. Physical properties and antioxidant activity of gelatin-sodium alginate edible films with tea polyphenols. *Int. J. Biol. Macromol.* **2018**, *118*, 1377–1383. [CrossRef] [PubMed]
140. Vetsch, J.R.; Paulsen, S.J.; Müller, R.; Hofmann, S. Effect of fetal bovine serum on mineralization in silk fibroin scaffolds. *Acta Biomater.* **2015**, *13*, 277–285. [CrossRef] [PubMed]
141. Villasante, A.; Robinson, S.T.; Cohen, A.R.; Lock, R.; Guo, X.E.; Vunjak-Novakovic, G. Human Serum Enhances Biomimicry of Engineered Tissue Models of Bone and Cancer. *Front. Bioeng. Biotechnol.* **2021**, *9*, 658472. [CrossRef] [PubMed]
142. Yilmaz, B.; Pazarceviren, A.E.; Tezcaner, A.; Evis, Z. Historical development of simulated body fluids used in biomedical applications: A review. *Microchem. J.* **2020**, *155*, 104713. [CrossRef]
143. van der Valk, J. Fetal bovine serum (FBS): Past-present-future. *ALTEX* **2018**, *35*, 99–118. [CrossRef]
144. Sangeetha, K.; Thamizhavel, A.; Girija, E.K. Effect of gelatin on the in situ formation of Alginate/Hydroxyapatite nanocomposite. *Mater. Lett.* **2013**, *91*, 27–30. [CrossRef]
145. Tomić, S.L.; Nikodinović-Runić, J.; Vukomanović, M.; Babić, M.M.; Vuković, J.S. Novel Hydrogel Scaffolds Based on Alginate, Gelatin, 2-Hydroxyethyl Methacrylate, and Hydroxyapatite. *Polymers* **2021**, *13*, 932. [CrossRef]
146. Pham Minh, D.; Rio, S.; Sharrock, P.; Sebei, H.; Lyczko, N.; Tran, N.D.; Raii, M.; Nzihou, A. Hydroxyapatite starting from calcium carbonate and orthophosphoric acid: Synthesis, characterization, and applications. *J. Mater. Sci.* **2014**, *49*, 4261–4269. [CrossRef]
147. Sans, J.; Arnau, M.; Sanz, V.; Turon, P.; Alemán, C. Hydroxyapatite-based biphasic catalysts with plasticity properties and its potential in carbon dioxide fixation. *Chem. Eng. J.* **2022**, *433*, 133512. [CrossRef]
148. Kameo, Y.; Sakano, N.; Adachi, T. Theoretical concept of cortical to cancellous bone transformation. *Bone Rep.* **2020**, *12*, 100260. [CrossRef] [PubMed]
149. Fan, Z.; Liu, H.; Ding, Z.; Xiao, L.; Qiang, L.ü.; Kaplan, D.L. Simulation of Cortical and Cancellous Bone to Accelerate Tissue Regeneration. *Adv. Funct. Mater.* **2023**, *33*, 2301839. [CrossRef] [PubMed]
150. Mohd Pu'ad, N.A.S.; Koshy, P.; Abdullah, H.Z.; Idris, M.I.; Lee, T.C. Syntheses of hydroxyapatite from natural sources. *Heliyon* **2019**, *5*, e01588. [CrossRef] [PubMed]
151. Sawada, M.; Sridhar, K.; Kanda, Y.; Yamanaka, S. Pure hydroxyapatite synthesis originating from amorphous calcium carbonate. *Sci. Rep.* **2021**, *11*, 11546. [CrossRef] [PubMed]
152. Mohd Pu'ad, N.A.S.; Alipal, J.; Abdullah, H.Z.; Idris, M.I.; Lee, T.C. Synthesis of eggshell derived hydroxyapatite via chemical precipitation and calcination method. *Mater. Today Proc.* **2021**, *42*, 172–177. [CrossRef]
153. Javadinejad, H.R.; Ebrahimi-Kahrizsangi, R. Thermal and kinetic study of hydroxyapatite formation by solid-state reaction. *Int. J. Chem. Kinet.* **2020**, *53*, 583–595. [CrossRef]
154. Takagi, S.; Chow, L.C.; Ishikawa, K. Formation of hydroxyapatite in new calcium phosphate cements. *Biomaterials* **1998**, *19*, 1593–1599. [CrossRef]
155. Lotsari, A.; Rajasekharan, A.K.; Halvarsson, M.; Andersson, M. Transformation of amorphous calcium phosphate to bone-like apatite. *Nat. Commun.* **2018**, *9*, 4170. [CrossRef]
156. Wu, Y.X.; Choi, E.J.; Vu, A.A.; Jiang, P.; Ali, S.N.; Patel, R.M.; Landman, J.; Clayman, R.V. Comparison of Ureteral Stent Biomaterials: Encrustation Profile in Lithogenic Artificial Urine Models. *ACS Omega* **2023**, *8*, 29003–29011. [CrossRef] [PubMed]

157. Doostmohammadi, A.; Monshi, A.; Salehi, R.; Fathi, M.H.; Seyedjafari, E.; Shafiee, A.; Soleimani, M. Cytotoxicity evaluation of 63s bioactive glass and bone-derived hydroxyapatite particles using human bone-marrow stem cells. *Biomed. Pap. Fac. Med. Palacký Univ. Olomouc Czech Repub.* **2011**, *155*, 323–326. [CrossRef] [PubMed]
158. Branzoi, I.V.; Iordoc, M.; Branzoi, F.; Vasilescu-Mirea, R.; Sbarcea, G. Influence of diamond-like carbon coating on the corrosion resistance of the NITINOL shape memory alloy. *Surf. Interface Anal.* **2010**, *42*, 502–509. [CrossRef]
159. Lee, D.Y.; Lee, S.Y.; Yun, S.H.; Jeong, J.W.; Kim, J.H.; Kim, H.W.; Choi, J.S.; Kim, G.-D.; Joo, S.T.; Choi, I.; et al. Review of the Current Research on Fetal Bovine Serum and the Development of Cultured Meat. *Food Sci. Anim. Resour.* **2022**, *42*, 775–799. [CrossRef] [PubMed]
160. Murray, E.; Provvedini, D.; Curran, D.; Catherwood, B.; Sussman, H.; Manolagas, S. Characterization of a human osteoblastic osteosarcoma cell line (SAOS-2) with high bone alkaline phosphatase activity. *J. Bone Miner. Res.* **2009**, *2*, 231–238. [CrossRef] [PubMed]
161. Saldaña, L.; Bensiamar, F.; Bore, A.; Vilaboa, N. In search of representative models of human bone-forming cells for cytocompatibility studies. *Acta Biomater.* **2011**, *7*, 4210–4221. [CrossRef]
162. Rodan, S.B.; Imai, Y.; Thiede, M.A.; Wesolowski, G.; Thompson, D.; Bar-Shavit, Z.; Shull, S.; Mann, K.; Rodan, G.A. Characterization of a human osteosarcoma cell line (Saos-2) with osteoblastic properties. *Cancer Res.* **1987**, *47*, 4961–4966. [PubMed]
163. Fernandes, R.J.; Harkey, M.A.; Weis, M.; Askew, J.W.; Eyre, D.R. The post-translational phenotype of collagen synthesized by SAOS-2 osteosarcoma cells. *Bone* **2007**, *40*, 1343–1351. [CrossRef] [PubMed]
164. Bilbe, G.; Roberts, E.; Birch, M.; Evans, D.B. PCR phenotyping of cytokines, growth factors and their receptors and bone matrix proteins in human osteoblast-like cell lines. *Bone* **1996**, *19*, 437–445. [CrossRef]
165. Kolan, K.C.R.; Semon, J.A.; Bromet, B.; Day, D.E.; Leu, M.C. Bioprinting with human stem cells-laden alginate-gelatin bioink and bioactive glass for tissue engineering. *Int. J. Bioprinting* **2019**, *5*, 3. [CrossRef]
166. Neufurth, M.; Wang, X.; Schröder, H.C.; Feng, Q.; Diehl-Seifert, B.; Ziebart, T.; Steffen, R.; Wang, S.; Müller, W.E.G. Engineering a morphogenetically active hydrogel for bioprinting of bioartificial tissue derived from human osteoblast-like SaOS-2 cells. *Biomaterials* **2014**, *35*, 8810–8819. [CrossRef]
167. Hu, C.; He, S.; Lee, Y.J.; He, Y.; Kong, E.M.; Li, H.; Anastasio, M.A.; Popescu, G. Live-dead assay on unlabeled cells using phase imaging with computational specificity. *Nat. Commun.* **2022**, *13*, 713. [CrossRef]
168. Serbanescu, M. Evaluation of cytotoxic activity and anticancer potential of indigenous rosemary (*Rosmarinus officinalis* L.) and oregano (*Origanum vulgare* L.) dry extracts on mg-63 bone osteosarcoma human cell line. *Rom. J. Morphol. Embryol.* **2021**, *62*, 525.
169. Měřička, P.; Janoušek, L.; Benda, A.; Lainková, R.; Sabó, J.; Dalecká, M.; Prokšová, P.; Salmay, M.; Špunda, R.; Pecha, O.; et al. Cell Viability Assessment Using Fluorescence Vital Dyes and Confocal Microscopy in Evaluating Freezing and Thawing Protocols Used in Cryopreservation of Allogeneic Venous Grafts. *Int. J. Mol. Sci.* **2021**, *22*, 10653. [CrossRef] [PubMed]
170. Müller, W.E.G.; Schröder, H.C.; Tolba, E.; Diehl-Seifert, B.; Wang, X. Mineralization of bone-related SaOS-2 cells under physiological hypoxic conditions. *FEBS J.* **2015**, *283*, 74–87. [CrossRef] [PubMed]
171. Sanginario, V.; Ginebra, M.P.; Tanner, K.E.; Planell, J.A.; Ambrosio, L. Biodegradable and semi-biodegradable composite hydrogels as bone substitutes: Morphology and mechanical characterization. *J. Mater. Sci. Mater. Med.* **2006**, *17*, 447–454. [CrossRef] [PubMed]
172. Yang, X.; Yang, K.; Yu, F.; Chen, X.; Wu, S.; Zhu, Z. Preparation of novel bilayer hydrogels by combination of irradiation and freeze-thawing and their physical and biological properties. *Polym. Int.* **2009**, *58*, 1291–1298. [CrossRef]
173. Xu, N.; Xu, J.; Zheng, X.; Hui, J. Preparation of Injectable Composite Hydrogels by Blending Poloxamers with Calcium Carbonate-Crosslinked Sodium Alginate. *ChemistryOpen* **2020**, *9*, 451–458. [CrossRef]

processes

MDPI

Article

Numerical Analysis on Mechanical Properties of 3D Five-Directional Circular Braided Composites

Weiliang Zhang [1], Chunlei Li [1,*], Liang Li [1], Wei Wang [2], Lei Yang [2], Chaohang Zhang [1] and Xiyue Zhang [1]

1 Institute of Mechanical Engineering, Baoji University of Arts and Sciences, Baoji 721016, China; zhangweiliang-243@163.com (W.Z.); liliang@bjwlxy.edu.cn (L.L.)
2 School of Mechanical and Precision Instrument Engineering, Xi'an University of Technology, Xi'an 710048, China
* Correspondence: 602lcl-602lcl@163.com; Tel.: +86-1589-130-1856

Abstract: Based on the analysis of the motion law of 3D five-directional circular transverse braided fibers, this paper obtains the angle calculation formula between fibers and the local polar coordinate system in various cell models by transforming the position coordinates of fiber nodes. The stress transformation matrix between the local coordinate system and the global coordinate system of any fiber in the circular braided single cell is derived without considering the physical force on the single-cell micro-hexahedron unit. The calculation formulas of braided parameters such as the overall stiffness matrix and fiber volume content of the circular braided composite material after considering the matrix are derived by using the volume average method; the length of braided knuckles is 2 mm, the inner diameter of inner cells is 7 mm, the number of radial and axial braided yarns is 80, the height of inner cells is 0.5 mm, and the filling coefficient is 0.61. Comparing the results of the numerical prediction model with the experimental results in reference, it is found that the error of the numerical prediction model deduced in this paper is small. Therefore, this model can be used to fully study the effects of braided parameters such as cell inner diameter, cell height, and node length on the mechanical properties of composites.

Keywords: circular braided; stress transformation matrix; stiffness matrix; braided parameters; mechanical properties of composites

1. Introduction

The 3D braided composites have a specific geometric position relationship between braided fibers and matrix by pre-setting the braiding mode of fibers, so as to ensure that the materials have excellent mechanical properties in a specific direction and show good performance with designable characteristics. In recent years, with the improvement in braiding technology and the reduction in manufacturing cost, 3D braided composites have been widely used in aviation, aerospace, navigation, transportation, medical care, sports, and civil engineering [1–3].

In order to obtain 3D braided composites with different structures, the braiding methods can be divided into rectangular braiding and circular braiding according to the structural differences of preforms. In the aspect of reasonable construction of rectangular braided material model, Tian et al. proposed a diversity-oriented particle swarm optimization (DPSO-MLS) with multilevel learning strategy, which can be used to optimize the composite structure [4]. Zhang and Xu, based on the analysis of braiding laws of 3D five-directional and all five-directional braided composites, divided the studied materials into three kinds of cell models: inner, surface, and corner. By introducing reasonable

 https://doi.org/10.3390/pr13030800

boundary conditions, they predicted the effective stiffness characteristics of the materials, and compared the detailed mechanical properties of the three cell models. In addition, the effects of braiding angle and fiber volume fraction on the elastic properties of the composites were also analyzed [5–7]. Inspired by the slub fiber structure, Wang et al. designed and prepared a new type of three-dimensional six-way braided composite pipe [8].

In the aspect of mechanical properties of rectangular braiding, Ji et al. [9] analyzed the geometric model of composite materials with the help of finite element analysis software Abaqus 15.0, and obtained the elastic modulus, shear modulus, and Poisson's ratio of composite materials corresponding to different braiding angles. The influence of fiber volume fraction on the elastic properties of composite element is also studied, and the bending properties of braided composite material are evaluated by the multi-scale model in finite element analysis. The experimental results show that the predicted results are basically consistent with the experimental results. In order to study the progressive damage behavior and strength characteristics under uniaxial tension, Kang et al. established a finite element and progressive damage analysis model, and compared the finite element analysis results with experimental data [10]. Sun et al. [11] studied the effects of fiber fracture defects and corrugation defects on compressive fatigue behavior and damage evolution process of three-dimensional multiaxial braided composites.

The above research focuses on rectangular braiding. Compared with rectangular braiding, the micro-geometric structure of circular braiding is more complicated than that of rectangular preform, and it is more difficult to analyze its structure and mechanical properties theoretically [12–14]. Zhao and Li et al. [15,16] selected a micro-geometric cell, and found that the curvature of annular pipe has great influence on braided composites. The elastic modulus of braided composites of annular pipe was predicted by mathematical models, and the predicted results were found to be reasonable by comparing with the literature data. Shen [17] tested the compression and torsion mechanical properties of 3D, 4D, and 5D braided tubular members through experiments, and found that the compression properties of 5D materials were greatly improved compared with 4D materials, but the shear modulus was not significantly improved. Comparing the experimental data with the prediction data of FGM model, it was concluded that the prediction results of FGM simulation were lower. Zhang [18] applied braided composite materials to two-stage gear transmission system, realizing the design goal of lightweight gear transmission system.

As can be seen from the above literature, most of the research on braided composites is focused on rectangular braiding and circular braiding, but there are few research papers on the prediction of mechanical properties of 3D five-way circular braiding. Because of the anisotropy of mechanical properties of 3D five-way circular braided composites, the research in this paper provides the possibility for the special performance requirements of materials, such as the application of composites subjected to radial force, which is of great value.

The paper is divided into seven sections. In Section 2, the motion law of 3D five-directional circular braiding fiber is analyzed, and the composite material is divided into three microscopic cells. In Section 3, the formulas for calculating the overall stiffness matrix and fiber volume content of 3D five-directional circular braided composites are derived, and the mechanical properties of the composites can be obtained by inverting the overall stiffness matrix of the composites. In Section 4, the prediction model derived in this paper is used to further numerically analyze the mechanical properties of cells. In Section 5, there is numerical analysis of the influence of braiding parameters on the mechanical properties of cells. In Section 6, on the basis of the above analysis, the parametric study is completed. In Section 7, we state the main conclusions. In this paper, the numerical prediction model for mechanical properties of 3D five-directional circular braided composites fully considers the braiding parameters such as cell inner diameter, cell outer diameter, cell height, and knuckle length, and can quantitatively analyze the influence of different braiding parameters on the mechanical properties of composites.

2. The 3D Circular Braided Three-Cell Structure Analysis

The micro-structure of 3D annular braided preform should have good periodicity. In order to improve the modeling efficiency, according to Figure 1, the smallest research unit of 3D five-way annular braided composites is divided into three forms: internal cell, upper surface cell, and lower surface cell [19,20].

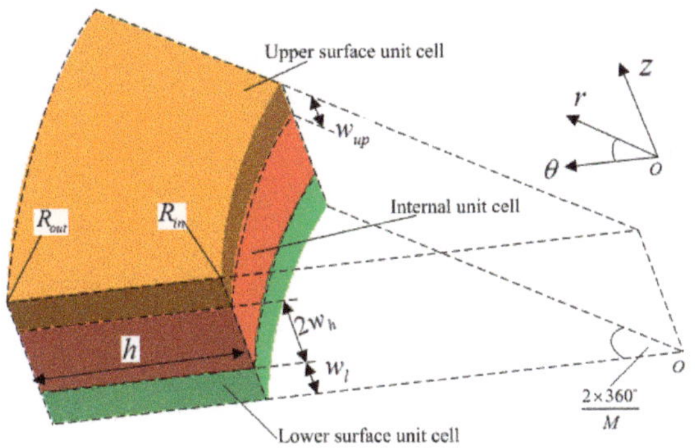

Figure 1. Three cell divisions.

As shown in Figure 2, the inner cell can be further subdivided into four sub-regions: A, B, C, and D, which extend periodically outward along the radial braiding direction. In Figure 2, h is the length of braiding knuckles, R_{in} is the inner diameter and R_{out} is the outer diameter of the inner cell, M and N are the number of radial and axial braiding fibers, respectively, and w_h is the height of the inner cell.

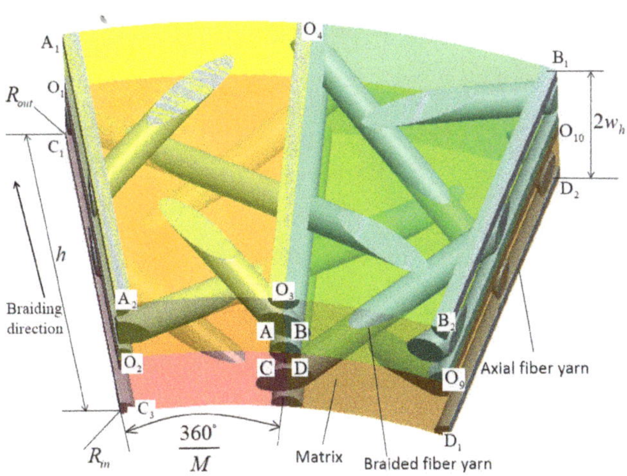

Figure 2. Internal unit cell matrix and fiber.

3. The Solving the Stress Transformation Matrix of Circular Braiding

In the process of 3D five-directional circular braiding, the fiber movement is carried out in the $\theta - r - z$ polar coordinate system in Figure 3, and the braiding angle is calculated under the X-Y-Z global coordinate rectangular system in Figure 3. The polar coordinate system $\theta_1 - r_1 - z_1$ is located on the inclined section of MNQ, and the θ_1 axis is parallel to the outer normal of plane MNQ, and the coordinate axis r_1 and z_1 are on plane MNQ.

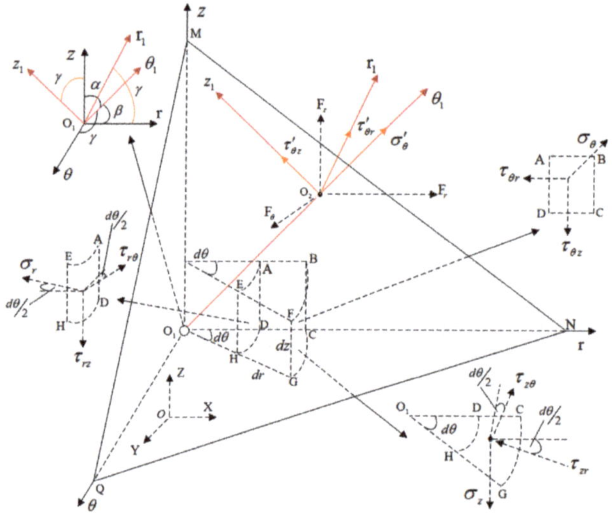

Figure 3. Three-direction stress on inclined section perpendicular to θ_1 axis.

In Figure 3, when M, N, and Q points approach infinitely to the origin O_1 of polar coordinate system $\theta - r - z$ along the opposite directions of z, r, and θ axes, respectively, the stress on the annular hexahedron element shown in Figure 3 is approximately equal to the stress on the inclined section MNQ.

Assuming that the force on the circular-shaped micro hexahedron element is not considered, the area of the inclined section $S_{MNQ} = 1$ will be projected on the three coordinate axes of the polar coordinate system $\theta - r - z$, and the relationship between the areas is obtained: $S_{MNO_1} \approx S_{ABCD} \approx \cos\gamma \times 1$, $S_{MQO_1} \approx S_{AEHD} \approx \cos\beta \times 1$, $S_{QNO_1} \approx S_{CDHG} \approx \cos\alpha \times 1$.

Assume that the total stress on the inclined section MNQ is F, and the components of F along the coordinate axis in polar coordinate system $\theta - r - z$ are expressed by, F_r and F_z respectively.

According to the equilibrium conditions $\sum F_\theta = 0$, $\sum F_r = 0$, and $\sum F_z = 0$, and in the annular hexahedron element in Figure 3, we can obtain:

$$\begin{cases} F_\theta = \sigma_\theta S_{ABCD} + \tau_{r\theta} S_{AEHD} \cos\frac{d\theta}{2} + \tau_{z\theta} S_{CDHG} \cos\frac{d\theta}{2} \\ F_r = \tau_{\theta r} S_{ABCD} + \sigma_r S_{AEHD} \cos\frac{d\theta}{2} + \tau_{zr} S_{CDHG} \cos\frac{d\theta}{2} \\ F_z = \tau_{\theta z} S_{ABCD} + \tau_{rz} S_{AEHD} + \sigma_z S_{CDHG} \end{cases} \quad (1)$$

The moment balance condition $\sum M_\theta = 0$, $\sum M_r = 0$, and $\sum M_z = 0$ are applied to the center of gravity coordinate axis of the annular hexahedron in Figure 3, we can obtain $\tau_{rz} = \tau_{zr}$, $\tau_{z\theta} = \tau_{\theta z}$ and $\tau_{\theta r} = \tau_{r\theta}$.

The force in Formula (1) is decomposed along the three coordinate axes of polar coordinate system $\theta_1 - r_1 - z_1$, and according to the angle relationship shown in Figure 3, the stress σ'_θ, $\tau'_{\theta r}$, and $\tau'_{\theta z}$ on the inclined section MNQ is:

$$\begin{cases} \sigma'_\theta = (F_\theta \cos\gamma + F_r \cos\beta + F_z \cos\alpha) \times S_{MNQ} \\ \tau'_{\theta r} = \left(F_\theta \cos^2\gamma \cos\beta + F_r \cos\gamma + F_z \cos\alpha \cos\gamma \cos\beta\right) \times S_{MNQ} \\ \tau'_{\theta z} = \left(F_\theta \cos^2\gamma \cos\alpha + F_r \cos\alpha \cos\gamma \cos\beta + F_z \cos\gamma\right) \times S_{MNQ} \end{cases} \quad (2)$$

$$\begin{cases} F_\theta = \sigma_\theta S_{abcd} + \tau_{r\theta} S_{aehd} \cos\frac{d\theta}{2} + \tau_{z\theta} S_{cdhg} \cos\frac{d\theta}{2} \\ F_r = \tau_{\theta r} S_{abcd} + \sigma_r S_{aehd} \cos\frac{d\theta}{2} + \tau_{zr} S_{cdhg} \cos\frac{d\theta}{2} \\ F_z = \tau_{z\theta} S_{abcd} + \tau_{rz} S_{aehd} + \sigma_z S_{cdhg} \end{cases} \tag{3}$$

In Figure 4, the z_1 axis is parallel to the outer normal of the plane mnq, δ is the angle between the axes z_1 and r; according to the angle relationship in Figure 4, we can obtain $\delta = \arccos[\cos\gamma \cos\alpha \cos\beta]$.

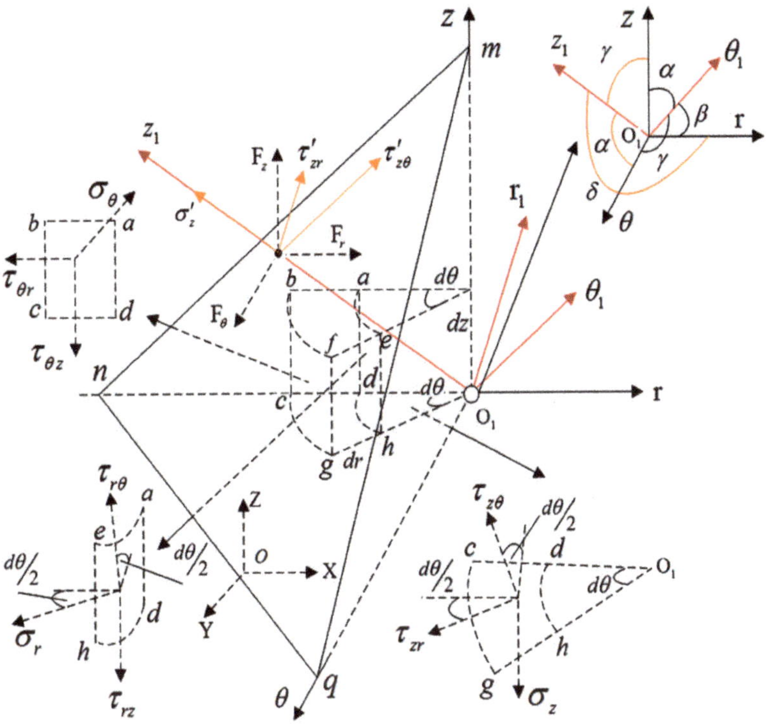

Figure 4. Three-direction stress on inclined section perpendicular to z_1 axis.

Let the area of inclined section mnq be $S_{mnq} = 1$, and project S_{mnq} on three coordinate axes plane of polar coordinate system $\theta - r - z$, and obtain the relationship between areas $S_{abcd} \approx \cos\alpha$, $S_{aehd} \approx \cos\alpha \cos\gamma \cos\beta$, $S_{cdhg} \approx \cos\gamma$. According to the equilibrium conditions in the annular hexahedron unit in Figure 4:

The force in Formula (3) is decomposed along three coordinate axes of polar coordinate system $\theta_1 - r_1 - z_1$, according to the angular relationship shown in Figure 4, the stresses σ'_z, τ'_{zr}, and $\tau'_{z\theta}$ on the inclined section mnq are obtained as follows:

$$\begin{cases} \sigma'_z = \left(F_\theta \cos^2\gamma \cos\alpha + F_r \cos\delta + F_z \cos\gamma\right) \times S_{mnq} \\ \tau'_{zr} = \left(F_\theta \cos^2\gamma \cos\beta + F_r \cos\gamma + F_z \cos\delta\right) \times S_{mnq} \\ \tau'_{z\theta} = \left(F_\theta \cos\gamma + F_r \cos\beta + F_z \cos\alpha\right) \times S_{mnq} \end{cases} \tag{4}$$

In the same way, σ'_r can be obtained when the inclined section is parallel to the outer normal of the r_1 axis. By using Formulas (2) and (4), the relationship matrix of stress components on the inclined section of the annular hexahedron element in polar coordinate system can be obtained as follows:

$$\begin{bmatrix} \sigma'_\theta \\ \sigma'_r \\ \sigma'_z \\ \tau'_{\theta r} \\ \tau'_{\theta z} \\ \tau'_{zr} \end{bmatrix} = T \begin{bmatrix} \sigma_\theta \\ \sigma_r \\ \sigma_z \\ \tau_{r\theta} \\ \tau_{r\theta} \\ \tau_{rz} \end{bmatrix} = \begin{bmatrix} \kappa_{11}\sigma_\theta + \kappa_{12}\sigma_r + \kappa_{13}\sigma_z + \kappa_{14}\tau_{r\theta} + \kappa_{15}\tau_{r\theta} + \kappa_{16}\tau_{rz} \\ \kappa_{21}\sigma_\theta + \kappa_{22}\sigma_r + \kappa_{23}\sigma_z + \kappa_{24}\tau_{r\theta} + \kappa_{25}\tau_{r\theta} + \kappa_{26}\tau_{rz} \\ \kappa_{31}\sigma_\theta + \kappa_{32}\sigma_r + \kappa_{33}\sigma_z + \kappa_{34}\tau_{r\theta} + \kappa_{35}\tau_{r\theta} + \kappa_{36}\tau_{rz} \\ \kappa_{41}\sigma_\theta + \kappa_{42}\sigma_r + \kappa_{43}\sigma_z + \kappa_{44}\tau_{r\theta} + \kappa_{45}\tau_{r\theta} + \kappa_{46}\tau_{rz} \\ \kappa_{51}\sigma_\theta + \kappa_{52}\sigma_r + \kappa_{53}\sigma_z + \kappa_{54}\tau_{r\theta} + \kappa_{55}\tau_{r\theta} + \kappa_{56}\tau_{rz} \\ \kappa_{61}\sigma_\theta + \kappa_{62}\sigma_r + \kappa_{63}\sigma_z + \kappa_{64}\tau_{r\theta} + \kappa_{65}\tau_{r\theta} + \kappa_{66}\tau_{rz} \end{bmatrix} \quad (5)$$

In Formula (5), T is the spatial transformation matrix of circular braiding stress, κ_{ij} is a 6×6 matrix, which is given in Appendix A.

4. The Equivalent Stiffness Matrix of Cells

4.1. Solving the Equivalent Stiffness Matrix of Internal Cells

Figure 5 shows the fiber angle relationship in the inner cell A, and the coordinates of J_1 and J_2 points in the cylindrical coordinate system θ-r-z in Figure 5 are:

$$\begin{cases} J_1^{\theta\text{-}r\text{-}z} = \left(R_{in} + \frac{3h}{4}, -\frac{180^\circ}{M}, w_h\right) \\ J_2^{\theta\text{-}r\text{-}z} = \left(R_{in} + \frac{h}{2}, 0, \frac{3w_h}{2}\right) \end{cases} \quad (6)$$

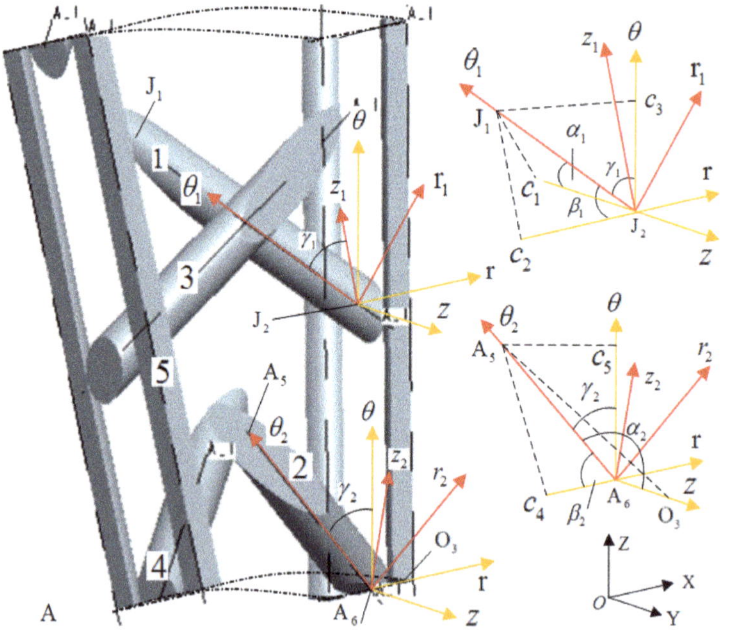

Figure 5. Angle relationship of fibers in inner cell A.

Therefore, the coordinates of the J_1 and J_2 points in the X-Y-Z rectangular coordinate system are:

$$\begin{cases} J_1^{\text{X-Y-Z}} = \left(-(R_{in} + \frac{3h}{4})\sin(\frac{180^\circ}{M}), w_h, (R_{in} + \frac{3h}{4})\cos(\frac{180^\circ}{M})\right) \\ J_2^{\text{X-Y-Z}} = \left(0, \frac{3w_h}{2}, R_{in} + \frac{h}{2}\right) \end{cases} \quad (7)$$

According to the relationship between cylindrical coordinate system θ-r-z and rectangular coordinate system X-Y-Z, in Figure 5, the distance between J_1 and J_2 points in rectangular coordinate system X-Y-Z is:

$$L_{J_1-J_2}^{\text{X-Y-Z}} = \sqrt{[(R_{in} + \frac{3h}{4})\sin(\frac{180^\circ}{M})]^2 + [\frac{w_h}{2}]^2 + [(R_{in} + \frac{3h}{4})\cos(\frac{180^\circ}{M}) - R_{in} - \frac{h}{2}]^2} \quad (8)$$

The distance between the c_1 and c_2 points in the cylindrical-coordinate system θ-r-z is:

$$\begin{cases} c_1^{\theta-r-z} = \left(R_{in} + \frac{h}{2}, 0, w_h\right) \\ c_2^{\theta-r-z} = \left(R_{in} + \frac{h}{2}, -\frac{180^\circ}{M}, \frac{3w_h}{2}\right) \end{cases} \quad (9)$$

Therefore, the coordinates of the c_1 and c_2 points in rectangular coordinate system X-Y-Z are:

$$\begin{cases} c_1^{\text{X-Y-Z}} = \left(0, w_h, (R_{in} + \frac{h}{2})\right) \\ c_2^{\text{X-Y-Z}} = \left(-\sin(\frac{180^\circ}{M})(R_{in} + \frac{h}{2}), \frac{3w_h}{2}, \cos(\frac{180^\circ}{M})(R_{in} + \frac{h}{2})\right) \end{cases} \quad (10)$$

Therefore, according to the positional relationship between the two coordinate systems, in Figure 5, the distance between the c_1 and J_2 points in the X-Y-Z rectangular coordinate system is $L_{c_1-J_2}^{X-Y-Z} = 0.5w_h$, the distance between c_2 and J_2 is [21]:

$$L_{c_2-J_2}^{\text{X-Y-Z}} = \sqrt{[(R_{in} + \frac{h}{2})\sin(\frac{180^\circ}{M})]^2 + [(R_{in} + \frac{h}{2})\cos(\frac{180^\circ}{M}) - R_{in} - \frac{h}{2}]^2} \quad (11)$$

According to the positional relationship shown in Figure 5, the formulas for calculating the angle between the θ_1 axis and the three coordinate axes θ-r-z on fiber 1 in inner cell A can be obtained:

$$\begin{cases} \alpha_1 = \arccos[\frac{0.5w_h}{\sqrt{[(R_{in}+\frac{3h}{4})\sin(\frac{180^\circ}{M})]^2+[\frac{w_h}{2}]^2+[(R_{in}+\frac{3h}{4})\cos(\frac{180^\circ}{M})-R_{in}-\frac{h}{2}]^2}}] \\ \beta_1 = \arccos[\frac{\sqrt{[(R_{in}+\frac{h}{2})\sin(\frac{180^\circ}{M})]^2+[(R_{in}+\frac{h}{2})\cos(\frac{180^\circ}{M})-R_{in}-\frac{h}{2}]^2}}{\sqrt{[(R_{in}+\frac{3h}{4})\sin(\frac{180^\circ}{M})]^2+[\frac{w_h}{2}]^2+[(R_{in}+\frac{3h}{4})\cos(\frac{180^\circ}{M})-R_{in}-\frac{h}{2}]^2}}] \\ \gamma_1 = \arccos[\frac{0.25h}{\sqrt{[(R_{in}+\frac{3h}{4})\sin(\frac{180^\circ}{M})]^2+[\frac{w_h}{2}]^2+[(R_{in}+\frac{3h}{4})\cos(\frac{180^\circ}{M})-R_{in}-\frac{h}{2}]^2}}] \end{cases} \quad (12)$$

In Formula (12), the angle between the inner braiding fiber and the braiding direction is defined as the inner braiding angle, and γ_1 is the inner braiding angle of the outer layer fiber of inner cell A.

In the same way, according to the positional relationship shown in Figure 5, the formulas for calculating the angle between the θ_1 axis on fiber 2 in inner cell A and the three coordinate axes of θ-r-z can be obtained:

$$\begin{cases} \alpha_2 = \arccos[\frac{0.5w_h}{\sqrt{[(R_{in}+\frac{h}{4})\sin(\frac{180^\circ}{M})]^2+[\frac{w_h}{2}]^2+[(R_{in}+\frac{h}{4})\cos(\frac{180^\circ}{M})-R_{in}]^2}}] \\ \beta_2 = \arccos[\frac{\sqrt{[R_{in}\sin(\frac{180^\circ}{M})]^2+[R_{in}-R_{in}\cos(\frac{180^\circ}{M})]^2}}{\sqrt{[(R_{in}+\frac{h}{4})\sin(\frac{180^\circ}{M})]^2+[\frac{w_h}{2}]^2+[(R_{in}+\frac{h}{4})\cos(\frac{180^\circ}{M})-R_{in}]^2}}] \\ \gamma_2 = \arccos[\frac{0.25h}{\sqrt{[(R_{in}+\frac{h}{4})\sin(\frac{180^\circ}{M})]^2+[\frac{w_h}{2}]^2+[(R_{in}+\frac{h}{4})\cos(\frac{180^\circ}{M})-R_{in}]^2}}] \end{cases} \quad (13)$$

In the same way, according to the positional relationship shown in Figure 6, the formulas for calculating the angle between θ_3 axis and θ-r-z three coordinate axes on fiber 1 in inner cell B can be obtained:

$$\begin{cases} \alpha_3 = \arccos\left[\dfrac{0.5w_h}{\sqrt{[X_{\alpha_3}]^2+[\frac{w_h}{2}]^2+[R_{out}\cos(\frac{360^\circ}{M})-(R_{in}+\frac{3h}{4})\cos(\frac{180^\circ}{M})]^2}}\right] \\ \beta_3 = \arccos\left[\dfrac{\sqrt{[R_{out}\sin(\frac{360^\circ}{M})-R_{out}\sin(\frac{180^\circ}{M})]^2+[R_{out}\cos(\frac{360^\circ}{M})-R_{out}\cos(\frac{180^\circ}{M})]^2}}{\sqrt{[X_{\alpha_3}]^2+[\frac{w_h}{2}]^2+[R_{out}\cos(\frac{360^\circ}{M})-(R_{in}+\frac{3h}{4})\cos(\frac{180^\circ}{M})]^2}}\right] \\ \gamma_3 = \arccos\left[\dfrac{0.25h}{\sqrt{[X_{\alpha_3}]^2+[\frac{w_h}{2}]^2+[R_{out}\cos(\frac{360^\circ}{M})-(R_{in}+\frac{3h}{4})\cos(\frac{180^\circ}{M})]^2}}\right] \end{cases} \quad (14)$$

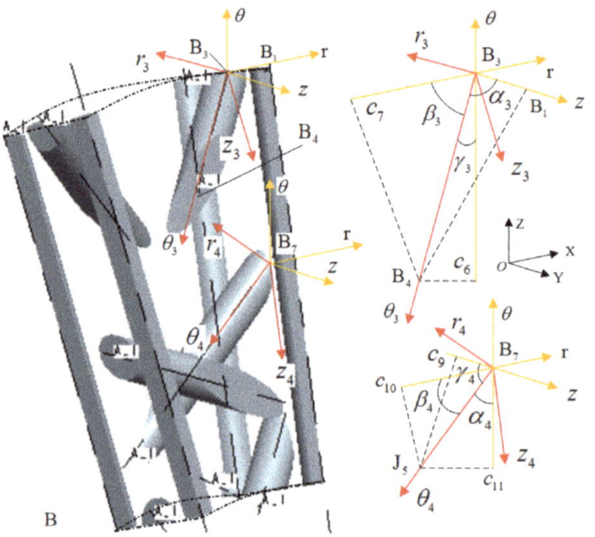

Figure 6. Angle relationship of fibers in inner cell B.

In the same way, the formulas for calculating the angle between θ_4 axis and θ-r-z three coordinate axes on fiber 2 in inner cell B can be obtained:

$$\begin{cases} \alpha_4 = \arccos\left[\dfrac{0.5w_h}{\sqrt{[X_{\alpha_4}]^2+[\frac{w_h}{2}]^2+[(R_{in}+\frac{h}{4})\cos(\frac{180^\circ}{M})-(R_{in}+\frac{h}{2})\cos(\frac{360^\circ}{M})]^2}}\right] \\ \beta_4 = \arccos\left[\dfrac{(R_{in}+\frac{h}{2})\sqrt{[2-2\sin(\frac{360^\circ}{M})\sin(\frac{180^\circ}{M})-2\cos(\frac{360^\circ}{M})\cos(\frac{180^\circ}{M})]}}{\sqrt{[X_{\alpha_4}]^2+[\frac{w_h}{2}]^2+[(R_{in}+\frac{h}{4})\cos(\frac{180^\circ}{M})-(R_{in}+\frac{h}{2})\cos(\frac{360^\circ}{M})]^2}}\right] \\ \gamma_4 = \arccos\left[\dfrac{0.25h}{\sqrt{[X_{\alpha_4}]^2+[\frac{w_h}{2}]^2+[(R_{in}+\frac{h}{4})\cos(\frac{180^\circ}{M})-(R_{in}+\frac{h}{2})\cos(\frac{360^\circ}{M})]^2}}\right] \end{cases} \quad (15)$$

In Formulas (14) and (15),

$$\begin{cases} X_{\alpha_3} = R_{out}\cos(\frac{360^\circ}{M}) - (R_{in}+\frac{3h}{4})\cos(\frac{180^\circ}{M}) \\ X_{\alpha_4} = (R_{in}+\frac{h}{4})\cos(\frac{180^\circ}{M}) - (R_{in}+\frac{h}{2})\cos(\frac{360^\circ}{M}) \end{cases}$$

There are five kinds of fibers with different orientations in inner cell A. According to the positional relationship shown in Figure 5, the angles between the five kinds of fibers in inner cell A and the coordinate axis are shown in Table 1.

Table 1. Fiber angle in inner cell A.

Axis ID	1	2	3	4	5
θ-θ_i	γ_1	γ_2	γ_1	γ_2	0
r-θ_i	$\pi-\beta_1$	$\pi-\beta_2$	β_1	β_2	0
z-θ_i	$\pi-\alpha_1$	α_2	α_1	$\pi-\alpha_2$	0

In the same way, there are five kinds of fibers with different directions in inner cell B. According to the positional relationship shown in Figure 6, the angles between the five kinds of fibers in inner cell B and the coordinate axis can be obtained, as shown in Table 2.

Table 2. Fiber angle in inner cell B.

Axis ID	1	2	3	4	5
θ-θ_i	$\pi-\gamma_3$	$\pi-\gamma_4$	$\pi-\gamma_3$	$\pi-\gamma_4$	0
r-θ_i	$\pi-\beta_3$	$\pi-\beta_4$	β_3	β_4	0
z-θ_i	α_3	$\pi-\alpha_4$	$\pi-\alpha_3$	α_4	0

Because of the symmetry of inner cell structure, the angle between fiber and coordinate axis in inner cell D is the same as that in inner cell A, and the angle between fiber and coordinate axis in inner cell C is the same as that in inner cell B.

Therefore, the stiffness matrix of inner cell A is:

$$C_{in}^{A} = 0.25[\left(V_{in-A}^{1} + V_{in-A}^{3}\right)T_{in-A}^{1}C_f\left(T_{in-A}^{1}\right)^T + \left(V_{in-A}^{2} + V_{in-A}^{4}\right)T_{in-A}^{2}C_f\left(T_{in-A}^{2}\right)^T + V_{in-A}^{5}C_f] \quad (16)$$

In Formula (16), V_{in-A}^{i} is the fiber volume ratio of fiber i in inner cell A to inner cell A, T_{in-A}^{i} is the stress transformation matrix of fiber i in inner cell A, and the angles in Table 1 are brought into Formula (5). According to Formulas (13) and (15), the stress transformation matrix T_{in-A}^{i} of inner cell A can be obtained. C_f is the stiffness matrix of fiber in local coordinates [16,17].

$$\begin{cases} V_{A-in}^{1} = V_{A-in}^{3} = \frac{h}{4V_{in}^{A}\cos\gamma_1} \\ V_{A-in}^{2} = V_{A-in}^{4} = \frac{h}{4V_{in}^{A}\cos\gamma_2} \\ V_{A-in}^{5} = \frac{h}{V_{in}^{A}} \end{cases} \quad (17)$$

In Formula (17), $V_{in}^{A} = \frac{h}{2\cos\gamma_1} + \frac{h}{2\cos\gamma_2} + h$, in the same way, the stiffness matrices C_{in}^{B}, C_{in}^{C}, and C_{in}^{D} of internal cells B, C, and D can be obtained.

$$\overline{C}_{in} = \frac{V_{A-in}}{V_{in}^{t}}C_{in}^{A} + \frac{V_{B-in}}{V_{in}^{t}}C_{in}^{B} + \frac{V_{C-in}}{V_{in}^{t}}C_{in}^{C} + \frac{V_{D-in}}{V_{in}^{t}}C_{in}^{D} \quad (18)$$

In Formula (18),

$$V_{A-in} = h(\frac{1}{2\cos\gamma_1} + \frac{1}{2\cos\gamma_2} + 1), V_{B-in} = h(\frac{1}{2\cos\gamma_3} + \frac{1}{2\cos\gamma_4} + 1)$$

$$V_{D-in} = h(\frac{1}{2\cos\gamma_1} + \frac{1}{2\cos\gamma_2} + 1)$$

$$V_{in}^{t} = \sum_{i=1}^{i=4}(\frac{h}{\cos\gamma_i}) + 4h$$

4.2. Solving the Equivalent Stiffness Matrix of Upper Surface Cells

Figure 7 shows the fiber angle relationship in the upper surface cell. Similarly, according to the position relationship shown in Figure 7, the calculation formulas of the angles between θ_5 axis and θ-r-z axis on the fiber in the upper surface cell can be obtained:

$$\begin{cases} \alpha_{top} = \arccos\left[\dfrac{0.5w_h}{\sqrt{\xi_{top}^2 + (\frac{w_h}{2})^2 + [(R_{in}+\frac{h}{4})\cos(\frac{360^\circ}{M}) - (R_{in}+\frac{3h}{4})\cos(\frac{180^\circ}{M})]^2}}\right] \\ \beta_{top} = \arccos\left[\dfrac{(R_{in}+\frac{h}{4})\sqrt{[2-2\sin(\frac{180^\circ}{M})\sin(\frac{360^\circ}{M})-2\cos(\frac{180^\circ}{M})\cos(\frac{360^\circ}{M})]}}{\sqrt{\xi_{top}^2 + (\frac{w_h}{2})^2 + [(R_{in}+\frac{h}{4})\cos(\frac{360^\circ}{M}) - (R_{in}+\frac{3h}{4})\cos(\frac{180^\circ}{M})]^2}}\right] \\ \gamma_{top} = \arccos\left[\dfrac{0.25h}{\sqrt{\xi_{top}^2 + (\frac{w_h}{2})^2 + [(R_{in}+\frac{h}{4})\cos(\frac{360^\circ}{M}) - (R_{in}+\frac{3h}{4})\cos(\frac{180^\circ}{M})]^2}}\right] \end{cases} \quad (19)$$

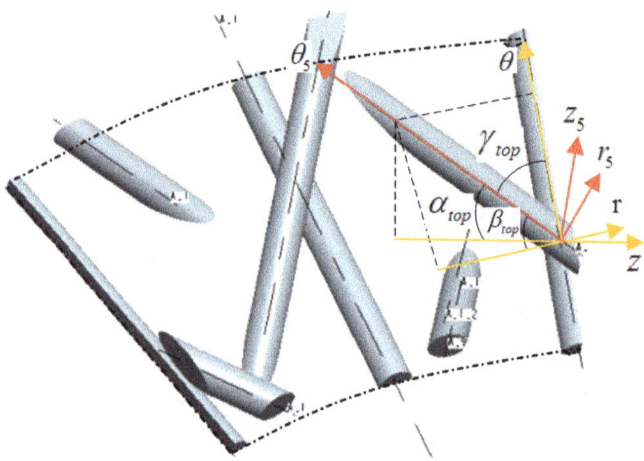

Figure 7. Angle relationship of fiber in upper surface cells.

In Formula (19), $\xi_{top} = (R_{in} + \frac{h}{4})\sin(\frac{360^\circ}{M}) - (R_{in} + \frac{3h}{4})\sin(\frac{180^\circ}{M})$.

In the same way, the equivalent stiffness matrix $\overline{C}_{top}$ of the upper surface cell is:

$$\overline{C}_{top} = V_{top}^1 T_{top}^1 C_f (T_{top}^1)^T + V_{top}^2 C_f \quad (20)$$

In Formula (20), $V_{top}^1 = \frac{h/\cos\alpha_{top}}{h/\cos\alpha_{top}+h/2}$, $V_{top}^2 = \frac{h/2}{h/\cos\alpha_{top}+h/2}$.

4.3. Solving the Equivalent Stiffness Matrix of Lower Surface Cells

Figure 8 shows the angle relationship of fibers in the lower surface cells. Similarly, according to the position relationship shown in Figure 8, the calculation formulas of the angle between θ_6 axis and θ-r-z axis on the fiber in the lower surface cells can be obtained:

$$\begin{cases} \alpha_{low} = \arccos\left[\dfrac{0.5w_h}{\sqrt{\xi_{low}^2 + (\frac{w_h}{2})^2 + [(R_{in}+\frac{3h}{4})\cos(\frac{360^\circ}{M}) - (R_{in}+\frac{h}{4})\cos(\frac{180^\circ}{M})]^2}}\right] \\ \beta_{low} = \arccos\left[\dfrac{(R_{in}+\frac{3h}{4})\sqrt{[2-2\sin(\frac{180^\circ}{M})\sin(\frac{360^\circ}{M})-2\cos(\frac{180^\circ}{M})\cos(\frac{360^\circ}{M})]}}{\sqrt{\xi_{low}^2 + (\frac{w_h}{2})^2 + [(R_{in}+\frac{3h}{4})\cos(\frac{360^\circ}{M}) - (R_{in}+\frac{h}{4})\cos(\frac{180^\circ}{M})]^2}}\right] \\ \gamma_{low} = \arccos\left[\dfrac{0.25h}{\sqrt{\xi_{low}^2 + (\frac{w_h}{2})^2 + [(R_{in}+\frac{3h}{4})\cos(\frac{360^\circ}{M}) - (R_{in}+\frac{h}{4})\cos(\frac{180^\circ}{M})]^2}}\right] \end{cases} \quad (21)$$

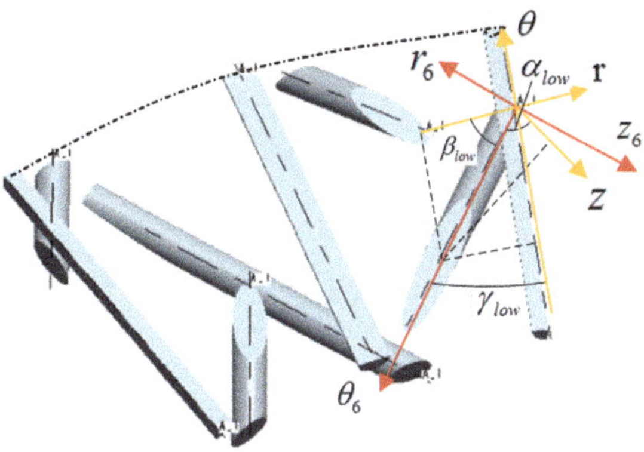

Figure 8. Angle relationship of fiber in lower surface cells.

In Formula (21), $\xi_{low} = (R_{in} + \frac{3h}{4})\sin(\frac{360^\circ}{M}) - (R_{in} + \frac{h}{4})\sin(\frac{180^\circ}{M})$.

In the same way, the equivalent stiffness matrix $\overline{C}_{low}$ of the lower surface cell is:

$$\overline{C}_{low} = V_{low}^1 T_{low}^1 C_f (T_{low}^1)^T + V_{low}^2 C_f \tag{22}$$

In Formula (22), $V_{low}^1 = \frac{h/\cos\alpha_{low}}{h/\cos\alpha_{low}+h/2}$, $V_{low}^2 = \frac{h/2}{h/\cos\alpha_{low}+h/2}$.

Therefore, the overall stiffness matrix of all fibers of the whole annular braided composite material is:

$$\overline{C}_B = \chi_{in} C_{in} + \chi_{top} C_{top} + \chi_{low} C_{low} \tag{23}$$

In Formula (23), $\chi_{in} = \frac{V_{in}^t}{V_{in}^t + h/\cos\alpha_{low} + h/\cos\alpha_{top} + h}$, $\chi_{top} = \frac{h/\cos\alpha_{top}+h/2}{V_{in}^t + h/\cos\alpha_{low} + h/\cos\alpha_{top} + h}$, $\chi_{low} = \frac{h/\cos\alpha_{low}+h/2}{V_{in}^t + h/\cos\alpha_{low} + h/\cos\alpha_{top} + h}$.

The overall stiffness matrix of that 3D and five-direction annular braided composite material considering the matrix is as follows:

$$\overline{C} = V_B \overline{C}_B + (1 - V_B) C_M \tag{24}$$

In Formula (24), V_B is the fiber volume fraction, C_M is matrix of matrix stiffness.

The calculation formulas of fiber volume fraction is as follows [21]:

$$\begin{cases} V_{in}^1 = \frac{4\pi(R_{out}^2 - R_{in}^2) w_h}{M} \\ V_1 = \frac{S_1 h}{\cos\gamma_1} + \frac{S_1 h}{\cos\gamma_2} + \frac{S_1 h}{\cos\gamma_3} + \frac{S_1 h}{\cos\gamma_4} + \frac{2S_1 h}{\cos\gamma_{top}} + \frac{2S_1 h}{\cos\gamma_{low}} + 6S_2 h \\ V_B = \frac{V_1}{V_{in}^1} \end{cases} \tag{25}$$

In Formula (25), V_{in}^1 is the total volume of internal cells, V_1 is the volume of fibers in the inner cell, S_1 is the cross-sectional area of braided fiber bundle, and the cross-sectional shape of braided fiber is oval, $S_1 = \pi ab$, the half of major axis of ellipse is a, the half of the minor axis of ellipse is b. S_2 is the cross-sectional area of the axial fiber, the cross-sectional shape of the axial fiber is square $S_2 = r^2 b^2$, the side length is rb, r is the cross-sectional dimension coefficient of axial fiber [17,18,21].

Table 3 gives the mechanical property parameters of fibers and matrix in carbon fiber braided composites. Because the 3D five-directional circular transverse braided materials approximately meet the conditions of transverse isotropy in a certain radial range in the plane perpendicular to the Z axis, the elastic modulus satisfies: $E_r > E_\theta \approx E_z$, the shear modulus meets the following requirements: $G_{\theta r} \approx G_{rz} > G_{\theta z}$, and Poisson's ratio meets the following requirements: $\mu_{\theta r} = \mu_{rz}$. Because the matrix is an isotropic material, the shear modulus of the matrix meets the following requirements: $G = \frac{E}{2(1+\mu)}$.

$$\begin{cases} E_{\theta}^{s} = E_{r}^{s} = \varepsilon E_{\theta} + (1-\varepsilon)E_{m} \\ E_{z}^{s} = \frac{E_{m}}{1-\sqrt{\varepsilon}(1-E_{m}/E_{z})} \\ G_{\theta z}^{s} = G_{rz}^{s} = \frac{G_{m}}{1-\sqrt{\varepsilon}(1-G_{m}/G_{\theta z})} \\ G_{\theta r}^{s} = \frac{G_{m}}{1-\sqrt{\varepsilon}(1-G_{m}/G_{\theta r})} \\ \mu_{\theta z}^{s} = \mu_{rz}^{s} = \varepsilon\mu_{\theta z} + (1-\varepsilon)\mu_{m} \\ \mu_{\theta r}^{s} = \frac{E_{z}^{s}}{2G_{\theta r}^{s}} - 1 \end{cases} \tag{26}$$

Table 3. Material parameters of fiber and matrix.

	E_θ/GPa	E_z/GPa	$G_{\theta r}$/GPa	$G_{\theta z}$/GPa	$\mu_{\theta r}$	$\mu_{\theta z}=\mu_{rz}$
T300-Carbon fiber	220	13.8	9.0	4.8	0.2	0.25
TDE85	4.5	4.5	1.68	1.68	0.34	0.34

According to the parameters of composite fibers and matrix in Table 3, the mechanical parameters of fiber bundles in composite materials in all directions can be calculated by using Chamis mechanical formulas shown in Formula (26) [21,22].

In Formula (26), E_i^s is the elastic modulus of fiber bundle in i direction, ε is the filling coefficient of matrix and fiber in fiber bundle, E_m is the elastic modulus of matrix, E_i is the elastic modulus of fiber in i direction, G_{ij}^s is the shear modulus of fiber bundle in i, j planes, G_{ij} is the shear modulus of fiber in i, j planes, μ_{ij}^s is Poisson's ratio of fiber bundle in i, j planes, μ_i is Poisson's ratio of fiber in i direction [18,21].

5. Numerical Analysis of the Influence of Braiding Parameters on the Mechanical Properties of Cells

The matrix parameters in Table 3 and the parameters of fiber bundles obtained by Formula (26) are brought into Formula (24), and the overall stiffness matrix of 3D five-directional annular braided composites can be obtained. By inverting the overall stiffness matrix, the engineering elastic constant of 3D five-directional annular braided composites can be obtained as shown in Formula (27).

$$[S] = [\bar{C}]^{-1} = \begin{vmatrix} 1/E_{\theta}^{s} & -v_{\theta r}^{s}/E_{\theta}^{s} & -v_{\theta z}^{s}/E_{\theta}^{s} & 0 & 0 & 0 \\ -v_{\theta r}^{s}/E_{\theta}^{s} & 1/E_{r}^{s} & -v_{rz}^{s}/E_{r}^{s} & 0 & 0 & 0 \\ -v_{\theta z}^{s}/E_{\theta}^{s} & -v_{rz}^{s}/E_{r}^{s} & 1/E_{z}^{s} & 0 & 0 & 0 \\ 0 & 0 & 0 & 1/G_{rz}^{s} & 0 & 0 \\ 0 & 0 & 0 & 0 & 1/G_{\theta z}^{s} & 0 \\ 0 & 0 & 0 & 0 & 0 & 1/G_{\theta r}^{s} \end{vmatrix} \tag{27}$$

In Formula (27), S is the overall flexibility matrix of composite materials, E_j is the elastic modulus of the composite material in the j direction, G_{ij} is the shear modulus of the composite in the i, j planes, v_{ij} is Poisson's ratio of composite materials in i, j planes.

The length of braiding knuckles h = 2 mm, inner cell diameter R_{in} = 7 mm, M and N are radial and axial braided fibers M = N = 80, respectively. The height of internal cells w_h = 0.5 mm, the filling factor $\varepsilon = 0.61$.

The numerical results calculated by Formula (27) are compared with the experimental results in reference [17], and the results are shown in Table 4. It is found that the relative error between the numerical prediction model deduced in this paper and the experimental results is 6.6%.

Table 4. Comparison of numerical results and experimental measurement results.

	Measured [17]	Predicted	Relative Error
$E_\theta = E_r$	30.8/GPa	32.98/GPa	6.6%
E_z/GPa		5.12/GPa	
$G_{\theta r}$/GPa		1.97/GPa	
$G_{\theta z} = G_{rz}$		1.82/GPa	
$\mu_{\theta z} = \mu_{rz}$		0.334	
$\mu_{\theta r}$		0.228	

In order to further analyze the influence of braiding parameters on the mechanical properties of 3D five-directional circular braided composites, the mathematical model described in this paper is used to numerically analyze the influence of braiding parameters on the mechanical properties of cells.

The relationship between that braiding angle γ and the knuckle length h of the 3D circular braiding material is shown in Figure 9, in which the cell inner diameter $R_{in} = 5$mm, the height of the inner cell $w_h = 0.5$mm, the number of radial and axial braiding fibers $M = N = 40$. It can be seen from Figure 9 that the braiding angles of the four internal cells are relatively close, that is, $\gamma_3 > \gamma_1 > \gamma_4 > \gamma_2$. With the increase in the length h of the knuckle, the braiding angles of each cell decrease, and the braiding angles of the four inner cells decrease the most; when that knuckle length increases from 1 mm to 3 mm, the braiding angle of the four inner cells is reduced by 38%~46%. With the increase in the length h of the knuckle, the braiding angle of the upper and lower surfaces decreases slightly, that is, $\gamma_{top} \approx \gamma_{low}$; when that knuckle length increases from 1 mm to 3 mm, the braiding angle of the upper and lower surfaces is reduced by 7.1%.

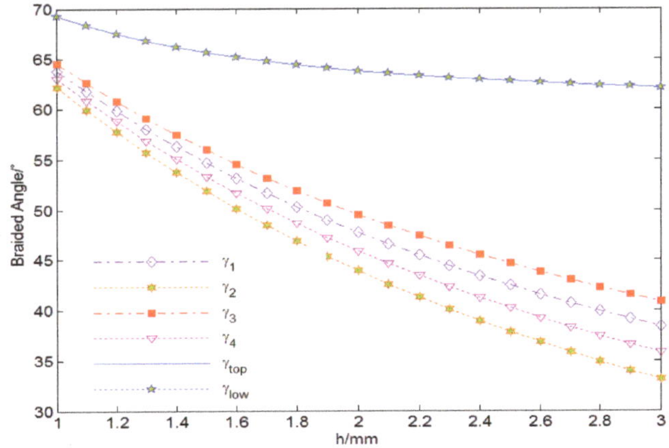

Figure 9. Relationship curve between braiding angle and knuckle length.

Figure 10 shows the influence curve of cell height w_h and knuckle length h on fiber volume content V_B and transverse elastic modulus E_r. It can be seen from Figure 10 that with the increase in knuckle length h and cell height w_h, the fiber volume content V_B of 3D five-directional annular braided composites decreases; when that knuckle length increase from 1 mm to 10 mm, the fiber volume content is reduced by 57.1%. With the increase in knuckle length h, the transverse elastic modulus E_r first increases and then decreases slowly, and with the increase in cell height w_h, the transverse elastic modulus E_r decreases. This is mainly because fiber volume content decreases due to the increase in knuckle length and cell height, and the number of carbon fibers per unit volume decreases, which, in turn, leads to a decrease in transverse elastic modulus.

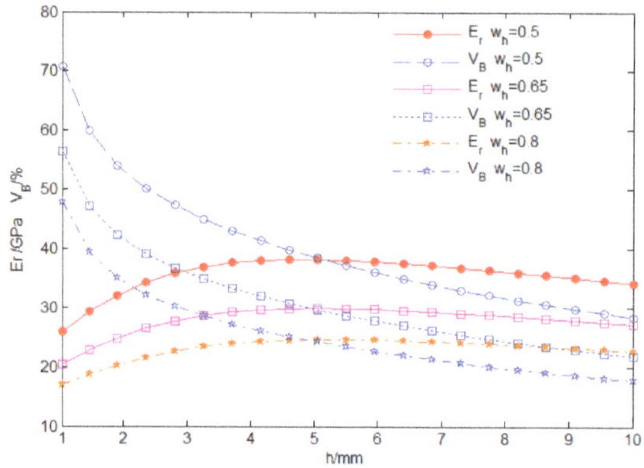

Figure 10. Influence curve of cell height and knuckle length on fiber volume content and transverse elastic modulus.

Figure 11 shows the influence curve of cell height w_h and knuckle length h on fiber volume content V_B and longitudinal elastic modulus E_z. It can be seen from Figure 11 that with the increase in knuckle length h and cell height w_h, the fiber volume content V_B of 3D five-directional annular braided composites decreases. The longitudinal elastic modulus E_z increases rapidly when the length of knuckle $h < 4$, and increases slowly when $h > 4$. When the knuckle length $h < 3$, the longitudinal elastic modulus E_z increases with the increase in the cell height w_h. When the node length $h > 3$, the longitudinal elastic modulus decreases with the increase in cell height w_h. The above results show that the cell height has less influence on longitudinal elastic modulus and more influence on fiber volume content.

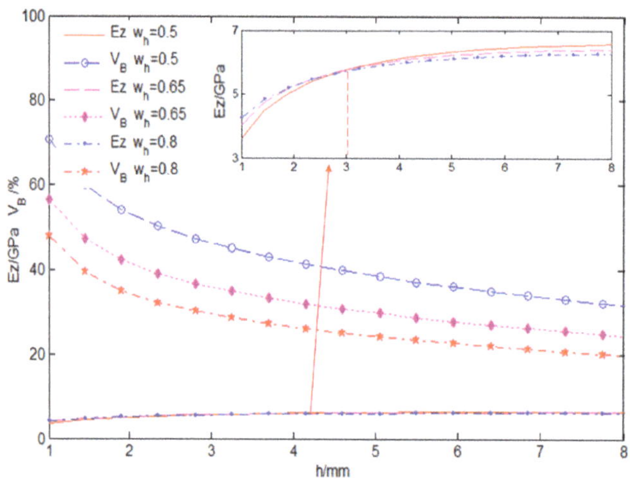

Figure 11. Influence curve of cell height and knuckle length on fiber volume content and longitudinal elastic modulus.

It can be seen from Figure 12 that with the increase in knuckle length h, the fiber volume content V_B decreases, and the shear modulus first increases and then decreases. With the increase in cell height w_h, the fiber volume content and elastic modulus decrease linearly. The above results show that the shear modulus $G_{\theta r}$ is greater than the shear modulus $G_{\theta z}$, and this is mainly because the mechanical properties of materials along the braiding direction are the best.

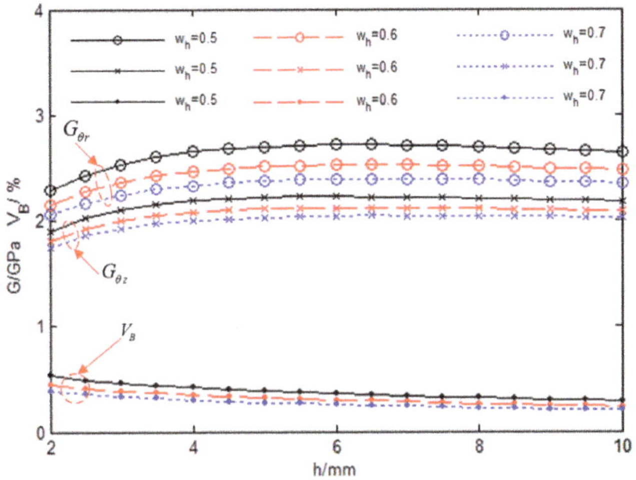

Figure 12. Relationship curve of cell height, knuckle length, fiber volume content, and shear modulus.

It can be seen from Figure 13 that with that knuckle length increase from 5 mm to 10 mm, the transverse elastic modulus is reduced by 8.6%; when the filling factor increases from 0.87 to 0.95, the transverse elastic modulus is increased by 7.5%; when the number of braiding fibers increases from 20 to 40, the transverse elastic modulus is increased by 114%.

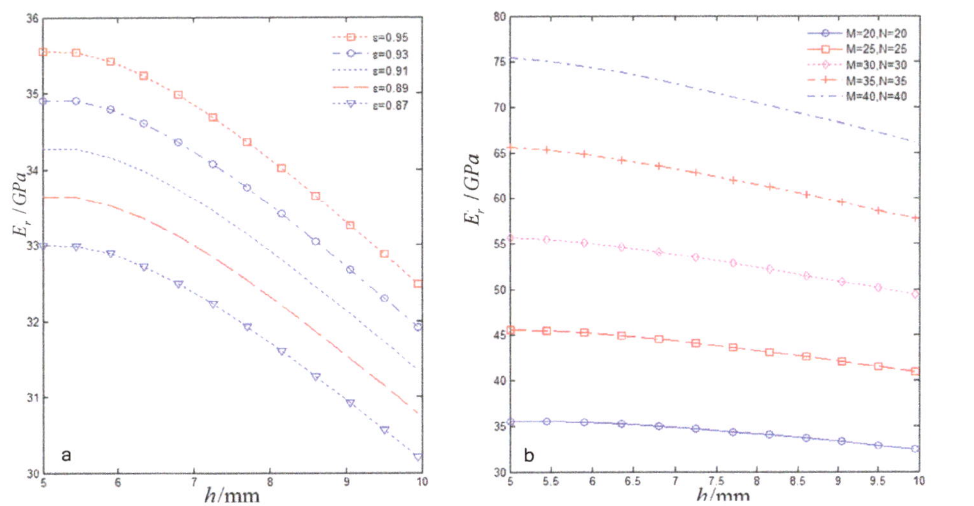

(**a**) The different length of knuckle and filling factor (**b**) The different length of knuckle and number of braiding fiber

Figure 13. Influence curve of braiding parameters and transverse elastic modulus.

It can be seen from Figure 14 that with that knuckle length increase from 5 mm to 30 mm, the shear modulus $G_{\theta r}$ is reduced by 33%, the shear modulus $G_{\theta z}$ is reduced by 31%, and when the number of braiding fibers increases, the shear modulus is increased.

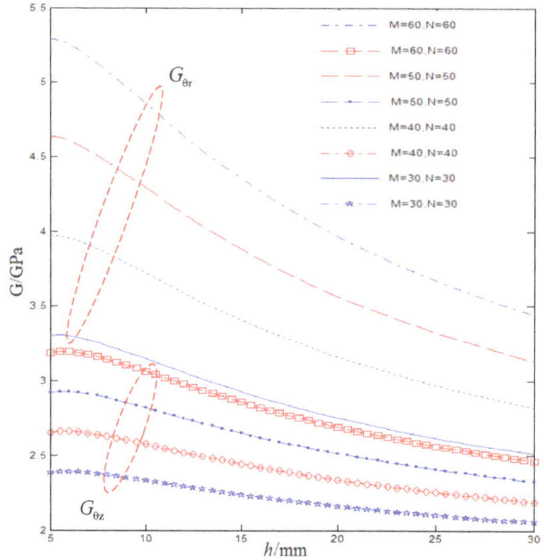

Figure 14. Relationship curve between knuckle length and shear modulus under different yarn numbers.

6. Future Work

Through the above analysis, it can be found that by changing the input design variables such as filling factor, the knuckle length, cell height, and number of braiding fibers, the braiding angle and fiber volume content will be changed, and finally, the mechanical properties such as elastic modulus, shear modulus, and Poisson's ratio will be affected, that is, the anisotropy index of mechanical properties of composites in different directions can be optimized by changing the input design variables, and the mechanical properties of braided composites can be designed, in order to meet the special requirements of industrial development for materials.

7. Conclusions

In this paper, the mathematical model of mechanical properties of 3D five-way circular braided composites is established, and the mechanical properties of materials are numerically analyzed. The research results show that the cell inner diameter is 5 mm, the cell height is 0.5 mm, and the number of radial yarns and axial yarns is 40. The numerical prediction results show that the braided angles of each cell decrease with the increase in the node length, and the braided angles of the four internal cells decrease the most. When that knuckle length increases from 1 mm to 3 mm, the braiding angle of the four inner cells is reduced by 38%~46%, and when that knuckle length increases from 1 mm to 3 mm, the braiding angle of the upper and lower surfaces is reduced by 7.1%. The braided angle of the upper and lower surfaces decreases slightly with the increase in the length of the knuckles. When the length of braided knuckles is 2 mm, the inner diameter of inner cells is 7 mm, the number of radial and axial braided yarns is 80, the height of inner cells is 0.5 mm, and the filling coefficient is 0.61. By comparing the numerical prediction model with the experimental results in reference [13], it is found that the error of the prediction model deduced in this paper is small.

Author Contributions: Final draft review, C.L.; overall scheme design L.Y.; writing, W.Z.; chart creation, W.W.; data analysis, L.L.; literature search, C.Z.; data collection, X.Z. All authors have read and agreed to the published version of the manuscript.

Funding: The research was supported by Key research and development projects in Shaanxi Province (grant No. 2024GX-YBXM-570), W.Z.; Key Project of Shaanxi Provincial Department of Science and Technology (Program No. 2024QY2-GJHX-38), L.L.

Data Availability Statement: The author provides the program code of Figure 12 in the paper. (DOCX).

Conflicts of Interest: The authors declare no conflicts of interest.

Nomenclature

R_{in}	The inner diameter of the inner cell.
R_{out}	The outer diameter of the inner cell.
M	The number of radial braiding fibers.
N	The number of axial braiding fibers.
w_h	The height of the inner cell.
V_{in-A}^{i}	The fiber volume ratio of fiber i in inner cell A to inner cell A.
T_{in-A}^{i}	The stress transformation matrix of fiber i in inner cell A.
T_{in-A}^{i}	The stress transformation matrix of the inner cell A.
C_f	The stiffness matrix of fiber in local coordinates.
V_{in}^{1}	The total volume of internal cells.
V_1	The volume of fibers in the inner cell.
E_i^s	The elastic modulus of fiber bundle in i direction.
ε	The filling coefficient of matrix and fiber in fiber bundle.
E_m	The elastic modulus of matrix.
E_i	The elastic modulus of fiber in i direction.
G_{ij}^{s}	The shear modulus of fiber bundle in i, j planes.
G_{ij}	The shear modulus of fiber in i, j planes.
μ_{ij}^{s}	Poisson's ratio of fiber bundle in i, j planes.
μ_i	Poisson's ratio of fiber in i direction.
V_B	The fiber volume content.
E_r	The transverse elastic modulus.
E_z	The longitudinal elastic modulus.
h	The knuckle length.

Appendix A

The stress transformation matrix.

$T = [\kappa_{ij}], (i = 1,2,\cdots,6; j = 1,2,\cdots,6)$.

$\kappa_{11} = \cos^2\gamma$, $\kappa_{12} = \cos^2\beta\cos\frac{d\theta}{2}$, $\kappa_{13} = \cos^2\alpha$, $\kappa_{14} = \cos\gamma\cos\beta(1+\cos\frac{d\theta}{2})$, $\kappa_{15} = \cos\gamma\cos\alpha(1+\cos\frac{d\theta}{2})$, $\kappa_{16} = \cos\beta\cos\alpha(1+\cos\frac{d\theta}{2})$, $\kappa_{21} = \cos^2\gamma\cos^2\beta$, $\kappa_{22} = \cos^2\gamma\cos\frac{d\theta}{2}$, $\kappa_{23} = \cos^2\alpha\cos^2\gamma\cos^2\beta$, $\kappa_{24} = \cos\gamma\cos\beta(1+\cos^2\gamma\cos\frac{d\theta}{2})$, $\kappa_{25} = \cos^2\beta\cos\gamma\cos\alpha(1+\cos^2\gamma\cos\frac{d\theta}{2})$, $\kappa_{26} = \cos^2\gamma\cos\beta\cos\alpha(1+\cos\frac{d\theta}{2})$, $\kappa_{31} = \cos^2\alpha\cos^2\gamma$, $\kappa_{32} = \cos^2\beta\cos^2\gamma\cos^2\alpha\cos\frac{d\theta}{2}$, $\kappa_{33} = \cos^2\gamma$, $\kappa_{34} = \cos^2\alpha\cos\gamma\cos\beta(\cos^2\gamma\cos\frac{d\theta}{2}+1)$, $\kappa_{35} = \cos\gamma\cos\alpha(1+\cos^2\gamma\cos\frac{d\theta}{2})$, $\kappa_{36} = \cos^2\gamma\cos\beta\cos\alpha(1+\cos\frac{d\theta}{2})$, $\kappa_{41} = \cos^3\gamma\cos\beta$, $\kappa_{42} = \cos\gamma\cos\beta\cos\frac{d\theta}{2}$, $\kappa_{43} = \cos^2\alpha\cos\gamma\cos\beta$, $\kappa_{44} = (\cos\frac{d\theta}{2}\cos^2\beta+1)\cos^2\gamma$, $\kappa_{45} = \cos^2\gamma\cos\alpha\cos\beta(\cos\frac{d\theta}{2}+1)$, $\kappa_{46} = \cos\alpha\cos\gamma(\cos\frac{d\theta}{2}+\cos^2\beta)$, $\kappa_{51} = \cos\alpha\cos^3\gamma$, $\kappa_{52} = \cos\alpha\cos\gamma\cos^2\beta\cos\frac{d\theta}{2}$, $\kappa_{53} = \cos\alpha\cos\gamma$, $\kappa_{54} = \cos^2\gamma\cos\alpha\cos\beta(\cos\frac{d\theta}{2}+1)$, $\kappa_{55} = (\cos\frac{d\theta}{2}\cos^2\alpha+1)\cos^2\gamma$, $\kappa_{56} = \cos\gamma\cos\beta(\cos^2\alpha\cos\frac{d\theta}{2}+1)$, $\kappa_{61} = \cos\alpha\cos\beta\cos^2\gamma$, $\kappa_{62} =$

$\cos\alpha\cos^2\gamma\cos\beta\cos\frac{d\theta}{2}$, $\kappa_{63}=\cos\alpha\cos^2\gamma\cos\beta$, $\kappa_{64}=\cos\alpha\cos\gamma(\cos\frac{d\theta}{2}\cos^2\beta\cos^2\gamma+1)$,
$\kappa_{65}=\cos\gamma\cos\beta(\cos^2\gamma\cos\frac{d\theta}{2}+\cos^2\alpha)$, $\kappa_{66}=\cos^2\gamma(\cos\frac{d\theta}{2}+\cos^2\alpha\cos^2\beta)$.

References

1. Zhang, W.L.; Wang, X.P.; Ji, X.M. Response analysis of 3D braided two-stage gear system excited by different frequency signals. *Adv. Mech. Eng.* **2021**, *13*, 1. [CrossRef]
2. Zhou, X. Construction of nanocomposite interphase with controllable thickness to relieve stress concentration and boost stress transfer from carbon fiber/epoxy resin interface. *Chem. Eng. J.* **2025**, *505*, 159542. [CrossRef]
3. Li, L.; Aliabadi, M.H. Elastic property prediction and damage mechanics analysis of 3D braided composite. *Theor. Appl. Fract. Mech.* **2019**, *104*, 102338. [CrossRef]
4. Tian, D.P.; Xu, Q.; Yao, X.H. Diversity-guided particle swarm optimization with multi-level learning strategy. *Swarm Evol. Comput.* **2024**, *86*, 101533. [CrossRef]
5. Zhang, C.; Xu, X.W.; Chen, K. Application of Three Unit-Cells Models on Mechanical Analysis of 3D Five-Directional and Full Five-Directional Braided Composites. *Appl. Compos. Mater.* **2013**, *20*, 803–825. [CrossRef]
6. Xu, K.; Xu, X.W. Finite element analysis of mechanical properties of 3D five-directional braided composites. *Mater. Sci. Eng. A* **2008**, *487*, 499–509. [CrossRef]
7. Zeng, T.; Fang, D.N.; Ma, L. Predicting the nonlinear response and failure of 3D braided composites. *Mater. Lett.* **2004**, *58*, 3237–3241. [CrossRef]
8. Wang, X.; Sun, J.; Pang, Z.; Qian, K.; Zhang, D. Design and fabrication of bioinspired bamboo-node three-dimensional six-directional braided composite tubes for superior axial compression capacity. *Constr. Build. Mater.* **2025**, *458*, 139593. [CrossRef]
9. Ji, X.B.; Khatri, A.M. Multi-scale simulation and finite element-assisted computation of elastic properties of braided textile reinforced composites. *J. Compos. Mater.* **2014**, *48*, 931–949. [CrossRef]
10. Kang, H.R. Progressive damage analysis and strength properties of fiber-bar composites reinforced by three-dimensional weaving under uniaxial tension. *Compos. Struct.* **2016**, *141*, 264–281. [CrossRef]
11. Sun, Y.; Zhang, Y.; Liu, T. Effect of fiber breakage defect and waviness defect on compressive fatigue behavior and damage evolution of 3D multiaxial braided composites. *Chin. J. Aeronaut.* **2025**, *38*, 103323. [CrossRef]
12. Wang, Q. Analysis of Vibration Characteristics for Rotating Braided Fiber-Reinforced Composite Annular Plates with Perforations. *Materials* **2024**, *17*, 5402. [CrossRef]
13. Gao, C.; Sun, H.; Gu, J. Dynamic modeling of a three-dimensional braided composite thin plate considering braiding directions. *Appl. Math. Mech.* **2025**, *46*, 123–138. [CrossRef]
14. Chen, Y.H. Micro-mechanical modelling of the overall response of plain woven polymer matrix composites. *Int. J. Eng. Sci.* **2019**, *145*, 103163. [CrossRef]
15. Zhao, Y.L. Analysis of structures and elastic properties of 3D tubular braided composites. *J. Tianjin Polytech. Univ.* **2010**, *29*, 15–18.
16. Li, D.S.; Lu, Z.X.; Chen, L. Microstructure Analysis and Prediction of the Elastic Properties of 3D and 5-D Tubular Braided Composites. *Acta Aeronaut. Astronaut. Sin.* **2007**, *28*, 123–129.
17. Shen, H.R. Investigation on Mechanical Properties for Three-dimensional Braided Circular Tubes and the Bearing Capacities for the Analogue Structure of a Rocket Linking Stage. *J. Natl. Univ. Defense Technol.* **1999**, *21*, 61–66.
18. Zhang, W.L. Structural design and dynamic characteristics analysis of braided composite two-stage gear transmission system. *Sci. Rep.* **2024**, *14*, 1–15. [CrossRef]
19. Zeng, J.H.; Zhang, Q.L.; Fu, Y.C. Heterogeneous ablation behavior of SiCf/SiC composite by nanosecond pulse laser. *Adv. Manuf.* **2025**, *13*, 196–210. [CrossRef]
20. Fang, G.D.; Liang, J.; Lu, Q. Investigation on the compressive properties of the three dimensional four-directional braided composites. *Compos. Struct.* **2011**, *93*, 392–405. [CrossRef]
21. Zhang, W.L.; Wang, X.P.; Ji, X.M. Parametric characteristics analysis of three cells in 3D five-directional annular braided composites. *PLoS ONE* **2021**, *16*, e0254691. [CrossRef] [PubMed]
22. Zhou, H.L.; Li, C.; Zhang, L.Q. Micro-XCT analysis of damage mechanisms in 3D circular braided composite tubes under transverse impact. *Compos. Sci. Technol.* **2018**, *155*, 91–99. [CrossRef]

processes

MDPI

Review

Polycrystalline Diamond Composite Bit Technology for Sustainable Work

Xiaoao Liu [1], Guodong Ji [1], Liyuan Yang [2], Haitao Ren [3], Qiang Wu [1], Jinping Yu [1] and Kuilin Huang [3,*]

[1] CNPC Engineering Technology R&D Co., Ltd., Beijing 102206, China; liuxadr@cnpc.com.cn (X.L.); jigddri@cnpc.com.cn (G.J.); wuqiangdri@cnpc.com.cn (Q.W.); yujpdr@cnpc.com.cn (J.Y.)
[2] CNPC Chuanqing Drilling Engineering Co., Ltd., International Engineering Company, Chengdu 610051, China; yangliyuan_sc@cnpc.com.cn
[3] College of Mechanical and Electrical Engineering, Southwest Petroleum University, Chengdu 610500, China; renhaitao781026@163.com
* Correspondence: huangklin@126.com

Abstract: Reducing the auxiliary drilling time and improving the drilling efficiency are not only important technical means to enhance the development and deep exploration of oil and gas reserves, but also have significant implications for ensuring energy security. Bottom hole self-renewal drill bit technology is a new technology that does not require the retrieval of drills from the well bottom, instead directly updating worn drill bits in place. Through the self-renewing structure of the drill bit, the sustainability of the drill bit can be increased. Conducting research on this technology is expected to overcome the technical issues related to the short useable life of drill bits, thus promoting single-trip drilling technology. This article summarizes the scientific progress made by researchers in self-renewal Polycrystalline Diamond Composite (PDC) drill bit technology; for example, the American National Oilwell Varco (NOV) company has developed a mobile self-renewal cone bit technology and introduced the structural characteristics and working principles of three types of rotating self-renewal drill bits: rack-driven, worm-driven, and ratchet-driven. This study compares and analyzes the advantages and disadvantages of these three rotary types of PDC drill bits, providing a detailed introduction to the working principles and modes of the triggering, transmission, limit and locking, and renewal structure devices of the drill bit. Relevant suggestions are proposed for the development of bottom hole self-renewal PDC drill bit technology; namely, strengthening research efforts relating to the intelligent judgment and recognition of cutting tooth wear of the drill bit, as well as the integration of multiple technologies. The sustainability of these novel bits can provide technical support for the development of single-trip drilling technology for deep formations, improve the service life of bits in deep formations that are difficult to drill, and contribute to the efficient development of deep-sea energy resources worldwide.

Keywords: sustainability; PDC bit; long life; self-renewal structure; transmission structure

1. Introduction

The low drilling efficiency of deep well sections is a bottleneck problem affecting the drilling cycle and the cost of drilling deep and ultra-deep wells [1–3]. Frequent tripping and the tripping of deep well sections due to poor formation drillability and the short life of bits are among the main reasons for this low drilling efficiency. Based on incomplete statistics, the tripping time associated with changing the drill bit is often longer than the pure drilling time in deep well sections with depths exceeding 6000 m; furthermore, the deeper the well, the higher the tripping time. The last 1246 m of the well Pengshen 6 (well depth: 9026 m; total drilling period: 561 days)—recorded as the deepest vertical well in Asia—took 200 days, accounting for 35.7% of the total drilling period; it required tripping 21 times, accounting for 34.4% of the total (61 times); and the total tripping time was 52.5 days, accounting for 26.1% of the total drilling time of this spud. The pure drilling time of this spud was only 24.9 days, and the tripping time was more than twice that of the pure drilling time. In deep formations, the combined action of the drilling fluid's column pressure and confining pressure makes the rock stratum more compact, thus worsening the rock's drillability and shortening the bit life; therefore, the tripping frequency associated with changing the bit is higher, and the single tripping time for drilling becomes longer in deeper well sections. Therefore, the high frequency of bit changes, the increase in the single tripping time, the problem of low drilling efficiency, and the issue of high cost caused by tripping and bit replacement in the context of deep drilling have been highlighted as key issues in the relevant literature [4,5].

Deep and ultra-deep wells have been increasing in number year by year. The exploration and development of deep oil and gas resources is the main way to achieve the strategic success of such resources [6]. Some achievements have been made through "deep engineering" approaches implemented in the field of oil and gas exploration and development. In this line, key technologies such as ultra-high-strength casings, drill pipes, drilling fluid, and measurement and control instruments have been put forward. More than 8000 m deep wells, such as the "bottom hole Mount Qomolangma" well, are emerging [7,8], and ultra-deep wells have been deployed and drilled in recent years [7]. However, it is still difficult to solve the problems related to the long drilling cycle and the high cost caused by frequent tripping and bit replacement when drilling deep well sections, which has directly affected the scale of the "bottom hole Mount Qomolangma" well and significantly hindered its implementation [9–11].

"Single-trip drilling" refers to drilling technology in which one drill bit and one set of downhole drilling tools are combined to drill one spud or the entire drilling footage, which has the prominent advantages of reducing the number of drill bits used, improving the drilling speed, shortening the drilling cycle, and reducing the drilling cost [12]. Single-trip drilling is a comprehensive technical method, the essence of which is to maximize the rock-breaking capacity of the drill bit through the comprehensive application of various means, thus completely avoiding the tripping associated with needing to change the drill bit. Single-trip drilling technology has gradually been developed due to the economic development of the shale and tight oil and gas (i.e., unconventional oil and gas) industry in North America, recognized in 2024 as one of the ten major scientific and technological advances in the international petroleum industry. In shale oil and gas horizontal well drilling in the United States, single-trip drilling of a single well section has gradually become normal, while the single-trip drilling of multiple well sections has not been conducted. The footage completed via single-trip drilling ranges from hundreds to thousands of meters, and the longest single-trip drilling footage exceeded 6000 m [13–16]. However, in long and ultra-long well sections, single-trip drilling is yet to be realized. At present, it is mainly performed in horizontal sections with a shallow burial depth (generally less than 2000 m), as this stratum has good homogeneity and drillability; in contrast, it is very difficult to realize single-trip drilling in strata characterized by a large burial depth, poor drillability, and heterogeneity.

Among many key tools, using a high-quality and high-efficiency bit is the most basic condition to realize single-trip drilling. When the bit life is short and the footage is deeper, single-trip drilling is impossible. The service life of the existing bit is affected by the material of the cutting teeth, and material technology is not likely to extend the wear life of the cutting teeth in a significant way. Therefore, a new structure for PDC bits that is suitable for integration into single-trip drilling technology needs to be developed.

2. Bottom Hole Self-Renewal PDC Bit New Technology

Bottom hole bit self-renewal technology is a new technical idea, where the old bit is directly renewed at the bottom hole without the need for tripping. Generally, when the rock-breaking efficiency of a bit is seriously reduced or drilling cannot be continued due to the wear of the bit, the service life of the bit ends and drilling can only be continued after the bit is pulled out and replaced. In contrast, it is not necessary to pull out and change the drill bit when using bottom hole bit self-renewal technology, as the worn cutting structure can be replaced with a preparatory drill bit cutting structure via ground control, thus completing the bit renewal process in place. The renewal bit is a new bit that can be directly used to drill, thus avoiding the tripping process to change the bit. The bottom hole bit self-renewal technology can be used to implement one or more bits without the need for tripping. This can extend the service life and continuous drilling time of the bit, greatly reduce the tripping time, increase the travel penetration rate (ROP), and reduce the drilling cycle and cost.

Schlumberger proposed indexable roller bit technology [17]. The left side of Figure 1 shows a sectional view of the assembly of the roller bit, while the right side of Figure 1 shows a schematic diagram of the roller engagement. The roller (133) on the bit is equipped with a driving mechanism (150) which is connected to the rotation mechanism (139). The driving mechanism can drive the rotation mechanism to rotate to a pre-determined position and fix it. The indexing mechanism (120) rotates the cone (121) and enables the teeth at different positions on the cone to break rock. When the teeth at a certain position are worn, the indexing mechanism can be driven to rotate and fix again to break rock with the teeth at a new position.

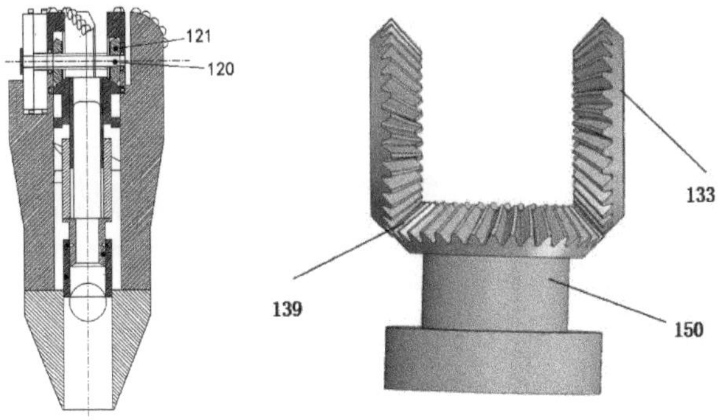

Figure 1. Technical scheme of rotary bit proposed by Schlumberger.

In 2018, the Drill Research Institute of Southwest Petroleum University proposed a long-life PDC bit with a translation blade [18]. The left side of Figure 2 shows a three-dimensional schematic diagram of a long-life PDC drill bit, the middle part of Figure 2 shows a cross-sectional view of the long-life drill bit, and the right side of Figure 2 shows a radial coverage tooth layout diagram for the long-life drill bit. The drill bit has a retractable translation blade (31), which is in a retracted state at the initial working stage of the drill bit (1), and does not participate in rock-breaking. When the cutting teeth (41a) on the fixed blade reach a certain degree of wear, the cutting structure (31a) of the translation blade stretches to the preset working position and is locked through the ground remote control, thus replacing the old cutting structures.

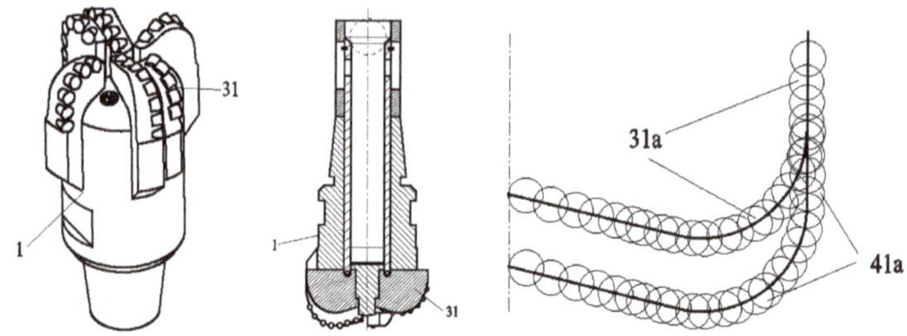

Figure 2. Long-life PDC bit with translation blade.

National Oilwell Varco (NOV) has developed a PDC drill bit that allows for the retraction of part of the cutter blades through a change in hydraulic parameters [19]. The retractable cutter wings of the drill bit are in a retraction state at the initial working stage. When the drill bit is worn, part of the cutter wings stretch out via hydraulic control, thus reinforcing the cutting structure. This scheme can compensate for the wear of the cutting structure of the drill bit once, extending the service life and continuous drilling capacity of the drill bit [20–24].

On the basis of a disc cutter bit and cross-scraping PDC bit technology, the bit research institute of Southwest Petroleum University has put forward a bit with a renewable cutting structure, comprising rotary disc cutter bit technology; that is, the rotary cutter wing (disc cutter) of the bit can be controlled to rotate [25]. The left side of Figure 3 shows a schematic diagram of a rotating renewable drill bit. The basic idea of this technology is that the disc cutter on the drill bit can be controlled to rotate to a preset working position and locked. When the bit is working in the bottom hole, if the initial cutting teeth are worn, the disc cutter can be rotated on the ground through remote control, such that the worn teeth are replaced with new cutting teeth to continue drilling. Therefore, the bit can complete multiple updates of the cutting structure in the bottom hole, achieving the effect of replacing the bit at the bottom hole without tripping. On the basis of the above ideas, the Drilling Research Institute of Southwest Petroleum University has continued to put forward technical schemes for non-disc cutter bits, where the disc cutter is designed instead as a curved polygon. The curved edge of the non-disc cutter can be designed according to the cutting contour of the drill bit, which matches the cutting contour of the drill bit. On the right side of Figure 3, the radial tooth layout diagram of the rotating renewable drill bit is shown. This scheme breaks the constraint of the idea of a circular disc cutter, and can be customized and flexibly designed according to the environment in which it is used and the conditions of the drill bit. The number of cutting structures on the drill bit is determined by the number of sides of the polygon [26–31].

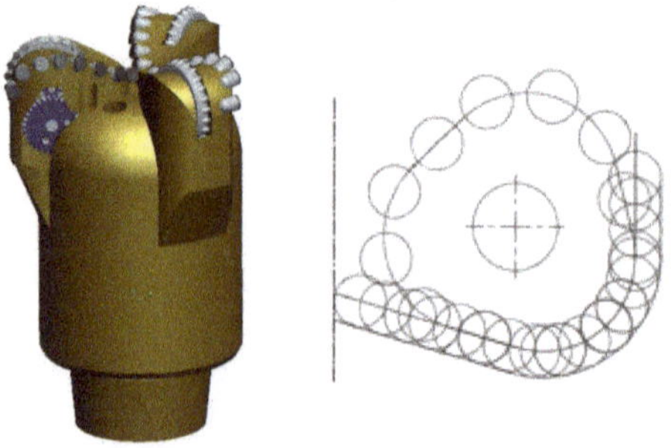

Figure 3. Technical principle of rotary table bit.

3. Scheme Design of Rotary Self-Renewal Bit

Translational self-renewal bits need to be provided with multiple cutter blades in a limited space; however, especially in the core area of the bit, it may be impossible to provide multiple cutter blades and, so, the translational self-renewal bit will be limited in actual application. However, the renewal structure of the rotary self-renewal bit has been updated through the circular disc cutter structure, occupying less bit space. The rotary self-renewal bit is composed of a trigger mechanism, a driving mechanism, a limiting mechanism, and a locking mechanism. According to the different driving structures [32,33], rotary self-renewal bits can be divided into rack drive self-renewal bits, worm drive self-renewal bits, and ratchet drive self-renewal bits.

3.1. *The Rack Drive Self-Renewal Drill*

The rotary rack drive self-renewal drill bit comprises a self-renewal wheel, a wheel shaft, a self-renewal gear, a triggering mechanism, a driving mechanism, a limiting and locking mechanism, and a spring reset mechanism [34–38]. The drill bit driving mechanism is a gear rack structure, and the triggering mode is triggered by a ball-throwing and pressure-holding mode. The limiting and locking mechanisms are triangular components. The three-dimensional model of a rack driving self-renewal drill bit is shown in Figure 4.

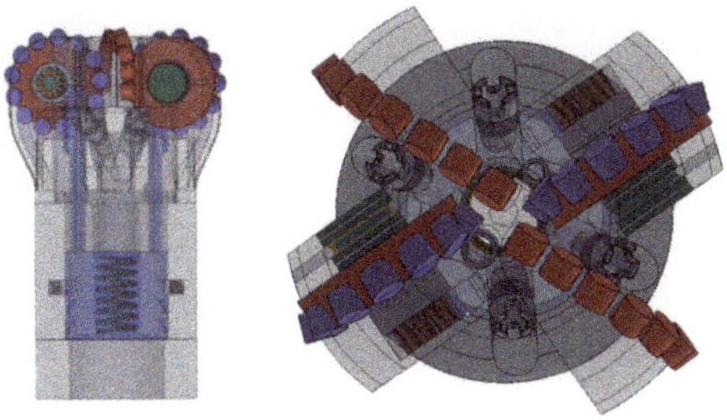

Figure 4. Three-dimensional view of rack-driven self-renewal bit.

The self-renewal wheel in the drill bit is connected with the drill bit body through the wheel shaft; the gear and the wheel shaft are connected through the key, the gear, and the rack mesh with each other; the rack and the piston member form a fixed connection; the spring reset mechanism is arranged between the piston member and the drill bit body; and a groove is arranged at the corresponding position of the piston member and the prism lock member. When the drill bit works normally, the self-renewal teeth are equivalent to the fixed teeth scraping the bottom rock and the spring reset mechanism is in a compression state, such that the piston member moves downward. Due to the unidirectional locking performance of the prism member, the piston is in a static state, thus locking the rack. Due to the meshing effect of the gear and rack, the gear is locked and, finally, the self-renewal wheel is locked. When the cutting teeth on the drill bit are worn, such that the drill cutting teeth need to be renewed, the drill bit is lifted away from the bottom of the well for a certain distance, and the displacement of drilling fluid is increased by throwing balls to hold the pressure. Figure 5 shows a cross-sectional view of the ball-type trigger mechanism. In this way, upward pressure is applied to the piston pressure-bearing surface. As the prism lock can only be locked downward, the piston pushes the rack to move upward under the pressure of the drilling fluid. Figure 6 shows the working diagram of the prism locking mechanism. The rack drives the gear to rotate, and the gear drives the self-regulating wheel to rotate. When the piston has moved a certain distance, the prism is locked under the action of the spring, such that the cutting teeth of the drill bit are renewed once [39,40].

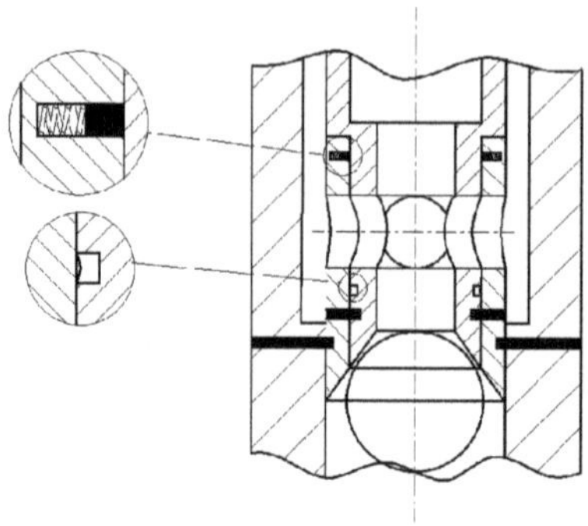

Figure 5. Ball-type trigger mechanism.

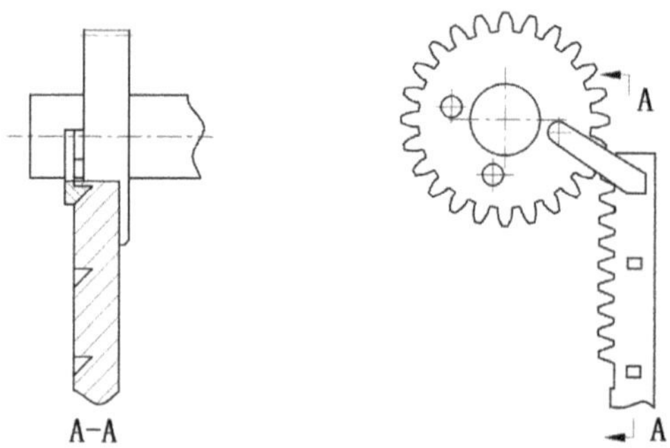

Figure 6. Working diagram of prism locking mechanism.

The key structural parameters of the rack-driven self-renewal drill bit mainly include the drill bit parameters, reset spring parameters, gear rack structural parameters, axle structural parameters, and so on. The specific values are given in Table 1.

Table 1. Key parameters of rack-driven self-renewal drill.

Drill bit parameter	Drill diameter (mm)	Drill length (mm)	Cutting tooth diameter (mm)	Cutting tooth length (mm)	Number of cutting teeth (grain)
	215.9	330	15.875	13	40
Parameters of return spring	Spring pitch diameter (mm)	Free height of spring (mm)	Spring diameter (mm)	Working Stroke (mm)	Coefficient of elasticity (KN·mm^{-1})
	20	60	4	35	37
Rack and pinion parameters	Gear tooth shape	Gear pressure angle (°)	Gear ratio	Rack movement distance (mm)	Gear module
	Involute	20	0.75	35	20
Structural parameters of axle	Diameter of main shaft (mm)	Length of main shaft (mm)	Small shaft diameter (mm)	Small shaft length (mm)	Bearing length (mm)
	37	20	14	13	35

3.2. The Worm-Driven Self-Renewal Drill

Three-dimensional models of a worm-driven self-renewal drill bit are shown in Figure 7. The driving mechanism is the worm gear mechanism, the trigger mechanism includes the piston component connected with the rack, the lower end of the piston component is provided with a bearing surface, the transmission structure is a ball screw, the

after 2–3 times the maximum conventional weight is applied to the bit, and the press-type rotating device under the safety spring begins to operate.

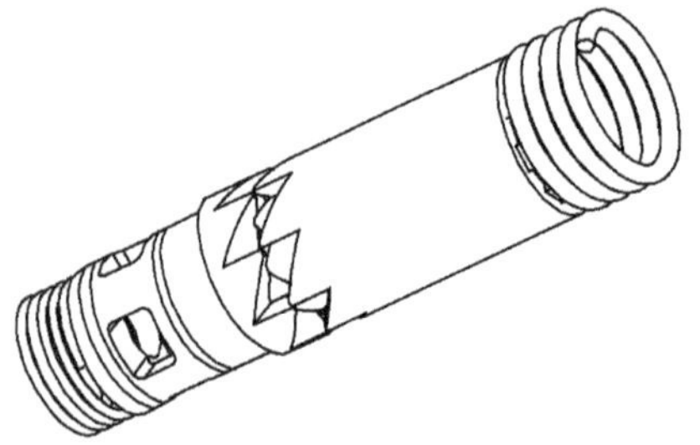

Figure 14. Bit pressure trigger device.

4.2. Transmission Structure Device

The transmission device comprises a gear rack transmission structure, a bevel gear structure, a worm gear and worm structure, and other different structures. In particular, the gear rack transmission structure comprises a rack and a gear set. When the rack is pushed downward by a trigger device, the rack and the gear set are engaged to drive the gear set to rotate. The gear set is composed of one or more gears which, when engaged, drive the indexing cutter blade to rotate. The bevel gear structure is composed of a pair of bevel gears, which are meshed with each other to realize 90° rotation transmission. The other bevel wheel is driven by the bevel wheel, in order to drive the rotation of the self-renewal wheel. The worm gear and worm structure consist of a worm and worm gear. This mechanism has a self-locking function, and the worm gear can only be driven by the worm gear to rotate. No separate locking mechanism is required.

4.3. Limit and Locking Device

The limiting and locking mechanism comprises a limiting pin and a limiting slot structure; for example, a ratchet structure. The limiting pin and limiting slot can be arranged on the transmission device or a trigger device. The elastic mechanism continuously applies a thrust, thus pushing the limiting pin toward the transmission device, and a limiting slot is arranged on the limiting pin of the corresponding transmission device, where the spacing between the limiting slots matches the angle between them. In this way, the limiting pin is inserted into the limiting slot. When the transmission device continues to move downward, the limiting pin is pushed to exit the limiting slot. Figure 15 shows the structure of the ratchet. The pawl and ratchet structure comprises a push rod, a rotating body, a pawl, and a ratchet wheel. The push rod is driven by a trigger device to move up and down in the drill bit body. The push rod and the rotating body are connected to drive the rotating body to rotate automatically. The rotating body has an inner cavity, which is provided with a pawl and a ratchet wheel. The pawl and the rotating body are elastically connected. A plurality of teeth are arranged on the outer edge of the ratchet wheel and, when the push rod moves up to drive the rotating body and its pawl to rotate, the pawl will freely pass through these teeth. When the push rod moves down to drive the rotating body and its pawl to rotate in a reverse direction, the pawl and a certain clamping wheel will engage to drive the ratchet wheel to rotate.

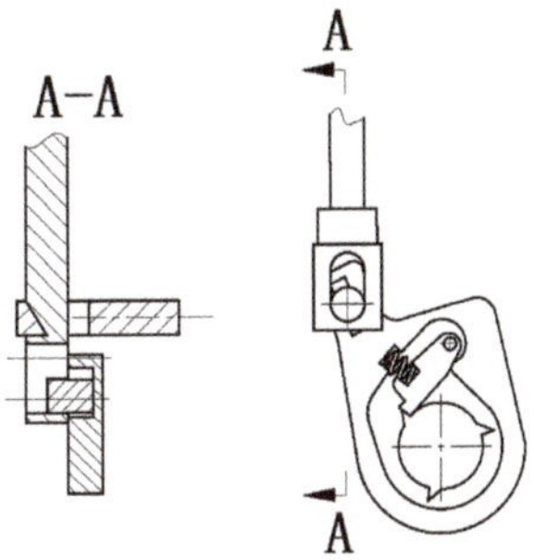

Figure 15. Structure of ratchet wheel.

4.4. Update of Structural Units

The rotary self-renewal structure device comprises two structures: a rotary disc cutter and a self-renewal wheel. Figure 16a shows a schematic diagram of a rotary drill bit, while Figure 16b shows a schematic diagram of a self-renewal wheel bit. The axis of the rotary disc cutter is perpendicular to the radial direction of the drill bit. The rotary disc cutter and the drill bit body can realize single- or double-side support connection. The axis of the self-renewal wheel is consistent with the radial direction of the drill bit (Figure 16b). The updating wheel is connected with the drill bit body through the shaft. The self-renewal wheel can realize multiple bottom hole renewals. While the bottom hole coverage area may be limited, especially the external area where the bit is easily worn, and the number of rotary disc cutter renewals is limited, the bottom hole coverage area of the bit is large, such that full coverage of the radial area of the bit can be achieved.

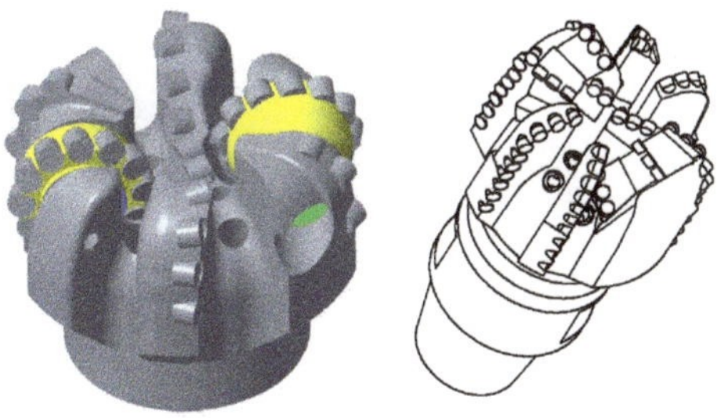

(a) (b)

Figure 16. Schematic diagrams of drills with different renewal structures: (**a**) schematic diagram of rotary disc bit; (**b**) schematic diagram of self-renewal wheel bit.

5. Prospects for Future Research

The real-time monitoring and identification of the working state (including wear state) of bottom hole bits has always been a difficult problem in the field of drilling engineering, which plays an important role in determining the timing of bit replacement. At present, the methods for identifying the working state of the bottom hole bit can be preliminarily divided into two types: indirect ground monitoring and identification (represented by the "mechanical specific energy principle") and direct monitoring and identification of the bottom hole bit (represented by the "intelligent bit").

The principle of mechanical specific energy is to observe its impact on the ROP by changing the bit wear parameters and bit design. Specific mechanical energy (MSE) models can be used to adjust the working parameters. Under the condition of not affecting the comprehensive wear, the ROP value can be maximized in the local drilling process; therefore, it is used as a global optimization tool and has been popularized for application.

cellulose (BC) material using the standard PCB manufacturing method [16]. Geczy et al. presented a PLA/flax composite and validated the technology by a two-sided blinker demonstrator [17]. Khrustalev et al. presented a PLA/glass-fiber composite PCB with components soldered using the Rose alloy (melting point 95 °C) [18]. Jiva Materials Ltd. developed a plant-based PCB base material (Soluboard) made from natural fibers enclosed in a non-toxic polymer that dissolves when immersed in hot water. They claimed up to a 67% carbon footprint reduction compared to FR4 [19].

Furthermore, studies have focused on using pure biopolymers including PLA, polyvinyl alcohol (PVA), Polyhydroxybutyrate (PHB), and Cellulose acetate (CA) as a PCB substrate. A study reported by Habib Ullah et al. developed aerogel poly (butylene succinate) (PBS) as a substrate for RF and microwave applications [20]. Schramm et al. showed the capability of using PHB, CA, and PLA with the standard process chain of the electronics manufacturing. However, they chose hot embossing instead of wet etching which is not widely used in large production volumes. Furthermore, the components were soldered with a low-temperature solder alloy (Sn42Bi58). They realized that PHB showed a high deformation due to the high temperature [21]. Henning et al. demonstrated the functionality of a humidity sensor on a PLA/CA composite. However, they encountered a warping of the substrate and consequently unreliable solder joints due to the soldering temperature [22].

PLA is a popular thermoplastic polyester derived from renewable sources such as corn and sugarcane. It has good mechanical strength and thermoplastic properties that can be used for many applications [23]. Using PLA in PCB production offers advantages including biodegradability and natural origin, reducing the carbon footprint of PCB technology. Moreover, PLA's processability at lower temperatures can save energy during production. These features make this material a perfect choice for sustainable electronics. Naji et al. reported the environmental contribution of the PCB substrate regarding the Global Warming Potential (GWP). They showed that replacing FR4 with PLA contributes to a 90% reduction in GWP impacts [24]. However, there are still challenges with using PLA in the industrial PCB production line. The PLA polymer exhibits a low thermal stability compared to conventional materials. It has a low melting temperature of 130–180 °C and cannot withstand the high temperature of the standard soldering process in PCB technology. Geczy et al. have focused on optimizing different low-temperature soldering profiles for PLA. They created a custom vapor phase soldering (VPS) liquid with a boiling point temperature of around 170 °C, specific for low-temperature solder paste (NC-SMQ81; composition: 58Bi/42Sn). However, the PLA-based PCBs still deformed despite the decreased temperature of the new VPS process [25,26].

PHB is a biodegradable polyester produced by the bacterial fermentation of sugars. This biopolymer can be an environmentally friendly alternative to traditional PCB materials, as well. However, PHB faces challenges such as its brittleness and low thermal stability which can limit its dimensional stability in the PCB manufacturing processes [27]. To the best of our knowledge, no study reported on the usage and evaluation of PHB in the different steps of standard PCB production.

Note that PLA and PHB exhibit limitations in their mechanical properties that must be addressed for a broader adoption in electronic applications. They are relatively brittle with a low impact resistance restricting their application in mechanically dynamic environments [28,29]. Ongoing research is focused on reinforcing these biopolymers using natural or synthetic fibers to enhance their mechanical resilience [17,18]. However, it is important to note that such reinforced variants are still under development and are not yet commercially available. In this study we focused on commercially available PLA and PHB to demonstrate feasibility within the constraints of current material accessibility, while acknowledging that further enhancements will be essential for broader industrial and user applications.

In this study, a technology is introduced to fabricate a single-sided rigid PCB by applying the PLA or PHB substrate at the end of the PCB production process using injection molding technology. The biopolymer is applied on a flexible polyimide (PI) foil. This approach avoids PLA or PHB undergoing harsh PCB process conditions such as soldering (high temperatures) and Cu patterning through wet etching (acidic and/or alkaline media) leading to the deformation and/or decomposition of these polymers. This stability is crucial for preserving the mechanical properties and functionality of the circuit. Furthermore, injection molding is a widely used process in the plastics industry, known for its efficiency and scalability. This makes it particularly advantageous for mass production, as the technology is already well integrated into manufacturing workflows, allowing for a seamless adoption in large-scale PCB production. Furthermore, overmolded polymers can be produced in a wide variety of shapes and thicknesses using this technology. Thus, manufacturers can design PCBs with tailored dimensions and configurations for various applications. This capability enhances the applicability of biopolymer-based PCBs and supports innovation in eco-friendly electronics design.

This paper investigates the feasibility and repeatability of the introduced technology. Moreover, the challenges encountered including adhesion between flexible PCB and biopolymers are addressed. Eventually, two end-of-life practices of debonding and quantitative composting are presented. Experimental results show that this approach is a reproducible, scalable, and reliable method for using biopolymers as base substrate of PCBs.

2. Materials

For the flexible PCB, three commercial Cu cladding PI foils have been used as follows: Upisel SR1220, Pyralux FR9120R, and Shengyi SF305R. The layer build-up of the three foils is illustrated in Table 1.

Table 1. The build-up of the three Cu cladding PI foils used in this work.

Flexible Foil Layers	Upisel-N SR1220	Pyralux FR9120R	Shengyi SF305R
Polyimide (PI)	50 μm	50 μm	25 μm
Adhesive	No adhesive layer	25 μm acrylic-based adhesive	20 μm epoxy-based adhesive
Copper (Rolled-annealed)	18 μm	35 μm	18 μm

The Upisel foil was purchased from UBE Europe GmbH, Düsseldorf, Germany, the Pyralux foil was supplied by DuPont, Wilmington, DE, USA, and the Shengyi foil was obtained from SYTECH, Dongguan, China.

Concerning sustainable polymers for the overmolding process, PLA was supplied by NatureWorks, Plymouth, MN, USA, in pellet form with the reference 3100HP. The other biopolymer, PHB, was purchased from Biomer, Schwalbach, Germany, in pellet form with the reference P263E. Both PLA and PHB are easy-to-process polymers suitable for injection molding and extrusion.

To improve adhesion between the PI foil and the overmolded biopolymer, the Bemis 3914 TPU tie layer with a 100 μm thickness was obtained from Bemis Associates Inc., Shirley, MA, USA.

core of the drill is provided with a reset spring mechanism, and the limiting and locking mechanisms have a ratchet groove structure.

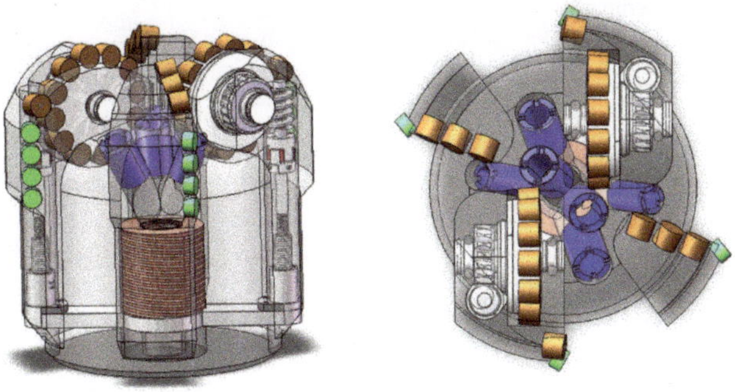

Figure 7. Three-dimensional views of worm-driven self-renewal bit.

The self-renewal wheel axis is vertical to the radial direction of the drill bit, and is connected with the drill bit through the wheel shaft to achieve support on both sides. The axial locking of the self-renewal wheel is realized through the ball. The self-renewal wheel, worm wheel, and wheel shaft are connected through the spline. The worm axis is parallel to the drill bit axis. The lower end of the worm is connected with the ball screw through a coupling. A bearing is arranged at the upper end of the worm and the lower end of the screw. The piston is connected with the ball screw through the connecting rod. A ratchet groove limiting and locking mechanism is arranged at the connecting rod, as shown in Figure 8 as a schematic diagram. When the drill bit works normally, the cutting teeth on the renewable wheel are fixed teeth. Due to the self-locking effect of the worm wheel and worm, the worm wheel cannot drive the worm to rotate, ensuring that the renewable wheel will not rotate reversely and the spring at the center of the drill bit is in a compressed state, causing the piston member to move downward. Due to the one-way locking performance of the ratchet groove, the piston is in a balanced state, thus locking the ball screw. When the drill bit needs to be updated, the pressure on the bearing surface of the piston increases, such that the piston moves upward, causing the ball screw to move, the worm to rotate and, finally, the renewable wheel to rotate. When the connecting rod moves to the ratchet groove position, limiting and locking can be completed.

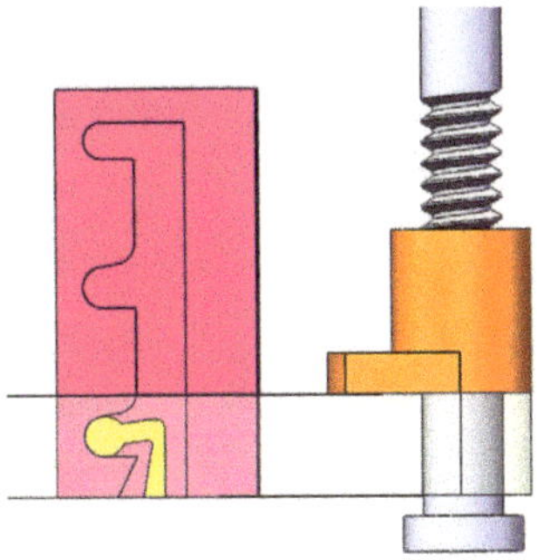

Figure 8. Schematic diagram of ratchet groove locking mechanism.

The key structural parameters of the worm-driven self-renewal drill bit mainly include drill bit parameters, reset spring parameters, worm gear and worm gear structural parameters, wheel shaft structural parameters, ball screw parameters, and other parameters. Specific values are provided in Table 2.

Table 2. Key parameters of worm-driven self-renewal drill.

Drill bit parameter	Drill diameter (mm)	Drill length (mm)	Cutting tooth diameter (mm)	Cutting tooth length (mm)	Number of cutting teeth (grain)
	215.9	221	15.875/13.44	13/8	43
Parameters of return spring	Spring pitch diameter (mm)	Free height of spring (mm)	Spring diameter (mm)	Working stroke (mm)	Coefficient of elasticity (KN·mm^{-1})
	45	60	5	19.2	367
Parameters of worm wheel and worm	Worm gear modulus	Number of threads	Gear ratio	Number of worm gear teeth	Worm reference circle diameter (mm)
	2	20	18	18	18
Structural parameters of axle	Diameter of main shaft (mm)	Length of main shaft (mm)	Small shaft diameter (mm)	Small shaft length (mm)	Bearing length (mm)
	37	20	30	35	35
Parameters of ball screw	Ball screw lead (mm)	Ball diameter (mm)	Working stroke (mm)	Diameter (mm)	Total length (mm)
	2	1.2	28.5	8	100

3.3. The Ratchet Drive Self-Renewal Drill

Three-dimensional models of a ratchet-driven self-renewal drill bit are shown in Figure 9. The drive mechanism is a ratchet mechanism, the trigger mechanism is a ratchet shaft, the lower end of the ratchet shaft is provided with a bearing surface, the transmission structure is a pair of bevel gears, a return spring mechanism is provided between the bevel gear and the ratchet shaft, and the limit and locking mechanisms are ratchet groove structures on the ratchet shaft.

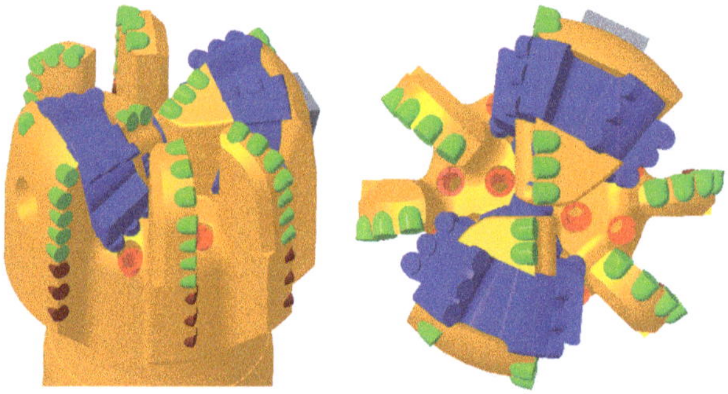

Figure 9. Three-dimensional views of ratchet drive self-renewal bit.

The assembly diagram of the drill parts is shown in Figure 10. A self-renewal wheel is provided on the support blade, while a self-renewal gear is provided on the self-renewal wheel. The self-renewal wheel is rotatably connected with the support blade through a wheel shaft, and a cone wheel is provided at one end of the self-renewal wheel. A drive shaft, a ratchet shaft, a spring, and a sliding pin are arranged inside the bit body. One end of the drive shaft is a tapered wheel, while the other end is a spline groove. The tapered wheel of the drive shaft and the tapered wheel of the self-regulating wheel mesh with each other. One end of the ratchet shaft is an upper spline. The ratchet shaft is provided with a ratchet groove, and the lower end of the ratchet wheel is provided with a bearing surface. Figure 11 shows a three-dimensional diagram of the ratchet shaft. The upper spline of the ratchet shaft and the spline groove of the drive shaft cooperate with each other, and the sliding pin and the ratchet groove form a sliding fit. The spring is arranged between the ratchet shaft and drive shaft, and the sliding pin is arranged between the bit and the joint. When the drill bit works normally, the drilling fluid pressure and spring elastic force are balanced mutually, the sliding pin locks the ratchet shaft's rotation in the ratchet groove, and the self-renewal wheel is locked by the transmission mechanism. When the cutting teeth at the bottom of the well are worn and need to be replaced, the drill bit is lifted to a certain distance from the bottom of the well and the drilling fluid pressure is increased using the ground equipment. Once the sliding pin reaches the lower dead center, the displacement of the drilling fluid is reduced. Driven by the spring, the ratchet moves downward in the axial direction, and the sliding pin gradually reaches the upper slope of the ratchet groove. This slope pushes the ratchet shaft to rotate continuously until it reaches the top dead center of the ratchet groove. The slope will push the ratchet shaft to continue to rotate until the sliding pin is located at the lower dead center of the ratchet groove. Under the action of the spring, the ratchet moves downward in the axial direction, and the sliding pin gradually reaches the upper slope of the ratchet groove; this pushes the ratchet shaft to rotate continuously under the action of the slope until it is located at the top dead center of the ratchet groove. In this way, the ratchet shaft moves and rotates at the same time. The upper end of the ratchet shaft is provided with an upper spline. The drive shaft spline groove, the spline, and the spline groove cooperate with each other, and the ratchet drives the drive shaft to rotate. The conical wheel at the upper end of the drive shaft meshes with the conical wheel on the self-renewal wheel to drive the self-renewal wheel to rotate, so that the self-renewal teeth are turned to the outer surface of the blade, forming a part of the cutting structure of the drill bit, thus providing the bottom hole self-renewal capacity and achieving the purpose of extending the service life of the bit.

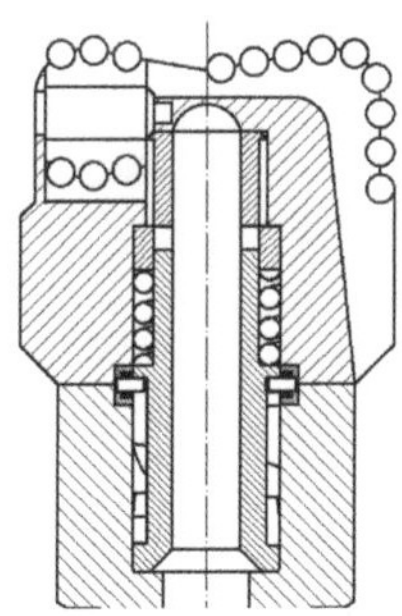

Figure 10. Assembly diagram of drill bit.

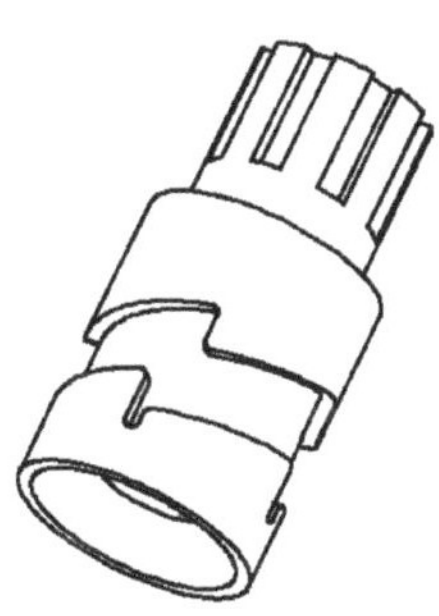

Figure 11. Three-dimensional diagram of ratchet shaft.

The key structural parameters of a ratchet-driven self-renewal bit mainly include bit parameters, reset spring parameters, cone wheel parameters, wheel shaft structural parameters, and ratchet wheel structure parameters, among others. The specific values are given in Table 3.

Table 3. Key parameters of ratchet-driven self-renewal bit.

Drill bit parameter	Drill diameter (mm)	Drill length (mm)	Cutting tooth diameter (mm)	Cutting tooth length (mm)	Number of cutting teeth (grain)
	215.9	413.5	15.875/13.44	13	45
Parameters of return spring	Spring pitch diameter (mm)	Free height of spring (mm)	Spring diameter (mm)	Working stroke (mm)	Coefficient of elasticity (KN·mm^{-1})
	62	120	16	50	37
Parameters of cone pulley	Inner hole diameter (mm)	Gear pressure angle (°)	Gear ratio	Tooth width (mm)	Gear module
	20/10	20	3	35	1
Structural parameters of axle	Diameter of main shaft (mm)	Length of main shaft (mm)	Small shaft diameter (mm)	Small shaft length (mm)	Bearing length (mm)
	28.5	20/18	15	20	45
Ratchet wheel parameters	Groove depth (mm)	Ratchet wheel diameter (mm)	Working stroke (mm)	Number of clips	Ratchet wheel length (mm)
	10	122	60	6	248

Through a comparison and analysis of these three self-renewal schemes, it can be seen that the advantage of the rack-and-pinion drill bit is that the rack-and-pinion structure is relatively simple and its reliability is good, while its disadvantages include that the control accuracy is low and the gear can easily buckle. The worm-driven drill bit structure has the advantages of strong self-locking capacity and high rotation control accuracy, as well as the disadvantages of a complex drill structure and need for the precise design of the pump pressure and spring rigidity. The advantage of ratchet-driven bits is that the bit is updated frequently, while the disadvantages are that the hydraulic structure of the bit is difficult to design and the internal transmission structure is complex.

4. Methodology

4.1. Triggering Device

Triggering devices include mechanical, electric, pneumatic, and hydraulic triggers. Common trigger devices include ball-throwing pressure-holding trigger devices, piston sliding trigger devices, and drilling pressure start trigger devices.

Figure 12 shows the pressure-holding processes of multiple pitching-type devices. The ball-throwing triggering device comprises a pressure-holding ball put into the ground and a multi-level one-way triggering mechanism arranged between the bit body and the transmission device or the indexing cutter wing. The multi-level one-way triggering mechanism comprises at least two layers of (inner and outer) nested sliding sleeves arranged in the drilling fluid flow channel of the bit body. The top surfaces of the inner and outer sliding sleeves are used to receive the pressure-holding ball, while the innermost layer of the inner sliding sleeve is used to drive the transmission device or the indexing cutter wing. The inner sliding sleeve of each layer can slide up and down in the inner cavity of the outer sliding sleeve, while the outermost outer sliding sleeve can slide up and down in the inner cavity of the drilling fluid flow channel. Shear pins are provided between the inner and outer sliding sleeves of each layer, as well as between the outermost outer sliding sleeve and the inner wall of the drilling fluid flow channel. The inner and outer sliding sleeves of each layer are provided with through-holes communicating inside and outside the inner cavity, the inner sliding sleeve of each layer is provided with a limit hole above the through-hole, and the outer sliding sleeve of the outer layer is correspondingly provided with a limit pin below the through-hole. The limit pin is continuously applied by the elastic mechanism, thus pushing it and causing a pushing force inside the limit hole. The pressure-holding processes of multiple ball-throwing devices are shown in Figure 12.

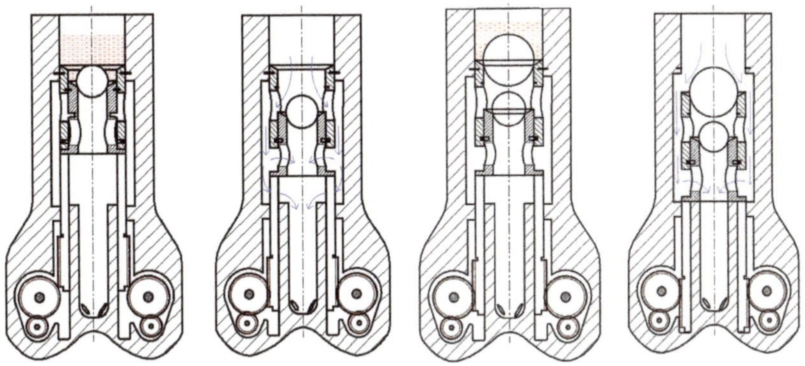

Figure 12. Pressure-holding process of multiple pitchers.

Figure 13 shows the piston sliding triggering device. The triggering device comprises a piston body, a return spring, and a transmission structure. When the drill bit works normally, the pressure of the drilling fluid on the piston and the pre-pressure of the spring counteract each other. When the self-renewal wheel needs to be rotated, the pressure of the drilling fluid is increased to push the piston compression spring to move upward, thus realizing the self-regulating trigger function of the drill bit.

Figure 13. Piston sliding trigger device.

Figure 14 shows the trigger device for starting the drill bit pressure. In the pressure triggering device, the elastic coefficient of the safety spring above the pressing member is greater than that of the starting spring below the rotating member, and the pressure at which the safety spring deforms is 2–3 times the maximum value of the normal working bit pressure. When the bit is under normal bit pressure, the press-type rotating device will not start; when the bit needs to be renewed, the upper safety spring will be deformed

Intelligent bit technology involves installing sensors and intelligent microchips on the bit; collecting the characteristics of the drilling stratum; measuring the bit weight, torque, rotation speed, and other data; identifying the working state of the bit at the bottom of the hole; and adjusting the drilling parameters of the bit. The use of these measurement data and interfaces helps to reduce or eliminate the low efficiency caused by bit design, bottom hole assembly, and parameter selection, and can support decisions regarding the dynamic performance of the bit.

To summarize, bit technology has developed rapidly, thus playing an important role in promoting the progress of drilling technology. In deep-sea and other high-cost drilling projects, bottom hole self-renewal bit technology has great application potential.

6. Conclusions

With the progress of resource exploration and development, there exist overall relationships between the response of the excavation bit, drilling parameters, and geological characteristics. Targeted tool/process design and drilling parameter optimization for speed and efficiency improvement have become urgently necessary for the promotion of deep and unconventional resource development. With the self-renewal bit as a representative, the self-renewal mode characterized by remote control, strong reliability, and high transfer efficiency has gradually become the key direction of research and development regarding long-life rock-breaking tools for oil and gas wells worldwide. Based on the technical idea of a rotary self-renewal structure, authors have studied the sustainability of self-renewal bits and explored some key problems relating to self-renewal bit triggers, transmission, limiting and locking, and renewal structure devices.

(1) The downhole renewal of cutting teeth provides a way to increase a bit's ability to drill sustainably; as opposed to increasing the wear resistance of cutting teeth to extend the bit life, the method of replacing cutting teeth at the bottom of the well is the most effective means to realize one-trip drilling in oil fields at present. The vigorous development and promotion of this technology is expected to overcome the technical issues related to low footage and the difficulty of increasing the speed of drilling in deep hard rock strata in China.

(2) Rotary bottom hole self-renewal PDC bit technology relies on the self-locking characteristics of the worm wheel and worm to achieve one-way locking, which is highly reliable and allows no less than three updates to be achieved. This significantly improves the bit sustainability; however, at present, there is a lack of research on this technology by both domestic and foreign scholars, and there is no systematic research conclusion regarding the wear judgment for the bit's cutting structure, the new control technology, the working mechanics of the update system, and the design method. Therefore, it is suggested to enhance the progress of this technology.

(3) The key to extending the sustainability of drill bits is to clarify their working mechanisms, starting from the driving mechanism kinematics and transmission dynamics; the blocking force and its source in the processes driving excitation, startup, and updating; the load in the driving process of updating; and the ultimate load of each component.

Author Contributions: Conceptualization, X.L. and L.Y.; methodology, G.J.; software, H.R.; validation, Q.W., J.Y. and X.L.; formal analysis, L.Y.; investigation, G.J.; resources, H.R.; data curation, Q.W.; writing—original draft preparation, K.H.; writing—review and editing, X.L.; visualization, L.Y.; supervision, G.J.; project administration, X.L. and J.Y.; funding acquisition, K.H. All authors have read and agreed to the published version of the manuscript.

Funding: This research was funded by the open fund project of the State Key Laboratory of Oil and Gas Reservoir Geology and Exploitation (Southwest Petroleum University) in 2022 [grant no. PLN2022-29], and Nanchong City-Southwest Petroleum University City School Science and Technology Strategic Cooperation Project [No. 23XNSYX0018].

Data Availability Statement: No new data were created or analyzed in this study.

Conflicts of Interest: Authors Xiaoao Liu, Guodong Ji, Qiang Wu and Jinping Yu were employed by CNPC Engineering Technology R&D Co., Ltd. Author Liyuan Yang was employed by CNPC Chuanqing Drilling Engineering Co., Ltd. The remaining authors declare that the research was conducted in the absence of any commercial or financial relationships that could be construed as a potential conflict of interest.

References

1. Deng, H.; Jia, L. Key Technology and Prospect of Deep and Ultra-deep Well Drilling in Sichuan Basin. *Gas Ind.* **2022**, *42*, 82–94.
2. Yin, H.; Zhao, X.; Li, Q. Drill bit selection method for deep well and ultra-deep well based on Vague set of entropy weight. In Proceedings of the 2022 China Oil and Gas Intelligent Technology Conference—The 5th High-End Forum on Artificial Intelligence in Petroleum and Petrochemical Industry and the 8th Open Forum on Intelligent Digital Oil Field Beijing Energy Association, China University of Petroleum (Beijing), National Key Laboratory of Oil and Gas Resources and Exploration, National Key Laboratory of Heavy Oil, Baidu Intelligent Cloud, Beijing, China, 6 February 2022; pp. 178–185.
3. Zuo, H.; He, F.; Yan, W.; Liu, W. PDC bit customization design and application in ultra-deep wells in East China Sea. *Technol. Superv. Pet. Ind.* **2020**, *36*, 17–22.
4. Wu, Z.; Lu, L.; Wang, Y. Rock-breaking characteristics and temperature field variation of cone-PDC hybrid bit. *Nat. Gas Ind.* **2020**, *40*, 99–106.
5. Song, D.; Ren, Z.; Yang, Y.; Chen, Y.; Nie, G.; Tan, L.; Peng, H.; Li, Z.; Chen, X.; Li, M.; et al. Drilling performance analysis of impregnated micro bit. *Mech. Sci.* **2022**, *13*, 867–875. [CrossRef]
6. Shi, Z.; Yang, Y. Research on Rock Breaking Mechanism with PDC Impact Composite Drill. *J. Front. Soc. Sci. Technol.* **2020**, *2*, 80–82.
7. Cheung, H.-D. Research status and key technologies of 3D printing technology. *Mod. Ind. Econ. Inf. Technol.* **2022**, *12*, 139–141.
8. Zhang, D.; Yang, Y.; Ren, H.; Niu, S. Failure resistance research of the dual-stage hybrid bit. *J. Eng. Fail. Anal.* **2023**, *145*, 107009. [CrossRef]
9. Wang, H.; Huang, H.; Bi, W.; Huang, K. Development and Prospect of Deep-well and Ultra-deep Well Oil and Gas Drilling Technology. *Gas Ind.* **2021**, *41*, 163–177.
10. Ji, H. Breakthroughs in "Bottom hole Zhufeng" Ultra-deep Oil and Gas Exploration. *China Sci. News* **2022**, *46*, 01-005.
11. Extend the wings Significant breakthrough in oil prospecting of "deep bottom hole engineering". *Guangming Dly.* **2022**, *26*, 11–16.
12. Wang, Z.; Wang, W.; Zhang, L.; Zhu, X. Current Situation and Prospect of Drilling and Completion of Wanmi Scientific Ultra-deep Well. *Sci. Technol. Bull.* **2022**, *40*, 27–35.
13. Liu, K. Status quo and development prospect of "one trip drilling" key tool technology. *Pet. Mach.* **2019**, *47*, 13–18.
14. Yuan, J. New Progress and Development Prospect of Sinopec Shale Gas Ultra-long Horizontal Well Drilling Technology. *Pet. Drill. Technol.* **2023**, *46*, 1–11.
15. Wang, H.; Zhou, B. Development and Prospect of Tight Sandstone Gas Drilling and Completion Technology. *Gas Ind.* **2022**, *42*, 159–169.
16. Yuan, G.; Fu, L.; Wang, Y.; Wang, W. Current Situation and Development Suggestions on Drilling and Completion Technology for Economic and Effective Development of Unconventional Oil and Gas in China. *Pet. Drill. Technol.* **2022**, *50*, 1–12.
17. Gao, D. Research Advances in Unconventional Oil and Gas Well Engineering Technology. *Gas Ind.* **2021**, *41*, 153–162.
18. Dahlgren, S.S.; Hall, D.R.; Hanson, J.J. Indexing Drill Bit: China. WO 2016160637 A1, 6 October 2016.
19. Yang, Y.; Qi, Q.; Lin, M.; Wang, W. A Long Life Pdc Bit with Indexing Cutter Blade: China. CN208137866U, 23 November 2018.
20. Vié, G. At-bit steerable system enables single-trip vertical, curve and lateral drilling in North American shales. *Drilling Contractor*, 6 March 2020.
21. Zhou, J.; Huang, Z.; An, X.; Tan, J. Failure Analysis and Improvement of Bimetal Seal of High Speed Cone Bit Bearing. *Oil Field Mach.* **2011**, *8*, 50–53.
22. Zhou, Y.; Huang, Z.; Tan, L.; Ma, Y.; Qiu, C.; Zhang, F.; Yuan, Y.; Sun, C.; Guo, L. Cone bit bearing seal failure analysis based on the finite element analysis. *Eng. Fail. Anal.* **2014**, *45*, 292–299. [CrossRef]
23. Li, Y. *Research on Mechanical Behavior of Hollow Cylindrical Roller Bearing with Cone Bit*; Southwest Petroleum University: Chengdu, China, 2015.

24. Shen, H. *Structural Strength Analysis and Parameter Research of Tricone Bit Bearing*; Southwest Petroleum University: Chengdu, China, 2015.
25. Huang, Z.; Li, G. Optimization of cone bit bearing seal based on failure analysis. *Adv. Mech. Eng.* **2018**, *3*, 168781401876748. [CrossRef]
26. Ji, W.; Cao, Z.; Zhong, L.; Li, G. Failure Analysis of Roller Bearing of Type KX732-250 Mining Roller Bit. *Min. Equip.* **2024**, *1*, 150–155.
27. Han, C.; Li, Y.; Zhang, J.; Yan, J. Thermal Coupling Analysis of Hollow Cylindrical Roller Bearing with Cone Bit. *Mech. Strength* **2016**, *6*, 1294–1299.
28. Huang, J.; Zeng, B.; He, Y.; Wang, X.; Qian, L.; Xia, C.; Yi, X. Numeric al study of rock- breaking mechanism in hard rock with full PDC bit model in compound impact drilling. *Energy Rep.* **2023**, *9*, 3896–3909. [CrossRef]
29. Ming, J.; Lei, C.; Bin, T.; Li, G. Experimental and numerical study on the dynamic characteristics of full-size PDC bit. *Mech. Syst. Signal Process.* **2023**, *200*, 123–134.
30. Liu, Y.; Dai, S.; Wei, J.; Li, G. Study on factors influencing the self -propelling capacity of self-propelled water jet drill bits. *Coal Geol. Explor.* **2023**, *5*, 198–206.
31. Cai, C.; Tan, Z.; Ling, C.; Li, G. Research on rock breaking Mechanism and Drilling rock breaking of separate impact scraping combined drill bit. *J. Vib. Shock.* **2022**, *16*, 232–241.
32. Zhu, X.; Tang, W.; Liu, W.; Liu, S. The influential factors on electric breakdown damage of insulator in electric pulse rock breaking drill bits. *Geoenergy Sci. Eng.* **2025**, *245*, 213504. [CrossRef]
33. You, Z.; Chen, L.; Li, C.; Liu, G.; Ye, B.; Shi, X. Cutting Migration and Core Disturbance Caused by the Hydraulic Structure of Pressure- and Gas-Maintaining Coring Bits. *Geofluids* **2022**, *36*, 16–29. [CrossRef]
34. Tao, W.; Jue, W.; Ping, Y. Research of electrical treeing characteristics in pmma under nanosecond pulses. *Adv. Technol. Electr. Eng. Energy* **2017**, *36*, 62–68.
35. Von Hippel, A. Electric breakdown of solid and liquid insulators. *J. Appl. Phys.* **1937**, *8*, 815–832. [CrossRef]
36. Wang, X.; Yang, S.; Li, C.; He, X. Nsga-ii algorithm-based structural parameters of electric pulse rock-breaking electrode bit multi-objective optimization. *J. Energy Resour. Technol.* **2024**, *146*, 33201. [CrossRef]
37. Yamano, Y. Control of electrical tree at initiation stage in ldpe by mixed addition of al2o3 nano-particle and azobenzoic compound. *IEEE Trans. Dielectr. Electr. Insul.* **2014**, *21*, 209–216. [CrossRef]
38. Yan, J.; Jia, P.; Huang, R.; Zhao, Y.; Li, L.; Zhou, Y. Investigation of Temperature dependent Dc Breakdown Characteristics of Nano-Al_2O_3/epoxy Composites at Low Temperatures. *Cryogenics* **2023**, *135*, 103744. [CrossRef]
39. Yao, T.; Chen, K.; Niu, T.; Yang, Y. Effects of frequency and thermal conductivity on dielectric breakdown characteristics of epoxy/cellulose/bn composites fabricated by ice-templated method. *Compos. Sci. Technol.* **2021**, *213*, 108945. [CrossRef]
40. Zarubin, V.S.; Kuvyrkin, G.N.; Savelyeva, I.Y. A variation model of thermal breakdown of a high-voltage dc cable electrical insulation. *Russ. Electr. Eng.* **2020**, *91*, 279–285. [CrossRef]

processes

MDPI

Article

Polylactic Acid and Polyhydroxybutyrate as Printed Circuit Board Substrates: A Novel Approach

Zahra Fazlali [1,*], David Schaubroeck [1], Maarten Cauwe [1], Ludwig Cardon [2], Pieter Bauwens [1] and Jan Vanfleteren [1]

[1] Centre for Microsystems Technology, Imec and Ghent University, Technology Park 126, B-9052 Gent-Zwijnaarde, Belgium; david.schaubroeck@ugent.be (D.S.); maarten.cauwe@ugent.be (M.C.); pieter.bauwens@ugent.be (P.B.); jan.vanfleteren@ugent.be (J.V.)

[2] Centre for Polymer and Material Technology (CPMT), Gent University, Technology Park 126, B-9052 Gent-Zwijnaarde, Belgium; ludwig.cardon@ugent.be

* Correspondence: zahra.fazlali@ugent.be

Abstract: This study presents a novel approach to manufacture a rigid printed circuit board (PCB) using sustainable polymers. Current PCBs use a fossil-fuel-based substrate, like FR4. This presents recycling challenges due to its composite nature. Replacing the substrate with an environmentally friendly alternative leads to a reduction in negative impacts. Polylactic acid (PLA) and Polyhydroxybutyrate (PHB) biopolymers are used in this study. These two biopolymers have low melting points (130–180 °C, and 170–180 °C, respectively) and cannot withstand the high temperature soldering process (up to 260 °C for standard SAC (SnAgCu, tin/silver/copper) lead free solder processes). Our approach for replacing the PCB substrate is applying the PLA/PHB carrier substrate at the end of the PCB manufacturing process using injection molding technology. This approach involves all the standard PCB processes, including wet etching of the Cu conductors, and component assembly with SAC solder on a thin flexible polyimide (PI) foil with patterned Cu conductors and then overmolding the biopolymer onto the foil to create a rigid base. This study demonstrates the functionality of two test circuits fabricated using this method. In addition, we evaluated the adhesion between the biopolymer and PI to achieve a durable PCB. Moreover, we performed two different end-of-life approaches (debonding and composting) as a part of the end-of-life consideration. By incorporating biodegradable materials into PCB standard manufacturing, the CO_2 emissions and energy consumption are significantly reduced, and installation costs are lowered.

Keywords: green electronics; printed circuit board; biodegradable materials; sustainability

1. Introduction

In recent years, sustainable and eco-friendly technologies and materials have gained prominence in many industries, including electronics. Electronics impact the environment in many ways, from raw material production to their end-of-life. Electronic waste (e-waste) is the fastest-growing waste stream in the world. Globally, 25 to 50 million tons of e-waste are accumulated each year. According to a UN report, less than one quarter (22.3%) of the year's e-waste mass was documented as properly collected and recycled in 2022, while the rest ended up buried in landfills or incinerated for the recovery of precious metals which generates a large amount of CO_2 gas [1]. Moreover, electronic circuits often contain toxic substances, including heavy metals and flame retardants like brominated bisphenol A. As a result, this industry introduces pollution to the air, water, and soil, being a threat to human health and the environment. In response to the growing concern about e-waste generation and greenhouse gas emissions, scientists have been exploring innovative materials and approaches to reduce the ecological footprint of electronic devices and to develop a shift towards green electronics [2,3].

Printed circuit boards (PCBs) are the backbone of almost every electronic device, enabling efficient and organized connections between electronic components. PCBs are the largest category of waste, accounting for 4–7% of the total weight of generated e-waste [4]. The current most frequently used PCB substrates (e.g., FR4) are composite materials based on brominated epoxy resin reinforced with fiberglass. The epoxy resins are derived from fossil fuel-based resources. This composite nature makes recycling a costly and challenging process [5]. Various recycling methods, including mechanical, chemical, and pyrolysis have been used to partially recycle the PCB waste, especially the (precious) metals. These methods are energy-intensive and can produce harmful and toxic byproducts which contribute to environmental pollution, an increased carbon footprint, and potential health risks [6]. Waste PCBs consist of a metallic (~30 wt.%) and a non-metallic fraction including the substrate, solder mask, and other organic or inorganic compounds (~70 wt.%) [7]. Therefore, PCB substrates occupy a significant volume and weight depending on the design, size, and number of layers. So, using an eco-friendly material as the base substrate reduces the harm by proposing alternative end-of-life options such as composting [8].

Employing bio-based and biodegradable materials as a PCB base material presents a promising solution to the environmental challenges associated with traditional PCBs [9]. Bio-based materials, which are derived from renewable natural resources such as plants, animals, or microorganisms, offer a sustainable alternative to petroleum-based materials. These materials contribute to reducing reliance on fossil fuels and lowering the carbon footprint. On the other hand, biodegradable materials are designed to break down naturally through the action of microorganisms, such as bacteria and fungi, at the end of their useful life. This decomposition process reduces the accumulation of electronic waste in landfills and minimizes environmental pollution. Examples of bio-based and biodegradable materials include polylactic acid (PLA), Polyhydroxyalkanoates (PHA), cellulose, and other polymers derived from plant starches [10]. Using bio-based and biodegradable materials in PCB technology aligns with the principles of circular economy, where products are designed with their entire lifecycle in mind from cradle to grave.

The main PCB manufacturing technology is the subtractive method. This method involves starting with a copper-clad laminate and removing unwanted copper using chemical etching to define the circuit traces. On the other hand, additive manufacturing such as inkjet printing or screen printing involves the selective deposition of conductive materials directly on a substrate to pattern the desired circuit. This method is advantageous for minimizing material waste, making them suitable for prototyping and low-volume production [11,12]. While subtractive processes generate significant material waste, they are the preferred choice for large-scale manufacturing due to their high precision, scalability, and cost efficiency. Additionally, the infrastructure supporting subtractive methods is well established, further lowering production costs in high-volume applications [13].

Several studies have explored the use of natural fibers as reinforcement in composite materials for PCB base material in conventional subtractive methods. Genc et al. developed Luffa fiber-reinforced/epoxy substrates and investigated the electrical properties for PCB applications [14]. Bharath et al. developed banana fiber-reinforced polymer composites and characterized mechanical, thermal, and electrical properties [15]. However, these studies used petroleum-based epoxies as the impregnating material which is not environmentally friendly. Geiger et al. produced rigid PCB with cellulose nanofibrils (CNF) and a bacterial

3. Experimental Procedure

The process steps are divided into four main sections: FPCB (flexible PCB) design and fabrication, the overmolding of the biopolymers, adhesion optimization, and end-of-life practices.

3.1. FPCB Design and Fabrication

The first circuit design shown in Figure 1a consists of two groups of 12 SMD resistors (0 Ω resistors) placed in parallel and linked over long Cu tracks to pads that are positioned left and right. Cu connection pads are patterned on the PI in a way that facilitates the measurements of resistance values.

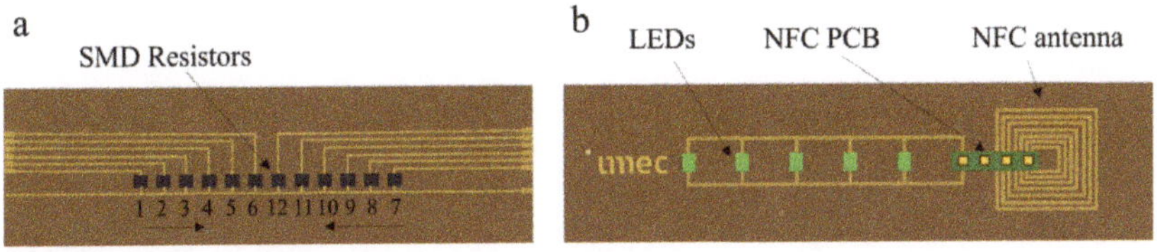

Figure 1. Two test circuit designs: (**a**) a zero-ohm resistors circuit, and (**b**) LED demonstrators powered by NFC antenna. The arrows show the direction of numbering the resistors.

As another circuit, a demonstrator was selected to power LEDs wirelessly using an integrated NFC antenna avoiding the need for an integrated battery. Doing so, the circuit functionality can be evaluated without connecting probes to the conductive traces. The circuit is therefore not exposed to the clamping force of the injection molding machine. The circuit design is shown in Figure 1b.

As mentioned, three commercial flexible foils are used as the substrate for patterning the copper traces. PI has a high thermal resistance, outstanding mechanical and chemical properties, and stable dielectric properties over high temperatures (up to 400 °C), which makes it a widely used substrate in FPCB industry. The FPCB was fabricated in-house through PCB manufacturing technology. The process flow includes patterning the copper structure by lithography and wet etching. Afterwards, the solder paste (lead-free SAC305) is manually applied to the soldering pads using a dispenser followed by the mounting of the components on the defined areas. Finally, the solder paste is reflowed by vapor phase soldering (VPS) at 230 °C (vapor phase reflow oven IBL SLC 300).

3.2. Rigid PCB Production Using Overmolding Technology

Overmolding generally refers to applying a material over another material or object using injection molding technology. To achieve the desired part quality, dimensional accuracy, and production efficiency, injection molding parameters must be optimized for biopolymers (PLA and PHB). Both biopolymers are hygroscopic; therefore, the biopolymer pellets were dried for 24 h at 60 °C in a FarrgTech dryer (Austria) before overmolding to avoid the presence of moisture and to prevent the generation of bubbles and air traps during the overmolding process.

An Engel 80T injection molding machine (Austria) is used to overmold the FPCB. As schematically shown in Figure 2, it is performed by inserting the FPCB into the mold cavity. The mold is closed on the foil and the injection screw pushes the molten biopolymer towards the mold cavity. The molten biopolymer flows and starts covering the FPCB. Following that, the biopolymer solidifies rapidly, and the mold opens again and ejects the part out from the machine. Figure 2 demonstrates the two main fabrication methods of FPCB and overmolding.

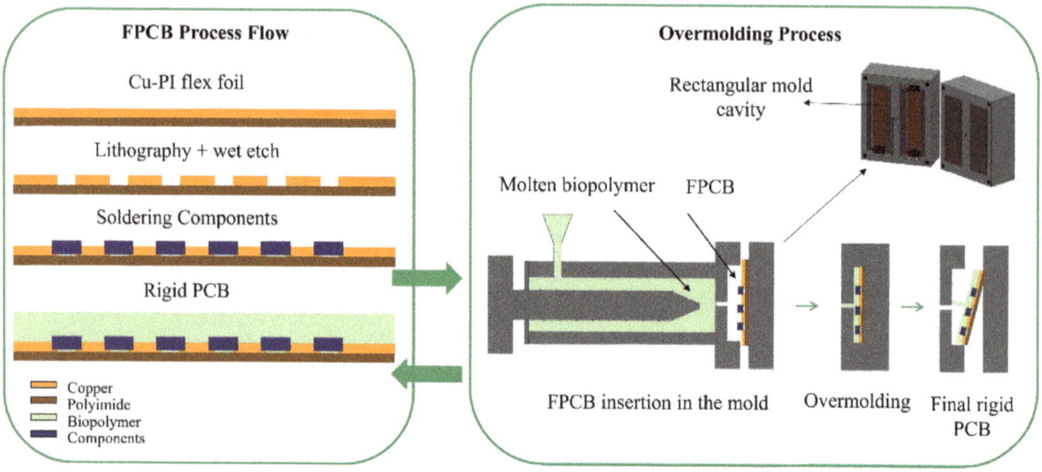

Figure 2. The two main fabrication steps of producing rigid PCB.

In this study, a rectangular aluminum mold with a dimension of $150 \times 40 \times 1.5$ mm^3 was used. One of the molds has one flow entry and the other one has two flow entries. The double-entry mold resulted in some air traps and shrinkage in the middle of the samples which is not desirable. Therefore, the single-entry mold was selected for this research.

During the injection molding process of polymers, various defects may arise if the process parameters are not properly adjusted. Some of the most common defects include warpage (bending due to uneven shrinkage), flash (excess material leaking from the mold cavity), short shots (incomplete mold filling), and bubbles (trapped air within the material). These defects can affect the structural integrity, functionality, and appearance of the final product [30].

To achieve a high dimensional accuracy and optimal part quality, it is crucial to optimize the injection molding parameters for each biopolymer. The key parameters affecting the final quality include material temperature, mold temperature, injection speed, cooling time, and holding time. Holding time refers to the duration of which pressure is applied to the molten material after the mold cavity is filled but before cooling begins. It ensures that material shrinkage is compensated, preventing voids and incomplete packing [30].

3.3. *Adhesion Optimization*

The adhesion between the PI foil and the PLA/PHB substrate is a crucial factor influencing the mechanical stability and reliability of the PCB. PI and the two selected biopolymers (PLA and PHB) show a very poor adhesion towards the three types of PI foils after overmolding. Therefore, two adhesion improvement methods were evaluated, and their effectiveness was investigated. Oxygen plasma surface treatment has been performed to activate the PI foil surface prior to overmolding. In addition to that, a thermoplastic polyurethane (TPU) tie layer was used as an adhesive to improve the bonding.

For these types of circuitry, the majority area of the flex circuit (Cu side) consists of exposed PI, resulting from the etching of the covering Cu layer. Hence, for the adhesion evaluation experiments, the surface treatment was conducted on completely etched PI foils. Note that for the Pyralux and Shengyi foils, the outer layers consist of adhesive layers (acrylic and epoxy based, respectively).

The oxygen plasma treatment was performed with a PICO—Diener electronic GmbH & Co. KG, Germany. The power and frequency were adjusted to 190 W and 40 kHz, respectively. The samples were treated for 2 min with an oxygen pressure of 0.8 mbar.

For the other improvement method, a TPU tie layer is applied on the etched PI foil using a vacuum lamination method (Lauffer RLKV 25) at a 120 °C temperature and 20 N/cm^2 pressure for 30 min. Afterwards, the TPU-laminated foil is overmolded by biopolymers.

To compare the effectiveness of each method, the adhesive strength is measured using the 180-degree peel test method. To measure the peel strength, an Instron 5543 universal testing machine equipped with a 500 N load cell is used. The bilayer samples were cut to a 3 cm width and fixed to the grips to measure the force (Figure 3). Three identical samples of each case were evaluated to ensure reproducibility.

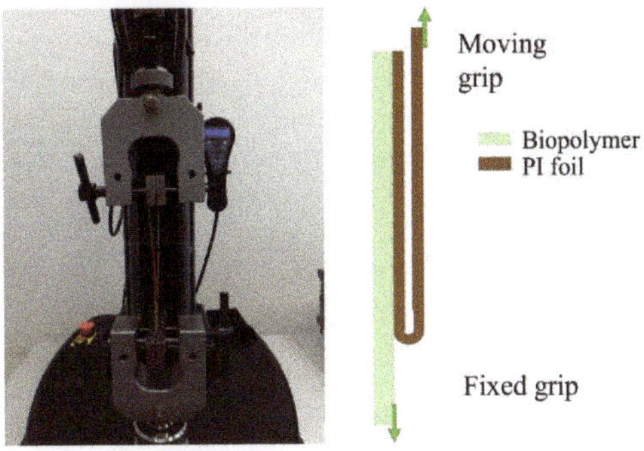

Figure 3. The 180-degree peel test setup and schematic.

3.4. End-of-Life Practices

In this study, as the end-of-life treatment, we analyzed debonding and composting methods separately. The rigid PCBs including the TPU layer are selected as a sample test in this section.

3.4.1. Debonding Strategy

The debonding approach suggests a separation of the biopolymer part from the FPCB part using a liquid. This allows the electrical components to be potentially reused or recycled.

The rigid PCBs consist of a PI foil with copper tracks and components, a laminated TPU tie layer, and an overmolded biopolymer. To separate the layers, the samples are immersed in an acetone solution and monitored over time. The debonding concept is demonstrated in Figure 4.

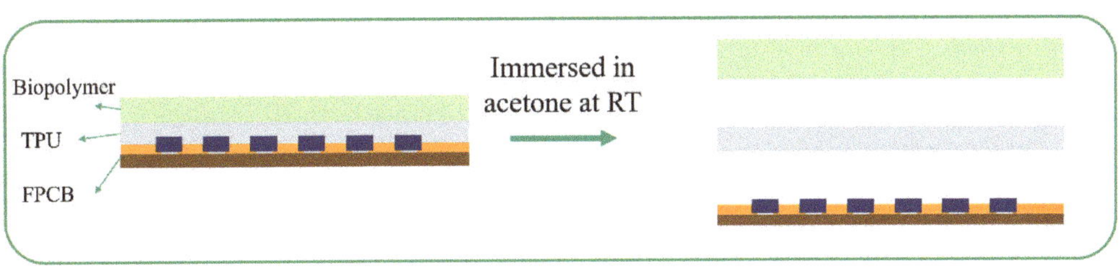

Figure 4. Debonding concept demonstration.

The samples were cut into sizes of 4 $\times$ 2.5 cm^2 and immersed in the analytical acetone solution (p. 99+% from Chemlab, Belgium) in a jar at room temperature.

3.4.2. Composting

On the other hand, a quantitative composting method is used to decompose the biopolymer part through microorganism digesting the nutrients of the biopolymers. Afterwards, the copper tracks and components are recovered from the compost.

Four small glass containers were filled with organic composting soil. The samples were cut into 4×2.5 cm^2 and buried within the compost. Four samples consisting of one pure PLA, one PCB with PLA, one pure PHB, and one PCB with PHB were selected. The pure biopolymers are considered as a reference. At the start of the experiment, the samples were visually inspected and weighed in a dry condition. Next, the containers were put in an oven at 58 ± 2 °C for 60 days. The medium was kept moist by adding deionized water every day and the soil was gently moved for better oxygen exchange. For checking the degradation behavior, the samples were removed from the compost, and brushed and washed gently with water to remove soil debris followed by drying at 40 °C for 5 h in the oven. Afterwards, the samples were visually analyzed for any color or morphological changes and then reweighed (Figure 5). The analyses were performed every 15 days.

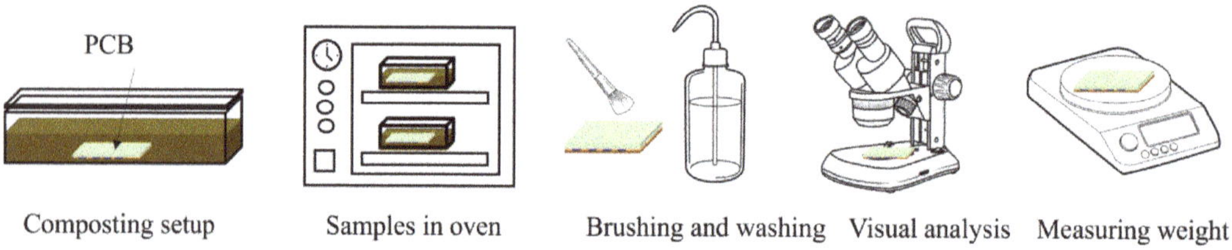

Figure 5. Composting evaluation process steps.

The weight loss was calculated with the following Formula (1):

$$weight\ loss\ (\%) = \frac{w_i - w_t}{w_i} \times 100 \quad (1)$$

where w_i is the initial weight of the samples, and w_t is the weight of the samples after they are removed from the compost at different time points.

4. Results and Discussion

4.1. Overmolding Process Parameters Optimization

Process optimization was initially conducted without inserting the flexible foil to fine-tune the injection molding parameters. The barrel temperature of the injection molding machine was set within the range recommended by the manufacturer and remained constant throughout the process, as illustrated in Figure 6. The nozzle temperature is crucial because it ensures that the polymer is fully melted before it is injected into the mold under applied pressure.

The optimization process began with PLA. The mold temperature, injection speed, and cooling time are maintained constant at 30 °C, 120 mm/s, and 45 s, respectively. The two main variables adjusted to optimize the molding process were the dosing length and switchover point, which directly affect the amount of material injected into the mold and the transition between the filling and packing phases.

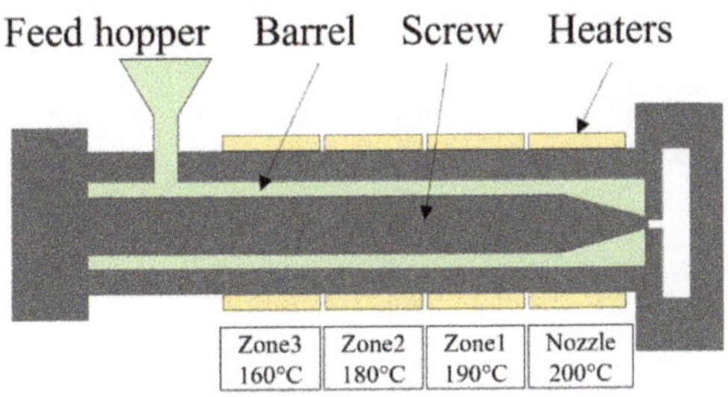

Figure 6. Injection molding barrel zones and temperatures for PLA and PHB.

The dosing length refers to the amount of molten plastic (shot size) that is prepared in the barrel before each injection cycle. The screw retracts by a set distance to accumulate the correct volume of molten material. The switchover point defines the position at which the injection phase switches from velocity-controlled filling to pressure-controlled packing. It is crucial to prevent defects such as flashing or short shots.

In the first cycle, the dosing length was set to 30 mm, and the switchover point was 5 mm, meaning that the screw initially retracts 30 mm to accumulate molten material, but moves forward 25 mm to inject it into the mold cavity. With these settings, excess material overflowed from the mold (flash and burr defects), indicating that too much material was injected.

To counteract this, the switchover point was increased to 15 mm, meaning the transition from filling to packing happens earlier. However, this adjustment led to an underfilled mold (short shot), because the material did not flow sufficiently due to its high viscosity.

Thus, the switchover point was adjusted to 12 mm allowing the mold to be completely filled. However, shrinkage and an uneven surface finish were observed. To address this issue, the holding time was increased to 2 s, ensuring that the material had sufficient time to settle and solidify properly, resulting in a smooth and well-defined final shape. Table 2 (Figure 7) summarizes the optimization cycles performed for PLA molding.

Table 2. The injection molding process parameters optimization for the PLA biopolymer.

Parameters	Cycle 1	Cycle 2	Cycle 3	Cycle 4
Switchover point (mm)	5	15	12	12
Dosing (mm)	30	30	30	30
Holding time (s)	0	0	0	2

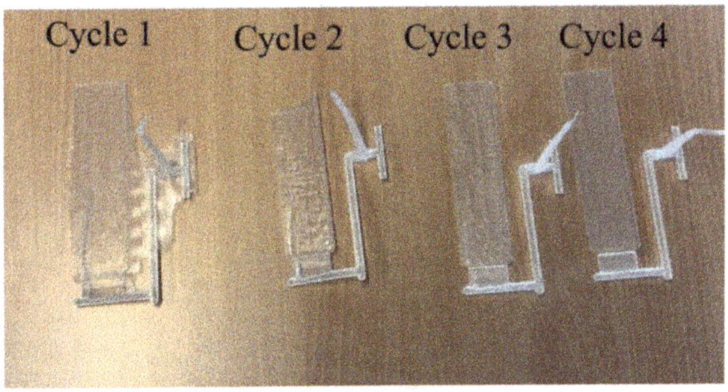

Figure 7. PLA molded pieces with different parameter settings.

The same parameters were applied to the PHB. The switchover point was initially set at 12 mm, and the holding time was set at 2 s. This setting resulted in some burrs and flashes. In the second stage, the switchover point was changed to 14 mm and the holding time was decreased to 0 s. The mold was not filled, resulting in a short shot. Therefore, the holding time was increased to 2 s, and the switchover point to 14 mm. The desired part was achieved in cycle 3 as shown in Table 3 (Figure 8).

Table 3. Injection molding process parameters for the PHB biopolymer.

Parameters	Cycle 1	Cycle 2	Cycle 3
Switchover point (mm)	12	14	14
Dosing (mm)	30	30	30
Holding time (s)	2	0	2

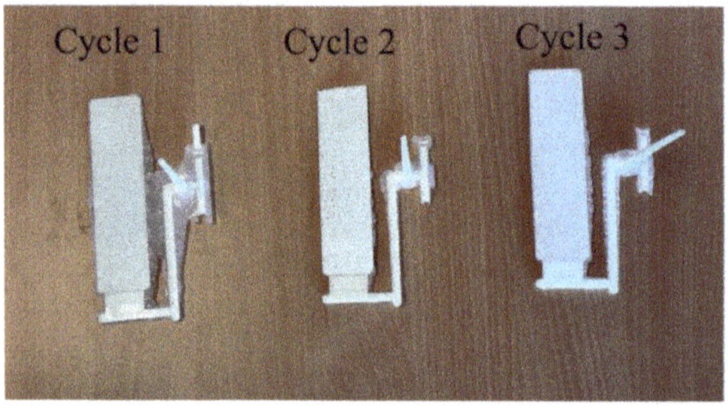

Figure 8. PHB molded pieces with different parameter settings.

Five tests confirmed the repeatability of these parameters, resulting in identical molded pieces. Optimized parameters were tested with PI foil inserted into the mold, and no defects were found.

4.2. The Validation of Circuit Functionality

The resistor circuits on three different PI foils after overmolding are demonstrated in Figure 9. They were all functional with a slight increment in the resistance value after overmolding.

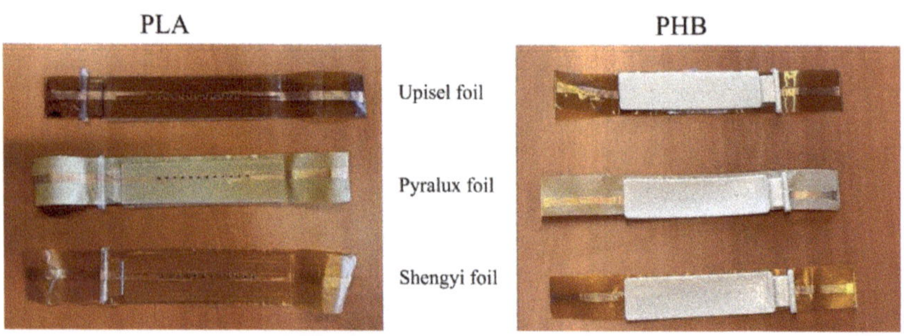

Figure 9. Rigid PCBs (FPCBs overmolded with biopolymers).

A component's measured resistance is the sum of its resistance, the Cu trace's resistance, the soldered connection's resistance, and the probe's contact resistance:

$$R_{measured} = R_{trace\ length} + R_{solder} + R_{contact} \quad (2)$$

The $R_{initial}$ of the component is the value before assembling the resistors on the foil. Each mounted 0 Ω resistor has a different track length (L) between the component and the connection pad. This length has an effect on the final value according to resistance formula ($R = \rho\, L/A$) with R being the resistance, ρ being the resistivity, L being the length, and A being the cross-sectional area) showing a direct relation to its length L [31].

Figure 10 presents average $R_{measured}$ values for Upisel foil before and after overmolding with PLA and PHB.

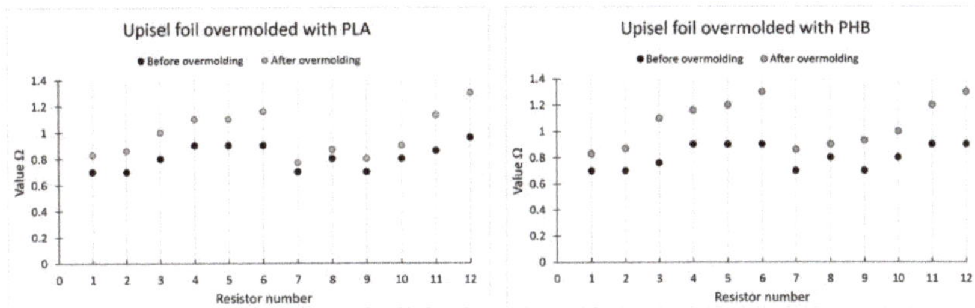

Figure 10. The average resistance values for Upisel foil before and after overmolding with PLA and PHB.

Overmolding slightly increased the measured resistance. This change could be attributed to the oxidation of Cu tracks because of heat, the contact resistance of the probe, or the local reduction in Cu track thickness due to mold clamping pressure [31].

Regarding the demonstrator circuit, a smart phone with an NFC reader application was used to power the LEDs wirelessly. By keeping the smart phone close to the NFC antenna, the voltage from the phone battery is induced to the antenna and transferred to the LEDs through an NFC chip. Figure 11 shows the functionality of the LEDs after the overmolding process.

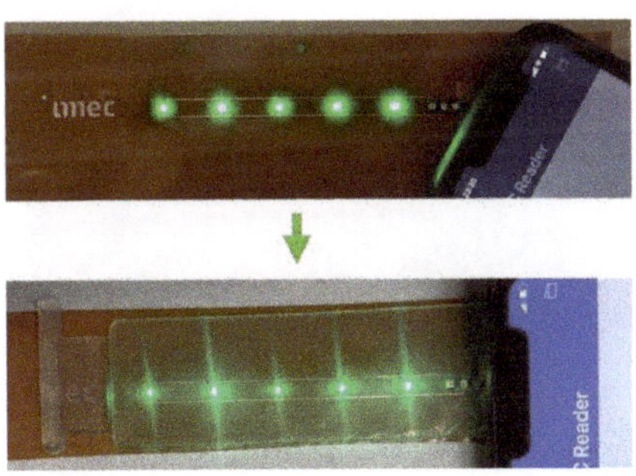

Figure 11. The functionality test of the LED demonstrator circuit before and after overmolding with PLA.

Functionality measurements of the fabricated test circuits demonstrated that the signal connection was maintained after overmolding with PLA and PHB.

4.3. Adhesive Strength Measurement

The peel test results of adhesion between the biopolymer (PLA or PHB) and three flexible foils in different conditions of a pristine (no-treatment), oxygen plasma, and TPU layer are shown in Figure 12.

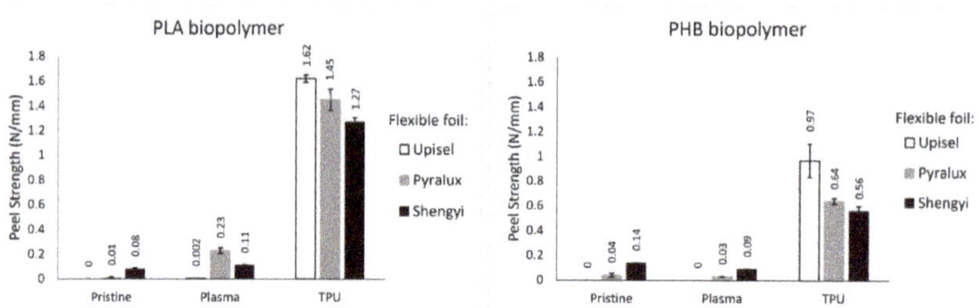

Figure 12. Peel strength results of bilayer samples before and after adhesion improvements.

As can be seen in the bar chart, the pristine Upisel foil shows no adhesion and peels off spontaneously from both biopolymers upon ejection from the mold. However, the adhesive strength related to the Pyralux and Shengyi samples were measurable but negligible (less than 0.1 N/mm). This could be attributed to the existence of adhesive layers (acrylic or epoxy) on the mentioned foils.

After treating the foils with oxygen plasma, oxygen functional groups such as hydroxyl (-OH) and carbonyl (C=O) groups are introduced to the foil surfaces. These polar functional groups can potentially improve the adhesion to overmolded PLA.

Using a TPU layer between PI foils and biopolymers resulted in significantly higher peel strengths. It was expected because this TPU has excellent adhesive properties to a large variety of polymer foils.

The approach of using the TPU layer shows a strong bond between the FPCB and biopolymer and guarantees the long-term durability of the rigid PCB. To ensure environmental sustainability, it would be appropriate to use biobased or biodegradable TPU, which is currently in the development phase [32].

4.4. End-of-Life

4.4.1. Debonding

The debonded layers after being recovered from the acetone solution are shown in Figure 13. As can be seen, the three fractions of the FPCB, TPU, and biopolymer are detached.

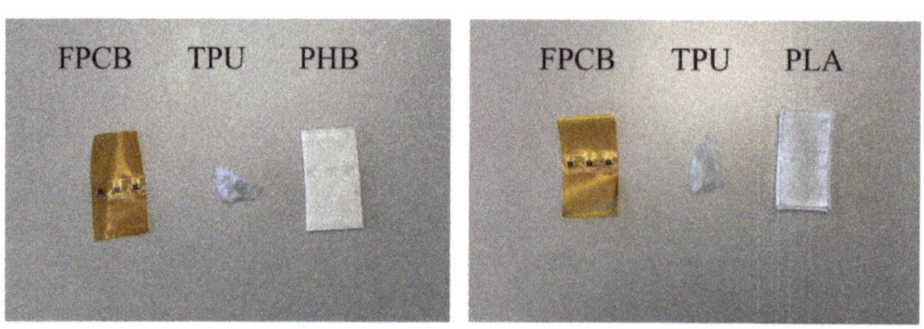

Figure 13. Debonded layers after being recovered from acetone (Shengyi foil).

For all the samples, the TPU started to wrinkle after about 1 h. After 2 h, the PI foil and biopolymer showed a small delamination from each other. After 24 h, the three parts were completely delaminated, and the fractions were recovered. Figure 14 illustrates the debonding phenomenon over time.

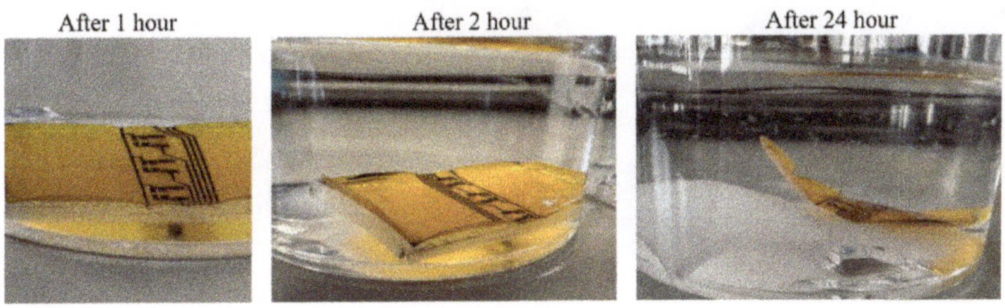

Figure 14. The debonding phenomenon over time for Shengyi foil overmolded with PLA.

This method allowed for the complete separation of the FPCB, TPU, and biopolymer under mild chemical treatment, without using additional mechanical and thermal processes. It is possible to use an ultrasonic bath or raise the temperature to enhance the separation rate; however, we explored the possibility of debonding with having the least environmental effects compared to other solvents. Alternative solvents can be used such as dimethyl formamide (DMF) and tetrahydrofuran (THF) for dissolving TPU; however, these chemicals bring more environmental and health hazards compared to acetone [33].

The electrical components and copper tracks remained intact (Figure 13) and retained their electrical properties. The debonded layers of biopolymers present a valuable opportunity for upcycling, where they can be repurposed into new products with an enhanced quality and value for different applications, such as advanced composites, and biodegradable packaging [34]. The debonding method could be an energy and time-saving approach by avoiding the production of virgin biopolymer.

4.4.2. Quantitative Composting Analysis

The weight change after 60 days of storage in the composting setup is represented in Figure 15. The weight loss shows a consistent trend over time. The pure PLA was degraded by more than 90% after 60 days, confirming biodegradability. However, the PCB with PLA showed a 37.3% weight loss after 60 days. It should be noted that metals could have negative impacts on the compost quality. PHB samples did not show any significant weight loss.

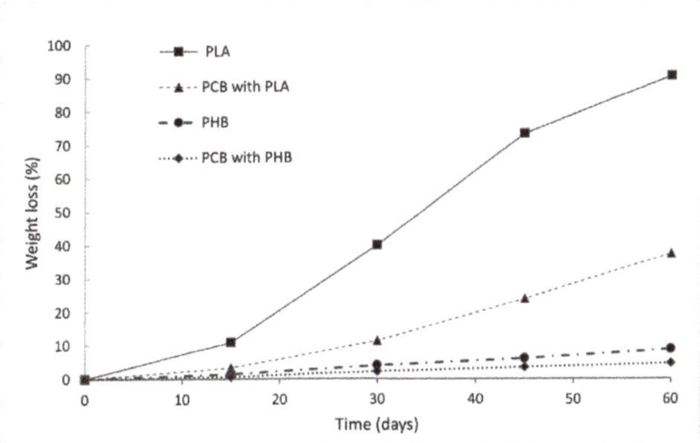

Figure 15. The weight loss in samples as a result of composting.

In samples with circuits, the degradation happened at the biopolymer side and no degradation was observed on the PI side. The biodegradation process of the four samples is shown in Figure 16. As can be seen, the pure PLA disintegrated into coarse parts after

15 days of storage in compost. PCB with PLA also showed some holes and surface change during the first 15 days. The PLA samples exhibited a significant color change during the composting. The amorphous regions of the polymers degrade faster than the crystalline regions. This selective degradation leads to an increase in the degree of crystallinity and opacity [35]. In addition, hydrolysis and oxidation reactions during biodegradation could result in the formation of shorter polymer chains. These chemical changes can alter the optical properties of PLA, resulting in color changes (e.g., yellowing or darkening) [36]. The blue-green patina color in the PCB sample is the result of the reaction of copper with oxygen and moisture [37].

Figure 16. Images of solid fractions after biodegradation in compost at 0, 15, 30, 45, and 60 days.

In the case of PHB samples, a slight deterioration showed up with minor holes after 30 days. The PCB with PHB almost maintained the same appearance as the initial sample. According to research [38], the degree of humidity plays a key role in the degradation of PHB. Moreover, PHB has a crystalline structure which makes it resistant to microbial attack since their tightly packed molecular structure hinders the penetration of enzymes. PHB degradation is also dependent on thermophilic microbes that produce PHB-depolymerase enzymes [39]. Therefore, the degradation process may be delayed if these microbes are absent or present in low concentrations in our compost media.

As a conclusion, PLA shows a faster degradation due to its higher amorphous fraction after molding compared to PHB. PCB samples exhibit a slower weight loss due to their smaller surface area exposed to microorganisms. It should be noted that the proposed PCBs showed no sign of natural degradation during the usage phase.

4.4.3. End-of-Life Strategies and Sustainability

The end-of-life phase is a critical aspect of evaluating the environmental sustainability of PCBs. In Section 4.4.1., a debonding strategy is discussed using a mild solvent treatment (acetone), enabling the separation of the biopolymer from the PI and copper layer without mechanical grinding or high-temperature processes. This separation method facilitates material recovery and recycling, while preserving the electrical and mechanical properties of the PI-copper layer for potential reuse [40].

Among biopolymers, industrial composting is a widely recognized end-of-life strategy. Under controlled conditions (temperature $\geq$58 °C, regulated humidity, and microbial activity), both PLA and PHB can be biodegraded into CO_2, water, and biomass offering a low-energy, low-impact alternative to landfill or incineration. However, improper disposal in uncontrolled environments (e.g., home composting) may result in an incomplete degradation and the formation of microplastics [41].

In addition, mechanical recycling is feasible for PLA and PHB-based PCBs. This method involves shredding the PCB and separating metallic and non-metallic parts. However, repeated processing can lead to a reducing of the mechanical and thermal performance of the biopolymer. Chemical recycling, including hydrolysis, or enzymatic depolymerization, enables the recovery of high-purity monomers like lactic acid or 3-hydroxybutyric acid. These processes allow closed-loop recycling and retain material quality. However, they require solvents, catalysts, and energy input, making them less favorable in terms of environmental cost [42,43].

Each end-of-life strategy has trade-offs. The possibility of material recovery through a debonding process as shown above, biodegradation under industrial composting, or chemical depolymerization gives biopolymer-based PCBs a more favorable environmental profile than conventional alternatives [44].

4.5. Life Cycle Assessment Perspective

From a life cycle assessment (LCA) perspective, the environmental impact of PLA and PHB substrates is significantly lower compared to FR4. In general, the production of FR4 involves high greenhouse gas emissions, energy-intensive manufacturing processes, and extensive resource extraction. In contrast, PLA and PHB production typically emit fewer greenhouse gases due to their bio-based origin.

A benchmarking analysis conducted by NatureWorks assessed the environmental impact of Ingeo PLA compared to traditional petroleum-based polymers such as polyethylene terephthalate (PET) and polystyrene (PS). The study reports that the cradle-to-gate nonrenewable energy use for PLA is approximately 42 MJ/kg, whereas petroleum-based polymers typically require 60–100 MJ/kg, depending on the energy source used. In terms of greenhouse gas (GHG) emissions, PLA emits approximately 1.3–1.5 kg CO_2-eq per kg, while PET and PS emissions range between 3.0–6.0 kg CO_2-eq per kg. The lower emissions of PLA are attributed to its bio-based origin, which makes it a more sustainable alternative [45].

Similarly, Naji et al. analyzed the cradle-to-gate LCA of PLA-based composite and FR4 PCBs, demonstrating that PLA-based substrate has a Global Warming Potential (GWP) of 0.63 kg CO_2-eq, whereas FR4 has a significantly higher GWP of 6.29 kg CO_2-eq due to high emissions during epoxy and fiber glass production. Additionally, FR4 exhibits a substantially higher Human Toxic Potential (HTP) of 19.5 kg DCB-eq, compared to 2.32 kg DCB-eq for PLA-based substrates. Since PLA is biodegradable under industrial composting conditions, its organic fraction decomposes into water, CO_2, and biomass, thereby minimizing waste accumulation [24].

A literature review by Fonseca et al. compared the environmental impacts of PLA with fossil-based plastics, such as polypropylene (PP), PET, and PS. Their study highlights that PLA has lower impacts in fossil fuel depletion and ozone layer depletion, but higher eutrophication and human toxicity impacts due to the use of fertilizers in agricultural feedstock production. However, utilizing agricultural byproducts (e.g., sugarcane bagasse) instead of food crops can significantly reduce the environmental toxicity of PLA production. One of the primary challenges of biopolymer production is its competition with food supply chains, necessitating the development of alternative biomass sources, such as agricultural residues or waste-derived feedstocks [46].

Regarding PHB production, Rostkowski et al. analyzed its life cycle impacts, comparing biogas-based PHB production with plant-derived PHB. Their study found that the total energy requirement for PHB production from waste biogas is 37.4 MJ/kg, whereas corn-based PHB production requires 41.9 MJ/kg. This suggests that biogas-derived PHB is more sustainable, as it requires less energy, water, and land, unlike corn-based PHB. This demands significant agricultural input. Despite this, PHB production remains less energy-intensive than petroleum-based polymers [47].

The sustainability benefits of our approach are primarily due to the reduced carbon footprint and energy consumption of biobased and biodegradable polymers, as well as the improved end-of-life options for PCBs. In our approach, we utilize PLA/PHB (1.5 mm) as the main substrate, but we also use a 50 μm PI layer and a 100 μm TPU layer to enhance performance and durability. While the addition of PI and TPU introduces some environmental impact, their contribution remains minimal compared to the significant reduction in CO_2 emissions, energy consumption, and material weight achieved by replacing FR4 with PLA/PHB. Furthermore, our approach facilitates the easier material separation and potential compostability of PLA/PHB, providing a more sustainable end-of-life solution compared to FR4, which typically ends up in landfills or requires high-energy recycling processes.

Comparing the weight of our PLA-based PCB with FR4 PCB provides a better insight into the lower environmental impacts of the proposed approach. The total weight is calculated using following Equation (3):

$$Weight = Length \times Width \times Thickness \times DensityWeight_{PLA-based} = Weight_{PLA} + Weight_{PI} + Weight_{TPU} \quad (3)$$

The PLA-based thickness will be 1.65 mm considering PI and TPU layers. The densities of each layer are D_{PLA} = 1.24 g/cm^3, D_{PI} = 1.42 g/cm^3, and D_{TPU} = 1.2 g/cm^3. The total weight for the dimension of 150 × 40 × 1.65 mm^3 results in 12.31 g. For a 150 × 40 × 1.6 mm^3 FR4-based PCB, the total weight is approximately 17.76 g. This shows in a 30.71% reduction in material weight. This lower mass contributes to resource savings and a reduced environmental impact.

According to Naji et al. data [24], we can estimate the GWP of the calculated weight of PCBs. Our PLA-based PCB results in a CO_2 footprint of approximately 0.00703 kg CO_2-eq (only PLA is considered), while an equivalent FR4 PCB substrate emits about 0.111 kg CO_2-eq, representing a 93.72% reduction in greenhouse gas emissions.

It should be noted that PI and TPU will increase the total CO_2 emission for PLA-based; however, in view of their low thickness, the total value is expected to remain lower than conventional FR4. Thus, while the PI and TPU layers introduce a minor environmental impact, it is more than compensated by using biopolymers. This demonstrates that biopolymer-based PCBs can serve as an eco-friendly alternative in electronic applications.

5. Conclusions

This study presents a novel potential approach for manufacturing PCBs using sustainable PLA and PHB biopolymers. The method involves overmolding the biopolymer onto a flexible PI circuit, allowing for a compatibility with standard PCB production in terms of Cu patterning, soldering, and component assembly. The results demonstrate that TPU interlayers significantly enhance adhesion, maintaining mechanical integrity while preventing delamination. Additionally, conductivity measurements confirm that PLA/PHB-based PCBs can sustain basic electronic functionalities.

In addition to fabrication and performance, the study also explores sustainable end-of-life pathways, which are essential for reducing PCB-related electronic waste. The successful debonding of the PI circuit from the PLA/PHB substrate enables material recovery and recycling. Furthermore, biodegradation testing confirms that PLA/PHB-based substrates break down under industrial composting conditions, offering a practical alternative to landfill disposal. These findings demonstrate the viability of biopolymer PCBs; however, further research is needed to improve thermal and mechanical stability and to provide the reliability of the electronics. By addressing these challenges, biopolymer-based PCBs could play an important role in sustainable electronics.

Author Contributions: Conceptualization, Z.F., D.S., M.C. and J.V.; formal analysis, Z.F., D.S., M.C. and J.V.; methodology, Z.F.; resources, D.S., M.C., L.C. and J.V.; supervision, D.S., M.C., P.B. and J.V.; validation, Z.F.; visualization, Z.F.; writing—original draft, Z.F.; writing—review and editing, Z.F., D.S., M.C., L.C., P.B. and J.V. All authors have read and agreed to the published version of the manuscript.

Funding: This research received no external funding.

Data Availability Statement: The original contributions presented in this study are included in the article. Further inquiries can be directed to the corresponding author.

Conflicts of Interest: The authors declare no conflicts of interest.

References

1. Available online: https://unitar.org/about/news-stories/press/global-e-waste-monitor-2024-electronic-waste-rising-five-times-faster-documented-e-waste-recycling (accessed on 1 March 2025).
2. Kiddee, P.; Naidu, R.; Wong, M.H. Electronic Waste Management Approaches: An Overview. *Waste Manag.* **2013**, *33*, 1237–1250. [CrossRef] [PubMed]
3. Ahirwar, R.; Tripathi, A.K. E-Waste Management: A Review of Recycling Process, Environmental and Occupational Health Hazards, and Potential Solutions. *Environ. Nanotechnol. Monit. Manag.* **2021**, *15*, 100409. [CrossRef]
4. Botelho Meireles de Souza, G.; Bisinotto Pereira, M.; Francisco dos Santos, L.; Gonçalves Alonso, C.; Jegatheesan, V.; Cardozo-Filho, L. Management of Waste Printed Circuit Boards via Supercritical Water Technology. *J. Clean. Prod.* **2022**, *368*, 133198. [CrossRef]
5. Kaya, M. *Printed Circuit Boards (PCBs)*; The Minerals, Metals & Materials Series; Springer Nature Switzerland AG: Cham, Switzerland, 2019; ISBN 9783030265922.
6. Jadhao, P.R.; Ahmad, E.; Pant, K.K.; Nigam, K.D.P. Advancements in the Field of Electronic Waste Recycling: Critical Assessment of Chemical Route for Generation of Energy and Valuable Products Coupled with Metal Recovery. *Sep. Purif. Technol.* **2022**, *289*, 120773. [CrossRef]
7. Marques, A.C.; Marrero, J.M.C.; de Fraga Malfatti, C. A Review of the Recycling of Non-Metallic Fractions of Printed Circuit Boards. *SpringerPlus* **2013**, *2*, 521. [CrossRef]
8. Dušek, K.; Koc, D.; Veselý, P.; Froš, D.; Géczy, A. Biodegradable Substrates for Rigid and Flexible Circuit Boards: A Review. *Adv. Sustain. Syst.* **2024**, *9*, 2400518. [CrossRef]
9. Soon, C.F.; Yee, S.K.; Nordin, A.N.; Rahim, R.A.; Ma, N.L.; Hamed, I.S.L.A.; Tee, K.S.; Azmi, N.H.; Sunar, N.M.; Heng, C. Advancements in Biodegradable Printed Circuit Boards: Review of Material Properties, Fabrication Methods, Applications and Challenges. *Int. J. Precis. Eng. Manuf.* **2024**, *25*, 1925–1954. [CrossRef]
10. Oliver-Cuenca, V.; Salaris, V.; Muñoz-Gimena, P.F.; Agüero, Á.; Peltzer, M.A.; Montero, V.A.; Arrieta, M.P.; Sempere-Torregrosa, J.; Pavon, C.; Samper, M.D.; et al. Bio-Based and Biodegradable Polymeric Materials for a Circular Economy. *Polymers* **2024**, *16*, 3015. [CrossRef]
11. Dhinesh, S.K.; Joshua Robert, J.; Tushar Nair, S.; Sharne Moni, D.H.; Sona Fowzeya, S.; Senthil Kumar, K.L.; Raghunath, M.; Nagarajan, P. Recent Trends in Additive Manufacturing of Electronics Devices. *Mater. Today Proc.* **2022**, *66*, 928–941. [CrossRef]
12. Egger, F.; Schiller, D.; Stockinger, T.; Pretschuh, C.; Müller, U.; Kaltenbrunner, M. Direct Fabrication of Electronic Circuits on Wooden Surfaces. *Adv. Sens. Res.* **2024**, *3*, 2400010. [CrossRef]

13. Coombs, C.F. Printed Circuits Handbook. McGraw-Hill. 2007. Available online: http://www.gbv.de/dms/hebis-darmstadt/toc/195066995.pdf (accessed on 1 March 2025).
14. Genc, G.; Sarikas, A.; Kesen, U.; Aydin, S. Luffa/Epoxy Composites: Electrical Properties for PCB Application. *IEEE Trans. Compon. Packag. Manuf. Technol.* **2020**, *10*, 933–940. [CrossRef]
15. Bharath, K.N.; Puttegowda, M.; Yashas Gowda, T.G.; Arpitha, G.R.; Pradeep, S.; Rangappa, S.M.; Siengchin, S. Development of Banana Fabric Incorporated Polymer Composites for Printed Circuit Board Application. *Biomass Convers. Biorefinery* **2024**, *14*, 12599–12612. [CrossRef]
16. Geiger, T.; Geiger, C.K.; Hoffmann, K.G.; Nyström, G. Printed circuit boards made from cellulose fibrils. In Proceedings of the 20th European Conference on Composite Materials: Composites Meet Sustainability, Lausanne, Switzerland, 26–30 June 2022; pp. 1457–1464.
17. Farkas, C.; Gál, L.; Csiszár, A.; Grennerat, V.; Jeannin, P.O.; Xavier, P.; Rigler, D.; Krammer, O.; Plachy, Z.; Dusek, K.; et al. Sustainable Printed Circuit Board Substrates Based on Flame-Retarded PLA/Flax Composites to Reduce Environmental Load of Electronics: Quality, Reliability, Degradation and Application Tests. *Sustain. Mater. Technol.* **2024**, *40*, e00902. [CrossRef]
18. Khrustalev, D.; Tirzhanov, A.; Khrustaleva, A.; Mustafin, M.; Yedrissov, A. A New Approach to Designing Easily Recyclable Printed Circuit Boards. *Sci. Rep.* **2022**, *12*, 22199. [CrossRef]
19. The World's First Fully Recyclable Printed Circuit Board Substrate. Available online: https://www.jivamaterials.com/ (accessed on 1 March 2025).
20. Habib Ullah, M.; Mahadi, W.N.L.; Latef, T.A. Aerogel Poly(Butylene Succinate) Biomaterial Substrate for RF and Microwave Applications. *Sci. Rep.* **2015**, *5*, 12868. [CrossRef]
21. Schramm, R.; Reinhardt, A.; Franke, J. Capability of Biopolymers in Electronics Manufacturing. In Proceedings of the 2012 35th International Spring Seminar on Electronics Technology, Bad Aussee, Austria, 9–13 May 2012; pp. 345–349. [CrossRef]
22. Henning, C.; Schmid, A.; Hecht, S.; Harre, K.; Bauer, R. Applicability of Different Bio-Based Polymers for Wiring Boards. *Period. Polytech. Electr. Eng. Comput. Sci.* **2019**, *63*, 1–8. [CrossRef]
23. Taib, N.A.A.B.; Rahman, M.R.; Huda, D.; Kuok, K.K.; Hamdan, S.; Bin Bakri, M.K.; Bin Julaihi, M.R.M.; Khan, A. *A Review on Poly Lactic Acid (PLA) as a Biodegradable Polymer*; Springer: Berlin/Heidelberg, Germany, 2023; Volume 80, ISBN 0123456789.
24. Nassajfar, M.N.; Deviatkin, I.; Leminen, V.; Horttanainen, M. Alternative Materials for Printed Circuit Board Production: An Environmental Perspective. *Sustainability* **2021**, *13*, 12126. [CrossRef]
25. Geczy, A.; Gal, L.; Hajdu, I.; Kovacs, B.; Nagy, D.; Ruszinko, M. Optimizing Solder Joints on Biodegradable PCBs with Vapour Phase Soldering. In Proceedings of the 2014 37th International Spring Seminar on Electronics Technology, Dresden, Germany, 7–11 May 2014; pp. 33–38. [CrossRef]
26. Geczy, A.; Garami, T.; Kovacs, B.; Nagy, D.; Gal, L.; Ruszinko, M.; Hajdu, I. Soldering Tests with Biodegradable Printed Circuit Boards. In Proceedings of the 2013 IEEE 19th International Symposium for Design and Technology in Electronic Packaging (SIITME), Galati, Romania, 24–27 October 2013; pp. 39–42. [CrossRef]
27. McAdam, B.; Fournet, M.B.; McDonald, P.; Mojicevic, M. Production of Polyhydroxybutyrate (PHB) and Factors Impacting Its Chemical and Mechanical Characteristics. *Polymers* **2020**, *12*, 2908. [CrossRef]
28. Carrasco, F.; Pagès, P.; Gámez-Pérez, J.; Santana, O.O.; Maspoch, M.L. Processing of Poly(Lactic Acid): Characterization of Chemical Structure, Thermal Stability and Mechanical Properties. *Polym. Degrad. Stab.* **2010**, *95*, 116–125. [CrossRef]
29. Cao, M.; Cui, T.; Yue, Y.; Li, C.; Guo, X.; Jia, X.; Wang, B. Preparation and Characterization for the Thermal Stability and Mechanical Property of PLA and PLA/CF Samples Built by FFF Approach. *Materials* **2023**, *16*, 5023. [CrossRef]
30. Valero, J.R.L. Part 3: Injection: Machines and Processes: CHAPTER 13: The Injection Molding Machine. In *Plastics Injection Molding*; Valero, J.R.L., Ed.; Plastics Injection Molding; Hanser: Munich, Germany, 2020; pp. 159–177, ISBN 9781569906897. [CrossRef]
31. Bakr, M.; Su, Y.; Rezaei, A.; Bossuyt, F.; Vanfleteren, J. Over-Molding of Flexible Polyimide-Based Electronic Circuits. *Flex. Print. Electron.* **2021**, *6*, 025007. [CrossRef]
32. Kaur, R.; Singh, P.; Tanwar, S.; Varshney, G.; Yadav, S. Assessment of Bio-Based Polyurethanes: Perspective on Applications and Bio-Degradation. *Macromol* **2022**, *2*, 284–314. [CrossRef]
33. Li, B.; Liu, Y.; Wei, S.; Huang, Y.; Yang, S.; Xue, Y.; Xuan, H.; Yuan, H. A Solvent System Involved Fabricating Electrospun Polyurethane Nanofibers for Biomedical Applications. *Polymers* **2020**, *12*, 3038. [CrossRef] [PubMed]
34. Sun, C.; Wei, S.; Tan, H.; Huang, Y.; Zhang, Y. Progress in Upcycling Polylactic Acid Waste as an Alternative Carbon Source: A Review. *Chem. Eng. J.* **2022**, *446*, 136881. [CrossRef]
35. Sedničková, M.; Pekařová, S.; Kucharczyk, P.; Bočkaj, J.; Janigová, I.; Kleinová, A.; Jochec-Mošková, D.; Omaníková, L.; Perd'ochová, D.; Koutný, M.; et al. Changes of Physical Properties of PLA-Based Blends during Early Stage of Biodegradation in Compost. *Int. J. Biol. Macromol.* **2018**, *113*, 434–442. [CrossRef]
36. Tor-Świątek, A.; Garbacz, T. Effect of Abiotic Degradation on the Colorimetric Analysis, Mechanical Properties and Morphology of PLA Composites with Linen Fibers. *Adv. Sci. Technol. Res. J.* **2021**, *15*, 99–109. [CrossRef]

37. FitzGerald, K.P.; Nairn, J.; Skennerton, G.; Atrens, A. Atmospheric Corrosion of Copper and the Colour, Structure and Composition of Natural Patinas on Copper. *Corros. Sci.* **2006**, *48*, 2480–2509. [CrossRef]
38. Kim, J.; Gupta, N.S.; Bezek, L.B.; Linn, J.; Bejagam, K.K.; Banerjee, S.; Dumont, J.H.; Nam, S.Y.; Kang, H.W.; Park, C.H.; et al. Biodegradation Studies of Polyhydroxybutyrate and Polyhydroxybutyrate-Co-Polyhydroxyvalerate Films in Soil. *Int. J. Mol. Sci.* **2023**, *24*, 7638. [CrossRef]
39. Thomas, G.M.; Quirk, S.; Huard, D.J.E.; Lieberman, R.L. Bioplastic Degradation by a Polyhydroxybutyrate Depolymerase from a Thermophilic Soil Bacterium. *Protein Sci.* **2022**, *31*, e4470. [CrossRef]
40. Brasse, Y.; Laguna Moreno, M.; Blum, S.; Horter, T.; Janek, F.; Gläser, K.; Emmerechts, C.; Clanet, J.M.; Verhaert, M.; Grymonprez, B.; et al. Recyclable In-Mold and Printed Electronics with Polymer Separation Layers. *RSC Sustain.* **2024**, *2*, 1883–1894. [CrossRef]
41. Haider, T.P.; Völker, C.; Kramm, J.; Landfester, K.; Wurm, F.R. Plastics of the Future? The Impact of Biodegradable Polymers on the Environment and on Society. *Angew. Chem. Int. Ed.* **2019**, *58*, 50–62. [CrossRef] [PubMed]
42. Spierling, S.; Venkatachalam, V.; Mudersbach, M.; Becker, N.; Herrmann, C.; Endres, H.J. End-of-Life Options for Bio-Based Plastics in a Circular Economy-Status Quo and Potential from a Life Cycle Assessment Perspective. *Resources* **2020**, *9*, 90. [CrossRef]
43. Sikorska, W.; Musioł, M.; Zawidlak-Węgrzyńska, B.; Rydz, J. End-of-Life Options for (Bio)Degradable Polymers in the Circular Economy. *Adv. Polym. Technol.* **2021**, *2021*, 6695140. [CrossRef]
44. Van Roijen, E.C.; Miller, S.A. A Review of Bioplastics at End-of-Life: Linking Experimental Biodegradation Studies and Life Cycle Impact Assessments. *Resour. Conserv. Recycl.* **2022**, *181*, 106236. [CrossRef]
45. Vink, E.T.H.; Davies, S.; Kolstad, J.J. The Eco-Profile for Current Ingeo® Polylactide Production. *Ind. Biotechnol.* **2010**, *6*, 212–224. [CrossRef]
46. Fonseca, A.; Ramalho, E.; Gouveia, A.; Figueiredo, F.; Nunes, J. Life Cycle Assessment of PLA Products: A Systematic Literature Review. *Sustainability* **2023**, *15*, 2470. [CrossRef]
47. Rostkowski, K.H.; Criddle, C.S.; Lepech, M.D. Cradle-to-Gate Life Cycle Assessment for a Cradle-to-Cradle Cycle: Biogas-to-Bioplastic (and Back). *Environ. Sci. Technol.* **2012**, *46*, 9822–9829. [CrossRef]

processes

MDPI

Review

Strategic Selection of a Pre-Reduction Reactor for Increased Hydrogen Utilization in Hydrogen Plasma Smelting Reduction

Bernhard Adami [1], Felix Hoffelner [2], Michael Andreas Zarl [1] and Johannes Schenk [2,*]

[1] K1-MET GmbH, Stahlstraße 14, A-4020 Linz, Austria; bernhard.adami@k1-met.com (B.A.); michael.zarl@k1-met.com (M.A.Z.)

[2] Chair of Ferrous Metallurgy, Montanuniversität Leoben, Franz Josef-Straße 18, A-8700 Leoben, Austria; felix.hoffelner@unileoben.ac.at

* Correspondence: johannes.schenk@unileoben.ac.at; Tel.: +43-3842-402-2214

Abstract: The hydrogen plasma smelting reduction process has the potential to drastically reduce the CO_2 emissions of the steel industry by using molecular, atomic and ionized hydrogen as a reducing agent for iron ores. To increase the hydrogen and thermal efficiency of the process, a pre-reduction and pre-heating stage should be incorporated in a future upscaling of an existing HPSR demonstration plant within the scope of the "SuSteel follow-up" project to a target capacity of 200 kg/h of iron ore. The determination of the optimal process parameters is followed by a review of possible reactor types. A fluidized bed cascade, a cyclone cascade and a rotary kiln are compared for this purpose. Their applicability for the hydrogen plasma smelting is discussed, based on their fundamental design and operational procedures. Additionally, critical features of the different reactor types are outlined. A cyclone cascade with at least 3 stages is proposed to be the optimal reactor for pre-heating and pre-reducing the input material for the upscaled hydrogen plasma smelting reduction demonstration plant, based on the assessment.

Keywords: hydrogen plasma smelting reduction (HPSR); hydrogen direct reduction (H_2-DR); green steel; iron ore pre-reduction; fluidized bed reactor cascade; cyclone cascade; rotary kiln; hydrogen utilization; Reh diagram

1. Introduction

The steel industry emits around 7% of the total annular CO_2 emissions worldwide [1]. To counteract climate change, those emissions need to be reduced [2]. One possible future technology for green steel production is the hydrogen plasma smelting reduction (HPSR) process, which has been under development since the beginning of the 1990s at the Chair of Ferrous Metallurgy at Montanuniversität Leoben, Austria [3]. It uses molecular, atomic and ionized hydrogen to directly produce molten crude steel from iron ore fines, thereby generating environmentally friendly water vapor [4]. The plasma, stabilized by argon, burns between the molten bath as anode and the hollow graphite cathode, through which the oxidic input material and the gases are charged continuously [5]. The schematics of the process are shown in Figure 1.

The detailed fundamentals of the process have already been described multiple times by other researchers, to which the reader is referred to [6–9]. Besides hardly emitting any CO_2, its main advantages include more favorable thermodynamics and faster reaction kinetics compared to conventional hydrogen-based direct reduction processes [10]. Furthermore, the direct processing of low-grade iron ore fines is possible since impurities like phosphorus or sulfur can be removed by operating certain process conditions [4]. A first study estimates the specific production costs for hot strip based on the HPSR process to be around 20% lower compared to conventional integrated steel mills [11]. The status of the process development is currently at technology readiness level 5, as a demonstration plant with a capacity for 200 kg ore per trial has been built [12]. Basically, it is a gas-tight direct-current electric arc furnace, operated at a slight overpressure, with hydrogen introduction through a hollow graphite cathode. A process feature is, that the injected hydrogen is heated to very high temperatures and hot unreacted hydrogen is still contained in the off-gas. This can be used for pre-reducing iron ores as well as pre-heating the charge. In principle, any hydrogen direct reduction plant can be used for this purpose.

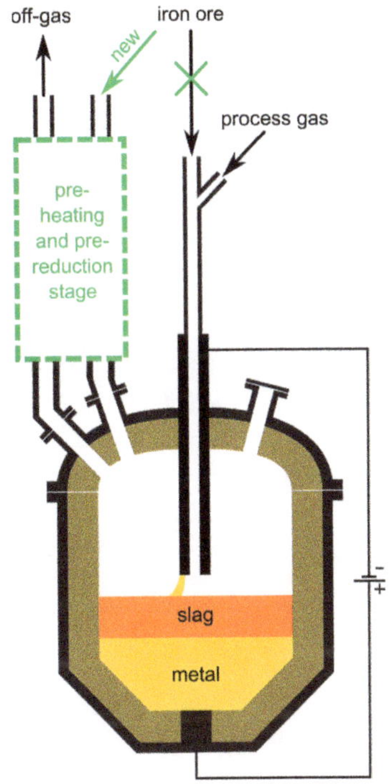

Figure 1. Overview of the HPSR process. The green marked area indicates the pre-heating and pre-reduction stage, for which its reactor selection is the topic of this work.

To be able to conduct experiments which such a setup, the previous batch-wise operation is currently being transformed within the scope of the "SuSteel follow-up" project into a fully continuous one with the ability to reduce 200 kg/h of ore [13]. The pre-heating and pre-reduction of the input material should be done solely with the off-gas of the HPSR reactor to increase the process' hydrogen utilization and thermal efficiency. Additionally, because of the lower residence time of pre-reduced ore in the plasma reactor, the refractory wear and electrode consumption also decrease [14]. This work aims to compare possible reactor types and propose the most suitable one for the pre-heating and pre-reduction of iron ores with HPSR off-gas. Firstly, a short summary of the ideal process parameters from thermodynamic and kinetics considerations is given, followed by the requirements for the reactor. After comparing different reactor types, the optimal one is selected based on a thorough assessment of their advantages and disadvantages.

2. Ideal Parameters for Iron Ore Pre-Reduction with Hydrogen

To select a suitable reactor for the desired process, the operating parameters need to be defined beforehand. Therefore, a study on the most optimal conditions for the pre-reduction of iron ores in hydrogen has to be conducted. Since a multitude of literature exists on this topic, including some review papers [15–17], only the most important findings are summarized in the following. For the optimal reduction of iron oxides with hydrogen containing gas in a pre-reduction reactor, the following conditions should be satisfied:

- The pre-reduction process should be operated at the highest possible temperature [15].
- The reducing gas should contain as much hydrogen as possible and few other components like H_2O, Ar, CO and CO_2 [18,19]. However, the water vapor concentration in the pre-reduction stage will be high at good hydrogen utilization in the proceeding HPSR process. For the benefit of the latter, the contents of carbon monoxide, carbon dioxide and argon in the off-gas are desired to be low. CO and CO_2 originate from the cathode wear [20], which ideally is minimized. While argon acts as plasma stabilizer [5], increased contents lead to a higher energy consumption for the heating of the reducing gas to reaction temperature.
- The input material should be very fine to facilitate fast reactions also at lower temperatures during heating [15]. A particle size of <500 μm is suggested for the reduction to FeO with carbon monoxide at 1000 °C [21]. Due to the better diffusivity of hydrogen [15], the reduction should also happen fast below this threshold. The grain sizes and distributions of iron ore concentrates vary between producers. Some provide the whole fraction below 0.1 mm, while others target the spectrum of 0.1–1 mm with only around 15% below 0.1 mm [22]. The typical pellet feed, however, consists of grains up to 0.15 mm [23]. This means that it can be used directly for pre-reduction in the HPSR process without further pre-treatment.
- Higher porosity of the iron ore benefits the reduction because of the decreased gas diffusion resistance and higher specific surface area [15].
- The effects of morphology are intertwined with the ones of porosity, since denser microscopical structures reduce the mass transfer in the gas phase. Therefore, the preferential input materials for the HPSR process are limonitic and porous hematitic ores, while magnetitic and dense hematitic ones are less beneficial regarding their slow reduction kinetics. From the viewpoint of carbon direct avoidance [24], siderite cannot be used because of its carbonatic nature, although it is easily reducible due to its high specific surface area [25].
- The presence of gangue elements can affect the reducibility positively and negatively. On the one hand, they can facilitate micro-cracking, which increases the reduction rate. However, dense compounds can also raise the resistance of the interfacial chemical reaction, thereby hindering the reduction process. Those effects are not necessarily specific for a given gangue element but depend on the actual conditions. For example, alumina forms hercynite, which can aid the reduction process by introducing micro-cracks. Nevertheless, this phase can also decrease the reduction rate for magnetite [16].

3. Definition of the Scenario Parameters

Before selecting a reactor type, its use case must be defined. First of all, it is operated as a pre-heating and pre-reduction step solely with the off-gas of the HPSR process. This means that the reducing gas flow rate is dictated by thermodynamic limits and increasing it would lead to decreased efficiencies in the HPSR process. Namely, raising the amount of hydrogen leads to a lower hydrogen utilization degree in the HPSR reactor and adding another gas lowers its partial pressure, yielding unfavorable reducing conditions. However, putting more hydrogen through the HPSR reactor than theoretically needed does not necessarily lower the overall hydrogen utilization degree of the HPSR process combined with the pre-reduction stage. If the excess hydrogen reacts in the pre-reduction facility accordingly, then the overall hydrogen efficiency is conserved and only the distribution of the reduction extend between the two stages is changed. For simplification, this behavior is not included in the following assessments. On the other hand, decreasing the hydrogen supply means that too little reducing agent is provided, which would lead to an accumulation of iron oxides in the HPSR reactor and is therefore not allowed for a continuous operation.

As input material a typical hematitic iron ore with around 95 wt.-% Fe_2O_3 (extracted from data provided in [22]) and a maximum grain size $d_{p,max}$ of 150 μm is used, so that the direct processing of pellet feed is enabled as stated above. In the pre-reduction step the hematite gets fully reduced to stochiometric FeO for simplification, which then is charged in the HPSR reactor, where the reduction process to metallic iron is finished. If the HPSR reactor operates at the thermodynamically maximum possible hydrogen utilization $\eta_{H_2}^{HPSR}$ of around 40 vol.-% (see the Baur-Glässner diagram shown in Figure 2 for the transformation of liquid wustite into liquid iron), then the mass balance reveals that the normal specific HPSR off-gas rate

$$\beta^0 = \frac{\dot{V}_f^0}{\dot{m}_s} = \frac{\dot{n}_{pg} V_m^0}{\dot{m}_s} = \frac{\dot{n}_{H_2,FeO\to Fe}}{\dot{m}_s\left(1 - x_{Ar}^{pg}\right)\eta_{H_2}^{HPSR}} = \frac{2\dot{n}_{Fe_2O_3}}{\dot{m}_s\left(1 - x_{Ar}^{pg}\right)\eta_{H_2}^{HPSR}} = \frac{2w_{Fe_2O_3}}{\left(1 - x_{Ar}^{pg}\right)\eta_{H_2}^{HPSR} M_{Fe_2O_3}} \quad (1)$$

is 0.9 Nm^3/kg iron ore charged. This is under the assumption that 25 vol.-% of argon need to be added to the plasma gas for arc stabilization [26] and neglects the cathode gasification for simplification. Therein, $\dot{V}_f^0$ means the volume flow rate of the HPSR off-gas at normal conditions, $\dot{m}_s$ the iron ore mass flow rate charged in the pre-reduction and pre-heating reactor, $\dot{n}_{pg}$ the molar flow rate of the plasma gas supply, V_m^0 the molar volume of an ideal gas at normal conditions, $\dot{n}_{H_2,FeO\to Fe}$ the molar flow rate of hydrogen for the complete reduction of FeO, x_{Ar}^{pg} the molar fraction of argon in the plasma gas input, $\dot{n}_{Fe_2O_3}$ the molar flow rate of hematite, $w_{Fe_2O_3}$ its mass fraction in the ore and $M_{Fe_2O_3} = 159.7$ g/mol [27] its molar mass. If the targeted value for the ore throughput is 200 kg/h for the purpose described in the introduction, a $\dot{V}_f^0$ of around 180 Nm^3/h is obtained.

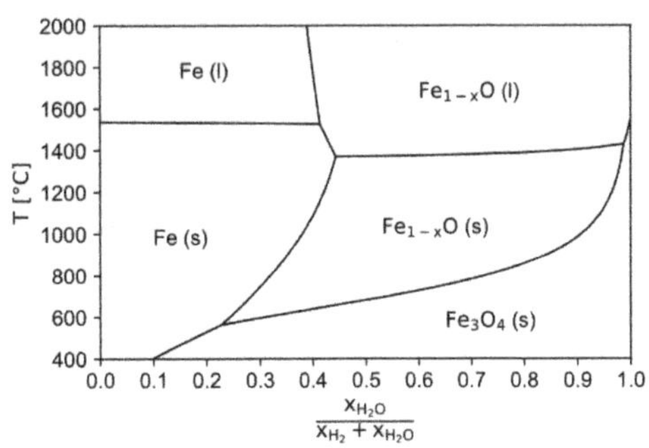

Figure 2. Baur-Glässner diagram, calculated with the software FactSage 8.3 [28], databases FactPS, FToxid and FSstel 2023.

Measurements during trials at the demonstration plant as currently configured give off-gas temperatures of usually around 1000 °C. However, since this reactor is basically a scaled-up version of the laboratory setup referred to in the introduction (Section 1), it is not efficiently insulated. Additionally, the upper vessel provides a relatively large surface for heat transfer until the gas reaches the exhaust port. As those problems have been identified, countermeasures will be implemented in the upgraded setup. Further, the cathode unit will be redesigned to feature lower heat losses. Therefore, it can be assumed that higher off-gas temperatures will be attainable, which benefits the pre-reduction and pre-heating stage.

Since the off-gas of the HPSR reactor is in contact with the melt, it can be estimated to leave the reactor at a similar temperature. To keep the produced iron liquid, a minimum temperature of around 1600 °C is required, accounting for local thermal imbalances and some heat losses in the reactor since the melting point of pure iron is 1536 °C [29] and there are not many impurities present [11]. The melting point of wustite is slightly below 1400 °C according to Figure 2. To safely prevent the melting of the material, the target operating temperature T_{op} is set to 1300 °C. The heat balance [30] of the input materials for a counter-current flow reactor

$$\frac{\dot{n}_{Fe_2O_3}}{w_{Fe_2O_3}}\Delta H^0_{s,T_{i,s}\rightarrow T_{op}} + \dot{n}_{Fe_2O_3}\Delta H^0_{Fe_2O_3+H_2\rightarrow 2FeO+H_2O,T_{op}} = \frac{\dot{V}^0_f}{V^0_m}\Delta H^0_{f,T_{e,f}\rightarrow T_{op}} \tag{2}$$

allows an estimation of whether this is theoretically possible. If heat losses in the reactor are neglected, an exhaust gas temperature of 240 °C is obtained. The exchanged heat energy is 1.2 MJ/kg solids. For this rough estimation, the initial temperature of the ore $T_{i,\,s}$ was set to 25 °C. The determination of the standard enthalpy changes [27] for the ore $\Delta H^0_{s,T_{i,s}\rightarrow T_{op}}$ is based on the assumption of pure Fe_2O_3, while the one for the reducing gas $\Delta H^0_{f,T_{e,f}\rightarrow T_{op}}$ was superimposed according to its composition, which is 45 vol.-% H_2, 30 vol.-% H_2O and 25 vol.-% Ar under the given scenario parameters. $\Delta H^0_{Fe_2O_3+H_2\rightarrow 2FeO+H_2O,T_{op}}$ denotes the standard enthalpy of reaction at temperature T_{op} and is obtained from the software FactSage 8.3 [28] with the database FactPS 2023 to 31,524.8 J/mol. If the absolute operating pressure p_{op} is assumed to be 1.1 bar_a, the adapted HPSR off-gas rate β is calculated with the ideal gas law [31]

$$\beta = \frac{\dot{V}_f}{\dot{m}_s} = \frac{\dot{V}^0_f T_{op} p^0}{\dot{m}_s T^0 p_{op}} \tag{3}$$

to 4.7 m^3/kg. The off-gas volume flow rate at operating conditions $\dot{V}_f$ is then 940 m^3/h.

For many of the calculations in the following sections, some gas and material properties need to be known. They are summarized here to provide a better overview of the assumed and calculated quantities. From the composition of the HPSR off-gas, a molar mass M_f of 16.299 g/mol and a density ρ_f at operating conditions of 0.137 kg/m^3 results with the treatment as an ideal gas. The kinematic viscosity ν_f of the off-gas mixture at operating conditions is calculated to $4.89 \cdot 10^{-4}$ m^2/s according to the mixing method of Wilke as presented in [30] together with the polynomial description of the temperature dependent properties of the components. From the data provided in [22], the apparent density of the ore ρ_s is assumed to be 5000 kg/m^3, while its bulk density ρ_{bulk} is set to 2200 kg/m^3.

The complete reaction of hematite to FeO corresponds to a reduction degree [32]

$$RD = \frac{\Delta m_O}{m_{O,i}} = 1 - \frac{w_{Fe^{2+}} + 1.5 w_{Fe^{3+}}}{1.5 w_{Fe,tot}} \tag{4}$$

of 33%. Therein, $m_{O,i}$ denotes the initial and Δm_O the change in the mass of oxygen. w_χ means the mass fraction of the iron species χ. The water vapor content in the gas entering the pre-reduction reactor is around $x^{bulk}_{H_2O} = 30$ mol.%. From the Baur-Glässner diagram (Figure 2) the equilibrium H_2O fraction at 1300 °C for the reaction of magnetite to wustite can be determined as approximately $x^{equ}_{H_2O} = 97$ mol.%. The following equation provides a rough empirical correlation between the time t_{RD} to reach a certain reduction degree for the reduction of iron oxides [33]:

$$t_{RD} = \frac{\left(1 - \sqrt[3]{1 - RD}\right) x_{H_2O}^{equ}}{25\ \frac{\mu m}{s} \left(x_{H_2O}^{equ} - x_{H_2O}^{bulk}\right)} d_p e^{\frac{17,600 \frac{J}{mol}}{RT}} \quad (5)$$

This allows the estimation of the necessary residence time $t_{33\%} = 4.2$ s of particles with $d_{p,max}$ in the pre-reduction reactor at T_{op} to reach the required 33% reduction degree. For simplification, the heating time of the ore to reaction temperature is neglected since the particles are small.

4. Overview of Reactor Types for Non-Catalytic Gas-Solid Reactions

There are several options for classifying chemical reactors. The most fundamental one is according to the operating mode into continuous and discontinuous reactors. For optimal productivity and due to the continuous operation of the HPSR vessel itself, a continuous pre-reduction of the input materials should be implemented. Another classification criterion is the energetic operating mode. The basic types are the isotherm, adiabatic and polytropic reactors. They feature a timely and locally constant temperature profile, the absence of heat transfer through the reactor walls, and the presence of heating or cooling facilities to control the temperature profile, respectively. The desired reactor ideally features adiabatic conditions because the iron ores should only be pre-heated by the sensible heat contained in the HPSR off-gas. Chemical reactors also can be classified according to the phases present. In this case, only two phases exist, namely the iron ores and the HPSR off-gas, which make a gas-solid reactor necessary. Since the chemical reaction is activated thermally, a catalysator is not needed. Non-catalytic gas-solid flow reactors can be classified based on their arrangement into co-current, counter-current and cross-current flow. The counter-current flow reactor is the most energy efficient, making it ideal for an optimal pre-heating stage [34].

To summarize, the optimal reactor for a pre-reduction and pre-heating stage in the HPSR process can be classified as a continuous, counter-flow, adiabatic, non-catalytic, gas-solid reactor.

4.1. The Reh Diagram

Before describing the specific reactor types, another classification system of continuous gas-solid reactors must be explained according to their fluid dynamics. Those are the moving (MBR) and fluidized bed (FBR) as well as the transport reactors (TR). In the MBR, the bed is moved through the reactor either mechanically or gravitationally with a limitation on the gas flow to retain a bulk. The FBR utilizes higher gas velocities to detach the particles from each other and obtain a fluidized state. In a TR, the reactions between the phases happen during the pneumatic conveying of the solids [35].

The operating range of those reactor types can be visualized with the Reh diagram, given in Figure 3. Its derivation according to the procedure described originally in [36] with minor modifications is summarized in Appendix A of this paper, together with an explanation and definition of the symbols used in the following formulas. The equations necessary for the description of the most important lines are [36,37]:

$$\epsilon = \frac{V_f}{V}, \quad (6)$$

$$n = \frac{F_d}{F_g - F_b}, \quad (7)$$

$$Re_p = \frac{\overline{u} d_p}{\nu_f}, \tag{8}$$

$$c_d = \frac{24}{Re_p}\left(1 + 0.1806 Re_p^{0.6459}\right) + \frac{0.4251}{1 + \frac{6880.95}{Re_p}}, Re_p < 2.6 \cdot 10^5, \tag{9}$$

$$Fr^* = \frac{4n\epsilon^{4.65}}{3c_d}, \begin{cases} n > 1, \ \epsilon \to 1 \\ n = 1, \ 0.4 \leq \epsilon \leq 1. \\ n < 1, \ \epsilon = 0.4 \end{cases} \tag{10}$$

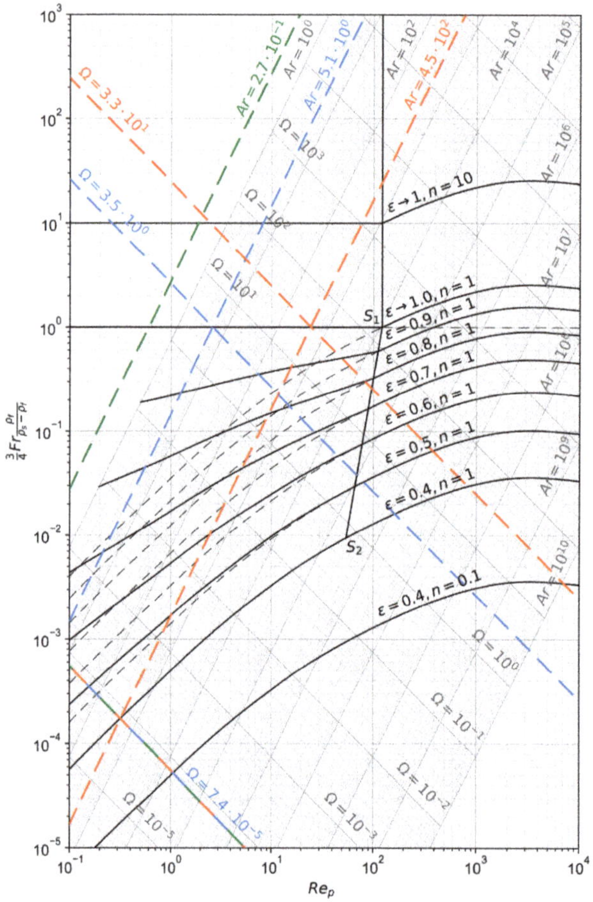

Figure 3. Reh diagram calculated as described in Appendix A according to the procedure presented in [36].

To aid the practical use of the Reh diagram, two auxiliary parameters can be defined, namely the Archimedes number

$$Ar = \frac{\left(\rho_s - \rho_f\right) g d_p^3}{\rho_f \nu_f^2}, \tag{11}$$

and the Ω number

$$\Omega = \frac{\rho_f \overline{u}^3}{\left(\rho_s - \rho_f\right) g \nu_f} \quad (12)$$

In contrast to Re_p and Fr^*, which both depend on the particle diameter as well as the superficial fluid velocity $\overline{u}$, besides material properties Ar is only a function of d_p and Ω only of $\overline{u}$. Their relationship to Re and Fr^* is as follows [38]:

$$Fr^* = \frac{Re_p^2}{Ar} = \frac{\Omega}{Re_p} \quad (13)$$

Depending on the load factor, three different cases are possible [39]:

- $n < 1,\ \epsilon = \epsilon_{min} = 0.4$: The fluid velocity is too small to lift the particles and the bulk remains packed at its minimal relative void volume ϵ_{min} of around 40 vol.-%.
- $n = 1,\ 0.4 < \epsilon < 1$: The fluid velocity is balanced with the bed weight to keep the bulk in a floating state and the bed fluidizes and expands.
- $n > 1,\ \epsilon \to 1$: The fluid velocity is too large and pneumatic conveying occurs due to the entrainment of the bulk by the flow.

4.2. Fluidized Bed Reactor Cascades

An FBR is a vertical retort, where the gas flows upwards through the bed of solids with a certain velocity at which fluid-like behavior of the whole charge is obtained. For illustrative purposes, a schematic drawing of an FBR for the drying of solids is shown in Figure 4. The gas velocity and particle size distribution need to be carefully balanced to maintain the material in a fluidized state, as can be seen in the derivations of the Reh diagram above. A gas distribution system is used to provide a constant stream of reducing gas over the whole reactor cross section. This can range from simple plates with regularly spaced holes to more complex bubble-cap tuyere distributors, which also prevent the backflow of solids in the gas supply pipes if the gas flow rate is too low. The gas outlet must be equipped with a gas-solid separator, like a cyclone or electrostatic precipitator, to minimize dust discharge. The solids outlet is usually designed as an overflow of the fluidized bed. Many variations of reactors which feature a fluidized bed exist, including circulating or multi-stage processes. The latter are necessary for the scenario assumed in this work to approximate a counter-current reactor system, since a single FBR alone behaves like a continuously stirred tank. However, for simplicity the following derivations of the reactor dimensions and critical parameters only consider the most basic FBR, consisting of one single fluidized bed in a cylindrical retort without solids recirculation [40].

The advantages of FBRs include the excellent heat and mass transfer due to the good mixing behavior of the phases and the adjustable solids retention time during engineering when defining the reactor dimensions [41]. Disadvantageous are the agglomeration and sintering of the solids due to the high temperatures, leading to larger particles which cannot be kept fluidized above a certain point. This results in the deposition of those lumps at the bottom of the reactor. The easiest ways to overcome this phenomenon are the decrease of the operating temperature and the increase in the gas velocity. However, the first measure usually lowers the conversion rate while the second one facilitates the discharge of fine particles [42].

Problems can occur also at the gas distributing system when using dust-loaded gas as input. Because of the high temperatures, sticking of particles at holes and baffles is possible [40].

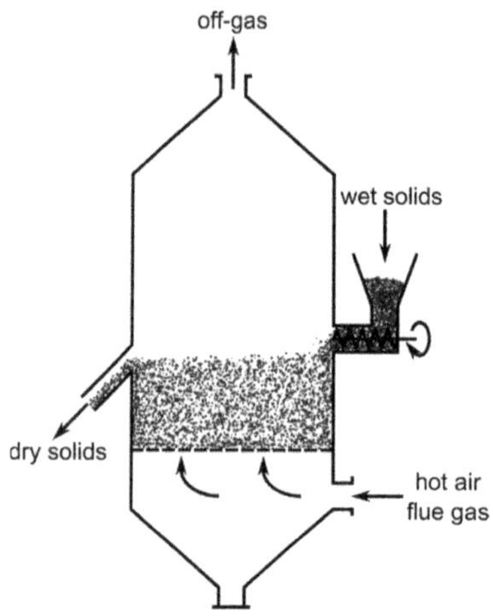

Figure 4. Schematic drawing of a fluidized bed reactor used for the drying of solids, modified with permission of Elsevier Science & Technology Journals from [40]; permission conveyed through Copyright Clearance Center, Inc.

One of the most important parameters of a chemical reactor is its reactant residence time. For an FBR, the average residence time

$$t_{s,d_{p,X}} = \frac{1}{\frac{\dot{m}_{fo} w^{fo}_{d_{p,X}}}{m_{bed} w^{bed}_{d_{p,X}}} + \frac{\dot{m}_{so}}{m_{bed}}} = \frac{1}{K_{el,d_{p,X}} + \frac{\dot{m}_{so}}{m_{bed}}} \quad (14)$$

of particles with constant size $d_{p,i}$ results from a mass balance of the solids over the bed and models elutriation of each particle size with a kinetic parameter $K_{el,d_{p,X}}$. This parameter is the ratio of the mass flow of particles of a specific size entrained in the off-gas to the mass of those particles in the bed. In the formula above, $\dot{m}_{fo}$ and $\dot{m}_{so}$ denote the mass flows through the gas and the product outlet respectively. The mass fractions of the particle size $d_{p,X}$ in the bed as well as in the flow through the gas outlet are denoted by $w^{fo}_{d_{p,X}}$ and $w^{bed}_{d_{p,X}}$. m_{bed} indicates the total mass of the bed. If the elutriation is neglected, the average residence time becomes the ratio of the bed's mass to the solids charging rate. Rearranging allows the calculation of the bed's mass to 0.23 kg for the given scenario parameters. Besides the average residence time its distribution is also important. Since a single fluidized bed is constantly mixed, each particle has the same possibility to leave it at every moment, assuming a uniform particle size distribution. This behavior is best described by an exponential function. However, when incorporating more stages, the residence time distribution approximates the one of a plug flow reactor, since the probabilities equalize [40].

The maximum reactor diameter can be obtained graphically from the Reh diagram (Figure 3). The Archimedes number $Ar_{d_{p,max}}$, which corresponds to the required maximum particle diameter $d_{p,max}$ of 150 μm, is calculated to 5.1. The crossing point of this line with the minimum fluidization curve ($\epsilon = 0.4,\ n = 1$) defines the lowest possible Ω number and, therefore, the necessary operating velocity as well as the maximum reactor diameter. However, since the particle size threshold is small, it lies outside the chart's validity range. When extrapolating, a minimum Ω number of $1.2 \cdot 10^{-8}$ is obtained, corresponding to a minimum fluidization velocity of 0.038 m/s. This allows for the calculation of the maximum reactor diameter from the principle of continuity to 3.0 m. Another method for the calculation of $d_{FBR,max}$ is based on the Ergun correlation [38], with which the fluidization starting line ($n = 1, \epsilon = \epsilon_{min} = 0.4$) in the Reh diagram can also be described, as explained during its derivation. It has to be noted, that the real design Ω number needs to be higher than the theoretical minimum to ensure the fluidization of small agglomerates with some margin. The quotient of two Archimedes numbers $Ar_{d_{p1}}/Ar_{d_{p2}}$ for a particle diameter ratio d_{p1}/d_{p2} grows cubically, meaning that if a particle agglomerates to a ratio of 10:1 its original size the corresponding Archimedes number increases 10^3:1. On the other hand, if the design Ω number gets too high, the entrainment of fines by the gas is increased. An FBR is therefore only applicable to narrow grain size distributions [43]. For further sizing of the reactor, the desired relative void volume needs to be defined. If it is set to 0.6, the design Ω number of $7.4{\cdot}10^{-5}$ is obtained graphically from the Reh diagram as the intercept between the fanned-out line for $\epsilon = 0.6$ and the one for $Ar_{d_{p,max}}$. This allows the calculation of the design gas velocity to 0.23 m/s which yields a reactor diameter of 1.2 m. The diameter of the smallest operatable particles is approximately 6 µm, which is obtained from the Archimedes number at the intercept between the design Ω number and the extrapolated fanned-out line for $\epsilon = 0.7$. Above this threshold the elutriation by the gas phase becomes more significant and hinders the operation of dense fluidized beds, introducing the need of a solids recirculation system [40]. Similarly, the maximum fluidizable particle diameter of around 670 µm can be obtained from the crossing with the line for $\epsilon = 0.4$ and $n = 1$. This means, that an agglomeration of up to 4.5-fold the maximum particle diameter would be possible for this proposed system.

The bed's height

$$h_{bed} = \frac{4m_{bed}}{\rho_s(1-\epsilon)d_{FBR}^2\pi} \tag{15}$$

can be obtained from the definition of the relative void fraction combined with the volume formula of a cylinder and the number of contained mono-sized particles. For the given scenario, the bed would have a height of just 0.1 mm, which of course, is not practical. This value results from the relatively low ore feed rate and solids retention time, which in turn requires only a small bed mass, together with the relatively large diameter necessary to prevent solids elutriation. This behavior shows that for small grain sizes, a solids separation and recirculation system is inevitable. Another possibility to obtain higher beds is to increase the actual residence time above the minimum required one for reduction. For every 10 cm of height, the residence time of the solids is around 4000 s, corresponding to a mass of approximately 220 kg.

The problem of exposure to high temperatures for elongated times is especially the formation of agglomerates, which can grow too large for fluidization [42]. As mentioned, this phenomenon can be counteracted to a certain degree by increasing the gas velocity. However, this is not desired since it increases the entrainment of solids by the reactor's off-gas. The operation of pure, homogeneous fluidized bed reactors under the required conditions is only possible with low gas velocities as well as a high dead bed volume and is prone to be influenced by sticking effects.

Alternatively, a cascade of circular fluidized bed reactors (CFBR) could be used to overcome some of the presented limitations. However, since a solids recirculation system is necessary, it is more complex [40].

4.3. *Transport Reactor Cascades*

When a high proportion of solids discharge is expected in FBR operation anyway and only short residence times are required, a logical step would be to increase the gas velocity vastly to eject all particles deliberately. With the pre-heating and pre-reduction then happening in a transport pipe, the reactor is transformed into a TR. An advantage over the FBR is the ability to operate without a gas distribution system [35]. This mitigates the sticking-problem with the hot, dust-loaded off-gas and allows particles or agglomerates to drop down instead of accumulating in the pre-reduction reactor. Therefore, the TR is best implemented in a vertical off-gas pipe directly over the gas outlet of the HPSR reactor, enabling the unhindered fall of unconveyable particles into the melt. However, the gas flow rate needs to be above a certain threshold to operate with reasonably sized pipes. If the reactor diameter is too small, sticking of particles to the walls can occur. Further disadvantages of using a TR include the low residence times, which lead to a high pre-reduction reactor and its inflexibility [35]. It must always be operated with the specified gas flow rate to maintain a constant velocity for the given reactor diameter, which cannot be changed after construction. Additionally, like a one-stage FBR, a single TR is also a co-current reactor. Therefore, to achieve counter-current flow behavior, it must be operated in multiple stages. The TR also needs some kind of gas-solid separation system, e.g., a cyclone, to separate the output stream into the pre-reduced solids and the off-gas [40].

While a TR provides good heat and mass transfer between phases, reaction kinetics need to be fast to reach the desired conversion because of the low particle residence time. These depend on the reactor's length and are highest in vertically upwards directed flows due to the effect of gravity at otherwise equal conditions [35].

Although higher heat transfer rates are achieved in TRs than in pure FBRs, both are inferior to CFBRs due to the effect of the solids recirculation [44].

The dimensions of the TR are calculated in the following paragraph according to the derivations presented in [35]. The equation

$$h_{TR,RD} = u_p t_{RD} = \frac{u_p d_p}{2K_{RD}} = \frac{\left(\overline{u} + u_{slip}\right) d_p}{2K_{RD}} \tag{16}$$

gives the height $h_{TR,RD}$ for a given reaction time necessary to reach a certain conversion t_{RD}. Therein, u_p indicates the velocity of the particles and K_{RD} the growth rate of the product layer into the inside of a dense spherical particle. The particle's velocity needs to be corrected by the slip velocity u_{slip} as follows:

$$u_p = \overline{u} + u_{slip} \tag{17}$$

It is positive in downwards directed, gravitational flows and negative in upwards directed, counter-gravitational ones. The terminal velocity u_{term} of a sphere in a non-moving atmosphere can be calculated for laminar flow with the Stokes' law

$$u_{term} = \frac{\left(\rho_s - \rho_f\right) g d_p^2}{18 \rho_f \nu_f} \tag{18}$$

and is equal to its slip velocity for dilute flows. For increased solids loadings, interactions between the particles themselves and with the wall can increase the slip velocity manyfold. When considering the two cases of co- and counter-gravitational flows the following equations are obtained for their minimum reactor heights h_{TR}^{co} and $h_{TR}^{counter}$, respectively:

$$h_{TR}^{co} = \frac{\left(\rho_s - \rho_f\right) g d_{p,max}^3}{18 \rho_f \nu_f K_r} \tag{19}$$

$$h_{TR}^{counter} = \frac{\sqrt{3}\left(\rho_s - \rho_f\right) g d_{p,max}^3}{9 \cdot 18 \rho_f \nu_f K_r} \tag{20}$$

In the first case, besides $d_p = d_{p,max}$, $\overline{u} = u_{term}$ is used for eased comparison between the two types. In the latter case the distribution of the reactor height for different particle diameters for a constant $\overline{u}$ must be considered. Small particles are converted fast, which leads to low reactor heights. Big particles need longer residence times for the reaction, but move slower with the flow, which also results in low reactor heights. The critical diameter $d_{p,crit}$ is obtained from

$$\frac{\mathrm{d}h_{TR,RD}\left(d_p = d_{p,c}\right)}{\mathrm{d}d_{p,c}} = 0 \tag{21}$$

and is described as

$$d_{p,crit} = \sqrt{\frac{6\overline{u}\rho_f\nu_f}{\left(\rho_s - \rho_f\right)g}} \tag{22}$$

Solving

$$h_{TR,RD}\left(d_p = d_{p,max}\right) = 0 \tag{23}$$

for $d_{p,max}$ gives

$$d_{p,max} = \sqrt{\frac{18\overline{u}\rho_f\nu_f}{\left(\rho_s - \rho_f\right)g}} \tag{24}$$

Combining the last two relations results in

$$d_{p,crit} = \frac{\sqrt{3}}{3}d_{p,max}. \tag{25}$$

Considering the reactor height for particles with the critical diameter in a flow with a velocity high enough to lift all particles yields the presented minimum reactor height for counter-gravitational flow (Equation (20)). It can be observed that counter-gravitational TRs can be by a factor of $\sqrt{3}/9 \approx 0.2$ shorter than co-gravitational ones due to the influence of gravity. A general disadvantage of the derived relations for the minimum reactor heights is that they assume chemical reaction control. The diffusion resistance of the reactants and products through the particle's shell is not considered in the presented kinetic model. However, this could be implemented by using modified expressions for t_{RD}.

The minimum Ω number for the entrainment of particles with $d_{p,max}$ by the gas flow can also be obtained graphically from Figure 3 as the intersection of the line for $Ar_{d_{p,max}}$ with the one for $\epsilon \to 1$, $n = 1$ and is 3.5. This results in a minimum gas velocity of 8.5 m/s, which equals a maximum reactor diameter of 0.2 m. Considering agglomerates up to 4.5 times the maximum particle size as in the FBR case, the maximum reactor diameter decreases to 0.14 m. The reactor heights for the two cases can be calculated with Equation (20) and the definition [33,35]

$$K_{RD} = 25\ \frac{\mu\mathrm{m}}{\mathrm{s}} e^{-\frac{17{,}600\,\frac{\mathrm{J}}{\mathrm{mol}}}{RT}} \frac{x_{H_2O}^{equ} - x_{H_2O}^{bulk}}{x_{H_2O}^{equ}} \tag{26}$$

to 6 m as well as 380 m, respectively. With the latter being unpractical, the severe influence of the grain size on the reactor height is clearly visible.

However, this actually only applies to a single TR. Since multiple stages are needed, the residence times in the gas-solid separators cannot be neglected. Vertically arranged cyclone cascades with four to six stages are widely used in the cement industry for preheating raw meal with the off-gas of the rotary kiln for clinker production, as shown schematically in Figure 5 [45]. They achieve high solids temperatures by approximating a counter-current heat exchanger with a stage-wise co-current operation [46]. Pendulum flaps are commonly used to prevent the short-circuiting of the gas flow up through the dust chambers [47]. The stages are usually numbered according to the flow direction of the solids, which are charged before the first separation stage. Since this stage is essential for the process efficiency, it consists of two parallel cyclones with a high separation degree. The other cyclones possess the same diameter and a low profile to minimize the pressure drop and increase the residence time. The separation efficiency is less important in the later stages since the entrained material gets recycled [48].

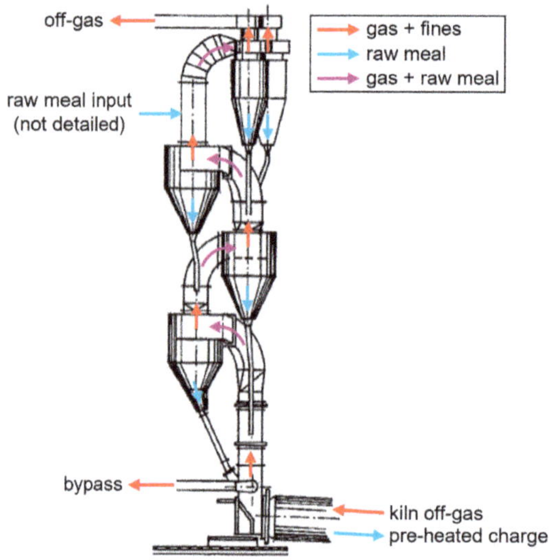

Figure 5. Schematic drawing of a four-stage cyclone cascade for pre-heating of raw meal in the cement industry, adapted from [48].

A typical cyclone is shown schematically in Figure 6. It consists of a vortex finder through which the cleaned gas exits, a cylindrical section in the vicinity of the tangential rectangular raw gas inlet duct, a conical separation chamber and a dust chamber where the solids are collected. Between the two chambers is an insert which prevents the pick-up of already separated solids by the gas vortex. The flow patterns inside cyclones are very complex, and the analytical description focuses mainly on separation efficiency and pressure drop, not retention times. In most cases, it is used as a gas-solid separator and not a reactor [49].

When considering the heat transfer, most of it will happen in the cylindrical part of the cyclone, because of the higher relative velocities between the solids and the gas than in the conical section, even at the longer solids residence times in the latter. Additionally, it is recommended that the solids mass loading of the gas

$$C_s = \frac{\dot{m}_s}{\dot{m}_f} = \frac{\dot{m}_s}{\dot{V}_f \rho_f} = \frac{1}{\beta \rho_f} \tag{27}$$

does not exceed 3, which applies for the given case. It is basically the inverse of the previously introduced coefficient β with the conversion of the gas volume flow into its mass equivalent $\dot{m}_f$ and equates to 1.5. At increased gas mass loadings the solids cannot be conveyed by the gas flow anymore, so they are deposited as strands at the cyclone's wall and slide thereon into the dust collection chamber [50].

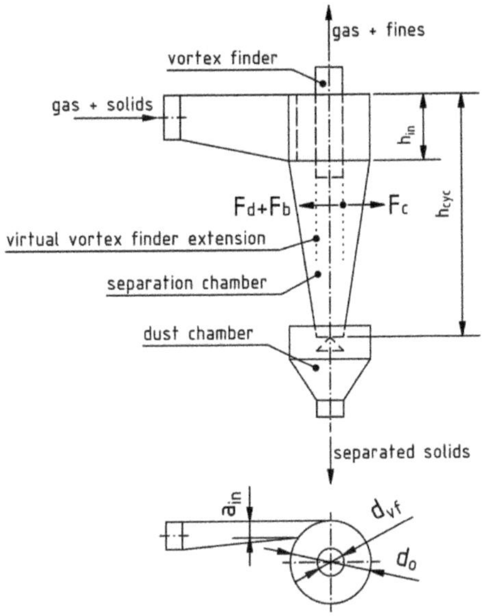

Figure 6. Dimensions of a cyclone separator and visualization of the force balance on an exemplary particle (Equation (29)), modified from [49] with permission from Elsevier.

An important measure for a gas-solid reactor is its solids retention time [34]. However, for the average particle residence time mainly empirical relations or computational fluid dynamics (CFD) models exist since the analytical description is very complex because of the three-dimensional flow pattern. The residence time of the solids in cyclone reactors is distributed relatively narrowly, like in a plug-flow reactor [51]. For rounded sand particles with a density of 2260 kg/m^3 and an average grain size of 100 µm t_s can be described as [52]:

$$t_s = \begin{cases} 0.1\ s + 1.26\ s \cdot C_s, & C_s \leq 15 \\ 19.45\ s - 0.03\ s \cdot C_s, & C_s > 15 \end{cases}. \tag{28}$$

This yields an average solids residence time t_s of 2 s for the given scenario, meaning that the material needs to pass 3 cyclones to reach the desired pre-reduction degree of the iron oxides. However, the time the solids are in contact with the gas in the transport pipes between the cyclones also must be considered. Therefore, a rough knowledge of the cyclone's dimensions is required.

In the following four paragraphs, the basic design factors of a cyclone separator after [49] with minor modifications to consider the system in this work are summarized. The dimensioning is based on the concept of optimal cyclones, which follow empirical guidelines for geometric ratios. An optimal cyclone balances the separation efficiency, expressed in terms of the critical size diameter, operating expenses, defined by the pressure drop, and construction expenses, described by the volume of the cyclone. The critical size diameter $d_{p,crit}$ is the theoretical cut-size and obtained by a force-balance

$$F_d = F_c - F_b, \tag{29}$$

on the vortex finder pipe diameter's d_{vf} virtual downwards extension. A visualization is given in Figure 6. The revolving particles with a tangential velocity $u_{t,vfe}$ at this location are pushed outwards by the centrifugal force

$$F_c = \frac{\rho_s \pi d_{p,crit}^3 u_{t,vfe}^2}{3d_{vf}}, \tag{30}$$

which gets countered by the buoyancy force

$$F_b = \frac{\rho_f \pi d_{p,crit}^3 u_{t,vfe}^2}{3d_{vf}} \tag{31}$$

as well as the drag force

$$F_d = \frac{c_d \pi d_{p,crit}^2 \rho_f u_{r,vfe}^2}{8} \tag{32}$$

of the inwards moving flow with a radial velocity of $u_{r,vfe}$. If either of these dominates, the particles leave their revolution path, which means that bigger particles move outwards and separate, while smaller ones are pushed further inwards and leave the cyclone with the cleaned gas through the vortex finder pipe. When considering that the small, inwards moving particles possess low Reynolds numbers, Stokes's law can be applied. Together with the knowledge that around 10% of the gas flow leave the cyclone directly and bypass the separator along its top boundary,

$$d_{p,crit} = \sqrt{\frac{0.9 \cdot 9 \rho_f \nu_f d_{vf} u_{r,vfe}}{\left(\rho_s - \rho_f\right) u_{t,vfe}^2}} \tag{33}$$

is obtained for the critical size diameter. Another important design parameter is the pressure drop

$$\Delta p = \frac{\xi_{tot} \rho_f u_{a,vf}^2}{2}. \tag{34}$$

It depends on the axial velocity in the vortex finder $u_{a,vf}$ and the cyclone's pressure drop coefficient ξ_{tot}. The latter is a combination of the pressure drop coefficient of the vortex finder with an auxiliary one considering the inlet and walls of the cyclone. $d_{p,crit}$ and Δp must be pre-defined to dimension the cyclone according to the optimization principle. Therefore, further explanations of their calculations for a given geometry are omitted in this work.

The relevant ratios for the dimensioning of an optimal cyclone are the following:

$$\mathrm{H}_{cyc} = \frac{2h_{cyc}}{d_{vf}}, \tag{35}$$

$$\mathrm{D}_{cyc} = \frac{d_o}{d_{vf}}, \tag{36}$$

$$U_{cyc} = \frac{u_{t,vfe}}{u_{a,vf}}. \tag{37}$$

Furthermore, two characteristic dimensionless numbers are introduced, namely the pressure drop coefficient

$$\xi_{cyc} = \frac{2\Delta p}{\rho_f u_o^{*2}} = \xi_{vf} \sqrt[3]{\mathrm{D}_{cyc}^8 \mathrm{H}_{cyc}^4} \tag{38}$$

and the Barth number

$$\mathrm{B}_{cyc} = \frac{u_{term,d_{p,crit}} u_o'}{d_o' g} = \frac{1}{4\left(\mathrm{D}_{cyc} \mathrm{H}_{cyc} U_{cyc}\right)^2}. \tag{39}$$

Those definitions use the velocity

$$u'_o = \frac{4\beta \dot{m}_{ore}}{\pi d'^2_o} \tag{40}$$

and the frictionless outer diameter

$$d'_o = 2\sqrt[3]{\frac{d_o^2 h_{cyc}}{4}}. \tag{41}$$

From the combination of the two dimensionless numbers and the terminal velocity of a particle with the critical diameter $u_{term,d_{p,crit}}$ (Equation (18)) the characteristic cyclone number

$$Z_{cyc} = B_{cyc}\sqrt[4]{\xi^3_{cyc}} = \frac{u_{term,d_{p,crit}}}{g}\sqrt{\frac{\pi}{\beta \dot{m}_s}}\sqrt[4]{\frac{(\Delta p)^3}{2\rho_f^3}} \tag{42}$$

can be determined. This first step in obtaining the cyclone dimensions from performance and operating conditions is followed by the determination of D_{cyc} and H_{cyc} from Figure 7 under an initial assumption of a friction coefficient λ, usually 0.2. The correlations plotted originate from a numerical study on optimal cyclones.

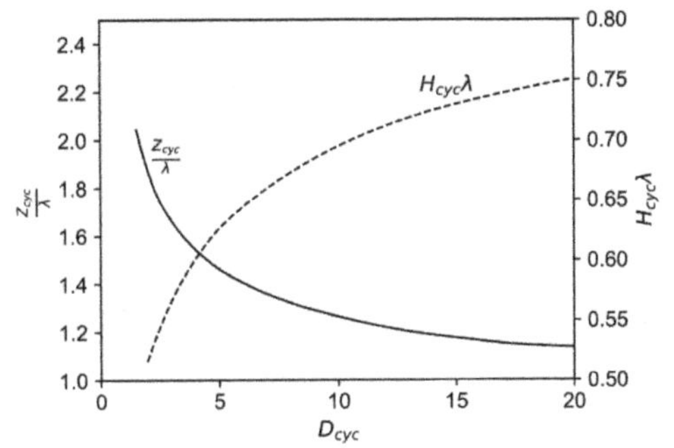

Figure 7. Correlation between the characteristic cyclone number as well as the diameter and height-diameter ratios for optimal cyclones, modified from [49] with permission from Elsevier.

With the two geometric ratios known, the area ratio

$$A_{cyc} = \frac{A_{in}}{A_{vf}} = \frac{4a_{in}h_{in}}{d^2_{vf}\pi} = 1.33(1 - \lambda H_{cyc})D_{cyc} - 0.16 \tag{43}$$

allows for the calculation of the inlet ratio

$$\frac{a_{in}}{h_{in}} = \frac{4}{\pi A_{cyc}}\left(\frac{D_{cyc}a_{in}}{d_o}\right)^2, \tag{44}$$

in which a_{in}/d_o is usually set to 0.15. Therin, A_{in} and A_{vf} indicate the cross-sectional area of the inlet as well as the vortex finder while a_{in} and h_{in} denote the width and height of the rectangular inlet duct, respectively. The optimal value of the Barth number to achieve a minimization of the pressure loss coefficient is always obtained for a velocity ratio $U_{cyc} = 1$. With this knowledge, B_{cyc} and ξ_{cyc} can now be obtained in this order from the Equations (39) and (42), respectively. Then, the definition of the vortex finder pressure drop

$$\xi_{vf} = \frac{2\Delta p}{\rho_f u^2_{a,vf}} \tag{45}$$

together with Equation (38) allows the calculation of the axial velocity

$$u_{a,vf} = \sqrt{\frac{2\Delta p}{\rho_f \xi_{cyc}}} \sqrt[3]{D_{cyc}^4 H_{cyc}^2}. \tag{46}$$

Its relationship to the vortex finder diameter is determined by the principle of continuity as

$$d_{vf} = 2\sqrt{\frac{\beta \dot{m}_s}{\pi u_{a,vf}}}. \tag{47}$$

This enables finally the calculation of the other cyclone dimensions with the definitions of the geometric ratios.

However, the wall friction coefficient was just assumed initially and, therefore, needs to be calculated using the obtained dimensions according to

$$\lambda = 0.005 + \frac{287.4}{Re_o'}. \tag{48}$$

The dependence on the Reynolds number Re_o'

$$Re_o' = \frac{u_{t,o}' d_o}{\nu_f} \tag{49}$$

with

$$u_{t,o}' = \frac{\beta \dot{m}_s}{h_{in} a_{in} \left(0.889 - \frac{0.408 a_{in}}{d_o}\right)} \tag{50}$$

accounts for changes in the gas properties at different temperatures. Therein, $u_{t,o}'$ denotes the frictionless tangential velocity at the wall. With the updated wall friction coefficient, the calculations presented above can be repeated. This procedure yields iteratively the main dimensions of an optimized cyclone.

Unfortunately, if the targeted critical size diameter is set to 2 μm and a pressure drop of 5 mbar is assumed according to measurements [49], the procedure does not converge to a single value for λ. It oscillates between 0.016 and 0.019, therefore, to obtain exemplary dimensions of the cyclone, their average is taken. This leads to the following values for the given scenario: $h_{cyc} = 1900$ mm, $d_o = 840$ mm, $d_{vf} = 95$ mm, $a_{in} = 130$ mm and $h_{in} = 200$ mm. As a simplification for the determination of the riser pipes lengths, all cyclones in the cascade are assumed to have the same dimensions. When assuming an arrangement like the one shown in Figure 5, the length of the riser pipe between two cyclones can roughly be approximated as the cyclone height. The cross-section of the riser pipe is limited on the one hand by the gas velocity necessary for transporting the particles upwards, and on the other hand must not exceed the vortex finder diameter to allow for loo large agglomerates to fall straight down through it and prevent accumulation below the particle charging section. In the given scenario, the vortex finder diameter is dominating the determination of the cross-section of the riser pipe. A time

$$t_{riser} = \frac{h_{cyc}}{u_{a,vf}} = \frac{h_{cyc} d_{vf}^2 \pi}{4 \beta \dot{m}_s} \tag{51}$$

of 0.05 s for conveying the particles through the riser is obtained. However, even with the adaption of the average residence time of a cyclone to 2.05 s, there are still 3 cyclones necessary to achieve the desired reduction time, resulting in a total height of 5.6 m for the cyclone cascade. In this calculation, the time for accelerating the particles to the gas velocity is neglected, meaning that the real conveying time is higher.

Since the presented method relies on many assumptions, simplifications and empirical correlations, it provides only rough estimates. In addition, the only factor that is included in the residence time correlation is the solids loading, thereby neglecting other influences like cyclone geometry, grain size distribution or gas flow rate. Furthermore, it was developed based on material properties different from those in this scenario. More exact values can either be obtained by using improved mathematical models or incorporating CFD simulations to gain knowledge about the complex three-dimensional flow patterns [53]. However, even with sophisticated CFD simulations, significant deviations can occur in some cases between the modeled and real cyclone behavior [54].

Additionally, it should be noted that there exists a multitude of different cyclone types with their own characteristics. These can range from different inlet geometries over varying inserts in the vortex finder for flow pattern optimization to the complete abandonment of a vortex finder in refractory lined cyclones and many more. The positioning of a scraper in the vortex finder is also possible to periodically remove material that stuck to the pipe's wall. It is mainly used when operating with sticky material at higher temperatures [55].

To summarize, a cascade of 3 cyclones in series is necessary for the required task, totaling around 5.6 m in height. For increased residence times and a decreased pressure loss, the cyclones after the first stage can be equipped with a wider cone for the required task, although this reduces their separation efficiency. A parallel arrangement is common for the first stage to increase the overall separation efficiency of the cascade. The solids are heated quickly and then react over some timespan with the reducing gas. The problem of sticking can be mitigated by using mechanical scraping devices. However, for small cyclones space constraints exist which limit or inhibit their usage. Additionally, larger particles and agglomerates charged from a lower staged cyclone can just fall through the vortex finder pipe of the next one into its dust collection chamber, repeatedly until the HPSR reactor is reached. The operation of the cyclone cascade at higher temperatures should therefore be possible, even at increased sticking effects. However, it has to be noted that those agglomerates then do not get pre-heated and pre-reduced to the desired degree due to the decreased residence time and limited heat and mass transfer. The application of this reactor type possesses the advantage of a relatively simple construction with few moving parts, the absence of a gas-distribution system, minimal heat losses due to its compact arrangement and longer residence times than in a vertical transport pipe alone.

4.4. Moving Bed Reactors

Another typical reactor type for gas-solid reactions is the moving bed reactor [56]. Due to the relatively small grain sizes and wide distributions thereof considered in this study, a setup with the gas flow through the bed is not possible because the bulk must possess a certain gas permeability, similar to a shaft furnace [57]. This means that the gas has to flow over the bed. Possible reactors therefore include the rotating hearth furnace (RHF) and the rotary kiln (RK) [58]. Since RHFs are typically heated by multiple burners positioned over the bed, the heat transfer to the bottom of the bulk is not optimal [59]. In directly heated RKs, the energy source is positioned at an end for counter-current operation at the material's outlet. A principal drawing of a RK is shown in Figure 8. The heating occurs not only by the burner directly, but also indirectly from the furnace's walls because of their rotation. The uncovered sections get heated by the burner and provide this energy to the bulk when positioned under it. Additionally, through the kiln's rotation the solids get mixed, thereby minimizing thermal gradients [56].

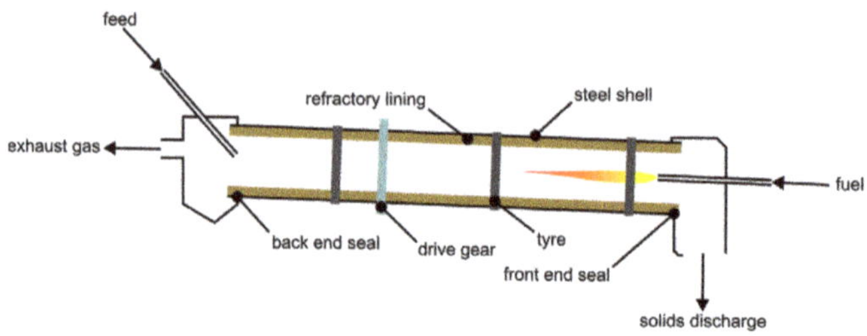

Figure 8. Basic schematic of a RK, re-illustrated after [60].

There are different movement patterns which the bulk can form, depending on characteristics like material properties, the friction between the charge and the wall, and the rotational speed of the kiln. These patterns include slipping, slumping, rolling, cascading, cataracting and centrifuging. The last two mentioned are avoided for the operation of RKs. Slumping means the periodical occurrence of avalanches when the gravitational forces exceed the frictional ones in an upwards-moving bulk near the wall. With increasing rotational speed, the slumping starts to transform into a continuous rolling motion. For low speeds, a constant incline of the bed's surface is obtained, while for even higher velocities a kidney-shaped cross-section forms. This is characterized as cascading movement. For the latter two, the bed can be split into two regions. In the active layer the particles move downwards on top of the passive layer. The material in the active zone mixes with the gas, increasing heat and mass transfer in this region. It gets thicker with rising rotational speed and deeper beds, thereby favoring reaction kinetics under these conditions [61].

The obtained transverse movement pattern can be visualized with the bed behavior diagram. It is a representation of the flow pattern depending on the bed depth or the filling degree Φ as well as the revolution frequency f_{RK} for a given material and kiln geometry. An exemplary one for nickel oxide can be seen in Figure 9. It can be clearly seen that for a constant filling degree there exists a range of Froude numbers

$$Fr_{RK} = \frac{\omega_{RK}^2 d_{RK}}{2g} \tag{52}$$

where rolling and cascading occur, which should be targeted for good heat and mass transfer. Therein, the kiln's angular velocity is defined as [62]

$$\omega_{RK} = 2\pi f_{RK}. \tag{53}$$

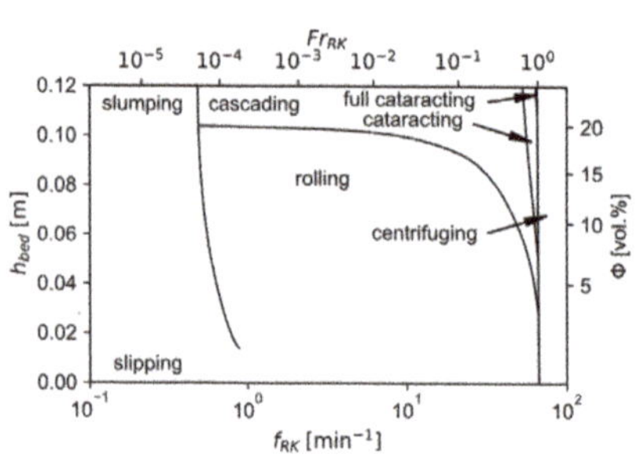

Figure 9. Exemplary bed behavior diagram for nickel oxide, modified with permission of Springer Nature BV from [62]; permission conveyed through Copyright Clearance Center, Inc.

During the transverse motion of the particles through the kiln segregation occurs between the finer and coarser fractions. This effect exists in three different kinds, namely in a radial, banding longitudinal and end longitudinal segregation. In the first one, the coarse fraction is at the bottom and the top of the bulk, enclosing the fines in its center. In the second one, bands of coarser and finer material move in an alternating pattern axially through the kiln. The last type is a variation of the radial segregation with fines accumulating at the kiln's ends. While the radial segregation always occurs to some extent, the other kinds require special criteria to be met [63].

In addition to the solids flow pattern, the effect of the gas velocity on the dust discharge must also be considered during the sizing of a RK. In industrial operations [64], off-gas dust loads of 0.01 kg/m^3 are often occurring, representing usually between 1–10% and in extreme cases up to 36% of the feed. Its most influential parameter is the gas velocity. The higher it gets, the more dust is discharged. Other values like the rotational speed and the concentration of fines in the bed also provoke dust discharge but are of minor importance, comparatively [65].

One problem which also occurs in RKs, as in all other presented reactors, is the effect of sticking. When heated, low-melting phases form rings normal to the rotation axis. Those buildups act basically as dams in the middle of the kiln and introduce back-mixing of the charge before them if they grow big enough so that the effective kiln incline in this region is too little. This leads to uncontrolled residence times and introduces variance in the process [66].

However, the overall process is relatively robust against the problem of grains growing many times their initial size. Those large particles are also simply transported through the kiln by its rotation and discharged at the end.

On the other hand, another major problem with RKs for the given task is the sealing of the rotating drum at its ends against the stationary housings. Industrial kilns usually operate at a slight underpressure, so that ambient air will leak in but no explosive or toxic furnace gases can get out [48]. This is not an option for hydrogen-containing kilns since the introduced oxygen could react explosively with the unused hydrogen [67]. For this reason, the sealing design gets very complicated since it needs to account for misalignment as well as movement of the kiln due to thermal elongation [68].

The design of RKs is mostly based on empirical correlations and engineering experience. The filling degree of the kiln, defined as the cross-sectional area taken up by the material to that of the whole drum, is usually between 5–17%. A higher filling degree increases the solids throughput at otherwise constant conditions. A rise in the filling degree can be achieved by reducing the kiln's slope. The incline of RKs is mostly in the range of 2–6%, whereas longer kilns are less steep. The latter require higher rotational speeds to convey the charge, which leads to better mixing of the solids. However, higher filling degrees reduce the heat transfer from the gas to the bed. It is therefore suggested that the filling degree does not exceed 13% in practice. Operational experience indicates that the corresponding slope for this filling degree should be 2.5%. Recommendations also provide estimates for the rotational speed. The circumferential velocity should be around 0.35 m/s, meaning that thinner kilns are rotating at higher revolutions than thicker ones [48].

The exact calculation of the solids' residence times in RKs is difficult since it depends on the flowing properties of the material, which itself can change during the process. The bed also gets thinner in the vicinity of the outlet, leading to the acceleration of the solids. Therefore, the residence times are best determined experimentally for the given operating conditions. However, there exist some models which use simplifications and empirical correlations to provide estimates of the real solids' residence times. The most basic ones use the principle of continuity for the bulk [56].

A simple formula for the estimation of the solids residence times in a RK for sponge iron production with length l_{RK} and incline α_{RK} is

$$t_{s,RK} = \frac{m_{bed}}{\dot{m}_s} = \frac{0.1026 l_{RK}^3}{\dot{V}_s} \left(\frac{\Theta_s}{\alpha_{RK}} \right)^{1.054} \left(\frac{\dot{V}_s}{l^3 f_{RK}} \right)^{0.981} \left(\frac{l_{RK}}{d_{RK}} \right)^{1.1}. \quad (54)$$

Therein, Θ_s indicates the dynamic angle of repose of the material and

$$\dot{V}_s = \frac{\dot{m}_s}{\rho_{bulk}}. \quad (55)$$

the volumetric charging rate. However, the use of exit dams severely influences the residence time by altering the filling degree for a constant mass flow rate, which is neglected in the provided correlation [69].

Such empirical correlations strongly depend on the material properties. Measurements with e.g., charcoal, olive nuts and sand yielded 0.38 instead of 0.1026 as proportionality coefficient. This nearly four-fold bigger value indicates that for these materials the average residence time is underestimated when using Equation (54) [70].

The solids residence times are distributed similarly to a plug-flow reactor [71], likewise to the cyclone cascade above. An increased feed rate narrows the residence time distribution, regarding the kiln's inclination and rotation rate are mixed behaviors observable for cohesive particles [72].

Practical insights reveal that exit gas velocities are as low as 2 m/s for industrial kilns to prevent excessive dust discharge [64]. If this value is targeted,

$$d_{RK} = 2\sqrt{\frac{\beta \dot{m}_s}{\pi u_f (1 - \Phi)}} \quad (56)$$

can be derived for the kiln's diameter when considering the principle of continuity as well as the definition of the filling degree [48]

$$\Phi = \frac{A_{bulk}}{A_{RK}} \quad (57)$$

as the ratio between the cross-sectional areas of the bulk and the empty kiln. When exit dams are present, A_{bulk} must consider the cross-section of the dam. However, since the mechanisms of particle pick-up by the gas from the charge are complicated, significant dust discharge may also occur at these low gas velocities due to the dominance of other effects like material properties, inserts and geometric relations [65]. The determination of the kiln's diameter for a desired filling degree would then allow for further design of the kiln. With the knowledge of the bed behavior diagram for the given material and geometric conditions, the rotational speed could then be estimated from the kiln's Froude number so that the targeted movement pattern is achieved. When determining the kiln's incline based on its filling degree empirically as mentioned above, Equation (54) allows the calculation of the length of the RK. Nevertheless, the obtained dimensions would only be initial estimates for further experimental investigations to find the optimal design parameters in the multitude of influential factors.

For the application of this sizing procedure to the given scenario, a filling degree of 10 vol.-% is assumed for good heat transfer conditions. With the gas velocity limited to 2 m/s a kiln diameter of 0.43 m is calculated. In the absence of better data, from the bed behavior diagram for nickel oxide a Froude number of 0.1 is selected to achieve a relatively fast-rolling pattern near the cascading border for a good mixing behavior. This yields an angular velocity of 2.1 s^{-1}, which corresponds to a revolution frequency of 20 min^{-1} and a maximum tangential velocity

$$u_{t,max} = \frac{\omega_{RK} d_{RK}}{2} \tag{58}$$

of 0.46 m/s. It shows good agreement with the recommended circumferential tangential velocity of 0.35 m/s stated above. Since the conditions in a RK are vastly different from the ones used for calculating the ore residence time in Section 3 a new estimation has to be made. Experiments show that the time required to reach the desired conversion degree of 33% for the reduction of hematite fines in a rotary drum furnace at 900 °C with a gas mixture of 50 vol.-% H_2 and 50 vol.-% N_2 is approximately 150 s [73]. The desired filling degree corresponds empirically to an incline of 2.3° [48]. Θ_s is estimated to be 30° according to measurements with ilmenite powder due to the unavailability of better data [74]. With this information, the length of the kiln is calculated to 8 m. This value can be decreased by incorporating an exit dam or other inserts to either increase the bed's mass or improve the mixing behavior [48].

5. Reactor Comparison and Selection

Based on the provided information in the sections above, in the following the reactor most suitable for the stated requirements shall be selected. To aid this process, Table 1 summarizes and compares the key facts of the respective reactor types.

Table 1. Comparison between the three different reactor types (summarized from Section 4).

Criteria	Fluidized Bed Reactor Cascade	Cyclone Cascade	Rotary Kiln
Solids residence time	Intermediate	Short	Long
Solids residence time distribution	Wide, but narrower, plug-flow like with more stages	Narrow, plug-flow like	Narrow, plug-flow like
Mixing of reactants	Excellent, in the whole bed volume	Good, mainly in the cyclones and less in the transport pipes	Bad, gas-solid mixing only in active layer, although mixing within the solids is good
Sensitivity to grain size enlargement	Strong, formation of a dead bed	Weak within some range, otherwise clogging of the vortex finder	Very weak, since the drum diameter is relatively large
Sensitivity to operation with sticky material	Strong, formation of a dead bed, clogging of the gas distributor and the hot gas cyclone	Weak within some margin, mechanical scrapers possible to prevent clogging	Weak, formation of rings
Sensitivity to variance in gas velocity	Strong, too low: defluidization; too high: more discharge, lower solids residence time	Weak, too low: less efficient separation; too high: increased pressure drop	Weak, too high: more dust discharge
Sensitivity to temperature fluctuations (sticking, gas velocity)	Strong	Weak	Weak
Engineering characteristics and key components	Hardly any moving parts, gas distributor, charging system, coarse material discharge, hot gas cyclone	Hardly any moving parts, charging system, dust chamber discharge system, possibly mechanical scrapers	Rotating drum, kiln end sealings, charging system, drivetrain system, motor
Overall complexity	High	Low	Medium

The proposed reactor for the implementation of a pre-reduction step in the HPSR process at an existing demonstration plant is the cyclone cascade because of its overall lowest complexity and sufficient flexibility regarding the process parameters. The rotary kiln performs well too but is more complex due to the presence of many moving parts. Since usually the simpler system is more cost-effective, the cyclone cascade is prioritized.

Sticking is also only a minor problem in this process. Although it must be designed for an operating point, slight deviations in the gas velocity just decrease its separation efficiency or alter the pressure drop, which is not a huge problem for the process at demonstration scale. Upscaling should also be possible since industrial plants generally do not vary their parameters as much compared to the development phase. There, the design for a specific operating point should be sufficient. Furthermore, the cyclone cascade is a wide-spread technology used in cement production.

Figure 10 shows schematically the proposed reactor concept. Future calculations with more detailed parameters might show the necessity of longer residence times than currently estimated. The most significant influencing factor is the implementation of more cyclones. To a certain extent, they can be stacked onto the planned ones. If the required time is not reachable within a reasonable number of stages, one of the other proposed technologies must be implemented due to the lack of alternatives. To keep the charge solid at all times during pre-reduction, some gas cooling is necessary after the HPSR-reactor's off-gas outlet. In the provided schematic, its location is indicated with the blue lines. Whether the length of the water-cooled pipe for sufficient cooling is within a reasonable limit needs to be calculated. For simplicity, this is omitted in this concept paper since other cooling options exist. To give an example, the mixing of the HPSR off-gas with another cold gas is also a possible solution. This can either be an inert gas like nitrogen or a reducing gas mixture with adjustable composition, in principal even cooled and recycled off-gas of the cyclone cascade. The advantage of the gas-injection method is that the temperature of the off-gas gets lowered but its enthalpy is conserved, while during water-cooling it is removed from the system. Furthermore, the cooling power can be easily adjusted by controlling the input flow rate of the cold gas. Due to the small dimensions of the vortex finder and riser pipes compared to industrial scale cyclone cascades, having a pipe-in-pipe charging system is susceptible to clogging. To mitigate this problem, the collection chamber's discharge pipe needs to end directly at the elbow of the next cyclones inlet duct instead of reaching downwards into it. In this case, the time the material remains in the riser is shortened, but for the given scenario there are still 3 cyclones sufficient, even when neglecting the residence time in the riser completely.

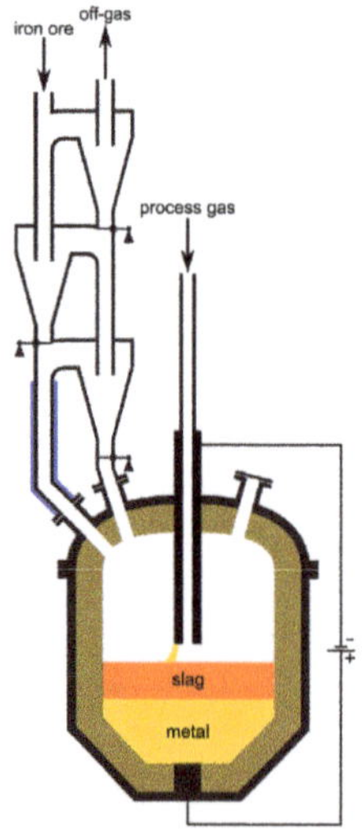

Figure 10. Schematic of the proposed concept for the HPSR process including a pre-heating and pre-reduction step.

6. Summary and Conclusions

The following key takeaways can be summarized:

- In principle, any hydrogen-based direct-reduction process can be used for a pre-heating and pre-reduction stage in the HPSR process, depending mainly on the grain size of the iron ores. In addition, the pre-reduction does not need to stop at wustite. The ores could be reduced nearly completely in the first stage and only melted for slag-metal separation in the HPSR reactor. In this case, the latter is used as an efficient electric hydrogen heating and smelting unit. The optimal reduction degree for the interlinked operation of the two processes needs to be determined in the future.
- The pre-reduction reactor operates with the off-gas of the HPSR process. Therefore, it must possess some robustness against dust-loaded reducing gas. This can lead to agglomeration and clogging when the solids get sticky because of the presence of some low-melting phases. Furthermore, the melting point of wustite must not be exceeded to keep the charge solid. Therefore, cooling of the off-gas from the HPSR reactor's exit temperature to the maximum allowable operating temperature before contacting the pre-heated and pre-reduced solids is necessary.
- The most critical component of an FBR is its gas distribution system. Additionally, the gas velocity needs to be carefully controlled to maintain a fluidized state. Agglomeration can further lead to a dead bed, since too big particles cannot be discharged and interfere with the gas distribution in the reactor. Furthermore, with the given mass flow ratios in combination with the desired particle sizes, an FBR is not operatable at the minimum solids residence time.
- The hydrogen-tight sealings of the rotating drum against the stationary housings are the most critical components of a rotary kiln, especially at the hot discharge end. Otherwise, it would be the least prone to sticking effects of the presented reactors.
- A cyclone cascade with at least 3 stages is proposed due to its overall lowest complexity, robustness of the process and sufficient flexibility. Agglomeration up to a certain degree will not hinder the operation, since the bigger particles can fall straight down from their charging point through the vortex finder pipe of the cyclone beneath. Additionally, there are no moving sealings with large diameters and only few moving parts required, lowering its complexity.

Finally, it needs to be emphasized that the stated sizing procedures are based on empirical correlations. Therefore, they only provide rough estimations for the design parameters. Practical experience is scarce, since to the knowledge of the authors, cyclone cascades are currently not used for the reduction of iron ores. First of all, the exact residence times necessary under the prevailing conditions need to be determined experimentally. On this basis, a show-case cyclone reactor should then be built to optimize it for its use case, in combination with more exact sizing calculations, CFD simulations or both methods. With the knowledge gained about the material behavior, the trial reactor can then be designed in detail, constructed, and incorporated into the existing HPSR demonstration plant.

Supplementary Materials: The following supporting information can be downloaded at https://www.mdpi.com/article/10.3390/pr13020420/s1: Figure S1: Reh diagram.

Author Contributions: Conceptualization, B.A. and M.A.Z.; methodology, B.A.; validation, J.S.; investigation, B.A.; resources, M.A.Z.; writing—original draft preparation, B.A.; writing—review and editing, M.A.Z., F.H. and J.S.; visualization, B.A. and F.H.; supervision, M.A.Z. and J.S.; project administration, M.A.Z.; funding acquisition, M.A.Z. All authors have read and agreed to the published version of the manuscript.

Funding: This research is part of the "Sustainable Steelmaking follow-up" project (no. 894616), funded by the Klima- und Energiefonds (www.klimafonds.gv.at, accessed on 28 January 2025) within the scope of the program "Vorzeigeregion Energie" through the Austrian Research Promotion Agency (www.ffg.at, accessed on 28 January 2025), K1-MET GmbH (www.k1-met.com, accessed on 28 January 2025), voestalpine Stahl GmbH (www.voestalpine.com/stahl/Gesellschaften/voestalpine-Stahl-GmbH, accessed on 28 January 2025), voestalpine Stahl Donawitz GmbH (www.voestalpine.com/stahldonawitz, accessed on 28 January 2025), Pirhofer-Automation e.U. (www.pirhofer-automation.at, accessed on 28 January 2025), the Chair of Ferrous Metallurgy at Montanuniversität Leoben (www.metallurgy.at/chair-of-ferrous-metallurgy, accessed on 28 January 2025), and WIVA P&G—Hydrogen Flagship Region Austria Power & Gas (www.wiva.at, accessed on 28 January 2025).

Data Availability Statement: No new data was created or analyzed in this study. Data sharing is not applicable to this article.

Conflicts of Interest: Author Bernhard Adami and Michael Andreas Zarl were employed by the company K1-MET GmbH. The remaining authors declare no conflicts of interest. The funders had no role in the design of the study; in the collection, analyses, or interpretation of data; in the writing of the manuscript; or in the decision to publish the results.

List of Symbols

Base

α ... inclination
A ... area ratio
β ... specific gas flow rate
γ ... drag coefficient of a particle collective
Δ ... difference
ϵ ... relative void volume
η ... utilization
H ... height ratio
Θ ... dynamic angle of repose
λ ... friction coefficient
ν ... kinematic viscosity
ξ ... pressure drop coefficient
ρ ... density
Φ ... filling degree
ψ ... Ergun parameter
ω ... angular velocity
Ω ... Omega number
a ... width
A ... cross-sectional area
Ar ... Archimedes number
B ... Barth number
c ... particle drag coefficient
C ... mass loading
d ... diameter
D ... diameter ratio
f ... revolution frequency
F ... force
Fr ... Froude number
Fr^* ... extended Froude number
g ... gravitational acceleration, $g = 9.81 \text{ m/s}^2$
h ... height

H ... enthalpy
K ... kinetic parameter
l ... length
m ... mass
$\dot{m}$ ... mass flow rate
M ... molar mass
n ... load factor
$\dot{n}$ ... molar flow rate
p ... absolute pressure
R ... ideal gas constant, $R = 8.314$ J/(mol K)
RD ... reduction degree
Re ... Reynolds number
t ... time
T ... absolute temperature
u ... velocity
$\overline{u}$ ... superficial velocity
U ... velocity ratio
V ... volume
$\dot{V}$ ... volume flow rate
w ... mass fraction
x ... molar fraction
Z ... characteristic cyclone number

Sub- and superscripts

$'$ ... frictionless
0 ... normal conditions ($T = 0$ °C and $p = 1$ atm$_a$)
a ... axial
b ... buoyancy
bed ... bed
$bulk$ ... bulk
c ... centrifugal
co ... co-gravitational flow
$counter$ ... counter-gravitational flow
$crit$ ... critical
cyc ... cyclone
d ... drag
e ... end
el ... elutriation
equ ... equilibrium
f ... fluid
fo ... gas outlet
FBR ... fluidized bed reactor
g ... gravitation
$HPSR$ ... hydrogen plasma smelting reduction
i ... initial
in ... cyclone inlet
m ... molar
max ... maximum
min ... minimum
o ... cylindrical part of the cyclone
op ... operating conditions
p ... particle

pg ... plasma gas
R ... reaction
r ... radial
RD ... reduction degree
$riser$ ... riser
RK ... rotary kiln
s ... solids
$slip$ ... slip
so ... solids outlet
t ... tangential
$term$ ... terminal
tot ... total
TR ... transport reactor
vf ... vortex finder
vfe ... virtual extension of the vortex finder pipe into the cyclone
χ ... indexing variable

Appendix A

This section describes the calculation of the Reh diagram stated in Figure 3. Additionally, since it might be useful to some reader, a high-quality version without annotations is provided as Supplementary Materials (Figure S1).

The Reh diagram is derived from basic fluid mechanical correlations of a bed with a specific height of solid spheres with constant size under an upwards directed fluid flow in a vertical pipe with cylindrical cross-section. Which contacting pattern is obtained depends on the superficial fluid velocity (i.e., the velocity in the empty pipe), the kinematic viscosity of the fluid, the density of the fluid, the density of the solids, the relative void volume

$$\epsilon = \frac{V_f}{V} = \frac{(1 - V_s)}{V}, \tag{A1}$$

and the grain size distribution of the particles. Hereby, V_f is the volume of the voids between the particles taken up by the fluid, V_s is the volume of the solid fraction, and V denotes the overall volume of the mixed phase. When starting with a packed bed and increasing the superficial fluid velocity, three states can be distinguished. At first, the bed remains as it is, then it begins to expand, and a fluidized bed is obtained. When further increasing the fluid velocity, it expands even more until the particles get entrained by the flow. The forces acting on the bed are the gravitational force

$$F_g = \rho_s V_s g = \rho_s (1 - \epsilon) A_{FBR} h_{bed} g, \tag{A2}$$

the buoyancy force

$$F_b = \rho_f V_s g = \rho_f (1 - \epsilon) A_{FBR} h_{bed} g, \tag{A3}$$

and the force of the flow by the pressure drop

$$F_{flow} = \Delta p A_{FBR}, \tag{A4}$$

with g denoting the gravitational acceleration. The force balance

$$F_{flow} = F_g - F_b \tag{A5}$$

becomes

$$\frac{\Delta p}{h_{bed}} = \left(\rho_s - \rho_f\right)(1 - \epsilon) g \tag{A6}$$

when all terms are inserted and rearranged. The load factor

$$n = \frac{F_{flow}}{F_g - F_b} = \frac{\Delta p}{\left(\rho_s - \rho_f\right)(1-\epsilon)h_{bed}g} \tag{A7}$$

is introduced as the ratio of the acting forces to describe the three different contacting patterns [36].

In the packed bed state, the relative void volume is at its minimum ϵ_{min} of around 0.4. The pressure drop over the height of a packed bed can be described with the Ergun correlation

$$\frac{\Delta p}{h_{bed}} = \psi \frac{(1-\epsilon_{min})\overline{u}^2\rho_f}{\epsilon_{min}^3 d_p}. \tag{A8}$$

Therein, the parameter ψ is defined as

$$\psi = 150\frac{1-\epsilon_{min}}{Re_p} + 1.75 \tag{A9}$$

with the particle Reynolds number

$$Re_p = \frac{\overline{u}d_p}{\nu_f}. \tag{A10}$$

Combined with the definition of the extended Froude number

$$Fr^* = \frac{\overline{u}^2\rho_f}{gd_p\left(\rho_s - \rho_f\right)}, \tag{A11}$$

$$Fr^* = \frac{n\epsilon_{min}^3}{\psi} \tag{A12}$$

is obtained for the region $n < 1$ [36,38].

The forces acting on a single particle in a fluid stream are the same as on the whole bulk, except that the force of the flow on the particle can also be expressed as the drag force

$$F_d = \frac{c_d A_p \rho_f \overline{u}^2}{2} = \frac{c_d d_p^2 \pi \rho_f \overline{u}^2}{8}, \tag{A13}$$

with c_d being the drag coefficient and A_p the cross-sectional area of the particle. To simplify the expression, instead of the relative velocity between the particle and the fluid, only the fluid's velocity is considered. Additionally, the virtual mass force of the entrained particle is also neglected. With the volume of the particle

$$V_p = \frac{d_p^3\pi}{6} \tag{A14}$$

inserted in the relations for the gravitational and buoyancy forces,

$$Fr^* = \frac{4n}{3c_d} \tag{A15}$$

for $n \geq 1$ is obtained from the definition of the load factor (Equation (A7)) [36].

The given derivations result in equations for the conditions necessary for the start of the fluidization as well as the beginning of the discharge of the solids. In the fluidized state

($0.4 < \epsilon < 1$), lines for a constant relative void volume can be derived when relating the drag coefficient of a single particle with the one of a particle collective

$$\gamma_d = \frac{c_d}{\epsilon^{4.65}}. \tag{A16}$$

This equation is valid for liquid-solid and low-expanded gas-solid fluidized beds ($\epsilon < 0.6$) with Reynolds numbers less than 10. As a result, for $n = 1$ and $0.4 < \epsilon < 1$

$$Fr^* = \frac{4n}{3\gamma_d} = \frac{4n\epsilon^{4.65}}{3c_d} \tag{A17}$$

is obtained. For the construction of the Reh diagram, this equation is extrapolated to higher Reynolds numbers and bigger bed expansions for simplicity [36].

When defining the drag coefficient as a function of the Reynolds number as in Formula (9), the Equations (A12), (A15) and (A17) can be plotted in terms of $3/4\ Fr^*$ over Re_p and the Reh diagram is obtained. Since the expression $3n\epsilon_0^3/(4\psi)$ provides approximately the same values as $n\epsilon_0^{4.65}/c_d$ for a constant n in the range of $10^{-1} < Re_p < 10^4$, instead of drawing Equation (A12) the validity area of Equation (A17) is expanded into the packed bed with $\epsilon = \epsilon_{min} = 0.4$. Furthermore, the calculation of the Reh diagram can be summarized by Equation (10) for the respective values of n and ϵ [36].

In contrast to liquid-solid fluidized beds, channels form at low particle diameters in gas-solid ones. The gas in those channels is not participating in fluidizing the solids, therefore, more gas is needed to obtain the same bed expansion. This results in a shift of the lines for a constant ϵ to higher Ω numbers for the same particle diameters, hence Ar numbers. It is constructed geometrically as a shift along the Ar lines according to the ratio between the fluidization and discharge line for liquid-solid fluidized beds and a new discharge line at

$$\frac{3}{4}Fr^* = 1 = c_d. \tag{A18}$$

Point S_1 is calculated from the intersection of the two discharge lines for the different types of fluidized beds. Re_p for Point S_2 is obtained from the comparison of Equations (A12) and (A15), which gives the relation

$$c_d = \frac{4}{3}\epsilon_0^{1.65}\psi = 1. \tag{A19}$$

The differences between the particulate and aggregative fluidization regimes occur to the left of the line connecting the points S_1 and S_2. In the given Reh diagram, the lines for particulate fluidization are drawn dashed and the ones for aggregative fluidization solid [36].

For aggregative fluidization at higher bed expansions, the solids are constantly expelled from the reactor. This is the operating region of a circulating fluidized bed reactor, where the solids are then separated from the gas and charged back into the reactor [75].

However, contrary to what was stated by Reh originally [36], the lines for a specific ϵ for the different fluidization regimes do not start deviating from each other at the S_1–S_2 line when carrying out the described scheme. Instead, they provide equal values regardless of their fluidization regime above the Archimedes number of point S_1. Therefore, to obtain a similar looking diagram, the curves between $Ar = 10^3$ and the line $S_1 - S_2$ are interpolated in Figure 3. This is done since the one described originally by Reh is still used widely [30], and the differences are relatively minor, considering the underlying assumptions and simplifications.

References

1. Somers, J. *Technologies to Decarbonise the EU Steel Industry*; JRC127468; European Union: Luxembourg, 2022. [CrossRef]
2. Cavaliere, P. *Clean Ironmaking and Steelmaking Processes: Efficient Technologies for Greenhouse Emissions Abatement*; Springer Nature: Cham, Switzerland, 2019. [CrossRef]
3. Bäck, E.; Badr, K.; Plaul, J.F.; Sormann, A. Überblick über die Entwicklung der Wasserstoff-Schmelzreduktion am Lehrstuhl für Metallurgie. *Berg Hüttenmännische Monatshefte* **2009**, *154*, 6–9. [CrossRef]
4. Adami, B. Untersuchung des Verhaltens von Phosphor, Schwefel und Kupfer Während des Wasserstoffplasma-Schmelzreduktionsprozesses. Master's Thesis, University of Leoben, Leoben, Austria, 2023. [CrossRef]
5. Zarl, M.A.; Ernst, D.; Cejka, J.; Schenk, J. A New Methodological Approach to the Characterization of Optimal Charging Rates at the Hydrogen Plasma Smelting Reduction Process Part 1: Method. *Materials* **2022**, *15*, 4767. [CrossRef] [PubMed]
6. Bäck, E. Schmelzreduktion von Eisenoxiden mit Argon-Wasserstoff-Plasma. Ph.D. Thesis, University of Leoben, Leoben, Austria, 1998.
7. Badr, K. Smelting of Iron Oxides Using Hydrogen Based Plasmas. Ph.D. Thesis, University of Leoben, Leoben, Austria, 2007.
8. Plaul, J.F. Schmelzreduktion von Hämatitischen Feinerzen im Wasserstoff-Argon-Plasma. Ph.D. Thesis, University of Leoben, Leoben, Austria, 2005.
9. Naseri Seftejani, M. Reduction of Hematite Using Hydrogen Plasma Smelting Reduction. Ph.D. Thesis, University of Leoben, Leoben, Austria, 2020.
10. Behera, P.R.; Bhoi, B.; Paramguru, R.K.; Mukherjee, P.S.; Mishra, B.K. Hydrogen Plasma Smelting Reduction of Fe_2O_3. *Metall. Mater. Trans. B* **2019**, *50*, 262–270. [CrossRef]
11. Hiebler, H.; Plaul, J.F. Hydrogen plasma smelting reduction—An option for steelmaking in the future. *Metalurgija* **2004**, *43*, 155–162.
12. Ernst, D. Strategies for Optimizing Process Parameters of Hydrogen Plasma Smelting Reduction Plants. Ph.D. Thesis, University of Leoben, Leoben, Austria, 2023.
13. Hydrogen Plasma Smelting Revolution: Green Steel for a Sustainable Future. Available online: https://www.k1-met.com/fileadmin/user_upload/Publications/Success_stories/2024_Success_Story_EN_Area_2.pdf (accessed on 6 September 2024).
14. Souza Filho, I.R.; Springer, H.; Ma, Y.; Mahajan, A.; Da Silva, C.C.; Kulse, M.; Raabe, D. Green steel at its crossroads: Hybrid hydrogen-based reduction of iron ores. *J. Clean. Prod.* **2022**, *340*, 130805. [CrossRef]
15. Spreitzer, D.; Schenk, J. Reduction of Iron Oxides with Hydrogen—A Review. *Steel Res. Int.* **2019**, *90*, 1900108. [CrossRef]
16. Heidari, A.; Niknahad, N.; Iljana, M.; Fabritius, T. A Review on the Kinetics of Iron Ore Reduction by Hydrogen. *Materials* **2021**, *14*, 7540. [CrossRef]
17. Zare Ghadi, A.; Valipour, M.S.; Vahedi, S.M.; Sohn, H.Y. A Review on the Modeling of Gaseous Reduction of Iron Oxide Pellets. *Steel Res. Int.* **2020**, *91*, 1900270. [CrossRef]
18. Kim, W.-H.; Lee, S.; Kim, S.-M.; Min, D.-J. The retardation kinetics of magnetite reduction using H_2 and H_2–H_2O mixtures. *Int. J. Hydrogen Energy* **2013**, *38*, 4194–4200. [CrossRef]
19. Zuo, H.; Wang, C.; Dong, J.; Jiao, K.; Xu, R. Reduction kinetics of iron oxide pellets with H_2 and CO mixtures. *Int. J. Miner. Metall. Mater.* **2015**, *22*, 688–696. [CrossRef]
20. Ernst, D.; Zarl, M.A.; Cejka, J.; Schenk, J. A New Methodological Approach on the Characterization of Optimal Charging Rates at the Hydrogen Plasma Smelting Reduction Process Part 2: Results. *Materials* **2022**, *15*, 4065. [CrossRef] [PubMed]
21. Corbari, R.; Fruehan, R.J. Reduction of Iron Oxide Fines to Wustite with CO/CO_2 Gas of Low Reducing Potential. *Metall. Mater. Trans. B* **2010**, *41*, 318–329. [CrossRef]
22. Legemza, J.; Findorák, R.; Fröhlichová, M.; Džupková, M. Advances in Sintering of Iron Ores and Concentrates. In *Iron Ores*; Shatokha, V., Ed.; IntechOpen: Rijeka, Croatia, 2021. [CrossRef]
23. Nomura, T.; Yamamoto, N.; Fujii, T.; Takiguchi, Y. Beneficiation Plants and Pelletizing Plants for Utilizing Low Grade Iron Ore. *Kobelco Technol. Rev.* **2015**, *33*, 8–15.
24. Kolbe, N.; Cesário, F.; Ahrenhold, F.; Suer, J.; Oles, M. Carbon Utilization Combined with Carbon Direct Avoidance for Climate Neutrality in Steel Manufacturing. *Chem. Ing. Tech.* **2022**, *94*, 1548–1552. [CrossRef]
25. Skorianz, M. Classification of Iron Ores Regarding Their Reduction Behavior in Fluidized Bed Technologies. Ph.D. Thesis, University of Leoben, Leoben, Austria, 2012.
26. Zarl, M.A.; Farkas, M.A.; Schenk, J. A Study on the Stability Fields of Arc Plasma in the HPSR Process. *Metals* **2020**, *10*, 1394. [CrossRef]
27. NIST Chemistry WebBook, NIST Standard Reference Database 69. Available online: https://webbook.nist.gov/chemistry/ (accessed on 6 September 2024).
28. FactSage Software and Databases. Available online: www.factsage.com (accessed on 7 November 2024).
29. Degner, M.; Fandrich, R.; Endemann, G.; Ghenda, J.T.; Letz, K.; Lüngen, H.B.; Steller, I.; Wieland, H.-J.; Winkhold, A.; Bartos, R.; et al. *Stahlfibel*; Verlag Stahleisen: Düsseldorf, Germany, 2011.

30. Stephan, P.; Kabelac, S.; Kind, M.; Mewes, D.; Schaber, K.; Wetzel, T. *VDI-Wärmeatlas*; Springer: Berlin/Heidelberg, Germany, 2019. [CrossRef]
31. Atkins, P.W.; de Paula, J. *Physical Chemistry*, 9th ed.; W. H. Freeman and Company: New York, NY, USA, 2010.
32. Qu, Y.; Yang, Y.; Zou, Z.; Zeilstra, C.; Meijer, K.; Boom, R. Reduction Kinetics of Fine Hematite Ore Particles with a High Temperature Drop Tube Furnace. *ISIJ Int.* **2015**, *55*, 952–960. [CrossRef]
33. Themelis, N.J.; Gauvin, W.H. Reduction of iron oxide in gas-conveyed systems. *Am. Inst. Chem. Eng. J.* **1962**, *8*, 437–444. [CrossRef]
34. Hertwig, K.; Martens, L.; Hamel, C. *Chemische Verfahrenstechnik: Berechnung, Auslegung und Betrieb Chemischer Reaktoren*, 3rd ed.; Walter de Gruyter: Berlin, Germany, 2018. [CrossRef]
35. Yannopoulos, J.C.; Themelis, N.J.; Gavvin, W.H. An evaluation of the pneumatic transport reactor. *Can. J. Chem. Eng.* **1966**, *44*, 231–235. [CrossRef]
36. Reh, L. Das Wirbeln von Körnigem Gut im Schlanken Diffusor als Grenzzustand Zwischen Wirbelschicht und Pneumatischer Förderung. Ph.D. Thesis, Technische Hochschule Karlsruhe, Karlsruhe, Germany, 1961.
37. Haider, A.; Levenspiel, O. Drag Coefficient and Terminal Velocity of Spherical and Nonspherical Particles. *Powder Technol.* **1989**, *58*, 63–70. [CrossRef]
38. Stieß, M. *Mechanische Verfahrenstechnik 2*; Springer: Berlin/Heidelberg, Germany, 1997. [CrossRef]
39. Reh, L. Verbrennung in der Wirbelschicht. *Chem. Ing. Tech.* **1968**, *40*, 509–515. [CrossRef]
40. Kunii, D.; Levenspiel, O. *Fluidization Engineering*, 2nd ed.; Butterworth-Heinemann: Boston, MA, USA, 1991. [CrossRef]
41. Reh, L. The circulating fluid bed reactor—Its main features and applications. *Chem. Eng. Process. Process Intensif.* **1986**, *20*, 117–127. [CrossRef]
42. Guo, L.; Bao, Q.; Gao, J.; Zhu, Q.; Guo, Z. A Review on Prevention of Sticking during Fluidized Bed Reduction of Fine Iron Ore. *ISIJ Int.* **2020**, *60*, 1–17. [CrossRef]
43. Sun, M.; Pang, K.; Jiang, Z.; Meng, X.; Gu, Z. Development and Problems of Fluidized Bed Ironmaking Process: An Overview. *J. Sustain. Metall.* **2023**, *9*, 1399–1416. [CrossRef]
44. Reh, L. Auswahlkriterien für Nichtkatalytische Gas/Feststoff-Hochtemperaturreaktoren. *Chem. Ing. Tech.* **1977**, *49*, 786–795. [CrossRef]
45. Chatterjee, A.K. *Cement Production Technology: Principles and Practice*; CRC Press: Boca Raton, FL, USA, 2018.
46. Alsop, P.A. *The Cement Plant Operations Handbook: The Concise Guide to Cement Manufacture*, 7th ed.; Tradeship Publications: Dorking, UK, 2019.
47. Pendulum Flap. Available online: https://atec-ltd.com/images/Pendulum_Flap_Produktblatt_05_2016_v2.pdf (accessed on 6 September 2024).
48. Duda, W.H. *Cement Data Book*, 3rd ed.; Bauverlag: Berlin, Germany, 1985.
49. Bohnet, M.; Gottschalk, O.; Morweiser, M. Modern design of aerocyclones. *Adv. Powder Technol.* **1997**, *8*, 137–161. [CrossRef]
50. Zhang, H.; Dewil, R.; Degrève, J.; Baeyens, J. The design of cyclonic pre-heaters in suspension cement kilns. *Int. J. Sustain. Eng.* **2014**, *7*, 307–312. [CrossRef]
51. Zhang, Y.; Wang, Z.; Jin, Y.; Li, Z.; Yi, W. CFD simulation and experiment of residence time distribution in short-contact cyclone reactors. *Adv. Powder Technol.* **2015**, *26*, 1134–1142. [CrossRef]
52. Chan, C.W.; Seville, J.P.K.; Fan, X.; Baeyens, J. Particle Motion in CFB Cyclones as Observed By Positron Emission Particle Tracking. *Ind. Eng. Chem. Res.* **2009**, *48*, 253–261. [CrossRef]
53. Wozniak, G.; Schade, K.-P.; Wozniak, K.; Shalaby, H.H. Über die Auslegung und den Entwicklungsstand von Zyklon-Abscheidern. *Forsch. Ingenieurwesen* **2007**, *71*, 171–180. [CrossRef]
54. Dimitrijević, D.; Schmid, M.; Harasek, M.; Bösenhofer, M. Comparison of experimental, empirical, and CFD pressure losses of lab-scale sampling cyclones. *Sep. Purif. Technol.* **2025**, *354*, 128992. [CrossRef]
55. Brauer, H. *Handbuch des Umweltschutzes und der Umweltschutztechnik: Band 3: Additiver Umweltschutz: Behandlung von Abluft und Abgasen*; Springer: Berlin/Heidelberg, Germany, 1996. [CrossRef]
56. Reschetilowski, W. *Handbuch Chemische Reaktoren*; Springer: Berlin/Heidelberg, Germany, 2020. [CrossRef]
57. Seetharaman, S.; McLean, A.; Guthrie, R.; Sridhar, S. (Eds.) *Treatise on Process Metallurgy: Industrial Processes*; Elsevier: Oxford, UK, 2014.
58. Runkana, V. Model-Based Optimization of Industrial Gas-Solid Reactors. *KONA Powder Part. J.* **2015**, *32*, 115–130. [CrossRef]
59. Kumari, V.; Roy, G.G.; Sen, P.K. Mathematical Model to Estimate the Rate Parameters and Thermal Efficiency for the Reduction of Iron Ore–Coal Composite Pellets in Multi-layer Bed at Rotary Hearth Furnace. *Trans. Indian Inst. Met.* **2015**, *68*, 109–116. [CrossRef]
60. Brough, D.; Jouhara, H. The aluminium industry: A review on state-of-the-art technologies, environmental impacts and possibilities for waste heat recovery. *Int. J. Thermofluids* **2020**, *1–2*, 100007. [CrossRef]

61. Henein, H.; Brimacombe, J.K.; Watkinson, A.P. Experimental study of transverse bed motion in rotary kilns. *Metall. Trans. B* **1983**, *14*, 191–205. [CrossRef]
62. Henein, H.; Brimacombe, J.K.; Watkinson, A.P. The modeling of transverse solids motion in rotary kilns. *Metall. Trans. B* **1983**, *14*, 207–220. [CrossRef]
63. Henein, H.; Brimacombe, J.K.; Watkinson, A.P. An experimental study of segregation in rotary kilns. *Metall. Trans. B* **1985**, *16*, 763–774. [CrossRef]
64. Tackie, E.N. Elutriation of Particles from Rotary Kilns. Ph.D. Thesis, University of British Columbia, Vancouver, BC, Canada, 1987. [CrossRef]
65. Tackie, E.N.; Watkinson, A.P.; Brimacombe, J.K. Experimental study of the elutriation of particles from rotary kilns. *Can. J. Chem. Eng.* **1989**, *67*, 806–817. [CrossRef]
66. Chatterjee, A.; Mukhopadhyay, P.K. Flow of materials in rotary kilns used for sponge iron manufacture: Part III. Effect of ring formation within the kiln. *Metall. Trans. B* **1983**, *14*, 393–399. [CrossRef]
67. Park, S.W.; Kim, J.H.; Seo, J.K. Explosion Characteristics of Hydrogen Gas in Varying Ship Ventilation Tunnel Geometries: An Experimental Study. *J. Mar. Sci. Eng.* **2022**, *10*, 532. [CrossRef]
68. EagleBurgmann. *DRO Drehrohrofen-Dichtsystem: Spezialprodukte*; EagleBurgmann: Wolfratshausen, Germany, 2024; Available online: www.eagleburgmann.com/de/produkte/spezialprodukte/dro-br-drehrohrofen-dichtsystem/datasheet_popup (accessed on 11 November 2024).
69. Chatterjee, A.; Sathe, A.V.; Mukhopadhyay, P.K. Flow of materials in rotary kilns used for sponge iron manufacture: Part II. Effect of kiln geometry. *Metall. Trans. B* **1983**, *14*, 383–392. [CrossRef]
70. Haeldermans, T.; Lataf, M.A.; Vanroelen, G.; Samyn, P.; Vandamme, D.; Cuypers, A.; Vanreppelen, K.; Schreurs, S. Numerical prediction of the mean residence time of solid materials in a pilot-scale rotary kiln. *Powder Technol.* **2019**, *354*, 392–401. [CrossRef]
71. Dinesh, L.; Sai, P.S.T. A Model for Residence Time Distribution of Solids in a Rotary Kiln. *Can. J. Chem. Eng.* **2004**, *82*, 392–398. [CrossRef]
72. Paredes, I.J.; Yohannes, B.; Emady, H.; Glasser, B.J.; Borghard, W.G.; Muzzio, F.; Cuitiño, A.M.; Beeckman, J.; Ilias, S.; Podsiadlo, P.; et al. The effect of operating conditions on the residence time distribution and axial dispersion coefficient of a cohesive powder in a rotary kiln. *Chem. Eng. Sci.* **2017**, *158*, 50–57. [CrossRef]
73. Wei, Z.; Zhang, J.; Qin, B.; Dong, Y.; Lu, Y.; Li, Y.; Hao, W.; Zhang, Y. Reduction kinetics of hematite ore fines with H_2 in a rotary drum reactor. *Powder Technol.* **2018**, *332*, 18–26. [CrossRef]
74. Lumay, G.; Boschini, F.; Traina, K.; Bontempi, S.; Remy, J.-C.; Cloots, R.; Vandewalle, N. Measuring the flowing properties of powders and grains. *Powder Technol.* **2012**, *224*, 19–27. [CrossRef]
75. Reh, L. Strömungs- und Austauschverhalten von Wirbelschichten. *Chem. Ing. Tech.* **1974**, *46*, 180–189. [CrossRef]

processes

MDPI

Review

Dry Machining Techniques for Sustainability in Metal Cutting: A Review

Shailendra Pawanr and Kapil Gupta *

Department of Mechanical and Industrial Engineering Technology, University of Johannesburg, Doornfontein Campus, Johannesburg 2028, South Africa; shailendrap@uj.ac.za
* Correspondence: kgupta@uj.ac.za; Tel.: +27-11-5599081

Abstract: Dry machining has gained significant importance in the last few years due to its promising contribution towards sustainability. This review study introduces dry machining, presents its benefits, and summarizes the recent technological developments that can facilitate dry machining. It aims to provide a concise overview of the current state of the art in dry machining to promote sustainability. This article synthesizes and emphasizes the useful information from the existing literature, and summarizes the methods and tools used to implement it. It also identifies some of the major problems and challenges and their potential solutions to make dry machining more viable and efficient. It concludes with some future research directions important for the scholars and researchers to establish the field further. From this review study, the major findings are: (1) tools with textures or patterns can enhance the cutting performance of dry machining for various materials, (2) tool coating is an effective way to lower the tool cost in dry machining and can achieve the required functionality for the cutting tool without affecting its core properties, (3) Alumina-based mixed ceramic tools with SiC whiskers have better fracture toughness, thermal shock resistance, and self-crack healing properties, (4) one effective method to improve the dry cutting of engineering materials is to apply external energy sources to assist the dry machining process, (5) by using microwave sintering, cutting tools with finer microstructures and higher densities can be produced, which improve their hardness, wear resistance, and thermal stability to perform well in dry machining conditions.

Keywords: cutting fluid; dry machining; environment; machinability; sustainability; textured tools; tool coatings; self-lubricating coatings; microwave sintering

1. Introduction

Energy and materials are vital resources for the expanding global economy, and their extraction from natural resources has negative environmental repercussions such as pollution, deforestation, biodiversity loss, and climate change. These environmental problems pose serious risks for life on the planet and are the drivers to adopt sustainable development. Sustainable development is defined as being able to meet current needs without endangering future generations' ability to meet their own [1,2]. As shown in Figure 1, it is a notion that refers to the ability to carry out a particular activity indefinitely encompassing environmental, social, and economic aspects of development [2]. The manufacturing sector is one of the key contributors to economic growth, a major consumer of natural resources including energy and materials, and can make a substantial contribution to sustainable development. The manufacturing industry, as the backbone of every industrialized economy, presents challenges in addressing concerns of sustainability [3,4]. In the present scenario, sustainable manufacturing is now not only an obligatory standard but an opportunity to create global competitiveness and benefit the community [3].

Machining is one of the most important and an absolutely essential manufacturing activity for producing different components [5,6] and undoubtedly a significant driving force behind global economic growth [7,8].

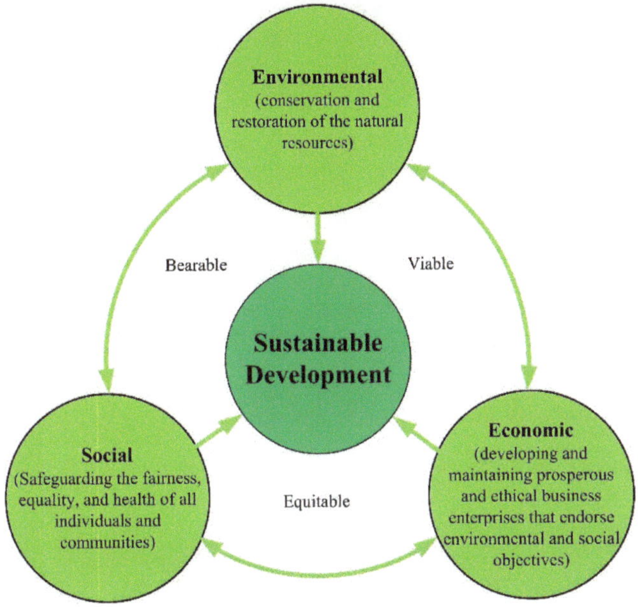

Figure 1. Sustainable development aspects.

Sustainable machining innovation is a key goal for the manufacturing sector, which faces increasing demand for eco-friendly methods and products. To achieve this goal, it is vital to design and implement systems that can combine research and industrial skills to save energy, resources, and the environment [9,10]. The manufacturing and machining industries in the United States have an annual value of over 100 billion dollars, demonstrating the relevance of sustainability in the practices of machining due to the scale of manufacturing processes involved and the related potential environmental dangers [11]. Machining is a process that involves a workpiece's material removal by the application of a cutting tool that produces chips, and the goal of machining is to attain the necessary shape and size of the workpiece according to the specifications [6]. A substantial volume of heat is produced through the machining process due to extreme metal deformation at the tip of tool and workpiece interaction, causing high temperature, build-up edge, and lower tool life, controlled using proper cooling agents or cutting fluids, a significant source of waste and adverse environmental impacts [12]. Although cutting fluids are favorable in many aspects of machining performance, such as heat removal, higher tool life, and better surface smoothness, they also incur financial costs and have a negative environmental impact due to their high fraction of total machining costs, i.e., around 17% [13], disposal issues [14], and rapid degradation when used at high temperatures [15]. Every year, over 30 million gallons of cutting fluid are released in the United States alone, and the cost of cutting fluid retreatment is around USD 15 per gallon [16]. Cutting fluid is a hazardous substance; when it comes into contact with the skin, it can cause irritation, inflammation, and even cancer, and when it evaporates at the cutting zone, it can release toxic vapors that can damage the respiratory system [16–18].

Furthermore, in certain cutting applications, such as high-cutting-speed machining, there is the possibility of insufficient cutting fluid penetration into the machining zone [19] and thermal shocks to the cutting tool, which reduces tool life [20], limiting the implementation of such a cooling strategy. An increase in global environmental awareness and competitive cost pressures on manufacturing industries led a critical consideration of cutting fluids in machining [4,21], and dry machining has gained popularity in the recent decade and has been extensively researched, particularly in the last five years [20]. Machining performed without the use of cutting fluids is referred to as "dry machining" [12]. Dry machining is a green and economical alternative to conventional machining that requires cooling and lubrication fluids. It is expected to become a crucial technique for the manufacturing industry in the coming decades, as it can reduce waste, energy consumption, and health risks.

The literature shows that various approaches have been proposed and implemented to achieve dry machining. This article provides a brief review of the current state of the art in dry machining, extracts and highlights the valuable information provided in the existing literature, summarizes the methods and tools used to implement it, and also discusses the future directions and opportunities for further research and development in this field.

The remaining structure of the papers is as follows (see Figure 2): Section 2 discusses the constraints and benefits of dry machining. The highlights of the existing review studies are discussed in Section 3. The different approaches for achieving dry machining are outlined in detail in Section 4. Section 5 concludes with a summary and future research directions.

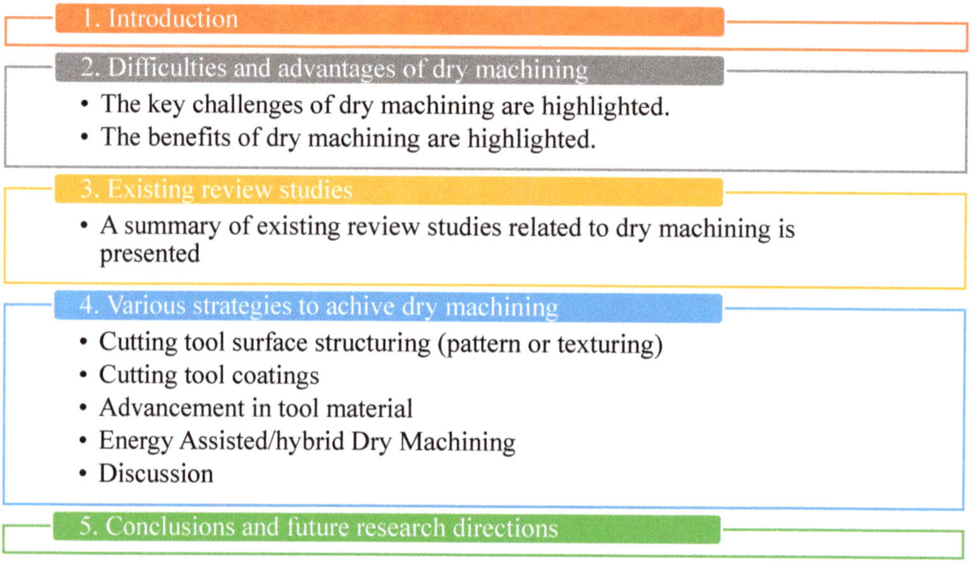

Figure 2. Structure of the review.

2. Difficulties with and Benefits of Dry Machining

The quest to eliminate the application of flooded fluids has been one of the key difficulties and goals in the context of machining in recent years, namely "Dry Machining", in order to meet the sustainable development standards and to reduce the machining cost [15,22,23]. Dry machining is an eco-friendly process that will become a vital requirement for the manufacturing sector in the near future. Whether voluntarily or by legal obligation, industries will have to adopt dry machining to abide by the environmental safeguarding guidelines and the regulations governing workplace safety and health [14,24]

Indeed, there are many benefits of dry machining, as follows and summarized in Figure 3a:

- No liquid waste or pollution [19];
- No exposure to hazardous substances that can cause skin or respiratory problems [14];
- No thermal shocks to the cutting tool [20];
- Delivers cost-effective machining, as it eliminates the need for purchasing, storing, and disposing of cutting fluids [19].

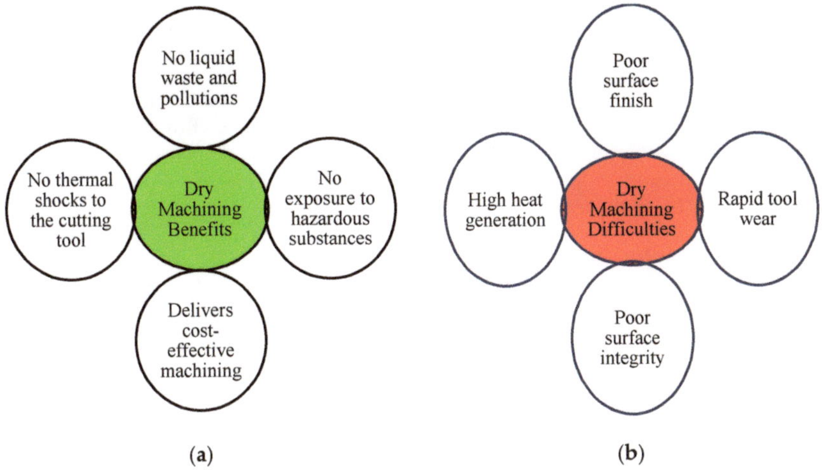

Figure 3. Dry machining (**a**) Benefits and (**b**) Difficulties.

However, in the absence of any cutting fluid, which can help with cooling, chip removal, and lubrication at the interaction point between the component being worked on and the cutting tool, the following are some of the major difficulties/challenges experienced in dry machining:

- Heat generated in the cutting zone: the tool and workpiece will interact more strongly due to the higher friction and adhesion forces, which causes high temperature at the cutting tool [12].
- At elevated temperature, the tool loses its mechanical strength and wear resistance, leading to faster deterioration of the tool and shorter tool life [25].
- Formation of tool wear causes deterioration in the surface roughness, dimensional accurateness, and surface integrity of the machined component [16].

Furthermore, without the use of a cutting fluid, dry machining may produce small particles of metals that remain in the air and can cause health issues if inhaled. These challenges worsen and intensify when dry machining has been performed at high speeds with materials that are susceptible to adhere to the tool surface, such as aluminum [20] and stainless steel [26], or materials that have low machinability, such as tool steel, titanium alloy [15,27], and nickel alloys [22,28], etc.

Aluminum, steel, stainless steel titanium alloy, and nickel base alloys are some of the most common materials used in engineering. They are suitable for different sectors, such as aerospace, automotive, construction and medical, because of their specific properties. Machining is a common process that shapes and modifies these materials to meet the desired specifications and requirements. Aluminum is a soft material, but despite its low shear strength, chip adherence to the tool edge and rake face is a serious issue in the absence of cutting fluid [14]. Stainless steel has a high strength that requires high cutting forces during machining but also a low thermal conductivity that leads to heat accumulation. Without the use of cutting fluids, the high friction at the tool interface and its sticky nature result in phenomena such as adhesive layer build-up edge (BUE) and rapid tool degradation [20]. Titanium and nickel base alloys are also hard to cut in the absence of cutting fluid due to their high hot strength and low heat conductivity, these properties cause high increases in temperature at the tool—chip interaction, which leads to premature tool failure [15,19,27].

Nowadays, dry machining is a topic of global interest due to its importance for sustainable manufacturing. However, it also poses some difficulties/challenges that hinder its widespread adoption. Figure 3b illustrates some of the major difficulties that need to be overcome to make dry machining more feasible and effective. Various researchers are investigating different methods and strategies to address these challenges and improve the performance of dry machining.

These strategies for research topics cover different aspects, such as new materials for cutting tools [29], coatings [28], surface structuring/texturing of cutting tools [30], and energy-assistive dry machining, which is a hybrid process that combines dry machining with other processes like laser [31] and ultrasonic [32]. The following section provides more details on these strategies.

3. Existing Review Studies

The literature search shows that a few review studies have addressed the topic of dry machining, either exclusively or in comparison with other techniques such as near-dry lubrication, i.e., minimum quantity lubrication (MQL). The review articles by Sreejith and Ngoi [24] and Goindi and Sarkar [12] focused solely on dry machining, while the articles by Weinert et al. [21], Zhang et al. [18], Chetan et al. [33], and Rajmohan et al. [34] discussed it along with other strategies such as MQL and liquid nitrogen. One of the first review articles on dry machining was presented by Sreejith and Ngoi [24]. They discussed the current trends and developments in dry machining and concluded that more technological innovation is needed to make dry machining feasible and effective. Weinert et al. [21] presented a review study on minimal quantity lubrication (MQL) supporting near-dry and dry machining. They provided a thorough discussion of MQL and its technological implications, as well as studies on cutting tool development, coating technologies, and machine tool optimization, emphasizing that they were critical to enabling dry machining. Zhang et al. [18] conducted a systematic analysis of the features and challenges of high-speed dry milling for hard-to-cut metal materials. They discussed the effects of different energy fields, such as laser, ultrasonic, and cryogenic minimum quantity lubrication, on the milling process and performance. The review by Goindi and Sarkar [12] identified the challenges and limitations of dry machining. They discussed recent advances in dry machining methods, like the utilization of advanced tool materials and coatings, tool geometry modifications, the incorporation of laser- or modulation-assisted machining, and MQL machining. A survey of different eco-friendly machining methods was conducted by Chetan et al. [33] and Rajmohan et al. [34]. They examined the use of MQL, liquid nitrogen, vegetable oil, or compressed air as alternatives to conventional coolant–lubricant fluids and found that these methods often achieved comparable results to wet machining in terms of machining performance.

In addition to these review studies, Bayraktar [35] explained the basic mechanics of dry machining and provided important information about tool geometry, material, and coatings, and discussed relevant case studies.

Although the existing significant review studies on dry machining offer valuable insights for the researchers, they also have the following limitations that need to be addressed:

- None of the recent review studies focused exclusively on the dry machining process.
- None of the studies provided any information on the application of textured cutting tools in dry machining.
- The technological advancements and strategies discussed in the four papers were limited in scope and did not cover all the aspects of dry machining.

Therefore, this study focuses solely on dry machining, extracts and highlights valuable information from the existing literature, and summarizes the strategies and tools used to implement it, including new cutting tool materials, coated cutting tools, surface structured/textured cutting tools, and energy-assistive conventional machining.

4. Various Dry Machining Techniques

The key challenges in dry machining are the higher coefficients of friction between the tooltip and the work material, as well as heat dissipation, while the presence of cutting fluid minimized these two, providing lubrication at the tooltip chip and carrying heat out from the cutting region [15,35]. To achieve comparable machining performance and explore alternatives to wet machining, researchers examined various dry machining methods such as textured/structured tools [30], coated tools [28], improved tool materials [29], and hybrid machining with energy assistance [31,32].

As a result of these facts, the current review focuses primarily on the following developments.

- Cutting tool surface structuring (pattern or texturing)
- Cutting tool coatings
- Advancement in tool material
- Energy-Assisted/hybrid Dry Machining

4.1. Cutting Tool Surface Structuring: Pattern or Texturing

Heat generation and plastic deformation of the workpiece depend on the friction at the tool–chip interaction, and that is mainly responsible for machinability. A possible way to enhance the machinability is to create a microscale or nanoscale texture that reduces the friction [36]. Recent improvements in micromanufacturing technologies have made it possible to build useful surfaces for cutting tools, dies, and molds that are utilized in crucial manufacturing procedures. Some common techniques for fabricating textured surfaces are chemical etching, ECM: electro chemical machining, LBM: laser beam machining, and EDM: electric discharge machining [37]. The literature review by Biermann et al. [38] provides a comprehensive overview of the benefits and drawbacks of these techniques; we have not included it here since it is outside the scope of this review. Surface textured/patterned cutting tools are a new generation of cutting tools that have improved tribological performance at the tool–chip interaction [39]. Numerous studies have been conducted on cutting tools that have patterns or textures on their surfaces, because they can reduce friction and wear during machining processes. One of the earliest works that presented the concept of textured tools was by Kawasegi et al. [36]. Kawasegi et al. [36] developed microscale or nanoscale textured cemented carbide tools using a laser beam. The textured tools showed a decrease in cutting force which significantly influenced the direction of the texture. Applying textures to the rake along with the flank surfaces of tools can significantly lower the friction ~13–26% during machining processes [30,40]. The texture reduces the contact between the cutting tool and the chip in cutting. It consequently leads to a lesser amount of friction, less heat generation, and changes in the cutting temperature distribution [41]. A typical representation can be seen in Figure 4.

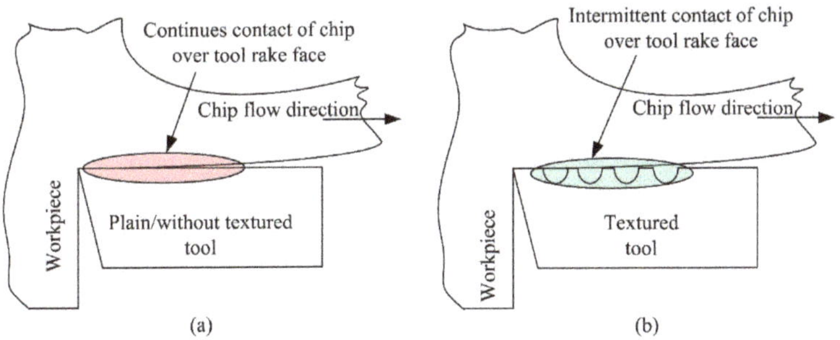

Figure 4. (**a**) Continuous contact of chip over tool rake face, (**b**) Intermittent contact of chip over tool rake face.

It is well established in previous studies that textured cutting tools can reduce the cutting force and feed force by 10–20% and 20–30%, respectively [42]. Thus, tool surface texturing/pattern is an innovative and effective method for improving machining process performance and quality. It entails making micro- or nano-scale patterns on the tool surface in order to change its tribological properties [43]. Because of the potential benefits for machinability and machining efficiency, the textured/patterned cutting tools have received a lot of attention in recent years.

Researchers have explored various types of texturing the surface of cutting tools to reduce friction. Some studies have focused on the rake surfaces [44–47], some on the flank surfaces [48,49], and some on both surfaces [42,50] simultaneously, forming dual-textured cutting tools. This study focused on the influences of texturing on the tools in dry cutting conditions and only included the relevant significant literature that met this criterion.

The performance of machining Inconel 718 alloy in dry turning with micro-textured dimple, micro-textured channel, and plain carbide cutting inserts was compared by Rajurkar and Chinchanikar [47]. The tool life of micro-textured channel cutting tools was 60% longer than that of micro-textured dimple tools at low-level speed cutting conditions. Nevertheless, the wear resistance was similar for both cutting tools (textured and plain) at high levels of cutting speeds. The primary mechanisms of wear were adhesion and abrasion, with some notch wear. The chips from micro-textured dimple tools were more segmented than the chips from micro-textured channel tools. A typical arrangement of texturing on the rake surface is presented in Figure 5.

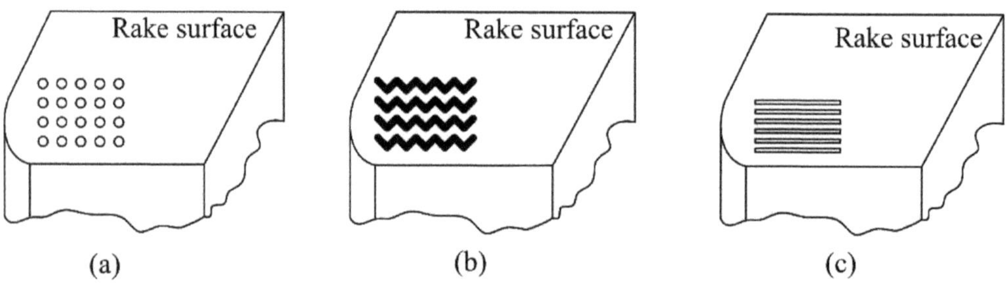

Figure 5. Micro-textured (**a**) dimples, (**b**) waves, and (**c**) channels on insert rake surface.

The machinability of SS316, an austenitic stainless steel, was investigated by Soni and Mashinini [46] using a tungsten carbide cutting tool with a straight-line texture pattern. The results suggest that 0.2 mm/rev of feed rate and 0.1 mm of cutting depth can achieve improved wear resistance and machinability when machining SS316 at higher speed. However, this low cutting depth may compromise the productivity, so further research is needed in this aspect.

Singh et al. [37] produced micro-textured pattern on a high-speed steel cutting tool by the application of micro-EDM and optimized the parameters of the process. They also examined the textured tool's effectiveness in cutting force reduction and determined that it lowered cutting force and prevented chip adherence on the textured tool surface during aluminum alloy dry machining. The textured tool performed better at lower speed, reducing the cutting force by 13% at lower speed and 5% at higher speed. Moreover, when the textured cavities of the cutting tool were filled with MoS_2 solid lubricant, the cutting force was reduced by 10% at high speed as well. Fatima et al. [51] examined the effectiveness of two kinds of textures (hole and straight line) on the cutting tool's rake surface on various cutting performance indicators, such as cutting forces, chip–tool contact length, wear, ratio of compression, and contact type. They found that textured rake faces enhanced the cutting performance in general, but the hole textures outperformed the straight-line textures in minimizing 14% in forces, 24% in tool–chip contact length, and 4% in wear. Moreover, the hole textures significantly decreased the sticking interaction from 52% to 25% at medium speed. Ma et al. [52] investigated how microbumps texturing on the rake surface of a cutting tool affect the AISI 1045 orthogonal machining. They changed the width, height, and position from the cutting edge of the microbumps and recorded the cutting forces. They discovered that microbumps can lower the cutting forces, but the influence of bump width on forces is complex.

Instead of the rake face, tool texturing can alternatively be done on the flank face [49]. The flank face is the tool's surface that comes into contact with the workpiece while cutting, and it can wear and deform over time. One of the most important regions to inspect for tool wear is the flank face, which frequently reveals when a cutting tool has reached the end of its service life and needs replacement [53]. The influences of the texturing on the tool flank surface on the machining of AISI 1045 steel was examined by Dhage et al. [54] using cemented tungsten carbide inserts. They found that the surface quality of the machined workpiece and machining forces were substantially reduced by the texturing on the flank face surface. Rajbongshi et al. [48] investigated how texturing on the surface of the coated cutting insert on the flank surface affects the machining performance. The studies were carried out on AISI D2 steel having HRC of 45 in a dry cutting environment using textured and plain (without texturing) cutting inserts. They found that the texturing on the inserts reduces flank wear as compared to the plain cutting inserts. They also report that the quality of the machined surface is better by the textured tool in most cases and concluded that flank face textured surface of the carbide tool with coatings improves the machinability. The effect of different micro-texturing forms (dimples, lines, and splines) on the cutting tool's flank face was investigated by Khan and Gupta [55] in a turning experiment of Inconel 600. They measured the tool life and wear under various conditions and found that the micro-texturing reduced the friction and heat significantly, enhancing the tool durability. The findings revealed that the lines and splines patterns performed superior to the dimples pattern in relation to tool life and wear. The SEM images of tool wear with different textures are shown in Figure 6.

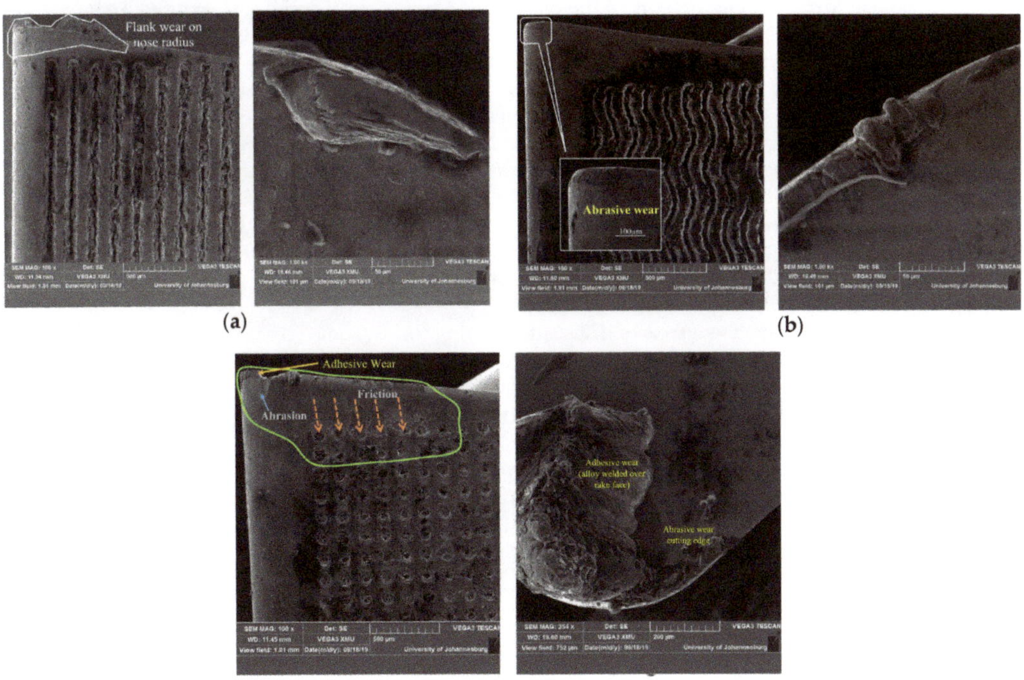

(c)

Figure 6. SEM images of tool wear with (**a**) line texture, (**b**) splines texture, and (**c**) dimple texture [55].

Some researchers investigated the effects of texturing on both faces of the cutting tools: rake and flank, as shown in Figure 7.

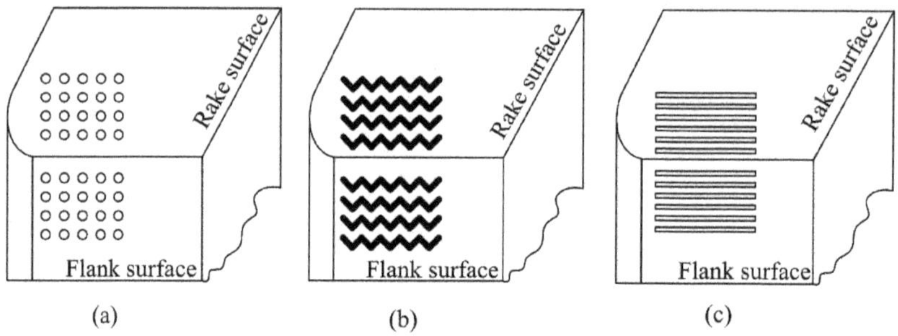

Figure 7. Rake and flank face textured tool (**a**) dimples, (**b**) waves, and (**c**) channels.

Fatima and Mativenga [42] used a micro-grooves laser-texturing flank surface (rake and flank) cutting tool for the turning of the AISI/SAE 4140 material workpiece. The findings revealed that the laser-texturing on the tool improved machining performance by a lowering of 14% in the friction coefficient, 10% in cutting forces, 23% in feed forces, 10% in compression ratio, thus enhancing the machining performance. Fatima and Mativenga [56] extended their study and compared the effectiveness of a plain cutting tool and three cutting tools with texturing. The textured tools had different surface patterns: only on the rake face, only on the flank face, or on both faces. The texturing on the flank face achieved the best performance in minimizing flank wear and tool temperature at high speeds. It also had lower cutting forces than the plain tool. The tool with both faces textured had the smallest friction coefficient but only reduced flank wear at low speeds. The rake face textured tool did not improve flank wear reduction significantly.

Bio-inspired textures on cutting tools can improve their performance by mimicking natural patterns. One example is the micro-crescent structure, which reduces friction and wear by creating gaps and channels between the tool and the workpiece. Ranjan and Hiremath [50] studied the effect of this structure on both surfaces of the cutting tools: rake and flank when machining hard martensitic steel AISI 420. They found that the micro-crescent structure on both surfaces was superior to other conventional and structured tools. Compared to a conventional tool, the crescent-shaped tool reduced the flank wear by 56.44%, the feed force by 28.91%, the tangential force by 19.04%, and the average surface roughness by 19.02%.

Given the positive effects of different texturing forms, some researchers examined the influences of dual-texturing on the cutting tools, as shown in Figure 8. Siju et al. [30] studied the effects of dual textures (grooves and dimple) on a cutting tool's rake surface during the machining on Ti6Al4V alloys. The machining forces were significantly reduced while using dual-textured tools. Reductions of 16.2% in cutting force, 31.2% in thrust force, and 17.8% in friction were observed as compared to the plain tool. Because the worn debris was contained inside the textures, the cutting tool with dual texturing had less adhesion for the workpiece and thus built-up-edge development. This impact could be seen by rake surface texturing and elemental dispersive spectroscopy. The chips formed by the dual-textured tools were smoother and had thin segments at low feed rates, while the non-textured tools created rougher chips [30].

Texturing cutting tools is a technique that can enhance their performance in relation to wear resistance, cutting forces reduction, and quality of surface improvement. Some studies also investigate how texturing influences chip segmentation during dry machining. Siju et al. [43] used carbide inserts with dual textures and analyzed the chip segmentation features quantitatively for machining Ti-6Al-4V titanium alloy. They report that texturing has a noticeable effect on chip segmentation compared to untextured inserts within a certain range of cutting velocity. Textured inserts have a relatively higher cutting performance than conventional plain without-textured inserts when the cutting velocities are low. Similar findings were reported by Rajurkar and Chinchanikar [47], that at greater cutting rates, there is no noticeable distinction in tool life between the two tools (textured and plain). However, some studies reported that the application of solid lubricants together with texturing provides promising results at higher cutting speeds as well [37], and this could be an aspect of future investigation.

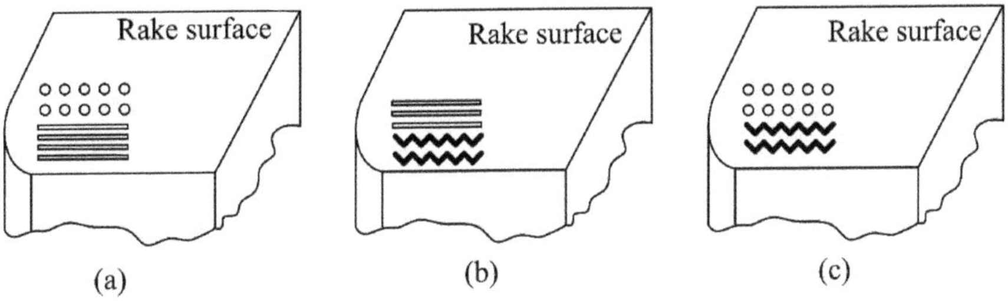

Figure 8. Dual-textured cutting inserts: (**a**) dimple and channel, (**b**) wave and channel, and (**c**) wave and dimple.

Some scholars have considered the performance of the cutting tool having texturing as well as coating and compared their results with conventional tools. Texturing creates micro- or nano-scale features on the tool surface, while coating adds a thin layer of hard material. Both techniques can reduce wear and friction and facilitate chip removal. A nanotextured and TiAlN-film-coated cutting tool was developed by Zhang et al. [44] and tested its machining performance on SLM 316 L stainless steel in a dry cutting environment. The nanotextured tool showed a 10–20% decline in cutting force, a 10–15% fall in cutting temperature, and an improvement in the surface quality of the workpieces at high cutting speeds as compared to the plain cutting tools.

Zhang et al. [45] produced different sizes of texturing on cemented carbide cutting tools and applied a thin layer of TiAlN on them. They applied the surface textures both prior to and following the PVD TiAlN coatings being applied and tested how these patterns and layer cause impacts on the cutting tool's wear resistance when cutting stainless steel made by SLM. Undoubtedly, the textured cutting tools had less wear than the ones without them, and the tool has both micro-patterns and nano-patterns that had the best performance, and the wear-resistive ability depended on the dimensions of the patterns. Further, TiAlN-coated tools' cutting performance is influenced by the texture sequence. The tools that are textured prior to the application of TiAlN coatings have better coating bond and anti-adhesive wear-resistive ability than the tools that are textured following the application of TiAlN coatings, irrespective of the texture dimension. This is because the pre-texture enhances the bond of the coating and the base material, i.e., substrate [45].

Besides texturing, some other approaches have been explored by scholars to enhance the machining performance in dry conditions, such as modifying the cutting tools geometry. This approach aims to reduce the friction and heat generation during the machining process, along with improving the chip evacuation and tool life.

Aslantas and Alatrushi [57] examined how the geometry of tool (angles: helix and axial rake and cutting edges) influences the forces in machining, surface roughness, and formation of burr in the micro-milling of nickel base superalloy grade Inconel 718. A helix angle of 45° was found to be optimal for reducing the cutting forces in the experiment. The findings also indicate that cutting edges with fewer numbers and rake angles with negative values result in smaller burr widths. The cutting tool with a fewer number of cutting edges can improve the surface roughness, and axial rake angles with negative values can enhance the tool durability. Zhang et al. [58] explored the application of wiper inserts as a substitute to conventional inserts for dry machining of laser-cladded components. They found that wiper inserts can enhance both the productivity and quality of the surface in dry turning.

A review of the literature shows that texturing on the rake and flank faces of the cutting tools has been investigated and assessed by various researchers in the dry machining of different materials. The texturing can be applied to either one or both faces of the tools, creating a dual texture such as circular cavities/holes [37,51], straight/linear/channel/grooves [46,56], dimple [47], splined/wave [55], and a combination of these [30,43,50]. The geometry of texturing, the area covered by the texture, and location on the surface of the tool are important factors for tool performance [47]. Different texture shapes such as dimples, grooves, channels, and waves work better in different situations. However, the literature review does not provide clear criteria for choosing the texture shape and size. Therefore, depending on the situation and the desired outcome, different textures and shapes may work better than others.

4.2. Cutting Tool Coatings

One of the challenges in designing cutting tools is to balance different properties. To perform well, cutting tools need different properties at the surface and the core. The tool surface should be hard, smooth, wear-resistant, and inert, while the tool core should be strong, heat-resistant, tough, and thermally conductive. These properties are hard to achieve in a single material, so coatings are used to enhance the tool surface. Coatings can provide the desired functionality for the cutting tool without compromising its core properties [59]. Cemented carbide tools, one of the widely used tool inserts in machining, are prone to premature tool failure when used under a dry cutting environment and at high cutting speeds. This is because the extreme temperature within the cutting zone causes the tool binder particles to diffuse into the chips that flow under the tool surface; this weakens the tool and reduces its durability [13]. Using costly harder tool materials like ceramics, CBN: cubic boron nitride, PCD: polycrystalline diamond, etc., is not economically viable. The investigation found that the presence of a coating is one of the most critical elements influencing the friction coefficient [60]. Tool coating has become widely popular in the last couple of decades as a more feasible alternative for costly tools in dry machining [13]. The cutting performance was improved by the coating properties, which include superior hardness, good wear resistance, and great heat stability, especially in a dry cutting environment. These properties made the coating more resistant to abrasion, deformation, and thermal degradation [28]. A coating over the substrate made of cemented carbide has the main objective of protecting it from mechanical, chemical, and thermal damage at high temperatures [61]. The coating is a rigid, thin layer that is chemically inert and has little attrition, and it also serves as a thermal insulation. Many research studies in the literature have investigated and reported on the performance of various types of coatings for machining different materials. Dry machining demands high-performance and durable cutting tools that can withstand high temperatures and friction. Coatings are one way to enhance the tool performance and life span by reducing wear and friction.

The type of coating can significantly affect the performance of a tool in machining processes [62]. In the literature, various coatings have been reported and evaluated for their performance in dry machining such as TiAlN [28,63,64], TiAlN [65], AlTiCrN/AlTiN coatings [62,66], TiN [63–65,67], Al_2O_3 [64,68], TiCN [64], and TiCN-Al_2O_3-TiOCN [19] and many more.

PVD: physical vapor deposition and CVD: chemical vapor deposition are the most widely used methods for depositing coating layers on carbide substrates [62]. These coatings can improve the hardness, wear-resistive ability, thermal stability, and resistance to oxidation of the cutting tools, thus increasing their tool life and cutting efficiency, which are discussed in the following paragraphs.

Superalloys based on nickel are extensively used in various strategic arenas, such as aviation, energy, and nuclear industries. These materials pose significant challenges for machining since they have high chemical susceptibility, low heat conductivity, and strength, and these challenges even worsen in a dry environment. Zhao et al. [28] carried out research on the impact of TiAlN coatings and the weightage of Al content ($Ti_{0.41}Al_{0.59}N$ and $Ti_{0.455}Al_{0.45}N$) regarding Inconel 718's dry machining performance. They revealed that the TiAlN-coated tools had advantages over the uncoated ones in relation to longer tool lifespan, lower cutting forces, and better thermal conductivity. Among the different TiAlN coatings, $Ti_{0.41}Al_{0.59}N$ displayed the most favorable outcomes with regard to extended tool lifespan, less friction, and a smaller heat resistance compared to the plain and $Ti_{0.455}Al_{0.45}N$ tools. Thakur et al. [64] carried out research on the high-speed machining on Inconel 718 using multicoated tools made of tungsten carbide: TiN/TiAlN (PVD-coated) and $TiN/Al_2O_3/TiCN$ (CVD-coated). They measured the forces and temperature during cutting and vibrations as indicators of machinability and found that the tool coated with CVD outperformed the tool coated with PVD in all aspects. Thakur et al. [63] performed machining on nickel base super alloy grade Incoloy 825 using a plain tool, tools coated with $TiCN/Al_2O_3$ (CVD) and TiAlN/TiN (PVD). The tools having PVD coating had improved machinability characteristics compared to the other tools excluding tool life with the CVD-coated tools. The tools having PVD coating show high antifriction and antisticking properties due to TiN and high durability as a result of the multilayer TiAlN phase in the structure.

The effectiveness of four types of cutting inserts: PCD-coated, TiN, TiAlN, and plain cemented carbide was evaluated by Uddin et al. [69] in a dry machining of aluminum base alloy grade Al 2024 alloy in relation to wear resistance and quality of machined surface. The results show that tool wear is minimal with TiAlN-coated inserts, whereas surface roughness is minimal with TiN-coated inserts. The coatings improved the cutting conditions compared to the uncoated insert. The coatings' adhesion and mechanical properties were observed as important factors for wear resistance and quality of surface. PCD coating has very low adhesion strength and breaks off at the first critical load, which is reflected in the wear behavior of PCD-coated tools as the abrupt failure of the coating due to adhesion loss, unlike other tools.

Sahoo and Datta [70] investigated the effects of carbide inserts: plain and coated by MT-CVD $TiCN–Al_2O_3$ on the dry machining of AA7075 T6 alloy. They found that the coated insert has lower temperature on the tool tip and wear resistance on the flank surface compared to the uncoated inserts. However, the coated insert exhibits higher cutting forces than the uncoated inserts during cutting of AA7075- T6 alloy. The performance of coated inserts and a PCD tool in the machining of rolled aluminum was investigated by Pattnaik et al. [67]. They observed that the coating layers on the inserts were susceptible to chipping and peeling off due to the adhesion of aluminum, whereas the PCD tool demonstrated improved wear resistance and surface finish.

Diamond possesses advantageous physical and chemical qualities, for example, low COF, highest hardness, wear, and chemical resistance. Thus, diamond coatings possess the capability to significantly improve the performance of micrometer-scale cutting tools. Heaney et al. [71] stated that diamond-coated tools lessen cutting forces and work adhesion, improving tool life and lowering breakage during the end milling of the aluminum-based alloy grade Al 6061-T6. Furthermore, the diamond coating allows for high-speed dry end milling without burr development.

To improve diamond coating performance, Gomez et al. [72] studied new tungsten carbide insert substrates that increase the interlocking effect of the diamond film while limiting cobalt's catalytic influence on the cemented carbide tool. They tested the performance in dry machining of aluminum alloys A390 and observed improved results. This study investigated the enhancement of cemented carbides' surface characteristics for more effective diamond coatings in the dry machining of aluminum alloys with high silicon.

Although the coatings described above, including CBN and diamond (extremely hard coatings) and TiAlN, TiC, TiN, and Al_2O_3, enhance the wear resistance and cutting performance of tools for various engineering materials, coatings with hard solid lubricant are a recent innovation in the field of machining that can promote the dry machining by providing lubrication without using fluids [73]. Solid lubrication, which reduces the environmental impact and enables dry machining, is becoming more popular nowadays. It relies on self-adaptive materials that can develop a low friction layer over the contact area during operation. Among the best examples of this approach are self-lubricating coatings that can produce low-friction phases during the cutting operation and lessen the friction that the tool has against the workpiece [29]. The performance and life of cutting tools in dry machining can be improved by applying self-lubricant coatings on them [15]. Cavaleiro et al. [15] used silver as a solid lubricant coating with different concentrations (6% and 10%) applied to TiSiN-coated inserts. They tested the performance of these inserts on conventional and CNC lathes during the machining of TiAl6V4 titanium alloy. The results showed that both TiSiNAg6 and TiSiNAg10 coatings increased the tool life by minimizing wear, friction, and the formation of BUE during cutting at higher speeds.

Aluminum is a soft material, but it can stick to the cutting tool and cause built-up edge (BUE) [74]. The BUE has a substantial impact on the tool lifespan and surface integrity. This is a typical machining issue of aluminum. Therefore, soft coating materials are increasingly used to prevent adhesion and improve machining performance [14]. Gr-X is a new solid lubricant coating made of graphite and polytetrafluoroethylene (PTFE). Kannan and Ghosh [14] tested its performance against MoS_2, a common solid lubricant. They found that Gr-X performed better than MoS_2 for cutting aluminum in humid conditions.

The self-lubricative composite coatings, i.e., TiN and MoSx, showed significant advantages in the study by Gangopadhyay et al. [73]. They observed that there were fewer cutting forces by about ~45%, the tool lifespan increased by two folds, and the quality of surface improved by 13–21% compared to the uncoated tools.

According to Kumar et al. [29], vanadium addition to transition metals nitride like AlCrSiN and TiSiN leads to fast oxidation at high temperatures, which produces V_nO_{3n-1} phases (V_2O_5) with low melting points and lubricating properties, resulting in lower friction coefficients. They investigated this under the dry machining of titanium alloy Ti6Al4V and found that the coated tool with TiSiVN showed less adhesion and chip deformation than the TiSiN-coated tool. The coating of TiSiVN achieved an overall decline of ~18–23% in adhesion as compared to the plain tool. Knowing the advantages of self-lubricative TiSiN and TiSiVN coatings, Kumar et al. [75] conducted a further analysis of how these coatings affect the machining process effectiveness in relation to cutting forces, temperature, wear resistive ability, and quality of surface. They found that the TiSiVN-coated tool had more consistent forces and less vibrations than the uncoated and TiSiN-coated tools. They also observed that the tool coated with TiSiVN generated better surfaces compared to the TiSiN-coated tool at cutting speeds of 75 m/min and above, while there was no significant difference at 50 m/min.

Si-doped TiAlxN super nitride nanocomposite (HSN_2) is a novel coating material for carbide tool inserts that improves their performance in machining by enhancing the hardness, wear resistive ability, and stability against oxidation of the carbide inserts [76]. Pandey and Datta [13] found that HSN_2-coated carbide inserts reduce cutting temperatures by ~17%, cutting forces by ~50%, and improve the tool lifespan by ~15% as compared to plain counterparts for the turning of Inconel 718. In another investigation, Sahoo and Datta [76] further tested these inserts during the machining of titanium-based alloy Ti–6Al–4V in dry environment and obtained similar promising results.

Li et al. [65] examined the effectiveness of nanocomposite coating on malleable cast iron machining. They employed tools of cemented carbide having three varieties of coatings: TiN/TiAlN: thick, TiN/TiAlN: thin, and TiAlSiN: nanocomposite (nc), as shown in Figure 9. They measured the cutting forces, temperature during cutting, wear, tool lifespan, and quality of surface. As seen in Figure 10, the results revealed that the coated tools with nc-TiAlSiN show the best result in all aspects. Further, they pointed out that the principal tool-wear mechanism in malleable cast iron machining was adhesive, which necessitated low friction coefficient and low affinity-coated tools.

Some researchers employed coatings with multiple layers to improve machining performance. The use of multilayers with a TiN/TiCN/Al_2O_3/ZrCN coatings tool revealed a substantial enhancement in tool lifespan and chip properties when compared to its plain corresponding insert for the complete range of cutting speed in the machining of stainless steel grade 17–4 PH under dry environment [59]. The performance of a coating is affected by a number of factors, including layer sequencing in multilayer coatings, compositional ratio, thickness of coating, and coating method. Sonawane et al. [62] evaluated the effectiveness of single-layer (AlTiCrN and AlTiN) and multilayer coatings (TiN/TiAlN- and AlTiN/TiAlN) for the machining of DSS2205 on steel tool wear and tool life. The AlTiCrN coating outperformed all of them.

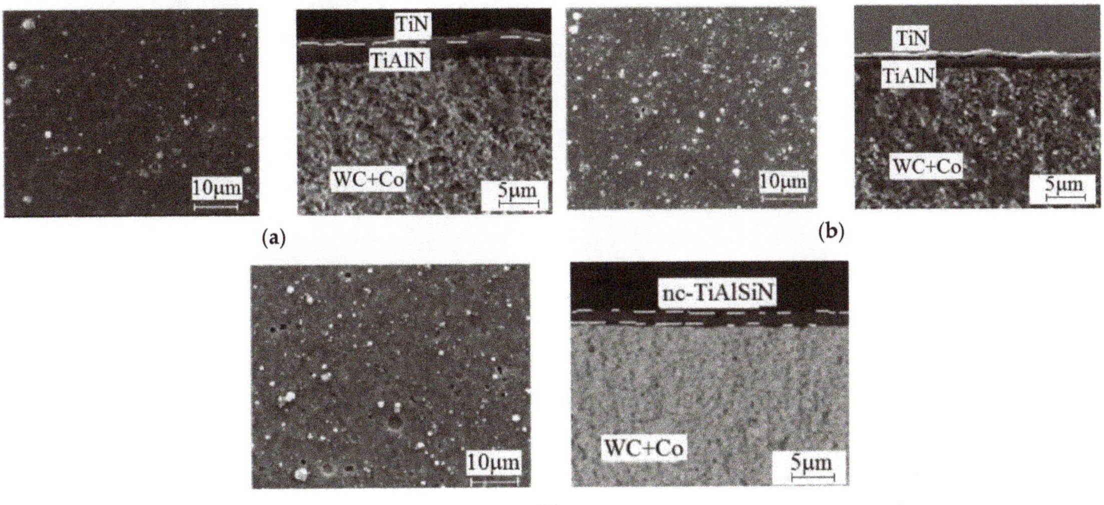

Figure 9. SEM images of coated tools surface and cross-sectional morphology: (**a**) thick TiN/TiAlN, (**b**) thin TiN/TiAlN, (**c**) (nc) TiAlSiN [65].

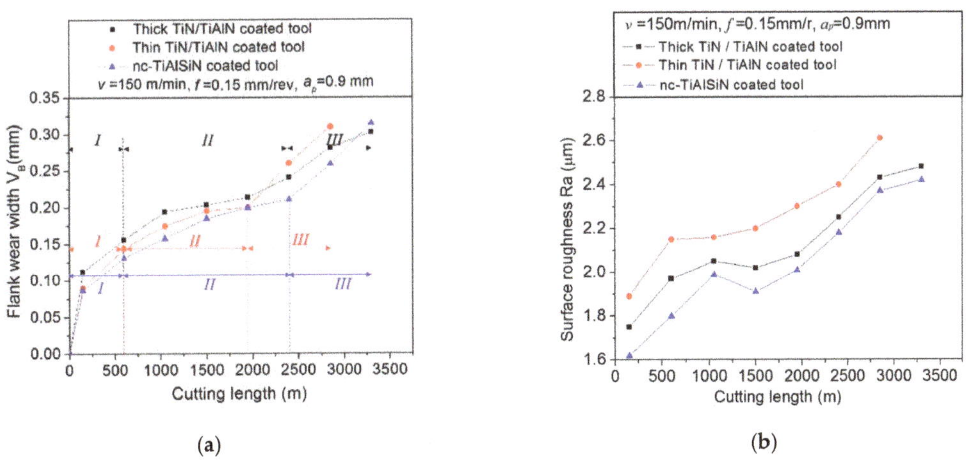

Figure 10. Coated tools performance comparision: (**a**) Flank wear, (I) initial wear, (II) stable wear, and (III) severe wear; (**b**) Surface roughness [65].

The coating can deteriorate and cause the tool to fail abruptly in processes that involve elevated mechanical and thermal stresses such as the turning process of hard materials. Boing et al. [61] studied the deterioration phenomena of tool coating materials TiCN/Al_2O_3/TiN using CVD and TiAlN using PVD, applied on the cemented carbide cutting substrate, during the machining of steels grade AISI 4340 which hardened 55 HRC. They found abrasion, crack initiation and growth, the development of crack grids, delamination, detachment of interlayer, and spalling as the prominent mechanisms for wear of the CVD-coated tool. For the PVD coating, the wear mechanism was associated with abrasion and high deformation at the cutting edge that led to coating cracks and weakening. In the present cutting conditions PVD inserts perform well compared to the CVD.

Sahoo and Datta [19] compared the dry machining performance of two types of coated inserts: CVD-produced multilayer coated: TiCN-Al_2O_3-TiOCN and PVD-produced single layer: TiN-coated composite ceramic (Al_2O_3/TiCN) on Ti-6Al-4V alloy. They found that the coated ceramic tool by the PVD method was more appropriate for machining speeds up to 80 m/min, whereas the CVD-coated carbide tool performed better at higher speeds. Sahoo and Datta [19] also suggested that adding a chip-breaker and a multilayered coating matrix could further enhance the ceramic tool's performance at high speeds.

A duplex stainless steel, DSS 2205 is a popular stainless steel because of its high strength and low cost [66,77]. However, due to its low ductility, thermal conductivity, and high toughness, machining in the dry environment is difficult. Sonawane and Sargade [66] examined the usage of AlTiCrN- and AlTiN-coated tools over uncoated tools and observed significant improvements in tool lifespan, cutting forces, and surface polish, showing the potential of these coatings in the dry environment. Parsi et al. [77] found that the PVD-coated AlTiN insert performed better than the coated-with-AlCrN and uncoated inserts in a low range of cutting speeds, resulting in lengthier tool lifespan, improved surface finish and reduced chip thickness. They also observed that AlCrN-coated and AlTiN-coated inserts formed Al–Cr oxides and Al–Ti oxides, respectively. These oxides increased the resistance for abrasion of AlCrN coatings more than AlTiN coatings. However, AlCrN coatings had more adhesive wear due to the chemical interaction with Cr in SDSS, which caused more crater wear on the cutting edge.

The thickness of coating is also an imperative influencing aspect on the machining effectiveness of coated inserts. Sargade et al. [78] investigated the effect of coating thickness on the effectiveness of inserts under dry cutting of steel grade C40. They found that the optimal coating thickness for carbide inserts coated by TiN during the dry machining of C40 steel is 4 μm, resulting in a significant tool life improvement, higher critical load, and better coating retention. This thickness also showed less wear and damage on both surfaces (rake and flank).

A review of the existing literature reveals that researchers have explored and assessed the performance of various kinds of coatings for machining different materials, such as conventional, nanocomposites, and self-lubricative coatings. The performance of the coatings depends on several factors, such as layer ordering in multilayer coatings, coating thickness, composition ratio, and deposition method. Moreover, there is a growing interest in developing multilayer and nanocomposite coatings that can improve the attributes of single-layer coatings and provide better performance in relation to tool lifespan, quality of surface, and cutting forces.

4.3. Advancement in Cutting Tool Material

The quality of tool materials is a key factor for the progress of metal-machining methods and the expansion of the manufacturing sector, especially machine tools. High-quality tool materials are essential for the progress of metal-machining strategies and the machining industry. Tool material technology is advancing rapidly, so novel cutting materials for tools are needed for different economic and technical situations. This would enable machining of materials which possess poor machinability and higher cutting parameters for better productivity [62]. For decades, cemented carbide has been a high-performance material for cutting tools, with good properties such as compression strength, modulus of Young's, bending endurance, conductivity of heat, and impact resistance, though it loses its hardness at elevated temperature, which restricts its use in dry and machining processes involving high speed [79]. A major challenge in dry machining is to improve the tool life, where the cutting tools are exposed to extreme heat and wear. Due to the dry machining, the tool and the workpiece experience higher levels of friction and adhesion, resulting in severe heat production and subsequently lessening in tool lifespan. To address this challenge, several researchers around the globe are exploring new types of cutting tools and related innovations that can withstand such harsh environments. A possible way to enable dry machining is to utilize/develop new/advanced tool materials that can lower heat generation and decrease tool wear by reducing friction coefficient and resist the severe conditions of high temperature and friction. Recent developments in cutting tool materials have shown promising results in reducing friction coefficient and improving machining performance. The following Sections 4.3.1 and 4.3.2 focus on such recent advancements in developing and evaluating novel cutting tools/tool materials for dry machining applications.

4.3.1. New Cutting Tool Material Advancement

Iron-based cermet has many benefits over Co or Ni, such as being non-toxic, cheap, and hardenable by heat treatment. This can result in high hardness with less ceramic phase. High-speed steel reinforcement through hard materials is a popular choice in commercial cutting tools. The M2 is a well-known high-speed steel with a prominent behavior of sintering and responsiveness to heat treatment. Canteli et al. [80] demonstrated the potential of the new base material, i.e., cermet (M2 + 50 vol.% TiCN), to improve the cutting processes during the machining of steel St 35 and compared to the without-reinforcement plain M2 and commercially available HSS M2 and found significant improvement in tool life. This indicates that HSS reinforced with hardened materials can be a suitable choice for dry machining and needs more investigations like optimum percentage of reinforcement, tool geometry, etc. Further, the cutting speed range of this study was only around 30–60 m/min, and more research on higher cutting speed ranges is required.

PCBN tools offers an impressive hardness and good thermal conductivity, while ceramic tools exhibit mechanical properties similar to PCBN [81]. Both tools are widely utilized in industries for hardened steel dry machining. However, PCBN tools are very expensive, and ceramic tools have low toughness. Another option is Ti (C, N)-based cermet, which offers various benefits like high-temperature hardness, wear-resistive ability, chemical stability, and a low friction coefficient. However, Ti (C, N)-based ceramic tools must enhance their hardness and abrasion resistance at temperatures above 1000 °C. The addition of Al_2O_3 particles to the cermet can enhance its high-temperature mechanical properties and wear resistance [81]. Microwave sintering is a novel technique for producing Ti (C, N)-based cermet, which is widely used in place of conventional hot-pressed sintering. Unlike the conventional sintering process, where heat originates from the heating source to the specimen through conduction and thermal stress, microwave sintering generates heat within the whole specimen by electromagnetic wave coupling [81,82]. This reduces the temperature gradient and improves the mechanical attributes of the ceramic material [82].

β-SiAlON is a type of Sialon that can dissolve β-Si3N4 in its structure. It has better resistance to thermal shock, high-temperature wear, and oxidation than β-Si3N4 because it has a lower thermal expansion coefficient. Therefore, β-SiAlON ceramics show their potential for cutting at high speeds for high-temperature alloys. A novel β-SiAlON ceramic tool with good machining qualities was developed by Yin et al. [82] using microwave sintering. The β-SiAlON ceramic tool includes 5 weights per cent Y_2O_3, and 0.5 weight per cent Yb_2O_3 sintered additives had a uniform microscopic structure and optimal mechanical attributes and can be utilized as cutting inserts regardless of being smoothed and polished. The productivity of the β-SiAlON ceramic tool was found to be 5-fold more as compared to the nearest companion TiAlN-coated tool during the machining of Inconel 718 at high speeds.

A new ceramic Al_2O_3-YAG ($Y_3Al_5O_{12}$) cutting tool was designed by de Sousa et al. [83] for the turning of nodular cast iron in the dry cutting environment. The tool consists of 85% Al_2O_3 and 15% YAG, which were mixed, pressed, and sintered at a temperature of 1600 °C for the duration of 2 h with a rate of heating of 5 °C/min. The tool exhibits high fracture toughness due to the fine YAG grains evenly dispersed along the Al_2O_3 grain boundaries.

The tool also shows good thermal compatibility with the Al_2O_3 ceramic matrix. The tool lifespan, quality of surface, and forces related to machining found with the Al_2O_3-YAG tool are superior compared to the commercially available carbide tools for all ranges of the cutting speeds, implying that the improved tool is more affordable and effective for the dry turning of nodular cast iron.

Machining of heat-treated AISI D3 tool steel in the absence of any cooling agents is difficult as it poses good hardness and abrasiveness. Conventional tool inserts such as CBN and ceramic experience rapid tool wear and high cutting temperatures at cutting speeds of more than 200 m per minute. Rath et al. [84] used a novel hybrid insert composed of Al_2O_3+Ti (C,N) with a TiN coating for this process. The experimental results show that the hybrid insert outperforms conventional inserts in relation to quality of machined surface, wear resistance ability, and cutting force at 175 m per minute cutting speed.

Ti(C, N)/Al_2O_3 composite cermet is a potential tool alternative for the dry machining of hardened steel. It has high hardness, good resistance against wear, and does not lose its chemical stability at elevated temperatures. Y. Zhang et al. [81] prepared this cermet by microwave sintering and tested its cutting effectiveness and wear mechanisms on 40Cr steel (AISI 5140). The cermet tool has a much smaller wear area than the commercial cemented carbide tools YS-8 and YT-15. The YS8 tool has a composition of 94%WC and 6%Co, while YT15 has 15%TiC, 79%WC, and 6%Co.

The outcomes revealed that the cermet tool had a longer tool life of 64.5 min, which was about twofold that of YS-8 and YT-15. The average quality of the machined surface was found to be 15.9% less compared to the cemented carbide tools.

By the addition of nano-sized Si3N4 and TiC elements into a micro-sized Si3N4 matrix, Lü et al. [85] created Si3N4/TiC micro-nanocomposite ceramic tool materials (ST28). The ST28 inserts show superior cutting performance compared to conventional inline SNM88 tools during the turning of hardened steel of 61 HRC. The Si3N4/TiC cutting tool materials have outstanding attributes that result from the combined influence of lengthened microscopic structure and nanoscale particle refinement on matrix grains. The worn morphology of ST-28 and SNM-88 cutting tools demonstrates the effects of these properties on enhancing the wear resistance and inhibiting the crack growth. These properties result from the addition of nano-sized particles and the formation of a composite structure.

Sun et al. [86] developed ceramic tools showing a remarkable improvement in tool life, which was 14-fold longer than the commercially carbide tool at 400 m/min cutting speed. Moreover, they achieved a maximum metal removal rate of 8.15 cm^3/min, which means that the ceramic tool could cut 1.5-fold faster compared to the carbide tool on average. The ceramic tool also reduced the cutting force by 13.5%, but the quality of surface of the machined part was slightly worse than the carbide tool.

High-speed machining enhances productivity but poses challenges for ceramic tools, which can reach temperatures above 1200 °C when cutting nickel-based superalloys [86,87].

The intense temperature exposes the ceramic tool to thermal cracking due to sharp thermal gradients and sometimes leads to tool fracture. The addition of SiC into ceramics as a healing agent can oxidize into SiO_2 at high temperature. The oxidation of SiC causes an 80% volume expansion that can fill the cracks [88]. Zhao et al. [88] developed a new Al_2O_3-ceremet ceramic tool (Al_2O_3/SiCw/TiCn) with whiskers by SiC and TiC nanoscale elements reinforcements and tested it in the cutting of Inconel 718 at high speeds and the dry cutting environment. The tool showed stable performance at 400 m/min speed and shows a tool life 50% longer than the commercial CC670 tool with SiC whisker toughening. They reported that the high cutting temperature (1200 °C) oxidized the SiC whisker to form a glassy segment that healed and passivated the tool cracks, preventing crack growth and tool failure.

Cemented carbide with Co and CaF_2 composition gradient, also known as self-lubricating FGCC, is a promising material in the absence of any coolant, i.e., dry machining as a cutting tool insert [89]. It has high hardness and solid lubricant on the surface, and high toughness in the core. The fabrication of self-lubricating FGCC is challenging due to the migration and homogenization of Co, which can be overcome by using the spark plasma sintering (SPS) technique [79]. The design and distribution of residual thermal stresses in the gradient structure are crucial for the performance of self-lubricating FGCC and need to be evaluated by accurate and efficient models. In the Singh Parihar et al. [79] study, a power law composition gradient model was used to design the self-lubricating FGCC with optimal configuration gradient and desired residual stress dispersal. The results indicated that the self-lubricating FGCC had compressive stress on the surface and core, which demonstrated the effectiveness of the proposed structural design and the SPS technique for developing FGCC with solid lubricant gradient.

4.3.2. New Technical/Design Advancements in Cutting Tools

The friction at the tool–chip interface during dry machining can result in high cutting temperatures and extreme tribological conditions, resulting in adhesive layer BUE and increased tool wear. This is among the most crucial variables influencing machining performance and quality. Pang et al. [26] designed the variable-length restricted contact tools (VRCT) to improve the effectiveness of the machining process and wear resistance in dry cutting of austenitic stainless-steel grade AISI 316L. The VRCTs limit the contact length to less than the normal extent, which improves the tribological attributes and tools durability. The VRCTs reduce the cutting force (~20%), cutting temperature (~17%), and flank wear (~44.18%) in contrast to traditional tools.

Metal machining is an energy-intensive process that generates a lot of heat in the cutting zone. The substantial percentage of the heat is carried out by the chips, but the rest of the heat affects the tool and the workpiece [12]. In dry machining, no coolant is used to cool down the cutting region. To enhance the cooling effect, a heat-pipe-embedded cooling arrangement can be implanted in the tool as a sustainable means of cooling the cutting zone and improving machining performance. A graphic illustration of the heat-pipe-embedded cutting tool system is shown in Figure 11.

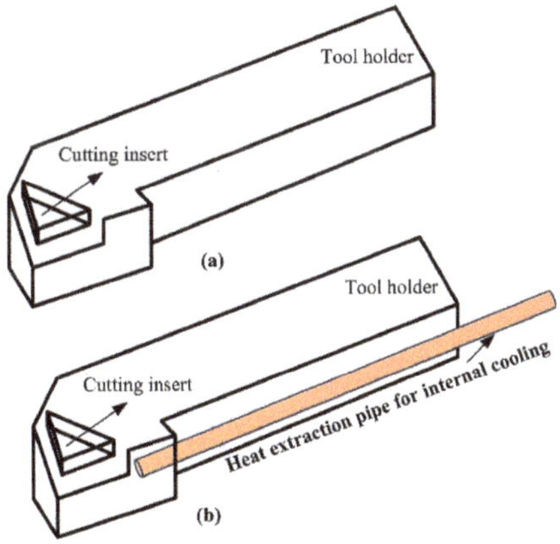

Figure 11. (**a**) Conventional cutting tool. (**b**) Schematic arrangement for the internally cooled cutting tool.

Quan and Mai [90] studied how this system works with different cutting tool–materials combinations and cutting parameters. They found that the heat pipe transferred about 40% of the tool heat, dropping the temperature about 50 °C at the rake surface and the maximum temperature of the tool in the range of 200 °C. However, the heat transfer efficiency and the cooling effect depended on the type of metal and the cutting conditions, as Quan and Mai [90] reported that M05-hardened steel AISI 5140 had the highest efficiency and cooling effect, followed by P10-AISI 1045, while K20-AlSi Alloy had the lowest.

One of the challenges of the machining of aluminum alloys in dry environment is the strong adhesion of chips to the cutting tool rake area, which interrupts the cutting effectiveness undesirably. To overcome this situation, Sugihara et al. [23] anticipated a novel method to remove the aluminum-embedded chips from the tool rake zone by using liquid metal embrittlement. This method involves applying liquid gallium to the tool surface when the tool remains in its position to weaken the bond between the aluminum and the tool substrate. The experiments showed that this method can efficiently eradicate the aluminum adhesion deposit without damaging the tool substrate and restore the cutting effectiveness of the tool to a level similar to its original state during dry machining of aluminum alloys. This approach works effectively in the machining of aluminum alloys under the dry cutting environment, and it can be utilized to resharpen the cutting tools during the dry machining of other sticky materials such as stainless steel.

The researchers have made significant efforts to develop new or advanced cutting tool materials that can enable dry machining. The cutting tool advancement work covers various aspects, and some studies have suggested that polycrystalline diamond [81,88] and ceramic tools [84,85] can replace cemented carbide tools in a sustainable way. These cutting tools form protective layers when they react with the workpiece material [29]. Another option is to use SiC-whiskered reinforced alumina-based mixed ceramic tools, which have better fracture toughness and resistance ability to thermal shocks [29,84,88]. Moreover, microwave-sintering-based cutting tools have shown promising performance in the dry cutting environment [81], and some researchers proposed an inbuilt cooling system for dry machining [90]. Some researchers have proposed novel methods to enhance the dry machining performance of iron-based cermet tools, such as removing the metal deposited chips from the tool rake zone by liquid metal embrittlement [23] and designing variable-length restricted contact tools (VRCT) that reduce the cutting heat generation and tool wear [26].

4.4. Energy-Assisted/Hybrid Dry Machining

In the last few years, dry cutting has gained popularity, as it can lower the manufacturing cost and avoid the negative impacts of cutting fluids. However, dry machining faces challenges like low tool life, intense cutting force, and deteriorated quality of surface. Therefore, new hybrid machining processes are emerging to overcome these difficulties and achieve cost-effective machining in the dry cutting environment [31]. The constraints of dry machining may be addressed by hybrid or aided machining technologies. Hybrid/assisted machining is a method of improving productivity, quality of surface, and tool lifespan by combining conventional machining methods with alternative energy sources such as vibration and/or heat [32,91]. Some examples of key energy-assisted dry machining are ultrasonic vibration-assisted (UVA) machining [32] and laser-assisted (LA) machining [92,93].

In UVAM, one-directional or bi-directional ultrasonic vibrations are applied to either the cutting tool or the workpiece [91]. It can be employed for drilling, turning, and milling to improve the machining performance. Figure 12 depicts a UVAM system with a piezoelectric transducer (PZT) embedded in the tool shank to stimulate the tool in any desired direction with respect to the machining axes. Exemplary, UVA milling achieves recurrent cutting by utilizing high-frequency vibration to the component being machined or the milling cutter. This creates high-frequency intermittent gaps between the cutter and the chip, which reduce tool–chip stickiness, cutting force, and heat generation. UVAM also lowers tool wear and enhances the quality of machining [32].

LA machining (LAM) is a method that uses a focused laser beam to locally preheat the component being machined before cutting [92]. Figure 13 shows a laser-assisted turning experimental setup illustration.

This softens the material and changes the characteristics of the shear zone, improving machining performance [18,92]. It is widely recognized that LAM of titanium-based alloys shows a significant lessening of cutting force, ranging up to 50% restricted on the alloy being machined and degree of heating [94]. The most common types of lasers utilized are carbon dioxide (CO_2) lasers, high-power semiconductor lasers (HPDL), and neodymium-doped yttrium aluminum garnet (Nd:YAG) lasers. The next paragraphs cover some important research conducted on UVAM and LAM.

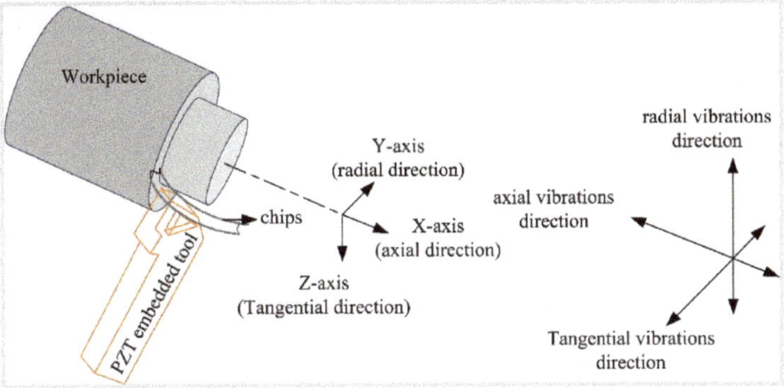

Figure 12. Ultrasonic vibration-assisted turning process schematic illustration.

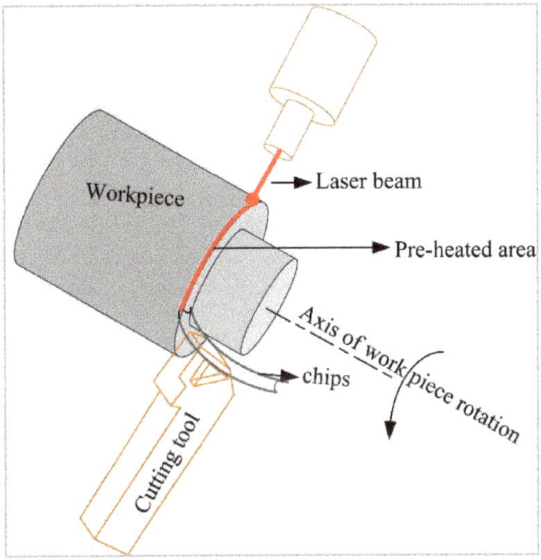

Figure 13. A schematic illustration of laser-assisted turning process.

A novel method of longitudinal bending hybrid ultrasonic vibration-assisted milling (LBVAM) of titanium-based alloy grade TC4 under the dry cutting environment was investigated by Hu et al. [32]. The LBVAM process is analyzed by simulated and analytical approaches and validated with experimental validations to study the kinematics and the material removal mechanisms. The results show that LBVAM can significantly improve the quality of surface and the cutting force compared to conventional milling. Moreover, LBVAM can achieve a surface roughness of 100 nm with optimal parameters, which is beneficial for clean dry manufacturing of titanium-based alloy grade TC4 in the absence of flooded coolants. A novel technology called UVA-elliptical in nature-assisted cutting (UEVC) was used by Tan et al. [95] to overcome the challenges of conventional high-precision cutting of titanium-based alloy grade Ti6Al4V, which is ideal but challenging due to the severe tool wear and surface damage produced by very high cutting temperature. Tan et al. [95] used a commercial PCD tool to attain cost-effective and sustainable machining. The UEVC reduced the frictional forces, cutting forces and dissipated the cutting heat faster, leading to a lesser temperature and less tool stress. The experiments showed that the UEVC technology improved the tool durability and quality of surface, and produced a flawless ultra-precise superior surface and achieved roughness value less than 30 nm. The amplitude and direction of UVA are critical for UVA machining because they affect the energy and cutting force supplied to the workpiece. The ultrasonic amplitude effects material removal chip formation indirectly through its relationship to cutting temperature. The cutting temperature influences chip formation because temperature changes the flow-softening behavior of the material. Gao et al. [91] investigated the impacts of amplitude on cutting force, temperature rise, and quality of surface for Ti-6Al-4V. When the amplitude is increased from 0 to 6 m, the cutting force, cutting temperature, and surface topography improve by 14.6, 20.3, and 35.1%, respectively.

Modulation-assisted machining (MAM) is a new type of energy-assisted process that uses a low-range frequency vibration (<500 Hz) of the tool to create intermittent cutting [96]. The vibration is employed to the cutter movement during the machining, which can be any continuous process like turning, boring, or drilling. By doing this, the continuous process is broken down into smaller segments of cutting. Sandoval et al. [96] examined the influence of MAM on wear resistance ability during the machining of graphite iron under dry environment. The outcomes indicates that the performance of MAM machining is better compared to the CM, and the wear lessening using MAM is negligible at lower cutting speeds, but it is outperformed at higher levels. This research shows the possibility of MAM in improving the efficiency of machining processes and can be explored in machining of other materials dry machining.

Attia et al. [92] investigated and compare the performance of laser-assisted machining (LAM) dry machining and conventional on Inconel 718 and found that the LAM decreases the cutting forces significantly, enhances the quality of surface by more than 25%, and allows for a much higher material removal rate with a significant improvement of 800%. Besides, the LAM attributed better surface characteristics to the lack of dispersed material and the enlarged plastic deformation region, which result in compressive residual stresses. Neeraj Deswal and Kant [97] investigated the machining effectiveness of aluminum alloy grade Al3003 LA-Turning (LAT). They found that LAT can improve the machinability of Al alloys in relation to reducing forces, temperature and improving the tool wear-resistive ability, morphology of chips, and quality of surface if the parameters are chosen carefully, which can otherwise result in poor machinability. Therefore, they emphasized the importance of selecting the optimal parameters for achieving a high performance. The Venkatesan et al. [31] investigation regarding LAM and CM reports that LAM improves the machining performance significantly. Further, the chip analysis indicates that the lower force, surface roughness, and flank wear are due to the increased chip thickness and reduced shear angle in LAM compared to CM. This is because the material is softer in LAM because of the higher temperature, reducing friction between the tool and the workpiece. Deswal and Kant [93] studied the LA-turning of AZ31B magnesium alloy and compared it with the CT. They compared the conventional turning (CT) and the LAT methods in terms of various parameters and found that LAT performed better than CT in all aspects, as it reduced the machining forces and improved wear-resistance ability by thermally softening the workpiece material.

Some researchers [98] have developed a hybrid Laser-Ultrasonic-assisted machining (LUAM) method that combines LAM with UVAM to capitalize on the benefits of both techniques. Figure 14 shows a novel hybrid arrangement of ultrasonic vibration and laser for machining assistance.

Deswal and Kant [98] conducted a comparison between traditional turning (CT), laser-assisted turning (LAT), ultrasonic vibration-assisted turning (UVAT), and the hybrid process ultrasonic vibration-laser-assisted turning (UVLAT). They reported that UVLAT achieves lower cutting forces and produces better quality of surface for a wider value of cutting parameters. This is due to the hybrid process's synergistic effects of thermal softening and force reduction. The chip morphology with different machining process CT,

LAT, UVAT, and UVLAT is illustrated in Figure 15. The UVLAT produced crack-free chip edges, outperforming the CT, LAT, and UVAT processes.

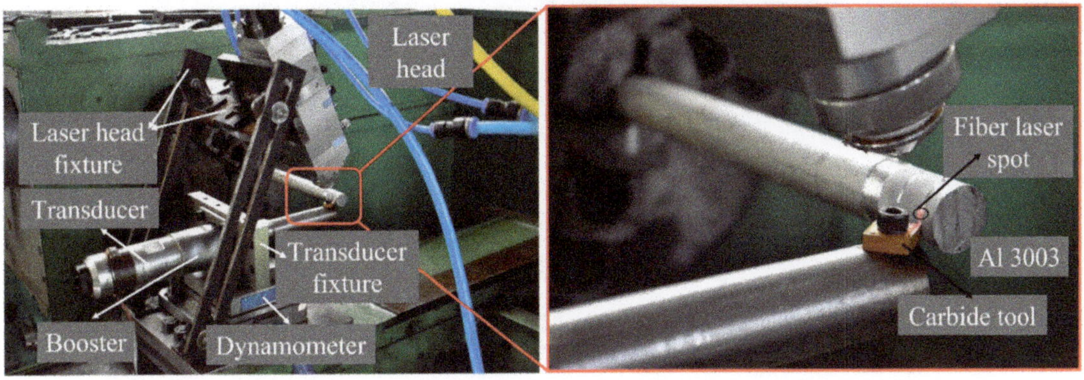

Figure 14. LUAM experimental setup [98].

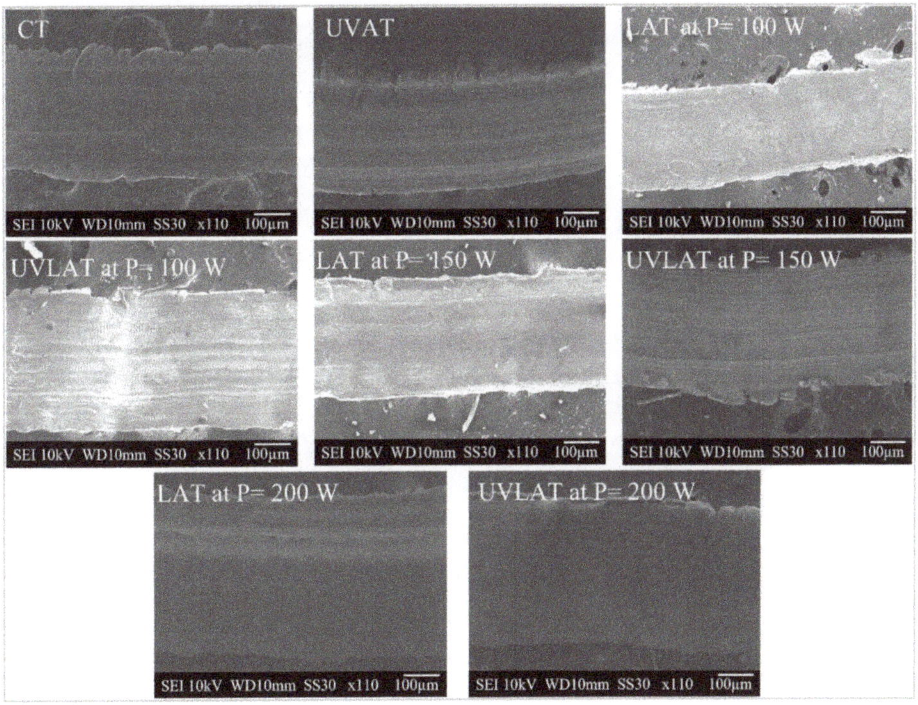

Figure 15. Chip morphology with CT, LAT, UVAT, and UVLAT [98].

To enhance the dry cutting of engineering materials, various methods have been developed that apply external energy to the machining process, i.e., assistive/hybrid machining process. In recent years, two of these methods, UVAM and LAM, have gained interest from researchers in the metal cutting field. The above assessment shows that the combination of energy-assisted machining and advanced cutting tools can improve machining performance and be a step towards overcoming the difficulties of dry cutting.

4.5. Discussion

The preceding sections demonstrate important advances in several dry cutting processes and techniques for achieving sustainability in metal cutting. However, each method has its own advantages and disadvantages in terms of their implementation ease, cost-effectiveness, and environmental impacts. A comparative analysis of the dry machining techniques considered in this study is presented in Table 1.

Table 1. A comparative analysis of different dry machining techniques.

Dry Machining Technique	Ease of Implementation	Cost-Effectiveness	Environmental Impacts
Cutting tool surface structuring	Easy to implement. There is adequate technical innovation and advanced machining processes that can engrave specific texturing on the tool surface [38].	Requires equipment such as electrical discharge machining, electro-chemical machining, and laser beam machine for engraving the tool surfaces. Although, these advanced machine tools are costly, yet texturing and patterning on these are affordable.	There are some environmental emissions and safety hazards while structuring by advanced machining processes. Furthermore, structured cutting tools minimize tool replacement and resharpening while machining and therefore minimize the resource consumption.
Cutting tool coatings	Easy to implement. The two most prevalent coating techniques are physical vapor deposition (PVD) and chemical vapor deposition (CVD).	The cost associated with tool coating is a bit high due to the use of expensive machines	The PVD process is environmentally clean, but the CVD process causes pollution from chemicals [99]. The coating deposition process requires a significant amount of power/energy [100]. But coated tools, while in use, minimize the consumption of energy and resources.
Advancement in tool material	Implementation is difficult and requires expertise. It involves a deep understanding of material science and the ability to conduct rigorous and innovative experiments.	The key costs associated with the development of new cutting tools includes the cost of materials, manufacturing, and testing and analysis of different alternatives [82,83].	The environmental implications of using advanced tool materials are not uniform and vary based on the kind and technique of the tool. For example, microwave-sintered tools and conventional sintered tools have various environmental effects. But the enhanced mechanical properties of the tools due to advanced materials minimize their wear and failure. It helps to reduce the machining cost specially while dealing with difficult-to-machine materials.
Energy-Assisted/hybrid Dry Machining	Implementation is moderately difficult. There is a need to build a novel setup that combines conventional machining methods with energy-assisted techniques.	The setup requires expensive equipment such as a laser energy field and an ultrasonic energy field.	The integrated laser energy fields have no direct harmful environmental impacts except safety hazard, although they only consume additional electricity.

5. Conclusions and Future Research Directions

This article has presented various techniques of dry cutting to achieve sustainability in metal cutting. A detailed review of the past work conducted in this area has also been discussed. Dry machining has some challenges, but its benefits are many that ensure the enhanced machinability to achieve overall sustainability in metal cutting. The following conclusions and future research directions can be drawn from this review:

- Textured or patterned tools significanlty improved the cutting performance of tools in dry machining for different materials. Previous studies have mainly focused on the rake surface texture, and few have explored the flank texture and the dual texture for dry machining. The most common textures used are dimple and straight grooves on the rake surface.
- However, there is no clear criterion for selecting the texture shape and size. This could be a future research topic to optimize the texture parameters and develop a selection guideline.
- Tool coating is a viable option for reducing the cost of tools in dry machining and can provide the desired functionality for the cutting tool without compromising its

core properties. Some of the commonly used coatings are TiN, AlTiN, TiAlN, TiCN, AlTiCrN, AlCrN, and Al_2O_3.

- Self-lubricative coatings, multilayered coatings, and nanocomposite coatings are attracting more attention in dry machining. There have been few investigations on these coatings, and further research is needed in light of dry machining at higher cutting speeds.
- The cutting tool advancement work covers various aspects:
  - Dry machining with polycrystalline diamond and ceramic tools rather than cemented carbide tools can be a more sustainable selection.
  - Alumina-based mixed ceramic tools reinforced with SiC whiskers offer better fracture toughness, thermal shock resistance, and self-crack healing properties. These tools can be an appropriate selection for machining in which temperatures exceed 1000 °C.
  - One possible way to enhance the performance of ceramic cutting tools is to create a micro-nanocomposite structure, e.g., nano-sized Si_3N_4 and TiC particles are dispersed in a micro-sized Si_3N_4 matrix. This micro-nanocomposite ceramic tool poses higher hardness, wear resistance, and thermal conductivity than conventional ceramic tools, but there is a lack of research on its properties and applications.
  - By using microwave sintering, it is possible to produce cutting tools with finer microstructures and higher densities, which improve their hardness, wear resistance, and thermal stability.
  - Cutting tools with an integrated cooling system and variable-length restricted contact tools (VRCT), which reduce cutting temperature and tool wear.
  - One way to improve the dry cutting of engineering materials is to use external energy sources to assist the dry machining process, i.e., energy-assistive or hybrid machining. Two examples of this technique are UVAM and LAM, which show potential to overcome the limitations of dry machining.

The effects of combining two different machining strategies, such as texturing and coating, texturing and self-lubricating coating, or texturing on advanced tool material, energy-assistive dry machining with these combinations, on the performance of dry machining may be interesting and deserve further exploration. These combinations may offer advantages in terms of obtaining friction reduction, wear resistance, and surface quality while machining various difficult-to-machine materials.

Author Contributions: Conceptualization S.P. and K.G.; methodology and review, S.P.; writing—first draft, S.P.; editing and supervision, K.G. All authors have read and agreed to the published version of the manuscript.

Funding: This research received no external funding.

Data Availability Statement: The research data will be made available upon request.

Acknowledgments: The authors would like to acknowledge the technicians and lab staff of the department of mechanical and industrial engineering technology at the University of Johannesburg.

Conflicts of Interest: The authors declare no conflicts of interest.

References

1. U.S. Department of Commerce. Introduction to Sustainable Manufacturing. International Trade Administration, 2011. Available online: https://legacy.trade.gov/green/sm-101-module.asp (accessed on 10 January 2024).
2. Gupta, K.; Laubscher, R.F.; Davim, J.P.; Jain, N.K. Recent developments in sustainable manufacturing of gears: A review. *J. Clean. Prod.* **2016**, *112*, 3320–3330. [CrossRef]
3. Peng, T.; Xu, X. Energy-efficient machining systems: A critical review. *Int. J. Adv. Manuf. Technol.* **2014**, *72*, 1389–1406. [CrossRef]
4. Halim, N.H.A.; Haron, C.H.C.; Ghani, J.A. Sustainable Machining of Hardened Inconel 718: A Comparative Study. *Int. J. Precis. Eng. Manuf.* **2020**, *21*, 1375–1387. [CrossRef]

5. Warsi, S.S.; Jaffery, S.H.I.; Ahmad, R.; Khan, M.; Ali, L.; Agha, M.H.; Akram, S. Development of energy consumption map for orthogonal machining of Al 6061-T6 alloy. *Proc. Inst. Mech. Eng. Part B J. Eng. Manuf.* **2018**, *232*, 2510–2522. [CrossRef]
6. Shokrani, A.; Dhokia, V.; Newman, S.T. Environmentally conscious machining of difficult-to-machine materials with regard to cutting fluids. *Int. J. Mach. Tools Manuf.* **2012**, *57*, 83–101. [CrossRef]
7. Pusavec, F.; Kramar, D.; Krajnik, P.; Kopac, J. Transitioning to sustainable production—Part II: Evaluation of sustainable machining technologies. *J. Clean. Prod.* **2010**, *18*, 1211–1221. [CrossRef]
8. Pawanr, S.; Garg, G.K.; Routroy, S. Development of an empirical model to quantify carbon emissions for machining of cylindrical parts. *Environ. Sci. Pollut. Res.* **2023**, *30*, 21565–21587. [CrossRef] [PubMed]
9. Gupta, K. A review on green machining techniques. *Procedia Manuf.* **2020**, *51*, 1730–1736. [CrossRef]
10. Khanna, N.; Wadhwa, J.; Pitroda, A.; Shah, P.; Schoop, J.; Sarıkaya, M. Life cycle assessment of environmentally friendly initiatives for sustainable machining: A short review of current knowledge and a case study. *Sustain. Mater. Technol.* **2022**, *32*, e00413. [CrossRef]
11. Nouzil, I.; Eltaggaz, A.; Pervaiz, S.; Deiab, I. Toxicity Analysis of Nano-Minimum Quantity Lubrication Machining—A Review. *Lubricants* **2022**, *10*, 176. [CrossRef]
12. Goindi, G.S.; Sarkar, P. Dry machining: A step towards sustainable machining—Challenges and future directions. *J. Clean. Prod.* **2017**, *165*, 1557–1571. [CrossRef]
13. Pandey, K.; Datta, S. Performance of Si-doped TiAlxN supernitride coated carbide tool during dry machining of Inconel 718 superalloy. *J. Manuf. Process.* **2022**, *84*, 1258–1273. [CrossRef]
14. Kannan, S.; Ghosh, A. Anti-friction and wetting behavior of a new polymer composite coating towards aluminium and dry machining of AA2024 alloy by coated end mills. *J. Mater. Process. Technol.* **2018**, *252*, 280–293. [CrossRef]
15. Cavaleiro, D.; Figueiredo, D.; Moura, C.W.; Cavaleiro, A.; Carvalho, S.; Fernandes, F. Machining performance of TiSiN(Ag) coated tools during dry turning of TiAl6V4 aerospace alloy. *Ceram. Int.* **2021**, *47*, 11799–11806. [CrossRef]
16. Ramachandran, K.; Yeesvaran, B.; Kadirgama, K.; Ramasamy, D.; Ghani, S.A.C.; Anamalai, K. State of art of cooling method for dry machining. *MATEC Web Conf.* **2016**, *90*, 01015. [CrossRef]
17. Dhar, N.R.; Paul, S.; Chattopadhyay, A.B. The influence of cryogenic cooling on tool wear, dimensional accuracy and surface finish in turning AISI 1040 and E4340C steels. *Wear* **2001**, *249*, 932–942. [CrossRef]
18. Zhang, J.; Huang, X.; Kang, X.; Yi, H.; Wang, Q.; Cao, H. Energy field-assisted high-speed dry milling green machining technology for difficult-to-machine metal materials. *Front. Mech. Eng.* **2023**, *18*, 28. [CrossRef]
19. Sahoo, S.P.; Datta, S. Comparative experimental study on application feasibiility of MTCVD TiCN-Al2O3-TiOCN multi-layer coated carbide and PVD TiN single layer coated composite ceramic inserts during dry machining of Ti-6Al-4V. *Sadhana Acad. Proc. Eng. Sci.* **2022**, *47*, 105. [CrossRef]
20. Yuan, J.; Fox-Rabinovich, G.S.; Veldhuis, S.C. Control of tribofilm formation in dry machining of hardened AISI D2 steel by tuning the cutting speed. *Wear* **2018**, *402*, 30–37. [CrossRef]
21. Weinert, K.; Inasaki, I.; Sutherland, J.W.; Wakabayashi, T. Dry machining and minimum quantity lubrication. *CIRP Ann. Manuf. Technol.* **2004**, *53*, 511–537. [CrossRef]
22. Umbrello, D. Investigation of surface integrity in dry machining of Inconel 718. *Int. J. Adv. Manuf. Technol.* **2013**, *69*, 2183–2190. [CrossRef]
23. Sugihara, T.; Nishimoto, Y.; Enomoto, T. On-machine tool resharpening process for dry machining of aluminum alloys employing LME phenomenon. *Precis. Eng.* **2015**, *40*, 241–248. [CrossRef]
24. Sreejith, P.S.; Ngoi, B.K.A. Dry machining: Machining of the future. *J. Mater. Process. Technol.* **2000**, *101*, 287–291. [CrossRef]
25. Song, W.; Deng, J.; Zhang, H.; Yan, P.; Zhao, J.; Ai, X. Performance of a cemented carbide self-lubricating tool embedded with MoS2 solid lubricants in dry machining. *J. Manuf. Process.* **2011**, *13*, 8–15. [CrossRef]
26. Pang, X.; Zhang, B.; Li, S.; Zeng, Y.; Liu, X.; Shen, P.; Li, Z.; Deng, W. Machining performance evaluation and tool wear analysis of dry cutting austenitic stainless steel with variable-length restricted contact tools. *Wear* **2022**, *504*, 204423. [CrossRef]
27. Jahaziel, R.B.; Krishnaraj, V.; Sudhagar, S.; Priyadarshini, B.G. Improving dry machining performance of surface modified cutting tools through combined effect of texture and TiN-WS2 coating. *J. Manuf. Process.* **2023**, *85*, 101–108. [CrossRef]
28. Zhao, J.; Liu, Z.; Wang, B.; Song, Q.; Ren, X.; Wan, Y. Effects of Al content in TiAlN coatings on tool wear and cutting temperature during dry machining IN718. *Tribol. Int.* **2022**, *171*, 107540. [CrossRef]
29. Kumar, C.S.; Urbikain, G.; de Lacalle, L.N.L.; Gangopadhyay, S.; Fernandes, F. Investigating the effect of novel self-lubricant TiSiVN films on topography, diffusion and oxidation phenomenon at the chip-tool interface during dry machining of Ti-6Al-4V alloy. *Tribol. Int.* **2023**, *186*, 108604. [CrossRef]
30. Siju, A.S.; Gajrani, K.K.; Joshi, S.S. Dual textured carbide tools for dry machining of titanium alloys. *Int. J. Refract. Met. Hard Mater.* **2021**, *94*, 105403. [CrossRef]
31. Venkatesan, K.; Ramanujam, R.; Kuppan, P. Investigation of machinability characteristics and chip morphology study in laser-assisted machining of Inconel 718. *Int. J. Adv. Manuf. Technol.* **2017**, *91*, 3807–3821. [CrossRef]
32. Hu, W.; Du, P.; Qiu, X.; Zhao, X.; Hu, Z.; Zhang, J.; Liu, Y. Enhanced dry machinability of TC4 titanium alloy by longitudinal-bending hybrid ultrasonic vibration-assisted milling. *J. Clean. Prod.* **2022**, *379*, 134866. [CrossRef]
33. Chetan, S.; Ghosh, P.; Rao, V. Application of sustainable techniques in metal cutting for enhanced machinability: A review. *J. Clean. Prod.* **2015**, *100*, 17–34. [CrossRef]

34. Rajmohan, T.; Chakravarthy, V.V.K.; Nandakumar, A.; Kumar, S.D.S. Eco Friendly Machining Processes for Sustainability—Review. *IOP Conf. Ser. Mater. Sci. Eng.* **2020**, *954*, 012044. [CrossRef]
35. Bayraktar, S. Dry cutting: A sustainable machining technology. In *Sustainable Manufacturing*; Elsevier: Amsterdam, The Netherlands, 2021. [CrossRef]
36. Kawasegi, N.; Sugimori, H.; Morimoto, H.; Morita, N.; Hori, I. Development of cutting tools with microscale and nanoscale textures to improve frictional behavior. *Precis. Eng.* **2009**, *33*, 248–254. [CrossRef]
37. Singh, B.; Sasi, R.; Subbu, S.K.; Muralidharan, B. Electric discharge texturing of HSS cutting tool and its performance in dry machining of aerospace alloy. *J. Braz. Soc. Mech. Sci. Eng.* **2019**, *41*, 152. [CrossRef]
38. Özel, T.; Biermann, D.; Enomoto, T.; Mativenga, P. Structured and textured cutting tool surfaces for machining applications. *CIRP Ann.* **2021**, *70*, 495–518. [CrossRef]
39. Mishra, S.K.; Ghosh, S.; Aravindan, S. 3D finite element investigations on textured tools with different geometrical shapes for dry machining of titanium alloys. *Int. J. Mech. Sci.* **2018**, *141*, 424–449. [CrossRef]
40. Xing, Y.; Deng, J.; Zhao, J.; Zhang, G.; Zhang, K. Cutting performance and wear mechanism of nanoscale and microscale textured Al_2O_3/TiC ceramic tools in dry cutting of hardened steel. *Int. J. Refract. Met. Hard Mater.* **2014**, *43*, 46–58. [CrossRef]
41. Sun, F.; Qu, S.; Pan, Y.; Li, X.; Yang, C. Machining performance of a grooved tool in dry machining Ti-6Al-4 v. *Int. J. Adv. Manuf. Technol.* **2014**, *73*, 613–622. [CrossRef]
42. Fatima, A.; Mativenga, P.T. A comparative study on cutting performance of rake-flank face structured cutting tool in orthogonal cutting of AISI/SAE 4140. *Int. J. Adv. Manuf. Technol.* **2015**, *78*, 2097–2106. [CrossRef]
43. Siju, A.S.; Jose, S.; Waigaonkar, S.D. Experimental analysis and characterisation of chip segmentation in dry machining of Ti-6Al-4V alloy using inserts with hybrid textures. *CIRP J. Manuf. Sci. Technol.* **2022**, *36*, 213–226. [CrossRef]
44. Zhang, K.; Guo, X.; Sun, L.; Meng, X.; Xing, Y. Fabrication of coated tool with femtosecond laser pretreatment and its cutting performance in dry machining SLM-produced stainless steel. *J. Manuf. Process.* **2019**, *42*, 28–40. [CrossRef]
45. Zhang, K.; Guo, X.; Wang, C.; Meng, X.; Sun, L.; Xing, Y. Effect of Scale and Sequence of Surface Textures on the Anti-adhesive Wear Performance of PVD Coated Tool in Dry Machining SLM-Produced Stainless Steel. *Int. J. Precis. Eng. Manuf. Green Technol.* **2021**, *8*, 1571–1586. [CrossRef]
46. Soni, H.; Mashinini, P.M. An Analysis on Tool-Chip Interaction During Dry Machining of SS316 Using Textured Carbide Tools. *Arab. J. Sci. Eng.* **2021**, *46*, 7611–7621. [CrossRef]
47. Rajurkar, A.; Chinchanikar, S. Experimental Investigation on Laser-Processed Micro-Dimple and Micro-Channel Textured Tools during Turning of Inconel 718 Alloy. *J. Mater. Eng. Perform.* **2022**, *31*, 4068–4083. [CrossRef]
48. Rajbongshi, S.K.; Singh, M.A.; Sarma, D.K. A comparative study in machining of AISI D2 steel using textured and non-textured coated carbide tool at the flank face. *J. Manuf. Process.* **2018**, *36*, 360–372. [CrossRef]
49. Anand, A.; Kumar, A.; Yadav, S.; Jayal, A.D. Influence of Cutting Tool Flank Surface Texture in Dry Metal Machining. *Int. J. Surf. Eng. Mater. Technol.* **2013**, *3*, 2249–7250.
50. Ranjan, P.; Hiremath, S.S. An Experimental Investigation on Bio-inspired Structure Position Variation on Tool Surface during Turning of Difficult to Machine Materials. *J. Mater. Eng. Perform.* **2023**, *30*, 1–20. [CrossRef]
51. Fatima, A.; Zaheer, A.; Fahad, M. Comparative performance analysis of micro-structured carbide inserts in machining of EN19 alloy steel. *J. Braz. Soc. Mech. Sci. Eng.* **2019**, *41*, 405. [CrossRef]
52. Ma, J.; Duong, N.H.; Lei, S. Numerical investigation of the performance of microbump textured cutting tool in dry machining of AISI 1045 steel. *J. Manuf. Process.* **2015**, *19*, 194–204. [CrossRef]
53. Fatima, A.; Mativenga, P.T. Performance of flank face structured cutting tools in machining of AISI/SAE 4140 over a range of cutting speeds. *Proc. Inst. Mech. Eng. Part B J. Eng. Manuf.* **2016**, *230*, 3–18. [CrossRef]
54. Dhage, S.B.; Goindi, G.; Jayal, A.D.; Sarkar, P. Influence of Cutting Tool Flank Surface Texture in Mechanical Micromachining of AISI 1045 Steel. In Proceedings of the National Conference on Manufacturing: Vision for Future, Guwahati, India, 23 April 2013.
55. Khan, M.A.; Gupta, K. A study on machinability of nickel based superalloy using micro-textured tungsten carbide cutting tools. *Mater. Res. Express.* **2019**, *7*, 016537. [CrossRef]
56. Fatima, A.; Mativenga, P.T. Experimental study on cutting comparison of structured cutting tools in dry cutting of EN19 alloy steel. *Proc. Inst. Mech. Eng. Part B J. Eng. Manuf.* **2019**, *233*, 505–514. [CrossRef]
57. Aslantas, K.; Alatrushi, L.K.H. Experimental Study on the Effect of Cutting Tool Geometry in Micro-Milling of Inconel 718. *Arab. J. Sci. Eng.* **2021**, *46*, 2327–2342. [CrossRef]
58. Zhang, P.R.; Liu, Z.Q.; Guo, Y.B. Machinability for dry turning of laser cladded parts with conventional vs. wiper insert. *J. Manuf. Process.* **2017**, *28*, 494–499. [CrossRef]
59. Mohanty, A.; Gangopadhyay, S.; Thakur, A. On Applicability of Multilayer Coated Tool in Dry Machining of Aerospace Grade Stainless Steel. *Mater. Manuf. Process.* **2016**, *31*, 869–879. [CrossRef]
60. Claudin, C.; Rech, J.; Grzesik, W.; Zalisz, S. Characterization of the frictional properties of various coatings at the tool/chip/workpiece interfaces in dry machining of AISI 4140 steel. *Int. J. Mater. Form.* **2008**, *1*, 511–514. [CrossRef]
61. Boing, D.; de Oliveira, A.J.; Schroeter, R.B. Evaluation of wear mechanisms of PVD and CVD coatings deposited on cemented carbide substrates applied to hard turning. *Int. J. Adv. Manuf. Technol.* **2020**, *106*, 5441–5451. [CrossRef]
62. Sonawane, G.D.; Bachhav, R.; Barnwal, A. Correlating the Characterization and Machining Performance of Advanced PVD Coatings for Dry Turning. *JOM* **2023**, *76*, 313–326. [CrossRef]

63. Thakur, A.; Gangopadhyay, S.; Maity, K.P.; Sahoo, S.K. Evaluation on Effectiveness of CVD and PVD Coated Tools during Dry Machining of Incoloy 825. *Tribol. Trans.* **2016**, *59*, 1048–1058. [CrossRef]
64. Thakur, D.G.; Ramamoorthy, B.; Vijayaraghavan, L. Some investigations on high speed dry machining of aerospace material inconel 718 using multicoated carbide inserts. *Mater. Manuf. Process.* **2012**, *27*, 1066–1072. [CrossRef]
65. Li, S.; Lin, H.; Zhang, T.; Sui, J.; Wang, C. High-Speed Machining of Malleable Cast Iron by Various Cutting Tools Coated by Physical Vapor Deposition. *Chin. J. Mech. Eng.* **2021**, *34*, 46. [CrossRef]
66. Sonawane, G.D.; Sargade, V.G. Machinability Study of Duplex Stainless Steel 2205 During Dry Turning. *Int. J. Precis. Eng. Manuf.* **2020**, *21*, 969–981. [CrossRef]
67. Pattnaik, S.K.; Bhoi, N.K.; Padhi, S.; Sarangi, S.K. Dry machining of aluminum for proper selection of cutting tool: Tool performance and tool wear. *Int. J. Adv. Manuf. Technol.* **2018**, *98*, 55–65. [CrossRef]
68. Gong, F.; Zhao, J.; Ni, X.; Liu, C.; Sun, J.; Zhang, Q. Wear and breakage of coated carbide tool in milling of H13 steel and SKD11 hardened steel. *SN Appl. Sci.* **2019**, *1*, 1111. [CrossRef]
69. Uddin, G.M.; Joyia, F.M.; Ghufran, M.; Khan, S.A.; Raza, M.A.; Faisal, M.; Arafat, S.M.; Zubair, S.W.H.; Jawad, M.; Zafar, M.Q.; et al. Comparative performance analysis of cemented carbide, TiN, TiAlN, and PCD coated inserts in dry machining of Al 2024 alloy. *Int. J. Adv. Manuf. Technol.* **2021**, *112*, 1461–1481. [CrossRef]
70. Sahoo, S.P.; Datta, S. Dry Machining Performance of AA7075-T6 Alloy Using Uncoated Carbide and MT-CVD TiCN-Al_2O_3-Coated Carbide Inserts. *Arab. J. Sci. Eng.* **2020**, *45*, 9777–9791. [CrossRef]
71. Heaney, P.J.; Sumant, A.V.; Torres, C.D.; Carpick, R.W.; Pfefferkorn, F.E. Diamond coatings for micro end mills: Enabling the dry machining of aluminum at the micro-scale. *Diam. Relat. Mater.* **2008**, *17*, 223–233. [CrossRef]
72. Gomez, H.; Durham, D.; Xiao, X.; Lukitsch, M.; Lu, P.; Chou, K.; Sachdev, A.; Kumar, A. Adhesion analysis and dry machining performance of CVD diamond coatings deposited on surface modified WC-Co turning inserts. *J. Mater. Process. Technol.* **2012**, *212*, 523–533. [CrossRef]
73. Gangopadhyay, S.; Bhaduri, D.; Chattopadhyay, A.K.; Paul, S. *Performance Evaluation of a Hard Composite Solid Lubricant Coating when Dry Machining of High-carbon Steel*; Taylor & Francis Group: Abingdon, UK, 2018. [CrossRef]
74. Kumar, R.; Pattnaik, S.K.; Minz, J.K.; Padhi, S.; Sarangi, S.K. Influence of cutting parameters on cutting forces and surface roughness in dry turning of Al using PCD and different coated tools. *Sadhana Acad. Proc. Eng. Sci.* **2019**, *44*, 186. [CrossRef]
75. Kumar, C.S.; Urbikain, G.; De Lucio, P.F.; De Lacalle, L.N.L.; Pérez-Salinas, C.; Gangopadhyay, S.; Fernandes, F. Investigating the self-lubricating properties of novel TiSiVN coating during dry turning of Ti6Al4V alloy. *Wear* **2023**, *532*, 205095. [CrossRef]
76. Sahoo, S.P.; Datta, S. Dry machining of Ti–6Al–4V using PVD Si-doped TiAlxN super nitride (HSN2) coated WC–Co tool. *J. Braz. Soc. Mech. Sci. Eng.* **2022**, *44*, 407. [CrossRef]
77. Parsi, P.K.; Kotha, R.S.; Routhu, T.; Pandey, S.; Dwivedy, M. Machinability evaluation of coated carbide inserts in turning of super-duplex stainless steel. *SN Appl. Sci.* **2020**, *2*, 1933. [CrossRef]
78. Sargade, V.G.; Gangopadhyay, S.; Paul, S.; Chattopadhyay, A.K. Effect of coating thickness on the characteristics and dry machining performance of Tin film deposited on cemented carbide inserts using cfubms. *Mater. Manuf. Process.* **2011**, *26*, 1028–1033. [CrossRef]
79. Parihar, R.S.; Sahu, R.K.; Setti, S.G. Novel design and composition optimization of self-lubricating functionally graded cemented tungsten carbide cutting tool material for dry machining. *Adv. Manuf.* **2021**, *9*, 34–46. [CrossRef]
80. Canteli, J.A.; Cantero, J.L.; Marín, N.C.; Gómez, B.; Gordo, E.; Miguélez, M.H. Cutting performance of TiCN-HSS cermet in dry machining. *J. Mater. Process. Technol.* **2010**, *210*, 122–128. [CrossRef]
81. Zhang, Y.; Cheng, Y.; Hu, H.; Yin, Z. Experimental study on cutting performance of microwave sintered Ti(C, N)/Al_2O_3 cermet tool in the dry machining of hardened steel. *Int. J. Adv. Manuf. Technol.* **2017**, *91*, 3933–3941. [CrossRef]
82. Yin, Z.; Hao, X.; Peng, H.; Yuan, J. A new β-SiAlON ceramic tool prepared by microwave sintering and its cutting performance in high-speed dry machining Inconel718. *Int. J. Adv. Manuf. Technol.* **2022**, *118*, 3105–3117. [CrossRef]
83. de Sousa, L.F.; Simba, B.G.; de Souza, J.V.C.; Ribeiro, M.V.; de Macedo Silva, O.M.; Santos, C.D. Dry machining of nodular cast iron using a YAG-reinforced alumina ceramic cutting tool. *Int. J. Adv. Manuf. Technol.* **2022**, *123*, 99–110. [CrossRef]
84. Rath, D.; Panda, S.; Pal, K. Performance Analysis of Hybrid Ceramic Insert in Dry Turning of Hardened Tool Steel. *Arab. J. Sci. Eng.* **2022**, *47*, 15455–15476. [CrossRef]
85. Lü, Z.; Deng, L.; Tian, Q.; Zhao, X. Cutting performance of Si3N4/TiC micro-nanocomposite ceramic tool in dry machining of hardened steel. *Int. J. Adv. Manuf. Technol.* **2018**, *95*, 3301–3307. [CrossRef]
86. Sun, H.; Zou, B.; Chen, W. Cutting performance of silicon-based ceramic end milling tools in high-efficiency machining of GH4099 under dry condition. *Int. J. Adv. Manuf. Technol.* **2022**, *118*, 1719–1732. [CrossRef]
87. Kitagawa, T.; Kubo, A.; Maekawa, K. Temperature and wear of cutting tools in high-speed machining of Inconel 718 and Ti-6Al-6V-2Sn. *Wear* **1997**, *202*, 142–148. [CrossRef]
88. Zhao, B.; Liu, H.; Huang, C.; Wang, J.; Wang, B.; Hou, Y. Cutting performance and crack self-healing mechanism of a novel ceramic cutting tool in dry and high-speed machining of Inconel 718. *Int. J. Adv. Manuf. Technol.* **2019**, *102*, 3431–3438. [CrossRef]
89. Parihar, R.S.; Setti, S.G.; Sahu, R.K. Effect of sintering parameters on microstructure and mechanical properties of self-lubricating functionally graded cemented tungsten carbide. *J. Manuf. Process.* **2019**, *45*, 498–508. [CrossRef]

90. Quan, Y.; Mai, Q. Investigation of the cooling effect of heat pipe-embedded cutter in dry machining with different thermal conductivities of cutter/workpiece materials and different cutting parameters. *Int. J. Adv. Manuf. Technol.* **2015**, *79*, 1161–1169. [CrossRef]
91. Gao, H.; Ma, B.; Zhu, Y.; Yang, H. Enhancement of machinability and surface quality of Ti-6Al-4V by longitudinal ultrasonic vibration-assisted milling under dry conditions. *Meas. J. Int. Meas. Confed.* **2022**, *187*, 110324. [CrossRef]
92. Attia, H.; Tavakoli, S.; Vargas, R.; Thomson, V. Laser-assisted high-speed finish turning of superalloy Inconel 718 under dry conditions. *CIRP Ann. Manuf. Technol.* **2010**, *59*, 83–88. [CrossRef]
93. Deswal, N.; Kant, R. Machinability and surface integrity analysis of magnesium AZ31B alloy during laser assisted turning. *J. Manuf. Process.* **2023**, *101*, 527–545. [CrossRef]
94. Bermingham, M.J.; Sim, W.M.; Kent, D.; Gardiner, S.; Dargusch, M.S. Tool life and wear mechanisms in laser assisted milling Ti-6Al-4V. *Wear* **2015**, *322*, 151–163. [CrossRef]
95. Tan, R.; Zhao, X.; Guo, S.; Zou, X.; He, Y.; Geng, Y.; Hu, Z.; Sun, T. Sustainable production of dry-ultra-precision machining of Ti-6Al-4V alloy using PCD tool under ultrasonic elliptical vibration-assisted cutting. *J. Clean. Prod.* **2020**, *248*, 119254. [CrossRef]
96. Sandoval, J.; Ali, A.; Kwon, P.; Guo, Y. Modulation-assisted machining of compacted graphite iron with coated carbide tool in dry condition. *Manuf. Lett.* **2022**, *33*, 452–460. [CrossRef]
97. Deswal, N.; Kant, R. Machinability Analysis During Laser Assisted Turning of Aluminium 3003 Alloy. *Lasers Manuf. Mater. Process.* **2022**, *9*, 56–71. [CrossRef]
98. Deswal, N.; Kant, R. Radial direction ultrasonic-vibration and laser assisted turning of Al3003 alloy. *Mater. Res. Express.* **2023**, *10*, 044004. [CrossRef]
99. Zhao, J.; Liu, Z.; Wang, B.; Hu, J.; Wan, Y. Tool coating effects on cutting temperature during metal cutting processes: Comprehensive review and future research directions. *Mech. Syst. Signal Process.* **2021**, *150*, 107302. [CrossRef]
100. Dabees, S.; Mirzaei, S.; Kaspar, P.; Holcman, V.; Sobola, D. Characterization and Evaluation of Engineered Coating Techniques for Different Cutting Tools—Review. *Materials* **2022**, *15*, 5633. [CrossRef]

processes

MDPI

Article

Parameter Optimization and Surface Roughness Prediction for the Robotic Adaptive Hydraulic Polishing of NAK80 Mold Steel

Dequan Shi [1], Xiongyawei Zeng [1], Xuhui Wang [1] and Huajun Zhang [2,*]

[1] School of Materials and Chemistry, University of Shanghai for Science and Technology, Shanghai 200093, China; shidequan@usst.edu.cn (D.S.); 223353199@st.usst.edu.cn (X.Z.); 233393111@st.usst.edu.cn (X.W.)
[2] School of Materials Science and Engineering, Shanghai Jiao Tong University, Shanghai 200240, China
* Correspondence: zhanghuajun@sjtu.edu.cn

Abstract: Pneumatic polishing tools are commonly used in traditional robot mold polishing systems, but they have problems with the stable control of mold surface roughness due to low precision and poor adaptability in polishing force adjustment. The integration of an adaptive hydraulic polishing (AHP) tool and robot system effectively solves the above problems, providing a robust solution for the high-precision polishing of various molds. This study systematically investigates the robotic polishing of NAK80 mold steel using an AHP-equipped robotic platform with 3M abrasive discs of progressively refined grit sizes (P180, P400, P800). Through single-factor experiments and response surface methodology, the effects of polishing force, rotational speed, and feeding speed on surface roughness were quantitatively analyzed. The relationship between surface roughness and the polishing parameters was derived to elucidate the roughness evolution before and after over-polishing. Orthogonal experiments combined with range analysis identified optimal parameter combinations for P180 (20 N polishing force, 5000 RPM rotational speed, and 5 mm·s^{-1} feeding speed) and P400 abrasives (10 N polishing force, 4000 RPM rotational speed, and 5 mm·s^{-1} feeding speed), achieving minimum surface roughness values of 0.08 μm and 0.044 μm, respectively. For P800 abrasives, a central composite design was used to develop a roughness prediction model with a ≤7.14% relative error, and the optimal parameters are a 20 N polishing force, a 5000 RPM rotational speed, and a 5 mm·s^{-1} feeding speed. The sequential application of the optimized parameters across all the grit sizes can reduce the surface roughness from an initial 0.4 μm to a final 0.017 μm, representing a 95.75% improvement in the surface finish.

Keywords: polishing parameter optimization; surface roughness prediction; robots; NAK80 mold steel; adaptive hydraulic polishing

1. Introduction

The mold constitutes the most pivotal component in the injection molding process, and its surface finish directly governs both the product quality and mold lifespan [1]. In order to eliminate tool marks and hardened layers resulting from machining operations on the mold cavity surface, polishing is indispensable for attaining the desired dimensions, shape, and surface roughness. Techniques for enhancing mold surface finish commonly include grinding, polishing, and lapping [2]. Traditional manual polishing suffers from unstable surface quality, low efficiency, and difficulty in meeting nanoscale surface requirements, leading to its gradual replacement by CNC polishing machines and robots, which effectively improve polishing efficiency and surface uniformity [3].

There are many ways to reduce the surface roughness of a mold. Currently, commonly employed methods include airbag polishing, magnetorheological polishing, and flexible disc polishing. Airbag polishing technology, initially proposed by Walker et al. [4], represents a flexible polishing technique. Through long-term research and development, it has demonstrated numerous advantages in the polishing domain due to its distinctive airbag structure. Nevertheless, it also gives rise to issues such as non-uniform stress distribution and unstable polishing trajectories, thereby rendering it less adaptable to the polishing of free-form surfaces [5]. Magnetorheological polishing, put forward by Kordonski et al. [6] based on the integration of electromagnetic and fluid dynamics theories, accomplishes the grinding and polishing of the workpiece surface via the frictional interaction between abrasive particles and the workpiece surface. However, this method is constrained by the size and curvature of the polishing wheel. It demands high surface shape accuracy of the workpiece, allows for only a small material removal amount, and suffers from low efficiency [7]. Flexible disc polishing is applicable to complex free-form surface workpieces, enabling uniform polishing and yielding a smooth and high-quality surface. This method chiefly depends on CNC centers to complete the automated polishing process [8]. Shiou et al. [9] accomplished the ultra-precision surface finishing of NAK80 mold steel by executing continuous ball polishing on CNC machining centers, optimizing parameters via orthogonal arrays and ANOVA to reduce average roughness from 0.06 μm to 0.016 μm. Huang et al. [10] developed a flexible polishing head chuck, achieving an R_a value of 0.006 μm on plastic mold steel. Zhao et al. [11] established a roughness prediction model for integrally bladed rotors utilizing response surface methodology, achieving a minimum roughness of 0.18 μm. Liu et al. [12] employed a nanosecond pulsed laser surface treatment to rectify polishing defects, offering new insights into laser polishing for roughness improvement.

The flexibility of high-precision mold polishing on CNC centers is relatively limited. In contrast, employing robots for mold polishing can effectively address the issue of the poor flexibility inherent in CNC machining. This offers a more adaptable solution for the complex requirements of mold polishing processes [13,14]. Xie et al. [15] optimized parameters for robotic mold steel polishing, achieving a roughness of 0.3 μm utilizing regression models. Qi et al. [16] proposed an integrated framework that integrates the L-BFGS-B optimization algorithm and image processing to predict the surface roughness of Inconel718. However, research on the roughness of molds polished by robots should not be limited solely to methods and experiments. Wang et al. [17] designed a contact motion planning strategy based on dynamic systems to generate trajectory routes during local polishing. This strategy ensures a constant contact force between the end effector of the polishing robot and the workpiece, enabling the robot to have flexible control. Wahballa et al. [18] put forward a robot polishing method for compound surfaces, including planes and curved surfaces, based on smooth trajectory planning, a constant force algorithm, and contour removal depth modeling. The method uses an online stiffness and inverse damping force impedance controller supported by a gravity compensation algorithm to achieve a constant polishing force. Mohsin et al. [19] proposed an effective approach for the robot polishing of complex curved surfaces, which relies on tool path planning with controlled forces and polishing parameter optimization using experiment design. Moreover, this theoretical modeling holds significant guidance for roughness reduction [20]. Verma et al. [21] investigated the contributions of various process parameters, materials, and geometric characteristics to the modeling of surface roughness obtained using pneumatic polishing processes. Based on Hertz's contact theory for elastic surfaces, a mathematical model for predicting the surface roughness of molds was proposed, which is a function of workpiece characteristics (such as hardness and tool properties) and key process parameters (such as the working gap, applied pneumatic pressure, spindle speed, abrasive slurry concentration, abrasive mesh size, and dwell time). Savio et al. [22] established a relationship between the workpiece under friction and polishing parameters and proposed a model to evaluate the change in the surface roughness of abrasively polished glass molds as a function of the volume of material removed according to the Reye wear hypothesis. Feng et al. [23] deduced a pressure distribution model within the contact area between the polishing disc and the workpiece during both flat- and curved-surface polishing and established a removal depth distribution model along the vertical feed direction for each polishing process to minimize the fluctuation of the removal depth. Qi et al. [24] utilized elastic and plastic contact theory to establish a model for the maximum cutting depth of abrasive particles and developed a predictive model for surface roughness in belt polishing based on a backpropagation neural network. Wang et al. [25] considered the shape and distribution of abrasives as well as the effect of pressure on surface roughness. Based on abrasive cutting and probability theory, a surface roughness prediction model was established by calculating the indentation depth of abrasive particles on the workpiece surface. There has been extensive research on polishing roughness modeling, with most theories focusing on the removal of material surface profiles by abrasive particles to reduce surface roughness.

Nevertheless, traditional robot polishing predominantly employs pneumatic flexible polishing tools. Owing to the challenges in stably controlling the polishing force, these tools lead to a reduction in polishing accuracy. In contrast, the hydraulic flexible polishing tool offers an effective solution for high-precision robot polishing, which can markedly enhance both polishing efficiency and precision [26]. Additionally, previous experimental studies on polishing have largely overlooked the determination of the ideal surface roughness level to which each abrasive grade should polish the workpiece, thereby ensuring the attainment of the minimal possible surface roughness in the final product [27]. In essence, the polishing process necessitates a methodical and incremental decrease in the workpiece's surface roughness, with specific and optimized polishing parameters for each abrasive grade.

In summary, this article focuses on research into the robot polishing of NAK80 mold steel using a self-developed adaptive hydraulic polishing (AHP) tool. Based on the fundamental principles of sandpaper polishing, this study calculated the cutting depth of abrasive grains under pressure, correlated this depth with surface roughness, derived a relationship between surface roughness and the parameters of polishing force, rotational speed, and feeding speed, and examined the variation patterns of roughness before and after over-polishing. Subsequently, AHP polishing experiments were conducted for each abrasive grain size, and regression models between the surface roughness and the above three polishing parameters were established using the response surface method. Ultimately, a series of process parameters, including polishing force, rotational speed, and feeding speed, were obtained to achieve the lowest surface roughness, and the prediction of surface roughness was realized.

2. Materials and Experimental

The experimental material was NAK80 mold steel with specimen dimensions of $300 \times 100 \times 30$ mm^3. The robotic polishing platform consisted mainly of a KR210 robot, a self-developed AHP tool, and a sandpaper tray. The key parameters of the AHP polishing tool included the following: an adjustable spindle speed from 2000 to 6000 RPM with a precision of 1 RPM, a spindle radial floating range of 0–20 mm with a precision of 0.01 mm, and an adjustable polishing force from 5 to 120 N with a precision of 1 N.

Before the experiment, the AHP polishing tool was mounted on the sixth axis of the KR210, with its feeding speed controlled by the KR210 robot and its rotational speed and polishing force regulated by the AHP. The sandpaper tray was installed on the output spindle of the AHP, with a piece of ceramic alumina round 3M sandpaper of 1-inch in diameter attached to its bottom for polishing. Three grit sizes of sandpaper, P180, P400, and P800, were used, and the surface roughness R_a of the sample was measured using a stylus-type surface roughness tester, JITAI810 (Jiahe Precision Instrument, Shenzhen, China). The tester has a measurement range for the roughness R_a from 0.005 to 16.000 μm, with an indication accuracy of 0.001 μm. The measuring stylus is diamond-tipped, characterized by a cone angle of 90° and a tip radius of 5 μm. During the measurement process, the speed is maintained at 1 mm·s^{-1}. The sampling length is set to 2.5 mm, and the measurement length (also known as the evaluation length) is 10 mm, which is exactly four times the sampling length.

Polishing experiments were conducted sequentially from coarse to fine grits (namely, P180→P400→P800) with varying parameters of polishing force, rotational speed, and feeding speed. The polishing direction is illustrated in Figure 1. To determine the optimal polishing parameters and minimum achievable surface roughness for each grit, single-factor experiments were first conducted to determine rough parameter ranges. Then, orthogonal experiments were conducted within this range to obtain the optimal parameters.

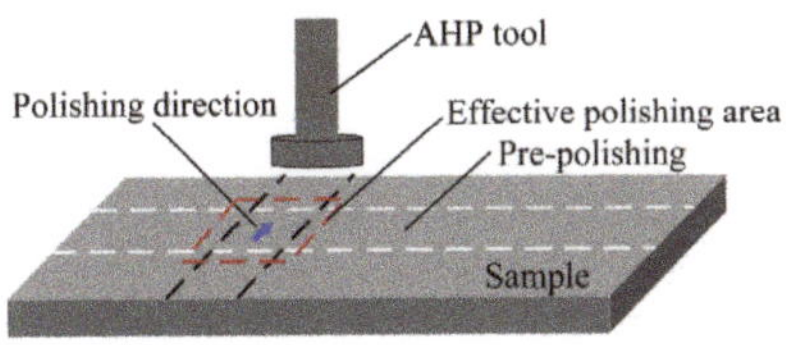

Figure 1. Schematic of the polishing direction.

The samples were pre-polished to Ra_1 = 0.35–0.40 μm as the initial roughness for the P180 grit tests. Subsequent experiments used the minimum roughness achieved by the preceding grits as the initial conditions; thus, the P400 experiments began at the minimum Ra_2 from P180 and P800 at the minimum Ra_3 from P400. Finally, the samples were polished with P800 to achieve a surface roughness of Ra_4. The experimental polishing process is shown in Figure 2.

The single-factor experiments were conducted iteratively without a predefined limit, ceasing when the roughness of the samples plateaued or began to increase. A new piece of sandpaper was used for each polishing iteration, the surface roughness of each experiment was measured five times, and the average value was determined as the effective R_a value. The single-factor experimental scheme is shown in Table 1.

Table 1. Single-factor experimental scheme for the influence of polishing parameters on surface roughness.

Polishing Force (N)	Rational Speed (RPM)	Feeding Speed (mm·s^{-1})
5,10, 15, 20, 25, 30	4000	10
10	2000, 3000, 4000, 5000	10
10	4000	5, 10, 15, 20, 25

Building on the single-factor results, the orthogonal experimental designs were implemented on the NAK80 samples with the P180 and P400 grits to obtain the optimal polishing parameters. The experimental scheme is shown in Table 2. A central composite design was adopted for the P800 grit, as shown in Table 3, including three replicates at central experimental points, to establish a mathematical relationship between the surface roughness and the polishing parameters and to obtain the optimal polishing parameters.

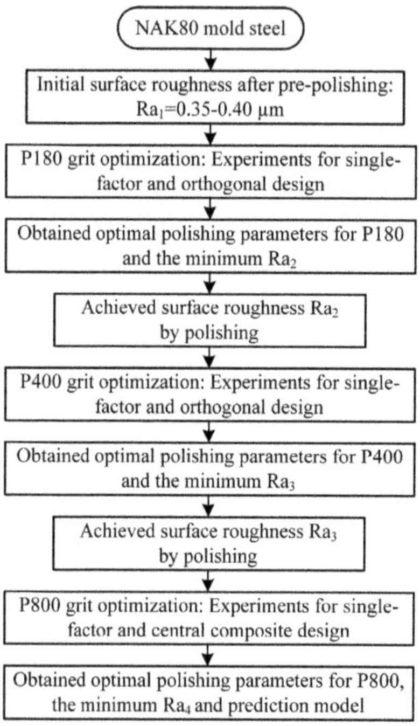

Figure 2. Flowchart of the experimental polishing process.

Table 2. Factors and levels of orthogonal experimental design for the P180 and P400 grits.

Levels	Factors		
	Polishing Force (N)	**Rotational Speed (RPM)**	**Feeding Speed (mm·s^{-1})**
	A_{180} I A_{400}	**B_{180} I B_{400}**	**C_{180} I C_{400}**
1	10	3000	5
2	20	4000	15
3	30	5000	25

Table 3. Factors and levels of central composite design for the P800 grit.

Levels	Factors		
	Polishing Force (N)	**Rotational Speed (RPM)**	**Feeding Speed (mm·s^{-1})**
	A_{800}	**B_{800}**	**C_{800}**
−1	5	3000	5
0	12.5	4000	15
1	20	5000	25

3. Results and Analysis

3.1. Influence of the Polishing Process Parameters on Surface Roughness

The initial surface roughness of the polished samples with the P180, P400, and P800 sandpaper was maintained at 0.35–0.40 μm, 0.07–0.08 μm, and 0.040–0.045 μm, respectively. Figures 3–5 demonstrate the influence of the polishing parameters on the minimum surface roughness and polishing iterations of the NAK80 samples.

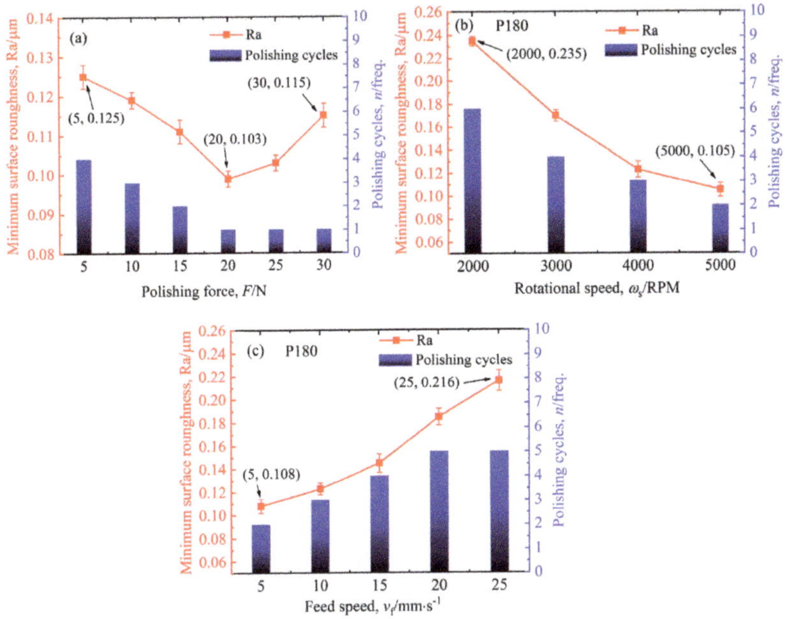

Figure 3. Variations in minimum surface roughness and polishing cycles with the parameters of (**a**) polishing force, (**b**) polishing speed, and (**c**) feeding speed for P180.

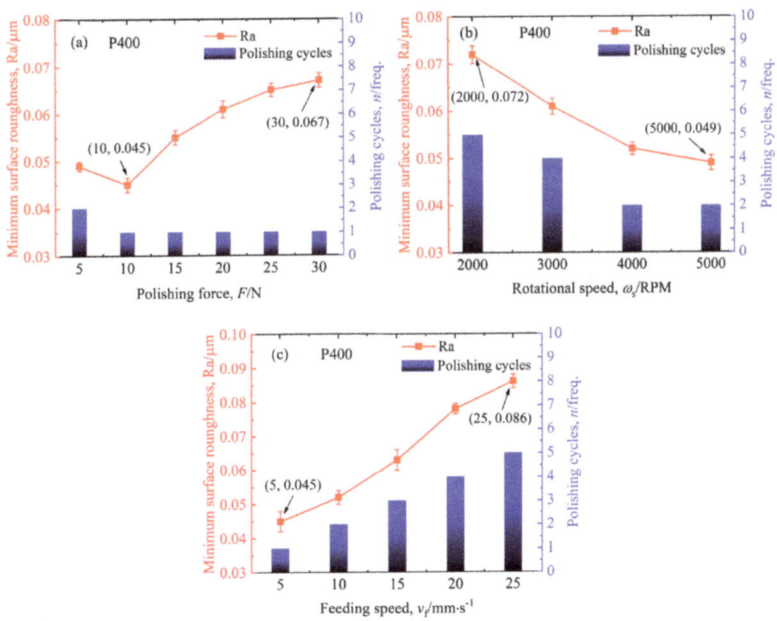

Figure 4. Variations in minimum surface roughness and polishing cycles with the parameters of (**a**) polishing force, (**b**) polishing speed, and (**c**) feeding speed for P400.

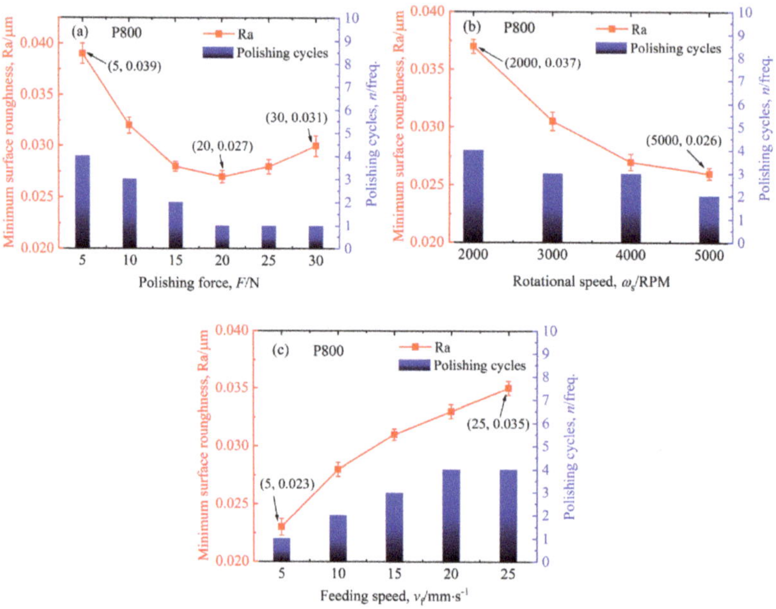

Figure 5. Variations in the minimum surface roughness and polishing cycles with the parameters of (**a**) polishing force, (**b**) polishing speed, and (**c**) feeding speed for P800.

As shown in Figures 3–5, the effects of polishing force, rotational speed, and feeding speed on the minimum surface roughness exhibit consistent trends across different grit sizes. With the increase in polishing force, the minimum surface roughness initially decreases and then increases. For the P180 sandpaper, when the polishing force is 20 N, the surface roughness of the sample decreases from 0.125 μm at 5 N to a minimum value of 0.103 μm, with a decrease of 82.4%. When the polishing force is increased to 30 N, the surface roughness of the sample actually increases to 0.115 μm. For the P400 sandpaper, when the polishing force is 10 N, the surface roughness reaches a minimum value of 0.045 μm. Afterward, as the polishing force increases, the surface roughness gradually increases. When the polishing force is 30 N, the surface roughness reaches a maximum value of 0.067 μm. For the P800 sandpaper, when the polishing force increases from 5 N to 20 N, the surface roughness of the sample decreases from the maximum value of 0.039 μm to the minimum value of 0.027 μm. As the polishing force continues to increase, the surface roughness increases, but the magnitude of the increase is not significant.

The surface roughness of the sample gradually decreases with the increase in AHP rotational speed. When the rotational speed of AHP increases from 2000 RPM to 5000 RPM, the surface roughness of samples polished with the P180, P400, and P800 sandpapers decreases from the maximum values of 0.235 μm, 0.072 μm, and 0.037 μm to the minimum values of 0.105 μm, 0.049 μm, and 0.026 μm, respectively, with a decrease of 44.7%, 68.1%, and 70.3%.

The surface roughness of the sample gradually increases with the increase in feeding speed. When the feeding speed is 5 mm$\cdot$s^{-1}, the surface roughness of the sample is the smallest. The P180, P400, and P800 sandpapers can achieve a minimum surface roughness of 0.108 μm, 0.045 μm, and 0.023 μm, respectively. When the feeding speed increases to 25 mm$\cdot$s^{-1}, the surface roughness of the sample reaches the maximum values, which are 0.216 μm, 0.086 μm, and 0.035 μm, respectively, increasing by 100%, 91.1%, and 52.2%.

When polishing the samples with different parameters, the number of polishing cycles to achieve the minimum surface roughness varies significantly, ranging from 1 to 6. For instance, Figure 3a reveals that the P180 sandpaper can attain its minimum roughness of 0.103 μm in one polishing cycle at 20 N while requiring four cycles at 5 N to reach a minimum roughness of 0.125 μm. As illustrated in Figure 3b, five polishing cycles are necessary for the rotational speed of 2000 RPM to reach a minimum surface roughness of 0.235 μm, whereas at a rotational speed of 5000 RPM, a minimum surface roughness of 0.105 μm is obtained after polishing twice. Figure 3c indicates that the feeding speed of 5 $mm{\cdot}s^{-1}$ enables a minimum roughness of 0.108 μm in two cycles, whereas a feeding speed of 25 $mm{\cdot}s^{-1}$ demands five cycles to achieve a minimum surface roughness of 0.216 μm; analysis of Figures 4 and 5 yields the same conclusion. Therefore, to obtain the minimum surface roughness, the number of polishing cycles varies for different parameters. These findings indicate that parameter selection critically determines polishing efficiency, with the optimal parameters enabling the single-polishing achievement of the minimum surface roughness. Further analysis of Figures 3–5 shows that when the P180, P400, and P800 sandpapers are used to polish the sample, the smallest number of polishing times achieves the minimum surface roughness for any of the polishing parameters of polishing force, rotational speed, and feeding speed.

To further elucidate the parametric dependence of surface roughness, a mathematical relationship for the polishing process was derived, which relates the surface roughness R_a of the sample to the polishing force, rotational speed, and feeding speed of the AHP tool and clearly explains the variation in surface roughness with polishing parameters.

The polishing process using sandpaper is actually the result of collective cutting by a large number of individual abrasive grains [28]. For analytical simplification, it is assumed that the sandpaper surface is a single layer of abrasives with uniform distribution, and each abrasive grain is perpendicular to the substrate, pointing outward at its tip, and has only one effective cutting edge. Thus, the number of effective cutting edges is equal to the number of abrasive grains on the surface.

According to the research results of Baidakova et al. [29], spherical grain geometry provides a more realistic representation of abrasive grains. Figure 6 schematically illustrates the indentation mechanism of a simplified spherical abrasive grain into the sample surface during polishing.

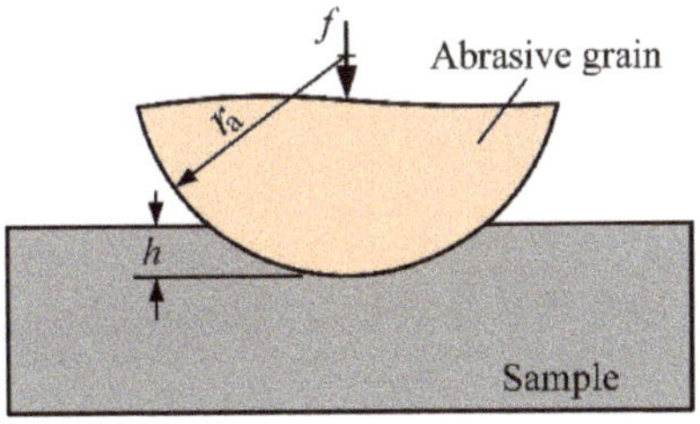

Figure 6. Contact model between a simplified spherical abrasive grain and a sample.

The indentation mechanism of a single spherical abrasive grain into the sample resembles Brinell hardness testing. Given the significant difference in Brinell hardness between the alumina abrasive and NAK80 mold steel, the Brinell test model was adopted to determine the indentation depth of a single abrasive grain. The Brinell hardness H_B of the sample can be calculated as the ratio of force f acting on the spherical grain to the projected area A of the spherical cap indentation:

$$H_{\mathrm{B}} = \frac{f}{A} = \frac{f}{2\pi r_{\mathrm{a}} h} \tag{1}$$

where r_{a} denotes the radius of the spherical grain, and h represents the indentation depth.

When external force F is applied to the abrasive tool, the force per grain f can be expressed as follows:

$$f = \frac{F}{N} = \frac{F}{\pi R^2 N_0} \tag{2}$$

where N indicates the total number of grains within the circular contact area of radius R, and N_0 represents the grain density per unit area, calculated by the following [30]:

$$N_0 = \frac{6V_{\mathrm{g}}}{\pi d_{\mathrm{m}}^2} \tag{3}$$

where V_{g} denotes the volume fraction of abrasive grains, which is related to the structural number S of the abrasive tool [31].

Combining Equations (1)–(3), the indentation depth h can be derived as follows:

$$h = \frac{F d_{\mathrm{m}}}{6\pi R^2 V_{\mathrm{g}} H_{\mathrm{B}}} \tag{4}$$

where d_{m} represents the mean grain diameter, $d_{\mathrm{m}} = 2r_{\mathrm{a}}$.

The relative linear velocity during polishing is governed by the rotational speed and feeding speed of the spindle. As shown in Figure 7, during linear feed polishing along the X-direction, the instantaneous relative velocity v_{t} at any point P(x, y) within the circular polishing zone with the radius of R can be expressed as follows:

$$v_{\mathrm{t}} = \sqrt{v_{\mathrm{s}}^2 + v_{\mathrm{f}}^2 - 2v_{\mathrm{f}} v_{\mathrm{s}} \cos(\pi - \theta)} \tag{5}$$

where v_{f} is the robotic feeding speed, and v_{s} denotes the tangential velocity at point P, given by the following:

$$\begin{cases} v_{\mathrm{s}} = \omega_{\mathrm{s}} \sqrt{x^2 + y^2} \\ \cos\theta = \frac{y}{\sqrt{x^2+y^2}} \end{cases} \tag{6}$$

where ω_{s} represents the rotational speed of the sandpaper.

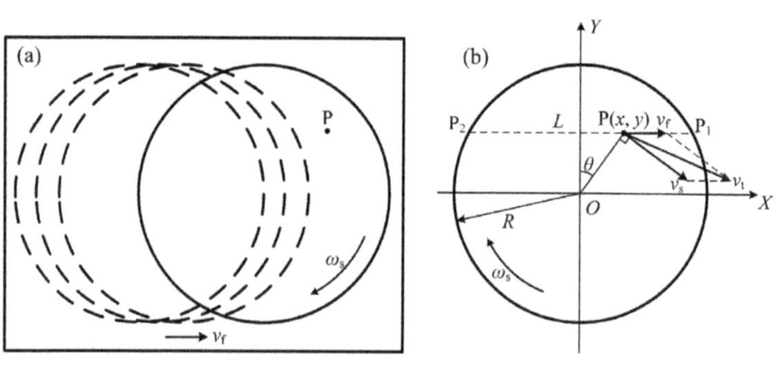

Figure 7. Polishing diagram: (**a**) polishing trajectory diagram; (**b**) schematic of the circular contact area.

Substituting Equation (6) into Equation (5) yields the following:

$$v_t = \sqrt{v_f^2 + \omega_s^2(x^2 + y^2) + 2v_f\omega_s y} \quad (7)$$

According to Figure 7, x can be represented by the product of feeding speed v_f and time t. Substituting this into Equation (7) gives the following:

$$v_t = \sqrt{(\omega_s y + v_f)^2 + \omega_s^2 v_f^2 t^2} \quad (8)$$

Considering $v_f \ll \omega_s$, Equation (8) simplifies to the following:

$$v_t \approx \omega_s\sqrt{y^2 + v_f^2 t^2} \quad (9)$$

As shown in Figure 7, the length L of the polishing path between entry point P_1 and exit point P_2 can be calculated as follows:

$$L = 2\sqrt{R^2 - y^2} \quad (10)$$

Thus, the polishing duration t_p at point P becomes the following:

$$t_p = \frac{L}{v_f} = \frac{2\sqrt{R^2 - y^2}}{v_f} \quad (11)$$

The actual polishing length s at point P is obtained by integrating the relative velocity v_t over t_p, and it is expressed as follows:

$$s = 2\int_0^{t_p} v_t dt = \frac{2\omega_s}{v_f}\left[\frac{\sqrt{R^2 - y^2}\cdot R}{2} + \frac{y^2}{2}\ln\left(\frac{R}{y}\right)\right] \quad (12)$$

The cumulative indentation depth $H(y)$, equivalent to the surface roughness, results from integrating single-grain indentation h over the polishing length s:

$$H(y) = \int_0^s h dy \quad (13)$$

Substituting Equations (4) and (12) into Equation (13) yields the following:

$$H(y) = \frac{F\omega_s d_m}{3\pi R^2 V_g H_B v_f}\left[\frac{\sqrt{R^2 - y^2}\cdot R}{2} + \frac{y^2}{2}\ln\left(\frac{R}{y}\right)\right] \quad (14)$$

Surface roughness fundamentally arises from the irregular topography generated by abrasive material removal [32]. In this study, there is a uniform initial surface roughness of $R_a^{\ominus}$. As shown in Figure 8a, it is assumed that the initial surface of the sample exhibits a semicircular contour with a height of $2H^0$, where $H^0 = R_a^{\ominus}$. When the sample is polished, the surface peaks progressively reduce under the cutting action of the abrasive grains, leading to a gradual decrease in surface roughness, as depicted in Figure 8b. Once the surface peaks are completely removed, subsequent polishing creates new surface irregularities, as illustrated in Figure 8c. Therefore, as the cutting depth increases, the surface roughness of the sample initially decreases and then increases.

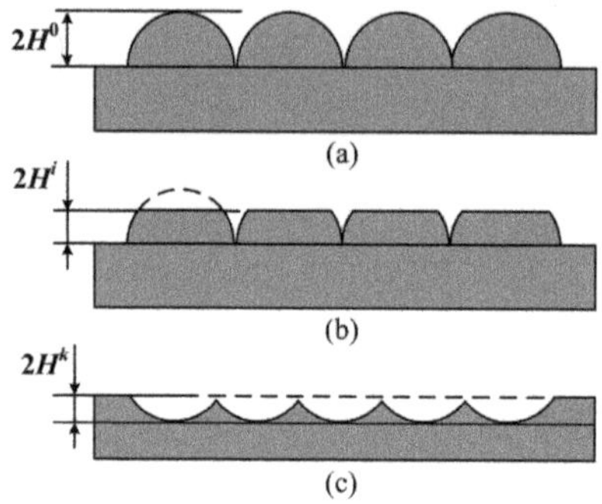

Figure 8. Sample surface profile: (**a**) with uniform semicircular peaks; (**b**) with the contour after polishing *i* times; (**c**) with the new surface contour.

After *n* polishing cycles, the surface profile height becomes the following [33]:

$$\begin{cases} 2H^i = 2H^0 - nH(y) & (\text{for } nH(y) < 2H^0) \\ 2H^k = nH(y) - 2H^0 & (\text{for } nH(y) > 2H^0) \end{cases} \quad (15)$$

where $2H^i$ and $2H^k$ represent the height of the surface profile of the sample after polishing, and they are equal to the surface roughness R_a of the sample.

Substituting Equation (14) into Equation (15) gives the evolution model of the surface roughness as follows:

$$\begin{cases} R_a = R_a^{\ominus} - \dfrac{nF\omega_s d_m}{6\pi R^2 V_g H_B v_f}\left[\dfrac{\sqrt{R^2-y^2}\cdot R}{2} + \dfrac{y^2}{2}\ln\left(\dfrac{R}{y}\right)\right] (a) \\ R_a = \dfrac{nF\omega_s d_m}{6\pi R^2 V_g H_B v_f}\left[\dfrac{\sqrt{R^2-y^2}\cdot R}{2} + \dfrac{y^2}{2}\ln\left(\dfrac{R}{y}\right)\right] - R_a^{\ominus} (b) \end{cases} \quad (16)$$

where F represents the polishing force, ω_s represents the rotational speed, and v_f denotes the feeding speed.

The model (16a) represents the process where the surface roughness gradually decreases before the peaks on the sample surface are completely removed, and the model (16b) expresses the process where the surface roughness gradually increases after the surface peaks are completely removed, also called over-polishing. Summarily, the model reveals that surface roughness decreases with increasing polishing force and rotational speed but increases with feeding speed before over-polishing. The inverse relationship holds after over-polishing. Notably, over-polishing initiates when the single-grain indentation depth exceeds the initial surface profile height.

In practical polishing, achieving the minimum surface roughness under given parameters typically requires multiple polishing cycles. When the sample undergoes repeated polishing, the surface roughness initially decreases, followed by stabilization or a subsequent increase, indicating the onset of over-polishing.

For Figures 3a, 4a and 5a, the surface roughness progressively decreases with the increasing polishing force until reaching a minimum value. This is consistent with Equation (16a), confirming the absence of over-polishing. After reaching a minimum value, the surface roughness becomes proportional to the polishing force, with minimum roughness achieved in a single polishing cycle. This aligns with Equation (16b), demonstrating irreversible over-polishing where further polishing cannot reduce roughness below this limit.

Figures 3b, 4b and 5b indicate that the surface roughness of the samples continuously decreases with the increase in rotational speed, and the number of polishing cycles is greater than 1, which remains consistent with Equation (16a), confirming no over-polishing occurrence. Similarly, Figures 3c, 4c and 5c demonstrate gradual roughness elevation with higher feeding speeds, also adhering to Equation (16a) without over-polishing.

3.2. Orthogonal Test Results and the Surface Roughness Prediction Model

The initial surface roughness of the samples for the P180, P400, and P800 sandpapers was controlled within 0.35–0.40 μm, 0.07–0.08 μm, and 0.040–0.045 μm, respectively. The orthogonal experimental results for the P180 and P400 sandpapers are summarized in Table 4.

Table 4. Orthogonal test results and range analysis for the P180 and P400 sandpapers.

No.	Polishing Force (N)	Rotational Speed (RPM)	Feeding Speed ($mm{\cdot}s^{-1}$)	R_a (μm)	
	A_{180} ǀ A_{400}	B_{180} ǀ B_{400}	C_{180} ǀ C_{400}	P180	P400
1	1	1	1	0.138	0.054
2	1	2	2	0.155	0.060
3	1	3	3	0.185	0.079
4	2	1	2	0.139	0.093
5	2	2	3	0.169	0.090
6	2	3	1	0.090	0.069
7	3	1	3	0.213	0.104
8	3	2	1	0.113	0.055
9	3	3	2	0.125	0.083
K_1	0.159 ǀ 0.064	0.163 ǀ 0.084	0.114 ǀ 0.059	—	—
K_2	0.133 ǀ 0.084	0.146 ǀ 0.068	0.140 ǀ 0.079	—	—
K_3	0.150 ǀ 0.081	0.133 ǀ 0.077	0.189 ǀ 0.091	—	—
R	0.027 ǀ 0.020	0.030 ǀ 0.015	0.189 ǀ 0.032	—	—

Range analysis was conducted to identify the influence of the process parameters, the optimal combinations, and the corresponding minimum roughness values. For the P180 sandpaper, the optimal parameter combination yielding the minimum R_a of 0.08 μm was determined as a 20 N polishing force, a 5000 RPM rotational speed, and a 5 $mm{\cdot}s^{-1}$ feeding speed, with factor significance ranked as C_{180} (feeding speed) > B_{180} (rotational speed) > A_{180} (polishing force). For the P400 sandpaper, the optimal parameters achieving R_a = 0.044 μm were a 10 N polishing force, a 4000 RPM rotational speed, and a 5 $mm{\cdot}s^{-1}$ feeding speed, with a significance order of C_{400} > A_{400} > B_{400}.

Analysis of variance (ANOVA) is primarily used to assess the fitting accuracy of a model. Larger F-values and smaller p-values indicate superior model significance. The ANOVA results for the orthogonal experiments using the P180 and P400 sandpapers are presented in Tables 5 and 6, respectively.

As shown in Tables 5 and 6, at a significance level of $\alpha = 0.05$, the models for P180 and P400 exhibit F-values of 84.129 and 106.167, respectively, which are greater than the critical value $F_{0.05} = 4.459$. Moreover, the p-values are 0.012 and 0.009, respectively, both of which are less than 0.05, confirming that the models have high significance. Furthermore, all three factors in both orthogonal experiments have F-values greater than $F_{0.05}$, indicating that all of them are also significant.

Table 5. ANOVA results of surface roughness for P180.

Source	Sum of Squares	df	Mean Square	F-Value	p-Value	
Model	0.011	6	0.002	84.129	0.012	significant
A_{180}	0.001	1	0.193	19.896	0.048	—
B_{180}	0.001	2	0.000	31.836	0.030	—
C_{180}	0.009	2	0.001	200.657	0.005	—
Pure error	4.467×10^{-5}	2	2.233×10^{-5}	—	—	—
Cor total	0.011	8	—	—	—	—

Significant factors at a 0.05 significance level. R^2 (coefficient of determination) = 0.996.

Table 6. ANOVA results of surface roughness for P400.

Source	Sum of Squares	df	Mean Square	F-Value	p-Value	
Model	0.003	6	0.000	106.167	0.009	significant
A_{400}	0.001	1	0.000	83.083	0.012	—
B_{400}	0.000	2	0.000	44.333	0.022	—
C_{400}	0.002	2	0.001	191.083	0.005	—
Pure error	8.000×10^{-6}	2	4.000×10^{-6}	—	—	—
Cor total	0.003	8	—	—	—	—

Significant factors at a 0.05 significance level. R^2 (coefficient of determination) = 0.997.

The results of the central composite design (CCD) for the P800 sandpaper are presented in Table 7, where the number of central experiments is three.

Table 7. Results of the central composite design for the P800 sandpaper.

No.	A_{800}	B_{800}	C_{800}	R_a (μm)
1	−1	−1	0	0.040
2	1	−1	0	0.036
3	−1	1	0	0.035
4	1	1	0	0.026
5	−1	0	−1	0.030
6	1	0	−1	0.021
7	−1	0	1	0.037
8	1	0	1	0.035
9	0	−1	−1	0.030
10	0	1	−1	0.024
11	0	−1	1	0.037
12	0	1	1	0.034
13	0	0	0	0.034
14	0	0	0	0.033
15	0	0	0	0.032

Given the coupled effects of the polishing parameters on surface roughness, a quadratic regression was used to derive a prediction model for surface roughness. To improve the reliability of the prediction model, it is necessary to eliminate nonsignificant factors in the model via stepwise selection, thus obtaining a refined prediction model. The ANOVA results for the P800 sandpaper is shown in Table 8.

Table 8. ANOVA results of the regression model for P800.

Source	Sum of Squares	df	Mean Square	F-Value	p-Value	
Model	4×10^{-4}	9	0.0000	23.21	0.0015	significant
A_{800}	10^{-4}	1	10^{-4}	40.00	0.0015	
B_{800}	10^{-4}	1	10^{-4}	40.00	0.0015	
C_{800}	2×10^{-4}	1	2×10^{-4}	100.28	0.0002	
$A_{800}B_{800}$	6.25×10^{-6}	1	6.25×10^{-6}	3.47	0.1214	
$A_{800}C_{800}$	0.0000	1	0.0000	6.81	0.0477	
$B_{800}C_{800}$	2.25×10^{-6}	1	2.25×10^{-6}	1.25	0.3144	
A_{800}^2	5.192×10^{-6}	1	5.192×10^{-6}	0.2885	0.6142	
B_{800}^2	2.827×10^{-6}	1	2.827×10^{-6}	1.57	0.2655	
C_{800}^2	0.0000	1	0.0000	14.13	0.0132	
Residual	9×10^{-6}	5	1.8×10^{-6}			
Lack-of-fit	7×10^{-6}	3	2.333×10^{-6}	2.33	0.3141	nonsignificant
Pure error	2×10^{-6}	2	10^{-6}			
Cor total	4×10^{-4}	14				

$R^2 = 0.976$, Pred $R^2 = 0.951$, Adeq-precision = 16.203, $F_{0.05} = 4.459$.

At a significance level of $\alpha = 0.05$, the regression model yields a statistical F-value of 23.21, which exceeds the critical value $F_{0.05} = 4.459$, indicating the model is highly significant. The lack-of-fit test produces an F-value of 2.33, confirming that the lack-of-fit of the model is not significant. Combined, these statistical evaluations demonstrate that the regression model not only aligns well with the experimental data at the tested points but also maintains good predictive capability across the entire experimental range. The coefficient of determination ($R^2 = 0.976$) signifies that 97.6% of the experimental variability is explained by the model, leaving only 2.07% unexplained. The predicted $R^2 = 0.951$ indicates that the model can account for 95.1% of variability when applied to new observations. Furthermore, the ratio of adequacy to precision is 16.203, which is much bigger than the threshold of 4. Therefore, these collectively validate the model's reliability for predicting surface roughness within the defined parameter space.

If the p-value of partial regression coefficients in the model exceeds 0.05, it is deemed statistically nonsignificant to the predictive model. As shown in Table 8, the partial regression coefficients for polishing force A_{800}, rotational speed B_{800}, feeding speed C_{800}, and the interaction term $A_{800}C_{800}$ and C_{800}^2 demonstrate that the p-values are below 0.05, indicating significant effects on the response variable. These terms were retained. Conversely, the interaction terms $A_{800}B_{800}$, $B_{800}C_{800}$, A_{800}^2, and B_{800}^2 demonstrate that the p-values are above 0.05, showing that they are not significant and should be excluded. The finalized predictive model after eliminating nonsignificant factors is expressed as follows:

$$R_a = 0.033 - 0.003A_{800} - 0.003B_{800} + 0.0048C_{800} + 0.0018A_{800}C_{800} + 0.0009B_{800}^2 \quad (17)$$

Based on the experimental results in Table 8 and the roughness prediction model, response surfaces for the polishing parameters were generated, as illustrated in Figure 9. Figure 9a demonstrates that surface roughness decreases with increasing feeding speed and rotational speed, though the rate of reduction diminishes progressively. These trends align with the single-factor experimental results. The minimum surface roughness of 0.017 μm is achieved at a 5000 RPM rotational speed and a 5 mm·s^{-1} feeding speed. Figure 9b reveals that the roughness decreases with a higher rotational speed and polishing force, reaching its minimum value of 0.017 μm at a 20 N polishing force and a 5000 RPM rotational speed. Consequently, the optimal polishing parameters for the P800 sandpaper are identified as a

20 N polishing force, a 5000 RPM rotational speed, and a 5 mm·s^{-1} feeding speed, yielding the minimum surface roughness of 0.017 μm.

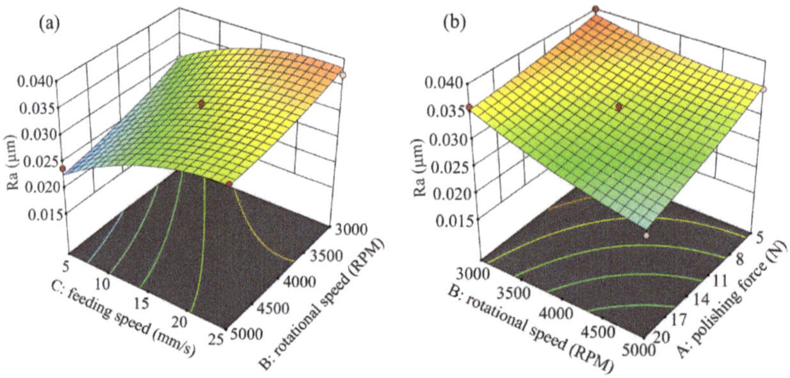

Figure 9. The response surfaces between surface roughness and the polishing parameters of (**a**) rotational speed and feeding speed, (**b**) polishing force and rotational speed.

To validate the accuracy of the surface roughness prediction model, five randomly selected polishing parameters were tested, and the results are shown in Table 9. The relative errors range from 3.70% to 7.14%, all within acceptable limits, demonstrating the reliability of the surface roughness prediction model.

Table 9. Results of the polishing experiments for model verification.

Polishing Force (N)	Rotational Speed (RPM)	Feeding Speed (mm·s^{-1})	R_a (μm)		Error (%)
			Predicted	Measured	
20	5000	5	0.017	0.018	5.56
15	5000	5	0.021	0.022	4.55
8	5000	5	0.026	0.027	3.70
20	5000	15	0.027	0.026	3.85
20	5000	20	0.030	0.028	7.14

4. Conclusions

In this study, the integration of an AHP polishing tool and a robot system was used to analyze the effects of polishing force, rotational speed, and feeding speed on the surface roughness of NAK80 steel, and a relationship model between surface roughness and the polishing parameters was derived. Before over-polishing, surface roughness exhibits direct proportionality to the polishing force and rotational speed but inverse proportionality to the feeding speed. These relationships were reversed after over-polishing. For the P180 sandpaper, the optimal polishing parameters are a 20 N polishing force, a 5000 RPM rotational speed, and a 5 mm·s^{-1} feeding speed, achieving a minimum surface roughness of 0.08 μm. For the P400 sandpaper, the optimal parameters of a 10 N polishing force, a 4000 RPM rotational speed, and a 5 mm·s^{-1} feeding speed yielded a minimum surface roughness of 0.044 μm. The optimal parameters for the P800 sandpaper are a 20 N polishing force, a 5000 RPM rotational speed, and a 5 mm·s^{-1} feeding speed. By sequentially applying the optimal parameters to all three grit sizes, the surface roughness of the sample can be reduced from 0.4 μm to 0.017 μm, representing a 95.75% improvement. In addition, the error of the surface roughness prediction model established according to the central composite method for the P800 sandpaper is less than 7.14%.

The above research provides a new polishing framework for the AHP high-precision polishing of mold steel, which helps to meet the high-precision polishing requirements of molds. However, this study only focuses on the optimal polishing parameters obtained for NAK80 using P180, P400, and P800 sandpaper. After replacing the polishing material and sandpaper, parameter optimization still needs to be carried out based on the obtained results.

Author Contributions: Conceptualization, X.Z.; methodology, D.S. and H.Z.; validation, X.Z. and H.Z.; formal analysis, X.Z. and X.W.; investigation, D.S. and X.W.; data curation, X.W.; writing—original draft, X.Z.; writing—review and editing, D.S.; supervision, D.S. and H.Z. All authors have read and agreed to the published version of the manuscript.

Funding: This study was partially supported by the Key Technology Research and Development Program of Heilongjiang Province (GA21A103).

Data Availability Statement: The original contributions presented in the study are included in the article; further inquiries can be directed to the corresponding author.

Conflicts of Interest: The authors declare that they have no known competing financial interests or personal relationships that could appear to influence the work reported in this paper.

References

1. Wang, K.; Ding, L.; Dailami, F.; Matthews, J. A Methodology to Support Robotic Polishing of Molds Integrated with CAD/CAM. *J. Adv. Manuf. Syst.* **2024**, *23*, 349–390.
2. Li, J.; Huang, W.; Xie, Y.; Yang, J.; Zhao, M. The Parameters Optimization of Robotic Polishing with Force Controlled for Mold Steel Based on Taguchi Method. *J. Braz. Soc. Mech. Sci.* **2024**, *46*, 313.
3. Li, C.L.; Nan, R.; Wei, Y.Y.; Li, L.; Liang, J.X.; Li, N. Application Research of Vision-Guided Grinding Robot for Wheel Hub Castings. *Processes* **2025**, *13*, 238. [CrossRef]
4. Walker, D.; Brooks, D.; King, A.; Freeman, R.; Kim, S.W. The 'Precessions' Tooling for Polishing and Figuring Flat, Spherical and Aspheric Surfaces. *Opt. Express* **2003**, *11*, 958–964. [PubMed]
5. Lee, H.; Kim, J.; Kang, H. Airbag Tool Polishing for Aspherical Glass Lens Molds. *J. Mech. Sci. Technol.* **2010**, *24*, 153–158.
6. Kordonski, W.I.; Golini, D. Fundamentals of Magnetorheological Fluid Utilization in High Precision Finishing. *J. Intell. Mater. Syst. Struct.* **1999**, *10*, 683–689.
7. Wang, W.; Ji, S.; Zhao, J. Review of Magnetorheological Finishing on Components with Complex Surfaces. *Int. J. Adv. Manuf. Technol.* **2024**, *131*, 3165–3191.
8. Wu, D.; Wang, H.; Zhang, K.; Lin, X. Research on Flexible Adaptive CNC Polishing Process and Residual Stress of Blisk Blade. *Int. J. Adv. Manuf. Technol.* **2019**, *103*, 2495–2513.
9. Shiou, F.J.; Cheng, C.H. Ultra-Precision Surface Finish of Nak80 Mould Tool Steel Using Sequential Ball Burnishing and Ball Polishing Processes. *J. Mater. Process. Tech.* **2008**, *201*, 554–559.
10. Huang, H.Y.; Fuh, K.H.; Wu, J.S.; Tai, Y.H. A New Integrated Polishing Process Design for Plastic Mold Steel to Mirror-Like Surface. *Int. J. Adv. Manuf. Technol.* **2014**, *73*, 1633–1645.
11. Zhao, T.; Shi, Y.; Lin, X.; Duan, J.; Sun, P.; Zhang, J. Surface Roughness Prediction and Parameters Optimization in Grinding and Polishing Process for IBR of Aero-Engine. *Int. J. Adv. Manuf. Technol.* **2014**, *74*, 653–663. [CrossRef]
12. Liu, Z.; Hu, Y.; Zhang, M.; Zhang, W.; Wang, J.; Lei, W.; Wang, C. Surface Morphology Evolution Mechanisms of Pulse Laser Polishing Mold Steel. *Int. J. Mech. Sci.* **2024**, *269*, 109039. [CrossRef]
13. Zou, L.; Liu, X.; Huang, Y.; Fei, Y. A Numerical Approach to Predict the Machined Surface Topography of Abrasive Belt Flexible Grinding. *Int. J. Adv. Manuf. Technol.* **2019**, *104*, 2961–2970. [CrossRef]
14. Yin, T.F.; Zhang, H.Q.; Hang, W.; To, S. A Novel Approach to Optimizing Grinding Parameters in the Parallel Grinding Process. *Processes* **2024**, *12*, 493. [CrossRef]
15. Xie, Y.; Chang, G.; Yang, J.; Zhao, M.; Li, J. Process Optimization of Robotic Polishing for Mold Steel Based on Response Surface Method. *Machines* **2022**, *10*, 283. [CrossRef]
16. Qi, B.; Huang, X.; Guo, W.; Ren, X.; Chen, H.; Chen, X. A Novel Comprehensive Framework for Surface Roughness Prediction of Integrated Robotic Belt Grinding and Burnishing of Inconel 718. *Tribol. Int.* **2024**, *195*, 109574. [CrossRef]
17. Wang, X.; Wang, X.; Yang, Z.; Zou, Y. An Investigation of Real-Time Robotic Polishing Motion Planning Using a Dynamical System. *Machines* **2024**, *12*, 278. [CrossRef]

18. Wahballa, H.; Duan, J.; Dai, Z. An Automatic Robot Polishing Control Method for Compound Surface Comprising Plane and Curved Surfaces. *Int. J. Adv. Manuf. Technol.* **2024**, *132*, 3801–3819. [CrossRef]
19. Mohsin, I.; He, K.; Li, Z.; Du, R. Path Planning Under Force Control in Robotic Polishing of the Complex Curved Surfaces. *Appl. Sci.* **2019**, *9*, 5489. [CrossRef]
20. Fernandez, A.; Dieste, J.A.; Javierre, C.; Santolaria, J. Surface Roughness Evolution Model for Finishing Using an Abrasive Tool on a Robot. *Int. J. Adv. Robot. Syst.* **2015**, *12*, 119. [CrossRef]
21. Verma, T.; Chawla, O.; Jha, S. Mathematical Modelling for Prediction of Surface Roughness in Pneumatically Configurable Polishing Process. *Wear* **2022**, *504*, 204434.
22. Savio, G.; Meneghello, R.; Concheri, G. A Surface Roughness Predictive Model in Deterministic Polishing of Ground Glass Moulds. *Int. J. Mach. Tools Manuf.* **2009**, *49*, 1–7.
23. Feng, D.; Sun, Y.; Du, H. Investigations on the Automatic Precision Polishing of Curved Surfaces Using a Five-Axis Machining Centre. *Int. J. Adv. Manuf. Technol.* **2014**, *72*, 1625–1637. [CrossRef]
24. Qi, J.; Zhang, D.; Li, S.; Chen, B. Modeling and Prediction of Surface Roughness in Belt Polishing Based on Artificial Neural Network. *Proc. Inst. Mech. Eng. Part B J. Eng. Manuf.* **2018**, *232*, 2154–2163.
25. Wang, G.; Zhou, X.; Meng, G.; Yang, X. Modeling Surface Roughness for Polishing Process Based on Abrasive Cutting and Probability Theory. *Mach. Sci. Technol.* **2018**, *22*, 86–98. [CrossRef]
26. Zhang, X.; Sun, Y. Development of Pneumatic Force-Controlled Actuator for Automatic Robot Polishing Complex Curved Plexiglass Parts. *Machines* **2023**, *11*, 446. [CrossRef]
27. Shang, Y.R.; Hu, S.B.; Qiao, H. Sensitivity Study of Surface Roughness Process Parameters in Belt Grinding Titanium Alloys. *Metals* **2023**, *13*, 1825. [CrossRef]
28. Yang, Z.Y.; Xu, X.H.; Zhu, D.H.; Yan, S.J.; Ding, H. On Energetic Evaluation of Robotic Belt Grinding Mechanisms Based on Single Spherical Abrasive Grain Model. *Int. J. Adv. Manuf. Technol.* **2019**, *104*, 4539–4548.
29. Baidakova, N.V.; Kryukov, S.A. Wear of Abrasive Grains. *Russ. Eng. Res.* **2024**, *44*, 1550–1552.
30. Doman, D.A.; Warkentin, A.; Bauer, R. A Survey of Recent Grinding Wheel Topography Models. *Int. J. Mach. Tools. Manuf.* **2006**, *46*, 343–352.
31. Han, Y.X.; Yang, Y.S.; Lyu, B.; Hang, W.; Wang, X.; Yuan, J.L. Optimization of Shear-Thickening Polishing Parameters for Optical Glass Based on Grey Relational Analysis. *Machines* **2025**, *13*, 40. [CrossRef]
32. Qiao, H.; Hu, S.B.; Xiang, Y.; Liu, S.S.; Zhang, L. Research on the Analysis and Prediction Model of Machining Parameters of Titanium Alloy by Abrasive Belt. *Metals* **2023**, *13*, 1085. [CrossRef]
33. Xi, F.; Zhou, D. Modeling Surface Roughness in the Stone Polishing Process. *Int. J. Mach. Tools Manuf.* **2005**, *45*, 365–372.

processes

MDPI

Article

Quality Control of Electro-Discharge Texturing of Rolls Through Six Sigma

Emil Evin [1], Vladimír Kokarda [2] and Miroslav Tomáš [1,*]

1 Department of Automotive Production, Faculty of Mechanical Engineering, Technical University of Košice, Mäsiarska 74, 040 01 Košice, Slovakia; emil.evin@tuke.sk

2 Carl Zeiss Slovakia, s.r.o, Račianska 12481/77/A, 831 02 Bratislava, Slovakia; vladimir.kokarda@zeiss.com

* Correspondence: miroslav.tomas@tuke.sk; Tel.: +421-55-602-3524

Abstract: This article presents the implementation of the Six Sigma methodology in the electro-discharge texturing process of cold mill work rolls. The final surface quality of sheet metal must meet the specific demands of car body part producers, which require a specific surface texture described by surface microgeometry parameters: the average roughness and the peak density. The requirements for the surface microgeometry of sheet metal are mainly related to improving the formability and adhesion of the paint in the body painting process. These microgeometry parameters can be controlled by the texture of work rolls: this texture is transferred onto the sheet metal surface. The electro-discharge texturing process allows for control of the average roughness and peak density according to individual customer specifications. In this study, a model is proposed to predict the average roughness based on the input parameters of the electro-discharge texturing process: current, voltage, and time. Compared to previous models, this model includes more input parameters. The process suitability was analyzed using control charts, capability indices, and Z scores. The modified weighted product method was used to create a purpose function describing the relationships between the input and output quality parameters. Based on the agreement of the target quality characteristics and the calculated values according to the models obtained, an algorithm to control the texturing process of the work rolls was designed. The proposed model was also validated on results published by other authors and demonstrated good agreement. This study should contribute to the philosophy of continuously improving the surface quality of cold-rolled sheet metal.

Keywords: roughness; peak density; sheet metal; electro-discharge texturing; six sigma; statistical method; process control

1. Introduction

Customers often perceive the quality of cars through the attractiveness of the design, color, and gloss of the body paint. These quality features are closely related to the sheet metal texture, the production quality of the car body surface parts made of this sheet metal, the accuracy of the assembly of these parts, etc. [1]. The rolling process used to make the sheet metal determines its manufacturing qualities, including its formability, glued joint strength, and suitability for paint application [2,3]. Automobile companies specify requirements for texture characteristics—roughness average (*Ra*), peak density (*Pc*), and waviness (*Wa*)—of the sheet metal surface for visible body parts [4–7]. For these car body parts, carmakers prefer the roughness (*Ra*) of the sheet metal surface to range from 1.0 μm to 1.6 μm; for the inner parts, it can be higher [2,3]. For example, Škoda Auto, in its specifications, states roughness values of *Ra* from 1.1 to 1.6 μm and peak density $Pc \geq 40$ cm^{-1}; Ford requires roughness *Ra* from 1.1 to 1.7 μm and peak density $Pc \geq 50$ cm^{-1}; Volkswagen requires roughness *Ra* from 1.1 to 1.6 μm and maximum peak density $Pc \geq 60$ cm^{-1}; and Renault requires roughness *Ra* from 0.7 to 1.3 μm and peak density $Pc \geq 110$ cm^{-1} [4,8].

The manufacturability of car body parts by cold forming—their formability—is greatly affected by the friction between the contact surfaces of the tool and the sheet metal. With a suitable sheet metal surface texture, lubricant accumulates in close pockets on the surface of the sheet metal. This lubricant is pushed out of these pockets during the movement of the material (during forming) and forms a continuous layer between the contact surfaces. The created lubricant layer prevents abrasion (sticking) of softer material (sheet metal) on the contact surfaces of the tool and contributes to reducing frictional forces and improving formability. On the contrary, with an inappropriate texture, the pressure on the contact surfaces increases several times when the sheet moves between the contact surfaces of the tool. As a result of a local increase in pressure on the contact surfaces, especially on the drawing edge of the drawbar, an insufficient amount of lubricant is captured in the shallow pockets. An adequate amount of lubricant is required to create a continuous layer. It has been found that the combination of smaller *Ra* and greater *Pc* values is not the most suitable because when the material is moved between the contact surfaces of the tool, the peaks on the surface of the sheet metal wear and are pushed into the depths on the surface of the sheet metal. On the other hand, a more significant increase in the roughness *Ra* will cause the surface of the tool to be matte after painting [2,3].

The sheet metal surface texture specified by individual car manufacturers can be achieved via controlled formation of the surface texture of cold mill work rolls and rolling parameters (rolling speed, thickness reduction, rolling force, etc.) [9,10]. Thus, the texture from the surface of the cold mill work rolls is transferred to the sheet metal surface [11,12].

Texturing of the work rolls' surface can be carried out using different methods. In shot blast texturing (SBT) with fine-grained granules, the roll surface quality depends on the shot shape and bulk, the kinetic energy of blast shots, and the surface quality before texturing. Even though the process is simple, highly productive, and relatively low-cost, the surface roughness that it produces is not homogeneous and has low roughness retention compared to other texturing methods [13,14]. Electric discharge texturing (EDT) has been proven to be a reliable method of providing the required surface microtopography [15]. The process consists of the formation of minute surface craters through the discharge of electrical energy between two metal electrodes that are separated by a dielectric fluid. The resultant surface integrity of the roll can be achieved through regulation of the amount of peak current, pulse on/off time, capacitance, polarity, tool electrode material, etc. [16]. Laser beam texturing (LBT) methods, such as Pomini digital texturing, allow good control of the surface texture matrix by means of software, thus resulting in a wide range of surfaces, ranging from deterministic to stochastic, that can be produced. Using this method, it is possible to alter the spacing parameters, density, crater overlap, and various combinations of effects such as fixed crater size and random placement (or variable crater sizes with fixed placement) [5,17]. Electron beam texturing (EBT), similar to laser beam texturing, creates craters with a solid collar when the roll turns into a vacuum chamber and moves along the axis. The distribution of the craters can be controlled by the roll speed and the frequency of beam irradiation; in addition, both may be synchronized. However, the texture obtained is not the most desirable in terms of tribological conditions on contact surfaces of the die since worsened sheet formability can be expected as a result [15]. The Topocrom process creates convex hemispheres on the roll surface that offer an opportunity for a wide range of different resulting topographies, thus achieving the required surfaces by adapting the mean radius, size distribution, and the number of these hemispheres per unit area. This hard chromium electroplating process improves the life of the surface textures created via conventional processes (SBT, EDT, EBT) [15,17]. As Gorbunov and Colak summarized, textures created with EDT, LBT, EBT, and Topocrom technologies show a good correlation between Ra and Pc. Even though the rolls produced via the Topocrom process were found to have a life that was roughly ten times longer than others, high Pc may occur on the surface rolls. The LBT process produces rolls with a very high roughness retention property, but the texturing process is time-consuming, which is a big disadvantage [15,17].

When applying *EDT*, the texture of the surface and the properties of the surface layer of the work rolls change due to electric discharges. The amount of material reduction in *EDT* depends on the erosion effect of the electrical discharge between two electrodes that are separated by a dielectric liquid and exposed to the electrical discharge E_{DP} [9–11,18]. The E_{DP} can be expressed by Relation (1):

$$E_{DP} = U_p.I_p.T_{on}, \tag{1}$$

while

$$Q_{DP} = I_p.T_{on}, \tag{2}$$

where U_p is the peak voltage of the discharge; I_p is the peak current; T_{on} is the pulse on-time; and Q_{DP} is the pulse charge.

It can be seen from Relation (1) that the amount of material reduction depends on the parameters of the *EDT* process, i.e., voltage U_p, current I_p, and discharge duration T_{on} [9–11,18]. To track non-conforming products, control tables are used during data collection to evaluate whether the product passes (OK) or fails (N-OK). Based on these data, the DPMO can be evaluated [19]. The number of defective parts (defects per million opportunities, DPMO) can be determined using Relation (3) [20–22]:

$$DPMO = \frac{D}{N.O}.10^6, \tag{3}$$

where *DPMO* is defects per million opportunities; *D* is the total number of defective parts; *N* is the total number of verified parts; and *O* is the number of opportunities per unit.

Another sign of process quality is the Six Sigma (Z) level. In the case of continuous data, Z can be calculated according to Relations (4) and (5):

$$Z_U = \frac{UCL - \overline{X}}{\sigma}, \tag{4}$$

$$Z_L = \frac{\overline{X} - LCL}{\sigma}, \tag{5}$$

where Z_U is the sigma level for the upper specification limit; Z_L is the sigma level for the lower specification limit; $\overline{X}$ is the measured average value of the quality characteristic; σ is the standard deviation of the quality characteristic; *UCL* is the upper control limit of the quality characteristic specified by the customer; and *LCL* is the lower control limit of the quality characteristic specified by the customer.

Requirements of car manufacturers for the quality of the surface of the body parts and the elimination of the waste of resources force the steel industry to produce sheet metals of high surface quality [20,21]. Therefore, steel companies are taking measures to continuously improve the surface texturing processes of work rolls. One possible measure is to reduce the variability of the input parameters of the roll texturing process. According to the International Organization for Standardization, one possibility is the application of Six Sigma principles. Six Sigma tools make it possible to identify the causes of variability and subsequently manage them in order to gradually eliminate the number of non-conforming parts (defective or N-OK parts) with the specification (see Figure 1) [22–26]. The Six Sigma level (Z) can also be calculated on the basis of *DPMO* using Relation (6):

$$Z = 1.5 * \sigma^{-1} * \left(\frac{DPMO}{10^6}\right), \tag{6}$$

or read from Table 1.

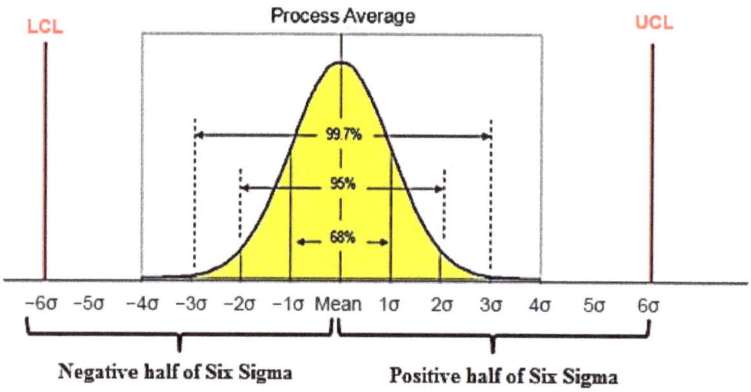

Figure 1. Probability of occurrence of non-conforming products.

Table 1. Six Sigma conversion table.

Capability Index $C_{pk\ min}$	1	1.13	1.33	1.49	1.667	1.75	1.83	2
Six Sigma level (Z_{min} score)	3	3.39	4	4.5	5	5.25	5.5	6
DPMO	660,807	30,396	6210	1350	233	88	32	3
Cost of poor quality COPQ	25% to 45%		15% to 25%			5% to 15%		

Reducing the number of defective products and deviations from the required value of the monitored quality parameter of a given product is the basis for achieving process stability. As mentioned previously, the level of process perfection can be expressed as the maximum number of defects per million opportunities (DPMOs). Using DPMOs as a quality measure in the product creation process can also sensitively reflect the economic aspects of the process. The occurrence of defects always represents, in the final sense, an increase in the cost of poor quality, which ultimately has a negative impact on profit generation. Table 1 shows the relationship between the number of defects, the Six Sigma level, and the cost of quality (cost of poor quality, COPQ) [27].

The aim of this study is to assess the ability of *EDT* to achieve the specified characteristics (*Ra* and *Pc*) of the surface texture of cold mill work rolls on the basis of control diagrams, C_{pk}, and Z scores. Furthermore, in this study, we aim to carry out a regression analysis of the dependence of the *Ra* and *Pc* quality characteristics on the *EDT* input parameters U_p, I_p, T_{on}, and T_{off} and propose an algorithm for gradual improvement of the output parameters of the *Ra* and *Pc*.

2. Materials and Methods

The Six Sigma method offers several techniques that can be used to achieve continuous process improvement. One of them is *DMAIC*. According to this methodology, the procedure needs to be divided into five phases: define, measure, analyze, improve, and control [21,22].

Define: When defining the process, the following assumptions were made:

- When texturing with *EDT* technology, there is a very good correlation between the roughness *Ra* and the peak density *Pc* [4,5,28];
- Prior results show that, depending on the rolling force, it is possible to achieve a 10% to 50% texture transfer from the rolls of the rolling mill to the sheet surface [2,4];
- On the EDT 2100/4500 device, it is possible to repeatedly create a texture on the surface of rolls with roughness *Ra* in the range from 1.5 μm to 3.8 μm and with peak density $Pc > 40$ cm^{-1}.

Measure: The rolls were textured on EDT 2100/4500 equipment in an oil dielectric BP250 with eight copper electrodes and using three different plans (*PE 1*, *PE 2*, and *PE 3*; see Table 2). The measurement of the output parameters of the work rolls' surface roughness was carried out with a Hommel Tester T 1000 roughness meter (JENOPTIK Industrial metrology, Villingen-Schwenningen, Germany) according to DIN EN ISO 4288 [29]. Roughness measurements *Ra* and *Pc* were taken three times on the left side, three times in the middle, and three times on the right side along the rolls when the rolls were turned at angles of 0°, 90°, 180°, and 270° (see Figure 2). The average *Ra* values of the work rolls are shown in Table 3.

Table 2. Plan of experiments—input values.

	PE 1 T_{Ra1} = 1.5 (μm)	PE 2 T_{Ra2} = 2.5 (μm)	PE 3 T_{Ra3} = 3.8 (μm)
Texture current *Ip* (A)	4	18	19
Lower voltage limit U_{PL} (V)	5	10	15
Upper voltage limit U_{PH} (V)	8	20	25
Pulse duration T_{on} (μs)	4	14	29
Technological break duration T_{off} (μs)	6	16	38
Electrode diameter *d* (mm)	8	8	8

Note: PE 1—plan of experiment 1; PE 2—plan of experiment 2; PE 3—plan of experiment 3; $T_{Ra,1}$—target value in PE 1; $T_{Ra,2}$—target value in PE 2; $T_{Ra,3}$—target value in PE 3.

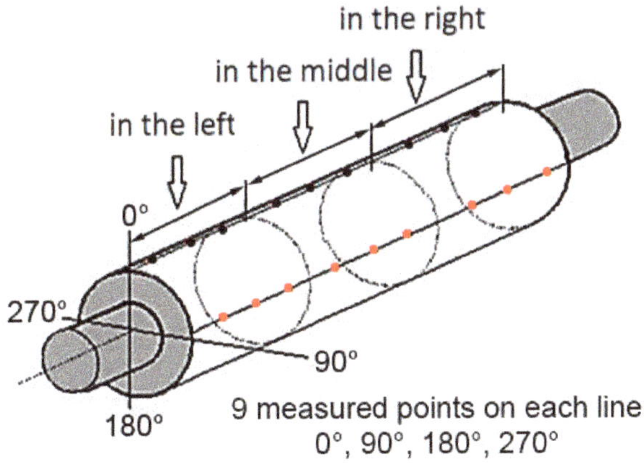

Figure 2. Representation of measuring points on the work rolls.

Table 3. Measured values in the variation range $VR_{Ra} = CL_{Ra,RR} \pm VR_{Ra,SS}$—first step.

Measure Value	PE 1		PE 2		PE 3	
	Ra (μm)	*Pc* (cm^{-1})	*Ra* (μm)	*Pc* (cm^{-1})	*Ra* (μm)	*Pc* (cm^{-1})
Average value $\overline{Ra}$	1.548	161	2.5	99	3.845	68
Standard deviation *SD*	0.06	7	0.06	3	0.04	3
Upper-level Six Sigma *ZU*	3.38	-	4.49	-	2.25	-
Lower-level Six Sigma *ZL*	4.98	18.6	4.48	21.1	7.53	10.0
C_{pKU}	1.13	-	1.50	-	1.75	-
C_{pKL}	1.66	6.2	1.49	7.05	2.51	-
DPMO	30,396	0	1350	0	88	3.36

3. Results and Discussion

Analysis: In the first phase, variability was analyzed based on the records of the number of non-conforming parts. Out of 175 inspected rolls, 17 were out of specification. Using Relation (3), the value of $DPMO \approx 30{,}396$ was calculated, and a Six Sigma level of 3.39 was subtracted from Table 1. Defects were identified on work rolls, such as burning of the rolls, shine-through of the textured surface of the rolls, strips around the perimeter of the rolls, strips from rolls (aluminations), and texture parameters *Ra* and *Pc*.

Burning of the work rolls can be caused by the short duration of the technological T_{off}, during which there is no interruption of the pulse. Roll burnout can also occur if the circumferential speed of the roll rotation is low. In this case, impurities are not removed from the dielectric liquid, and as a result, the electric arc between the electrode and the workpiece (roll) recurs.

Defects on the textured surface of the roll can be caused by imperfect grinding of the roll. When taking over the rolls from the supplier, precise grinding of the rolls is required. Another cause of defects in the roll texture is damage to the roll pins.

Bands around the circumference of the roll can be caused by an incorrect setting of the texturing I_p. The occurrence of this error is permissible within a certain interval; otherwise, the input values of the texturing current I_p need to be adjusted. Other causes of these strips are poorly ground pins, a worn bearing, an unseated support on the base, and bad centering in the chrome plating bath. These errors can be eliminated by taking measures related to the acceptance procedure from the producer, following technological procedures when setting the rolls in the device, and regular maintenance of the electrospark device.

When analyzing the stability of the process, control diagrams are used, in which it is necessary to mark the following control limits of quality indicators:

- *CL*, the central line, or the target value ($CL \approx TV$) of the regulated roughness characteristic.
- *UCL*, the upper control limit, is calculated from the range of variation (*VR*) of the regulated quality indicator ($UCL = CL + VR/2$). If the measured values are above the *UCL*, the process is not stable.
- *LCL*, the lower control limit, is calculated from the *VR* range of the regulated quality indicator ($UCL = CL - VR/2$). If the measured values are below the *LCL*, the process is not stable.
- *UWL*, the upper warning limit, is calculated from twice the *VR* ($UWL = CL + 2.VR/6$).
- *LWL*, the lower warning limit, is calculated from twice the *VR* ($UWL = CL - 2.VR/6$).
- *USL*, the upper safe limit, is calculated from a multiple of the *VR* ($USL = CL + VR/6$).
- *LSL*, the lower safe limit, is calculated from a multiple of the *VR* ($USL = CL - VR/6$).

Regulation diagram creation was based on the requirement specifications of the Volkswagen car company for the roughness of the sheet metal surface (Ra_{SS}) for the external parts of the car body (Ra_{max} = 1.6 μm and Ra_{min} = 1.1 μm). The variation range ($VR_{Ra,SS}$) of the sheet metal surface roughness was determined by Relation (7):

$$VR_{Ra,SS} = Ra_{max} - Ra_{min}, \quad (7)$$

For the work rolls, $UCL_{Ra,FR}$ and $VR_{Ra,FR}$ were determined based on the variation range of roughness $VR_{RA,SS}$ specified by the automobile company Volkswagen for the sheet surface according to Relations (8) and (9):

$$UCL_{Ra,FR} = CL_{Ra,FR} + \frac{VR_{Ra,SS}}{2}, \quad (8)$$

$$LCL_{Ra,FR} = CL_{Ra,FR} - \frac{VR_{Ra,SS}}{2}, \quad (9)$$

where $CL_{Ra,FR} = TV_{Ra,FR}$ is the target value of the quality sign *Ra* of the finishing rolls, and $VR_{Ra,SS}$ is the variation range of the *Ra* quality sign of the sheet metal.

In the first step of the analysis phase, the following were indicated in the control diagram: $CL_{Ra,FR} = TV_{Ra,FR}$, $UCL_{Ra,FR}$, $LCL_{Ra,FR}$, $UW_{Ra,FR}$, $LWL_{Ra,FR}$, $UCL_{Ra,FR}$, and $LCL_{Ra,FR}$. Attention was focused first on identifying data that were out of the regulatory field $CL_{Ra,FR} \pm 0.25$ μm; see Figure 3 (samples from 1 to 20, which are outside the regulatory boundaries, represent an unstable process). Data on these parts were excluded from further analysis because they are unstable processes. In the second phase, the *Ra* roughness data were analyzed; these were within the specified control limits (these are samples at *x* from 20 to 40 in Figure 3). It is assumed that the variability arose as a result of common (predictable) events, which the regulation of texturing process parameters can retain within the specified limits of $UCL_{Ra,FR}$ and $LCL_{Ra,FR}$. If the statistical process is stabilized, i.e., definable causes are identified and analyzed, it is necessary to propose measures preventing their recurrence and, subsequently, propose measures for improvement. To assess the stability of the production process, it is possible to use the capability index C_{pk}, which is calculated according to Relations (10) and (11):

$$C_{Pk,U} = \frac{UCL - \overline{X}}{3.\sigma} > 1.33, \quad (10)$$

$$C_{Pk,L} = \frac{\overline{X} - LCL}{3.\sigma} > 1.33, \quad (11)$$

where $C_{pk,U}$ is the process capability index calculated for the *j*-th texture quality sign UCL_j; $\overline{X}$ is the average value from the measured data of the *j*-th sign of texture quality; σ is the standard deviation of the measured data of the *j*-th sign of texture quality; and $C_{pk,L}$ is the process capability index calculated for the *j*-th texture quality sign LCL_j.

If the C_{pk} value is less than 1.0, the process is not capable. If the C_{pk} value is between 1.0 and 1.33, the process is considered barely capable, and if the C_{pk} value is greater than 1.33, the process is considered capable. If possible, it is necessary to aim for the highest possible C_{pk} value (greater than 2). If both the $C_{pk,U}$ and $C_{pk,L}$ values are greater than 1.33, the process is capable, and the process is under statistical control. In cases where one or both of the $C_{pk,U}$ and $C_{pk,L}$ values are smaller than 1.33, the process is not capable, and the process is not under statistical control.

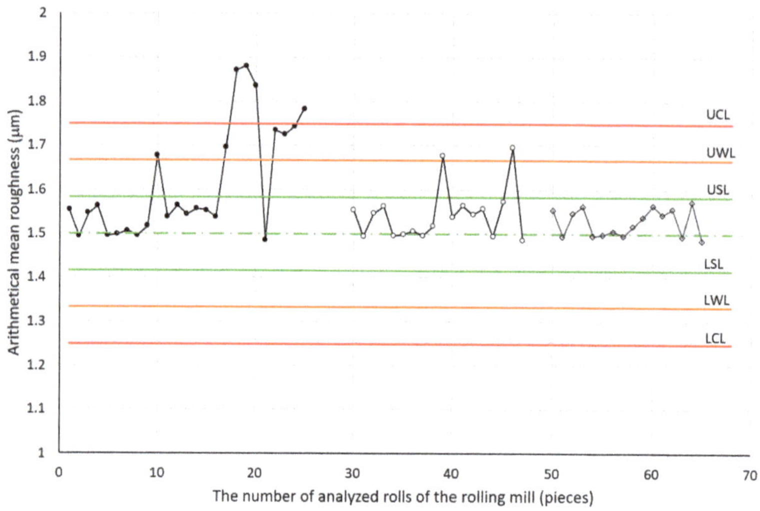

Figure 3. Detection of a gradual change in roughness *Ra*—PE1.

Table 3 shows the *Ra* values measured in the zone between UCL_{Ra} and LCL_{Ra} (in the edge zone). From Table 3, it can be seen that the capability indexes $C_{pk,URa}$ and C_{pkLRa} for both the *PE2* and *PE3* experiment plans are greater than 1.33. This means that the *PE2* and *PE3 EDT* processes are capable. For the *EDT* according to plan PE1, the value of C_{pkU} = 1.13. This means that the process under this plan is not capable. However, measures need to be taken to improve it, as it is less than 1.33, and the projected number of non-compliant $DPMO_{L,Ra}$ products would be approximately 30,396 out of 1 million possibilities.

Table 4 shows the results after filtering out the data that were in the marginal area. The capability indexes' values were greater than 1.33 in all three plans. This means that the processes in plans *PE1*, *PE2*, and *PE3* are capable after filtering out data from the marginal area, and the predicted number of $DPMO_{Ra}$ of non-conforming products would be 1350 in the case of *PE2*, and ≈0 in the cases of *PE1* and *PE3*. For illustration, a stepwise data filtering procedure is shown in Figure 3 for *PE1*. Data within the range 0–25 are from before the filtering out was performed (raw data), data within the range 30–45 are from after filtering out the data outside the UCL and LCL, and data in the range 50–65 are from after filtering out the data outside UWL-UCL and LWL-LCL.

Table 4. Measured values in the variation range $VR_{Ra} = CL_{Ra,RR}/3 \pm VR_{Ra,SS}$—second step.

Measure Value	PE 1		PE 2		PE 3	
	Ra (μm)	*Pc* (cm^{-1})	*Ra* (μm)	*Pc* (cm^{-1})	*Ra* (μm)	*Pc* (cm^{-1})
Average value $\overline{Ra}$	1.526	158	2.495	99	3.83	68
Standard deviation *SD*	0.03	7	0.05	3	0.02	3
Upper-level Six Sigma *ZU*	7.19	-	4.73	-	10.0	-
Lower-level Six Sigma *ZL*	8.88	18.1	4.56	21.1	12.91	9.59
C_{pKU}	2.40	-	1.58	-	3.34	-
C_{pKL}	2.96	6.0	1.52	7.05	4.31	3.20
DPMO	0	0	1350	0	0	0

In the case of Pc, only the lower limit value (LCL_{Pc} = 40 cm^{-1}) is defined. Assuming that the lower control limit (LCL_{Pc}) is determined using Relation (12):

$$LCL_{Pc} = CL_{Pc} - k.\sigma_{PC}, \tag{12}$$

Then, after adjusting Equation (12), we obtain Equation (13):

$$CL_{Pc} = LCL_{Pc} + k.\sigma_{PC}, \tag{13}$$

where k is a multiple of the standard deviation of the distance of the upper control limit from the center line (k = 3).

In our case, the maximum value of the directional deviation is equal to 7. This means that CL = 40 + 3 × 7 = 61 cm^{-1} (see Figure 4). It follows from this figure that the values of Pc determined from records $PE1$, $PE2$, and $PE3$ are higher than the lower LCL_{Pc} limit value. This means that the EDT processes $PE1$, $PE2$, and $PE3$ are capable of producing Pcs according to specifications without the need to take corrective action.

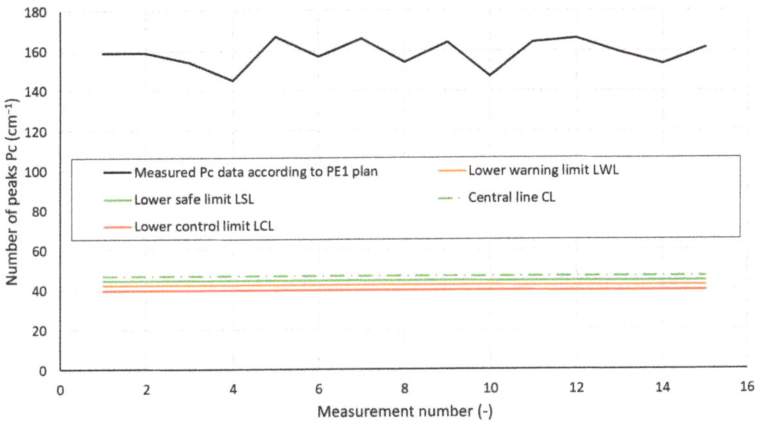

Figure 4. Detection of a gradual change in the peak density Pc—PE1.

Control: Managing the process (object) by means of trial and error is ineffective. By purposefully controlling the process aimed at achieving a specified goal, it is possible to gradually set the input parameters using the purpose function [30]. The objective function of multicriteria optimization describing the relationships between input parameters Xi and output quality features S_{Qj} [22] was created by modifying the weighted product method (WPM) in the form of Equation (14):

$$S_{Qi} = \frac{1}{n}\sum_{1}^{n}(x_1.w_1 + x_2.w_2 \ldots + x_n.w_n) \pm \Delta_{Qj}, \tag{14}$$

where S_{Qj} represents the output quality feature; $x_1, x_2, \ldots x_n$ represent the input independent factors of the process; w_i represents the criterion weight; j represents a sign of quality; n represents the number of process input parameters; and Δ_{Qj} represents the accuracy or uncertainty.

It can be seen from Figures 5–8 that the dependences of Ra on the discharge energy, or the texturing process parameters I_p, U_p, and T_{on}, show a good correlation, and the regression equations (Equations (15)–(17)) describe them with very high accuracy:

$$Ra_{Ip} = 1.476 * \ln(I_p) - 0.5262; \ R^2 = 0.99 \tag{15}$$

$$Ra_{Up} = 0.1244 * U_p;\ R^2 = 0.99 \quad (16)$$

$$Ra_{Ton} = 0.0915 * T_{on} + 1.19;\ R^2 = 0.99 \quad (17)$$

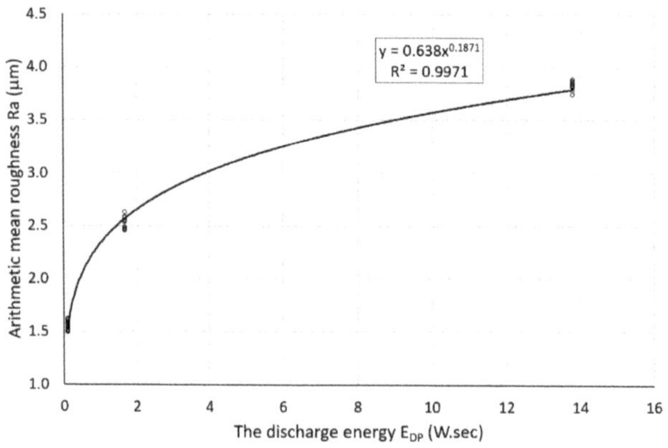

Figure 5. Dependence of roughness *Ra* on discharge energy DE_p.

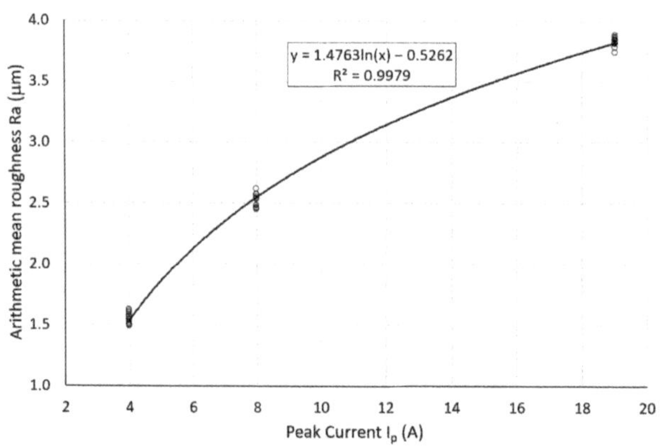

Figure 6. Dependence of roughness *Ra* on discharge current I_p.

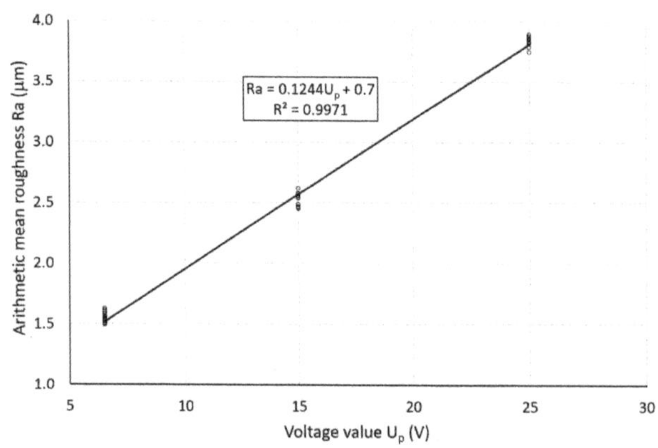

Figure 7. Dependence of roughness *Ra* on discharge voltage U_p.

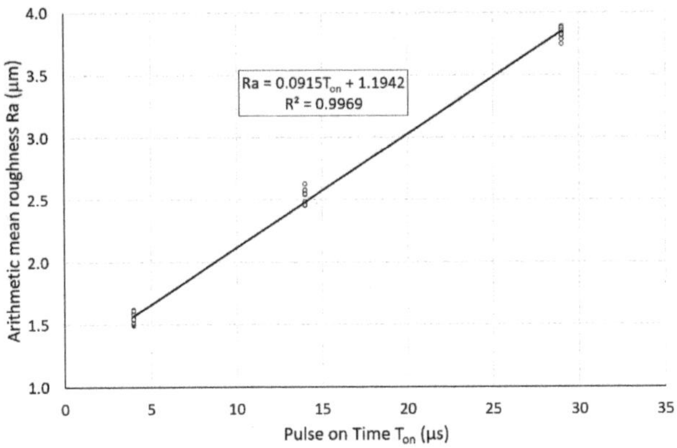

Figure 8. Dependence of roughness *Ra* on pulse on-time T_{on}.

One of the problems of multicriteria decision-making tasks is that individual criteria (characteristics of quality) or input parameters are expressed in different measurement units [22]. For this reason, it is convenient to transform them into a common unit so that they can be added. The method of relative parameters—planning assistance through technical evaluation of relevance number, PATTERN—is used in the technical, technological, and economic fields. This means that the change in the input parameters (I_p, U_p, and T_{on}) and the quality sign Q_j can be expressed as a ratio of the value of the $Q_{j\text{-th}}$ parameter to the value of the basic parameter Q_{j0} according to Relation (18):

$$IQ_{ji} = \frac{Q_{ji}}{Q_{ji0}}, \quad (18)$$

where IQ_{ji} is the degree of change in quality signs (*Ra* and *Pc*) according to the *i*-th process parameter; w_i is the weight (significance) of the *i*-th process parameter; and *n* is the number of *i*-th process parameters.

If we substitute the individual indices of the change in quality *Ra* into Equation (14), we obtain

$$S_{QRa} = \frac{1}{n}\sum_1^3 \left(IQ_{Ra(Ip)} \cdot w_{Ra(Ip)} + IQ_{Ra(Up)} \cdot w_{Ra(Up)} + IQ_{Ra(Ton)} \cdot w_{Ra(Ton)}\right), \quad (19)$$

After substituting into Equation (18), the change indices, depending on the input parameters of the texturing process, take the following form:

$$IQ_{Ra(Ip)} = \frac{Q_{Ra(Ip)}}{Q_{Ra0(Ip)} = 4} = \frac{(1.4763 * \ln(I_p) - 0.0526)}{(1.4763 * \ln(4) - 0.0526)}, \quad (20)$$

$$IQ_{Ra(up)} = \frac{Q_{Ra(Up)}}{Q_{Ra(Up)} = 6.5} = \frac{\left(0.1244 * U_{p)} + 0.7\right)}{(0.1244 * 6.5) + 0.7)}, \quad (21)$$

$$IQ_{Ra(Ton)} = \frac{Q_{Ra(Ton)}}{Q_{Ra0(Ton)} = 6.5} = \frac{(0.0915 * T_{on} + 1.1942)}{(0.0915 * 6.5) + 1.1942)}, \quad (22)$$

In the scalar multicriteria objective function, the approximation function (Figures 6–8) with weight value w_i is expressed in Equations (15)–(17) by coefficients for individual variables ($w_{Ra(Ip)}$ = 1.476, $w_{Ra(Up)}$ = 0.1244, $w_{Ra(Ton)}$ = 0.0915) so that no influence of the considered process parameters is preferred. After substituting Equations (20)–(22) into

Equation (19), we obtain the overall effect of the texturing process parameters on the *Ra* quality sign:

$$S_{QRa} = \frac{1}{3}[\ln(I_{I,Ra} + I_{U,Ra} + I_{Ton,Ra})] \pm \Delta_{QRa}, \quad (23)$$

Figure 9 shows the dependence between the total value of the quality feature S_{QRa} and the calculated roughness values Y_{QRa} according to Relation (23), which is described by regression Equation (24):

$$Y_{QRa} = 1.526 * S_{QRa} + 0.0091 \pm \Delta_{QRa}, \quad (24)$$

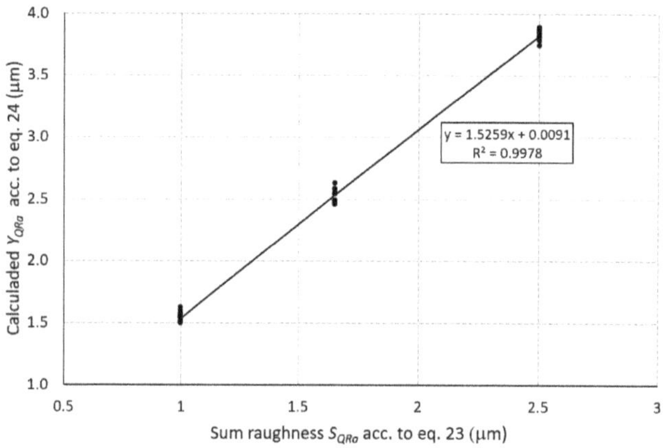

Figure 9. Dependence of roughness Y_{QRa} on S_{QRa}.

After modifying Model (23) into a simpler form, we obtain

$$Y_{QRa} = 0.494 * \ln(I_p) + 0.0424 * U_p + 0.0298 * T_{on} + 0.437 \pm \Delta_{QRa}, \quad (25)$$

Model (25) was subsequently tested for target roughness values TV_{Ra1} = 1.5 μm, TV_{Ra2} = 2.5 μm, and TV_{Ra3} = 3.8 μm. The presented model allows one to predict the roughness Y_{QRra} based on the pulse current I_p, the gap voltage U_p and the pulse on-time T_{on}. Table 5 compares the calculated Y_{QRa} values from Model (25) with the target values TV_{Ra1} = 1.5 μm, TV_{Ra2} = 2.5 μm, and TV_{Ra2} = 3.8 μm.

The comparison shows that there was good agreement between the calculated values according to Model (25) and the target values: for target value TV_{Ra1} = 1.5 μm, ΔRa = 0.014 μm and capability index C_{pkRa1} = 1.311; for target value TV_{Ra2} = 2.5 μm, ΔRa = 0.011 μm and capability index C_{pkRa2} = 1.325; and for target value TV_{Ra3} = 3.8 μm, ΔRa = 0.006 μm and capability index C_{pkRa3} = 1.355. This means that the texturing process for plans PE1 and PE2 is ineligible because the capability index is C_{pk} < 1.33. When T_{on} was changed by 1 μs, the C_{pkRa} capability indices were assumed to be greater than 1.33 when PE 1, PE2, and PE 3 plans were applied. This implies that texturing with these plans should be eligible. Reducing the Six Sigma level by one level, from four to five, will reduce the cost of poor quality by 10%. Thus, it is necessary to change the process parameters to achieve a change in the standard deviation from 0.06 to 0.048.

Table 5. Results of prediction Model (25) and its testing using data from ref. [28].

	Input Data					Calculated Data			
Plan	I_P **(A)**	U_P **(V)**	T_{on} **(µs)**	TV_{Ra} **(µm)**	Y_{QRa} **(µm)**	ΔRa **(µm)**	$S.D._{Ra}$ **(µm)**	C_{pk} **(µm)**	***DPMO*** **(Pieces)**
PE 1	4	6.5	4	1.5	1.514	0.014		1.311	≈6200
PE 2	8	15	14	2.5	2.511	0.011	0.06	1.325	≈6200
PE 3	19	25	29	3.8	3.806	0.006		1.355	≈4300
After improvement T_{on}-1									
PE 1	4	6.5	3	1.5	1.484	−0.016		1477	≈4300
PE 2	8	15	13	2.5	2.482	−0.018	0.06	1491	≈4300
PE 3	19	25	28	3.8	3.776	−0.024		1522	≈4300
Verification of Model (25) acc. to input parameters of [28]									
[28]	20	-	12	3.0	3.103	0.103	0.06	0.818	≈190,000
After improvement T_{on}-2 and I_p-1									
[28]	19	-	10	3.0	2.979	−0.003	0.06	1.373	≈4300

The proposed model was also tested using the parameters presented in [28], as they are closely related to the results in this paper. Since the authors of [28] only investigated the effect of pulse current I_p and pulse on-time T_{on}, Model (23) was modified as follows:

$$Y_{QRa} = 0.745 * \ln(I_p) + 0.045T_{on} + 0.331 \pm \Delta_{QRa}, \tag{26}$$

The difference is that the authors of [28] tested their prediction model for the target roughness value TV_{Ra} = 3 µm only for the input texturing parameters I_p = 20 A and T_{on} = 12 µs. In this case, the capability index C_{pkRa} was 0.818. When T_{on} was changed by 2 µs and I_p by 1 A, the capability index was 1.373.

The deviation (inaccuracy) ΔRa was calculated from the difference between the calculated YQ_{Ra} values and the specified target values $CL_{Ra,FR} = TV_{Ra,FR}$:

$$\Delta_{Ra} = Y_{QRa} - CL_{Ra,FR}, \tag{27}$$

In the analyzed interval of texturing parameters I_p, U_p, and T_{on}, the values of ΔR were ±0.015 µm. Similar to [4,28], the assumption was confirmed in our case that a very good correlation (R^2 = 0.989) was recorded between the roughness *Ra* and the peak density *Pc* at the output of the EDT rolls (see Figures 9 and 10). This dependence was described by the regression equation, Equation (28):

$$Y_{QPc} = 233.6 * Y_{QRa}^{-0.923}, \tag{28}$$

$$\Delta_{Pc} = Y_{QPc} - CL_{Pc}, \tag{29}$$

The decision process (process control) of texturing was described by the following algorithm (see Figure 11):

The sequence of steps followed in this algorithm is as follows:

1. *Step:* defining the target values of quality characteristics TV_{Ra} and TV_{Pc}.
2. *Step:* Assignment of input parameters of the *EDT* process (I_p, U_p, T_{on}, and T_{off}) to defined target values TV_{Ra} and TV_{Pc} according to Formulas (30)–(33):

$$Pc_{Ip} = 322.24\, I_p^{-0.539};\ R^2 = 0.97 \quad (30)$$

$$Pc_{Up} = 512.74U_p^{-0.621};\ R^2 = 0.98 \quad (31)$$

$$Pc_{Ton} = 287.87T_{on}^{-0.421},\ R^2 = 0.98 \quad (32)$$

$$Pc_{Toff} = 357.33T_{off}^{-0.459},\ R^2 = 0.99 \quad (33)$$

3. *Step:* Calculation of the values of qualitative characteristics Y_{QRa} and Y_{QPc} according to Formulas (25) and (28), calculation of ΔRa and ΔPc according to Formulas (27) and (29), and calculation of indices C_{pkL} and C_{pkU} according to Formulas (10) and (11). Thus, for $C_{pk} \geq 1.33$, the standard deviation after adjustment of the equations is

$$\sigma = \frac{UCL - \overline{\Delta Ra}}{3 * 1.33} = \frac{0.25 - 0}{3 * 1.33} = 0.06 \quad (34)$$

4. *Step:* Comparison of calculated values of C_{pkRa} and C_{pkPC} with the limit value 1. If the calculated values of C_{pkRa} and C_{pkPc} are less than 1, the process is not capable, and it is necessary to adjust the input parameters.
5. *Step:* If the calculated values of C_{pkRa} and C_{pkPC} are in the interval from 1 to 1.33, the process is capable, but the compliance rate is low, so it is recommended to take measures to improve the stability of the process.
6. *Step:* If the calculated C_{pkRa} and C_{pkPC} values are greater than 1.33, the compliance rate is high, and no process improvement measures need to be taken.
7. *Step:* Printing of output parameters of the quality characteristics *Ra*, *Pc*, and the input parameters of the *EDT* process I_p, U_p, T_{on}, and T_{off}.

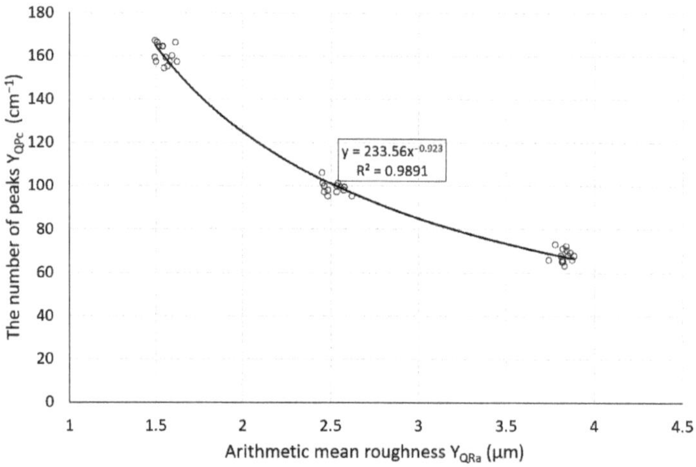

Figure 10. Dependence the number of peaks *Pc* on mean roughness Y_{QRa}.

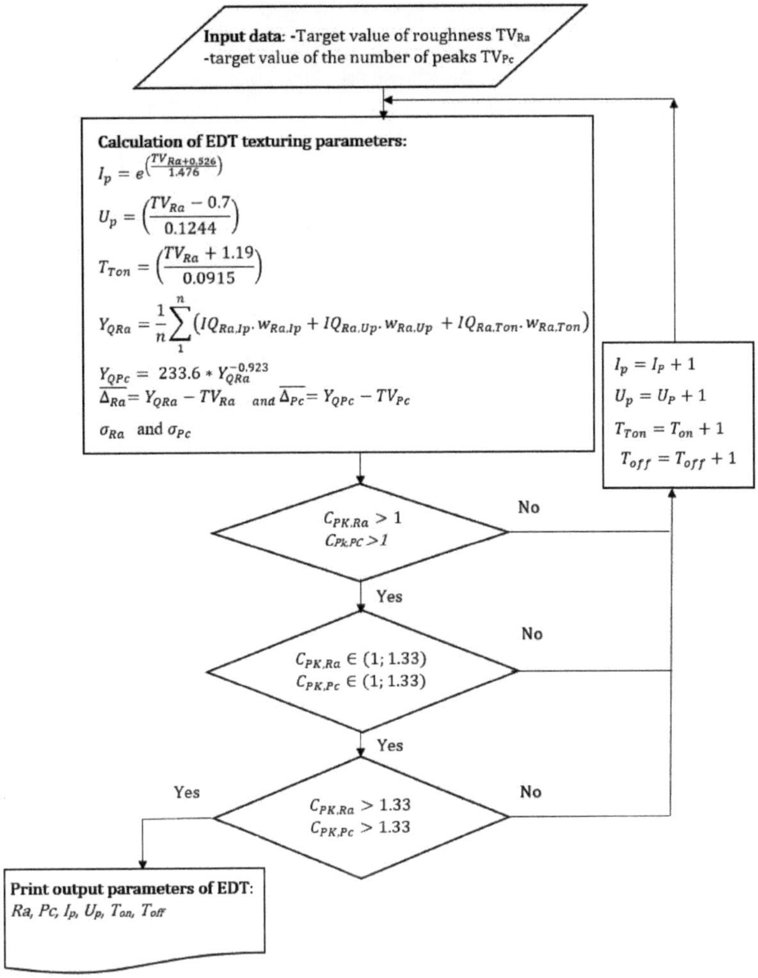

Figure 11. Process control of EDT.

4. Conclusions

Based on the results of the analyses, the following conclusions were formulated:

- On the basis of control diagrams, C_{pk} capability indices, and Z scores, it is possible to take measures to improve the quality characteristics *Ra* and *Pc* of the surface textures of the rolls in rolling mills.
- The results obtained by means of regression analysis indicate that the roughness *Ra* created during *EDT* is strongly dependent on the discharge energy. The coefficient of determination R^2 reached values greater than 0.99.
- The parameters I_p, U_p, and T_{on} of the *EDT* process have a major influence on the roughness *Ra*. An increase in these parameters of the texturing process leads to an increase in *Ra* values.
- Using regression analysis, the dependences of *Ra* on I_p were described by a logarithmic regression equation, and the dependences of *Ra* on both U_p and T_{on} were described by linear regression equations. The regression models showed a strong correlation ($R^2 > 0.9$).

- Using the obtained dependencies, a purpose function was created and verified. This function allows one to predict the roughness *Ra* with an accuracy of $\pm$ 0.02 μm in the analyzed interval of texturing parameters I_p, U_p, and T_{on}.
- The regression equation of the dependence of the peak density *Pc* on T_{off} was used to predict the peak density. The R^2 coefficient was greater than 0.98.
- Based on the comparison of the conformity of the C_{pk} capability indices with the limit values, an algorithm was designed, which allows step-by-step decisions to be made to improve the *EDT* process of rolling mill rolls.

The design of a procedure for taking corrective actions to improve the roughness Ra of EDT rolls based on the C_{pkRa} capability index also allows the prediction of the amount of non-conforming products out of one million DPMO cases compared to other procedures. Thus, by controlling the texturing process on the cold mill work roll surface based on the proposed algorithm, the number of non-conforming products can be minimized. Future efforts will be focused on verification of the roughness *Ra* of work rolls ranging from 1.1 μm to 1.6 μm, as presented in a previous work [31]. Furthermore, future studies will also verify whether both the required values of TV_{Ra} and TV_{Pc} are achieved on the surface of the sheet metal.

Author Contributions: Conceptualization, E.E. and V.K.; methodology, E.E.; formal analysis, E.E. and V.K.; investigation, E.E.; resources, E.E. and V.K.; data curation, V.K.; writing—original draft preparation, E.E. and M.T.; writing—review and editing, E.E. and M.T.; visualization, E.E.; supervision, E.E.; project administration, E.E.; funding acquisition, E.E. All authors have read and agreed to the published version of the manuscript.

Funding: This research and the APC were funded by Vedecká grantová agentúra MŠVVaM SR a SAV, grant number VEGA 1-0238-23.

Data Availability Statement: Data are available on request due to restrictions of the funder.

Acknowledgments: The work was accomplished under project VEGA 1-0238-23 "Implementation of CAx systems and virtual engineering techniques in the redesign of car-body parts for deformation zones".

Conflicts of Interest: Author Vladimír Kokarda was employed by Carl Zeiss Slovakia. The remaining authors declare that the research was conducted in the absence of any commercial or financial relationships that could be construed as a potential conflict of interest.

References

1. Sakuramoto, C. Automotive Skin Panels Quality Improvement by Means of Finite Element Method. In Proceedings of the Conference SAE Commercial Vehicle Engineering Congress (COMVEC), Rosemont, IL, USA, 1–3 October 2013. [CrossRef]
2. Schulte, C.; Li, X.; Abel, D.; Hirt, G. Model-based Control of the Strip Roughness in Cold Rolling. *FAC Pap.* **2021**, *54*, 109–114. [CrossRef]
3. Çolak, B.; Zahoor, A.; Özakin, B.; Kurgan, N. An Experimental Investigation into Roughness Transfer in Asymmetrical Rolling of Steel Strips. *Karadeniz Fen Bilim. Derg.* **2023**, *13*, 1070–1089. [CrossRef]
4. Çolak, B.; Kurgan, N. An experimental investigation into roughness transfer in skin-pass rolling of steel strips. *Int. J. Adv. Manuf. Technol.* **2018**, *96*, 3321–3330. [CrossRef]
5. Cavallari, M.; Gaboardi, P.; McWhirter, R.; Perassolo, M.; Trevisan, C. New Surface Finishes for Cold Rolling Mills. In Proceedings of the METEC & 2nd ESTAD 2015, European Steel Technology and Application Days, Düsseldorf, Germany, 15–19 June 2015.
6. Elkoca, O. A study on the characteristics of electrical discharge textured skin pass mill work roll. *Surf. Coat. Technol.* **2008**, *202*, 2765–2774. [CrossRef]
7. Rodriguez-Vidal, E.; Matthews, D.T.A.; de Viteri, V.S.; Korver, F.; Wentink, D.J.; Quintana, I. Surface design and texturing of strip steel using nanosecond pulsed lasers for simulated roughness transfer and paint appearance. *J. Mater. Process. Technol.* **2020**, *275*, 116365. [CrossRef]
8. Evin, E.; Tomáš, M.; Kmec, J. Optimization of Electro-Discharge Texturing Parameters for Steel Sheets' Finishing Rollers. *Materials* **2020**, *13*, 1223. [CrossRef]

9. Bartkowiak, T.; Mendak, M.; Mrozek, K.; Wieczorowski, M. Analysis of Surface Microgeometry Created by Electric Discharge Machining. *Materials* **2020**, *13*, 3830. [CrossRef] [PubMed]
10. Nieslony, P.; Wojciechowski, S.; Gupta, M.K.; Chudy, R.; Krolczyk, J.B.; Maruda, R.; Krolczyk, G.M. Relationship between energy consumption and surface integrity aspects in electrical discharge machining of hot work die steel. *Sustain. Mater. Technol.* **2023**, *36*, e00623. [CrossRef]
11. Hourmand, M.; Farahany, S.; Sarhan, A.A.D.; Noordin, M.Y. Investigating the electrical discharge machining (EDM) parameter effects on Al-mg 2 Si metal matrix composite (MMC) for high material removal rate (MRR) and less EWR–RSM approach. *Int. J. Adv. Manuf. Technol.* **2015**, *77*, 831–838. [CrossRef]
12. Straka, Ľ.; Piteľ, J.; Čorný, I. Influence of the main technological parameters and material properties of the workpiece on the geometrical accuracy of the machined surface at wedm. *Int. J. Adv. Manuf. Technol.* **2021**, *115*, 3065–3087. [CrossRef]
13. Bidulska, J.; Kvackaj, T.; Bodak, V.; Bidulsky, R. The Microgeometry Parameters of Uncoated and Zinc-Coated Cold Rolled Steel Strips. *J. Met. Mater. Miner.* **2007**, *17*, 1–7.
14. Çolak, B.; Kurgan, N. EDT ve Sbt Yöntemleriyle Pürüzlendirilmiş Soğuk Hadde Merdanelerinin Yüzey Pürüzlülük Etkilerinin Karşılaştırılmasına Dair Bir Çalışma. In Proceedings of the 1st International Iron and Steel Symposium (IISS'12), Karabuk, Turkey, 2–4 April 2012; Karabuk University: Karabuk, Turkey, 2012; p. 861.
15. Gorbunov, A.V.; Belov, V.K.; Begletsov, D.O. Texturing of Rollers for the Production of Auto-Industry Sheet. *Steel Transl.* **2009**, *39*, 696–699. [CrossRef]
16. Simao, J.D.; Aspinwall, K.; Wise, M.L.H.; El-Menshawy, M.F. Electrical discharge texturing of cold mill work rolls using different tool electrode materials. *Iron Steel Eng.* **1996**, *73*, 42–47.
17. Colak, B.; Basoglu, F.; Kurgan, N. Texturing methods for cold mill work rolls. In Proceedings of the Fourth International Iron and Steel Symposium, Karabuk, Turkey, 4–6 April 2019; pp. 424–431.
18. Świercz, R.; Holubek, R. Experimental investigation of influence electrical discharge energy on the surface layer properties after EDM. *Weld. Technol. Rev.* **2020**, *92*, 7–13. [CrossRef]
19. Munro, R.A.; Ramu, G.; Zrymiak, D.J. *The Certified Six Sigma Green Belt Handbook*, 2nd ed.; Quality Press: Welshpool, Australia, 2015.
20. Giampieri, A. A review of the current automotive manufacturing practice from an energy perspective. *Appl. Energy* **2020**, *261*, 114074. [CrossRef]
21. Coleman, S. Six Sigma: An opportunity for statistics and for statisticians. *Significance* **2008**, *5*, 94–96. [CrossRef]
22. Tyzdek, T.; Keller, P.A. *The Six Sigma Handbook*, 3rd ed.; The McGraw-Hill Companies: New York, NY, USA, 2010.
23. Schindlerova, V.; Marcik, J.; Čada, R.; Sajdlerova, I. Analysis of Non-Conforming Production in an Engineering Company. *Tech. Gaz.* **2024**, *31*, 296–302. [CrossRef]
24. Mittal, A.; Gupta, P.; Kumar, V.; Ali Al Owad, A.; Mahlawat, S.; Sumanjeet Singh, S. The performance improvement analysis using Six Sigma DMAIC methodology: A case study on Indian manufacturing company. *Heliyon* **2023**, *9*, e14625. [CrossRef]
25. Gijoa, E.V.; Scaria, J.; Antony, J. Application of Six Sigma Methodology to Reduce Defects of a Grinding Process. *Qual. Reliab. Eng. Int.* **2011**, *27*, 1221–1234. [CrossRef]
26. Ganguly, K. Improvement process for rolling mill through the DMAIC six sigma approach. *Int. J. Qual. Res.* **2012**, *6*, 221–231.
27. Laddha, S.A.; Desai, G. Quality improvement in building construction using six sigma. *Int. Res. J. Eng. Technol.* **2019**, *6*, 7711–7713.
28. Terpák, L.; Dorčák, J.; Revaj, J. Quality Control of the Electro-Discharge Texturing. *Metalurgija* **2010**, *49*, 19–22.
29. *DIN EN ISO 4288*; Geometrical Product Specifications (GPS)—Surface Texture: Profile Method—Rules and Procedures for the Assessment of Surface Texture. International Organization for Standardization (ISO): Geneva, Switzerland, 2009.
30. Červeňanská, Z.; Važan, P.; Juhás, M.; Juhásová, B. Multi-Criteria Optimization in Operations Scheduling Applying Selected Priority Rules. *Appl. Sci.* **2021**, *11*, 2783. [CrossRef]
31. Kundracik, V. Strip surface microgeometry for automotive surface car-body parts. Galvanized material for automotive industry. In Proceedings of the 66th Workshop of the STEEL STRIP Company, Košice-Jahodná, Slovakia, 12–14 October 2009; pp. 28–33. (In Slovak)

processes

MDPI

Article

TopADDPi: An Affordable and Sustainable Raspberry Pi Cluster for Parallel-Computing Topology Optimization

Zhi-Dong Zhang [1,*], Dao-Yuan Yu [1], Osezua Ibhadode [2], Liang Meng [1,*], Tong Gao [1], Ji-Hong Zhu [1] and Wei-Hong Zhang [1,*]

1 State IJR Center of Aerospace Design and Additive Manufacturing, School of Mechanical Engineering, Northwestern Polytechnical University, Xi'an 710072, China

2 Multi-Scale Additive Manufacturing Lab, University of Waterloo, 200 University Avenue West, Waterloo, ON N2L 3G1, Canada

* Correspondence: zhidong.zhang@nwpu.edu.cn (Z.-D.Z.); liang.meng@nwpu.edu.cn (L.M.); zhangwh@nwpu.edu.cn (W.-H.Z.)

Abstract: Parallel-Computing Topology Optimization (PCTO) has gained importance, especially with the advancement of additive manufacturing (AM), due to its ability to tackle high-dimensional, high-resolution challenges. PCTO is highly relevant to sustainable manufacturing processes and technologies, enabling resource-efficient designs, reduced emissions, and advancements in Industry 4.0 integration. However, PCTO poses difficulties for newcomers or researchers, mainly because of its reliance on non-traditional computing environments and the limited availability of high-performance computing (HPC) resources. Addressing this, the study introduces TopADDPi, a Raspberry Pi-based cluster system, which has been purpose-built to facilitate learning and research in PCTO. It provides detailed instructions for assembling and configuring a Raspberry Pi cluster, with a focus on cost-effectiveness and ease of use. The study thoroughly investigates how different hardware and software configurations affect computing efficiency. In addition, through extensive numerical testing, the performance, energy consumption, and environmental impact of the Raspberry Pi cluster are benchmarked against conventional computing systems. The findings demonstrate the cluster's advantages in handling parallel computing, its indispensable role in debugging, its remarkable energy efficiency, and its significantly reduced carbon footprint compared to conventional systems. These attributes establish the Raspberry Pi cluster as an invaluable tool for both educational and research applications in structural engineering, offering an affordable, sustainable, and indispensable solution for PCTO.

Keywords: topology optimization; raspberry pi cluster; parallel computing; arbitrary design domain; sustainability; 3D printing

1. Introduction

Topology optimization (TO) has become essential in designing high-performance structures, especially with the rise in additive manufacturing (AM) [1–4]. This method optimizes material distribution within a design space and is highly relevant to modern manufacturing processes, transportation, and technologies, enabling resource-efficient designs, reduced greenhouse gases emissions, and advancements in Industry 4.0 integration [5]. The detrimental impact of greenhouse gasses on the environment has attracted the attention of researchers in both industry and academia [6,7]. However, TO faces challenges when it comes to addressing high-dimensional, high-resolution problems [8]. High

 https://doi.org/10.3390/pr13030633

resolutions are preferred for detailed features and smoother surfaces are suitable for AM, a key aspect of advanced manufacturing and precision engineering. In contrast, low resolutions can lead to suboptimal designs due to artificial length-scale constraints from the coarse grids [9]. Therefore, managing high-resolution TO is critical, despite substantial computational demands.

The need for increased computational resources due to larger numbers of finite elements has led to the widespread adoption of parallel computing in TO [10]. Parallel computing divides large problems into smaller, concurrently solved segments, enabling an efficient data exchange and faster iterations. Parallel-Computing Topology Optimization (PCTO) has been under investigation for over two decades, with Borrvall and Petersson [11] being pioneers in applying parallel computing to large-scale TO. Subsequent studies, such as that of Aage et al. [9], have developed flexible, open-source frameworks for PCTO, contributing significantly to the field of sustainable processes and green manufacturing. Researchers have expanded these frameworks for various applications, including designing high-resolution support structures [12,13], and incorporating features like minimum and maximum feature size [14] and arbitrary design domain initialization [15], all of which are available as open-source code.

These open-source programs are invaluable for newcomers and researchers focused on PCTO, offering hands-on experience with code and the opportunity to obtain optimized structures [9]. However, the predominance of C++ code and the requirement for a Linux operating system (OS), e.g. Ubuntu, with application software installations can pose challenges, especially for those who are only familiar with Windows environments. In addition, debugging these programs often requires a distributed-memory computational environment like an in-house cluster or high-performance computing (HPC).

Raspberry Pi (RPi) clusters present an affordable solution for learning and research in parallel computing, cluster management, and distributed systems [16]. These clusters have been used for cost-effective experimentation in computing, offering an efficient and eco-friendly option for multi-node system applications [17–20]. However, there is limited research on using Raspberry Pi clusters specifically for TO.

This study presents TopADDPi, an RPi-based cluster system specifically developed to support the open-source TopADD code [15] and other codes to enhance the learning, programming, and mastery of PCTO. Firstly, this study will demonstrate TopADDPi's usefulness to users with limited resources or experience in high-performance computing environments. Then, it will offer detailed guidelines on constructing an RPi cluster tailored for topology optimization tasks. Additionally, the study will thoroughly evaluate the cluster's effectiveness and efficiency. This evaluation will focus on how TopADDPi stands as a practical and eco-friendly solution for handling the computational demands of PCTO. Furthermore, the TopADDPi cluster is suitable for additive technologies like 3D printing, enabling exploration of sustainable supply chains and process designs.

2. Motivation

The increasing complexity of large-scale topology optimization and corresponding computational tasks have underscored the critical need for a dedicated learning and research platform in the field of parallel computing. As modern applications leverage parallelism for the purpose of performance enhancement, mastering the intricacies of parallel computing becomes imperative for structural engineers. The need for an in-house Raspberry Pi (RPi) cluster is demonstrated, followed by two specific requirements for the cluster.

2.1. Need for an In-House Cluster to Learn PCTO

The development of PCTO program necessitates access to an actual computing cluster. While debugging in a serial or single-node setup is feasible, it is crucial to thoroughly familiarize newcomers or researchers with the parallel-computing environment. Newcomers or researchers often face limited computational resources; they are typically restricted to a single computer, frequently one running a Windows operating system. These users would need to install a Linux system or a virtual Linux machine on their existing Windows machines to perform parallel computing tasks. This setup presents several challenges: dual-boot complexity, loss of Windows functionality, and resource limitations in virtual machine.

Firstly, installing Linux alongside Windows often requires setting up a dual-boot system, which can be technically challenging for users who are unfamiliar with system partitioning and bootloaders. Secondly, when running Linux, users lose access to their familiar Windows environment, which they may rely on for tasks such as searching for solutions to technical problems, accessing documentation, or using Windows-specific tools. Thirdly, virtual machines are often used as an alternative to dual-boot setups; however, they are not native systems and suffer from performance overheads, limited hardware access, and incomplete functionalities. These hurdles can significantly impede users' learning and research process, leading to frustration and potential abandonment of their educational and research journey, as shown in Figure 1.

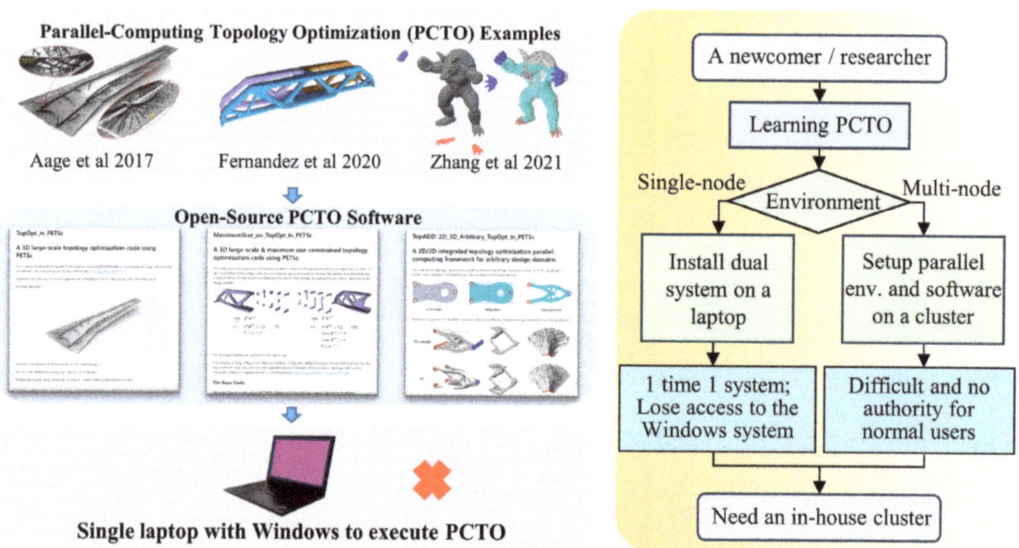

Figure 1. Schematic depicting the complexities and challenges of mastering parallel-computing topology optimization and operating its related open-source software, culminating in the need for an in-house cluster. Adapted with permission from Refs. [9,14,15].

Additionally, accessing high-performance computing (HPC) resources can be daunting for newcomers or researchers, especially when faced with an unfamiliar HPC environment and the lack of skilled technicians for support. The process of setting up complex application software like PETSc—a parallel numerical software library for partial differential equations and sparse matrix computations [21]—can be overwhelming, even for a normal programmer, let alone a beginner.

Therefore, having an affordable and accessible real cluster is vital for newcomers or researchers. A Raspberry Pi-based cluster can be an excellent solution. It not only prompts the learning and research process for PCTO but also provides hands-on experience with parallel computing with particular focus on distributed-memory systems. This practical exposure is instrumental in helping people gain a deeper and more realistic understanding of parallel computing, thereby enhancing their ability to develop more effective PCTO programs.

2.2. Need for ARM Computers

The Advanced RISC Machines (ARM) architecture is renowned for its power efficiency, simplicity, and flexibility, and is particularly suitable for energy-efficient mobile devices like smartphones, tablets, and laptops [22]. Given the pronounced architectural differences between ARM and common platforms like ×86, which are prevalent among desktop computers and laptops, it is crucial to conduct thorough testing on both ×86 and ARM platforms for software development targeted at various devices. The Raspberry Pi is a widely acclaimed single-board computer leveraging ARM architecture. It serves as an exemplary tool for debugging and testing code tailored for ARM Central Processing Units (CPUs) [23]. Figure 2 illustrates the process of utilizing a Raspberry Pi to identify a bug that is challenging to detect on an ×86 platform. The bug, arising from casting negative float to an unsigned int in C++ [24], results in architecture-dependent behavior. The corresponding code is provided in Appendix A.

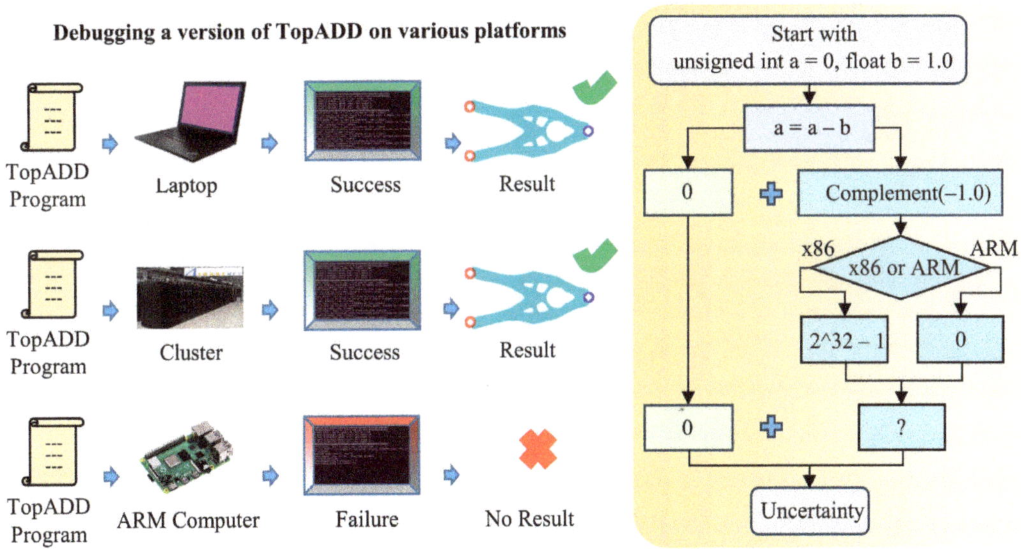

Figure 2. Schematic of utilizing a Raspberry Pi for undefined-behavior bug identification, showing the necessity of bug detection in ARM platforms.

2.3. Need for Distributed-Memory Systems

The escalating demand for computational power and scalability underscores the imperative need for distributed-memory systems. Achieving efficient communication and data sharing among processes is essential. For example, techniques like MPI_Win which can be used for communication on the same node can enhance overall performance and scalability in parallel applications [25]. However, transitioning a program designed for shared-memory parallel computing to a distributed-memory system presents challenges.

Despite successful execution on laptops with ×86 architecture and Raspberry Pis with ARM architecture, issues may arise on clusters, as shown in Figure 3. This can be attributed to the malfunctioning of MPI_Win on the cluster without meticulous configuration, which emphasizes the crucial role of a distributed-memory cluster in code debugging. Therefore, a convenient distributed-memory testbed emerges as indispensable in the development of parallel-computing programs. The corresponding code is provided in Appendix B.

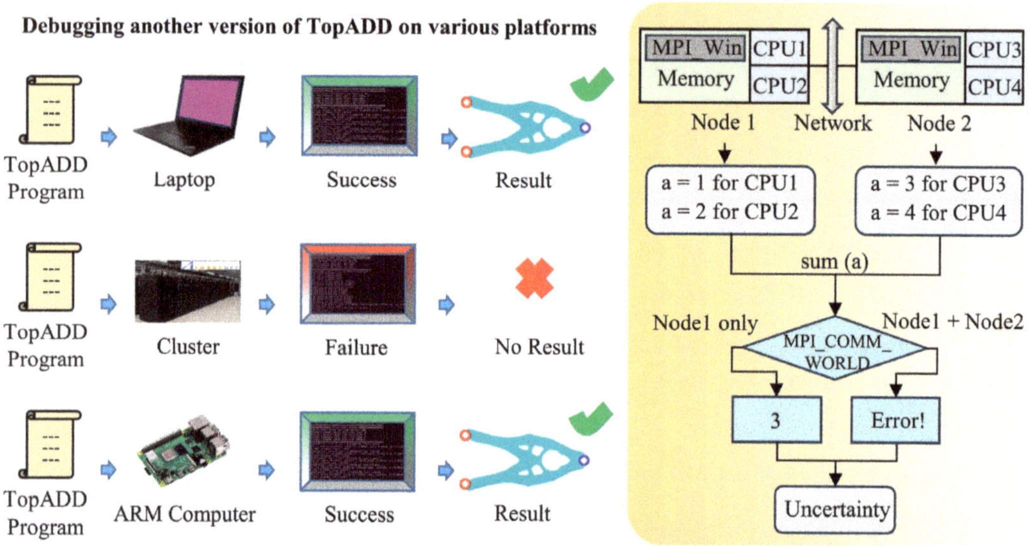

Figure 3. Schematic of utilizing a distributed-memory system for distributed-memory bug identification, showing the necessity of bug detection in multi-node platforms.

3. Building the Cluster for PCTO

This section introduces a Raspberry Pi cluster specifically designed for learning and researching PCTO. Comprehensive details on configuring and setting up this cluster have been made publicly available on GitHub at: https://github.com/wonderfulzzd/Pi_Cluster_for_TopADD (accessed on 1 January 2024).

3.1. Existing Raspberry Pi Clusters

Several projects have previously assembled RPis into a variety of clusters. Table 1 lists nine existing Raspberry Pi clusters developed at different universities and institutions. During the initial release of the first-generation Raspberry Pi Model B (RPi Model B) in 2012, several pioneering projects utilized it to construct RPi clusters. These early endeavors, now over a decade old, were based on the earlier Raspberry Pi models. With the evolution and advancement of Raspberry Pi technology, the exploration of newer models has become increasingly relevant. Consequently, Table 1 includes several recent projects that employ the latest Raspberry Pi 4 Model B (RPi 4 Model B) for building parallel-computing clusters, which was the latest version available during the time of this study.

These contemporary projects delve into various performance aspects of the clusters, examining factors such as cost-efficiency, input/output (I/O) performance, and capabilities in scientific computing. The examples presented in Table 1 represent just a fraction of the myriad projects that have been developed around Raspberry Pi clusters in the literature. This subset highlights an active community of developers and researchers. The ongoing interest in constructing and utilizing Raspberry Pi clusters underscores their potential as a valuable tool in both research and educational spheres. While these initiatives have contributed to the field, it should be noted that none of them were specifically designed for PCTO.

Table 1. Existing Raspberry Pi clusters.

	Year	RPi Version	RPi Number	Purpose
University of Southampton [17]	2012	RPi Model B	64	Testing MPI functionality
Boise State University [18]	2013	RPi Model B	32	Testing MPI functionality
University of Glasgow [19]	2013	RPi Model B	56	Testbed for cloud computing
Free University of Bozen-Bolzano [20,26]	2013	RPi Model B	300	Testbed for cloud computing; Container-based edge cloud
Prometheus Research [27]	2013	RPi Model B	2	Book on building a cluster and exploring application software
Daejin University [28]	2022	RPi 4 Model B	16	Testing cost-performance
Deakin University [29]	2022	RPi 4 Model B	12	Testing performance of scientific computing
Federal University of Sergipe [30]	2022	RPi 4 Model B	9	Testing performance of big data Input/Output (I/O)
Hellenic Mediterranean University [31]	2023	RPi 4 Model B	24	Testing performance of scientific computing

3.2. System Specification of Raspberry Pi 4 Model B

Raspberry Pi 4 Model B will be employed as building blocks of the RPi cluster. Figure 4 shows the most prominent components on a Raspberry Pi 4 Model B board. The heart of the RPi is an integrated circuit Broadcom BCM2711 (System on Chip, SoC), combining a Quad core ARM Cortex-A72 1.5 GHz CPU and a Broadcom VideoCore VI Graphics Processing Unit (GPU). The Random Access Memory (RAM) can be 4 GB/8 GB LDDR4 SDRAM, depending on the choice of users. Moreover, there is an SD card slot for storage, as well as I/O units such as Universal Serial Bus (USB), USB-C as power supply port, a Gigabit ethernet port, audio, and 2 micro-High-Definition Multimedia Interface (HDMI) ports. Table 2 summarizes the hardware specifications of the RPi [32].

Figure 4. Illustration on the most prominent components on the Raspberry Pi 4 model B board.

For the system software, the Raspberry Pi Foundation typically recommends Raspbian as the OS of choice. Raspbian is a tailored version of the well-known Debian Linux distribution. However, in this case, Ubuntu 20.04 has been installed on the Raspberry Pis to facilitate a more consistent comparison with desktop and laptop environments. Ubuntu is a widely recognized Linux distribution that is also based on Debian and offers a more familiar environment for those accustomed to traditional computer systems. This approach allows for a more direct and relevant comparison of performance and usability between Raspberry Pi devices and standard computers.

Table 2. Raspberry Pi 4 Model B hardware specifications.

Component Name	Specifications
Integrated circuit	Broadcom BCM2711
CPU	Quad core ARM Cortex-A72 1.5 GHz processor
GPU	Broadcom VideoCore VI 0.5 GHz processor
Memory	4 GB/8 GB LDDR4 SDRAM
Ethernet	Gigabit ethernet port
USB	2 USB 2.0 ports, 2 USB 3.0 ports
Video output	2 micro-HDMI ports
Audio output	3.5 mm port, HDMI
Storage	SD card slot

Details concerning the installation of Ubuntu and other application software on the Raspberry Pis will be provided. This information will include step-by-step guidance for setting up the Raspberry Pi devices with Ubuntu, ensuring the devices are well prepared for the tasks and comparisons intended in this study.

3.3. Cluster Hardware Configurations

The multitude of projects utilizing Raspberry Pis inspired the authors to set up a new Raspberry Pi cluster for PCTO. In this work, an RPi with 8 GB RAM is used as a login node and 4 RPis with 4 GB RAM are employed as compute nodes. Figure 5 illustrates the components of this Raspberry Pi cluster as well as its final assembly.

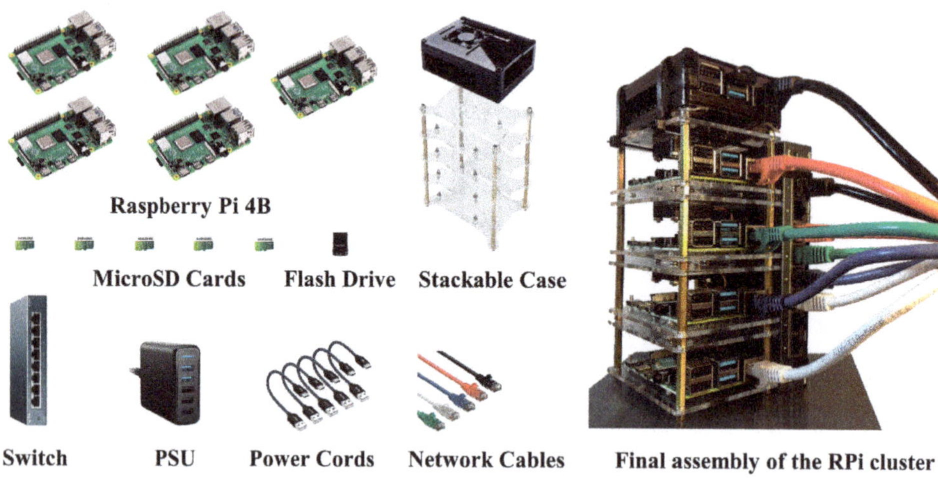

Figure 5. Components of the RPi cluster and its final assembly outlook.

The cluster is designed with the capability to support graphical output, allowing for the connection of a monitor, as depicted in Figure 6a. Such graphical interface support is particularly advantageous for newcomers or researchers who may prefer interacting with graphical operations rather than command-line interfaces. It is important to note that a a valuable tool in both research and educational spheres. While these initiatives have contributed to the field, it should be noted that none of them were specifically designed for PCTO.

Table 1. Existing Raspberry Pi clusters.

	Year	RPi Version	RPi Number	Purpose
University of Southampton [17]	2012	RPi Model B	64	Testing MPI functionality
Boise State University [18]	2013	RPi Model B	32	Testing MPI functionality
University of Glasgow [19]	2013	RPi Model B	56	Testbed for cloud computing
Free University of Bozen-Bolzano [20,26]	2013	RPi Model B	300	Testbed for cloud computing; Container-based edge cloud
Prometheus Research [27]	2013	RPi Model B	2	Book on building a cluster and exploring application software
Daejin University [28]	2022	RPi 4 Model B	16	Testing cost-performance
Deakin University [29]	2022	RPi 4 Model B	12	Testing performance of scientific computing
Federal University of Sergipe [30]	2022	RPi 4 Model B	9	Testing performance of big data Input/Output (I/O)
Hellenic Mediterranean University [31]	2023	RPi 4 Model B	24	Testing performance of scientific computing

3.2. System Specification of Raspberry Pi 4 Model B

Raspberry Pi 4 Model B will be employed as building blocks of the RPi cluster. Figure 4 shows the most prominent components on a Raspberry Pi 4 Model B board. The heart of the RPi is an integrated circuit Broadcom BCM2711 (System on Chip, SoC), combining a Quad core ARM Cortex-A72 1.5 GHz CPU and a Broadcom VideoCore VI Graphics Processing Unit (GPU). The Random Access Memory (RAM) can be 4 GB/8 GB LDDR4 SDRAM, depending on the choice of users. Moreover, there is an SD card slot for storage, as well as I/O units such as Universal Serial Bus (USB), USB-C as power supply port, a Gigabit ethernet port, audio, and 2 micro-High-Definition Multimedia Interface (HDMI) ports. Table 2 summarizes the hardware specifications of the RPi [32].

Figure 4. Illustration on the most prominent components on the Raspberry Pi 4 model B board.

For the system software, the Raspberry Pi Foundation typically recommends Raspbian as the OS of choice. Raspbian is a tailored version of the well-known Debian Linux distribution. However, in this case, Ubuntu 20.04 has been installed on the Raspberry Pis to facilitate a more consistent comparison with desktop and laptop environments. Ubuntu is a widely recognized Linux distribution that is also based on Debian and offers a more familiar environment for those accustomed to traditional computer systems. This approach allows for a more direct and relevant comparison of performance and usability between Raspberry Pi devices and standard computers.

Table 2. Raspberry Pi 4 Model B hardware specifications.

Component Name	Specifications
Integrated circuit	Broadcom BCM2711
CPU	Quad core ARM Cortex-A72 1.5 GHz processor
GPU	Broadcom VideoCore VI 0.5 GHz processor
Memory	4 GB/8 GB LDDR4 SDRAM
Ethernet	Gigabit ethernet port
USB	2 USB 2.0 ports, 2 USB 3.0 ports
Video output	2 micro-HDMI ports
Audio output	3.5 mm port, HDMI
Storage	SD card slot

Details concerning the installation of Ubuntu and other application software on the Raspberry Pis will be provided. This information will include step-by-step guidance for setting up the Raspberry Pi devices with Ubuntu, ensuring the devices are well prepared for the tasks and comparisons intended in this study.

3.3. Cluster Hardware Configurations

The multitude of projects utilizing Raspberry Pis inspired the authors to set up a new Raspberry Pi cluster for PCTO. In this work, an RPi with 8 GB RAM is used as a login node and 4 RPis with 4 GB RAM are employed as compute nodes. Figure 5 illustrates the components of this Raspberry Pi cluster as well as its final assembly.

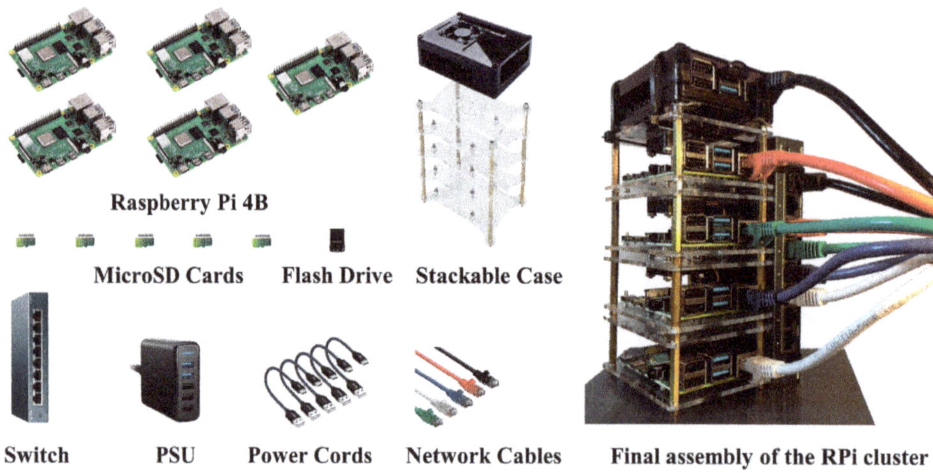

Figure 5. Components of the RPi cluster and its final assembly outlook.

The cluster is designed with the capability to support graphical output, allowing for the connection of a monitor, as depicted in Figure 6a. Such graphical interface support is particularly advantageous for newcomers or researchers who may prefer interacting with graphical operations rather than command-line interfaces. It is important to note that a monitor is not included in the overall expense of the cluster, as its desktop environment can be accessed remotely via a laptop or other computers, eliminating the necessity for a dedicated monitor. Figure 6b presents a schematic of the cluster setup, illustrating the connections between the monitor, compute nodes, login node, switch, and Network File System (NFS).

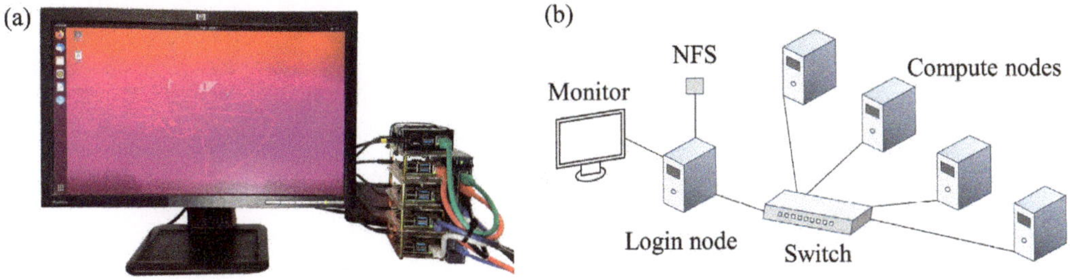

Figure 6. RPi cluster connected with a monitor, (**a**) final assembly outlook, and (**b**) its corresponding setup schematic.

The cost breakdown of the five-node RPi cluster is illustrated in Figure 7. All the components were purchased from Amazon.ca. The prices listed reflect the cost as of December 2019, inclusive of taxes and shipping fees. The total expenditure for the setup is presented in Canadian dollars (CAD), amounting to CAD 730, which is equivalent to approximately USD 550.

	Product	Qty	Unit Price (CAD)	Total Price (CAD)
Raspberry Pi		5	92	460
MicroSD Cards		5	15.4	77
Flash Drive		1	25	25
Switch		1	30	30
PSU		1	50	50
Cable and Cases		1	88	88
	Sum			**730**

Figure 7. Cost breakdown of the five-node RPi cluster.

To investigate the impact of hardware configurations on the performance of the RPi cluster, three key aspects will be examined:

(a) Internet cable influence: The study will compare two types of internet cables: CAT6 and CAT8. The CAT6 cable supports data transfer rates of up to 10 gigabits per second (Gbps), whereas the CAT8 cable can handle up to 40 Gbps. This comparison aims to understand how these different capabilities affect the cluster's data transfer efficiency.

(b) NFS influence: The NFS, where all application files are installed and stored, is a critical component. This research will explore the performance differences between two types of NFS hardware: a flash drive with a capacity of 128 GB and a Non-Volatile Memory express (NVMe) solid-state drive (SSD) with 256 GB. The flash drive offers read and write speeds of 130 MB/s and 50 MB/s, respectively, while the NVMe SSD provides significantly faster speeds of 2400 MB/s for reading and 1950 MB/s for writing.

(c) Switch influence: The network switch plays a vital role in the communication among the cluster's nodes, thereby impacting the overall performance of the cluster. This study will investigate the performance differences between switches with 100-Megabit and 1-Gigabit capacities to gauge the extent of their influence on the cluster's efficiency.

3.4. Cluster Software Configurations

Comprehensive details on configuring and setting up this cluster have been made publicly available on GitHub at https://github.com/wonderfulzzd/Pi_Cluster_for_TopADD (accessed on 1 January 2024). Readers interested in constructing their own Raspberry Pi cluster are encouraged to refer to these step-by-step procedures whether for research or educational purposes.

The OS installed on all Raspberry Pis in this project is Ubuntu 20.04, sourced from the official Raspberry Pi website. Each Raspberry Pi's SD card is prepared with this OS image using the Raspberry Pi Imager. Following the image flashing, the systems require configuration, including various application software installation, such as MPI and PETSc software.

Furthermore, the study will briefly delve into the impact of software configuration on the cluster's performance. This includes examining the effects of software's different optimization levels, installation locations, and CPU overclocking settings.

Lastly, a flowchart describing the assembly and configuration process is plotted in Figure 8 to enhance clarity and reproducibility.

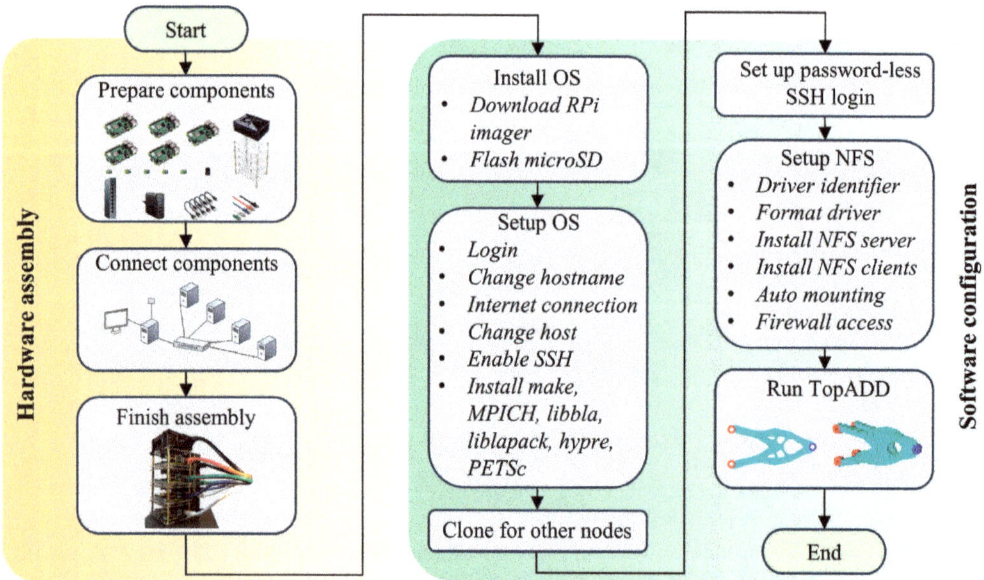

Figure 8. Flowchart describing the hardware assembly and software configuration process.

4. Topology Optimization Problem

The TopADD program [15] encompasses three distinct topology optimization (TO) problems: minimum compliance, a compliant mechanism, and heat conduction. These problems are schematically represented in Figure 9. The primary goal in each TO problem is to determine the optimal distribution of material that minimizes the objective function. This could be the compliance of structures, negative displacement, or heat compliance. They are all subjected to predefined boundary and loading conditions.

$$\begin{aligned} \underset{\rho}{Min.} \quad & c(\rho) = \boldsymbol{L}^T\boldsymbol{U}(\rho) \\ s.t. \quad & \boldsymbol{K}(\rho)\boldsymbol{U}(\rho) = \boldsymbol{F} \\ & g(\rho) = \frac{V(\rho)}{V_0} = \frac{1}{V_0}\sum_{i=1}^{n}\rho_i \leq \overline{V}_{\mathrm{f}}. \\ & 0 < \rho_i \leq 1,\ i = 1,\ 2, \cdots,\ N_e \end{aligned} \tag{1}$$

where $\boldsymbol{U}$ and $\boldsymbol{F}$ are the solution and load vectors, respectively. The vector $\boldsymbol{L}$ serves as an auxiliary constant vector. In the context of minimum compliance and heat conduction problems, $\boldsymbol{L}$ is equivalent to $\boldsymbol{F}$. However, for the compliant mechanism problem, $\boldsymbol{L}$ is a unique constant vector characterized by zeros at all degrees of freedom except at the output point's degree of freedom, where it is set to -1.

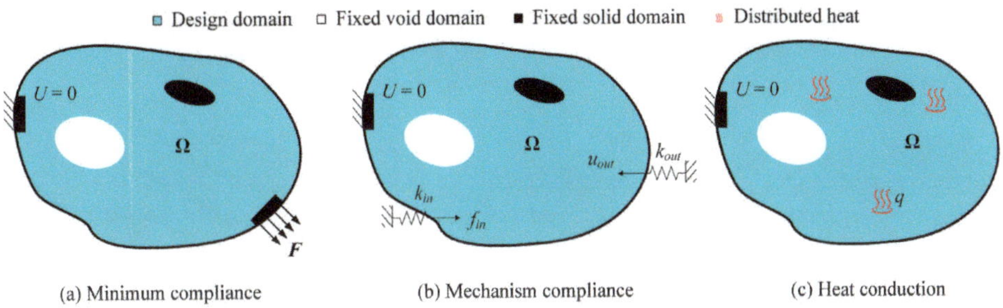

Figure 9. Schematic of three representative topology optimization problems.

The term $\boldsymbol{K}(\rho)$ denotes the global stiffness matrix in the case of minimum compliance and compliant mechanism problems, whereas it signifies the global heat conductivity matrix in the heat conduction problem. The matrix is dependent on the design variable ρ. The variable ρ_i indicates the density of the material in the ith element, where $\rho_i = 1$ represents a solid element and $\rho_i = 0$ indicates an empty element.

Lastly, g(ρ) symbolizes the volume constraint function. Within this function, $V(\rho)$ is the volume of material used, $\overline{V}_{\mathrm{f}}$ is the volume fraction, and V_0 is the total volume of the design domain. This formulation effectively encapsulates the constraints and objectives of the topology optimization problems incorporated by the TopADD program.

5. Results and Discussion

This section will conduct numerical tests on the Raspberry Pi cluster to assess its performance. The impact of hardware and software configurations on the cluster's efficiency will be explored to determine the most effective components for the Raspberry Pi cluster. Additionally, the speedup and efficiency gained through parallelization will be evaluated and compared with standard computing systems. Unless specified otherwise, the default test case will be the 2D arbitrary geometry problem in the TopADD program.

5.1. Hardware Configurations' Influences

5.1.1. Cable Influence

Figure 10 compares the computational time differences when utilizing CAT6 versus CAT8 network cables. Interestingly, the disparity in performance is minimal, despite CAT8 cables being reputed to offer triple the speed of CAT6 cables. To evaluate network connection bandwidth, the "iperf" tool is employed. The procedure involves two primary steps. The first step is the installation and server setup: Install the "iperf" program on the first node (e.g., rpi1) and initiate it in server mode. Secondly, on a second node (e.g., rpi2), after installing "iperf", start the program in client mode, which connects to the server node. The syntax is as follows:

```
ubuntu@rpi1:~$ sudo apt install iperf   #installation on rpi1
ubuntu@rpi1:~$ iperf -s                 #start server on rpi1
ubuntu@rpi2:~$ sudo apt install iperf   #installation on rpi2
ubuntu@rpi2:~$ iperf -c rpi1            #start client on rpi2
```

Network bandwidth tests conducted between various nodes reveal that the connection speed approximates 900 Mbits/s. This speed aligns closely with the 1 gigabit switch's maximum capacity, as anticipated. Consequently, it is evident that the cables do not impact the cluster's performance.

However, CAT8 cables have superior shielding and reduced crosstalk [33], which can improve signal integrity and reliability in high-density cluster environments. While the bandwidth advantage of CAT8 is not utilized here, its enhanced durability and interference resistance provide a more robust infrastructure for consistent performance.

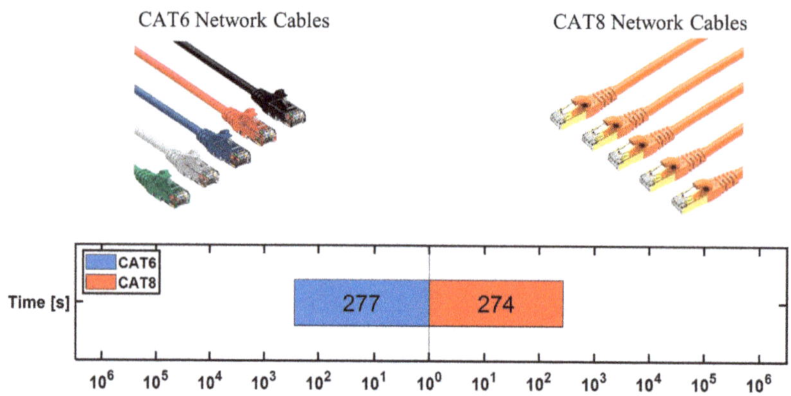

Figure 10. Comparison of computational time using different network cables: CAT6 versus CAT8 network cables.

5.1.2. Network File System Influence

Figure 11 presents a comparative analysis of computational times using different storage solutions, specifically a 128 GB flash drive and a 256 GB NVMe SSD. Despite the NVMe SSD's reported superiority in terms of speed—it is alleged to be over 17 times faster at reading and 38 times faster at writing compared to the flash drive—the differences in computational times are surprisingly marginal. To quantify this, a program named "disk" is employed to offer a user-friendly interface for testing real reading and writing speeds. The results show average reading speeds of 57 MB/s for the flash drive and 333 MB/s for the SSD, with corresponding writing speeds of 48 MB/s and 303 MB/s. While these testing results indicate significant disparities in the reading and writing speeds of the two drives, the final runtime of the TopADD application was nearly identical for both storages. This suggests that the NFS drives are not the bottleneck in the cluster's performance.

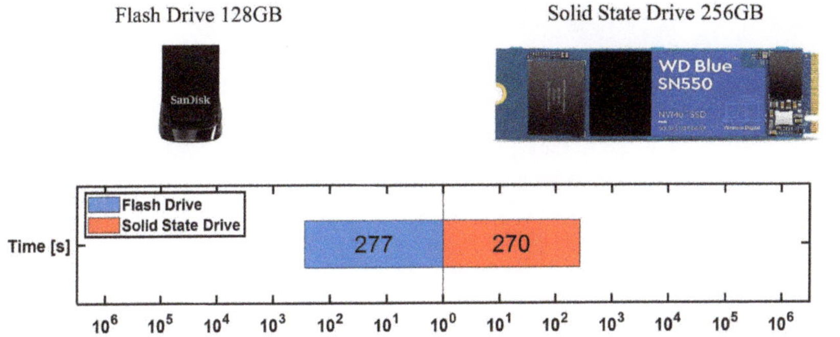

Figure 11. Comparison of computational time using different NFSs: flash drive versus solid state drive network.

5.1.3. Network Switch Influence

Figure 12 compares computational times when using different network switches: a 1-gigabit switch versus a 100-megabit switch. The configuration with the 100-megabit switch shows a 16% slower performance compared to the 1-gigabit one. Network bandwidth was tested using the program "iperf", which has been previously discussed. These tests show that the network bandwidth hovers around 95 Mbits/s, aligning with the maximum capacity of the 100-megabit switch, as expected. However, this comparison reveals an interesting observation: even though the network speed associated with the 100-megabit switch is only a tenth of that achieved with the 1-gigabit switch, the execution time of the TopADD program increases by only 16%. This suggests that although switches do impact the cluster's performance, the effect is less substantial than expected, especially considering the 90% decrease in network speed.

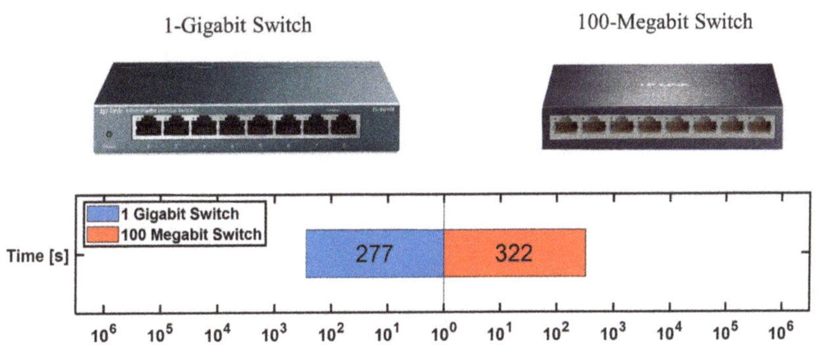

Figure 12. Comparison of computational time using different network switches: a 1-gigabit switch versus a 100-megabit switch.

As noted in the above discussion, peripheral devices, such as network connections and NFSs, have a marginal impact on the cluster's execution time. These devices, including NFSs and network switches, are essentially I/O devices that are connected to the bus of the RPis. However, a more critical component connected to this bus is the RPis' RAM. The RAM's bandwidth plays a significant role in influencing the cluster's performance. To understand this better, the RAM bandwidth has been examined using open-source code form McCalpin [34], available at https://www.cs.virginia.edu/stream/FTP/Code/stream.c (accessed on 1 January 2024). This code can be compiled and executed as follows:

```
ubuntu@rpi1:~$ mpicc -o stream -O3 stream.c   #compiling on rpi1
ubuntu@rpi1:~$ mpirun -n 4 ./stream           #run on rpi1
```

The resulting average memory bandwidth per core is 900 MB/s, with a total bandwidth of 3.6 GB/s. This is notably low, especially when considering the reported total memory bandwidth of 5.5 GB/s and the CPU core frequency of 1.5 GHz [32]. Consequently, it appears that memory bandwidth is a primary limiting factor in the cluster's performance. This topic will receive more in-depth analysis in Section 5.3. Therefore, it is reasonable to infer that investing in high-performance peripheral devices, like advanced network cables or SSD drives, is unnecessary for most users. The identified bottleneck in memory bandwidth suggests that basic versions of these peripherals should suffice.

5.2. *Software Configurations' Influences*

5.2.1. Compiler Optimization Level Influence

Figure 13 presents a comparative analysis of computational time in relation to the level of optimization in the TopADD program, specifically contrasting the release version against the debug version. In the release version, both the TopADD program and its underlying software, PETSc 3.16.3, were compiled with an optimization level of O3. On the contrary, the debug version is compiled with no level of optimization. The results demonstrate a marked difference: the debug version required over six times the execution time of the release version. This significant disparity underscores the importance of flexibility in configuring a cluster, such as the ability to reinstall optimized versions of software. Such flexibility is crucial for enhancing the performance of parallel computing.

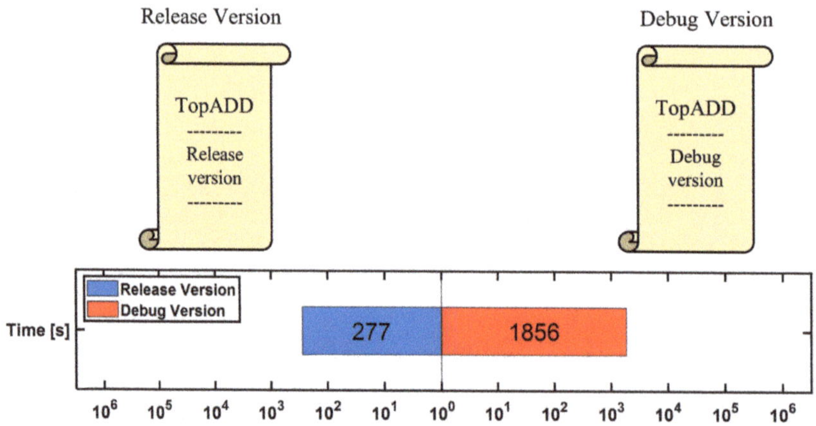

Figure 13. Comparison of computational time using different software versions: the release version versus the debug version.

In addition, the single and double float performance has been compared using the code presented in Appendix D. Single float (12.9455 s) outperforms double float (19.4908 s) in terms of execution time. This is due to the smaller memory footprint and faster memory access associated with float, which allows for better cache utilization and reduced computational overhead. On the other hand, while double float provides higher precision, it incurs additional memory and computation costs, leading to slower performance for large data sizes.

5.2.2. Software Installation Location Influence

Figure 14 illustrates a comparison of computational time when installing the PETSc software 3.16.3 in different locations: on each compute node versus on an NFS. Notably, the execution times are identical in both cases, demonstrating that the installation location does not influence the cluster's performance. This outcome is surprising, given that the NFS-installed version requires loading PETSc via the network each time, and network speeds are typically slower than flash drive speeds. This could be attributed to PETSc being loaded from the network only once, at the beginning of execution, minimizing the impact of loading speed. This demonstrates that conducting practical experiments on an in-house cluster is instrumental in gaining a more comprehensive understanding of parallel computing mechanisms.

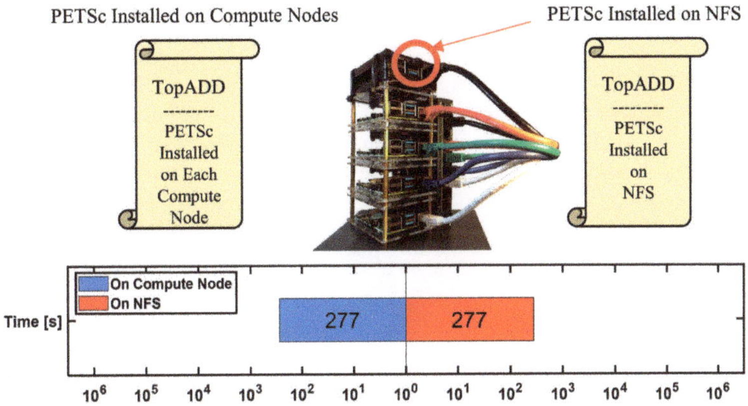

Figure 14. Comparison of computational time using different PETSc installation locations: on each compute node versus on NFS.

5.2.3. Overclocking Level Influence

Figure 15 depicts a comparison of computation times between scenarios with and without overclocking settings. Overclocking refers to the process of increasing the clock speed of a computer's CPU beyond the manufacturer's specified rate. Overclocking can be implemented using the specified syntax, though it is important to note that this is carried out at the user's own risk.

```
sudo nano /boot/firmware/config.txt
# add the following at the end of config.txt
over_voltage=6
arm_freq=2000
```

The results indicate that the difference in computational times due to overclocking is negligible. This aligns with the findings in Section 5.1 which stated that RAM bandwidth has been identified as the primary bottleneck. As the CPU frequency increases due to overclocking, the limiting effect of RAM bandwidth becomes more pronounced. Despite a 25% increase in CPU frequency, the execution time only shows a marginal improvement of 3%. This underscores the minimal impact of overclocking under the current RAM bandwidth constraints.

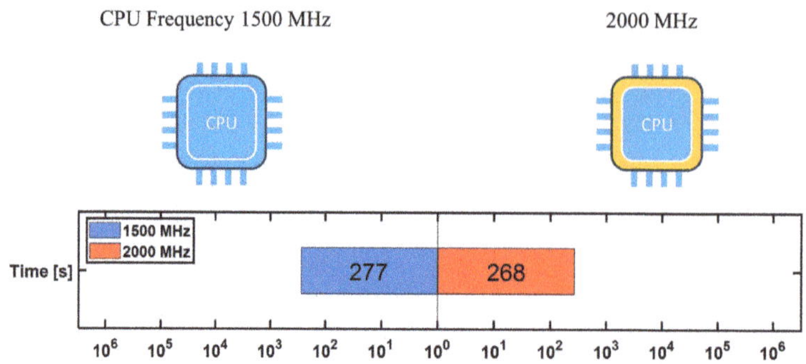

Figure 15. Comparison of computational time using a different CPU frequency: 1500 MHz versus 2000 MHz.

5.3. Parallelization Speedup and Efficiency

The investigation of parallelization speedup and efficiency is conducted. The computational time from a single core was used as the reference. The speedup factor, denoted as S_N, is calculated based on the ratio of the computational time for one core (T_1) to the computational time using N cores (T_N). This is represented by Equation (2):

$$S_N = \frac{T_1}{T_N} \quad (2)$$

Furthermore, the efficiency of parallelization, named as E_p, is a measure of the proportion of time during which a processor is effectively utilized. It is defined as follows in Equation (3):

$$E_P = \frac{S_N}{N} \quad (3)$$

To evaluate these metrics, two benchmarks have been utilized. The first benchmark involves an implementation of parallel computing of pi in mathematics.

It should be mentioned that the core allocation rule in the cluster evenly distributes cores across nodes. For example, when using 4 cores, each node contributes 1 core; for 8 cores, each node contributes 2 cores, and so on. This ensures balanced utilization of resources across all nodes.

5.3.1. Pi Calculation via Integral Method

In the first case study, a method of calculating Pi using the integral method was explored. The integral is represented by

$$\pi = 4\int_0^1 \frac{1}{1+x^2}dx \quad (4)$$

where x varies from 0 to 1, incrementing by an infinitesimally small value dx. The relevant C++ code for this computation is provided in Appendix C.

Table 3 displays the speedup and efficiency results obtained by parallelizing this Pi calculation. Figure 16a,b graphically illustrate these results, showing both the speedup and efficiency achieved through parallelization. The performance of RPi cluster in this context is noteworthy. The observed speedup closely aligns with the theoretical ideal for various core counts. Similarly, the efficiency maintains a high level, reaching up to 0.99 in different scenarios involving varying numbers of cores. This demonstrates the RPi cluster's robust capability in handling parallel computations effectively.

Table 3. Speedup and efficiency of the pi code parallelization.

Core Number	1	2	3	4	8	12	16
Wall clock time (s)	9.70	4.88	3.25	2.44	1.22	0.82	0.61
Speedup (S_N)	1	1.99	2.98	3.98	7.95	11.83	15.90
Efficiency (E_P)	1	0.99	0.99	0.99	0.99	0.99	0.99

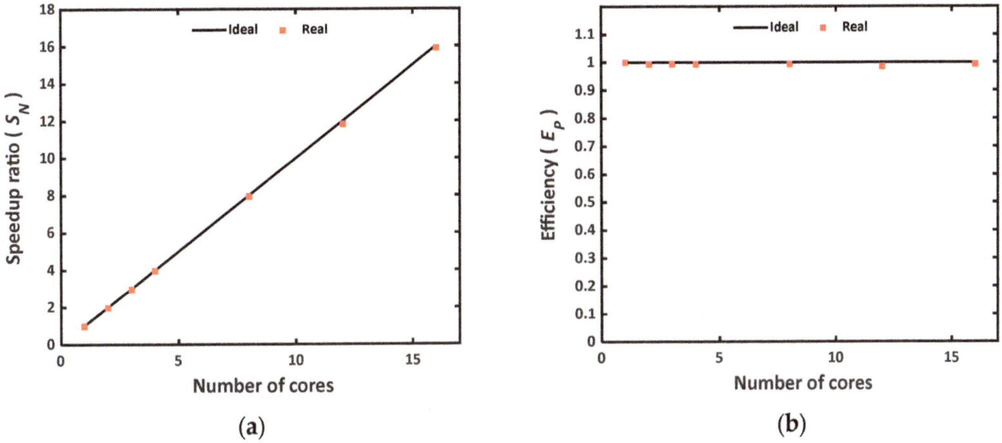

Figure 16. Visualization of code parallelization for the pi calculation: (**a**) speedup and (**b**) efficiency.

5.3.2. TopADD Default 2D and 3D Models

In another analysis, the performance of TopADD executed with varying numbers of cores is examined. Figure 17 illustrates the initial arbitrary design domain and the corresponding optimized results in TopADD, showing both the default 2D and 3D models. The 2D model features a 240 × 120 mesh, while the 3D model is constructed with a 64 × 32 × 32 mesh. Table 4 provides detailed information about the speedup and efficiency observed when running TopADD on the cluster. Furthermore, Figure 18a,b visually represent these metrics, displaying the speedup and efficiency achieved through parallelization. The data reveal that employing more cores enhances the speedup ratio up to a certain point, beyond which additional cores only marginally reduce the computation time. This trend aligns with findings reported in the literature [29].

Table 4. Speedup and efficiency of the TopADD parallelization on the RPi cluster.

Core Number	1	2	3	4	8	12	16
		2D					
Wall clock time (s)	386	283	219	198	207	221	277
Speedup (S_N)	1	1.36	1.76	1.95	1.86	1.75	1.39
Efficiency (E_P)	1	0.68	0.59	0.49	0.23	0.15	0.09
		3D					
Wall clock time (s)	6593	3844	2662	2119	1948	1808	1853
Speedup (S_N)	1	1.72	2.48	3.11	3.38	3.65	3.56
Efficiency (E_P)	1	0.86	0.83	0.78	0.42	0.30	0.22

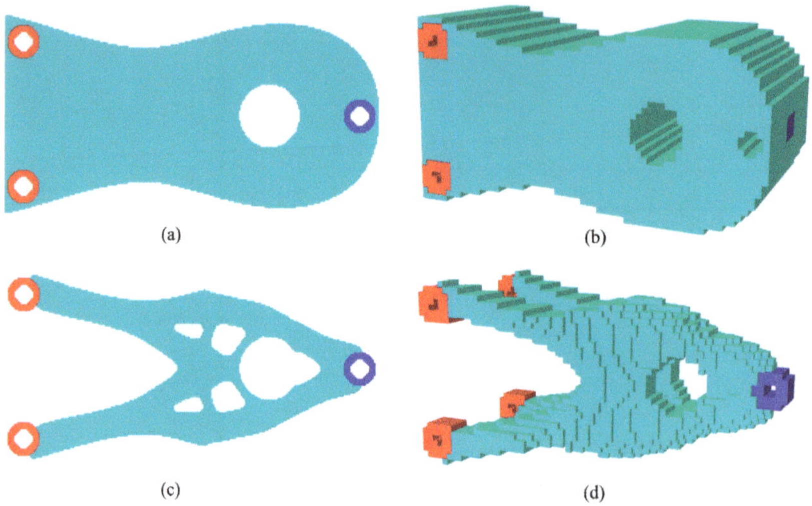

Figure 17. Default domain initialization and optimized results in TopADD, (**a**) 2D design domain with a mesh of 240 × 120, (**b**) 3D design domain with a mesh of 64 × 32 × 32, (**c**) 2D-optimized result, and (**d**) 3D-optimized result. The red color areas are the fixed locations, and the blue areas are the loading positions.

The observed results highlight superior speedup and efficiency in the 3D model compared to the 2D model. This variance is likely a consequence of the increased network communication involved in the 2D scenario, which adds substantial overhead time. Conversely, the 3D model necessitates a larger allocation of CPU time for computation. This shift in resource utilization diminishes the proportion of time spent on network communication, thereby reducing overhead and enhancing overall efficiency and speedup in the parallel computing environment.

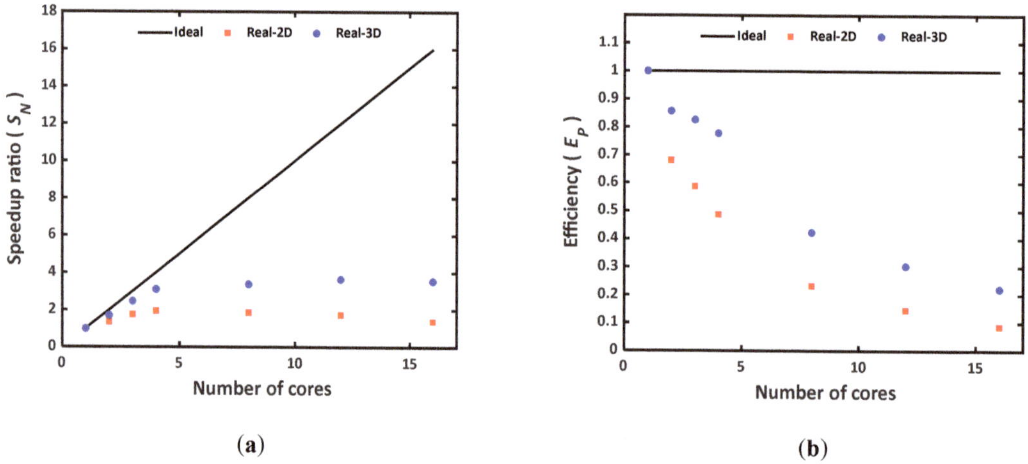

Figure 18. The visualization of the code parallelization for the TopADD default 2D and 3D models: (**a**) speedup and (**b**) efficiency.

It is important to note that performance begins to decline significantly when more than four cores are utilized. This trend may be attributable to the RAM bandwidth limitations of the RPis, as discussed earlier. Consequently, the RAM bandwidth testing code by [34] remains relevant and will be employed again, this time across the entire cluster. The setup involves creating a machine file containing the names of the nodes. The test is then compiled and executed using a varying number of cores, ranging from 1 to 16, as demonstrated in the following example.

```
ubuntu@rpi0:~$ mpicc -o stream -O3 stream.c          #compile on rpi0
ubuntu@rpi0:~$ mpirun -n 16 -f machinefile ./stream #run on rpi1
```

Table 5 displays the total memory bandwidth results. Notably, a decline in bandwidth is observed at 8, 12, and 16 cores. When up to 4 cores are used, each core fully utilizes the bandwidth available on a Raspberry Pi node. However, adding more cores per node leads to a reduction in overall performance. This suggests that the Raspberry Pi 4 Model B is highly efficient at saturating the bus when a single core is used. The addition of more cores does not enhance performance; in fact, it results in a decrease due to increased idle cycles from RAM. This is caused by contention leading to more frequent opening and closing of memory pages.

Table 5. RAM bandwidth with different number of cores used for the RPi cluster.

Core Number	1	2	3	4	8	12	16
Total RAM bandwidth (GB/s)	5.46	10.92	15.82	21.81	26.90	14.68	14.36
Bandwidth per core (GB/s)	5.46	5.46	5.27	5.45	3.36	1.22	0.90

Since parallelization efficiency tends to decrease with an increasing number of cores, a further analysis of network latency, MPI communication overhead, and workload balancing strategies is necessary.

Firstly, the network speed and latency were tested using the method described in Section 5.1.1. The results show a maximum network speed of approximately 900 Mbits/sec, which is close to the theoretical limit of the 1 Gigbits Ethernet on the Raspberry Pi 4 model B. Additionally, the transmission latency between two nodes is around 0.2 ms when the "ping" software is used.

Secondly, the MPI communication overhead was evaluated using the OSU benchmark [35]. The measured latency for transferring 64-byte message is 0.056 ms, indicating good MPI communication performance.

Lastly, the TopADD [15] utilizes the DMDA function in PETSc [21] to generate structured grids, which are evenly distributed across all cores. As a result, factors such as network latency, MPI communication overhead, and workload balancing strategies are unlikely to significantly impact the parallelization efficiency.

From Sections 5.3.1 and 5.3.2, it can be observed that the parallel-computing performance of Raspberry Pi can indeed vary depending on the nature of the application being tested. CPU-bound tasks (e.g., Pi calculation) scale well, while memory-intensive tasks (e.g., TopADD) degrade when more cores are added due to shared memory bus contention, as shown in Table 5. This highlights the value of affordable RPi clusters for testing diverse workloads.

The scalability test on the 4-node Raspberry Pi cluster (4 cores per node) shows parallel efficiency drops beyond 4 cores due to shared-memory bandwidth saturation. However, this is unnecessarily negative, as single-core saturation maximizes performance for non-multithreaded apps, which is necessary for Raspberry as a general-purpose computing device. In the future, scalability can be improved by adding more nodes with one core each, and further multi-node tests will be necessary.

5.4. Comparison to Typical Computers

To benchmark the RPi cluster's performance against the standard computing hardware typically accessible to newcomers or researchers, a laptop and a desktop computer were selected as test systems. The specifications of these devices are detailed in Table 6. Both the laptop and desktop were set up with the same OS to ensure consistency in the testing environment. The following sections will provide a concise evaluation of their performance and energy consumption in comparison to the RPi cluster. This comparison aims to offer a clearer perspective of the relative capabilities and efficiency of the RPi cluster in common computational tasks.

Table 6. Test computers' specifications.

Platform	Cost (CAD)	OS	CPU	Cores	MHz
RPi Cluster	730	Ubuntu 20.04	4 × BCM2711 ARM Cortex-A72	16	1500
Laptop	1500	Ubuntu 20.04	Core i5-8350U	4	1900
Desktop	2000	Ubuntu 20.04	Core i7-11700	8	2500

Figure 19 provides a comparative analysis of the computing and energy performance among three different types of computers: an RPi cluster, a Core i5 laptop (4 cores), and a Core i7 desktop (8 cores). The comparison reveals distinct differences in their computing times and energy consumption. In terms of computing time, as depicted in Figure 19a, the RPi cluster had the longest recorded duration for task completion, whereas the Core i7 desktop demonstrated the shortest computing time. However, when it comes to power consumption, the scenario shifts. As shown in Figure 19b, the Core i7 desktop consumes the most power when operating at full capacity. In contrast, the RPi cluster exhibits remarkably low power consumption. When in an idle state, which is typical during program development and debugging, the RPi cluster can operate with just the login node active, significantly reducing its power usage to merely 6 W. However, the Core i7 desktop consumes approximately 12 times more power than the RPi cluster, as presented in Figure 19c.

The annual energy consumption calculation considers both full-power and idle power states. For the RPi cluster, the calculation assumes an equal split between these two states. For the laptop and desktop, their full-power duration is based on their respective computing time ratios compared to the RPi cluster. For instance, the desktop's full-power time is calculated as $0.5 \times 365 \times (19/198)$, resulting in a total consumption of 763 kWh of electricity annually. Figure 19d summarizes these energy calculations, clearly showing that the RPi cluster is the most energy-efficient option, incurring the lowest annual energy costs among the three.

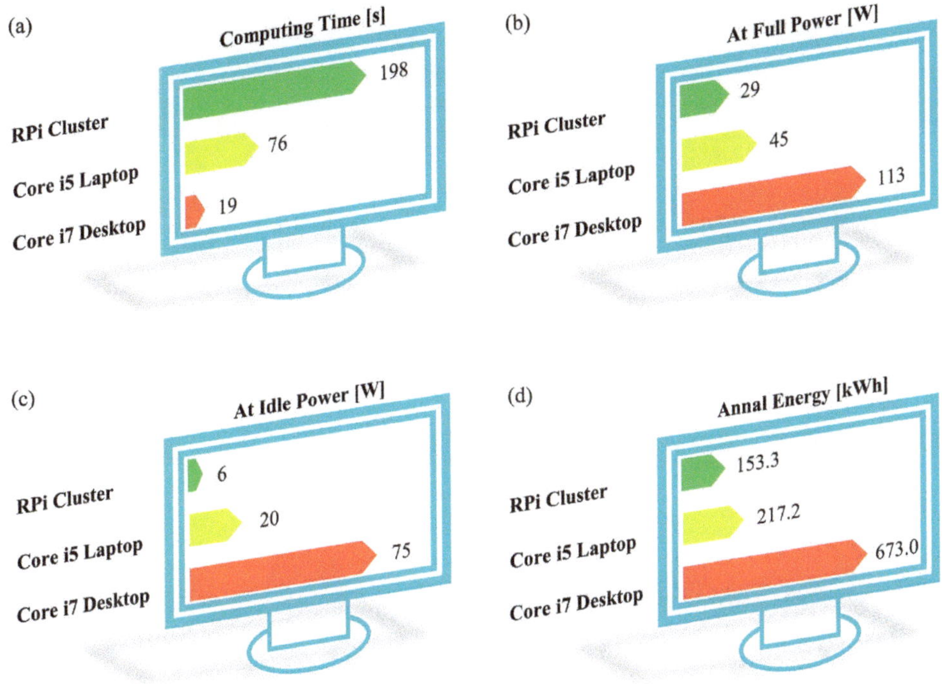

Figure 19. Comparison of computing and energy performance among the three different computers: (**a**) computing time, (**b**) full power consumption, (**c**) idle power consumption, and (**d**) annual energy.

It should be noted that the RPi cluster could be compared to other low-cost clusters in the future. However, comprehensive comparisons between Raspberry Pis, ODROID-XU4, Jetson Nano and other low-cost single-board computers (SBCs) have already been presented in the literature [36], as depicted in Table 7. Table 7 shows the RPi cluster stands out in terms of cost-effectiveness and energy efficiency, although the RPi's shared memory architecture limits its scalability for memory-heavy tasks compared to other SBCs. Furthermore, the Raspberry Pi is one of the best-selling SBCs worldwide, with cumulative sales exceeding 60 million units as of 2024, which demonstrates user satisfaction [37]. Its vibrant ecosystem, including active forums (e.g., raspberrypi.com/forums accessed on 1 January 2024), rich documentation, and a large community, ensures timely problem resolution, making it a trusted platform for both hobbyists and professionals. Finally, although this work uses a Raspberry Pi cluster as the experimental platform, the performance analysis and optimization techniques (e.g., memory bandwidth testing and parallel efficiency evaluation) are universal and can be extended to other SBCs (e.g., ODROID, Jetson Nano, etc.).

Table 7. Low-cost single board computers' comparisons.

Computer	Cost per Node	Power Draw (W/Node)	Peak RAM Bandwidth (GB/s)	MHz
RPi 4	92 CAD	2–3	4.3	1500
ODROID-XU4	197 CAD	4–6	6.1	1900
Jetson Nano	260 CAD	5–10	12.8	2000

5.5. Environmental Sustainability

The year 2023 has been identified as the hottest on record, according to [38], highlighting the growing urgency of sustainability and carbon-neutral initiatives. With an increasing global focus on reducing carbon footprints, electricity savings have become a critical aspect of sustainable practices. As noted by [39], each kilowatt-hour (kWh) of electricity is equivalent to approximately 0.3 kg of CO_2 emissions.

Given this relationship, the annual carbon footprints of three different types of computers have been calculated and are presented in Figure 20. The results reveal a stark contrast in environmental impact: the RPi cluster requires the carbon absorption capacity of approximately 2 mature trees to offset its emissions, compared to around 9 trees needed for a Core i7 desktop. For context, a mature tree can absorb about 22 kg of CO_2 annually, as noted by the United States Department of Agriculture [39].

The significant difference underscores the sustainability of the RPi cluster, whose carbon footprint is over 70% lower compared to that of the desktop. Its lower power consumption and consequent reduced CO2 emissions make it an attractive choice for those seeking to minimize the environmental impact in the realm of computing.

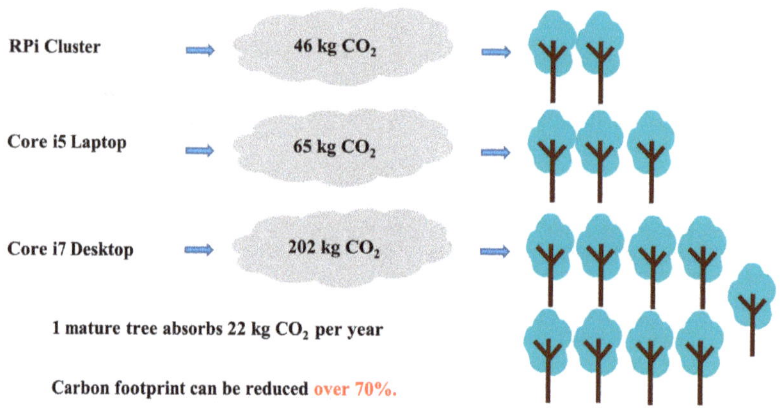

Figure 20. Comparison of the annual carbon footprint among the three different computers.

5.6. Maximum Capability

The cluster's proficiency in high-resolution 2D topology optimization was rigorously evaluated through three physical TO problems, which are outlined in Section 4. The outcomes, as shown in Figure 21, were executed in 2D to optimize efficiency, achieving enhanced resolution while maintaining manageable computational costs. The first scenario involved a compliance minimization problem, utilizing a 2048 × 1024 mesh that comprised 2 million elements. For the second scenario, the compliant mechanism problem, a smaller 1280 × 640 mesh was used, resulting in 0.8 million elements. This reduction in mesh size was necessary, as a larger mesh exceeded the cluster's memory capacity. The potential crashes are because of the increased memory requirements of the self-adjoint problem calculation in the compliant mechanism problem. The third scenario addressed a heat conduction problem with a 1648 × 2048 mesh, accounting for 3.4 million elements. The computational durations for these problems were recorded at 11.9 h, 1.5 h, and 3.6 h, respectively. Notably, the results from these 2D optimizations can be transformed into 3D geometries through extrusion, as shown in Figure 21b. All the procedures were performed exclusively on the login node of the cluster. Furthermore, these optimized designs are suitable for 3D printing, such as Fused Deposition Modeling (FDM), as demonstrated in Figure 21c.

(a)

(b)

(c)

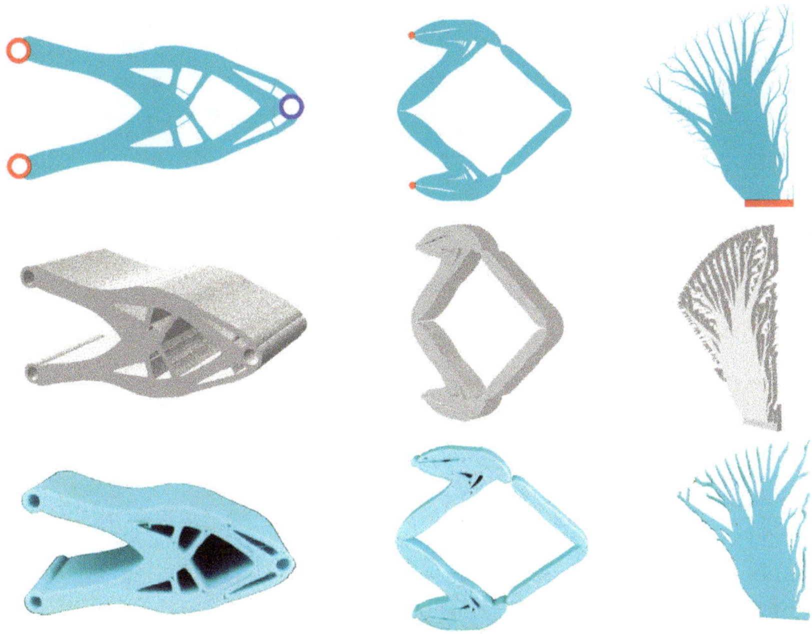

Figure 21. RPi cluster's maximum capacity for the three physical problems in TopADD, from left to right: compliance minimization, compliant mechanism, and heat conduction: (**a**) 2D TO results, (**b**) extruded 3D geometries, and (**c**) parts which were 3D printed via FDM.

6. Conclusions

A Raspberry Pi-based cluster system, TopADDPi, is proposed for the field of PCTO with an arbitrary design domain. TopADDPi addresses the need for cost-effective, accessible distributed-memory computational resources in PCTO. As an affordable, sustainable, and indispensable tool, it offers new opportunities for newcomers and researchers, particularly those with limited access to conventional high-performance computing resources. The main conclusions are as follows:

1. TopADDPi establishes a model in distributed computing and educational resources, presenting a holistic framework that enhances programming and learning capabilities in PCTO.
2. The detailed analysis of hardware and software setups in the study reveals that peripheral devices like internet cables and Network File Systems (NFSs) exert minimal influence on performance. The primary constraint, however, lies in the RAM bandwidth of the Raspberry Pis, which is identified as a critical bottleneck.
3. In tests, the cluster achieves remarkable computational efficiency, approaching 99% parallel efficiency for moderate workloads such as the Pi calculation. For more demanding tasks, the parallel efficiency markedly drops below 10%. To optimize performance and maintain parallel efficiency above 40%, the study recommends using only one core per node, a strategy to mitigate the limitations imposed by RAM bandwidth saturation.
4. When compared with traditional computing systems, TopADDPi emerges as both a cost-effective and environmentally sustainable solution. The study highlights that has a significantly reduced carbon footprint, offering a reduction of over 70% annually compared to standard desktop computers.

5. The proposed method achieves high-resolution topology optimization design results while balancing computational costs and memory constraints. All the procedures were performed exclusively on the cluster. The design results can be processed and manufactured by 3D printing, enabling the exploration of sustainable and digitalized manufacturing processes.

Author Contributions: Conceptualization, Z.-D.Z. and O.I.; methodology, Z.-D.Z.; software, Z.-D.Z.; validation, Z.-D.Z. and D.-Y.Y.; formal analysis, Z.-D.Z. and D.-Y.Y.; investigation, Z.-D.Z. and O.I.; resources, L.M., T.G., J.-H.Z. and W.-H.Z.; data curation, Z.-D.Z.; writing—original draft preparation, Z.-D.Z.; writing—review and editing, all authors; visualization, Z.-D.Z.; supervision, W.-H.Z.; project administration, T.G. and J.-H.Z.; funding acquisition, Z.-D.Z., T.G., J.-H.Z. and W.-H.Z. All authors have read and agreed to the published version of the manuscript.

Funding: This research was funded by the National Natural Science Foundation of China (12202359, 12032018), the National Key R&D Program of China (2022YFB3402200, 2022YFB4603101) and the Key Project of NSFC (92271205).

Data Availability Statement: The configuration information of Raspberry Pi cluster and further details have been provided at: https://github.com/wonderfulzzd/Pi_Cluster_for_TopADD (accessed on 1 January 2024).

Acknowledgments: The authors would like to thank all the specialists from SHARCNET for providing valuable tutorials and assistance.

Conflicts of Interest: The authors declare that they have no conflicts of interest.

Appendix A. Code Needs a Raspberry Pi to Debug

```
/*
Authors: Zhidong Brian Zhang in Dec. 2023
Copyright (C) 2023-2023

Disclaimer:
The authors reserve all rights but do not guaranty that the code is
free from errors. Futhermore, we shall not be liable in any event
caused by the use of the program.

Usage:
 1. g++ main.cc
 2. ./a.out
*/

#include <iostream>
#include <fstream>

int main () {

  float val = 1.0;
  std::ofstream fs("example.bin", std::ios::out | std::ios::binary);
  fs.write(reinterpret_cast<const char*>(&val), sizeof(float));
  fs.close();

  float buf;
  std::ifstream in ("example.bin", std::ios::binary); // reload the
file stream as binary
  in.read (reinterpret_cast<char*> (&buf), sizeof(float));
  in.close();

  unsigned int a = 0;
  float b = buf;

  a = a - b;
  std::cout << "a - b: " << a << std::endl;

  return 0;
}
```

Appendix B. Code Needs a Cluster to Debug

```
/*
Authors: Zhidong Brian Zhang in Dec. 2023
Copyright (C) 2023-2023

Disclaimer:
The authors reserve all rights but do not guaranty that the code is
free from errors. Futhermore, we shall not be liable in any event
caused by the use of the program.

Usage:
 1. mpic++ main.cc
 2. Create machinefile containing computer name(s)
 3. mpiexec -n 4 --hostfile machinefile ./a.out
*/

#include <mpi.h>
#include <iostream>

int main (int argc, char *argv[]) {

  // Initialize MPI
  MPI_Init (&argc, &argv);

  // Sum up a
  int rank;
  int rankSize;
  MPI_Comm_rank(MPI_COMM_WORLD, &rank);
  MPI_Comm_size(MPI_COMM_WORLD, &rankSize);
  int a = rank;

  int *shared;
  MPI_Win win;
  MPI_Win_allocate_shared(1*sizeof(int), sizeof(int), MPI_INFO_NULL,
                          MPI_COMM_WORLD, &shared, win);
  shared[0] = a;
  a = 0;
  MPI_Barrier(MPI_COMM_WORLD);

  MPI_Aint queried_size;
  int disp_unit;
  MPI_Win_shared_query(win, 0, &queried_size, &disp_unit, &shared);

  for (int i = 0; i < rankSize; ++i)
  {
    a += shared[i];
  }

  std::cout << "sum(a): " << a << std::endl;

  MPI_Win_free(&win);
  MPI_Finalize();

  return 0;
}
```

Appendix C. Code for Pi Calculation Using Integral Method

```
/*
Authors: Zhidong Brian Zhang in Dec. 2023
Copyright (C) 2023-2023

Disclaimer:
The authors reserve all rights but do not guaranty that the code is
free from errors. Futhermore, we shall not be liable in any event
caused by the use of the program.

Usage:
 1. mpic++ main.cc
 2. Create machinefile containing computer name(s)
 3. mpiexec -n 4 --hostfile machinefile ./a.out
*/

#include "mpi.h"

int main( int argc, char *argv[]) {
    int rank, np, i, nLoc, nTot = 1024000000 ;
    double mypi, pi, detx, sum, x;

    MPI_Init(&argc,&argv);
    MPI_Comm_size(MPI_COMM_WORLD,&np);
    MPI_Comm_rank(MPI_COMM_WORLD,&rank);

    double t1, t2;
    // start timer
    t1 = MPI_Wtime ();

    if (rank == 0) { nLoc= nTot / np; }

    MPI_Bcast(&nLoc, 1, MPI_INT, 0, MPI_COMM_WORLD);

    detx = 1.0 / (double) nTot;
    sum = 0.0;
    for (i = rank + 1; i <= nTot; i += np) {
      x = detx * ((double)i -0.5);
      sum += (4.0 / (1.0 + x*x));
    }
    mypi = detx * sum;

    MPI_Reduce(&mypi, &pi, 1, MPI_DOUBLE, MPI_SUM, 0, MPI_COMM_WORLD);

    t2 = MPI_Wtime ();
    if (rank == 0) { printf("pi: %1.15f, time: %f\n", pi, t2-t1);}

    MPI_Finalize();
    return 0;
}
```

Appendix D. Code for Single and Double Float Comparison

```
/*
Authors: Zhidong Brian Zhang in Feb. 2025
Copyright (C) 2023-2025

Disclaimer:
The authors reserve all rights but do not guaranty that the code is
free from errors. Futhermore, we shall not be liable in any event
caused by the use of the program.

Usage:
 1. mpic++ floatVSdouble floatVSdouble.cc
 2. Create machinefile containing computer name(s)
 3. mpiexec -n 4 --hostfile machinefile ./floatVSdouble
*/
#include <iostream>
#include <vector>
#include <mpi.h>
#include <cstring>
#include <ctime>
using namespace std;

template <typename T>
void compute_sum(vector<T>& array, int chunk_size, MPI_Datatype
mpi_type, int rank, int size, const string& precision) {
    vector<T> locArray(chunk_size);
    MPI_Scatter(array.data(), chunk_size, mpi_type, locArray.data(),
chunk_size, mpi_type, 0, MPI_COMM_WORLD);
    double locSum = 0, totSum = 0;
    for (int i = 0; i < chunk_size; ++i) locSum += static_cast<dou-
ble>(locArray[i]);
    MPI_Reduce(&locSum,&totSum,1,MPI_DOUBLE, MPI_SUM,0,MPI_COMM_WORLD);
    if (rank == 0) cout << "TotSum " <<precision<<":"<<totSum << endl;
}

int main(int argc, char *argv[]) {
    MPI_Init(&argc, &argv);
    int rank, size;
    MPI_Comm_rank(MPI_COMM_WORLD, &rank);
    MPI_Comm_size(MPI_COMM_WORLD, &size);
    int N = 200000000, CHUNK_SIZE = N / size;
    double t1start = MPI_Wtime(); vector<float> aF(N);
    if (rank==0) for(int i=0;i<N;++i) aF[i]=static_cast<float>(i+1);
    compute_sum(aF, N/size, MPI_FLOAT, rank, size, "float");
    double t1 = MPI_Wtime() - t1start;
    double t2start = MPI_Wtime(); vector<double> aD(N);
    if (rank==0) for(int i=0;i<N;++i) aD[i]=static_cast<double>(i+1);
    compute_sum(aD, N/size, MPI_DOUBLE, rank, size, "double");
    double t2 = MPI_Wtime() - t2start;
    if (rank == 0) {cout <<"T1: "<<t1<< " s\n"<<"T2: "<<t2<<" s\n";}
MPI_Finalize();
return 0;
}
```

References

1. Zhu, J.; Zhou, H.; Wang, C.; Zhou, L.; Yuan, S.; Zhang, W. A review of topology optimization for additive manufacturing: Status and challenges. *Chin. J. Aeronaut.* **2021**, *34*, 91–110. [CrossRef]
2. Meng, L.; Zhang, W.; Quan, D.; Shi, G.; Tang, L.; Hou, Y.; Breitkopf, P.; Zhu, J.; Gao, T. From Topology Optimization Design to Additive Manufacturing: Today's Success and Tomorrow's Roadmap. *Arch. Comput. Methods Eng.* **2020**, *27*, 805–830. [CrossRef]

3. Plocher, J.; Panesar, A. Review on design and structural optimisation in additive manufacturing: Towards next-generation lightweight structures. *Mater. Des.* **2019**, *183*, 108164. [CrossRef]
4. Ibhadode, O.; Zhang, Z.; Sixt, J.; Nsiempba, K.M.; Orakwe, J.; Martinez-Marchese, A.; Ero, O.; Shahabad, S.I.; Bonakdar, A.; Toyserkani, E. Topology optimization for metal additive manufacturing: Current trends, challenges, and future outlook. *Virtual Phys. Prototyp.* **2023**, *18*, e2181192. [CrossRef]
5. Bendsøe, M.P.; Sigmund, O. *Topology Optimization: Theory, Methods and Applications*; Springer Science & Business Media: New York, NY, USA, 2003.
6. Eslamipoor, R. Direct and indirect emissions: A bi-objective model for hybrid vehicle routing problem. *J. Bus. Econ.* **2024**, *94*, 413–436. [CrossRef]
7. Eslamipoor, R. An optimization model for green supply chain by regarding emission tax rate in incongruous vehicles. *Model. Earth Syst. Environ.* **2023**, *9*, 227–238. [CrossRef]
8. Aage, N.; Andreassen, E.; Lazarov, B.S.; Sigmund, O. Giga-voxel computational morphogenesis for structural design. *Nature* **2017**, *550*, 84–86. [CrossRef] [PubMed]
9. Aage, N.; Andreassen, E.; Lazarov, B.S. Topology optimization using PETSc: An easy-to-use, fully parallel, open source topology optimization framework. *Struct. Multidiscip. Optim.* **2015**, *51*, 565–572. [CrossRef]
10. Mukherjee, S.; Lu, D.; Raghavan, B.; Breitkopf, P.; Dutta, S.; Xiao, M.; Zhang, W. Accelerating Large-scale Topology Optimization: State-of-the-Art and Challenges. In *Archives of Computational Methods in Engineering*; Springer Nature: Berlin, Germany, 2021; pp. 4549–4571. [CrossRef]
11. Borrvall, T.; Petersson, J. Large-scale topology optimization in 3D using parallel computing. *Comput. Methods Appl. Mech. Eng.* **2001**, *190*, 6201–6229. [CrossRef]
12. Mezzadri, F.; Bouriakov, V.; Qian, X. Topology optimization of self-supporting support structures for additive manufacturing. *Addit. Manuf.* **2018**, *21*, 666–682. [CrossRef]
13. Zhang, Z.D.; Ibhadode, O.; Ali, U.; Dibia, C.F.; Rahnama, P.; Bonakdar, A.; Toyserkani, E. Topology optimization parallel-computing framework based on the inherent strain method for support structure design in laser powder-bed fusion additive manufacturing. *Int. J. Mech. Mater. Des.* **2020**, *16*, 897–923. [CrossRef]
14. Fernández, E.; Yang, K.; Koppen, S.; Alarcón, P.; Bauduin, S.; Duysinx, P. Imposing minimum and maximum member size, minimum cavity size, and minimum separation distance between solid members in topology optimization. *Comput. Methods Appl. Mech. Eng.* **2020**, *368*, 113157. [CrossRef]
15. Zhang, Z.D.; Ibhadode, O.; Bonakdar, A.; Toyserkani, E. TopADD: A 2D/3D integrated topology optimization parallel-computing framework for arbitrary design domains. *Struct. Multidiscip. Optim.* **2021**, *64*, 1701–1723. [CrossRef]
16. Johnston, S.J.; Cox, S.J. The raspberry Pi: A technology disrupter, and the enabler of dreams. *Electronics* **2017**, *6*, 51. [CrossRef]
17. Cox, S. Steps to make Raspberry Pi Supercomputer. *Univ. Southampt.* **2013**. Available online: https://www.southampton.ac.uk/~sjc/raspberrypi/pi_supercomputer_southampton_web.pdf (accessed on 1 January 2024).
18. Kiepert, J. Creating a raspberry pi-based beowulf cluster. *Boise State* **2013**. Available online: https://ogervasi.unipg.it/OpSysNet/4Students/Clusters/Creating.a.Raspberry.Pi-Based.Beowulf.Cluster_v2.pdf (accessed on 1 January 2024).
19. Tso, F.P.; White, D.R.; Jouet, S.; Singer, J.; Pezaros, D.P. The glasgow raspberry pi cloud: A scale model for cloud computing infrastructures. In Proceedings of the 2013 IEEE 33rd International Conference on Distributed Computing Systems Workshops, Philadelphia, PA, USA, 8–11 July 2013; pp. 108–112.
20. Abrahamsson, P.; Helmer, S.; Phaphoom, N.; Nicolodi, L.; Preda, N.; Miori, L.; Angriman, M.; Rikkilä, J.; Wang, X.; Hamily, K.; et al. Affordable and energy-efficient cloud computing clusters: The Bolzano Raspberry Pi cloud cluster experiment. In Proceedings of the 2013 IEEE 5th International Conference on Cloud Computing Technology and Science, Bristol, UK, 2–5 December 2013. [CrossRef]
21. Balay, S.; Abhyankar, S.; Adams, M.; Brown, J.; Brune, P.; Buschelman, K.; Dalcin, L.; Dener, A.; Eijkhout, V.; Gropp, W.; et al. PETSc Users Manual. 2019. Available online: https://publications.anl.gov/anlpubs/2019/12/155920.pdf (accessed on 1 January 2024).
22. Seal, D. *ARM Architecture Reference Manual*; Pearson Education: London, UK, 2001.
23. Upton, E.; Halfacree, G. *Raspberry Pi® User Guide*; John Wiley & Sons: Hoboken, NJ, USA, 2016. [CrossRef]
24. Stroustrup, B. *The C++ Programming Language*; Pearson Education: London, UK, 2013.
25. Potluri, S.; Wang, H.; Dhanraj, V.; Sur, S.; Panda, D.K. Optimizing MPI one sided communication on multi-core infiniband clusters using shared memory backed windows. In *Recent Advances in the Message Passing Interface, Proceedings of the 18th Eur. MPI Users' Gr. Meet. EuroMPI 2011, Santorini, Greece, 18–21 September 2011*; Springer: Berlin/Heidelberg, Germany; pp. 99–109.
26. Pahl, C.; Helmer, S.; Miori, L.; Sanin, J.; Lee, B. A container-based edge cloud PaaS architecture based on raspberry Pi clusters. In Proceedings of the 2016 IEEE 4th International Conference on Future Internet of Things and Cloud Workshops (FiCloudW), Vienna, Austria, 22–24 August 2016; pp. 117–124. [CrossRef]
27. Dennis, A.K. *Raspberry Pi Super Cluster*; Packt Publishing Ltd.: Birmingham, UK, 2013.

28. Han, H.; Kim, K.; Jung, S.; Chang, Y. Performance Evaluation on the Parallel Processing System with the Raspberry Pi 4. In Proceedings of the 2022 International Conference on Computational Science and Computational Intelligence (CSCI), Las Vegas, NV, USA, 14–16 December 2022; pp. 1372–1374. [CrossRef]
29. Warade, M.; Schneider, J.G.; Lee, K. Measuring the Energy and Performance of Scientific Workflows on Low-Power Clusters. *Electronics* **2022**, *11*, 1801. [CrossRef]
30. Neto, A.J.A.; Neto, J.A.C.; Moreno, E.D. The development of a low-cost big data cluster using Apache Hadoop and Raspberry Pi. A complete guide. *Comput. Electr. Eng.* **2022**, *104*, 108403. [CrossRef]
31. Papakyriakou, D.; Barbounakis, I.S. High Performance Linpack (HPL) Benchmark on Raspberry Pi 4B (8GB) Beowulf Cluster. *Int. J. Comput. Appl.* **2023**, *185*, 11–19. [CrossRef]
32. Raspberry Pi. Raspberry Pi 4 Tech Specs. 2022. Available online: https://www.raspberrypi.com/products/raspberry-pi-4-model-b/specifications/ (accessed on 1 January 2024).
33. Foroughi, T. Evaluation of 10G Ethernet Under Various Environments, Unitec Institute of Technology. 2024. Available online: https://www.researchbank.ac.nz/items/97dab53a-7950-4bc1-8244-0dc3f00db5f6 (accessed on 1 January 2024).
34. McCalpin, J. STREAM: Sustainable Memory Bandwidth in High Performance Computers. 2006. Available online: https://www.cs.virginia.edu/stream/ftp/code/stream.c (accessed on 1 January 2024).
35. OSU Micro-Benchmarks, Ohio State University. 2025. Available online: http://mvapich.cse.ohio-state.edu/download/mvapich/osu-micro-benchmarks-7.1.tar.gz (accessed on 1 January 2024).
36. Kaiser, T. sbc-bench. 2025. Available online: https://github.com/ThomasKaiser/sbc-bench.git (accessed on 1 January 2024).
37. Pounder, L. Raspberry Pi Celebrates 12 Years as Sales Break 61 Million Units, Tom's Hardw. 2024. Available online: https://www.tomshardware.com/raspberry-pi/raspberry-pi-celebrates-12-years-as-sales-break-61-million-units (accessed on 1 January 2024).
38. Copernicus Climate Change Service (C3S), Record Warm November Consolidates 2023 as the Warmest Year. 2023. Available online: https://climate.copernicus.eu/record-warm-november-consolidates-2023-warmest-year (accessed on 1 January 2024).
39. Stancil, J.M. The Power of One Tree–The Very Air We Breathe, USDA (Blog). Entry Posted. 2019. Available online: https://www.usda.gov/media/blog/2015/03/17/power-one-tree-very-air-we-breathe (accessed on 17 March 2015).

processes

Article

Development of a Value Stream Map to Optimize the Production Process in a Luxury Metal Piece Manufacturing Company

Beatriz Costa [1], José Varejão [1] and Pedro Dinis Gaspar [1,2,*]

[1] Department of Electromechanical Engineering, University of Beira Interior, Rua Marquês de D'Ávila e Bolama, 6201-001 Covilhã, Portugal; b.costa@ubi.pt (B.C.); jose.eduardo.v7@gmail.com (J.V.)
[2] C-MAST—Centre for Mechanical and Aerospace Science and Technologies, 6201-001 Covilhã, Portugal
* Correspondence: dinis@ubi.pt

Abstract: The current market is highly competitive, and customers are increasingly demanding. In this context, organizations need to adopt tools that enhance process efficiency to ensure competitiveness. This report aims to implement Lean tools, specifically Value Stream Mapping (VSM) and complementary tools, to optimize the production process in the metal treatment industry. A case study was conducted, beginning with a brief sector and process recognition, followed by an analysis of production stages using VSM. Value-added activities, non-value-added activities, and waste were identified. The current VSM revealed a Lead Time (LT) of approximately 336 h (14 days), value-added activities (VA) of 21 h, and a process cycle efficiency (PCE) of 6.29%. Improvement actions were proposed to reduce waste and increase competitiveness. After implementation, the LT decreased to approximately 318 h (13 days), VA increased to 23 h, and process efficiency improved to 7.15%. Despite the limitations of VSM in discontinuous flows, its use increased process efficiency, demonstrating its applicability in complex industrial contexts.

Keywords: lean manufacturing; value stream mapping; waste; process optimization; metal parts industry

1. Introduction

In the global context of increasingly competitive and dynamic markets, operational efficiency and customer satisfaction have become essential for companies wishing to stand out. The adoption of techniques and philosophies such as Lean Manufacturing is a strategy being used around the world to address challenges and optimize production processes [1]. Introduced by Ohno in *Toyota Production System: Beyond Large-Scale Production* in 1978, this revolutionary approach has demonstrated its ability to improve efficiency and reduce costs by eliminating waste, becoming a global benchmark for operational excellence [2,3]. The successful implementation of Lean tools, such as 5S, Kaizen, Kanban, Value Stream Mapping (VSM), and Poka-Yoke, has significantly improved the performance of companies in various industrial sectors [4]. VSM, in particular, makes it possible to visualize information and material flows, identify and analyze waste and non-value-added activities, and help suggest improvements [5].

This study focuses on the application of VSM to optimize the production process in the metal processing industry for luxury fashion brands. Currently, the turnaround time is relatively high, indicating that it can be improved by eliminating waste. By conducting observations and interviews and collecting data, the aim is to build a VSM of the process, identifying waste and suggesting improvements for optimization. Key issues addressed include the identification of production steps, associated tasks, information and material flows, time metrics, process bottlenecks, and waste elimination strategies. Given the complexity and variability inherent in the industrial production of metal parts, this optimization is crucial to meeting quality requirements and reducing delivery times [6].

The main objective is to address the themes of Lean Manufacturing, in particular VSM, to identify industrial challenges and opportunities for improvement. By defining the process through employee interviews and process observation, followed by time analysis, the aim is to create and follow the VSM methodology. The subsequent analysis will lead to proposals for improvement and the development of an optimized future state. This study aims to contribute by reducing time, standardizing work, and eliminating waste, thereby making the process more efficient. The methodology comprises four implementation phases: product selection, current state map, future state map, and action plan. This approach aims to improve the ability to deliver high-quality products efficiently, demonstrating the effectiveness of Lean Manufacturing, particularly VSM, in improving production processes.

2. Literature Review

This chapter provides a comprehensive overview of Lean Manufacturing, with a special focus on Value Stream Mapping (VSM) and complementary tools for reducing time and waste. The literature review was conducted based on a selection of articles relevant to the topic, covering everything from the fundamentals of Lean Manufacturing to the evolution and practical application of Value Stream Mapping.

2.1. *Toyota Production System and Lean Manufacturing*

The Toyota Production System, or TPS, was conceived and implemented shortly after the Second World War, but it was only later that it came to prominence. The TPS gained recognition for revolutionizing the traditional approach to mass production, with the essential focus being on the human factor and product quality. In contrast to Taylor and Ford's methods, TPS seeks to eliminate waste, enabling efficient production without excessive stocks (Just-in-Time) and maintaining high-quality standards [2].

The Toyota Production System (TPS) integrates methodologies such as Kanban, Heijunka, standardized work, Jidoka, and Kaizen to enhance efficiency and facilitate continuous improvement. These methodologies emphasize the importance of fostering a culture of perpetual enhancement and engaging all employees in the process. The foundational principles of TPS are articulated through the 4P model: Philosophy, Process, People and Partners, and Problem Solving. This model highlights the critical role of nurturing a culture of continuous improvement and ensuring the active involvement of all employees in achieving organizational excellence [7].

Lean Manufacturing, which derives from the Toyota Production System, is a more production-oriented approach. Still, all terminologies have a common goal: to optimize the use of resources by reducing waste as much as possible [8].

The Lean Manufacturing approach, adopted by several companies globally, seeks to reduce costs by eliminating activities that do not add value. The key areas of Lean Manufacturing include flexibility, eliminating waste, optimization, monitoring processes, and involving people [3].

2.1.1. The Origins of Lean

The origins of Lean Manufacturing can be traced back to the Toyota Production System (TPS), developed in Japan in the 1940s and 1950s, just after the Second World War. This system gained prominence, especially after the 1973 oil crisis, which fueled the search for greater production efficiency [2]. In the 1990s, James P. Womack and Daniel T. Jones introduced the concept to the West, popularizing the term 'Lean Manufacturing' with their book *The Machine That Changed the World* [9].

Ohno emphasized the importance of eliminating waste and implemented Just-in-Time. In his 1978 book *Toyota Production System: Beyond Large-Scale Production*, he identified seven types of waste: transport, stocks, waiting, overproduction, defects, unnecessary movements, and rework. Eliminating these wastes is fundamental to improving productivity and reducing costs by looking at the Gemba (shop floor) [2].

In 1990, the term Lean Manufacturing gained prominence in the West through the book *The Machine That Changed the World* [9], which analyzed American, Japanese, and

European car production. Womack et al. [9] argued that organizations should eliminate waste in operations to promote more efficient production, highlighting the importance of reducing Lead Time and adapting production systems to deal with small-quantity and large-variety orders [9,10].

In 1996, Womack and Jones [11] wrote 'Lean Thinking', identifying five fundamental principles of Lean: specifying the value, identifying the value stream, creating continuous flow, allowing customers to 'pull' value, and seeking continuous perfection. They emphasized that Lean is not limited to production, but involves the entire organization [11].

2.1.2. The Eight Wastes

Waste is an action that results in something unnecessary, as explained by Taichii Ohno, who initially identified seven types of waste and later expanded to eight. The categories are [8–13]:

i. Overproduction: Producing more than demand, resulting in excess stock and loss of resources.
ii. Waiting: Time spent waiting, whether for work, orders, or accessories, jeopardizes the flow of value.
iii. Transport: Moving materials between locations consumes time and money.
iv. Over-processing: Doing more than is necessary, generating additional costs and wasting resources.
v. Inventory: Keeping excessive stocks, increasing waiting time and space, and making it difficult to identify problems.
vi. Defects: Inadequate final products or services, resulting in rework and additional costs.
vii. Unnecessary movements: Movements of people or equipment that do not add value, resulting in wasted movement.
viii. Waste of talent and skills: Under-utilization of workers' skills, resulting in loss of motivation and ideas. This category was initially defined by [11], with the eighth waste being identified as employee underutilization, indicating that people's skills and capabilities are not properly used.

2.1.3. Lean Tools

A collection of techniques and tools enriches the Lean approach, the varied application of which is aimed at optimizing processes and eliminating waste. These tools include Just-In-Time (JIT), Single-Minute Exchange of Dies (SMED), Total Quality Management (TQM), work standardization, visual management, Kaizen, Kanban, VSM, and 5S. These techniques, when employed properly, lead organizations to identify and eliminate inefficiencies, improve the quality of products and services, and promote a culture of continuous improvement [12,14].

Lean tools can be categorized in different ways, ranging from analyzing processes to planning, executing changes, and monitoring performance [15]. This subsection will provide a detailed description of the tools that will be practically applied to implement improvements and achieve the desired future state.

Kaizen stands out as a fundamental technique for eliminating waste and promoting continuous improvement at all levels of the organization. It emphasizes not only constant improvement but also addressing problems and inefficiencies along the way, promoting active employee participation and implementing visual management systems to empower workers in the workplace [8].

The 5S method, developed to establish organized and efficient work environments, consists of five steps: Sort, Put in order, Shine, Standardize, and Sustain. Its application results in benefits such as reducing wastes of time and space and improving quality, safety, and hygiene [16].

Kanban, designed to visualize work and optimize flows, contributes to Lean practices and operational efficiency, indicating the right time to replenish materials and facilitating workflow management [16].

Just-in-Time (JIT), introduced by Toyota, aims for efficient production, depending on a significant reduction in tool changes and the practice of leveling production to avoid bottlenecks and safety buffers [2,11].

Standard Work, developed by Taiichi Ohno, seeks to eliminate variations and inconsistencies in results, promoting a culture of responsibility and innovation. Its implementation results in stable and measurable operations, reduced costs, and improved quality [17].

SMED (Single-Minute Exchange of Dies), an essential tool of Lean production, aims to reduce waste by enabling the rapid exchange of tools in less than 10 min. This technique gives companies greater flexibility by minimizing the time needed to change between tasks or equipment configurations. The effective implementation of SMED depends on the precise classification of activities, distinguishing between internal setup (carried out while the machine is stopped) and external setup (carried out while the machine is running) [18].

Time study, which originated with Taylor (1881) and was accompanied by a movement study introduced by Gilbreths in 1885, is used to find the most economical way to carry out a job, standardize methods, and define the time needed to complete a task. To carry out a time study, it is necessary to observe and understand the process, divide the operation into elements, determine the sample size, classify the operator's pace, and calculate normal and standard time [18].

The PDCA Cycle (Plan–Do–Check–Act), also known as the Deming Cycle or Continuous Improvement Cycle, is a management approach that is based on four stages. The Planning stage involves setting goals and allocating resources appropriately. The Execute stage consists of implementing what has been planned. The Verify stage monitors processes and communicates results, while the Act stage undertakes actions to improve performance if necessary [19].

Quality tools play a key role in the efficient management and continuous improvement of organizations. The 5W2H tool, derived from the PDCA cycle, offers an organized structure for carrying out tasks by answering seven fundamental questions. The GUT Matrix, meanwhile, is a qualitative and subjective tool for prioritizing problems, taking into account severity, urgency, and trend to identify and address critical issues within an organization [20].

Combining these tools with the VSM allows for a complete approach to optimizing processes, reducing waste, and achieving a desired future state, promoting efficiency and operational excellence.

2.2. *Value Stream Mapping*

Value Stream Mapping (VSM) is a tool that has been widely used since the late 1990s, mainly in the automotive industry, but also in other industries and the service sector [21]. This subsection will look at the evolution of the VSM tool, as well as its main objectives and application methods.

2.2.1. Historic Evolution

Value Stream Mapping (VSM) originated in the value chain analysis developed by Taiichi Ohno, mentor of the Toyota Production System. The first formal VSM approach was developed by Rother and Shook in 1999. It is a methodology that analyzes an organization's value stream, providing a comprehensive view of the activities related to a product family [21]. Various authors have contributed to the evolution and improvement of VSM, addressing its limitations and proposing improvements. For example, Hines and Rich [22] described seven tools that, used together, create an effective methodology for identifying and reducing waste. These tools include VSM, the supply chain response matrix, the production funnel, quality filter mapping, amplified demand mapping, decision point analysis, and physical structure. Meanwhile, Brunt [23] provided an in-depth look at current and future state mapping, highlighting the application of VSM in implementing the Lean Management philosophy. This approach was expanded to create a representation of the value flow of a product between three different companies—steel producer, steel service center, and first-line component supplier—to provide a visual understanding of the value flow and identify opportunities for improvement. In 2002, Jones and Womack [24] introduced Extended Value Stream Mapping (EVSM) to complement traditional value stream maps by extending the view to the entire value chain, from suppliers to customers, thereby raising the shared awareness of all participants along a given value chain. Another study by Lian and Landeghem [25] investigated simulation and value stream mapping, visualizing impacts before implementation and enabling organizational transformation with minimal costs. They built simulation models for push and pull (Kanban) systems, emphasizing physical and computer-assisted simulations to validate the current state map and explore future scenarios.

Further advancements include Improved Value Stream Mapping (IVSM), presented by Braglia et al. [26], which deals with products with a complex Bill of Materials (BOM) using a recursive approach to identify opportunities for improvement. Additionally, the Value Stream Mapping Paradigm by Lian and Landeghem [27] integrated VSM with simulation, developing a model generator that automatically creates VSM simulation models. VSM has also been used by Clifford and Golmohammadi [28] to unite the workflow of two companies with distinct processes into a single entity, improving operational metrics. This study highlighted challenges in implementing changes due to significant differences between the current state and the ideal future but emphasized the importance of organizational commitment to the success of the approach. In another example, the implementation of Lean Manufacturing in a small company in the USA highlighted the cost challenges associated with hiring specialized engineers. They used Kaizen events and methods such as 5 Whys to identify and solve problems, resulting in improvements in productivity and operational efficiency [29]. The Value Stream Mapping technique has also been applied to small companies by [30], showing that significant changes are expected in terms of productivity improvement if the proposed future states are implemented. This research suggests that VSM can substantially benefit companies of all sizes. A review by Pagliosa et al. [31] of the connection between Industry 4.0 and Lean Manufacturing highlighted that implementing technologies such as IoT and cyber-physical systems can produce rapid and synergistic results with Lean practices. The application of digitized VSM in manufacturing has been addressed by Horsthofer-Rauch et al. [32], emphasizing that manual VSM analysis becomes inefficient with increasing production complexity. This study explored the potential of process mining to model and analyze production processes, identifying challenges in data requirements, methodology, and visualization. Additionally, various perspectives have been analyzed by [33] to help adopt sustainable development through VSM, concluding that implementing VSM requires strengthening its applications with sustainability and Industry 4.0.

The evolution of the VSM reflects the continuous search for improvements and adaptations to meet the specific needs of different industries and production contexts. Various approaches have been developed to make VSM a more comprehensive and effective tool for managing and improving processes.

2.2.2. VSM Objectives and Associated KPI

VSM (Value Stream Mapping) is a technique that aims to identify and eliminate waste throughout the system, minimize resources used, and optimize organizational performance. It also acts as a proactive tool for analysing and selecting technological approaches right from the start of a project [34].

Silva et al. [20] point out that the VSM provides a detailed 'door-to-door' visualization of the production process, identifies waste, establishes a common language for processes, integrates Lean concepts, identifies material and information flows, and underpins decision-making and implementation plans.

The VSM includes concepts such as the pull system, which starts production in response to customer demand. Distinguishing between value-adding activities (VA), necessary non-value-adding activities (NNVA), and non-value-adding activities (NVA) is essential for eliminating waste. NVA is pure waste and should be eliminated, while NNVA is necessary within current procedures [8,21].

Work in Progress (WIP) refers to the inventory between the start and end points of a process. Lead Time is the total period a customer waits to receive a product after placing an order. If demand exceeds the system's capacity, the waiting time increases beyond the production time [24].

In terms of KPI, there is Takt time, calculated as available time divided by demand [1,8,21]:

$$\text{Takt time} = (\text{Available Time})/\text{Demand} \tag{1}$$

Process cycle efficiency (PCE) compares the value-added time (VA) with the total cycle time (Lead Time). A process is considered Lean if the value-added time is more than 25% of the total cycle time [35].

$$\text{PCE} = \text{VA}/\text{LT} \tag{2}$$

VSM is an integral tool for optimizing operational efficiency, aligning production with market needs and eliminating activities that do not add value to the end product [8].

2.2.3. VSM Symbols

Value Stream Mapping (VSM) uses a set of universal symbols to graphically represent production processes, providing a common language for value stream analysis and a deeper understanding of the connections in the flow of materials. Symbols adapted from the approach in [22] will be used as they are shown in Figure 1.

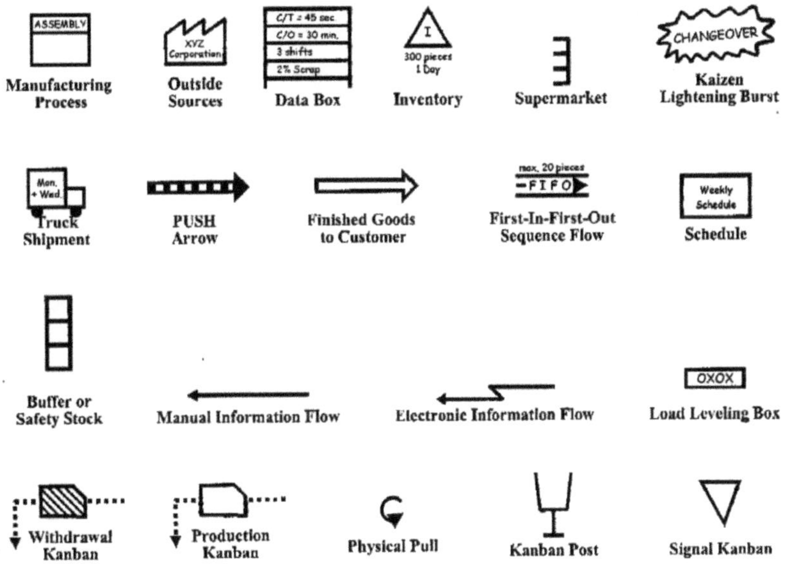

Figure 1. VSM symbols.

According to Nash and Poling [36], VSM consists of three main sections:

- Process or production flow: Shows the flow of materials throughout the production process.
- Communication or information flow: Identifies all communication within the value stream.
- Timeline and distance traveled: Represents the execution time of the process, the total cycle time, and the physical distance traveled by the product or people within the process.

2.2.4. Implementation Method

To effectively apply Value Stream Mapping (VSM), four main steps are followed, according to Oliveira et al. [3] and Hines and Rich [22]:

1. Product or Product Family Selection: A product or product family with common processing steps and equipment is chosen. The chosen product family, the number of parts, customer demand, and delivery frequency are documented. Value Stream Mapping involves touring the factory and outlining the processing steps from input to output.
2. Drawing the Current State Representation: A map is created representing the current state of the value stream, identifying and mapping each step of the process from raw materials to the final product. This detailed drawing reveals areas of waste, inefficiencies, and bottlenecks. The steps are categorized and associated with metrics such as time, distance, and involvement of people. The process includes gathering customer requirements, detailing the physical flow, mapping the supply of materials, and identifying information flows and push/pull systems
3. Value Stream Design (VSD): Once the current state is understood, the desired future state is designed, eliminating inefficiencies and developing an optimized model. This process aims to create a clear and improved vision of the value stream. It explores Takt time, continuous flow, pull systems, production scheduling, and process improvements to optimize operational efficiency and effectively meet customer demands. Strategic questions help to create a production environment aligned with Lean principles, promoting the consistent delivery of value to the customer.
   Questions include [21]:
   - What is the Takt time, considering the available labor time?
   - Should you produce for finished goods stock or directly for dispatch?
   - Where should you implement continuous flow?
   - Where should you introduce pull systems with supermarkets?
   - Where should you plan production in the production chain?
   - How should you level the production mix in the pulling process?
   - What increment of work will be released evenly by the pulling process?
   - What process improvements are needed to guarantee the flow of value according to the future state of the project?
4. Drawing up a Work Plan to Achieve the Future State: A detailed work plan is defined to implement the proposed improvements, covering the actions needed to move from the current state to the future state, with clearly defined responsibilities and deadlines.

2.2.5. Advantages and Limitations of VSM

Value Stream Mapping (VSM) is widely recognized as a powerful tool for analyzing and optimizing processes, but it has both advantages and limitations that must be carefully considered. Among the main advantages of VSM is its ability to serve as the basis for implementing Lean Production. Organizations can identify and eliminate waste by mapping and analyzing value flows, promoting operational efficiency. In addition, VSM establishes a valuable connection between internal manufacturing processes and the entire supply chain, providing an integrated view that facilitates a comprehensive understanding of processes and ensures that decisions are informed by detailed production and inventory data [25,26,37]. This tool also directly links product planning and demand forecasting to production scheduling and shop floor control, promoting a coordinated approach to integrated optimization.

VSM is particularly effective in continuous production layouts, such as assembly lines, where it can favor the fluidity and efficiency of process data [25,26,37]. However, VSM also has significant limitations. Its application in high-variety, low-volume manufacturing systems can be challenging, as it is less suitable for these settings due to their specific characteristic data [25,26,37]. In MTO (Make to Order) environments, the application of VSM can be even more complex. Conventional VSM is primarily designed for MTS (Make to Stock) environments and may not adequately address the logistics objectives of an MTO environment, such as schedule reliability. In these environments, production is more flexible, and customer demand is more variable, which can make it challenging to implement Lean principles such as FIFO (First In, First Out) [38]. Thus, the application of VSM in MTO environments can be hampered by the variable and dynamic nature of production, jeopardizing its effectiveness [39].

To address these limitations, alternatives have been developed such as the Waste Identification Diagram (WID), the Manufacturing Cost Deployment (MCD), the Manufacturing Critical-Path Time (MCT), and the Project Time Deployment (PTD). The WID aims to overcome some of the shortcomings of the VSM by allowing a more comprehensive representation of production units. WID can represent not only the flow of a product family but also entire production units, visualizing the layout and assessing all types of waste visually and intuitively. In addition, it offers detailed quantitative information on the performance of workstations, such as cycle time, changeover time, and the utilization of the workforce in value-adding and non-value-adding activities. However, WID still faces challenges, such as the lack of representation of the flow of information from Production Planning and Control (PPC) and the absence of links to suppliers and customers [40].

MCD focuses on systematic cost reduction through a rigorous and structured approach. This tool uses five matrices (A-matrix, B-matrix, C-matrix, D-Matrix, and E-matrix) to classify losses, clarify cause and effect relationships, convert losses into manufacturing costs, suggest improvement actions, and present the efficiency of the investment associated with each action. The MCD is effective in identifying and eliminating inefficiencies, highlighting opportunities to increase efficiency and effectiveness in reducing losses. Its implementation has been successful in various industries, especially in the automotive sector, where it has been integrated into the World-Class Manufacturing (WCM) model [41].

MCT is a time-based mapping approach that graphically represents the flow of an order and calculates the corresponding delivery time. It highlights the longest critical path of order-fulfilled activities, helping to identify non-value-adding activities that consume significant time and offering a robust metric for continuous improvement projects [41].

Finally, PTD is an iterative structured approach specifically designed for Engineer-to-Order (ETO) production environments. Combining MCT with specific features of MCD, PTD identifies the losses affecting an ETO project and facilitates planning on how to reduce them. It classifies, analyzes, and eliminates losses to reduce production time, focusing on business processes where the causal losses are located [41].

Despite the enhanced capabilities of WID, MCD, MCT, and PTD in terms of detailing and evaluating losses, the choice of VSM in the context of discontinuous flows is justified by its simplicity and practical effectiveness. VSM, although limited in its representation of complex production routes and various types of waste, provides a clear and direct view of the value stream, making it easier to quickly identify areas for improvement in terms of material flow and information flow. The adaptability of the VSM is a crucial factor. It allows the stages of the production process to be visualized understandably and intuitively, which is essential for implementing continuous improvements holistically and simply [40,41].

Therefore, the choice of VSM, even with its limitations, is based on its effectiveness in representing the value stream simply and straightforwardly, ensuring that visualization of processes and identification of waste can be carried out efficiently and practically in a short amount of time and with accessible materials.

3. Case Study

This study was conducted at a prominent entity in the metalworking and metal-mechanics industry, specializing in the processing of metal items such as jewelry and metal components/accessories for bags, belts, and scarves. This entity oversees the entire production process from machining to final assembly, with the finished products exported to other companies within its group.

Originally founded in a garage with just 12 employees, the organization underwent significant changes, and in 2013, it adopted its current name. Initially, the company focused exclusively on polishing metals with around 50 employees. In 2016, the entity expanded by acquiring another company, increasing its workforce to 80 employees. Over the years, it has consistently invested in employee training, new machinery, and equipment to enhance its capabilities and expand its activities to include machining, vibratory polishing, and galvanization, among others, with the primary objective of consolidating the entire production chain within Portugal.

This continuous investment has allowed the organization to grow and adapt, ensuring it remains at the forefront of the industry.

3.1. Selection of the Product Family

To develop the Value Stream Mapping (VSM), the initial step involved selecting the product family. Product E was chosen due to its significant order volume, making it a high runner in the company. This product family includes several variants, differing in stem length, hole configuration, and semicircle diameter.

Product E is considered a semi-finished product, consisting of components such as the *Ecrou* and *Pontet* (assembled onto the piece) and other parts included in the kitting delivered to the customer.

Initially, various production processes and areas within the company were observed and understood to gain a general overview of the entire process. To construct the VSM, a batch was tracked from machining to shipment to the customer. It is important to note that the polishing process was simulated in this analysis, as this step is currently outsourced by the company for Product E.

3.2. Productive Process and Flow

The manufacturing process for the Product E family begins with the receipt of stamped pieces from a company within the same group, identified by supplier code RN. This receiving process includes material reception and the issuance of a manufacturing order for the machining section (Usinage). After being stored in the receiving stock, a manufacturing order labeled USI is issued, and the batch is moved to advanced stock in the machining area. According to the weekly planning schedule, the batch proceeds to the CNC machine for machining until the end of each operator's shift. Once machining is completed, the pieces are placed in the control stock area. After inspection, the batch awaits transportation by logistics operators responsible for moving batches between sections. In this instance, the batch is collected by logistics and sent to a group company for polishing. However, to complete the VSM, the selected sample remained in stock until it was transported to the designated polishing production unit.

The next phase is polishing, which includes *Lapidage* preparation (of the tongue, sides, and "tongue tip" and the semicircle faces), washing, semi-automatic polishing, washing, manual polishing (around the piece, pre-*Avivage*, complete *Avivage*), washing (automatic), and finally a complete inspection. Transporting the batches is managed by two employees responsible for washing, each covering half of the layout.

After polishing, logistics transports the batch to the material reception area to add the *Ecrou*, a component pre-assembled onto the piece. The *Ecrou* is partially added to Product E to facilitate full plating in the subsequent galvanoplasty process. The galvanoplasty section includes an assembly area where pieces, previously transported in trays, are placed on a support called *Bouclard*, followed by the bath area and control area.

After this process, logistics retrieves the batch and performs the initial kitting to assemble all components into the final product. Some components are not assembled onto Product E but are included in the kit for customer assembly into the final product.

Once the kits are completed, they undergo inspection and are transported to the shipping area (with quality control in this area and final inspection by picking) for delivery to the customer. In Figure 2 there is a spaghetti diagram showing the flow of product E.

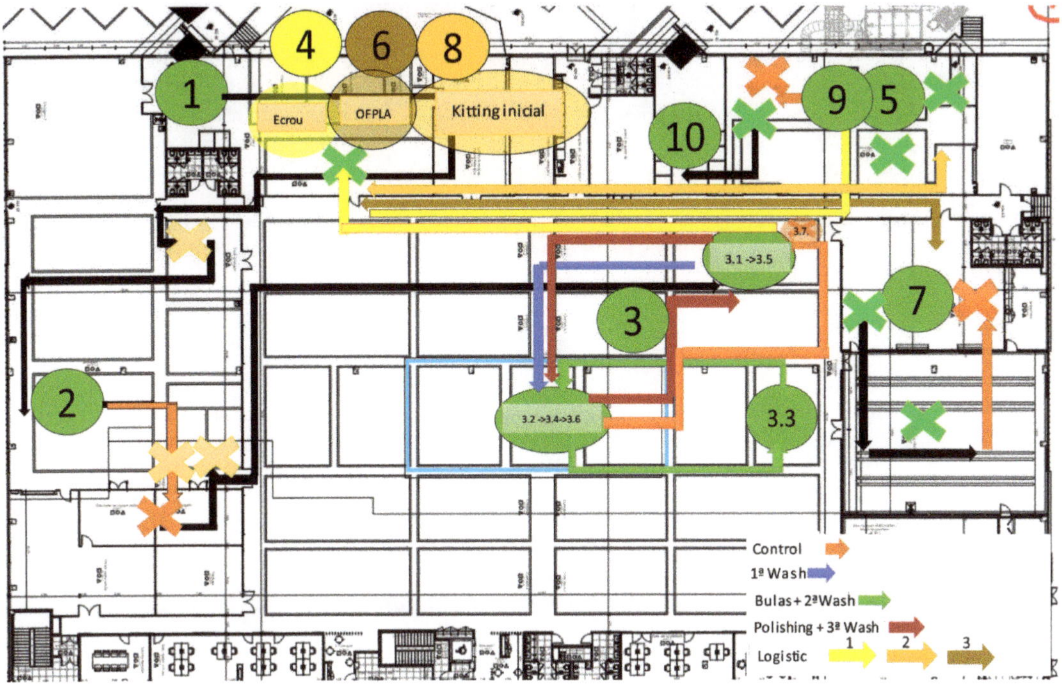

Figure 2. Spaghetti diagram of family of products E.

3.3. Data Collection and Analysis

This section presents the data collected and explains each section and production process, considering that the current production flow can be characterized as push and Make to Stock. To develop the VSM, data such as setup time, cycle time, available time, and actual operation time were collected. A batch of 50 pieces was tracked through the entire process, and a table was created with timed metrics (second/piece; minute/piece; hour/piece; minute/lot; hour/lot; and day/lot). The table also indicates the type of process each activity represents according to the ASME standard, with administrative processes added as needed. From the collected data, Uptime, Takt time, and PCE/efficiency were calculated.

The OEE (Overall Equipment Effectiveness) indicator was not mentioned in the VSM (Value Stream Mapping) analysis, even though it was calculated for the machining process, for various reasons. OEE is relevant for machining processes, but most of the other processes in the company, except electroplating, are manual and highly variable. Electroplating, being a complex electrochemical process, is difficult to control and standardize. The main aim of implementing the VSM was to identify waste and optimize the production process holistically, given that the time available to implement the tool was somewhat limited. Focusing exclusively on OEE could divert attention away from global improvements, concentrating instead on the efficiency of individual pieces of equipment.

3.3.1. Reception

The process starts with material reception, which takes about 3 days for stamping and 120 h for delivery. It is unloaded, verified (4 h), and then organized in stock. The reception area includes a control area, vertical storage modules, shelving areas, and a central logistics area for various tasks like issuing manufacturing orders and kitting. The vertical storage modules are distributed for machining materials, consumables, assembly components, and polishing.

The material arrives already stamped in batches of 2000 from a group supplier, so no reception control is performed. The material is received, counted, and stocked using FIFO for supplying the machining section. Typically, the material stays in stock for about a week. Logistics staff select the material and issue a weekly manufacturing order (USI), which is then divided into batches. The logistics transport worker takes the batch to advanced machining stock. This process takes about 5.2 days, though this can vary depending on stock levels and customer orders. The activities that add necessary value (VA) are material reception and order issuance (4.5 h/batch), while non-value-added activities (NVA) total 120.5 h per batch.

3.3.2. Machining

The machining area consists of three machine sets named B, H, and C. Material processing is performed on CNC machines in set B (machines B4, B5, and B6). Each CNC machine has a rotating panel that performs specific operations in sequence. The first operation (cycle time: 3 min and 14 s) includes shaft roughing, drilling with three bits, chamfering, hole finishing, and code engraving. The second operation (cycle time: 2 min and 50 s) involves creating channels, corner finishing, and additional drilling and chamfering.

Each operator handles three machines, and unplanned stoppages or micro-stoppages are minimal as the CNCs can operate continuously. Quality control is supposed to be hourly, but operators often spend time cleaning burrs, causing unplanned stops and leading to 78% availability (5.75 h). Setup time is 3 h, varying with adjustments and tuner availability. The batch advances by shift, with the operator leaving the production in stock at the end of their shift. The inspection takes about a day, followed by a 10% batch sampling for visual control and stock waiting (40 min) before transport to the polishing unit.

The planned time per shift is 7.33 h with a daily demand of 110 pieces (55 per shift), resulting in a Takt time of 8 min per piece. Considering cleaning and unplanned stops, the efficiency is reduced due to stock waiting times, totaling 163 h of NVA and 6.03 h of VA, leading to 54.48% efficiency.

3.3.3. Polishing

Polishing involves several phases: preparation, washing, semi-automatic polishing, manual polishing, and control. Each product family has different polishing flows. For Product E, initial preparation (*Lapidage*) takes 80 s per piece. Following *Lapidage*, the batch is transported for washing (29 min), semi-automatic polishing (33 min per pass, with 2 passes), and further washing and waiting. Manual polishing follows (95 s per piece) and final washing before quality control.

Efficiency is 45%, with transport and waiting times being significant. Activities adding value total 5.36 h, and non-value-added activities total 6.67 h, resulting in a 12 h Lead Time.

3.3.4. Pre-Assembly

Product E requires pre-assembly before precious metal deposition. Logistics transports the polished pieces for pre-assembly, where an Ecrou is partially added (32 min for 50 pieces, 15 s per piece). Following pre-assembly, the batch waits (33 min) before logistics transports it to galvanoplasty. In this process the VA is 0.53 h, with 4.81 h of NVA, resulting in 10% efficiency.

3.3.5. Galvanoplasty

The batch is mounted on Bouclards for a series of ultrasonic, degreasing, activation, and plating baths (0.97 h total). Post-bath, the pieces are rinsed, dried, and inspected before initial kitting. The total VA time is 1.89 h, and the NVA is 10.96 h, with a significant wait for plated components that need to be assembled into the main product. Efficiency is 15%.

3.3.6. Final Assembly

Final assembly involves visual inspection, complete Ecrou insertion, and Pontet attachment (75 s per piece). The batch is checked, kitted, and prepared for shipping. Total VA time is 6.33 h, with various waiting times reducing efficiency.

3.3.7. Shipping

Upon reaching shipping, the batch undergoes final quality checks and kit verification (10 min), is declared in the management software, packaged (1.5 min), and shipped (30 h to the customer). VA activities total 21.17 h, and NVA activities total 315.17 h, resulting in a Lead Time of 367 h and a cycle efficiency of 6.29%.

Overall, the process efficiency is low due to a push-based and discontinuous flow, with significant waiting times and transportation delays. Improving these aspects and eliminating waste can enhance process efficiency and effectiveness.

4. Discussion

4.1. Value Stream Mapping Actual State

Based on the data collected and described in the previous section, the VSM (Value Stream Mapping) of the current state of the product E was constructed.

The production flow is characterized as push, with production made to stock, to always have products available for each production process. This method results in the creation of safety stock to minimize the risk of interruptions. In addition, the flow is discontinuous, with different manufacturing orders for each section, resulting in a fragmented flow in logistical terms. Each section works independently, with a discontinuous flow, generating inefficiencies.

When receiving raw materials, the waiting time is significant (120 days) due to the use of FIFO (First In, First Out) and weekly consumption in machining. The raw material received is kept in stock until it is needed. The receiving process includes unloading, checking, and organizing stock. Receiving time is divided into value-added (VA) and non-value-added (NVA) activities.

In machining, setup time is relatively high due to the fine-tuning and adjustments required for each machine. The production cycle includes two main operations with several steps, from roughing to chamfering and drilling. Waiting time for control is significant (73 percent of total time), varying according to shifts and production priorities. Efficiency is reduced due to stock waits and unplanned stops, representing 163 h of NVA against 6.03 h of VA, resulting in an efficiency of 54.48 percent. There are opportunities for improvement, such as reducing setup time through tools such as 5S/SMED to minimize time and effort in fine-tuning and adjustments (See Figure 3).

In polishing, the process includes several phases such as *Lapidage*, washing, semi-automatic polishing, manual polishing, and control. Each phase has associated waiting and transport times. Lead Time is 8.97 h, with 3.61 h dedicated to waiting and transport. This process is carried out externally for strategic reasons and takes 432 h to complete. Internally, polishing efficiency is 45 percent, with value-added activities accounting for 5.36 h and NVA 6.67 h (See Figure 4).

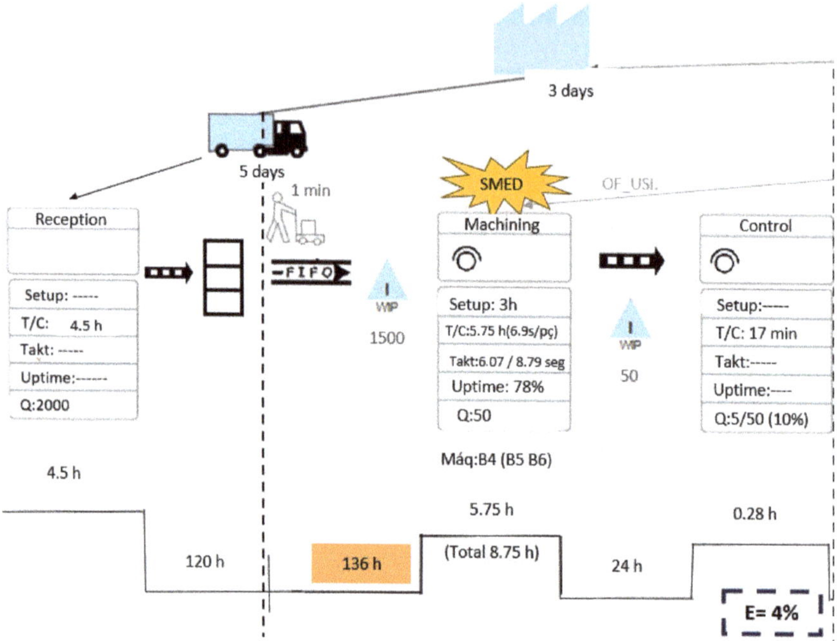

Figure 3. VSM, reception to machining.

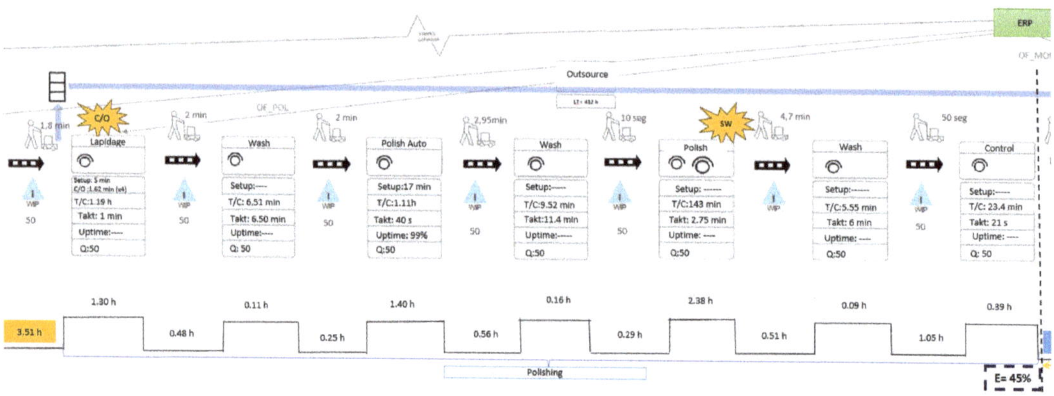

Figure 4. VSM—polishing process.

Before pre-assembly, logistics adds the component needed for assembly and transports the parts. Assembly time is 0.53 h, with the main difference between actual and theoretical time due to burr cleaning and visual control. It represents 10% efficiency, with 4.81 h of NVA.

In galvanoplasty, the process includes assembly in the *Bouclard*, plating, and control, with a total time of 1.89 h and 0.97 h of waiting. Process efficiency is 15%, with value-added activities totaling 1.89 h and NVA 10.96 h, including a significant 9 h wait for components (See Figure 5).

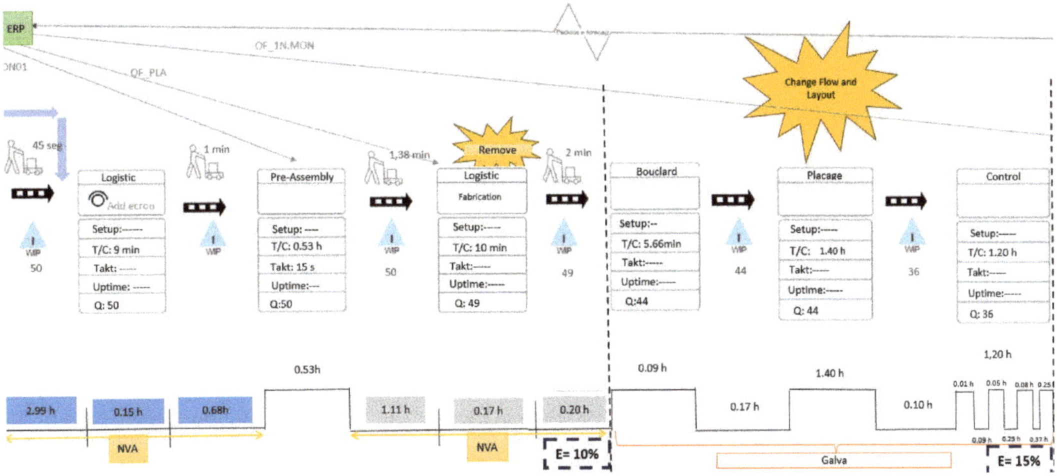

Figure 5. VSM—pre-assembly and galvanoplasty processes.

In the final assembly, the process includes visual control, final assembly of the *Ecrou*, and adding components. Part conformity caused a significant delay. The assembly time is 6.33 h, longer than expected due to doubts about quality parameters. Value-added activities totaled 1.75 h, while NVA accounted for 0.42 h.

The final dispatch process includes a final sample check, kit verification, computerized closing, packing, and wrapping. Transport time takes around 30 h for the batch to reach the customer after dispatch (See Figure 6).

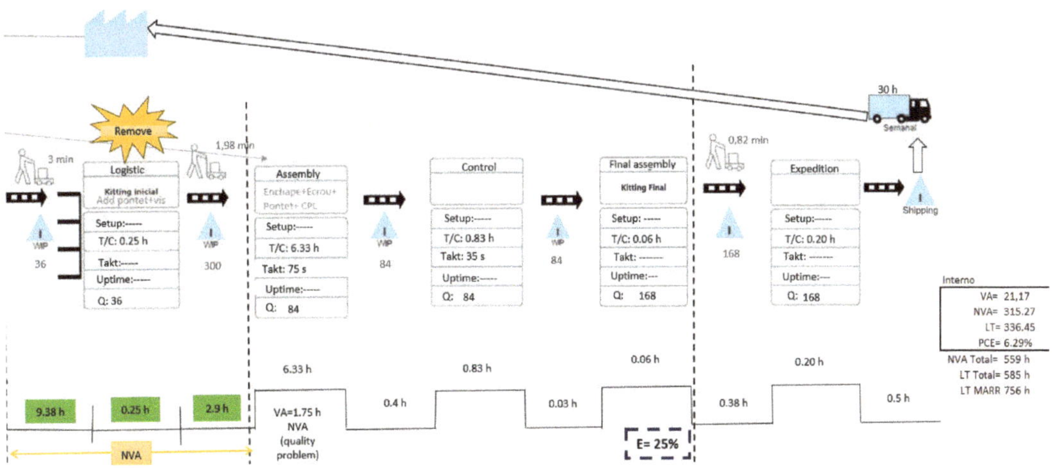

Figure 6. VSM—galvanoplasty to expedition.

The VSM shows that there are relatively long waits between each of the processes. Internally, in each of the processes, there are also NVA activities, indicating that the times represented as VA on the timeline are not entirely value-added. The main causes of waste identified include waiting and transport between the different phases, as well as inefficiencies in the setup and quality control processes.

4.2. Improvements

After analyzing the current state and considering the guidelines for the future state proposed by Rother and Shook [21], several areas for improvement and actions to be taken in each section of the process were identified.

The main problem identified was in final assembly, where the current cycle time exceeds the Takt time; it is essential to reduce the cycle time to match customer demand. The most appropriate approach is to maintain the push flow in machining but create a pull system between electroplating and final assembly. In addition, it is proposed to implement a Kanban system to control the stock of final products.

4.2.1. Machining

In machining, SMED needs to be implemented to reduce setup times. To identify opportunities for optimization, the setup process was monitored and timed. As shown in Table 1, the setup consists of 10 operations.

Table 1. SMED. Operations classification.

#	Operation	Time (min)	Cumulative Time	Internal	External	Observations
1	Check Tools	6.78	6.78	x		Team Leader
2	Removing Tools	1.20	7.98	x		Team Leader
3	Select Tools in the Cabinet	2.00	9.98	x		Team Leader
4	Preparing the tool	16.00	25.98	x		Team Leader
5	Measure Tool	2.67	28.65	x		Team Leader
6	Place Tool	2.00	30.65	x		Team Leader
7	Insert parameters (x,y,z)	10.13	40.78	x		Team Leader
8	Zero point	11.48	52.26	x		Shift Leader/Tuner
9	Insert Program	7.75	60.01	x		Shift Leader/Tuner
10	Make 1st Part-Validation	58.00	118.01		x	Shift Leader/Tuner
7–8	Wait Time	61.58	179.59	x		Wait for tuner. One part of the process is done by the team leader, another by the tuner or shift leader. This wait is also for the shift change. Tuner busy.
	Total	180				

It was observed that all the activities are internal, except for the validation of the first part, which generally requires adjustments. This means that they are carried out with the machine stopped. In addition, a waiting time of around 1 h was identified due to waiting for the tuner, who was busy, and due to the shift change.

Although some SMED mechanisms are already in place, it is possible to transform some of the internal steps into external ones. For example, operations 3, 4, and 5 in Table 1 can be carried out while production is taking place, allowing the team leader to prepare the tools; it is suggested that a tool shelf be installed like what already exists on the H needed for the next batch. In addition, machines are used to reduce tool changeover time. By implementing these improvements and maintaining the waiting time for the tuner, it is possible to reduce the internal setup time from 3 h to 2 h and 40 min.

4.2.2. Polishing

In polishing, a quick change of sandpaper is suggested, switching to self-adhesive sandpaper, which significantly reduces the changeover time and associated costs.

The sandpaper change process involves several steps, the time for which has been calculated as a standard 90 s. In the production of E, each batch of 50 units requires four sandpaper changes, totalling 36 min per day. Adopting self-adhesive sandpaper reduced the changeover time to 51 s, saving 15.6 min a day and resulting in monthly savings of 327.6 min, representing a time reduction of 43%. Financially, this change meant a monthly saving of around EUR 3 per operator and glue booth. In addition to the financial benefits, self-adhesive sandpaper contributes to a more sustainable approach, improving air quality and freeing up space in the industrial environment.

Additionally, it is proposed to standardize and speed up changing *Brosses* (groups of fabric assembled to create a rotating wheel for polishing). This can be achieved by creating kits of pre-prepared *Brosses* and establishing standards for the use of cloths to optimize the process.

The *Brosse* changeover process revealed a lack of standardization, both in the number of cloths and the arrangement of the tools. After analyzing the data and interviewing operators, a standard number of cloths was established for each type of *Brosse* configuration, resulting in kits designated as S, M, and L. These kits were organized in a specific station with a shelf for each configuration, also including the necessary key and reference tables. The standard *Brosse* changeover time was reduced to 2 min, saving 3 min and 24 s. This standardization not only improves efficiency but also guarantees better quality and controlled use of resources.

4.2.3. Assembly and Galvanoplasty

In assembly and galvanoplasty, it is essential to reorganize the layout to create a continuous flow between the two areas, eliminating non-value-adding activities such as transport and waiting. This involves implementing a Kanban production system for final products and creating shelves for pre-assembly and PP assembly, speeding up the workflow and ensuring more efficient production. To optimize the value flow of the product family under study, it is proposed to reorganize the layout between the assembly and galvanoplasty areas. This involves the elimination of logistical steps, the creation of stock between these areas, and the implementation of a pull-type system to create a continuous flow, with the aim of reducing non-value-added activities such as transport and waiting. The new layout will group assembly stages and associate the final assembly with control. A system will be implemented to control the entry and exit of parts, such as the use of a conveyor to guarantee FIFO during electroplating. The flow between assembly and electroplating will be continuous, supported by a Kanban of final products pulled by final kitting. The walls between the sections will be removed to promote a fluid connection. These changes are aimed at improving efficiency, reducing waste, and promoting a continuous and agile flow in the production process.

These actions aim to reduce non-value adding activities, optimize production processes and improve the overall efficiency of the system, in line with the principles of Lean Manufacturing.

4.3. Future State

Based on the improvements identified, the future state VSM was drawn up, which includes various optimizations in the setup, sandpaper changeover, and *Brosse* standardization processes. Through SMED, it was possible to reduce setup time by 20 min, while changing from traditional sandpaper to self-adhesive sandpaper reduced changeover times, resulting in more efficient cycle times. The standardization of *Brosses* led to a reduction in changeover times, as well as ensuring greater control and stability in production.

The most significant proposed improvement involves changing the production flow and layout between the assembly and galvanoplasty sections (See Figures 7 and 8). This includes eliminating non-value-added activities, such as transport and waiting, and creating a continuous flow of work. Through this reorganization, it is hoped not only to reduce idle times but also to promote better use of resources and a more agile response to customer demands.

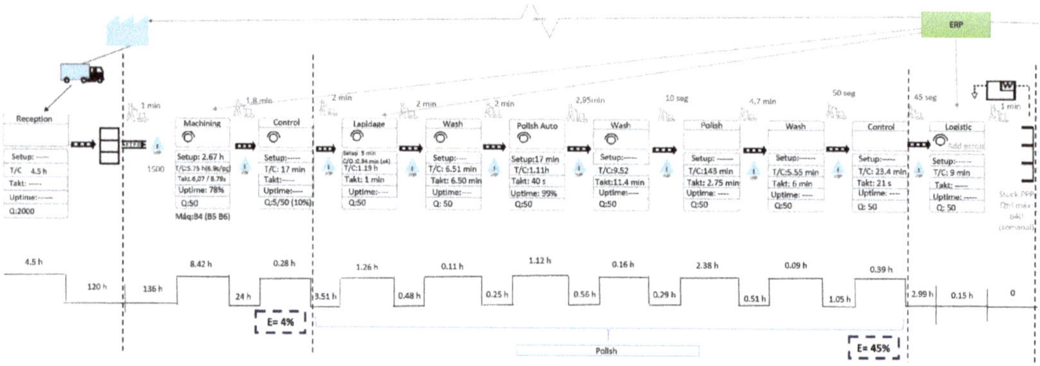

Figure 7. Future state VSM—reception to polish.

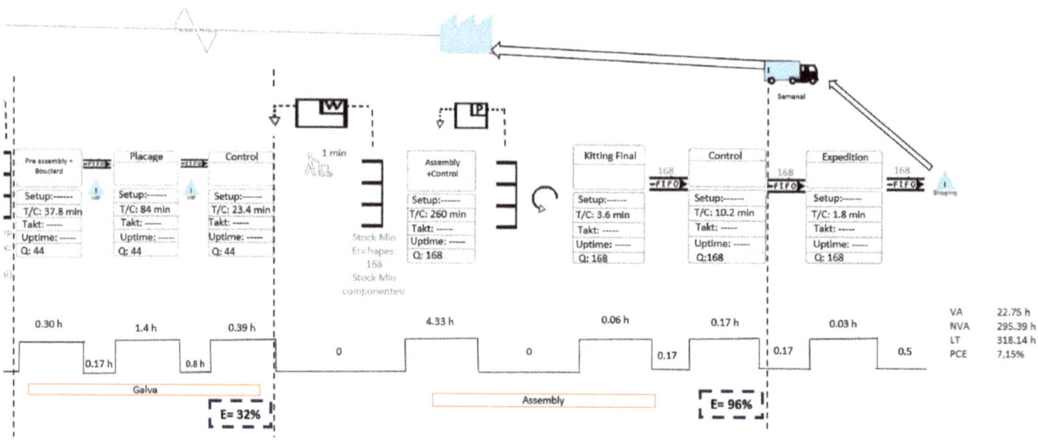

Figure 8. Future state VSM—galvanoplasty to expedition.

The flow of parts between sections has been redefined, starting with logistics reception and followed by PPP assembly and *Bouclard* assembly, now carried out sequentially at the same workstation. The flow continues with the bath and quality control stages, before returning to final assembly and control. This reconfiguration allows for smoother integration between the processes and the elimination of waste, such as unnecessary handling and waiting between stages.

With these improvements, it is hoped to eliminate waiting times between batches, guaranteeing a constant and efficient flow.

Once the main sources of waste have been eliminated, the future state shows an increase in efficiency of 0.86% and a reduction in Lead Time of 18.31 h with the following values:

- VA = 22.75 h
- NVA = 295.39 h
- LT = 318.14 h

- PCE = 7.15%

The future state VSM reflects these improvements, providing a clear view of the optimized value flow and the activities required at each stage of the process. These changes not only increase operational efficiency but also improve product quality and customer satisfaction.

5. Conclusions

By applying Value Stream Mapping (VSM) and other Lean tools, waste was identified and proposals for improvement were developed. The VSM allowed for a comprehensive visualization of the entire process, identifying waste and suggesting changes for the future state. Key wastes identified included waiting, unnecessary transport, and over-processing. Improvement proposals included the implementation of self-adhesive sandpaper, standardization of brushes, and reorganization of the production flow.

The current state was characterized by (1) Lead Time: 336.45 h; (2) value-added time: 21.17 h; (3) process cycle efficiency (PCE): 6.29%.

After implementing the proposed improvements, the following changes were observed:

- SMED reduced setup times by 20 min.
- Self-adhesive sandpaper reduced unnecessary movements by 0.04 h.
- The most significant improvement was the restructuring of the layout and flows, with the creation of component stock, which eliminated approximately 14 h of non-value-added (NVA) time.

The new state improved the process by achieving: (1) Lead Time: 318.14 h; (2) value-added time: 22.75 h; and (3) PCE: 7.15%.

This analysis resulted in significant operational improvements and cost reduction (details undisclosed due to institutional confidentiality). Thus, this research work stands out for its specific application in a discontinuous process context within a sector characterized by independent sections operating with entirely different processes. VSM focus is on continuous production processes, and this study offers an innovative approach by applying VSM in an industrial environment where the variable nature of processes presents unique challenges. Despite the challenges, VSM was successfully implemented in a complex industrial setting, achieving the initial objectives. By analyzing how VSM can effectively be applied in discontinuous processes. This study could fill a gap in the existing literature and establish standards for the practical application of conventional VSM, providing accessible insights for dynamic industrial environments.

Future research is suggested to assess VSM's applicability in the complex context of other surface treatment process industry discontinuous flow and to evaluate the effectiveness of the implemented improvements. Future work should include applying SMED actions, analyzing the processes of different product families to create FIFO or Kanban systems in centralized areas like washing. Additionally, conducting detailed time studies for each VSM activity is necessary to ensure the replicability of this research.

Author Contributions: Conceptualization, B.C., J.V. and P.D.G.; methodology, B.C., J.V. and P.D.G.; validation, B.C., J.V. and P.D.G.; formal analysis, B.C., J.V. and P.D.G.; investigation, B.C.; resources, J.V.; data curation, B.C. and J.V.; writing—original draft preparation, B.C., J.V. and P.D.G.; writing—review and editing, B.C., J.V. and P.D.G.; supervision, J.V. and P.D.G.; project administration, J.V. and P.D.G.; funding acquisition, P.D.G. All authors have read and agreed to the published version of the manuscript.

Funding: This work was partially supported by Fundação para a Ciência e Tecnologia (FCT), C-MAST (Centre for Mechanical and Aerospace Science and Technologies), under the projects UIDB/00151/2020 (https://doi.org/10.54499/UIDB/00151/2020; https://doi.org/10.54499/UIDP/00151/2020, accessed on 3 January 2024).

Data Availability Statement: The data presented in this study are available on request from the corresponding author due to company privacy concerns.

Conflicts of Interest: The authors declare no conflicts of interest.

References

1. Valencia, E.; Garcia, Y.; Santivanez, G.; Castaneda, E. Production Management Model to Reduce Cycle Times Using Lean Manufacturing in a Company in the Metalworking Sector. In Proceedings of the 2022 8th International Conference on Information Management (ICIM 2022), London, UK, 25–27 March 2022; pp. 182–185.
2. Ohno, T. *O Sistema Toyota de Produção-Além da Produção em Larga Escala*, 1st ed.; Bookman: Porto Alegre, Brazil, 1997; pp. 1–276. Available online: http://books.google.com.br/books?id=dg4_3tlM8EYC (accessed on 25 May 2024).
3. Oliveira, J.; Sá, J.C.; Fernandes, A. Continuous improvement through 'Lean Tools': An application in a mechanical company. *Procedia Manuf.* **2017**, *13*, 1082–1089. [CrossRef]
4. De-La-Cruz-Arcela, F.; Martinez-Castillo, J.; Altamirano-Flores, E.; Alvarez-Merino, J. Application of Lean Manufacturing Tools to Reduce Downtime in a Small Metalworking Facility. In Proceedings of the IEEE International Conference on Industrial Engineering and Engineering Management, Macao, China, 15–18 December 2019; pp. 551–555.
5. Tyagi, S.; Choudhary, A.; Cai, X.; Yang, K. Value stream mapping to reduce the lead-time of a product development process. *Int. J. Prod. Econ.* **2015**, *160*, 202–212. [CrossRef]
6. Dias, A.; Silva, M.T.; Nascimento, P.A.; Gonçalves, L.S.; Nunes, M.L.; Oliveira, P.S.; Fernandes, R. Utilization of the Arena simulation software and Lean improvements in the management of metal surface treatment processes. *Procedia Comput. Sci.* **2022**, *204*, 140–147. [CrossRef]
7. Liker, J.K. *The Toyota Way: 14 Management Principles from the World's Greatest Manufacturer*; McGraw-Hill: New York, NY, USA, 2004.
8. Kumar, N.; Hasan, S.S.; Srivastava, K.; Akhtar, R.; Yadav, R.K.; Choubey, V.K. Lean manufacturing techniques and its implementation: A review. *Mater. Today Proc.* **2022**, *64*, 1188–1192. [CrossRef]
9. Womack, J.P.; Jones, D.T.; Roos, D. *The Machine That Changed the World*; Macmillan Publishing Company: New York, NY, USA, 1990.
10. Azevedo, J.; Sá, J.C.; Ferreira, L.; Santos, G.; Lima, V. Improvement of production line in the automotive industry through lean philosophy. *Procedia Manuf.* **2019**, *41*, 1023–1030. [CrossRef]
11. Womack, J.P.; Jones, D.T. *Lean Thinking: Banish Waste and Create Wealth in Your Corporation, Revised and Updated*; Free Press: New York, NY, USA, 2003.
12. Perdomo-Verdecia, V.; Sacristán-Díaz, M.; Garrido-Vega, P. Lean management in hotels: Where we are and where we might go. *Int. J. Hosp. Manag.* **2022**, *104*, 103250. [CrossRef]
13. Klein, L.; Tonetto, M.; Avila, L.; Moreira, R. Management of lean waste in a public higher education institution. *J. Clean. Prod.* **2021**, *286*, 125386. [CrossRef]
14. Rahani, A.; Al-Ashraf, M. Production flow analysis through Value Stream Mapping: A lean manufacturing process case study. *Procedia Eng.* **2012**, *41*, 1727–1734. [CrossRef]
15. Proença, A.P.; Gaspar, P.D.; Lima, T.M. Lean optimization techniques for improvement of production flows and logistics management: The case study of a fruits distribution center. *Processes* **2022**, *10*, 1384. [CrossRef]
16. Antoniolli, I.; Guariente, P.; Pereira, T.; Ferreira, L.; Silva, F. Standardization and optimization of an automotive components production line. *Procedia Manuf.* **2017**, *13*, 1120–1127. [CrossRef]
17. Shingo, S. *A Study of the Toyota Production System from an Industrial Engineering Viewpoint*, Revised ed.; Productivity Press: Cambridge, MA, USA, 1989.
18. Barnes, R. *Motion and Time Study Design and Measurement of Work*, 7th ed.; Wiley India Pvt. Limited: New Delhi, India, 2009.
19. Chua, C.K.; Wong, C.H.; Yeong, W.Y. Quality Management Framework in Additive Manufacturing. In *Standards, Quality Control, and Measurement Sciences in 3D Printing and Additive Manufacturing*; Kumar, A., Zhang, Y., Eds.; Academic Press: Cambridge, MA, USA, 2017; pp. 213–239.
20. Silva, D.; Delai, I.; Castro, M.; Ometto, A. Quality tools applied to Cleaner Production programs: A first approach toward a new methodology. *J. Clean. Prod.* **2013**, *47*, 174–187. [CrossRef]
21. Rother, M.; Shook, J. *Learning to See: Mapping the Value Stream to Add Value and Eliminate Waste: Working Manual for a Lean Tool*; The Lean Enterprise Institute, Inc.: Cambridge, MA, USA, 2003.
22. Hines, P.; Rich, N. The seven value stream mapping tools. *Int. J. Oper. Prod. Manag.* **1997**, *17*, 46–64. [CrossRef]
23. Brunt, D. From Current State to Future State: Mapping the Steel to Component Supply Chain. *Int. J. Logist. Res. Appl.* **2000**, *3*, 259–271. [CrossRef]
24. Jones, D.; Womack, J. *Seeing the Whole: Mapping the Extended Value Stream*; Lean Enterprise Institute: Cambridge, MA, USA, 2002.
25. Lian, Y.H.; Landeghem, H. An Application of Simulation and Value Stream Mapping in Lean Manufacturing. In Proceedings of the 14th European Simulation Symposium, Dresden, Germany, 23–26 October 2002; pp. 300–307. Available online: https://www.scs-europe.net/services/ess2002/PDF/log-11.pdf (accessed on 25 May 2024).
26. Braglia, M.; Carmignani, G.; Zammori, F. A new value stream mapping approach for complex production systems. *Int. J. Prod. Res.* **2006**, *44*, 3929–3952. [CrossRef]
27. Lian, Y.H.; Landeghem, H. Analysing the effects of Lean manufacturing using a value stream mapping-based simulation generator. *Int. J. Prod. Res.* **2007**, *45*, 3037–3058. [CrossRef]
28. Clifford, B.; Golmohammadi, D. Using Lean Value Stream Mapping to Align the Work Streams of Two Companies. *Ind. Syst. Eng. Manag.* **2012**, *1*, 1–9.
29. Chen, J.; Li, Y.; Shady, B. From value stream mapping toward a lean/sigma continuous improvement process: An industrial case study. *Int. J. Prod. Res.* **2010**, *48*, 1069–1086.

30. Singh, H.; Bahl, A.; Kumar, A.; Mann, G.S. Materials and Information Flow Analysis and Optimization of Manufacturing Processes in MSMEs by the Application of Value Stream Mapping (VSM) Technique. *Mater. Today Proc.* **2018**, *5*, 28420–28426.
31. Pagliosa, M.; Tortorella, G.; Ferreira, J. Industry 4.0 and Lean Manufacturing: A systematic literature review and future research directions. *J. Manuf. Technol. Manag.* **2019**, *32*, 543–569. [CrossRef]
32. Horsthofer-Rauch, J.; Schumann, M.; Milde, M.; Vernim, S.; Reinhart, G. Digitalized value stream mapping: Review and outlook. *Procedia CIRP* **2022**, *112*, 244–249. [CrossRef]
33. Batwara, A.; Sharma, V.; Makkar, M.; Giallanza, A. Towards smart sustainable development through value stream mapping—A systematic literature review. *Heliyon* **2023**, *9*, e15852. [CrossRef] [PubMed]
34. Kumar, S.; Dhingra, A.; Singh, B. Process improvement through Lean-Kaizen using value stream map: A case study in India. *Int. J. Adv. Manuf. Technol.* **2018**, *96*, 2687–2698. [CrossRef]
35. George, M.L. *Lean Six Sigma: Achieve Major Cost, Inventory and Lead Time Reductions in Less Than a Year*; McGraw-Hill: New York, NY, USA, 2002; p. 89.
36. Nash, M.; Poling, S. *Mapping the Total Value Stream: A Comprehensive Guide for Production and Transactional Processes*; Productivity Press: New York, NY, USA, 2008.
37. Khaswala, Z.; Irani, S. Value Network Mapping (VNM): Visualization and Analysis of Multiple Flows in Value Stream Maps. In Proceedings of the Lean Management Solutions Conference, St. Louis, MO, USA, 4–5 October 2001; pp. 1–18. Available online: https://lib.asprova.com/images/stories/lean/lean-manufacturing/pdf/Value_Network_Mapping_VNM_.pdf (accessed on 25 May 2024).
38. Slomp, J.; Bokhorst, J.A.C.; Germs, R. A lean production control system for high-variety/low-volume environments: A case study implementation. *Prod. Plan. Control* **2009**, *20*, 586–595. [CrossRef]
39. Stamm, M.; Neitzert, T. Value Stream Mapping (VSM) in a Manufacture-To-Order Small and Medium Enterprise. In Proceedings of the 3rd World Conference on Production and Operations Management, Tokyo, Japan; 2008; pp. 1–15.
40. Dinis-Carvalho, J.; Moreira, F.; Bragança, S.; Costa, E.; Alves, A.; Sousa, R. Waste identification diagrams. *Prod. Plan. Control* **2015**, *26*, 235–247. [CrossRef]
41. Bertolini, M.; Braglia, M.; Marrazzini, L.; Neroni, M. Project Time Deployment: A new lean tool for losses analysis in Engineer-to-Order production environments. *Int. J. Prod. Res.* **2022**, *60*, 3129–3146. [CrossRef]

processes

MDPI

Article

Mathematical Model of the Electronic Cam in Terms of Application in a Dosing Machine

Karol Wójkowski [1], Krzysztof Talaśka [2] and Dominik Wilczyński [2,*]

[1] TREPKO S.A. Company, Roosevelta Str. 116, 62-200 Gniezno, Poland; karol.wojkowski21@gmail.com
[2] Faculty of Mechanical Engineering, Institute of Machine Design, Poznan University of Technology, Piotrowo Str. 3, 61-138 Poznań, Poland; krzysztof.talaska@put.poznan.pl
* Correspondence: dominik.wilczynski@put.poznan.pl

Abstract: The article analyses the use of servomotors in the control systems of industrial equipment, focusing on the alternative offered by position and speed synchronization in relation to classical mechanical mechanisms. A complete methodology is presented to determine the dynamic parameters of the adopted kinematic system using electronic motion profiles. The results obtained constitute a mathematical model of the execution chain and an analysis of the basic quantities for linear motion, supported by actual measurements of the drive parameters. The merit of the article is to show that the servomotors can significantly simplify the design of the device, make it more flexible in adaptation to different assortments, and allow integration with systems predicting the technical condition of the device. The analysis of the results revealed significant differences in the constant rotational speed of the servomotor, which do not align with previous findings. The results suggest that changing the angular working range of the assembly to the range (205°;270°) could significantly affect the generated linear acceleration, reducing the risk of stalling. The calculations and graphs conducted allowed for the accurate representation of the actual mechanical system, considering its dynamic characteristics. The key conclusion is that precise mathematical modelling is essential to ensure the stability and durability of engineering components.

Keywords: synchronization; real-time control system; electronic motion profile; electronic cam; cam coupling

1. Introduction

The economic starting point of any production device is its performance and the quality of the final product. In order to meet the wide range of requirements of the current market, newly designed devices must meet stringent requirements for production stability, consumption of process media, environmental safety and operation, minimizing the ecological footprint [1–3].

An increasingly common criterion in the design of industrial machines is their versatility and simplification of design. This is to ensure quick adaptation of the device to the production of various formats of the output product and to simplify the process of maintaining the proper technical condition of the device. As a result, operating costs are significantly reduced, while reliability is increased [4,5]. Newly designed industrial equipment is very often characterized by proven design solutions, supported by the experience of construction engineers and new technologies offered by the dynamically developing drive market [2,6]. The use of servomotors avoids the kinematic and dynamic nonlinearities of conventional mechanisms. On this basis, it can be assumed that the process of selecting components can be much simpler [2].

Because of the above criteria, designers are increasingly reaching for servomotors, appreciating their high potential. Servomotors are used to synchronize the position or speed of the slave axes with the real or virtual master axes, control the torque, and implement the point-to-point (p-to-p) trajectory [7,8]. They enable the generation of dynamic rotary and reciprocating movements of components with a significant mass and moment of inertia, while maintaining a high precision in position control [1,9,10].

The use of servomotors significantly simplifies the construction of the device. The kinematic chain is limited to the servomotor, gearbox, and executive apparatus, i.e., the pull-lever system or the cam mechanism [11]. This eliminates the need for complex torque transmission, bearing, and lubrication systems. Simplifying the design by eliminating unnecessary mechanisms allows to reduce the risk of failure and maintenance costs, including the costs of purchasing parts, their storage, and providing qualified personnel for service and repair activities. Compared to mechanical mechanisms, the servomotor control method, including the electronic cam, provides unparalleled flexibility and the ability to compensate for errors in the part manufacturing process [1,5,9,11].

In recent years, an innovative approach to the design of industrial processes, including processing equipment, has become noticeable. In the first place, an abstract approach to the activity has been suggested, focusing on the main elements and the effects achieved. To implement multiple layouts, you need to create interactions between them, with respect to the parent system [6]. Basing the design of devices on servomotors perfectly reflects the presented idea of creating a device model. The presented methodology in obtaining a mathematical model gives a general view of the activity performed. And the synchronization property satisfies the reference criterion of the parent system [6,7].

The use of the position synchronization functionality ensures constant positioning of the equipment, which is especially important in precise processing processes [3,10,12]. These processes often apply to objects where the geometry is irregular and the surface is easily deformable [1,9]. The dynamic capabilities offered by servomotors have made them an unrivalled solution in relation to classic actuators, i.e., pneumatic and hydraulic actuators [4,6,12].

High dynamic properties, precise position control, and the ability to design motion trajectories are widely used in the construction of processing machines, i.e., pouring and capping. An interesting example is the control of liquid sloshing by optimizing motion profiles, as a function of speed efficiency of the processing device [13]. Sloshing is a vibration or swaying that is most often in the form of asymmetric sine waves, which differ in frequency and modulus. The phenomenon of sloshing, classified as interference, is often amplified by the resonant vibrations of the device. These disturbances are compensated for by generating torque or changing velocity to suppress the defined disturbance. The whole process is carried out by the mathematical modelling of the disturbance, which requires a fast computational process to suppress the resulting disturbance in time [14].

In the era of Industry 4.0, a particularly important direction of development in the construction of machines based on servomotors is the monitoring of the condition of executive mechanisms. This research has focused on predicting the end of the life of parts, diagnostics, and determining the source of faults. Detailed diagnostics of the technical condition of the executive units has become possible thanks to advanced microprocessor systems that are an integral part of servo amplifiers–motion controllers. Precise measurements of speed, torque, etc., allow for quick reference to the adopted mathematical models of newly designed devices and taking the appropriate remedial steps [5,6].

The Motion Curve (Function) Editor allows tabular and graphical configurations of the start and end points of the disintervals into which the full course is divided. This makes it possible to precisely program the trajectory of motion and its derivatives, such as speed, acceleration, and spurt, which allows for the optimization of the production process. The graphical representation of the position and its subsequent derivatives after time are calculated for the defined constant rotational speed or linear master axis. The separated intervals are described by a function of the form [7,8,15]:

- linear interpolation;
- interpolation with third-degree spline functions;
- interpolation with Bezier curves;

- interpolation of a 5th degree polynomial;
- interpolation of a 7th degree polynomial;
- interpolation of a quadratic function.

The number of available interpolations may vary depending on the hardware platform used to implement the electronic motion profile. On the basis of the motion function, editors calculate successive derivatives of the function, which describe quantities, such as the following:

- velocity;
- acceleration;
- spurt.

Motion curve editors dedicated to the hardware platform do not allow you to define the kinematic layout of the executive apparatus. A tabular or graphical representation of the electronic motion profile is limited to the defined gear ratio and the mechanical characteristics of the motor [4,15].

Hardware platforms providing drive solutions enable integration with most industrial computers or PLCs. Interface solutions and dedicated visualization applications, e.g., in HMI, enable editing motion profiles directly from the workstation, bypassing the connection through a dedicated hardware platform IDE. Thanks to such functionalities, qualified construction staff can easily adjust the operating parameters of processing equipment to constantly changing production requirements, thus optimizing the manufacturing process in terms of efficiency and precision [7,8,11].

By simplifying the editing of motion profiles and being able to transfer it partially to a completely different hardware platform, it is possible to edit recipes, which allows for immediate and simple preparation of the machine for the production of a completely different end product. Recipe change can include all parameters directly related to the movement profile, e.g., the position of the master axis, the position of the slave axis, the type of characteristic describing the discontinuity interval, the speed of crossing the discontinuity interval boundaries, and other critical parameters that are required to start the production of a new product. This enables a quick response to dynamically changing market demand, reducing downtime, including changeover time [7,11].

The combination of these advanced solutions allows for intuitive management and real-time monitoring of production, which translates into better quality control and increased production flexibility. The result is increased automation and optimization, which has become the domain of modern production centers, constantly striving to maximize their position on the market [1,9,10,12].

Behind the whole range of advantages offered by servomotors and their implementation in industrial equipment, one should also be aware of their disadvantages and limitations. Servomotors are characterized by higher costs compared to other, classic drive solutions. The total cost of the drive unit is not only influenced by the cost of the motor, but also by advanced motion controllers, cables that are loaded with the maximum length, and the electrical installation that is designed and compatible with the latest EMC guidelines. These are the elements necessary for their proper functioning. Servomotors operating in harsh conditions, e.g., explosion hazard zones, marine environments, etc., are characterized by increased resistance to operating conditions and thus a much higher price [7].

The installation and configuration of complex servo drive units is more complex and complicated than that of simpler drives, such as inverters or stepper motors. This requires skilled knowledge backed by experience to fine-tune speed and position loop controllers based on position error over time. Achieving a satisfactory effect of the drive unit may lead to long production downtimes [7,8,11].

Behind a number of superlatives resulting from the use of electronic motion profiles, they also bring a number of limitations to the design of the device. The complexity of the software, which often differs significantly between hardware platforms, makes creating, editing, and visualizing complex and requires advanced technical knowledge. Without proper experience and knowledge, there is a significant risk of making mistakes that may result in damage to the device [8,11].

Often, motion profiles created at the beginning of the device's operation need to be optimized. This is a very time-consuming process, during which parameters related to positioning and generated torque are analysed, interpreted, and finally modified [7,8,11]. Such activities require advanced controllers that provide adequate computing power. This process significantly extends the time of implementation of new devices for operation, despite constantly developing methods of self-optimization of systems synchronized by electronic motion profiles [1,6,9].

The subject of this article is to determine the function of the motion of the executive apparatus and quantities, i.e., speed, acceleration, and spurt as a function of the position of the main axis. The seemingly basic parameters of movement allows for the precise dynamic analysis and selection of the right servomotor in terms of speed and generated torque. The results obtained create a precise mathematical model of the designed assembly, which is necessary for the implementation of advanced diagnostic systems and predicting the end of the life of parts.

2. Materials and Methods

2.1. Description of the Tested System

The pull-and-lever system, driven by a servomotor, was analysed in the context of replacing conventional mechanical systems. The position of the system was calculated on the basis of the electronic motion profile, the boundaries of which are shown in Table 1. The adopted division results from the need to synchronize with the other equipment installed in the production device. The intervals of discontinuities were described by a polynomial of the 5th degree and a linear function. The position synchronization took place in relation to the parent system, which was the virtual axis. The position of the virtual axis was in the range of $(0°;360°)$, and the maximum rotational speed was 150 rpm.

Table 1. Motion profile table.

No.	ΔX [Deg]	ΔY [Deg]
0	20	−5
1	60	0
2	120	−60
3	20	5
4	60	0
5	10	5
6	60	50
7	10	5

Based on the data contained in Table 1, we generated an electronic traffic profile presented in Figure 1. A graphical representation of the item and the next row of derivatives using Cam Builder Software IndraWorks Engineering ML 14V22 P17 can be found in the appendix.

The software IndraWorks Engineering ML 14V22 P17 was from Bosch Rexroth drives, the IndraDrive series. It offers sophisticated technological functions for the comprehensive implementation of drive issues in industrial automation. Users can parameterize and tune drives efficiently and precisely, easily adapting their operation to the dynamically changing requirements of the application being designed.

This software allows you to configure the topology of the drive communication network, including operating and synchronization modes. A clear and intuitive user interface simplifies the process of configuration, monitoring, and diagnostics. Advanced diagnostics allow the real-time monitoring of operating parameters and direct access to a descriptive list of all available parameters, error codes, and warnings. This allows for a quick diagnostic process and equally quick restoration of the full functionality of the application in the event of a failure.

The functionality also includes designing communication with other hardware platforms, i.e., PLCs or industrial computers. Thanks to communication based on most of the available communication protocols, the integration process with existing industrial automation applications is smooth and uncomplicated. This makes the drives of the IndraDrive series easy to integrate into complex industrial systems.

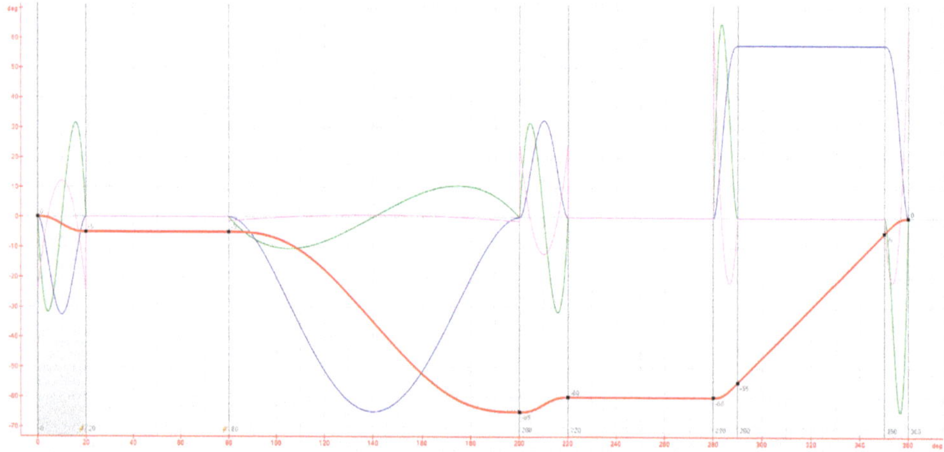

Figure 1. Electronic motion profile: red: synchronized axis position [deg], blue: synchronized axis speed [rpm], green: acceleration [rad/s^2], pink: synchronized axis spurt [rad/s^3].

The main processing material are pasted products, i.e., butter, milk fats with the addition of vegetable oils, margarines, cottage cheese, minced meat, lard, and yeast. The device in question has two working tracks that can carry out the packaging process in the range of 100–300 mL and with a capacity of up to 220 pieces/min [16].

The draw-lever system, analysed in this article and presented in Figure 2, is an integral part of the metering piston assemblies. The unit is made of two identical draw-lever systems, which are mirror-oriented to each other. The unit is driven by two separate servomotors that ensure synchronized movement and precise position control. Each system consists of the following:

- The L1 lever mounted on the gear shaft. The length of the L1 lever is crucial, as it is the distance between the piston rod axis and the axis of rotation of the gear shaft.
- The L2 link is the link between the lever and the piston rod of the metering piston. The role of the link is to transfer the rotary movement of the lever to the reciprocating movement of the piston rod.

The dotted line indicates the maximum angular position of the lever, which translates into a linear stroke. At point B—which is the end of the piston rod—a hygienic piston was installed. Tight seating in the steel cylinder ensured proper tightness of the system in the production cycle and precise dosing of liquids (74.2 mm).

The 822 HS devices allowed product dispensing in the range of 100–300 mL [16]. The changing dose was implemented by a modifiable electronic motion profile, which allowed precise adjustment of the amount of material dispensed in real time. Such flexible adaptation to the changing conditions of the production process would not be possible with the use of mechanical mechanisms. Classic solutions are limited by fixed movement parameters and the lack of possibility of quick adaptation.

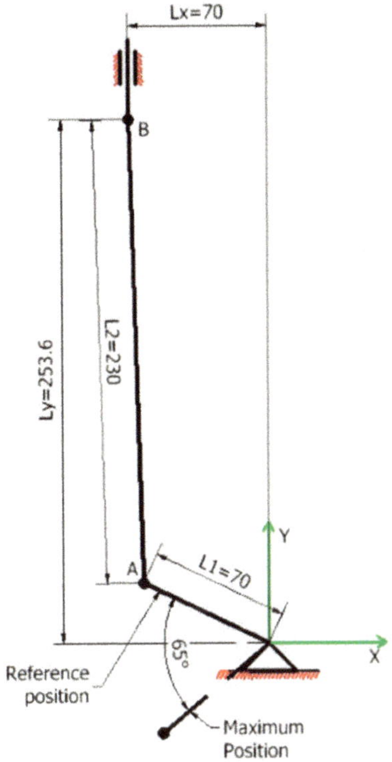

Figure 2. Execution chain kinematic system.

Changing the dose is often necessary from cycle to cycle of the device. Changes in product temperature, viscosity, density, and production line speed may result in fluctuations in the weight of the final product. The flexibility provided by the electronic cam allows for immediate and precise adaptation to these constantly changing conditions, which is crucial to maintaining high efficiency and quality of production.

The essence of this article is to determine the function of motion of the executive apparatus and to determine the key quantities, i.e., speed, acceleration, and spurt as a function of the position of the main axis. Seemingly basic motion parameters are necessary for further stages of design work, including the selection of the servomotor and gearbox, strength analysis, and the selection of bearing arrangements.

Based on the obtained mathematical model of the movement, it is possible to optimally select drive solutions, analyse various operating scenarios of the device, and accurately monitor the operating conditions. As shown in the research [1,5,17], comparing direct measurements from drives with the adopted mathematical model sets a completely new direction in the industry regarding the real-time diagnostics of actuator parts. As a result, we may acquire another level of efficiency and reliability with a device adapted to the realities of the modern manufacturing market.

Real-time monitoring of the operating status allows one to accurately track the technical condition of the equipment, identify deviations from accepted standards, and immediately respond to potential downtime resulting from equipment failure. This level of technological advancement not only increases efficiency, but also reduces maintenance and repair costs, increasing production uptime. As a result, companies can maximize their unrivalled position in the manufacturing market.

Determination of the Motion Function of the Kinematic System

Equation of motion for point B depicted in Figure 2. It takes the following form:

$$y_B = [\cos(\alpha_1) * L_1] + [\cos(\text{asin}(\frac{L_X - |\sin(\alpha_1) * L_1|}{L_2})) * L_2] - L_Y$$
$$x_B = 0 \tag{1}$$

The working movement of the unit was divided into 8 compartments presented in Table 2. The division was due to the cooperation of the team in question with regard to the other equipment included in the device. Modifying the width of the compartment may lead to the desynchronization of the equipment or damage to it (Δx).

Table 2. Table of coefficients of the position function of point B.

Interval	A	B	C	D	And	F
No. 0	−0.00009375318850	0.000468766060870	−0.006250222355561	0.000000207088533	−0.000000388970875	−0.000000048458332
No. 1	0	−5	0	0	0	0
No. 2	−0.000000014468037	0.000010127625759	−0.002662118786617	0.324084025380577	−18.5190869605351	397.975169692502
No. 3	0.000009375310839	−0.009844076379006	4.13138698156651	−866.278725275157	90753.0097407992	−3800191.05369414
No. 4	0	−60	0	0	0	0
No. 5	0.000050000248613	−0.071667031186851	41.0835468530585	−11774.062460404	1686915.79185127	−96662566.0039355
No. 6	0.83336	−296.67583	0	0	0	0
No. 7	0.000050002046225	−0.088336953382979	62.4192282221552	−22050.9062159613	3894640.2872562	−275126219.563166

The compartments of the electronic traffic profile are described by the following functions:

- Compartment No. 0: $y_0 = Ax^5 + Bx^4 + Cx^3 + Dx^2 + Ex + F$
- Compartment No. 1: $y_1 = Ax + B$
- Compartment No. 2: $y_2 = Ax^5 + Bx^4 + Cx^3 + Dx^2 + Ex + F$
- Compartment No. 3: $y_3 = Ax^5 + Bx^4 + Cx^3 + Dx^2 + Ex + F$
- Compartment No. 4: $y_4 = Ax + B$
- Compartment No. 5: $y_5 = Ax^5 + Bx^4 + Cx^3 + Dx^2 + Ex + F$
- Compartment No. 6: $y_6 = Ax + B$
- Compartment No. 7: $y_7 = Ax^5 + Bx^4 + Cx^3 + Dx^2 + Ex + F$

The values of the function coefficients were determined using the trend line—the functionality of the software Microsoft Excel 365 ver. 2407. The accuracy of coefficients to accurately represent the position curve shown in Figure 1 amounted to a minimum of 10 significant places.

2.2. Mathematical Analysis

The axis position diagram shown in Figure 1 is presented in the following units:

$$\frac{pozycja\ osi\ slave\ [^\circ]}{pozycja\ osi\ master\ [^\circ]} \tag{2}$$

The unit shown in Equation (2) is sufficient only if the position of the actuator in question is presented. In order to continue the calculation, it is necessary to change the unit in which the time occurs $[s]$. This is due to the fact that the basic multiplicities characterising the system, i.e., speed, acceleration, and spurt, are calculated on the basis of the change in the antiderivative function per unit of time.

In the available cam editors, the main axis is assumed to convert units. This assumption also results in the constant time of a single team cycle. This relationship allowed us to determine the positions of the main axis as a function of time. For the calculations, it was assumed that the speed of the main axis was 150 [rpm]. Based on this, the total cycle time was determined as follows:

$$\omega = 2\pi * \frac{n[\text{rpm}]}{60[\text{s}]} = 2\pi * \frac{150}{60} = 15.70796323\left[\frac{\text{rad}}{\text{s}}\right] \tag{3}$$

$$t_C = \frac{2\pi}{\omega} = \frac{2\pi}{15.70796323} = 0.4[\text{s}] \tag{4}$$

In order to determine the next order of the time derivative, it is necessary to specify the time step *dt*. To achieve this, it is necessary to define the number of points into which we could divide the entire time period. When determining this step, we took into account the precision of calculations and the accuracy of our results. Accepting the appropriate number of points is important for the reliability and accuracy of the analysis of the dynamics of time phenomena. The following value was used for the calculation: $2^{10} = 1024$.

$$dt = \frac{t_C}{p-1} = \frac{0.4}{1024-1} = 3.91007 * 10^{-4}[\text{s}] = 0.000391007\ [\text{s}] \tag{5}$$

Using the following relation, the position of the master axis in a unit of time was determined, the graph of which is shown in Figure 3.

$$\alpha = \frac{\omega * t * 360^{\circ}}{2\pi}\ [^{\circ}] \tag{6}$$

where

t [s] $\in (0; t_c)$—cycle time incremented by the value of dt.

By obtaining the positions of the main axis with high resolution as a function of time, it is possible to determine the *angular position of the synchronized axis*, the graph of which is shown in Figure 3. The calculations were made on the basis of the general form of the function, which was used to describe successive intervals of discontinuities, the table of coefficients presented in Table 2, and the widths of the intervals given in Table 1.

The determination of the angular position of the synchronized axis was calculated on the basis of the following relationships:

Compartment No. 0:

$$y_0 = A_0x^5 + B_0x^4 + C_0x^3 + D_0x^2 + E_0x + F \tag{7}$$

where $x \in (0^{\circ}; 20^{\circ})$.

Compartment No. 1:

$$y_1 = A_1x + B_1 \tag{8}$$

where $x \in (20^{\circ}; 60^{\circ})$.

Compartment No. 2:

$$y_2 = A_2x^5 + B_2x^4 + C_2x^3 + D_2x^2 + E_2x + F_2 \tag{9}$$

where $x \in (80^{\circ}; 200^{\circ})$.

Compartment No. 3:

$$y_3 = A_3x^5 + B_3x^4 + C_3x^3 + D_3x^2 + E_3x + F_3 \tag{10}$$

where $x \in (200^{\circ}; 220^{\circ})$.

Compartment No. 4:

$$y_4 = A_4x + B_4 \tag{11}$$

where $x \in (220^{\circ}; 280^{\circ})$.

Compartment No. 5:

$$y_5 = A_5x^5 + B_5x^4 + C_5x^3 + D_5x^2 + E_5x + F_5 \tag{12}$$

where $x \in (280°; 290°)$.

Compartment No. 6:

$$y_6 = A_6 x + B_6 \tag{13}$$

where $x \in (290°; 350°)$.

Compartment No. 7:

$$y_7 = A_7 x^5 + B_7 x^4 + C_7 x^3 + D_7 x^2 + E_7 x + F_7 \tag{14}$$

where $x \in (350°; 360°)$.

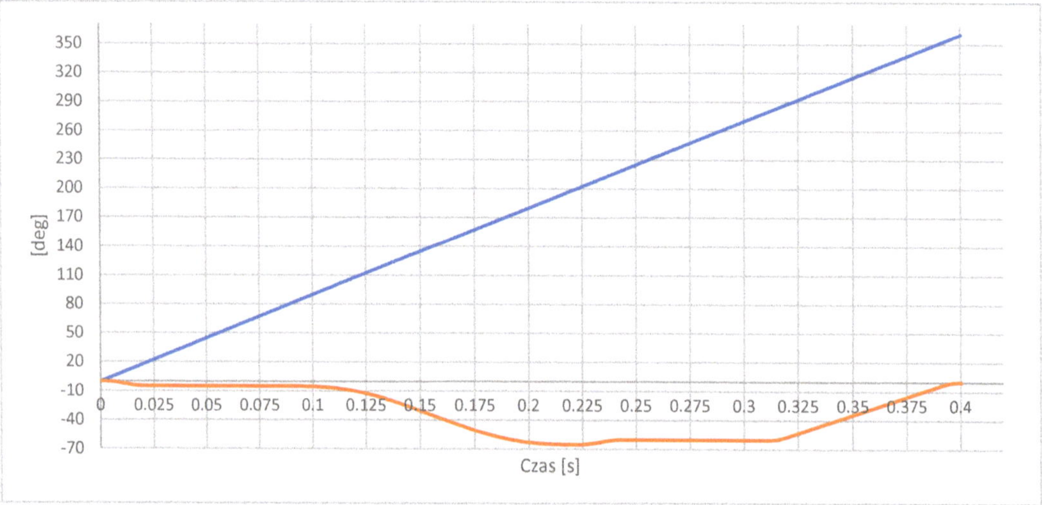

Figure 3. Chart of the position of the main and synchronized axis: blue: master axis position [deg], orange: synchronized axis position [deg].

The angular position, the position of the gear shaft, and its subsequent derivatives over time are only approximate conditions that prevail at the end of the execution chain. The inaccuracy resulted from the trigonometric functions describing the position of the final point of the executive system.

In order to describe the system in detail, a determination of the position of the kinematic system in mm is made. This is possible thanks to the determined angular position of the high-resolution synchronised axis substituted into (1). As a result of this arithmetic operation, the position of the synchronised axis at time t is obtained.

In order to determine the parameters, i.e., speed, acceleration, and spurt, the equation should be differentiated (1). Motion functions that describe the position of objects in two- or three-dimensional space are inherently complex, and their differentiation is time-consuming and complicated. To simplify the calculations, the following basic relation describing the antiderivative function was used:

$$f'(x) = \frac{dy}{dx} => f'(x) = \frac{dy}{dt} \tag{15}$$

According to the presented relation, it is possible to determine subsequent derivatives, constituting the basic parameters of linear motion on the basis of successive, instantaneous values after time—the time step determined in Equation (5). Based on the above, the following relationships are recorded:

The equation for the velocity of point B.

$$v_{y_B} = \frac{dy_B}{dt}\left[\frac{\text{mm}}{\text{s}}\right] \tag{16}$$

The acceleration equation of point B.

$$a_{y_B} = \frac{dv_{y_B}}{dt}\left[\frac{\text{mm}}{\text{s}^2}\right] \tag{17}$$

The equation for the breakaway of point B.

$$j_{y_B} = \frac{da_{y_B}}{dt}\left[\frac{\text{mm}}{\text{s}^3}\right] \tag{18}$$

In order to thoroughly analyse and obtain a mathematical model of the system under consideration, a number of calculations were performed, the results of which are presented in the above diagrams. In Figures 4–7, position, speed, acceleration, and spurt as a function of time are presented.

The analysis of the diagrams allowed for a detailed understanding of the dynamic behaviour of the kinematic system under consideration, which ultimately translated into an efficient and precise design process. An important aspect of this analysis is the fact that all units are expressed in millimetres. This is important from the point of view of precise representation of the actual operation of the executive module.

It is worth noting that generic motion curve editors often present data in degrees, which can lead to inaccuracies and inaccuracies in the interpretation of results. In our case, using millimetres as the unit of measurement provides greater accuracy and a better representation of the real operating conditions of the system.

The above diagram is representative; the values of individual derivatives were scaled in order to reflect the phenomena occurring in the considered complex. Detailed values of the parameters, i.e., position, speed, acceleration, and spurt, are presented in the following graphs.

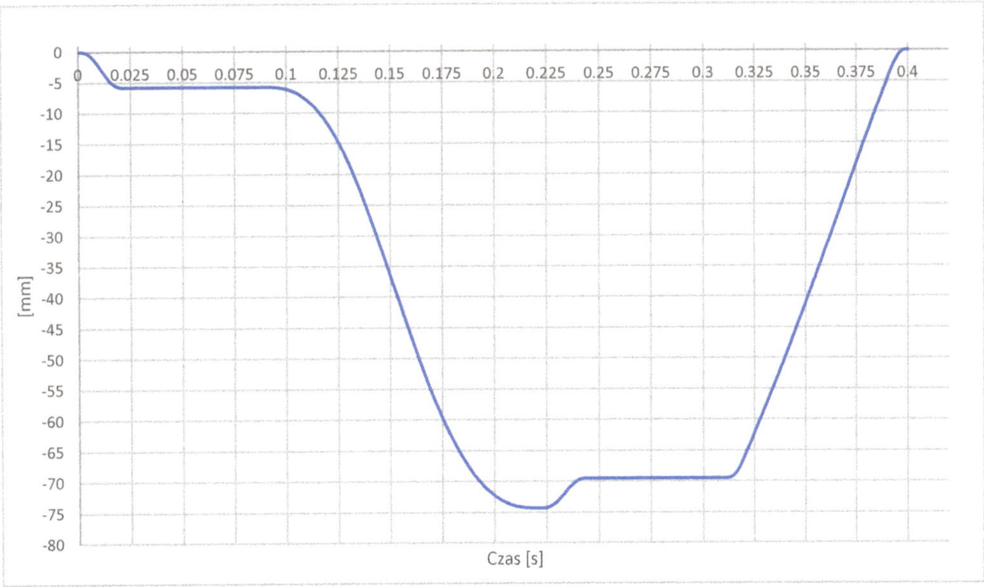

Figure 4. Position chart of point B.

In Figure 8, the complete kinematics of the considered executive system is presented. The purpose of this diagram is to demonstrate similarity to the diagram obtained in a dedicated electronic motion profile editor, designed for the drive Bosch Rexroth. Compare the charts in Figures 1 and 8 and it allows one to identify and better understand the differences between the calculations performed without taking into account the kinematic system and calculations with it included.

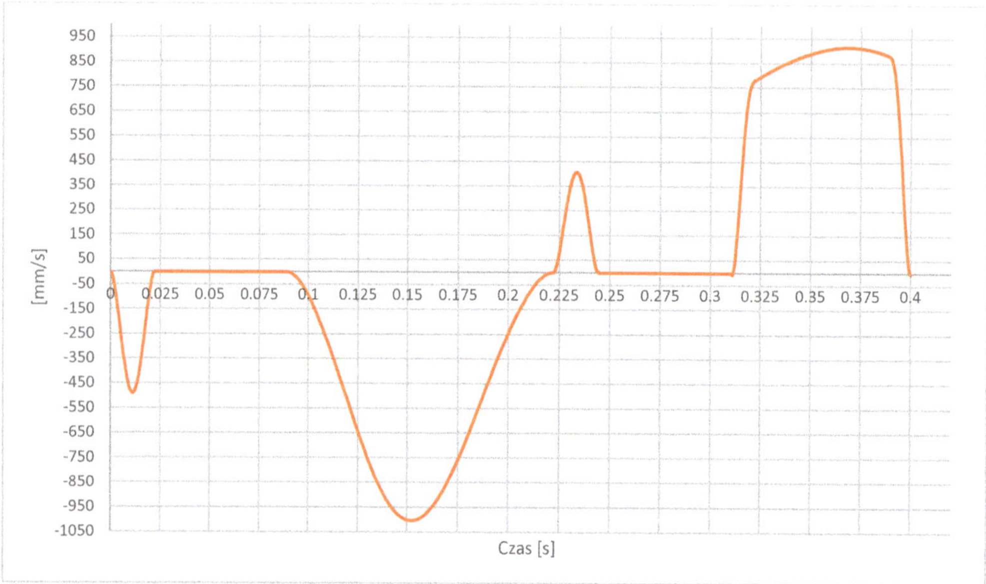

Figure 5. Velocity chart of point B.

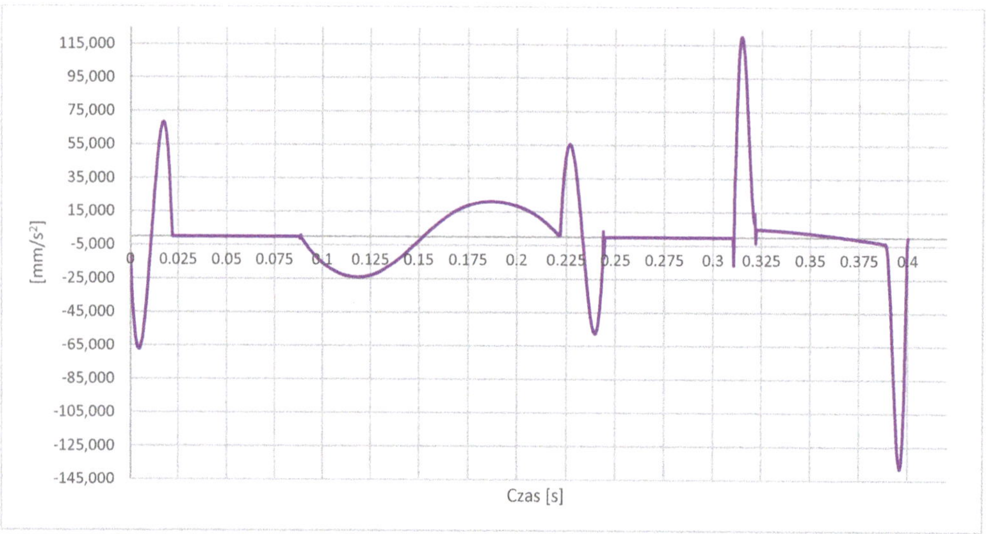

Figure 6. Acceleration chart of point B.

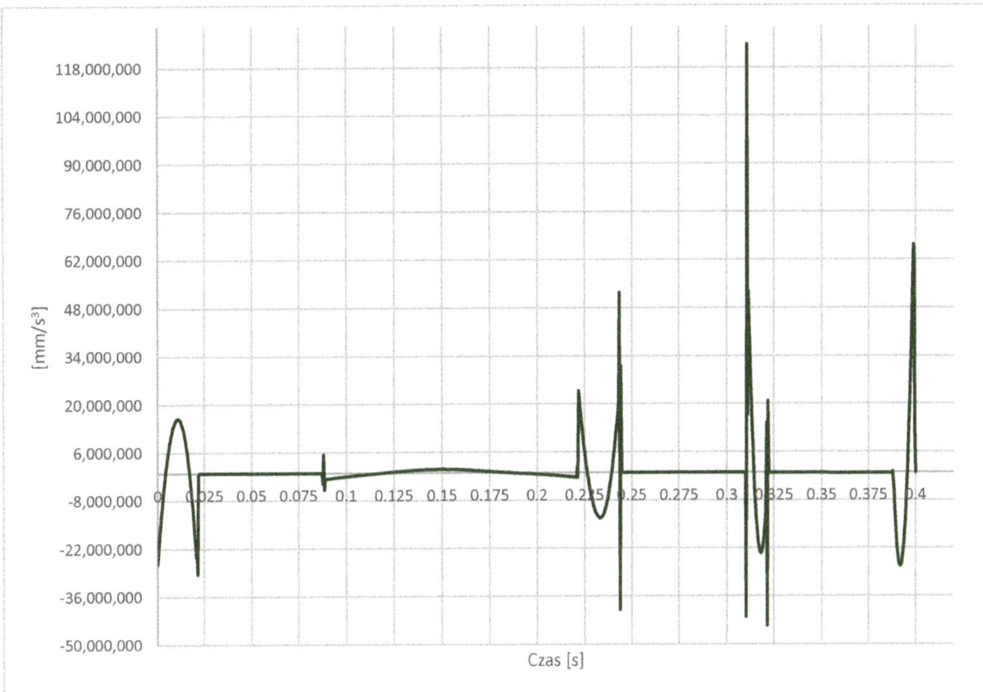

Figure 7. Jerk chart of point B.

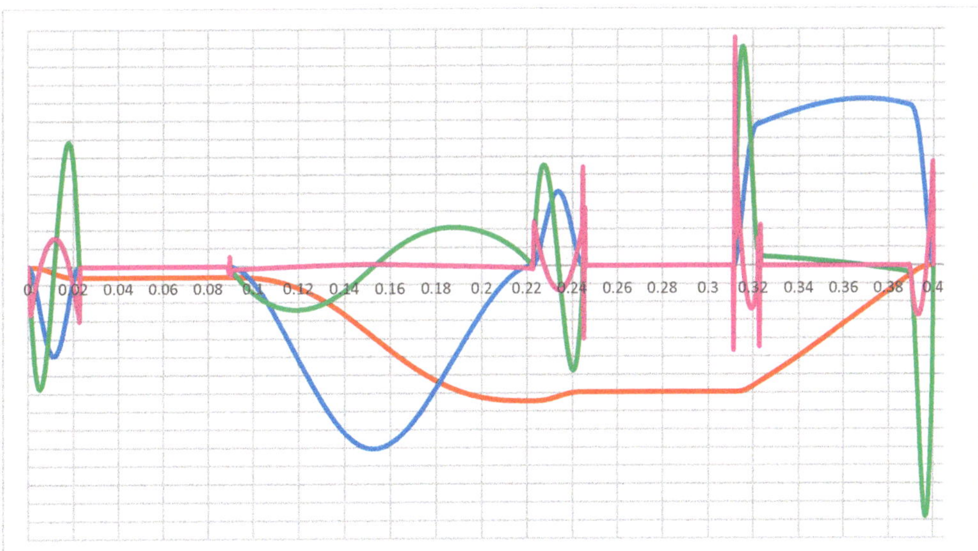

Figure 8. Actuator chart: red: synchronized axis position [mm], blue: synchronized axis speed [mm/s], green: acceleration [mm/s^2], pink: synchronized axis spurt [mm/s^3].

3. Results and Discussion

On the basis of the calculations performed, a mathematical model of the actuator system was obtained. The obtained number of results were an introduction to further design activities, i.e., the selection of construction materials, strength analysis, consideration of frictional wear, static, and dynamic load capacity of bearing arrangements, rod heads, the selection of drive and gears, etc. The obtained results were compared with the measurements of the drive parameters of the kinematic system in question.

The graphs show all units in millimetres (mm), which allows for a precise representation of the actual actuator. The adopted unit allowed for greater accuracy and better visualization of the system's operation, in contrast to general editors of electronic traffic profiles, which represent data in degrees, which may result in inaccuracies and inaccuracies in the interpretation of received messages. The calculations were limited to a single cycle of the assembly. This increased transparency and simplified the interpretation of the data and the identification of the key steps in the dispensing process.

By analysing the trajectory of motion depicted in Figures 4 and 9 (purple line), the most important and at the same time the most strenuous stage of movement is the dosing of the product accumulated in the space of the dosing cylinder (Figure 10). A product with a density of $900\frac{kg}{m^3}$, and a high kinematic viscosity offers a high resistance at the initial stage of dosing. This resistance is assumed to be related to the intermolecular cohesive forces of the product. The greater the bond strength, the greater the resistance, as evidenced by the rapid and momentary jump in torque generated by the servomotor.

It should be noted that the spurt diagram presented in Figure 7 reflects the shape of the measured torque generated by the drive shown in Figure 9. In the context of the spurt characteristics generated in the electronic motion profile editor, shown in Figure 1, it is clear that it did not reflect the operating conditions of the system at all. Such a discrepancy can lead to incorrect design assumptions and serious design errors.

The presented analyses show that motion profiles generated in the dedicated software can ignore important dynamic factors of the real system. In particular, the imprecise reproduction of the spurt may result in the improper selection of actuator components, which in turn affects the stability and reliability of the entire device.

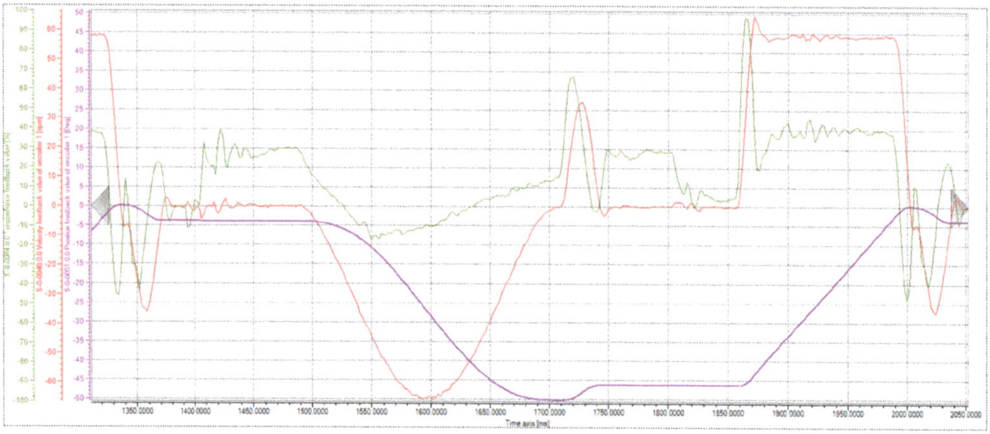

Figure 9. Measurement of actuator drive parameters: purple—position [deg], red—speed [rpm], green—torque [%] (Servomotor Max Torque Percentage).

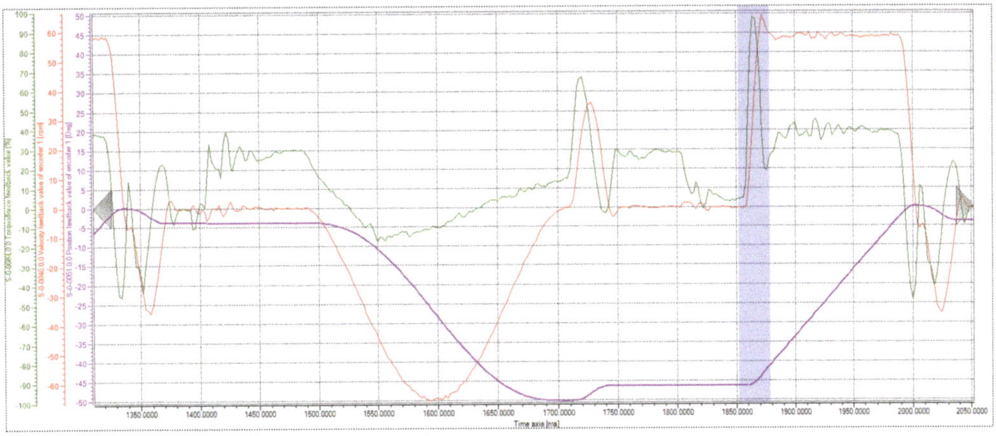

Figure 10. Start of product dispensing. The blue area indicates the start of the dosing cycle.

In practice, underestimating the spurt leads to excessive wear of bearing arrangements, rod heads, noise, and vibration of the equipment. The occurrence of a spurt resulting from rapid changes in acceleration causes the formation of large dynamic forces, which have a particularly negative impact on the executive system. The equation of the force acting on the system takes the following form:

$$F = ma = m\frac{dv}{dt} = \frac{dp}{dt} \tag{19}$$

From this equation, it is clear that the formation of large changes in acceleration has a direct effect on the executive system.

Continuing the analysis of the results obtained, another important difference is the constant rotational speed captured in the blue area in Figure 11 (red line), the course of which did not coincide with the results obtained in Figure 5. The constant rotational speed of the servomotor was the effect of the linear function described in interval No. 6 presented in Table 2. The obtained calculations showed that the course of the function within the range of No. 5 to No. 7 had a course similar to that of a parabola.

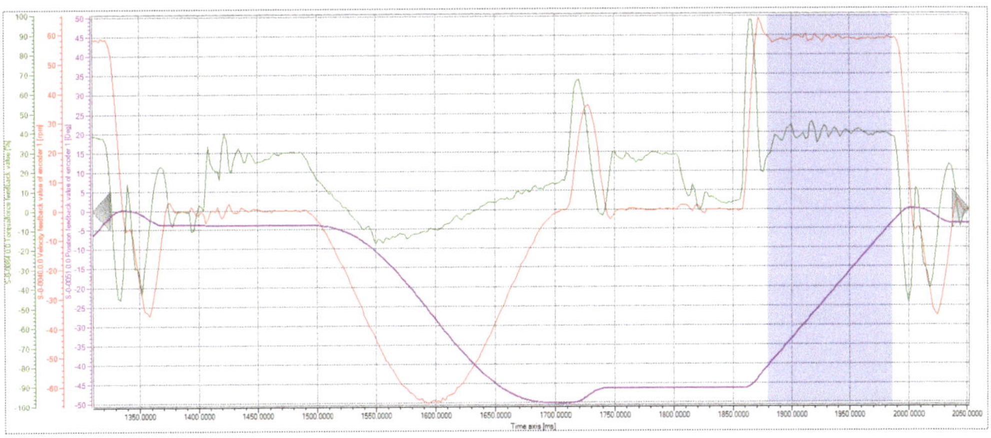

Figure 11. Constant speed section.

The difference shown is a consequence of the trigonometric functions with which the position of the executive member and its range of angular work were sown. The range of the angular work within the range of absolute measure $(220^\circ;285^\circ)$ caused an increase in the linear velocity in the Y axis—the axis of work in the piston rod of the system, the maximum of which falls for 270°.

On the basis of the obtained mathematical model and the analysis performed, it is recommended to transfer the range of the angular work of the assembly to the interval. This change could significantly affect the generated linear acceleration, thus reducing the risk of the occurrence of the spurt phenomenon, the negative effects of which were described earlier $(205^\circ;270^\circ)$.

This was due to the fact that the range of angular work directly affected the dynamic characteristics of the system. Moving the working range to the indicated angle range will ensure much smoother acceleration changes, which in the case of pasted products with high kinematic viscosity, will significantly reduce the wear of mechanical parts.

This paper presents an analysis of the results obtained from the mathematical model of the actuator system, which is a key step in the engineering design process. The calculations carried out and the generated diagrams made it possible to accurately map the real system, taking into account the basic parameters of linear motion. The analysis of the results focused on the problem of spurt and dynamic forces that particularly negatively affect the stability and durability of components.

The analysis showed that precise mathematical modelling is essential to accurately represent the actual behaviour of a mechanical system. Cam editors dedicated to drive solutions are only useful for general outlines of the motion trajectory, but do not take into account the key parameters resulting from the kinematic system. This can lead to serious design errors, which can disrupt the assumed reliability and durability of equipment that the modern process market expects.

4. Conclusions

Based on the above research and discussion it can be concluded that the primary scientific problem to be addressed is the optimization of kinematics and the efficiency of cable-lever systems used in industrial equipment. These systems offer an alternative to traditional cam mechanisms, potentially providing benefits in terms of precision and performance. Future research will focus on the following two key aspects:

- Kinematics Analysis and Modelling: There is a lack of sufficient research on the precise analysis of kinematics and the synchronization of movements in cable-lever systems. The goal is to develop advanced mathematical models and analytical methods that enable the precise modelling and optimization of these systems. The research will encompass both theoretical aspects of kinematics and practical approaches to movement synchronization, aiming to enhance the precision and efficiency of system performance.
- Durability and Reliability: It is crucial to investigate the durability and reliability of components used in cable-lever systems. This analysis will aim to assess the impact of these components on the stability and safety of the equipment, as well as to develop methods for improving their durability and reliability. Specifically, the research will focus on evaluating material wear, dynamic loads, and the effects of long-term operation on system performance and safety.

The planned research aims not only to advance the existing knowledge of cable-lever systems but also to provide practical solutions that can significantly impact the design and optimization of industrial equipment. The outcome of the doctoral dissertation is expected to include innovative approaches and recommendations that can be utilized in future engineering and research endeavours.

Author Contributions: Conceptualization, K.W.; methodology, K.W., D.W. and K.T.; validation, K.W., D.W. and K.T.; formal analysis, K.W.; investigation, K.W.; resources, K.W.; data curation, K.W.; writing—original draft preparation, K.W., D.W. and K.T.; writing—review and editing, K.W., D.W. and K.T.; visualization, K.W.; supervision, K.T. and D.W.; project administration, K.T.; funding acquisition, K.T. All authors have read and agreed to the published version of the manuscript.

Funding: This research received no external funding.

Data Availability Statement: The data presented in this study are available upon request from the corresponding author.

Conflicts of Interest: Author Karol Wójkowski was employed by the company TREPKO S.A. Company. The remaining authors declare that the research was conducted in the absence of any commercial or financial relationships that could be construed as a potential conflict of interest. The TREPKO S.A. Company had no role in the design of the study; in the collection, analyses, or interpretation of data; in the writing of the manuscript, or in the decision to publish the results.

References

1. Holownenko, O.; Troll, C.; Ihlenfeldt, S.; Majschak, J.-P. A Novel Adaptive Interpolation Approach for the Implementation of Operating-Speed-Dependent Motion Profiles in Processing Machines. *Eng. Sci. Technol. Int. J.* **2021**, *24*, 1308–1317.
2. Corves, B. Servo Drives, Mechanism Simulation and Motion Profiles. In *Mechanisms and Machine*; Ceccarelli, M., Ed.; Springer: Berlin/Heidelberg, Germany, 2010; pp. 147–155.
3. Dyllong, E.; Visioli, A. Planning and real-time modifications of a trajectory using spline techniques. *Robotica* **2003**, *21*, 475–482. [CrossRef]
4. Li, X.; Guo, L.; Liang, L. In That Motion control of guide rolls in intelligent simulation for profiled ring rolling process. *Procedia Manuf.* **2018**, *15*, 97–104. [CrossRef]
5. Barbers, M.; Several, R. A Language Condition monitoring of electric-cam mechanisms based on Model-of-Signals of the drive current higher-order differences. *IFAC-Pap.* **2020**, *53*, 802–807.
6. Troll, C.; Schebitz, B.; Majschak, J.-P.; Döring, M.; Holowenko, O.; Ihlenfeldt, S. Commissioning New Applications on Processing Machines: Part I–Proces Modelling. *Adv. Mech. Eng.* **2018**, *10*, 1687814018754955. [CrossRef]
7. Biagiotti, L.; Melchiorri, C. *Trajectory Planning for Automatic Machines and Robots*; Springer: Berlin/Heidelberg, Germany, 2008; p. 514.
8. Xiao, Y.; Zhu, K.; Liaw, H.C. Generalized synchronization control of multi-axis motion systems. *Control. Eng. Pract.* **2005**, *13*, 809–819. [CrossRef]
9. Balula, S.; Liao-McPherson, D.; Rupenyan, A.; Lygeros, J. Data-driven reference trajectory optimization for precision motion systems. *Control. Eng. Pract.* **2024**, *144*, 105834. [CrossRef]
10. Bianco, C.G.L.; Piazzi, A. Minimum-time trajectory planning of mechanical manipulators under dynamic constraints. *Int. J. Control.* **2002**, *75*, 967–980. [CrossRef]
11. Angeles, J.; Lopez-Cajun, C.S. *Optimization of Cam Mechanisms*; Kluwer Academic Publishers: Norwell, MA, USA, 1991.
12. Biagiotti, L.; Melchiorri, C. Smooth trajectories for high-performance multiaxes automatic machines. *IFAC Proc. Vol.* **2006**, *39*, 175–180. [CrossRef]
13. Troll, C.; Tietze, S.; Majschak, J.-P. Investigation on the Application of Operating Speed Dependent Motion Profiles in Processing Machines at the Example of Controlling Liquid Slosh. In *Advances in Mechanism and Machine Science*; Springer: Berlin/Heidelberg, Germany, 2019; pp. 1909–1918.
14. Rexroth Bosch Group. How Adaptive Systems Unlock Big Productivity Gains. Available online: https://www.semanticscholar.org/paper/How-Adaptive-Systems-unlock-big-productivity-gains/2ee259cf4f5f2266c1104a782259fc6367cb334e (accessed on 23 May 2024).
15. Siemens. Motion Control. Available online: https://publikacje.siemens-info.com/pdf/125/Motion%20Control%20-%20Sprz%C4%99%C5%BCenie%20krzywkowe.pdf (accessed on 21 May 2024).
16. Trepko. Seria 800. Available online: https://www.trepko.com/wp-content/uploads/2021/01/seria_800_automaty_do_pakowania_w_kostki.pdf (accessed on 8 June 2024).
17. Holowenko, O.; Kauschinger, B.; Ihlenfeldt, S. Increasing the Performance of Processing Machines by Executing Output Rate Dependent Motion Profiles. *Int. J. Autom. Technol.* **2017**, *11*, 165–170. [CrossRef]

MDPI AG
Grosspeteranlage 5
4052 Basel
Switzerland
Tel.: +41 61 683 77 34

Processes Editorial Office
E-mail: processes@mdpi.com
www.mdpi.com/journal/processes

www.ingramcontent.com/pod-product-compliance
Lightning Source LLC
Chambersburg PA
CBHW061500220926
51850CB00004B/135